美国数学会经典影印系列

出版者的话

近年来，我国的科学技术取得了长足进步，特别是在数学等自然科学基础领域不断涌现出一流的研究成果。与此同时，国内的科研队伍与国外的交流合作也越来越密切，越来越多的科研工作者可以熟练地阅读英文文献，并在国际顶级期刊发表英文学术文章，在国外出版社出版英文学术著作。

然而，在国内阅读海外原版英文图书仍不是非常便捷。一方面，这些原版图书主要集中在科技、教育比较发达的大中城市的大型综合图书馆以及科研院所的资料室中，普通读者借阅不甚容易；另一方面，原版书价格昂贵，动辄上百美元，购买也很不方便。这极大地限制了科技工作者对于国外先进科学技术知识的获取，间接阻碍了我国科技的发展。

高等教育出版社本着植根教育、弘扬学术的宗旨服务我国广大科技和教育工作者，同美国数学会（American Mathematical Society）合作，在征求海内外众多专家学者意见的基础上，精选该学会近年出版的数百种专业著作，组织出版了“美国数学会经典影印系列”丛书。美国数学会创建于 1888 年，是国际上极具影响力的专业学术组织，目前拥有近 30000 会员和 580 余个机构成员，出版图书 3500 多种，冯·诺依曼、莱夫谢茨、陶哲轩等世界级数学大家都是其作者。本影印系列涵盖了代数、几何、分析、方程、拓扑、概率、动力系统等所有主要数学分支以及新近发展的数学主题。

我们希望这套书的出版，能够对国内的科研工作者、教育工作者以及青年学生起到重要的学术引领作用，也希望今后能有更多的海外优秀英文著作被介绍到中国。

高等教育出版社

2016 年 12 月

AMS
AMERICAN MATHEMATICAL SOCIETY
美国数学会经典影印系列

Basic Complex Analysis:

A Comprehensive Course in Analysis, Part 2A

基础复分析：分析综合教程（第 2A 部分）

Barry Simon

中国教育出版传媒集团
高等教育出版社 · 北京

To the memory of Cherie Galvez

extraordinary secretary, talented helper, caring person

and to the memory of my mentors,

Ed Nelson (1932-2014) and Arthur Wightman (1922-2013)

who not only taught me Mathematics
but taught me how to be a mathematician

Contents

Preface to the Series

Young men should prove theorems, old men should write books.

—*Freeman Dyson*, quoting G. H. Hardy[1]

Reed–Simon[2] starts with "Mathematics has its roots in numerology, geometry, and physics." This puts into context the division of mathematics into algebra, geometry/topology, and analysis. There are, of course, other areas of mathematics, and a division between parts of mathematics can be artificial. But almost universally, we require our graduate students to take courses in these three areas.

This five-volume series began and, to some extent, remains a set of texts for a basic graduate analysis course. In part it reflects Caltech's three-terms-per-year schedule and the actual courses I've taught in the past. Much of the contents of Parts 1 and 2 (Part 2 is in two volumes, Part 2A and Part 2B) are common to virtually all such courses: point set topology, measure spaces, Hilbert and Banach spaces, distribution theory, and the Fourier transform, complex analysis including the Riemann mapping and Hadamard product theorems. Parts 3 and 4 are made up of material that you'll find in some, but not all, courses—on the one hand, Part 3 on maximal functions and H^p-spaces; on the other hand, Part 4 on the spectral theorem for bounded self-adjoint operators on a Hilbert space and det and trace, again for Hilbert space operators. Parts 3 and 4 reflect the two halves of the third term of Caltech's course.

[1]Interview with D. J. Albers, The College Mathematics Journal, **25**, no. 1, January 1994.

[2]M. Reed and B. Simon, *Methods of Modern Mathematical Physics, I: Functional Analysis*, Academic Press, New York, 1972.

While there is, of course, overlap between these books and other texts, there are some places where we differ, at least from many:

(a) By having a unified approach to both real and complex analysis, we are able to use notions like contour integrals as Stietljes integrals that cross the barrier.
(b) We include some topics that are not standard, although I am surprised they are not. For example, while discussing maximal functions, I present Garcia's proof of the maximal (and so, Birkhoff) ergodic theorem.
(c) These books are written to be keepers—the idea is that, for many students, this may be the last analysis course they take, so I've tried to write in a way that these books will be useful as a reference. For this reason, I've included "bonus" chapters and sections—material that I do not expect to be included in the course. This has several advantages. First, in a slightly longer course, the instructor has an option of extra topics to include. Second, there is some flexibility—for an instructor who can't imagine a complex analysis course without a proof of the prime number theorem, it is possible to replace all or part of the (non-bonus) chapter on elliptic functions with the last four sections of the bonus chapter on analytic number theory. Third, it is certainly possible to take all the material in, say, Part 2, to turn it into a two-term course. Most importantly, the bonus material is there for the reader to peruse long after the formal course is over.
(d) I have long collected "best" proofs and over the years learned a number of ones that are not the standard textbook proofs. In this regard, modern technology has been a boon. Thanks to Google books and the Caltech library, I've been able to discover some proofs that I hadn't learned before. Examples of things that I'm especially fond of are Bernstein polynomials to get the classical Weierstrass approximation theorem, von Neumann's proof of the Lebesgue decomposition and Radon–Nikodym theorems, the Hermite expansion treatment of Fourier transform, Landau's proof of the Hadamard factorization theorem, Wielandt's theorem on the functional equation for $\Gamma(z)$, and Newman's proof of the prime number theorem. Each of these appears in at least some monographs, but they are not nearly as widespread as they deserve to be.
(e) I've tried to distinguish between central results and interesting asides and to indicate when an interesting aside is going to come up again later. In particular, all chapters, except those on preliminaries, have a listing of "Big Notions and Theorems" at their start. I wish that this attempt to differentiate between the essential and the less essential

didn't make this book different, but alas, too many texts are monotone listings of theorems and proofs.

(f) I've included copious "Notes and Historical Remarks" at the end of each section. These notes illuminate and extend, and they (and the Problems) allow us to cover more material than would otherwise be possible. The history is there to enliven the discussion and to emphasize to students that mathematicians are real people and that "may you live in interesting times" is truly a curse. Any discussion of the history of real analysis is depressing because of the number of lives ended by the Nazis. Any discussion of nineteenth-century mathematics makes one appreciate medical progress, contemplating Abel, Riemann, and Stieltjes. I feel knowing that Picard was Hermite's son-in-law spices up the study of his theorem.

On the subject of history, there are three cautions. First, I am not a professional historian and almost none of the history discussed here is based on original sources. I have relied at times—horrors!—on information on the Internet. I have tried for accuracy but I'm sure there are errors, some that would make a real historian wince.

A second caution concerns looking at the history assuming the mathematics we now know. Especially when concepts are new, they may be poorly understood or viewed from a perspective quite different from the one here. Looking at the wonderful history of nineteenth-century complex analysis by Bottazzini–Grey[3] will illustrate this more clearly than these brief notes can.

The third caution concerns naming theorems. Here, the reader needs to bear in mind Arnol'd's principle:[4] *If a notion bears a personal name, then that name is not the name of the discoverer* (and the related Berry principle: *The Arnol'd principle is applicable to itself*). To see the applicability of Berry's principle, I note that in the wider world, Arnol'd's principle is called "Stigler's law of eponymy." Stigler[5] named this in 1980, pointing out it was really discovered by Merton. In 1972, Kennedy[6] named Boyer's law *Mathematical formulas and theorems are usually not named after their original discoverers* after Boyer's book.[7] Already in 1956, Newman[8] quoted the early twentieth-century philosopher and logician A. N. Whitehead as saying: "Everything of importance has been said before by somebody who

[3] U. Bottazzini and J. Gray, *Hidden Harmony—Geometric Fantasies. The Rise of Complex Function Theory*, Springer, New York, 2013.

[4] V. I. Arnol'd, *On teaching mathematics*, available online at `http://pauli.uni-muenster.de/~munsteg/arnold.html`.

[5] S. M. Stigler, *Stigler's law of eponymy*, Trans. New York Acad. Sci. **39** (1980), 147–158.

[6] H. C. Kennedy, *Classroom notes: Who discovered Boyer's law?*, Amer. Math. Monthly **79** (1972), 66–67.

[7] C. B. Boyer, *A History of Mathematics*, Wiley, New York, 1968.

[8] J. R. Newman, *The World of Mathematics*, Simon & Schuster, New York, 1956.

did not discover it." The main reason to give a name to a theorem is to have a convenient way to refer to that theorem. I usually try to follow common usage (even when I know Arnol'd's principle applies).

I have resisted the temptation of some text writers to rename things to set the record straight. For example, there is a small group who have attempted to replace "WKB approximation" by "Liouville–Green approximation", with valid historical justification (see the Notes to Section 15.5 of Part 2B). But if I gave a talk and said I was about to use the Liouville–Green approximation, I'd get blank stares from many who would instantly know what I meant by the WKB approximation. And, of course, those who try to change the name also know what WKB is! Names are mainly for shorthand, not history.

These books have a wide variety of problems, in line with a multiplicity of uses. The serious reader should at least skim them since there is often interesting supplementary material covered there.

Similarly, these books have a much larger bibliography than is standard, partly because of the historical references (many of which are available online and a pleasure to read) and partly because the Notes introduce lots of peripheral topics and places for further reading. But the reader shouldn't consider for a moment that these are intended to be comprehensive—that would be impossible in a subject as broad as that considered in these volumes.

These books differ from many modern texts by focusing a little more on special functions than is standard. In much of the nineteenth century, the theory of special functions was considered a central pillar of analysis. They are now out of favor—too much so—although one can see some signs of the pendulum swinging back. They are still mainly peripheral but appear often in Part 2 and a few times in Parts 1, 3, and 4.

These books are intended for a second course in analysis, but in most places, it is really previous exposure being helpful rather than required. Beyond the basic calculus, the one topic that the reader is expected to have seen is metric space theory and the construction of the reals as completion of the rationals (or by some other means, such as Dedekind cuts).

Initially, I picked "A Course in Analysis" as the title for this series as an homage to Goursat's *Cours d'Analyse*,[9] a classic text (also translated into English) of the early twentieth century (a literal translation would be

[9]E. Goursat, *A Course in Mathematical Analysis: Vol. 1: Derivatives and Differentials, Definite Integrals, Expansion in Series, Applications to Geometry. Vol. 2, Part 1: Functions of a Complex Variable. Vol. 2, Part 2: Differential Equations. Vol. 3, Part 1: Variation of Solutions. Partial Differential Equations of the Second Order. Vol. 3, Part 2: Integral Equations. Calculus of Variations*, Dover Publications, New York, 1959 and 1964; French original, 1905.

"of Analysis" but "in" sounds better). As I studied the history, I learned that this was a standard French title, especially associated with École Polytechnique. There are nineteenth-century versions by Cauchy and Jordan and twentieth-century versions by de la Vallée Poussin and Choquet. So this is a well-used title. The publisher suggested adding "Comprehensive", which seems appropriate.

It is a pleasure to thank many people who helped improve these texts. About 80% was TEXed by my superb secretary of almost 25 years, Cherie Galvez. Cherie was an extraordinary person—the secret weapon to my productivity. Not only was she technically strong and able to keep my tasks organized but also her people skills made coping with bureaucracy of all kinds easier. She managed to wind up a confidant and counselor for many of Caltech's mathematics students. Unfortunately, in May 2012, she was diagnosed with lung cancer, which she and chemotherapy valiantly fought. In July 2013, she passed away. I am dedicating these books to her memory.

During the second half of the preparation of this series of books, we also lost Arthur Wightman and Ed Nelson. Arthur was my advisor and was responsible for the topic of my first major paper—perturbation theory for the anharmonic oscillator. Ed had an enormous influence on me, both via the techniques I use and in how I approach being a mathematician. In particular, he taught me all about closed quadratic forms, motivating the methodology of my thesis. I am also dedicating these works to their memory.

After Cherie entered hospice, Sergei Gel'fand, the AMS publisher, helped me find Alice Peters to complete the TEXing of the manuscript. Her experience in mathematical publishing (she is the "A" of A K Peters Publishing) meant she did much more, for which I am grateful.

This set of books has about 150 figures which I think considerably add to their usefulness. About half were produced by Mamikon Mnatsakanian, a talented astrophysicist and wizard with Adobe Illustrator. The other half, mainly function plots, were produced by my former Ph.D. student and teacher extraordinaire Mihai Stoiciu (used with permission) using Mathematica. There are a few additional figures from Wikipedia (mainly under WikiCommons license) and a hyperbolic tiling of Douglas Dunham, used with permission. I appreciate the help I got with these figures.

Over the five-year period that I wrote this book and, in particular, during its beta-testing as a text in over a half-dozen institutions, I received feedback and corrections from many people. In particular, I should like to thank (with apologies to those who were inadvertently left off): Tom Alberts, Michael Barany, Jacob Christiansen, Percy Deift, Tal Einav, German Enciso, Alexander Eremenko, Rupert Frank, Fritz Gesztesy, Jeremy Gray,

Leonard Gross, Chris Heil, Mourad Ismail, Svetlana Jitomirskaya, Bill Johnson, Rowan Killip, John Klauder, Seung Yeop Lee, Milivoje Lukic, Andre Martinez-Finkelshtein, Chris Marx, Alex Poltoratski, Eric Rains, Lorenzo Sadun, Ed Saff, Misha Sodin, Dan Stroock, Benji Weiss, Valentin Zagrebnov, and Maxim Zinchenko.

Much of these books was written at the tables of the Hebrew University Mathematics Library. I'd like to thank Yoram Last for his invitation and Naavah Levin for the hospitality of the library and for her invaluable help.

This series has a Facebook page. I welcome feedback, questions, and comments. The page is at `www.facebook.com/simon.analysis`.

Even if these books have later editions, I will try to keep theorem and equation numbers constant in case readers use them in their papers.

Finally, analysis is a wonderful and beautiful subject. I hope the reader has as much fun using these books as I had writing them.

Preface to Part 2

Part 2 of this five-volume series is devoted to complex analysis. We've split Part 2 into two pieces (Part 2A and Part 2B), partly because of the total length of the current material, but also because of the fact that we've left out several topics and so Part 2B has some room for expansion. To indicate the view that these two volumes are two halves of one part, chapter numbers are cumulative. Chapters 1–11 are in Part 2A, and Part 2B starts with Chapter 12.

The flavor of Part 2 is quite different from Part 1—abstract spaces are less central (although hardly absent)—the content is more classical and more geometrical. The classical flavor is understandable. Most of the material in this part dates from 1820–1895, while Parts 1, 3, and 4 largely date from 1885–1940.

While real analysis has important figures, especially F. Riesz, it is hard to single out a small number of "fathers." On the other hand, it is clear that the founding fathers of complex analysis are Cauchy, Weierstrass, and Riemann. It is useful to associate each of these three with separate threads which weave together to the amazing tapestry of this volume. While useful, it is a bit of an exaggeration in that one can identify some of the other threads in the work of each of them. That said, they clearly did have distinct focuses, and it is useful to separate the three points of view.

To Cauchy, the central aspect is the differential and integral calculus of complex-valued functions of a complex variable. Here the fundamentals are the Cauchy integral theorem and Cauchy integral formula. These are the basics behind Chapters 2–5.

For Weierstrass, sums and products and especially power series are the central object. These appear first peeking through in the Cauchy chapters (especially Section 2.3) and dominate in Chapters 6, 9, 10, and parts of Chapter 11, Chapter 13, and Chapter 14.

For Riemann, it is the view as conformal maps and associated geometry. The central chapters for this are Chapters 7, 8, and 12, but also parts of Chapters 10 and 11.

In fact, these three strands recur all over and are interrelated, but it is useful to bear in mind the three points of view.

I've made the decision to restrict some results to C^1 or piecewise C^1 curves—for example, we only prove the Jordan curve theorem for that case.

We don't discuss, in this part, boundary values of analytic functions in the unit disk, especially the theory of the Hardy spaces, $H^p(\mathbb{D})$. This is a topic in Part 3. Potential theory has important links to complex analysis, but we've also put it in Part 3 because of the close connection to harmonic functions.

Unlike real analysis, where some basic courses might leave out point set topology or distribution theory, there has been for over 100 years an acknowledged common core of any complex analysis text: the Cauchy integral theorem and its consequences (Chapters 2 and 3), some discussion of harmonic functions on $\mathbb{R}^2$ and of the calculation of indefinite integrals (Chapter 5), some discussion of fractional linear transformations and of conformal maps (Chapters 7 and 8). It is also common to discuss at least Weierstrass product formulas (Chapter 9) and Montel's and/or Vitali's theorems (Chapter 6).

I also feel strongly that global analytic functions belong in a basic course. There are several topics that will be in one or another course, notably the Hadamard product formula (Chapter 9), elliptic functions (Chapter 10), analytic number theory (Chapter 13), and some combination of hypergeometric functions (Chapter 14) and asymptotics (Chapter 15). Nevanlinna theory (Chapter 17) and univalents functions (Chapter 16) are almost always in advanced courses. The break between Parts 2A and 2B is based mainly on what material is covered in Caltech's course, but the material is an integrated whole. I think it unfortunate that asymptotics doesn't seem to have made the cut in courses for pure mathematicians (although the material in Chapters 14 and 15 will be in complex variable courses for applied mathematicians).

Chapter 1

Preliminaries

> You know also that the beginning is the most important part of any work.
>
> —*Plato's Republic* [**436**]

This chapter sets notation and reviews various subjects that, it is hoped, the reader has seen before. For some, like calculus, the basic material is really a prerequisite. Others, like covering spaces, are only used in a few places, and enough background is given to perhaps tide over the student with less preparation. Essentially, Sections 1.1–1.4 are used often and the rest sporadically.

1.1. Notation and Terminology

> A foolish consistency is the hobgoblin of little minds ... Is it so bad, then, to be misunderstood? Pythagoras was misunderstood, and Socrates, and Jesus, and Luther, and Copernicus, and Galileo, and Newton, and every pure and wise spirit that ever took flesh. To be great is to be misunderstood.
>
> —*Ralph Waldo Emerson* [**164**]

For a real number a, we will use the terms positive and strictly positive for $a \geq 0$ and $a > 0$, respectively. It is not so much that we find nonnegative bad, but the phrase "monotone nondecreasing" for $x > y \Rightarrow f(x) \geq f(y)$ is downright confusing so we use "monotone increasing" and "strictly monotone increasing" and then, for consistency, "positive" and "strictly positive." Similarly for matrices, we use "positive definite" and "strictly positive definite" where others might use "positive semi-definite" and "positive definite."

Basic Rings and Fields.

$$\mathbb{R} = \textit{real numbers} \qquad \mathbb{Q} = \textit{rationals} \qquad \mathbb{Z} = \textit{integers}$$
$$\mathbb{C} = \textit{complex numbers} = \{x + iy \mid x, y \in \mathbb{R}\}$$

with their sums and products. For $z = x + iy \in \mathbb{C}$, we use $\operatorname{Re} z = x$, $\operatorname{Im} z = y$. $|z| = (x^2 + y^2)^{1/2}$.

Products. X^n = n-tuples of points in X with induced vector space and/or additive structure; in particular, $\mathbb{R}^n$, $\mathbb{Q}^n$, $\mathbb{Z}^n$, $\mathbb{C}^n$.

Subsets of $\mathbb{C}$ (and one superset).

- $\widehat{\mathbb{C}} = \mathbb{C} \cup \{\infty\}$ as a one-point compactification; in Section 7.2, we'll put a complex structure near ∞.
- $\mathbb{C}_+$ = upper half-plane = $\{z \mid \operatorname{Im} z > 0\}$;
  $\mathbb{H}_+$ = right half-plane = $\{z \mid \operatorname{Re} z > 0\}$
- $\mathbb{Z}_+ = \{n \in \mathbb{Z} \mid n > 0\} = \{1, 2, 3, \dots\}$;
  $\mathbb{N} = \{0\} \cup \mathbb{Z}_+ = \{0, 1, 2, \dots\}$
- $\mathbb{D} = \{z \in \mathbb{C} \mid |z| < 1\}$; $\partial\mathbb{D} = \{z \in \mathbb{C} \mid |z| = 1\}$
- $\mathbb{D}_\delta(z_0) = \{z \in \mathbb{C} \mid |z - z_0| < \delta\}$ for $z_0 \in \mathbb{C}$, $\delta > 0$
- $\mathbb{C}^\times = \mathbb{C} \setminus \{0\}$; $\mathbb{D}^\times = \mathbb{D} \setminus \{0\}$; $\mathbb{R}^\times = \mathbb{R} \setminus \{0\}$
  (in general, $Y^\times$ = invertible elements of Y)
- $\mathbb{A}_{r,R}$ = annulus = $\{z \mid r < |z| < R\}$ for $0 \le r < R \le \infty$
- $\mathbb{S}_{\alpha,\beta}$ = sector = $\{r(\cos\theta + i\sin\theta) \mid r > 0;\ \alpha < \theta < \beta\}$ for $\alpha < \beta$ and $|\beta - \alpha| \le 2\pi$; for example, $\mathbb{S}_{-\pi,\pi} = \mathbb{C} \setminus (-\infty, 0]$

Miscellaneous Terms.

- K a compact Hausdorff space, $C(K)$ = continuous complex-valued functions on K
- For $x \in \mathbb{R}$, $[x]$ = greatest integer less than x, that is, $[x] \in \mathbb{Z}$, $[x] \le x < [x] + 1$
- $\{x\} = x - [x]$ = fractional parts of x
- $\sharp(A)$ = number of elements in a set A
- $\operatorname{Ran} f$ = range of a function f
- $\log(z)$ = natural logarithm, that is, logarithm to base e; if z is complex, put the cut along $(-\infty, 0]$, i.e., $\log(z) = \log(|z|) + i\arg(z)$ with $-\pi < \arg(z) \le \pi$
- Region = open, connected subset of $\mathbb{C}$

- For sets A, B subsets of X, $A \cup B$ = union, $A \cap B$ = intersection, A^c = complement of A in X, $A \setminus B = A \cap B^c$, $A \triangle B = (A \setminus B) \cup (B \setminus A)$
- For matrices, M_{12} means row one, column two
- $f \restriction K$ = *restriction* of a function to K, a subset of the domain of f

Integer Notation (see Section 1.3 for modular arithmetic). This is mainly relevant in Chapter 13 of Part 2B. All letters in this subsection are in $\mathbb{Z}$

- $d \mid n$ means d is a divisor of n, that is, $n = dk$ for $k \in \mathbb{Z}$
- $a \equiv b \pmod{m} \Leftrightarrow m \mid a - b$
- $\mathbb{Z}_k \equiv$ conjugacy classes mod k
- $\mathbb{Z}_k^\times$ = relatively prime conjugacy classes

We use "big oh, litttle oh" notation, e.g., $f(x) = O(|x|)$ or $f(x) = o(1)$. If the reader is not familiar with it, see Section 1.4.

1.2. Complex Numbers

> Because all possible numbers which one can think of are either greater or less than 0, or else 0 itself, it is clear that the square roots of negative numbers cannot be counted among the possible numbers. Consequently we must say that these are impossible numbers.
>
> —*L. Euler*, as translated by U. Bottazzini [**68**]

We've already said something about complex numbers and their rectangular coordinates. Here we want to say something about the polar representation. We suppose the reader knows the geometric basics of $\sin x$ and $\cos x$ as well as the fact that they both solve $f''(x) = -f(x)$ and, indeed, that any solution has the form $f(x) = \alpha \cos x + \beta \sin x$ where $\alpha = f(0)$, $\beta = f'(0)$. In particular, $\sin\theta$ and $\cos\theta$ have the Taylor expansions (see Section 2.3 for more)

$$\sin\theta = \sum_{n=0}^{\infty} \frac{(-1)^n \theta^{2n+1}}{(2n+1)!} \qquad \cos\theta = \sum_{n=0}^{\infty} \frac{(-1)^n \theta^{2n}}{(2n)!} \tag{1.2.1}$$

$$e^{i\theta} \equiv \cos\theta + i \sin\theta \tag{1.2.2}$$

so $z = re^{i\theta}$. Later, in Section 2.3, we define e^z by a power series, and (1.2.2) will become a derived equality rather than a definition. The geometric picture of complex numbers in a plane (see Figure 1.2.1) will play an important role in many places in this volume. Arithmetic operations have geometric meaning. Addition is via parallelogram. Multiplication by $z = re^{i\theta}$ is uniform scaling by a factor r, followed by rotation by angle θ about 0.

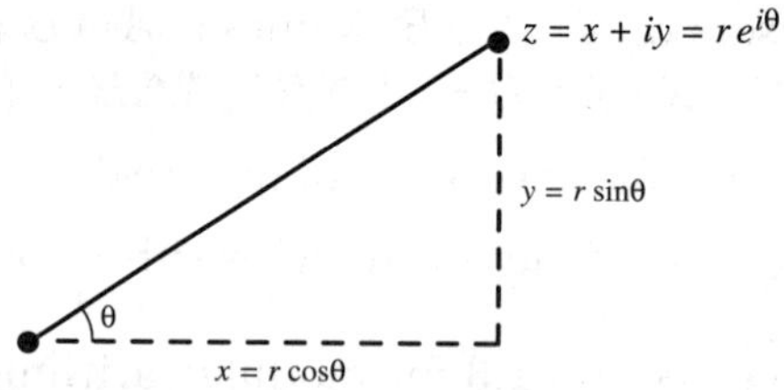

Figure 1.2.1. Polar and rectangular coordinates.

Notes and Historical Remarks. As the terms "imaginary" and "complex" attest, acceptance of complex numbers was a difficult process. One might have guessed, given the formula for solutions of the quadratic equation, that square roots of negative numbers would have gone back to antiquity. But scholars were content to think that some equations, like $x^2+1=0$, have no solutions. Rather, complex numbers came through cubic equations, where $x^3 - ax - b = 0$ was first studied. In 1515, Scipione del Ferro (1465–1526) found but did not publish the solution

$$x = \sqrt[3]{\frac{b}{2} + \sqrt{\frac{b^2}{4} - \frac{a^3}{27}}} + \sqrt[3]{\frac{b}{2} - \sqrt{\frac{b^2}{4} - \frac{a^3}{27}}} \tag{1.2.3}$$

Even if $\frac{b^2}{4} - \frac{a^3}{27} < 0$, the equation has real solutions for a, b real. The formula was finally published by Girolamo Cardano (1501–76) in 1545. Interestingly enough, while Cardano's book has examples where $\frac{b^2}{4} - \frac{a^3}{27} < 0$, he never applied the formula in such cases.

In 1572, Rafael Bombelli (1526–72) published a book [**62**] which spelled out rules of arithmetic for complex numbers and used them in Cardano's formula for finding real solutions of cubics. Key later work is by Wallis and Euler [**178**] (see the Notes to Section 9.2 for a capsule biography). In particular, Euler clarified complex roots of unity and found the multiple roots represented by (1.2.3). It took almost another century before mathematicians as a community fully accepted complex numbers. Thus much of Cauchy's work was done in a milieu where many, even sometimes Cauchy himself, were not totally comfortable with complex numbers.

The geometric view of complex numbers as a plane was introduced by Jean-Robert Argand (1768–1822) and Caspar Wessel (1745–1818) and championed by Gauss. Neither Argand nor Wessel were professional mathematicians. Wessel was a Danish surveyor and presented the geometric interpretation to the Royal Danish Academy in 1797. It was published in 1799 and promptly forgotten until rediscovered and publicized by two Danish mathematicians in 1895, Sophus Christian Juel and the famous Sophus Lie. So Wessel's work had no impact.

Argand was an accountant and bookkeeper, and in 1806 self-published his work in a book which appeared without his name! In 1813, a French mathematician, Jacques Français, published a follow-up paper asking the author of the book to identify himself. So Argand got some recognition, enough that the geometric picture like that in Figure 1.2.1 is often called an Argand diagram.

An interesting realization of $\mathbb{C}$ given $\mathbb{R}$ is to look at 2×2 real matrices

$$M_{a+ib} = \begin{pmatrix} a & b \\ -b & a \end{pmatrix} \tag{1.2.4}$$

with matrix addition and matrix multiplication realizing the operations on $\mathbb{C}$.

1.3. Some Algebra, Mainly Linear

We'll need some basic facts about finite matrices and, in Chapter 13 of Part 2B, some simple facts about modular arithmetic. We assume the reader knows about finite-dimensional vector spaces, linear maps between them, bases, inner products (see Section 3.1 of Part 1), and matrices. A *Jordan block* is a finite matrix of the form

$$J = \begin{pmatrix} \lambda & 1 & & & 0 \\ 0 & \lambda & 1 & & \\ 0 & 0 & \ddots & \ddots & \\ & & & \ddots & 1 \\ 0 & & & & \lambda \end{pmatrix} \tag{1.3.1}$$

that is, a constant complex number along the diagonal, 1's directly above and 0 elsewhere. Notice that $J(1\,0\ldots0)^{\mathrm{t}} = \lambda(1\,0\ldots0)^{\mathrm{t}}$, so λ is an *eigenvalue* of J. A fundamental result says:

Theorem 1.3.1 (Jordan Normal Form). *Let A be a linear map of a finite-dimensional complex vector space to itself. Then there is a basis in which the matrix form of A is*

$$\begin{pmatrix} J_1 & & \\ & \ddots & \\ & & J_\ell \end{pmatrix}$$

where each J_j is a Jordan block.

For a complex variable proof of this, see Problems 2, 3, and 4 in Section 3.9 and Section 2.3 of Part 4.

Generically, the Jordan blocks are 1×1, that is, the matrix is diagonal. A matrix or operator on a finite-dimensional space for which there is a basis in which its matrix is diagonal is called *diagonalizable*. If some block is not

1×1, we say A has a *Jordan anomaly.* One can show that the blocks are uniquely associated to A up to order. In particular, the λ's that arise are precisely all the *eigenvalues* of J.

There is a natural complex-valued function, *determinant*, denoted by det, from all linear maps of a finite-dimensional space to itself. In a matrix representation,

$$\det(A) = \sum_{\pi \in \Sigma_n} (-1)^\pi a_{1\pi(1)} \dots a_{n\pi(n)} \tag{1.3.2}$$

where Σ_n is the group of permutations and $(-1)^\pi$ is their sign. One has

$$\det(AB) = \det(A)\det(B) \tag{1.3.3}$$

and A is invertible if and only if $\det(A) \neq 0$. Writing out A in Jordan normal form shows that if $\{\lambda_j\}_{j=1}^n$ are the diagonal elements in the Jordan blocks (counting multiplicity), then

$$\det(\lambda \mathbb{1} - A) = \prod_{j=1}^n (\lambda - \lambda_j) \tag{1.3.4}$$

The multiplicity of the distinct λ's in this product is called their *algebraic multiplicity.* The number of distinct Jordan blocks associated to λ_i is $\dim\{v \mid (A - \lambda_0)v = 0\}$, called the *geometric multiplicity* of the eigenvalues.

We suppose the reader is familiar with the notion of inner product. Our convention, discussed in Section 3.1 of Part 1, is that $\langle x, y \rangle$ is linear in y and conjugate linear in x.

When one has an inner product, $\langle\ ,\ \rangle$, on a finite-dimensional complex vector space, V, there is an additional important structure on the operators from that space to itself. One defines the *adjoint*, A^*, of $A \colon V \to V$ by

$$\langle A^* x, y \rangle = \langle x, Ay \rangle \tag{1.3.5}$$

There is a unique linear operator obeying (1.3.5); for example, in a matrix representation, (1.3.5) is equivalent to

$$(a^*)_{ij} = \bar{a}_{ji} \tag{1.3.6}$$

One defines A to be

normal $\Leftrightarrow A^*A = AA^*$

self-adjoint $\Leftrightarrow A = A^*$

positive $\Leftrightarrow \langle x, Ax \rangle \geq 0$ for all x

unitary $\Leftrightarrow A^*A = AA^* = \mathbb{1} \Leftrightarrow \langle Ax, Ay \rangle = \langle x, y \rangle$ for all x, y

Positive operators on a complex space are self-adjoint. Self-adjoint and unitary operators are normal.

Every operator has a *polar decomposition*:

Theorem 1.3.2 (Polar Decomposition). *Any operator, A, on a finite-dimensional complex, inner product space can be uniquely written*

$$A = U|A| \tag{1.3.7}$$

where $|A|$ is positive, $U\varphi = 0$ if $\varphi \in (\text{Ran}(A))^\perp$ and $\|U\varphi\| = \|\varphi\|$ for $\varphi \in \text{Ran}(A)$.

One finds $|A|$ as $\sqrt{A^*A}$, where square root is defined by a power series for $\sqrt{1-x}$. U is then determined by its conditions. This is discussed in Section 2.4 of Part 4 even in the infinite-dimensional case.

Normal operators do not have Jordan anomalies; more is true—their eigenvectors can be chosen to be orthogonal. Recall that an orthonormal basis of V, a finite-dimensional inner product space of dimension λ, is a set $\{\varphi_j\}_{j=1}^d$ with $\langle\varphi_j, \varphi_k\rangle = \delta_{jk}$. Any φ can then be written

$$\varphi = \sum_{j=1}^d \langle\varphi_j, \varphi\rangle\varphi_j \tag{1.3.8}$$

Theorem 1.3.3 (Spectral Theorem for Finite-Dimensional Spaces). *For any normal operator, A, on a finite-dimensional vector space, there is an orthonormal basis of eigenvectors, $A\varphi_j = \lambda_j\varphi_j$. Equivalently, if a fixed basis is picked, there is a unitary U so that UAU^{-1} is a diagonal matrix in this fixed basis.*

Given that Theorem 1.3.1 implies that any operator has an eigenvector, the proof isn't difficult. The following, which we'll need again below, is crucial:

Proposition 1.3.4. *If A, B are two linear operators with $AB = BA$, then for any $\lambda \in \mathbb{C}$, B maps $\{v \mid Av = \lambda v\}$ to itself.*

This is immediate from $Av = \lambda v \Rightarrow ABv = BAv = \lambda Bv$. Given the proposition, if A is normal, A and A^* have common eigenvectors, that is, φ so $A\varphi = \lambda\varphi$ and $A^*\varphi = \mu\varphi$ (it is easy to see that $\mu = \bar{\lambda}$). This implies A and A^* leave $\varphi^\perp = \{\psi \mid \langle\psi, \varphi\rangle = 0\}$ invariant, so Theorem 1.3.3 follows inductively.

Proposition 1.3.4 also implies, by an inductive argument,

Theorem 1.3.5. *If $A_1, \dots, A_n$ are normal operators on a complex vector space, and $A_jA_k = A_kA_j$, $A_jA_k^* = A_k^*A_j$, then there is an orthonormal basis $\{\varphi_\alpha\}_{\alpha=1}^n$ so $A_j\varphi_\alpha = \mu_{j,\alpha}\varphi_\alpha$.*

If

$$B = \begin{pmatrix} \beta_1 & & \\ & \ddots & \\ & & \beta_n \end{pmatrix}, \quad \beta_j \neq 0, \qquad S = \begin{pmatrix} |\beta_1|^{-1/2} & & \\ & \ddots & \\ & & |\beta_n|^{-1/2} \end{pmatrix}$$

then S^*BS has every diagonal matrix element ± 1. Combing this with Theorem 1.3.3 yields

Theorem 1.3.6. *If A is an invertible symmetric matrix, there exists T so T^*AT is diagonal with every diagonal entry ± 1.*

* * * * *

We suppose that the reader is familiar with the language of groups: homomorphism, subgroups, cosets, quotient groups. We will also very occasionally refer to rings and ideals and the fact that quotient rings have an inherited product.

In particular, in Chapter 13 of Part 2B, we'll look at $n\mathbb{Z} \subset \mathbb{Z}$ where $n\mathbb{Z} = \{kn \mid k \in \mathbb{Z}\}$. This is an ideal. The quotient $\mathbb{Z}/n\mathbb{Z} \equiv \mathbb{Z}_n$ is a ring of n elements. Its elements viewed as subsets of $\mathbb{Z}$ are called *conjugacy classes.* $a \equiv b$, called a is *conjugate* to b if and only if $n \mid (a - b)$.

$p \in \mathbb{Z}_+$ is *prime* if $p > 1$ and its only divisors are 1 and p. Every $a \in \mathbb{Z}_+$ can be uniquely written as $a = p_1^{k_1} \dots p_\ell^{k_\ell}$, where the p_j are prime. a, b are called *relatively prime* if their only common divisor is 1. The *Euclidean algorithm* asserts that this is true if and only if there are integers m and n so that $ma + nb = 1$. In turn, this is true if and only if in $\mathbb{Z}_b$, the conjugacy class of a has an inverse (namely, the conjugacy class of m). $\mathbb{Z}_b^\times$ is the set of invertible elements of $\mathbb{Z}_b$; equivalently, those classes relatively prime to b.

Notes and Historical Remarks. For a textbook presentation of linear algebra, see Lang [**345**], and for the basics of groups and rings, see Dummit–Foote [**151**].

1.4. Calculus on $\mathbb{R}$ and $\mathbb{R}^n$

The key to the Cauchy approach to complex variables is the differential and integral calculus of functions of a region $\Omega \subset \mathbb{C}$ to $\mathbb{C}$. Such functions can be viewed as maps of $\mathbb{R}^2$ to $\mathbb{R}^2$ by forgetting the complex structure, so the calculus of such functions is important background. To set notation, we review the main elements in this section.

We begin with "big oh, little oh" notation. Let $f, g \colon U \to Y$ be functions defined on an open set $U \subset X$, some normed linear space to Y, another normed linear space. We say $f = O(g)$ at $x_0 \in U$ if there is $\delta > 0$ and C so that

$$\|f(x)\|_Y \le C\|g(x)\|_Y \tag{1.4.1}$$

if $\|x - x_0\|_X < \delta$. We say $f = o(g)$ at $x_0 \in U$ if and only if, for all $\varepsilon > 0$, there is $\delta > 0$ so that

$$\|x - x_0\|_X < \delta \Rightarrow \|f(x)\|_Y \le \varepsilon\|g(x)\|_Y \tag{1.4.2}$$

We say f is *differentiable* at x_0 if and only if there is a bounded linear map $T\colon X \to Y$, so at x_0,

$$f(x) - f(x_0) - T(x - x_0) = o(\|x - x_0\|) \tag{1.4.3}$$

T is called the *derivative* of f at x_0, written $D_{x_0}f$. Clearly, since T is a bounded linear map, (1.4.3) implies

$$f(x) - f(x_0) = O(\|x - x_0\|) \tag{1.4.4}$$

so differentiability at $x_0 \Rightarrow$ continuity at x_0. A function is called C^1, aka *continuously differentiable*, on U if and only if it is differentiable at all $x_0 \in U$ and $x \to D_x f$ is continuous. $\mathbb{C}^k$, $k = 2, 3, \dots$, is defined inductively by requiring f to be C^{k-1}, and $D_x^{k-1}f = D_x(D_x \dots D_x f)$ ($k-1$ times) is C^1 as a function in $\hom(X, \hom(X, \dots, \hom(X, Y)))$. $D_x^k f$ can be viewed as a multilinear functional of k variables in x.

Riemann integrals of continuous X-valued functions on $[a, b]$ can be defined so long as X is a Banach space, that is, complete; see Chapter 5 of Part 1. The *fundamental theorem of calculus* asserts that if f is C^1 on (a, b) to X, a Banach space, then for any $a < c < d < b$,

$$f(d) - f(c) = \int_c^d f'(x)\,dx \tag{1.4.5}$$

where $f'(x) = (D_x f)(1) \equiv \frac{df}{dx}(x)$. Conversely, if g is continuous on (a, b) and $c \in (a, b)$,

$$f(x) = \int_c^x g(u)\,du \tag{1.4.6}$$

is C^1 and $\frac{df}{dx}(x) = g(x)$.

A simple induction starting with (1.4.5) and using (1.4.5) at each step allows one to show

Theorem 1.4.1 (Taylor's Theorem with Remainder). *If f is C^k on (a, b) with values in a Banach space, Y, then for every $x, x_0 \in (a, b)$,*

$$f(x) = \sum_{j=1}^{k-1} \frac{f^{(j)}(x_0)}{j!}(x - x_0)^j + \frac{(x - x_0)^k}{(k-1)!}\int_0^1 t^{k-1} f^{(k)}(tx + (1-t)x_0)\,dt \tag{1.4.7}$$

One of the deepest results in differential calculus that will illuminate some of the theorems in this book is:

Theorem 1.4.2 (Implicit Function Theorem). *Let X and Y be Banach spaces, $U \subset X \times Y$ open, and $F\colon U \to X$, a C^1 function. Suppose for $(x_0, y_0) \in U$, $F(x_0, y_0) = 0$. Let $D_{(x_0,y_0)}F(x,y) = D^{(1)}_{(x_0,y_0)}F(x) + D^{(2)}_{(x_0,y_0)}F(y)$ and suppose that $D^{(1)}_{(x_0,y_0)}F$ is invertible. Then there is V, an open neighborhood of y_0, W, an open neighborhood of x_0, and g, a C^1 map of V to W, so that $W \times V \subset U$, and for $(x,y) \in W \times V$, $F(x,y) = 0 \Leftrightarrow y = g(x)$.*

As a corollary (take $F(x,y) = f(x) - y$),

Corollary 1.4.3 (Inverse Function Theorem). *Let X be a Banach space, U open in X, $f\colon U \to X$, a C^1 function. Let $x_0 \in U$, $y_0 = f(x_0)$, and suppose $D_{x_0}f$ is invertible. Then there exist neighborhoods V of y_0 and W of x_0 so that f is a bijection of W onto V, and so the inverse, g, on V is C^1.*

A second theme that will occur, especially in connection with the Weierstrass approach to complex analysis, involves infinite series. One subtle issue involves rearrangements. Given a sequence $\{a_n\}_{n=1}^\infty$ of real numbers, a *rearrangement* is a sequence of the form $b_n = a_{\pi(n)}$, where π is a bijection of $\mathbb{Z}_+$ to itself. Here the main result is:

Theorem 1.4.4. *Let $\{a_n\}_{n=1}^\infty$ be a sequence of real numbers. If $\sum_{n=1}^\infty |a_n| < \infty$, then for any rearrangement, $\lim_{N\to\infty} \sum_{n=1}^N b_n \equiv \sum_{n=1}^\infty b_n$ exists in $\mathbb{R}$ and*

$$\sum_{n=1}^\infty a_n = \sum_{n=1}^\infty b_n \tag{1.4.8}$$

If $\sum_{n=1}^\infty a_n$ exists in $\mathbb{R}$ but $\sum_{n=1}^\infty |a_n| = \infty$, then for any $\alpha \in [-\infty, \infty]$, there is a rearrangement with

$$\lim_{N\to\infty} \sum_{n=1}^N b_n = \alpha \tag{1.4.9}$$

Remarks. 1. By looking at $\operatorname{Re} a_n$ and $\operatorname{Im} a_n$, we see the rearrangement invariant result holds for complex a_n.

2. The idea behind the proof that absolute convergence implies rearrangement invariance is the following: Given ε, find N so $\sum_{N+1}^\infty |a_j| < \varepsilon$. Given b and π, find M so $\pi(j) < M$ for $j = 1, \dots, N$. Then for $k > N, M$, $|\sum_1^k (a_j - b_j)| \le 2\varepsilon$ since all the "large" a's are cancelled.

Closely related to rearrangement is Fubini's theorem for sums:

Theorem 1.4.5. *If $\sum_{i=1}^\infty \sum_{j=1}^\infty |a_{ij}| < \infty$, then*

$$\sum_{i=1}^\infty \biggl(\sum_{j=1}^\infty a_{ij} \biggr) = \sum_{j=1}^\infty \biggl(\sum_{i=1}^\infty a_{ij} \biggr) \tag{1.4.10}$$

The following "real analysis" result is occasionally useful in complex analysis:

Theorem 1.4.6 (C^∞ Partitions of Unity). *Let $K \subset \mathbb{R}^\nu$ be compact and $\{U_\alpha\}_{\alpha\in I}$ an open cover of K. Then there exist C^∞ functions, $\{j_k\}_{k=1}^n$, of compact support so that each j_k has support in some U_{α_k} and so that $\sum_{k=1}^n j_k(x) = 1$ for all $x \in K$.*

The idea of the proof is to first use products of functions of the form

$$g(x) = \begin{cases} \exp(-(x-x_0)^{-1}), & x \geq x_0 \\ 0, & x \leq x_0 \end{cases} \tag{1.4.11}$$

to get, for any $V \subset \mathbb{R}^\nu$ and $x \in V$, a function $q_{x,V}$, C^∞ of compact support so that $\sup q_{x,V} \subset V$, $0 \leq q_{x,V} \leq 1$, and $q_{x,V}(y) = 1$ for y in some neighborhood, $N_{x,V}$, of x.

Then, for each $\alpha \in I$ and $x \in U_\alpha$, pick such a q_{x,U_α} and N_{x,U_α}. Finitely many N_{x,U_α} cover K, so let $q_1, \dots, q_n; N_1, \dots, N_n$ be the corresponding functions and sets.

Finally, define $j_1, \dots, j_n$ by

$$j_1 = q_1, \dots, j_{k+1} = q_{k+1}(1-q_1)\dots(1-q_k), \dots \tag{1.4.12}$$

Then $\sum_{k=1}^n j_k = 1 - \prod_{k=1}^n (1-q_k)$ is 1 on K.

Finally, we note results on ODEs. We look at $\mathbb{R}^n$-valued functions. There is no loss in restricting to first order at least for equations of the form

$$u^{(n)}(t) = F(t; u(t), \dots, u^{(n-1)}(t)) \tag{1.4.13}$$

since this is equivalent to

$$v'(t) = A(t; v(t)) \tag{1.4.14}$$

if $v(t) = (u(t), \dots, u^{(n-1)}, t)$ and A has components

$$A_j(t) = -v_{j+1}(t);\ j = 0, \dots, n-2; \quad A_{n-1}(t) = F(t; v_0(t), \dots, v_{n-1}(t)) \tag{1.4.15}$$

The basic existence and uniqueness result (see Theorem 5.12.5 of Part 1) is local.

Theorem 1.4.7. *If $A(t; v)$ is continuous in t for v fixed and obeys*

$$|A(t; v) - A(t; \tilde{v})| \leq C\|v - \tilde{v}\| \tag{1.4.16}$$

for a fixed C and all $t \in (t_0 - \delta, t_0 + \delta)$, then there is $\delta' < \delta$ so that for any v_0, (1.4.14) has a unique solution in $(t_0 - \delta', t_0 + \delta')$ with $v(t_0) = v_0$.

If the equation is linear, we have global solutions (see Theorem 5.12.8 of Part 1).

Theorem 1.4.8. *If $A(t;v) = A(t)v$ where $A(t)$ is a linear map of $\mathbb{R}^n$ to $\mathbb{R}^n$ and continuous in t on (a,b), then for any $c \in (a,b)$ and $v_0 \in \mathbb{R}^n$, there is a unique solution of (1.4.14) on (a,b) with $v(c) = v_0$. Moreover,*

$$\|v(t)\| \leq \exp\left(\left|\int_c^t \|A(s)\|\, ds\right|\right)\|v_0\| \tag{1.4.17}$$

in the Euclidean norm, $\|v\| = (\sum_{j-1}^n |v_j|^2)^{1/2}$.

Notes and Historical Remarks. Big oh/little oh notation was introduced by Bachmann [**29**] in 1894 and popularized especially by Landau [**338**] and Hardy [**238**]. They are sometimes called Landau symbols. Hardy used $f \lesssim g$ for $O(\,\cdot\,)$ and $f \ll g$ for $o(\,\cdot\,)$, but by his book on divergent series [**239**], he had shifted to O and o.

For a textbook presentation of calculus on Banach spaces, including the implicit function theorem, see Choquet-Bruhat et al. [**115**], Lang [**346**], Loomis–Sternberg [**367**], and Section 5.12 of Part 1. For a discussion of theorems on series, see Knopp [**318**]. Many books discuss the details of the construction of a partition of unity, for example, DiBenedetto [**141**].

Theorem 1.4.7 is obtained by rewriting (1.4.14) as an integral equation and using the contraction mapping theorem (see Theorem 5.12.4 of Part 1) on a suitable metric space (see Theorem 5.12.5 of Part 1). Uniqueness can fail if A is only assumed continuous, for example, the scalar equation $v'(t) = 2\sqrt{v(t)}$ has the solutions $v(t) = 0$ and also $v(t) = (t-t_0)^2$ ($t \geq t_0$; 0 if $t \leq t_0$), all with $v(0) = 0$. This result is called the method of successive approximation (since the contraction mapping theorem can be proven by iteration) or Picard iteration, after its inventor, or sometimes the Picard–Lindelöf theorem, after [**434, 435, 355**].

(1.4.17) comes from $\frac{d}{dt}\|v(t)\| \leq \|A(t)\|\|v(t)\|$, so $\frac{d}{dt}\log\|v(t)\| \leq \|A(t)\|$, which can be integrated. With (1.4.17) as an a priori bound, one can see that the local can be continued indefinitely; hence, there are global solutions.

Details of the proofs of these basic ODE theorems can be found, for example, in Hille [**263**] or Ince [**275**].

1.5. Differentiable Manifolds

The language of differentiable manifolds is ideal to describe surfaces and hypersurfaces. We'll discuss in this book Riemann surfaces which have more structure, but both the structures and analogies to manifolds are useful. Here is a lightning summary.

Definition. (Preliminary) A C^∞ *manifold*, aka *differentiable manifold*, (manifold for short), is a metric space, M, with a collection, called an *atlas*, $\{(U_\alpha, f_\alpha)\}_{\alpha\in I}$ of pairs $U_\alpha \subset M$, an open set, and $f_\alpha\colon U_\alpha \to \mathbb{R}^\nu$, a homeomorphism of U_α, and an open set, $\mathrm{Ran}(f_\alpha)$, so that

(i) $\bigcup_{\alpha\in I} U_\alpha = M$

(ii) If $U_\alpha \cap U_\beta \neq \emptyset$, then $f_\beta \circ f_\alpha^{-1}\colon \mathrm{Ran}(f_\alpha) \to \mathrm{Ran}(f_\beta)$ is a C^∞ map with C^∞ inverse.

ν is called the *dimension* of M.

Two atlases, $\{(U_\alpha, f_\alpha)\}_{\alpha\in I}$ and $\{(V_\beta, g_\beta)\}_{\beta\in J}$, on M are called *compatible* if $U_\alpha \cap V_\beta \not\equiv \emptyset$ implies $f_\alpha \circ g_\beta^{-1}\colon g_\beta[U_\alpha \cap V_\beta] \to f_\alpha[U_\alpha \cap V_\beta]$ is a C^∞ map with C^∞ inverse.

Definition. A *manifold* is a metric space, M, with an equivalence class of atlases.

Definition. On a manifold, (U, f), an element of some atlas is called a *local coordinate* or *coordinate patch.* If $f(x) = (x^1, \dots, x^\nu)$, $\{x^j\}_{j=1}^\nu$ are called the coordinates.

Definition. If M, N are two manifolds, a map $f\colon M \to N$, is called C^∞ if for any local coordinates (U_α, f_α) on M and (V_β, g_β) on N, we have $g_\beta \circ f \circ f_\alpha^{-1}$ is C^∞. $C^\infty(M)$ is the family of C^∞ maps of M to $\mathbb{R}$.

Theorem 1.4.6 extends even to a noncompact setting.

Theorem 1.5.1. *Let $\{U_\alpha\}_{\alpha\in I}$ be an open cover of a manifold, M. Then there is a refinement, $\{V_\beta\}_{\beta\in J}$, which is locally finite, that is, for any x, there is a neighborhood, N, so that $\{\beta \mid N \cap V_\beta \neq \emptyset\}$ is finite and C^∞ functions, $j_\beta\colon M \to \mathbb{R}$ with $0 \le j_\beta \le 1$ so that $\mathrm{supp}(j_\beta) \subset V_\beta$, and for all $x \in M$,*

$$\sum_\beta j_\beta(x) = 1 \tag{1.5.1}$$

Remarks. 1. Refinement means every V_β is a subset of some U_α and $\cup V_\beta = M$.

2. By the local finiteness, the sum in (1.5.1) only has finitely many terms for each $x \in M$.

Here are the basic facts about differential and integral calculus on manifolds:

(1) A *point derivation* at $x_0 \in M$ is a linear map, $\ell\colon C^\infty(M) \to \mathbb{R}$, so that

$$\ell(fg) = f(x_0)\ell(g) + \ell(f)g(x_0) \tag{1.5.2}$$

The set of point derivations at x_0 is denoted $T_{x_0}(M)$.

(2) If $\gamma\colon [0,1] \to M$ is a C^∞ curve,

$$\ell(f) = \frac{d}{ds} f(\gamma(s))\Big|_{s_0} \tag{1.5.3}$$

is a point derivation at $\gamma(s_0)$, called the *tangent* to γ at $\gamma(s_0)$.

(3) If $x_0 \in M$ and (U, f) a coordinate patch, then every point derivation at x_0 has the form

$$\ell(f) = \sum_{i=1}^{\nu} a^i \frac{\partial f}{\partial x^i}\Big|_{x=x_0} \tag{1.5.4}$$

so that $\dim(T_{x_0}(M)) = \nu$.

(4) $T(M) \equiv \bigcup_{x\in M} T_x(M)$ is called the *tangent bundle.* For each coordinate patch, (U, f) for M, define $\widetilde{f}\colon \bigcup_{x\in U} T_x(M) \equiv T_U(M) \to \mathbb{R}^{2\nu}$ by $\widetilde{f}(x_0, \ell) = (f(x_0), \{a^j\}_{j=1}^{\nu})$, where the a^j's are given by (1.5.1). $T(M)$ can be given a metric topology so that these $(T_U(M), \widetilde{f})$ turn $T(M)$ into a 2ν-dimensional manifold. $\pi\colon (x,\ell) \to x$ from $T(M) \to M$ is called the *natural projection.*

(5) A *vector field* is a map $X\colon M \to T(M)$ so that $\pi \circ X$ is the identity. X is a general first-order differential operator on M. We say X is a *cross-section* of $T(M)$. $\Gamma(T(M))$ will denote the set of all vector fields.

(6) A *flow* is a C^∞ map, Φ, to M from a subset, $S \subset M \times \mathbb{R}$, with the property that $\{(x,s) \mid s \in I_x = (\alpha_x, \beta_x)\}$ is an interval, I_x, of $\mathbb{R}$ containing 0 so that

$$\Phi(x, 0) = x \tag{1.5.5}$$

and so that if $s \in I_x$ and $t \in I_{\Phi(x,s)}$, then $s + t$ is an I_x and (*the flow equation*)

$$\Phi(x, s+t) = \Phi(\Phi(x,s), t) \tag{1.5.6}$$

(7) If Φ is a flow, then there is a unique vector field, X, so that for any $f \in C^\infty$, we have

$$\frac{d}{dt} f(\Phi(x,t))\Big|_{t=0} = X(x)(f) \tag{1.5.7}$$

(8) Conversely (a variant of Theorem 1.4.8), given any vector field, X, there is a flow obeying (1.5.7). There is a unique such maximal flow, where maximal means S is maximal. For this maximal flow, for any x, either β_x is infinite or $\Phi(x,s)$ has no limit point in M as $s \uparrow \beta_x$ (and similarly for α_x).

(9) Elements of the dual space, $T_x(M)^*$, are *covectors* at x. If dx^i is the dual basis to $\frac{\partial}{\partial x^i}$ (i.e., $dx^j(\frac{\partial}{\partial x^i}) = \delta_{ij}$), then any $\omega \in T_x(M)^*$ has the form $\omega = \sum b_j dx^j$.

(10) As with the tangent bundle, $\bigcup_x T_x^*(M) \equiv T^*(M)$, the *cotangent bundle* is a 2ν-dimensional manifold with local coordinates $\langle x, b\rangle$.

(11) If $f \in C^\infty(M)$, then df is defined by $df(\ell) = \ell(f)$. The map $f \mapsto df$ is a map of $C^\infty(M)$ to $\Gamma(T^*(M))$, the *one-forms* which are cross-sections of $T^*(M)$. We also write $\wedge(M)$ for $\Gamma(T^*(M))$.

(12) (Exterior algebra is discussed in Section 3.8 of Part 1.) One defines $\wedge^k(T^*_x(M))$, $k = 1, 2, \dots, \nu$, to be the exterior spaces over $T^*_x(M)$. $\bigcup_x \wedge^k(T^*_x(M))$ is again a manifold and its cross-sections, denoted $\wedge^k(M)$, are called k-*forms*. The map d extends to $\wedge^k(M)$ into $\wedge^{k+1}(M)$ by requiring $d(f\, dx^1 \wedge \cdots \wedge dx^k) = df \wedge dx^1 \cdots \wedge dx^k$. One has $d^2 = 0$ because of the antisymmetry of $dx^1 \wedge \cdots \wedge dx^k$ and symmetry of $\partial^2 f/\partial x^i \partial x^j$.

(13) A manifold, M, is called *orientable* if there is a nowhere vanishing m-form. Picking one gives one a notion of positivity, called an *orientation.*

(14) ν-forms transform under coordinate transformations as determinants, that is, Jacobians of the coordinate change. So long as one only considers orientation-preserving coordinate transformations, we can define integrals of ν-forms by using a partition of unity supported on coordinate patches and defining $\int_{U_\alpha} j_\alpha \omega = \int_{U_\alpha} g j_\alpha \, dx_1 \dots dx_n$ if $\omega = g\, dx_1 \wedge \cdots \wedge dx_\nu$ in local coordinates on U_α. The Jacobian transformation means this definition is coordinate-independent.

(15) A C^∞ map $f\colon M \to N$ between manifolds induces maps on the bundles (or at least their fibers), which we've discussed. $f^*\colon C^\infty(N) \to C^\infty(M)$ by

$$f^*(g) = g \circ f \tag{1.5.8}$$

This then induces $f_*\colon T_m(M) \to T_{f(m)}(N)$ (the *push forward*) via

$$f_*(X)g = X(f \circ g) \tag{1.5.9}$$

Its transpose is then a map $f^*\colon T^*_{f(m)}(N) \to T^*_m(M)$ (the *pullback*). $\wedge^k(f^*)$, which we also denote f^*, then defines a pullback of $\wedge^k(T^*_{f(m)}(N)) \to \wedge^k(T^*_m(M))$.

If $\{x^\alpha\}$ and $\{y^\beta\}$ are local coordinates on M and N, respectively, and $f^\beta \equiv y^\beta \circ f$ as a function of $(x_1, \dots, x_m)$, then

$$f^*\biggl(\sum_\alpha a^\alpha \frac{\partial}{\partial x^\alpha}\biggr) = \sum_{\alpha,\beta} a^\alpha \frac{\partial f^\beta}{\partial x^\alpha} \frac{\partial}{\partial y^\beta} \tag{1.5.10}$$

and

$$f_*\biggl(\sum_\beta b_\beta(y)\, dy^\beta\biggr) = \sum_{\alpha,\beta} b_\beta(f(x)) \frac{\partial f^\beta}{\partial x^\alpha}\, dx^\alpha \tag{1.5.11}$$

(16) One cannot use f_* to push forward vector fields on M to ones on N since $n = f(m)$ may have no solution or many (i.e., f_* is purely local), but f^* can be used to push a one-form (or k-form) since $n = f(m)$ has a unique

solution for each m (we wrote (1.5.11) for one-forms). It can be seen that for any smooth f and k-form ω on N,

$$d(f_*(\omega)) = f_*(d\omega) \tag{1.5.12}$$

The proof of this depends on the symmetry of $\partial^2 f^\beta/\partial x^\alpha \partial x^\gamma$ in α and γ and antisymmetry of $dx^{\alpha_1} \wedge \cdots \wedge dx^{\alpha_k}$.

(17) Let X be a k-dimensional submanifold of M, that is, X is a subset of M and a k-dimensional manifold in a set of local coordinates that are restrictions of a coordinate system of M, that is, there exist local coordinates for M, $(x^1, \dots, x^\nu)$, in a neighborhood U, $p \in X \subset M$, so that $U \cap X = \{x \in U \mid (x^{k+1}, \dots, x^\nu) = (0, \dots, 0)\}$ and $(x_1, \dots, x^k)$ are the local coordinates on N. Let $f\colon X \to M$ be the identity map. If X is a k-dimensional orientable submanifold and ω is a k-form on M, then $f^*(\omega)$ is a k-form on X and its integral over X is defined to be the integral of ω over X.

(18) One can define manifolds, M, with a boundary ∂M. M is of dimension n and ∂M is an $(n-1)$-dimensional manifold. For points, m, in the boundary, a local coordinate path $(U_\alpha, \varphi_\alpha)$ has $\varphi_\alpha\colon U_\alpha \to V \in \mathbb{R}^\nu_+ = \{x \mid x_1 \geq 0\}$ with $\varphi_\alpha(m) \in \{x \mid x_1 = 0\}$. Consistency requires maps C^∞ in $(x_2, \dots, x_n)$ and C^∞ up to boundary in x_1. One can then define f^* and the various structures. If M^{int} has an orientation, there is an induced orientation on ∂M.

(19) If X is a k-dimensional submanifold of M so that $f\colon M \to N$ is one-one in a neighborhood of X (so $f[X]$ is a submanifold of N), then for any k-form, ω, on N,

$$\int_{f[X]} \omega = \int_X f^*(\omega) \tag{1.5.13}$$

because the (Jacobian) change of variables in local coordinates is precisely why f^* acts on k-forms.

(20) One has

Theorem 1.5.2 (Abstract Stokes' Theorem). *If N is a k-dimensional orientable compact submanifold with boundary, ∂N, and ω is a $(k-1)$-form, then*

$$\int_N d\omega = \int_{\partial N} \omega \tag{1.5.14}$$

Remark. Integrals depend on an orientation. There is an induced orientation on ∂N, given one on N.

Special cases of this abstract theorem are called Green's, Gauss', or Stokes' theorem.

Notes and Historical Remarks.

> Clearly Thomson knew of Green's theorem in 1845, but in a postscript to a letter (dated July 2, 1850) that he wrote to an academic friend at Cambridge, George Stokes (1819–1903), Thomson mentioned the theorem but neither gave a proof of it nor mentioned Green's authorship. In February 1854 Stokes made the proof of the theorem an examination question at Cambridge—on a test, it is amusing to note, taken by a youthful James Clerk Maxwell. Maxwell (1831–79) later developed the mathematical theory of electromagnetics in his 1873 masterpiece *Electricity and Magnetism* where, in a footnote to article 24, he attributes the theorem to Stokes. So, today, the theorem is often called—you guessed it—Stokes' theorem[1].
>
> —*Paul J. Nahin* [**399**].

The Thomson referred to is William Thomson, later Lord Kelvin (a biographical note can be found in Section 3.5 of Part 3), who while at Cambridge had obtained a copy of Green's privately published paper.

For texts on manifold theory, see [**367, 401, 530, 576, 578**].

The more common definition of C^∞ manifold doesn't require a priori that M be a metric space, but rather paracompactness (see the Notes to Section 2.3 of Part 1). Since all our examples are metric spaces (and metric spaces are paracompact), for simplicity we use the simpler requirement.

A *vector bundle*, E, over a manifold, M, is another manifold and map $\pi\colon E \to M$, so that there is an atlas $\{(U_\alpha, f_\alpha)\}_{\alpha\in I}$ of M and fixed vector space, $V \cong \mathbb{R}^\ell$ (S will be an explicit linear bijection of V and $\mathbb{R}^\ell$), so that

(i) $\widetilde{U}_\alpha \equiv \pi^{-1}[U_\alpha]$ is homeomorphic to $U_\alpha \times V$ under a map $g_\alpha\colon U_\alpha \times V \to \pi^{-1}[U_\alpha]$, with $\pi(g_\alpha(x,v)) = x$.

(ii) If $\widetilde{f}_\alpha\colon \pi^{-1}[U_\alpha] \to \mathbb{R}^{\nu+\ell}$ is given by

$$\widetilde{f}_\alpha(\eta) = \langle f_\alpha(\pi(\eta)), \tilde{\pi}_\alpha \circ g_\alpha^{-1}(\eta)\rangle$$

where $\tilde{\pi}_\alpha\colon U_\alpha \times V \to \mathbb{R}^\ell$ by $\tilde{\pi}_\alpha(x,v) = Sv$, then $\{(\widetilde{U}_\alpha, \widetilde{f}_\alpha)\}_{\alpha\in I}$ is an atlas for M.

(iii) Each $f_\alpha^{-1} \circ f_\beta$ is linearly restricted to each $\pi^{-1}(\{x\})$ for $x \in U_\alpha \cap U_\beta$.

The tangent, cotangent, and $\bigcup_x \wedge^k(T_x^*(M))$ manifolds are all vector bundles.

A *cross-section* of a vector bundle $\pi\colon E \to M$ is a C^∞ function $f\colon M \to E$ so that $\pi \circ f(x) = x$ for all x.

Given the notion of tensor products of vector spaces (see Section 3.8 of Part 2), one defines tensor bundles over M by taking tensor products of multiple copies of $T_x(M)$ and $T_x^*(M)$ and gets a coordinate-free formulation of the tensor analysis used in many applications in physics and engineering.

---

[1] We note that while we have lumped these all in as an abstract Stokes' theorem, until recently Green's theorem which involves $\mathrm{div}(\vec{A})$ and Stokes' theorem which involves $\mathrm{curl}(\vec{A})$ were thought of as distinct.

1.6. Riemann Metrics

While Riemann surfaces are a frequent figure in this volume, Riemannian manifolds, a distinct notion, are not—but they do appear as the major player in Chapter 12 of Part 2B, so we recall some of the basics in this section.

A *Riemann metric* on a ν-dimensional C^∞ manifold, M, provides local notions of distance and angle, namely, a real inner product $\langle\ ,\ \rangle_m$ on each tangent space, $T_m(M)$, which is smooth in the sense that if X and Y are vector fields, then $m \mapsto \langle(X(m), Y(m)\rangle_m$ is a C^∞ function. A connected manifold with Riemann metric is called a *Riemannian manifold.* Given a local coordinate system, x^i, near $m \in M$, we define the *metric tensor*, g_{ij}, by

$$g_{ij}(n) = \left\langle \frac{\partial}{\partial x^i}, \frac{\partial}{\partial x^j} \right\rangle_m \tag{1.6.1}$$

so if $X(m) = \sum a^i \frac{\partial}{\partial x^i}$, $Y(m) = \sum b^i \frac{\partial}{\partial x^j}$, then

$$\langle X(m), Y(n)\rangle_m = \sum_{i,j=1}^n g_{ij}(m) a^i b^j \tag{1.6.2}$$

The functions $g_{ij}(m)$ are C^∞ and transform as follows: If $\tilde{g}_{ij}$ is the metric tensor in the y coordinate system, then

$$\tilde{g}_{ij} = \sum_{k,\ell} g_{k\ell} \frac{\partial x^k}{\partial y^i} \frac{\partial x^\ell}{\partial y^j} \tag{1.6.3}$$

The positivity of the inner product implies $g \equiv \det(g_{ij})$ is nonzero. We denote the inverse to g_{ij} as g^{ij}. In the dual inner product, it is $\langle dx^i, dx^j\rangle_{\text{dual}}$. As usual,

$$\|X\|_m = \langle X, X\rangle_m^{1/2} \tag{1.6.4}$$

Riemann metrics have a natural ν-form which, in local coordinates, is given by

$$\omega = g^{1/2} dx_1 \wedge \cdots \wedge dx_\nu \tag{1.6.5}$$

Because (1.6.3) implies

$$\tilde{g} = g \left[\det\left(\frac{\partial x}{\partial y}\right)\right]^2 \tag{1.6.6}$$

this is independent of coordinate system. There is a question of sign due to reordering, but if M is orientable, there is a global form ω which defines a natural volume measure on M.

A Riemann metric allows one to define the length of a smooth curve, $\gamma(t)$, mapping $[t_0, t_1]$ to M by

$$L(\gamma) = \int_{t_0}^{t_1} \|\gamma'(t)\|_{\gamma(t)}\, dt \tag{1.6.7}$$

where γ' is the tangent to the curve, as defined in (1.5.3). The geodesic distance between $m_0, m_1 \in M$ is defined by

$$\begin{aligned} &d(m_0, m_1) \\ &\quad = \inf\{L(\gamma) \mid \text{ smooth curves, } \gamma, \text{ with } \gamma(t_0) = m_0,\ \gamma(t_1) = m_1\} \end{aligned} \tag{1.6.8}$$

A *geodesic* is a minimizing curve, if one exists. Such a minimizer is highly degenerate since length is unchanged under reparametrization (reparametrization is discussed in Section 2.2). Geodesic parametrization is one where $t_0 = 0$, $t_1 = 1$, and $\|\gamma(s)\|$ is constant, so $L(\gamma)$. It is not hard to see that for a given curve, there is a unique geodesic reparametrization.

The *energy* of a curve defined on $[0, 1]$ is given by

$$E(\gamma) = \int_0^1 \|\gamma'(s)\|^2_{\gamma(s)}\, ds \tag{1.6.9}$$

By the Cauchy–Schwarz inequality, $L(\gamma)^2 \le E(\gamma)$, with equality if and only if $\|\gamma(s)\|_{\gamma(s)}$ is constant. It follows that the minimum energy among all curves from m_0 to m_1 is $d(m_0, m_1)^2$ and the minimizers for E are precisely those for L when given geodesic reparametrization. There can still be multiple minimizers of E between points (think of the north and south pole on a sphere) but they must be distinct curves.

The energy is susceptible to study via the methods of the calculus of variations. The Euler–Lagrange equations in local coordinates, where $\gamma^i(s)$ are the coordinates of $\gamma(s)$ take the form

$$\frac{d^2\gamma^i(s)}{ds^2} + \Gamma^i_{jk} \frac{d\gamma^j(s)}{ds} \frac{d\gamma^k(s)}{ds} = 0 \tag{1.6.10}$$

called the *geodesic equation*. Here Γ^i_{jk} is the *Christoffel symbol* given by

$$\Gamma^i_{jk} = \frac{1}{2} \sum_m g^{im} \left[\frac{\partial g_{mj}}{\partial x^k} + \frac{\partial g_{mk}}{\partial x^j} - \frac{\partial g_{jk}}{\partial x^m} \right] \tag{1.6.11}$$

As a second-order equation with smooth coefficients, (1.6.10) has a local solution for any initial condition. That is, (1.6.10) has a unique solution for small s for any $m = \gamma(0)$ and $\frac{d\gamma'}{ds}\Big|_{s=0} = v \in T_m(M)$. If this solution can be defined up to $s = 1$, we set

$$\exp_m(v) = \gamma(1) \tag{1.6.12}$$

The local solubility and reparametrization implies $\exp_m(v)$ is defined at least for $\|v\|_m$ small.

It is not hard to see then for s small, the solutions of (1.6.10) are what we have called geodesics, that is, minimizers of $L(\tilde{\gamma})$ among $\tilde{\gamma}$'s with $\tilde{\gamma}(t_0) = m$, $\tilde{\gamma}(t_1) = \gamma(s)$. But for s larger, this may fail. Think of two close points on the sphere—there are two great circles between them: one is a minimizer and the other is a minimizer over short distances but not globally. We call solutions of (1.6.10) *local geodesics.*

Here is a criterion for existence of geodesics for all pairs of points:

Theorem 1.6.1 (Hopf–Rinow Theorem). *Let M be a Riemannian manifold. Then the following are equivalent:*

(1) *M is complete as a metric space in the metric d given by* (1.6.8).
(2) $\exp_m$ *is defined on all of $T_m(M)$ for one m in M.*
(3) $\exp_m$ *is defined on all of $T_m(M)$ for all m in M.*
(4) *Any closed, bounded* (*i.e.*, $\sup_{m' \in S} d(m_0, m') < \infty$ *for any* $m_0 \in M$) *subset of M is compact.*

If these conditions hold, a geodesic exists between any pair of points.

Remarks. 1. If M obeys these conditions, it is called *geodesically complete.*

2. If M is compact, (4) holds, so compact Riemannian manifolds are geodesically complete.

3. (3) is equivalent to saying solutions of the geodesic equations on an interval can be extended to all $s \in \mathbb{R}$.

The other aspect of Riemannian manifolds we need to discuss is curvature. This is a beautiful but involved subject. Since we only need it once (in Section 12.1 of Part 2B) and in two dimensions, we'll restrict ourselves to Gaussian curvature, K. That it has intrinsic geometric meaning is seen by the formula for the curvature $K(m)$, at m,

$$\operatorname{vol}\{m' \mid d(m, m') < r\} = \pi r^2 - \frac{\pi K(m) r^4}{12} + O(r^5) \qquad (1.6.13)$$

The formula for K in terms of g or Γ is complicated, but in a coordinate system where $g = \left(\begin{smallmatrix} g_{11} & 0 \\ 0 & g_{22} \end{smallmatrix}\right)$, it is somewhat simpler:

$$K = \frac{1}{2\sqrt{g}} \left[\frac{\partial}{\partial x} \frac{1}{\sqrt{g}} \frac{\partial g_{22}}{\partial x} + \frac{\partial}{\partial y} \frac{1}{\sqrt{g}} \frac{\partial g_{11}}{\partial y} \right], \quad \text{if } g_{12} = 0 \qquad (1.6.14)$$

Notes and Historical Remarks. The Hopf–Rinow theorem was proven by Heinz Hopf (1894–1971) and his student, Willi Rinow (1907–79), in 1931 [**267**].

Gaussian curvature was discovered by Gauss in 1827 [**205**]. Curvature of curves in space had earlier been defined depending on the embedding in 3-space. What Gauss discovered (his *Theorema egrigium,* Latin for "remarkable theorem") was that the product of the maximum and minimum curvatures of curves through a point on a surface (now called the Gaussian curvature) was intrinsic to the surface, that is, only dependent on angles and distances on the surface. It is the K of (1.6.14).

In generalizing Gauss' ideas to arbitrary dimensions, Riemann [**482**] essentially formulated the notions of Riemann metric, geodesic equation, and curvature (of course, without the formal manifold language of the twentieth century).

For books that expose the subject introduced in this section and, in particular, for proofs of (1.6.11), (1.6.14), and Theorem 1.6.1, see [**31, 112, 225, 291, 320, 332, 428, 458, 531, 539**].

1.7. Homotopy and Covering Spaces

The notion of homotopic curves and topologically simply connected regions of $\mathbb{C}$ will enter in many places. The related notion of covering space will come up more rarely, but will play a central role in Sections 7.1, 8.5–8.7, 10.7, and 11.3. This section discusses the main ideas in this subject.

A *curve* in a topological space, X, is a continuous map, $\gamma\colon [0,1] \to X$. X is called *arcwise connected* if for every $x, y \in X$, there is a curve with $\gamma(0) = x$, $\gamma(1) = y$. Any arcwise connected set is connected. Conversely, if X is connected and locally arcwise connected (i.e., every $x \in X$ has a neighborhood which is arcwise connected), it is arcwise connected. In particular, open subsets of $\mathbb{C}$ are topologically connected if and only if they are arcwise connected; this is discussed in Section 2.1 of Part 1; see Theorem 2.1.16 of Part 1.

Two curves, $\gamma, \tilde{\gamma}$ with $\gamma(0) = \tilde{\gamma}(0) = x$ and $\gamma(1) = \tilde{\gamma}(1) = y$, are called *homotopic* if and only if there is a continuous map $\Gamma\colon [0,1] \times [0,1] \to X$ called a *homotopy* with

$$\Gamma(t,0) = \gamma(t), \qquad \Gamma(t,1) = \tilde{\gamma}(t), \qquad \Gamma(0,s) \equiv x, \qquad \Gamma(1,s) \equiv y$$

Being homotopic is easily seen to be an equivalence relation; equivalence classes are called *homotopy classes.*

Equivalently, if X is a metric space with metric ρ, one puts a metric, d, on $\Omega_{x,y} = \{\text{curves } \gamma \text{ with } \gamma(0) = x,\, \gamma(1) = y\}$ by $d(\gamma, \tilde{\gamma}) = \sup_t \rho(\gamma(t), \tilde{\gamma}(t))$. A homotopy is then a curve in $\Omega_{x,y}$ (with $\Gamma_s(t) = \Gamma(t,s)$) so that $\Gamma_0 = \gamma$, $\Gamma_1 = \tilde{\gamma}$. Homotopy classes are thus arcwise connected components of $\Omega_{x,y}$.

A topological space, X, is called *simply connected* if it is arcwise connected, and for any $x, y \in X$, any two curves are homotopic. It is easy to see this is true for all pairs x, y if and only if it is true for one pair.

One defines the composition of two curves in $\Omega_{x,x}$ by

$$\tilde{\gamma} * \gamma(t) = \begin{cases} \gamma(2t), & 0 \le t \le \frac{1}{2} \\ \tilde{\gamma}(2t-1), & \frac{1}{2} \le t \le 1 \end{cases} \tag{1.7.1}$$

It can be shown that compositions of homotopic curves are homotopic and $(\gamma*\tilde{\gamma})*\gamma^\sharp$ is homotopic to $\gamma*(\tilde{\gamma}*\gamma^\sharp)$ so that composition defines an associative product on homotopy classes. The class of $\gamma_0(t) \equiv x$ is an identity and $\tilde{\gamma}(t) = \gamma(1-t)$ is an inverse (i.e, $\tilde{\gamma} * \gamma$ and $\gamma * \tilde{\gamma}$ are homotopic to γ_0). The *fundamental group* or *first homotopy group*, $\pi_1(X, x)$, is the homotopy classes in $\Omega_{x,x}$ with this group structure. x is called the base point. For $x, y \in X$, $\pi_1(X, x)$ and $\pi_1(X, y)$ are isomorphic as groups. Indeed, extending $*$ to any curves $\gamma, \tilde{\gamma}$ with $\gamma(1) = \tilde{\gamma}(0)$, the map $\gamma \mapsto \gamma^\sharp\gamma(\gamma^\sharp)^{-1}$ of $\Omega_{x,x}$ to $\Omega_{y,y}$ for a curve $\gamma^\sharp$ with $\gamma^\sharp(0) = x$, $\gamma^\sharp(1) = y$ sets up an isomorphism.

Given arcwise connected spaces, X and Y, a *covering map* is a continuous map $f\colon Y \to X$ so that for any x in X, there is a neighborhood N_x or X so that $f^{-1}[N_x]$ is a countable union (finite or infinite) of sets, $M_x^{(j)}$, on which f is a homeomorphism of each $M_x^{(j)}$ to N_x. The canonical example is $f\colon \mathbb{R} \to \partial\mathbb{D}$ by $f(x) = e^{ix}$. Y is called a *covering space* of X.

Two covering maps $f\colon Y_1 \to X$ and $g\colon Y_2 \to X$ are called isomorphic if there is a homeomorphism $h\colon Y_1 \to Y_2$ so that $gh = f$. Given a covering map $f\colon Y \to X$ and a continuous map $g\colon Z \to X$, a *lifting* of g is a map $h\colon Z \to Y$ so $fh = g$.

Theorem 1.7.1 (Path Lifting Theorem). *Let X, Y be arcwise connected spaces and $f\colon Y \to X$ a covering map. Let $\gamma\colon [0,1] \to X$ be a curve and $y \in Y$ with $f(y) = \gamma(0)$. Then there exists a unique lift $\tilde{\gamma}\colon [0,1] \to Y$ with $\tilde{\gamma}(0) = y$ and $f \circ \tilde{\gamma} = \gamma$.*

The idea of the proof is that $[0,1]$ can be covered by finitely many intervals $J_1, \dots, J_\ell$ with $J_j \cap J_{j+1} \neq \emptyset$, $0 \in J_1$, $1 \in J_\ell$, and so that $\gamma[J_j]$ lies in an arcwise connected open set on which f^{-1} is a union of disjoint sets on which f is a homeomorphism. The unique lift on each J_j is immediate and they can be pieced together. This result leads to:

Theorem 1.7.2 (Fundamental Theorem of Covering Spaces). *Let X, Y be arcwise connected spaces and $f\colon Y \to X$ a covering map. Let Z be an arcwise connected and simply connected space, and $g\colon Z \to X$ continuous. Then given any $z_0 \in Z$ and $y \in Y$ with $f(y) = g(z_0)$, there exists a unique lift, G, of g to Y with $G(z_0) = y$ and $F \circ G = g$.*

The idea of the proof is that given $z_1 \in Z$, find $\gamma\colon [0,1] \to Z$ with $\gamma(0) = z_0$, $\gamma(1) = z_1$. By Theorem 1.7.1, we can lift $g \circ \gamma$ to $\tilde{\gamma}$ to Y and define $G(z_1) = \tilde{\gamma}(1)$. If γ_1 is a homotopic curve from z_0 to z_1, a continuity argument shows $\tilde{\gamma}_1(1) = \tilde{\gamma}(1)$, so G is well-defined, and thus, continuous.

This pair of results has many consequences. First, given $x_0 \in X$ and y_0 so $f(y_0) = x_0$, we claim there is a natural action of $\pi_1(X, x_0)$ on Y. It is defined as follows. Given $y_1 \in Y$, pick a curve γ in Y with $\gamma(0) = y_0$,

$\gamma(1) = y_1$, Given an element $[\tilde{\gamma}] \in \pi_1$, consider the curve $(f \circ \gamma) * \tilde{\gamma}$. It goes from x_0 to $f(y_1)$, and so has a unique lift $\gamma^\sharp$ to Y with $\gamma^\sharp(0) = y_0$. One defines $\tau_{[\tilde{\gamma}]}(y_1) = \gamma^\sharp(1)$. This defines an action of π_1 on Y. The maps are called *deck transformations*. It is not hard to see the action is transitive on the fibers of f, that is, $f(y_1) = f(y_2) \Leftrightarrow \exists g \in \pi_1$ with $y_2 = \tau_g(y_1)$. The fibers are thus in one-one correspondence with π_1/I_f, where $I_f = \{g \in \pi_1 \mid \tau_g = \mathbb{1}\}$ and, in particular, the cardinality of the fibers is fixed.

We are particularly interested in the case where $I_f = \{1\}$. For the fibers to be countable, π_1 needs to be countable and that is true so long as X isn't too big; for example, a separable metric space. Here's the key result:

Theorem 1.7.3 (Universal Covering Spaces). *Let X be an arcwise connected separable metric space. Then there exists a covering space Y (i.e., Y and $f\colon Y \to X$) so that Y is arcwise connected and simply connected. Moreover, Y is essentially unique in that, given $x_0 \in X$ and $(Y, y_1), (Y_1, y_1)$, where $f\colon Y \to X$, $f_1\colon Y_1 \to X_1$ are covering maps, $f(y_0) = x_0$, $f_1(y_1) = x_1$, and Y and Y_1 are arcwise connected and simply connected, there is a unique homeomorphism, $h\colon Y \to Y_1$ so $h(y_0) = y_1$ and $f_1 \circ h = f$.*

The space Y guaranteed by the theorem is called the *universal covering space* of X.

Existence and uniqueness are separate arguments. To get existence, one puts a sup metric on all curves $\gamma\colon [0,1] \to X$ with $\gamma(0) = x_0$, a base point in X. One passes to a quotient topology on the homotopy classes, Y, of curves. One maps Y to X by $f([\gamma]) = \gamma(1)$. Y constructed in this manner is simply connected and is the universal covering space. Uniqueness follows from Theorem 1.7.2.

Given a universal cover Y of X and a subgroup $G \subset \pi_1$, we say $y \sim_G y_1$ for $y, y_1 \in Y$ if $y_1 = \tau_g(y)$ for some $g \in G$. This defines an equivalence relation on Y and the equivalence classes Y_G and inherited projection f_G yields a covering space with $I_{f_G} = G$. This is the existence part of

Theorem 1.7.4 (Classification of Covering Spaces). *For any $G \subset \pi_1(X, x_0)$, there is a covering map $f\colon Y \to X$ with $I_f = G$. Two such covering spaces are isomorphic if and only if the associated G's are conjugate, that is, $G = gG_1g^{-1}$ for some $g \in \pi_1$.*

Notes and Historical Remarks. For textbook presentations of homotopy and the theory of covering spaces, see Fulton [**197**] and Hatcher [**244**]. It was in the context of problems in complex analysis (branched covers of Riemann surfaces and later uniformization) that the notion of covering space was invented by Poincaré and Klein; see the Notes to Section 8.7.

1.8. Homology

> The computability of homology groups does not come for free, unfortunately. The definition of homology groups is decidedly less transparent than the definition of homotopy groups, and once one gets beyond the definition, there is a certain amount of technical machinery to be set up before any real calculations and applications can be given.
>
> —*A. Hatcher* [**244**]

The fundamental group, aka homotopy, measures holes in a region, Ω, of $\mathbb{C}$. Homology, which we reprise in this section, provides a different way of measuring such holes, one of greater relevance to the Cauchy integral theorem (CIT). In a sense, simple connectivity is relevant to that theorem because it implies the homology is trivial. We emphasize that while homology is relevant to the CIT, homotopy is critical in the theory of Riemann surfaces.

Here we'll focus on what is called singular homology and the first homology group—we'll thereby miss exact sequences, cohomology, and all the deeper parts of homology theory, but parts not relevant to complex analysis.

A *curve* is a continuous map, γ, of $[0,1]$ onto a topological space, X. We'll also call it a 1-simplex and points in X we'll call a 0-simplex. The *canonical 2-simplex* is the set $\Delta^2 = \{(x,y,z) \in \mathbb{R}^3 \mid x,y,z \geq 0,\ x+y+z = 1\}$ endowed with the induced topology from $\mathbb{R}^3$ (i.e., the topology of the Euclidean metric). A *2-simplex* in X is a continuous map $\varphi\colon \Delta^2 \to X$.

An *n-chain* for $n = 0,1,2$ is a formal sum $\sum_{j=1}^k m_j\varphi_j$, where $m_j \in \mathbb{Z}$ and φ_j is an n-simplex. The set of chains we'll denote $\Delta_n(X)$. The *boundary*, $\partial\gamma$, of a 1-simplex, γ, is the 0-chain

$$\partial\gamma = \gamma(1) - \gamma(0) \tag{1.8.1}$$

Extend ∂ to all of $\Delta_1(X)$ by

$$\partial\biggl(\sum_{j=1}^k m_j\varphi_j\biggr) = \sum_{j=1}^k m_j\partial\varphi_j \tag{1.8.2}$$

In $\Delta_1(\Delta^2)$, define three 1-simplexes by

$$\tilde\gamma_1(t) = (0, 1-t, t), \quad \tilde\gamma_2(t) = (t, 0, 1-t), \quad \tilde\gamma_3(t) = (1-t, t, 0) \tag{1.8.3}$$

so if p_j has coordinates δ_{ji} for $j = 1,2,3$, then

$$\partial\tilde\gamma_1 = p_3 - p_2, \quad \partial\tilde\gamma_2 = p_1 - p_3, \quad \partial\tilde\gamma_3 = p_2 - p_1 \tag{1.8.4}$$

and then,

$$\partial(\tilde\gamma_1 + \tilde\gamma_2 + \tilde\gamma_3) = 0 \tag{1.8.5}$$

Define $\partial\colon \Delta_2(X) \to \Delta_1(X)$, first on 2-simplexes, φ, by

$$\partial\varphi = \varphi\circ\tilde\gamma_1 + \varphi\circ\tilde\gamma_2 + \varphi\circ\tilde\gamma_3 \tag{1.8.6}$$

Figure 1.8.1. Homologous chains.

and extended by (1.8.2).

Thus,

$$\Delta_2(X) \xrightarrow{\partial} \Delta_1(X) \xrightarrow{\partial} \Delta_0(X) \tag{1.8.7}$$

and (1.8.5) implies, for any $\varphi \in \Delta_2(\mathrm{X})$, we have

$$\partial^2\varphi = 0 \tag{1.8.8}$$

Thus, if we define

$$Z_1(X) = \operatorname{Ker}(\Delta_1 \xrightarrow{\partial} \Delta_0), \qquad B_1(X) = \operatorname{Ran}(\Delta_2 \xrightarrow{\partial} \Delta_1)$$

called *cycles* and *boundaries*, respectively, then

$$B_1(X) \subset Z_1(X) \tag{1.8.9}$$

The quotient group, $Z_1(X)/B_1(X)$, is called the *first homology group*.

Geometrically, H_1 measures the extent to which closed curves (because the sum curves in Z_1 can be combined into closed curves since they have zero boundary) can be "filled" in. One thing that makes it different from homotopy is that we can break the curves in pieces (by adding and dividing lines in each direction; see Figure 1.8.1) and fill them in separately. Note that H_1 is an abelian group and π_1 may not be (Figure 4.1.1 can be written as a commutator in $\pi_1(\mathbb{C}\setminus\{0,1\})$).

There is a natural map of $\pi_1(X)$ to $H_1(X)$. If γ is a closed curve in X, it defines a 1-chain $h(\gamma)$ with boundary zero, that is, lies in Z_1. If γ_1 is homotopic to γ_2, the homotopy provides a map of $[0,1]\times[0,1]$ to X. Dividing the square into two triangles shows $\gamma_1-\gamma_2$ is a boundary, that is, h is actually a map of π_1 to H_1. One has that

Theorem 1.8.1 (Hurewicz's Theorem). *Let X be an arcwise connected topological space. The kernel of $h\colon \pi_1(X)\to H_1(X)$ is exactly $[\pi_1(X),\pi_1(X)]$, the subgroup of $\pi_1(X)$ generated by $\{xyx^{-1}y^{-1}\mid x,y\in\pi_1\}$, and h is onto. Thus,*

$$H_1(X) \cong \frac{\pi_1(X)}{[\pi_1(X),\pi_1(X)]} \tag{1.8.10}$$

Notes and Historical Remarks. Homology classes were first defined by Poincaré in 1895 [**442**]. The modern theory came in two waves: work in the late 1920s on the group and algebraic structure, of which key contributors are Noether [**411**], Vietoris [**563**], and Hopf [**266**]; and then two critical papers by Eilenberg [**161, 162**], the first on singular homology and the other, with Steenrood, on an axiomatic approach to homology theory.

Before singular homology, the subject was combinatorial, involving decomposing spaces into triangles and their higher-dimensional analogs. The pieces were homeomorphic to simplexes. What made Eilenberg's theory "singular" is that the images of simplexes need not be under injective maps.

Hurewicz's theorem is from 1935 [**269**]. Witold Hurewicz (1904–1956), was a Polish mathematician, student of Hahn and Menger and, in turn, his students include Felix Browder. He spent his last ten years at MIT. It was his work that established the importance of higher homotopy groups. He is also known for his work on dimension theory. Hurewicz's life had a strange end. In his early fifties, after attending a conference in Mexico City, he visited the Yucatan and fell to his death while exploring a Mayan step pyramid.

If one forms $\sum_{j=1}^{k} \alpha_j \varphi_j$, where φ_j are n-simplexes but now $\alpha_j \in \mathbb{Q}$ or $\alpha_j \in \mathbb{R}$, one gets homology groups $H_k(X;\mathbb{Q})$ and $H_k(X;\mathbb{R})$. There is also a dual theory, *cohomology theory*, $H^k(X;\mathbb{R})$. In this regard, we should mention deRham's theorem [**138**]. If X is a manifold, $\wedge^\ell \xrightarrow{d} \wedge^{\ell+1}$ and $d^2 = 0$, deRham's theorem asserts that $H^\ell(X;\mathbb{R})$ is then isomorphic to $\operatorname{Ker}(\wedge^\ell \to \wedge^{\ell+1}) / \operatorname{Ran}(\wedge^{\ell-1} \to \wedge^\ell)$. In particular, in case $H_1(X) = \mathbb{Z}^m$, the set of 1-forms, ω, with $d\omega = 0$ modulo the set of df's, is an n-dimensional vector space.

Hatcher [**244**] has a lovely presentation of homology theory, including Hurewicz's theorem. deRham's theorem is discussed in some of the references in Section 1.5.

1.9. Some Results from Real Analysis

Some parts of this volume need results from standard real analysis texts such as Part 1. For example, the chapter on spaces of analytic functions (Chapter 6) will assume that the reader knows the basics of Frechét space theory (Section 6.1 of Part 1) and the Ascoli–Arzelà theorem on equicontinuity implying compactness (see Theorem 2.3.14 of Part 1). Occasionally we'll need some basics in Hilbert space theory (the Riesz representation on the dual of a Hilbert space—Section 3.3 of Part 1) or Banach space theory (Hahn–Banach theorem—Section 5.5 of Part 1) and uniform boundedness principle (Section 5.4 of Part 1). In this section, we want to focus on two topics that are not found in all books on the subject.

1.9.1. Some Connected Results. A topological space, X, is *connected* if the only sets $A \subset X$ which are both open and closed are $\emptyset$, the empty set and X, the whole space. $Y \subset X$ is connected if it is connected in the relative topology. X is called *arcwise connected* if for all $x, y \in X$, there is a continuous map $\gamma : [0,1] \to X$ so $\gamma(0) = x$ and $\gamma(1) = y$. Any arcwise connected space is connected. An important fact is that if $\Omega \subset \mathbb{R}^\nu$ is open and connected, then it is arcwise connected.

It is a basic fact (see Theorem 2.1.13 in Part 1) that if $\{A_\alpha\}_{\alpha\in I}$ is a collection of nonempty connected subsets of a topological space X so that for all $\alpha \neq \beta$, $A_\alpha \cap A_\beta = \emptyset$, then $\bigcup_{\alpha\in I} A_\alpha$ is connected. It follows that any $x \in X$ is contained in a unique maximal connected set called the *connected component* of x. X is the disjoint union of its connected components. Since the closure of a connected subset is connected, connected components are closed but need not be open.

Here is a result we'll need in describing different equivalent definitions of simply connected regions in $\mathbb{C}$ (we'll need $\nu = 2$ below) in Section 4.5 below:

Theorem 1.9.1. *Let $\Omega \subset \mathbb{R}^\nu$ be an open connected set. Let $\mathbb{R}^\nu \cup \{\infty\}$ be the one point compactification of $\mathbb{R}^\nu$. Then $(\mathbb{R}^\nu \cup \{\infty\}) \setminus \Omega$ is connected if and only if every component of $\mathbb{R}^\nu \setminus \Omega$ is noncompact.*

This is a special case of Theorem 5.4.22 of Part 1. The proof is somewhat involved (although not much more than a page).

1.9.2. The Banach Indicatrix Theorem. An important tool in complex analysis is the rectifiable curve. A basic fact about them that we'll need in Section 4.6 below is that almost all coordinate lines intersect a given rectifiable curve in only finitely many points. When we define such curves, we'll see their vectorial components are of bounded variation so we begin by recalling that notion.

Let $\alpha : [0,1] \to \mathbb{R}$. Given $y_0 = 0 < y_1 < y_2 < \ldots < y_\ell = 1$, we define

$$V_{(y_0,\ldots,y_\ell)}(\alpha) = \sum_{j=1}^{n} |\alpha(y_j) - \alpha(y_{j-1})| \tag{1.9.1}$$

The *total variation* of α is

$$\operatorname{Var}(\alpha) = \sup_{(y_0,\ldots,y_\ell);\ell} V_{(y_0,\cdots,y_\ell)}(\alpha) \tag{1.9.2}$$

If $\operatorname{Var}(\alpha) < \infty$, we say α has bounded variation. Monotone functions have bounded variation since, in that case, for any $\vec{y}$, $V_{\vec{y}}(\alpha) = |\alpha(1) - \alpha(0)|$. $\alpha(x) = \sin\left(\frac{1}{x}\right)$ $\big(x > 0; \alpha(0) = 0\big)$ is an example of a function not of bounded variation. While we defined the notion for $[0,1]$, it makes sense for $[a,b]$ and for (a,b) if we require $a < y_0 < \ldots < y_\ell < b$.

If $f:(a,b)\to\mathbb{R}$ is a bounded continuous function, one defines its *indicatrix*, $N_f:\mathbb{R}\to\mathbb{N}\cup\{\infty\}$ by

$$N_f(s)=\sharp\{x\mid f(x)=s\} \tag{1.9.3}$$

the cardinality of $f^{-1}\big[\{s\}\big]$. Theorem 4.15.7 of Part 1 says

Theorem 1.9.2 (Banach Indicatrix Theorem). *Let f be a bounded continuous function on (a,b). Then N_f is measurable and*

$$\operatorname{Var}(f)=\int_{-\infty}^{\infty}N_f(s)\,ds \tag{1.9.4}$$

In particular, if f has bounded variation, then $N_f(s)<\infty$ for Lebesgue a.e. s.

Chapter 2

The Cauchy Integral Theorem: Basics

> The integral along two different paths will always have the same value if it is never the case that $\phi(x) = \infty$ in the space between the curves representing the paths. This is a beautiful theorem, whose not-too-difficult proof I will give at a suitable opportunity.
>
> —*C. F. Gauss* in a letter to F. Bessel, December 18, 1811,[1]

Big Notions and Theorems: Holomorphic Functions, Cauchy–Riemann Equations, Conformality, Contour Integral, Analytic Functions, Cauchy Radius Formula, Exponential and Logarithm, Euler's Formula, Goursat's Argument, Holomorphically Simply Connected, Fractional Powers, Cauchy Integral Theorem and Formula for the Disk

This chapter will focus on holomorphic ($\equiv$ complex differential) and analytic ($\equiv$ convergent Taylor series) functions. The central result of complex analysis, proven in Section 3.1, is that these notions are equivalent, but for this chapter we will keep them distinct. Once they are proven to be the same, we'll use the terms interchangeably—and many books do the same.

Section 2.1 will discuss complex differential calculus, including the equivalence to Cauchy–Riemann equations, and, for locally invertible maps, to conformal mapping. Section 2.2 will discuss complex integral calculus and the Cauchy integral theorem for derivatives. Section 2.3 will discuss Taylor

[1]as translated by Jeremy Gray in [**155**]. This was several years before Cauchy's work. Gauss never published this result.

series and analyticity and the important examples of exponential, trigonometric, and logarithmic functions. Section 2.4 provides the elegant argument of Goursat to prove the Cauchy integral theorem for triangles, which we boost in Sections 2.5 and 2.6 to all curves in star-shaped regions and their conformal images. Finally, Section 2.7 proves the Cauchy integral formula for the disk—that if f is holomorphic in a neighborhood of $\overline{\mathbb{D}}$, then for $z_0 \in \mathbb{D}$,

$$f(z_0) = \frac{1}{2\pi i} \oint_{|z|=1} \frac{f(z)}{z - z_0}\, dz$$

This formula will yield a cornucopia of consequences in the next chapter. We'll also prove the Cauchy integral formula for annuli.

In this chapter and throughout the book, we'll focus almost entirely on functions from $\Omega \subset \mathbb{C}$ to $\mathbb{C}$, but one can consider generalizing $\mathbb{C}$ in either domain or range to some kind of complex vector space. If we consider $\mathbb{C}^n$ valued functions rather than $\mathbb{C}$ valued, almost no change is needed—one need only demand that each component is holomorphic or analytic. We will note in Section 2.1 that it is easy to extend the notion of analyticity to functions with values in a complex Banach space. There is one deep result about such functions—Dunford's theorem on weak analyticity implying norm analyticity—which will appear at the end of Section 3.1.

As to changing the dimension of the domain of f, we'll say something about functions of several complex variables in Section 11.5, mainly two variables with some discussion of the Banach space case in the Notes to that section. Unlike the change of the dimension of the range, change of the dimension of the domain from one to several variables has profound effect. In these results, z^2 in a Taylor expansion is replaced by a bilinear functional. A rather different subject arises if the domain is an algebra, that is, has a product, in which case z^2 again means a square! The theory of analytic functions on a Banach algebra will play a major role in Chapter 6 of Part 4.

2.1. Holomorphic Functions

This section, roughly speaking, has three subsections on the basic definition, the Cauchy–Riemann equations, and conformality.

Recall that a region, Ω, is an open, connected subset of $\mathbb{C}$.

Definition. Let f be a complex-valued function on a region, Ω, and let $z_0 \in \Omega$. We say that f is *holomorphic at* z_0 if there exists a complex number, $f'(z_0)$, so that

$$f(z) - f(z_0) - f'(z_0)(z - z_0) = o(|z - z_0|) \tag{2.1.1}$$

that is, if and only if for all ε, there is a δ so that

$$|z - z_0| < \delta,\ z \in \Omega \Rightarrow |\text{LHS of (2.1.1)}| \le \varepsilon |z - z_0| \tag{2.1.2}$$

If f is holomorphic at each $z_0 \in \Omega$, we say that f is *holomorphic on* Ω.

The following is straightforward:

Theorem 2.1.1. (a) *If f is holomorphic at z_0, it is continuous at z_0.*

(b) *If f, g are holomorphic at z_0, so are $f + g$, fg, and if $g(z_0) \neq 0$, so is f/g and*

$$(f+g)'(z_0) = f'(z_0) + g'(z_0) \tag{2.1.3}$$

$$(fg)'(z_0) = f(z_0)g'(z_0) + f'(z_0)g(z_0) \tag{2.1.4}$$

$$(f/g)'(z_0) = [g(z_0)f'(z_0) - f(z_0)g'(z_0)]/g(z_0)^2 \tag{2.1.5}$$

(c) (Chain rule) *Suppose that f is defined on Ω, that $\widetilde{\Omega}$ is an open set containing $f[\Omega]$, and that g is defined on $\widetilde{\Omega}$ so $g \circ f$ is defined on Ω. If f is holomorphic at $z_0 \in \Omega$ and g is holomorphic at $f(z_0)$, then $g \circ f$ is holomorphic at z_0 with*

$$(g \circ f)'(z_0) = g'(f(z_0))f'(z_0) \tag{2.1.6}$$

(d) (Local inverse) *If f is holomorphic at z_0, $f'(z_0) \neq 0$, and for some $\delta > 0$, f is a bijection from $\mathbb{D}_\delta(z_0)$ to a region Ω, then the inverse function g is holomorphic at $f(z_0)$ and*

$$g'(f(z_0)) = \frac{1}{f'(z_0)} \tag{2.1.7}$$

Remarks. 1. We'll see in Section 3.4 that if f is holomorphic in Ω and $f'(z_0) \neq 0$, there is a δ, so f is a bijection on $\mathbb{D}_\delta(z_0)$.

2. Section 3.4 will also have an explicit formula for g in terms of a contour integral.

Proof. The proofs are the same as in the case of one real variable. For example, here is the chain rule which illustrates the use of little oh notation:

$$\begin{aligned} g(f(z)) - g(f(z_0)) &\equiv g'(f(z_0))(f(z) - f(z_0)) + o(f(z) - f(z_0)) \\ &= g'(f(z_0))f'(z_0)(z - z_0) + o(z - z_0) \end{aligned} \tag{2.1.8}$$

The error in (2.1.8) has two parts. Since $f(z) - f(z_0) = O(z - z_0)$, the $o(f(z) - f(z_0))$ is $o(z - z_0)$, and clearly, $g'(f(z_0))o(z - z_0)$ is $o(z - z_0)$. $\square$

Before analyzing (2.1.1) more closely, we want to note that if f is holomorphic on Ω and $z_0 \to f'(z_0)$ is continuous, then (2.1.1) holds uniformly on compact subsets of Ω.

Theorem 2.1.2. *Let f be holomorphic on a region Ω and suppose f' is continuous. Then for any compact subset $K \subset \Omega$ and any $\varepsilon > 0$, there is a $\delta > 0$ so that for all $z_0 \in K$ and z with $|z - z_0| < \delta$, we have $z \in \Omega$ and*

$$|f(z) - f(z_0) - f'(z_0)(z - z_0)| \le \varepsilon |z - z_0| \tag{2.1.9}$$

Proof. Since K is compact,

$$\delta_0 \equiv \tfrac{1}{2} \min\{|z - w| \mid z \in K,\, w \in \mathbb{C} \setminus \Omega\} > 0$$

Thus, if $z_0 \in K$ and $|z - z_0| < \delta_0$, then for all $t \in [0,1]$, $tz + (1-t)z_0 \in \Omega$ and so,

$$h(t) = f(z_0 + t(z - z_0)) \tag{2.1.10}$$

is a C^1 function on $[0,1]$ with

$$h'(t) = (z - z_0) f'(z_0 + t(z - z_0)) \tag{2.1.11}$$

(the $'$ on the left is an ordinary calculus derivative and on the right, the holomorphic derivative).

By the fundamental theorem of 1D calculus,

$$f(z) - f(z_0) = (z - z_0) \int_0^1 f'(z_0 + t(z - z_0))\, dt \tag{2.1.12}$$

so

$$|f(z) - f(z_0) - (z - z_0) f'(z_0)| \le |z - z_0| \sup_{0 \le t \le 1} |f'(z_0 + t(z - z_0)) - f'(z_0)| \tag{2.1.13}$$

Note next that

$$K_{\delta_0} = \{z \mid \text{dist}(z, K) \le \delta_0\} \subset \Omega \tag{2.1.14}$$

and is compact, so f' is uniformly continuous on it. Thus, for any ε, we can find $\delta < \delta_0$ so that if $|z - z_0| < \delta$, the sup in (2.1.13) is less than ε. □

We turn to the Cauchy–Riemann equations. A map $f \colon \Omega \to \mathbb{C}$ can be viewed as a map of $\Omega \subset \mathbb{R}^2$ to $\mathbb{R}^2$ and so, susceptible to multivariable real calculus. Specifically, write

$$f(x + iy) = f_r(x, y) + i f_i(x, y) \tag{2.1.15}$$

with f_r, f_i real-valued. We define $F \colon \mathbb{R}^2 \to \mathbb{R}^2$ by

$$F\begin{pmatrix} x \\ y \end{pmatrix} = \begin{pmatrix} f_r(x, y) \\ f_i(x, y) \end{pmatrix} \tag{2.1.16}$$

As explained in Section 1.4, F has a derivative at (x_0, y_0) in the $\mathbb{R}^2$ sense if there is a linear transformation $DF_{(x_0, y_0)}$ so that

$$F\begin{pmatrix} x \\ y \end{pmatrix} = F\begin{pmatrix} x_0 \\ y_0 \end{pmatrix} + DF_{(x_0, y_0)} \begin{pmatrix} x - x_0 \\ y - y_0 \end{pmatrix} + o(|(x, y) - (x_0, y_0)|) \tag{2.1.17}$$

If we note that on $z = x + iy$,

$$\text{multiplication by } a + ib = \begin{pmatrix} a & -b \\ b & a \end{pmatrix} \tag{2.1.18}$$

and that $(z - z_0) \mapsto f'(z_0)(z - z_0)$ is multiplication by $f'(z_0)$ and that

$$DF_{(x_0,y_0)} = \begin{pmatrix} \partial f_r/\partial x & \partial f_r/\partial y \\ \partial f_i/\partial x & \partial f_i/\partial y \end{pmatrix} \tag{2.1.19}$$

we see that

Theorem 2.1.3 (Cauchy–Riemann Equations). *f is holomorphic at $z_0 = x_0 + iy_0$ if and only if*

(i) *F given by* (2.1.16) *is $\mathbb{R}^2$-differentiable at* (x_0, y_0).

(ii) *The partial derivatives obey*

$$\frac{\partial f_r}{\partial x} = \frac{\partial f_i}{\partial y}, \qquad \frac{\partial f_i}{\partial x} = -\frac{\partial f_r}{\partial y} \tag{2.1.20}$$

In that case,

$$f'(z_0) = \frac{\partial f_r}{\partial x} + i\,\frac{\partial f_i}{\partial x} \tag{2.1.21}$$

Remarks. 1. (2.1.20) are called the *Cauchy–Riemann equations* (aka CR equations).

2. Rather that phrasing things in terms of 2D derivatives, the CR equations are often derived by noting that the existence of $\lim_{z\to z_0}[f(z)-f(z_0)]/(z-z_0)$ implies

$$\lim_{\varepsilon\downarrow 0} \frac{[f(z_0+\varepsilon) - f(z_0)]}{\varepsilon} = \lim_{\varepsilon\downarrow 0} \frac{[f(z_0+i\varepsilon) - f(z_0)]}{i\varepsilon}$$

3. By the Cauchy–Riemann equations and (2.1.21), we see that the Jacobian of F as a map of $\mathbb{R}^2$ to $\mathbb{R}^2$ is given by

$$\det(DF_{(x_0,y_0)}) = |f'(z_0)|^2 \tag{2.1.22}$$

There is a useful radial version of the CR equations. If $f(z)$ is nonzero near $f(z_0)$, we can write for z near z_0,

$$f(z) = |f(z)|e^{iA(z)} \tag{2.1.23}$$

with $A(z)$ real-valued and continuous (determined up to single $2\pi\mathbb{Z}$ ambiguity). It is easy to see (Problem 1) that if also $z_0 \neq 0$, the CR equations in terms of $z = re^{i\theta}$ coordinates become

$$\frac{1}{|f|}\frac{\partial}{\partial r}|f| = \frac{1}{r}\frac{\partial}{\partial \theta}A \tag{2.1.24}$$

$$\frac{1}{r|f|}\frac{\partial |f|}{\partial \theta} = -\frac{\partial A}{\partial r} \tag{2.1.25}$$

Before leaving the subject of CR equations, we introduce a useful shorthand, sometimes called $\bar\partial$ notation (pronounced "del bar"). For this discussion, suppose that $f(x+iy)$, complex-valued, is C^∞ in x and y for $x+iy \in \Omega$. Ω can be viewed as a real two-dimensional manifold with a single coordinate patch and (x,y) as local coordinates. Thus, dx, dy and $\partial/\partial x, \partial/\partial y$ are global bases for differential forms and vector fields, respectively. To accommodate complex functions, we want to allow complex functions and vector fields. In particular, if $z, \bar z$ are thought of as functions, we get the forms

$$dz = dx + i\,dy, \qquad d\bar z = dx - i\,dy \tag{2.1.26}$$

While, as functions, z and $\bar z$ are not "independent," $dz, d\bar z$ as forms are at every point and form a basis for the cotangent space at any point. Indeed,

$$dx = \frac{1}{2}\,(dz + d\bar z), \qquad dy = \frac{1}{2i}\,(dz - d\bar z) \tag{2.1.27}$$

The dual basis, $(\partial, \bar\partial)$, is defined by

$$\langle dz, \partial\rangle = \langle d\bar z, \bar\partial\rangle = 1, \qquad \langle dz, \bar\partial\rangle = \langle d\bar z, \partial\rangle = 0 \tag{2.1.28}$$

and is thus given by

$$\partial = \frac{1}{2}\left(\frac{\partial}{\partial x} - i\,\frac{\partial}{\partial y}\right), \qquad \bar\partial = \frac{1}{2}\left(\frac{\partial}{\partial x} + i\,\frac{\partial}{\partial y}\right) \tag{2.1.29}$$

The key point is that the CR equations have the compact form

$$\bar\partial f = 0 \tag{2.1.30}$$

as is easy to see (Problem 2). Moreover, (2.1.21)/(2.1.20) show

$$f' = \partial f \tag{2.1.31}$$

Of course, for any function,

$$df = \partial f\, dz + \bar\partial f\, d\bar z \tag{2.1.32}$$

Thus,

$$\begin{aligned} d(f\,dz) = df \wedge dz &= \partial f\, dz \wedge dz + \bar\partial f\, d\bar z \wedge dz \\ &= \bar\partial f\, d\bar z \wedge dz \end{aligned} \tag{2.1.33}$$

so the CR equations say (for functions C^∞ in x and y)

$$f \text{ holomorphic on } \Omega \Leftrightarrow \bar\partial f = 0 \Leftrightarrow d(f\,dz) = 0 \tag{2.1.34}$$

which, we will see, can be viewed as the Cauchy integral theorem for smooth curves and functions.

Notice also that

$$d^2 z \equiv dx \wedge dy = \frac{i}{2}\,(dz \wedge d\bar z) \tag{2.1.35}$$

Thus, Stokes' theorem and (2.1.33) plus (2.1.35) imply that if Ω is a smooth region with boundary $\partial\Omega$, then

$$\int_{\partial\Omega} f\,dz = \int_\Omega d(f\,dz) = -\frac{2}{i}\int_\Omega \bar\partial f\,d^2z \tag{2.1.36}$$

Note also that the usual Laplacian is given by

$$\Delta \equiv \frac{\partial^2}{\partial x^2} + \frac{\partial^2}{\partial y^2} \tag{2.1.37}$$

$$\equiv 4\,\partial\bar\partial \tag{2.1.38}$$

In particular, if f is holomorphic, then

$$\partial(\bar\partial f) = 0 \Rightarrow \Delta f = 0 \tag{2.1.39}$$

As the reader probably knows, functions with zero Laplacian are called *harmonic*. Since Δ is a real differential operator, (2.1.39) says $\operatorname{Re} f$ and $\operatorname{Im} f$ are harmonic. Later, we'll see that any real harmonic function is locally the real part of a holomorphic function. This and other aspects of the relation of holomorphic and harmonic functions are the subject of Section 5.4 and Chapter 3, especially Section 3.8, of Part 3. While it may not be clear, (2.1.38) uses the equality of the mixed partials, that is, $\partial^2/\partial x\partial y = \partial^2/\partial y\partial x$, which is the essence of the more prosaic proof that CR $\Rightarrow$ f harmonic (see Problem 3).

Finally, we turn to conformal maps. This involves a shift of point of view of what $z \mapsto f(z)$ is. Rather than a function whose values just happen to lie in $\mathbb{C}$, the same vector space which contains Ω, the domain of f, we think of it as a map from Ω to $f[\Omega] = \widetilde\Omega$. Eventually, we'll see (Theorem 3.6.1) that for any nonconstant holomorphic f, $\widetilde\Omega$ is open, so f is a map between regions. For this point of view, geometry comes to the fore.

Again, for this discussion, we consider maps of Ω to $\widetilde\Omega$ which are C^∞ as functions of $x = \operatorname{Re} z$ and $y = \operatorname{Im} z$. We also focus on points $z_0 \in \Omega$ where f is locally one-one which, by the inverse function theorem (see Corollary 1.4.3), is equivalent to Df_z, the 2×2 real matrix derivative, being invertible.

Definition. A C^∞ map, $f\colon \Omega \to \widetilde\Omega$, is called *conformal at* z if it is locally one-one near z_0, and if $\gamma, \tilde\gamma$ are two smooth curves emanating from z_0, then the angle between the tangents at z_0 of γ and $\tilde\gamma$ is the same as the angle at $f(z)$ between $f\circ\gamma$ and $f\circ\tilde\gamma$. If f is conformal at every $z_0 \in \Omega$, we say f is a *conformal map*.

We will later refine this slightly (to also require orientation preserving). Tangents are local (i.e., involve derivatives), so this is equivalent to saying Df preserves angles, so we care about linear angle-preserving maps.

Proposition 2.1.4. *An invertible linear map, T, on $\mathbb{R}^n$ is angle-preserving if and only if it is a product of rotations, reflections, and uniform scaling.*

Proof. Any such product preserves angles, so we need only prove the converse. If $T = U|T|$ is the polar decomposition (see Theorem 1.3.2), where U is orthogonal, so a product of rotations and reflections, and $|T|$ real self-adjoint, then $|T| = U^{-1}T$. So we need only prove that any angle-preserving positive matrix, T, is a multiple of the identity matrix, $\mathbb{1}$. T has an orthogonal basis $e_1, \ldots, e_n$ of eigenvectors $Te_j = \lambda_j e_j$ with $\lambda_j > 0$. $e_j + e_k$ is orthogonal to $e_j - e_k$. Thus, T angle-preserving implies $\langle \lambda_j e_j + \lambda_k e_k, \lambda_j e_j - \lambda_k e_k \rangle = 0$ or $\lambda_j^2 - \lambda_k^2 = 0$. Since $\lambda_j, \lambda_k > 0$, we have $\lambda_j = \lambda_k$, that is, T is a multiple of $\mathbb{1}$. $\square$

We refine our definition of conformal slightly to not allow reflection: a map is conformal if angle-preserving and orientation-preserving and *anti-conformal* if angle-preserving and orientation-reversing.

Multiplication on $\mathbb{D}$ by $r > 0$ is a scale factor on $\binom{x}{y}$ and multiplication by $e^{i\theta}$ is a rotation, so multiplication by $z = re^{i\theta}$ is conformal. Conversely, every conformal linear map on $\mathbb{C}$ is multiplication by some z. Therefore, we have proven:

Theorem 2.1.5. *A C^∞ map, $f\colon \Omega \to \Omega'$, is conformal at z_0 if and only if f is holomorphic at z_0 with $f'(z_0) \neq 0$.*

Even more important to us later (see Theorem 5.6.2) is that conformal maps change scales uniformly in all directions so that areas scale as the square of lengths; we'll make this precise in (2.2.20) and (2.2.22).

Notes and Historical Remarks. For a general comprehensive reference on the history of complex analysis in the nineteenth century, see Bottazzini–Gray [**69**]. We've emphasized two points of view for holomorphic functions: as functions on Ω with values in $\mathbb{C}$ and as maps. There is a third view implicit in our discussion of the CR equations—as a vector field on Ω—that is, at each point $(x, y) \in \Omega$, we consider a two-component vector $\vec{u} = \binom{u_1}{u_2}$. If $\vec{u}$ is smooth, standard vector calculus defines two scalar functions on Ω:

$$\operatorname{div}(\vec{u}) = \frac{\partial u_1}{\partial x} + \frac{\partial u_2}{\partial y}, \qquad \operatorname{curl}(\vec{u}) = \frac{\partial u_1}{\partial y} - \frac{\partial u_2}{\partial x} \tag{2.1.40}$$

It was realized in studies of hydrodynamics in the eighteenth century, where $\vec{u}$ was the velocity of a small piece of fluid near (x, y), that $\operatorname{div}(\vec{u}) = 0$ expressed the lack of buildup of fluid and $\operatorname{curl}(\vec{u}) = 0$ expressed a lack of vortices that was often valid. Thus, it was realized that the study of sourceless, irrotational fluid motion means one wants to understand vector fields that obey

$$\operatorname{div}(\vec{u}) = 0, \qquad \operatorname{curl}(\vec{u}) = 0 \tag{2.1.41}$$

The reader will notice that (2.1.41) are the same as CR equations for $f = u_1 - iu_2$. The CR equations first appeared in this format in work of Jean le Rond d'Alembert [**128**] in 1752 although he didn't realize the connection to complex differentiability. Gauss, in one of his 1816 proofs of the fundamental theorem of algebra, essentially used, albeit implicitly, the radial CR equations. Cauchy published them in 1827 and Riemann included them as a central part of his fundamental work on complex analysis [**480**] in 1851. It wasn't until about then that Cauchy realized their central importance.

Jean-Baptiste le Rond d'Alembert (1717–1783) was found abandoned in the church of Saint-Jean-le Rond in Paris named after John the Baptist. He had been abandoned by his mother, Claudine Guérin de Tencin, whose literary salon was a social center during the reign of Louis XV. Her many lovers included Richelieu and Louis-Camus Destouches, an army officer who was d'Alembert's father. While neither parent officially acknowledged d'Alembert, his father did arrange a foster home where d'Alembert lived for almost 50 years and, when Destouches died, d'Alembert was left an income that allowed him to pursue mathematics rather than the more mundane law that he'd studied. d'Alembert discovered the wave equation as describing plucked strings and found the general one-dimensional solution. He was an editor with Diderot of the Encyclopédie which led to his being made a member of Académie Française (the immortals). Laplace was his student. For more on the life of d'Alembert (and also Abel, Cauchy, and others, see Alexander [**11**]).

$\bar\partial$ notation was introduced by Poincaré about 1900 and later extensively developed by Wilhelm Wirtinger (1865–1945) [**594**]. It is sometimes called Wirtinger calculus. It came into extensive use only with the flourishing of the theory of several complex variables and differentiable operators on vector bundles in the 1940s and 1950s, especially in work of Kähler, Hodge, Kodaira, and Spencer.

In modern approaches to complex variables, the so-called *del-bar* problem, that is, given $g \in C^\infty(\mathbb{C})$, the existence of a C^∞ f with

$$\bar\partial f = g \tag{2.1.42}$$

is often important, especially the analog in higher complex dimension. Typically, it enters in some way of constructing analytic functions. First, one finds h which is close to the function we want, then solves $\bar\partial f = -\bar\partial h$ using general methods, and then $f + h$ is analytic since $\bar\partial(f + h) = 0$ (of course, one doesn't want to take $f = -h$). In Problem 2 of Section 9.3, the reader will use this idea to prove the Mittag-Leffler theorem. For $g \in C_0^\infty(\mathbb{C})$, an explicit solution to (2.1.42) is found in Problem 11 of Section 5.4. For general g, a solution is constructed in Problem 1 of Section 9.3.

Conformality was studied in the eighteenth century as part of geographic map making. Distortions like one has with a Mercator projection are only acceptable for the kind of accuracy sailors need if they locally preserve angles. It was Riemann [**480**] who realized the connection between complex function theory and conformality.

A C^∞ function (in the real variables x and y), f, on Ω is called *quasi-conformal* if and only if for a $Q < 1$,

$$|\bar\partial f| < Q|\partial f|$$

For an introduction to the theory, see Bers [**46**] and references therein. If f is conformal, then $\partial f = 0$, $\bar\partial f \neq 0$, so f is then quasi-conformal.

For general functions f which are C^2 in the $\mathbb{R}^2$ sense, the directional derivatives

$$(D_\theta f)(z_0) \equiv \lim_{r \downarrow 0} \frac{f(z_0 + re^{i\theta}) - f(z_0)}{re^{i\theta}} \tag{2.1.43}$$

$$= (\partial f)(z_0) + e^{-2i\theta}(\bar\partial f)(z_0) \tag{2.1.44}$$

so $(\bar\partial f)(z_0)$ measures the lack of constancy. As θ varies, this traces out the Kasner circle (after [**298**]), centered at $(\partial f)(z_0)$ of radius $|\bar\partial f(z_0)|$.

August-Louis Cauchy (1789–1859) was born and died in Paris but lived in interesting times ("May you live in interesting times" is a Chinese curse), which greatly impacted his life. He was first exiled from Paris at age four when his family left because of the turmoil of revolution. His family were then neighbors of Laplace who had also fled Paris. After studying engineering, his first job was building harbor fortifications in Cherbourg for Napoleon's plans to launch a naval invasion of England. He returned to Paris in 1812 and began his research and teaching in mathematics. He was a conservative royalist and devout Catholic, and this helped him get appointed to the Academy of Sciences in 1816 to fill one of the slots left open by the removal of some republicans. Throughout his career, Cauchy was shunned by many of his colleagues whose left-wing politics and atheism clashed with Cauchy's strong conservatism. That his appointment to the Academy was by royal intervention and not election by his peers was resented. Disappointed by the accession of a liberal monarch after the revolution of 1830, Cauchy went into exile and returned to Paris in 1838 at which point he could not resume his teaching positions because of his refusal to take an oath to the new king, an oath introduced after his self-exile. Ironically, he only regained teaching positions after the revolution of 1848 brought in an even more liberal government but one that dropped the need for an oath.

Cauchy's complex analysis grew out of his interest in the basics of real variable calculus. He was the first to carefully define limits, continuous functions, and a version of what we now call Riemann sums. (This history is discussed in the Notes to Sections 4.1 and 4.4 in Part 1.) He started developing his ideas on complex calculus around 1815, and in 1825 published a mémoire [**104**] that had what we now call the Cauchy integral theorem. He found the Cauchy integral formula in 1831 and Cauchy estimates in 1835. (See Smithies [**526**] for a detailed history of the development of Cauchy's ideas.)

We tend to look back with a view that Cauchy made one of the great discoveries of Western civilization and assume that he must have been lionized. But as late as 1843, he lost out on a chair at Collège de France to someone you've never heard of named Libri (the third candidate was Liouville!). No doubt, politics (both Cauchy's royalist views and Libri's notorious use of connections) played a role, but it still seems surprising until one realizes that it took a while before the subject of complex analysis was appreciated. Belhoste [**39**] has a scientific biography of Cauchy.

Many of the fundamental ideas of complex analysis were known to Gauss but never published—rather, they appeared in some of his letters and notebooks. In particular, he understood the importance of contour integrals contemporaneously with Cauchy and of the relation of complex differentiability and conformality before Riemann did.

Problems

1. Verify the radial CR equations, (2.1.24) and (2.1.25).

2. Using $f = f_r + if_i$ and (2.1.29) for $\bar\partial$, verify that $\text{Re}(\bar\partial f) = \text{Im}(\bar\partial f) = 0$ are equivalent to the CR equations.

3. If f_r and f_i are C^2 and the CR equations hold, write Δf_r in terms of derivatives of f_i, and so relate Δf_r to equality of mixed partials.

4. For $a \in \mathbb{C}$, $a \neq 0$, let $\widetilde{D}_a f$ be the $\mathbb{R}^2$ directional derivative (slightly different from (2.1.43))

$$\widetilde{D}_a f(z) = \lim_{t \downarrow 0} \frac{f(z+at) - f(z)}{t} \tag{2.1.45}$$

Prove that the CR equations are equivalent to

$$\widetilde{D}_{ia} f(z) = i\,\widetilde{D}_a f(z) \tag{2.1.46}$$

for all $a \in \mathbb{C} \setminus \{0\}$.

5. Let

$$f(z) = \begin{cases} \exp(-\frac{1}{z^4}), & z \in \mathbb{C}^\times \\ 0, & z = 0 \end{cases}$$

Prove that the partial derivatives, defined as directional derivatives along the axes, obey

$$\frac{\partial f}{\partial x}(z=0) = \frac{\partial f}{\partial y}(z=0) = 0$$

so the CR equations hold, but that f is not only not holomorphic at $z=0$, it is not continuous there!

Note. Theorem 2.1.3 does not apply because F is not $\mathbb{R}^2$ differentiable.

6. Let $f(z) = |z|^2$. Prove that f is holomorphic at exactly one point.

7. Prove that a C^∞ map in x, y coordinates is anticonformal at $z_0 + iy_0$ if and only if $\partial f = 0$, $\bar\partial f \neq 0$ at z_0.

8. Let $f(z) = \log|z|$. Prove that

$$\bar\partial f(z) = \frac{1}{2\bar z}, \qquad \partial f(z) = \frac{1}{2z}$$

(*Hint*: $\log|z| = \frac{1}{2}\log(z) + \frac{1}{2}\log(\bar z)$.)

2.2. Contour Integrals

A *curve* in Ω, a region in $\mathbb{C}$, is a continuous function $\gamma\colon [0,1] \to \mathbb{C}$. Its *endpoints* are $\gamma(0)$ and $\gamma(1)$ and it is called a *closed curve* if $\gamma(0) = \gamma(1)$. It is a basic fact (discussed in great generality in Theorem 2.1.16 of Part 1; see also Problem 1) that Ω as an open, connected set is also arcwise connected, that is, for any z_0, z_1, there is a curve γ with endpoints z_0 and z_1.

By allowing arbitrary continuous functions, we get into some phantom difficulties where γ sits still for a while, that is, where γ is constant on some interval, $[a,b] \subset [0,1]$. If γ is not constant on any interval, we will call the curve "proper."

Two curves, γ and $\tilde\gamma$ are called *reparametrizations* of each other if and only if there is a monotone function $g\colon [0,1] \to [0,1]$ with $g(0) = 0$, $g(1) = 1$ so that $\tilde\gamma = \gamma \circ g$. If g is discontinuous, say $g(t_0 - 0) \neq g(t_0 + 0)$, we require that γ is constant on $[g(t_0 - 0), g(t_0 + 0)]$. It is not too hard to see that reparametrization is an equivalence relation. Among proper γ's, it is very easy to see this since g is continuous and a bijection. For general γ's, you need to go through some contortions (or should I say contourtions); see Problem 2. It is not hard to see that any curve has a reparametrization that is proper.

A *contour* is an equivalence class of curves under reparametrizations. We'll also use the symbol γ for a contour. While $\gamma(t)$ is dependent on parametrization, it is not for $t = 0, 1$.

A *Jordan arc* is one for which γ is one-one. A *Jordan curve* is a closed curve which is one-one on $[0, 1)$, that is, for which $\gamma(t) = \gamma(s)$ only if $t = s$ or if $t = 0$, $s = 1$ (or vice-versa). A *Jordan contour* or *closed Jordan contour* is one with a Jordan arc or Jordan curve among its equivalence class. Some use *simple* or *non–self-intersecting* instead of "Jordan." Basically, Jordan arcs, curves, and contours capture the notion of no self-intersection. The Jordan curve theorem says that any Jordan curve, γ, in $\mathbb{C}$ has $\mathbb{C} \setminus \operatorname{Ran}(\gamma) = S_1 \cup S_2$, where S_1, S_2 are both connected and open (and so the connected components of $\mathbb{C}$) and where S_2 is unbounded, S_1 is bounded. Moreover, in terms of a notion of "winding number" that counts the number of times a curve winds around a point, the winding number of points in S_2 is 0, and in S_1, either all $+1$ or all -1. Colloquially speaking, this theorem says a Jordan curve separates the plane into a connected inside and a connected outside. If ever a theorem called out for proof by Goldberger's method,[2] this is it—but the proof is surprisingly subtle and basically topology and not analysis. We settle for proving it for C^1 Jordan curves (as defined immediately below) in Section 4.8. This special case relies on the implicit function theorem, and so this proof is analytical.

We call a curve C^1 if and only if γ is a C^1 function (including one-sided derivatives at the ends) with $|\gamma'(t)| \neq 0$ for all t. This last condition is needed to not allow getting around sharp corners smoothly by slowing down (see Problem 5). For closed curves to be C^1, we demand one-sided derivatives exist and at the end points, $\gamma'(0+) = \gamma'(1-)$. γ is called piecewise C^1 if there is $t_0 = 0 < t_1 < \cdots < t_n = 1$ so that γ is C^1 restricted to each $[t_j, t_{j+1}]$ (including one-sided derivatives at each endpoint so that at $t_1, \dots, t_{n-1}$, there are perhaps unequal derivatives from both the left and right). We can also speak of smooth (aka C^∞), piecewise smooth, and even analytic and piecewise analytic curves. A contour is C^1, etc. if and only if the equivalence class contains a C^1 curve.

If $\gamma, \tilde{\gamma}$ are curves with $\gamma(1) = \tilde{\gamma}(0)$, we can define their *sum* by

$$\tilde{\gamma} * \gamma(t) = \begin{cases} \gamma(2t), & 0 \le t \le \frac{1}{2} \\ \tilde{\gamma}(2t-1), & \frac{1}{2} \le t \le 1 \end{cases} \tag{2.2.1}$$

This behaves the right way under reparametrization so that we can refer to sums of contours. It is easy to see that for three curves, $\gamma_1, \gamma_2, \gamma_3$, with

[2]Murph Goldberger is a theoretical physicist who invented Goldberger's method: "The proof is by the method of *reductio ad absurdum*. Suppose the result is false. Why that's absurd!"

$\gamma_1(1) = \gamma_2(0)$, $\gamma_2(1) = \gamma_3(0)$, $\gamma_3 * (\gamma_2 * \gamma_1)$ is a reparametrization of $(\gamma_3 * \gamma_2) * \gamma_1$.

A *subdivision* of $[0,1]$ is a finite ordered set $t_0 = 0 < t_1 < \cdots < t_n = 1$. $\mathbf{t} = (t_n)_{n=1}^N$ is called a *refinement* of $\mathbf{s} = (s_j)_{j=1}^J$ (we then write $\mathbf{t} \rhd \mathbf{s}$) if the set of values $\{s_j\}_{J=1}^J$ is a subset of the set of values $\{t_n\}_{n=1}^N$ and the orderings of the points are consistent. We think of a subdivision as a decomposition of $[0,1]$ into intervals $[t_0, t_1], [t_1, t_2], \dots, [t_{n-1}, t_n]$, which overlap only at their ends. Refinement corresponds to breaking up each interval into subintervals.

The subdivisions are a partially ordered set under $\rhd$ and are a lattice, that is, $\mathbf{t}, \mathbf{s}$ has a least upper bound, namely, one takes the union of their values and reorders this union. Given a curve, γ, and subdivision, $\mathbf{t}$, we define

$$\ell_{\mathbf{t}}(\gamma) = \sum_{j=0}^{n-1} |\gamma(t_{j+1}) - \gamma(t_j)|$$

By the triangle inequality, $\mathbf{t} \rhd \mathbf{s} \Rightarrow \ell_{\mathbf{t}}(\gamma) \geq \ell_{\mathbf{s}}(\gamma)$, so under the order $\lim \ell_{\mathbf{t}}(\gamma)$ exists and is defined to be the *length*

$$\ell(\gamma) \equiv \sup_{\mathbf{t}} \ell_{\mathbf{t}}(\gamma) \tag{2.2.2}$$

If $\tilde{\gamma} = \gamma \circ g$ for a reparametrization, $\{g(t_j)\}_{j=1}^n$ is a subdivision if $\mathbf{t}$ is (at least if γ and $\tilde{\gamma}$ are proper so g is continuous and strictly monotone) and it is thus easy to see that $\ell(\gamma) = \ell(\tilde{\gamma})$. Therefore, we can speak of the length of a contour. If $\ell(\gamma) < \infty$, we say γ is *rectifiable.* If γ is a piecewise C^1 curve, then it is easy to see that γ is rectifiable with

$$\ell(\gamma) = \int_0^1 |\gamma'(t)|\, dt \tag{2.2.3}$$

It is also not hard (see Problem 6) to see that

$$\ell(\gamma) = \lim_{n\to\infty} \sum_{j=0}^{n-1} \left|\gamma\left(\frac{j+1}{n}\right) - \gamma\left(\frac{j}{n}\right)\right| \tag{2.2.4}$$

The condition $\ell(\gamma) < \infty$ is exactly the condition that γ is a function of bounded variation, which allows us to define Riemann–Stieltjes integrals (see Section 4.1 of Part 1 or Problem 8). If $f(z)$ is continuous on $\operatorname{Ran}(\gamma)$, then

$$\oint_\gamma f(z)\, dz \equiv \int f(\gamma(t))\, d\gamma(t) \tag{2.2.5}$$

$$= \lim_{n\to\infty} \sum_{j=0}^{n-1} f\left(\gamma\left(\frac{j}{n}\right)\right)\left[\gamma\left(\frac{j+1}{n}\right) - \gamma\left(\frac{j}{n}\right)\right] \tag{2.2.6}$$

If γ is piecewise smooth, then

$$\oint f(z)\,dz = \int_0^1 f(\gamma(t))\gamma'(t)\,dt \tag{2.2.7}$$

By the fact that in (2.2.6) one can replace $\{\frac{j}{n}\}_{j=0}^{n-1}$ by any partition $\mathbf{t}$ and take the limit of the net, we see that $\oint_\gamma f(z)\,dz$ is only dependent on the contour, that is, it is reparametrization invariant, and so it is called a *contour integral.*

Theorem 2.2.1 (Change of variables in contour integrals). *Let $g\colon \Omega \to \widetilde{\Omega}$ be a holomorphic bijection with g' continuous and g^{-1} holomorphic. Let γ be a curve in Ω and f a continuous function on* $\operatorname{Ran}(\gamma)$. *Then*

$$\oint_{g\circ\gamma} (f\circ g^{-1})(z)\,dz = \oint_\gamma f(z)g'(z)\,dz \tag{2.2.8}$$

Remark. Eventually we'll see that g a holomorphic bijection implies that g' is continuous and g^{-1} is holomorphic.

Proof. The left side is the limit of

$$\sum_{j=0}^{n-1} f\biggl(g^{-1}\circ g\circ\gamma\Bigl(\frac{j}{n}\Bigr)\biggr)\Bigl[g\circ\gamma\Bigl(\frac{j+1}{n}\Bigr) - g\circ\gamma\Bigl(\frac{j}{n}\Bigr)\Bigr]$$

By Theorem 2.1.2, this is the sum of

$$\sum_{j=0}^{n-1} f\biggl(\gamma\Bigl(\frac{j}{n}\Bigr)\biggr)g'\biggl(\gamma\Bigl(\frac{j}{n}\Bigr)\biggr)\Bigl[\gamma\Bigl(\frac{j+1}{n}\Bigr) - \gamma\Bigl(\frac{j}{n}\Bigr)\Bigr]$$

plus an error bounded by $\varepsilon\|f\|_\infty\ell(\gamma)$ where $\varepsilon\to 0$ as $n\to\infty$. □

Example 2.2.2. (2.2.9) below is perhaps the most celebrated integral in complex analysis. Let $\gamma(t) = e^{2\pi i t}$, a closed contour around $\partial\mathbb{D}$. Since $e^{2\pi i t}$ is shorthand for $\cos(2\pi t) + i\sin(2\pi t)$, we see

$$\gamma'(t) = 2\pi i\gamma(t)$$

Thus, by (2.2.5),

$$\oint_{|z|=1} \frac{dz}{z} = 2\pi i\int_0^1 \frac{\gamma(t)}{\gamma(t)}\,dt = 2\pi i \tag{2.2.9}$$

□

By using the triangle inequality in the sums approximating contour integrals, one immediately sees the useful bound (sometimes called Darboux's theorem)

$$\left|\oint_\gamma f(z)\,dz\right| \le \|f\|_\infty\ell(\gamma) \tag{2.2.10}$$

where

$$\|f\|_\infty = \sup_{0\le t\le 1} |f(\gamma(t))| \tag{2.2.11}$$

Clearly,

$$\oint_{\tilde\gamma * \gamma} f(z)\,dz = \oint_\gamma f(z)\,dz + \oint_{\tilde\gamma} f(z)\,dz \tag{2.2.12}$$

and if $\gamma^{-1}(t) = \gamma(1-t)$, then

$$\oint_{\gamma^{-1}} f(z)\,dz = -\oint_\gamma f(z)\,dz \tag{2.2.13}$$

Combining these lets us decompose one contour into smaller ones, as can be illustrated with polygonal paths.

If $x, y \in \mathbb{C}$, we use $[xy]$ for the contour with

$$[xy](t) = (1-t)x + ty \tag{2.2.14}$$

the straight line from x to y. Given $x_1, \dots, x_m \in \mathbb{C}$, $[x_1 \dots x_m]$ is the contour $[x_m x_{m-1}] * \cdots * [x_2 x_1]$. For closed polygons, $[x_1 \dots x_m x_1]$, by breaking into segments and rearranging, we have equality of contour integrals under cyclic rearrangement, that is, integrals over $[x_j x_{j+1} x_n x_1 \dots x_j]$ are independent of j. And one can inject intermediate points, that is, if $y \in [x_j x_{j+1}]$, then $[x_1 \dots x_j y x_{j+1} \dots x_n]$ gives the same contour integrals as $[x_1 \dots x_n]$.

Using $[xy]^{-1} = [yx]$ and (2.2.13), we can decompose a polygon into two by adding a segment, that is, integrals over $[x_1 \dots x_n x_j]$ are the same as over $[x_1 \dots x_j x_1]$ plus $[x_j x_{j+1} \dots x_n x_1 x_j]$. Combining this with subdivision proves, for example, the contour integral over the single triangle on the left in Figure 2.2.1, is the same as over the four triangles on the right.

Given that we've seen piecewise C^1 curves have some nice formulae, it is useful that any rectifiable contour integral is a limit of contour integrals along polygons:

Proposition 2.2.3. *Let γ be a rectifiable curve. Let γ_n be the polygon $[\gamma(0)\gamma(\frac{1}{n})\gamma(\frac{2}{n})\dots\gamma(1)]$. Let f be a continuous function in a neighborhood,*

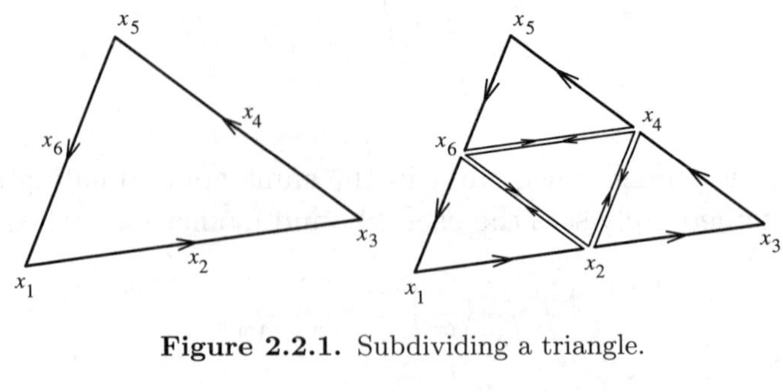

Figure 2.2.1. Subdividing a triangle.

N, of Ran(γ). *Then* $\gamma_n(t) \to \gamma(t)$ *uniformly and*

$$\int_{\gamma_n} f(z)\,dz \to \int_\gamma f(z)\,dz \tag{2.2.15}$$

Proof. Uniform convergence of $\gamma_n \to \gamma$ is a consequence of uniform continuity of γ. Pick 2δ so $\bigcup_t D_{2\delta}(\gamma(t)) \subset N$ and let $K = \bigcup_t \overline{D_\delta(\gamma(t))}$ so f is uniformly continuous on K. It follows that for any ε, there is an n so that for all j,

$$\int_{s=0}^1 \left| f\left((1-s)\gamma\left(\frac{j}{n}\right) + s\gamma\left(\frac{j+1}{n}\right)\right) - f\left(\gamma\left(\frac{j}{n}\right)\right)\right| ds \le \varepsilon$$

which implies that

$$\left|\int_{\gamma_n} f(z)\,dz - \sum_{j=0}^{n-1} f\left(\gamma\left(\frac{j}{n}\right)\right)\left(\gamma\left(\frac{j+1}{n}\right) - \gamma\left(\frac{j}{n}\right)\right)\right| \le \varepsilon\ell(\gamma)$$

Thus, (2.2.15) follows from (2.2.6). □

The fundamental theorem of calculus holds for contour integrals:

Theorem 2.2.4 (Fundamental Theorem of Calculus). *Let F be holomorphic on Ω with $F' = f$ continuous. Let γ be a rectifiable curve in Ω. Then*

$$\oint_\gamma f(z)\,dz = F(\gamma(1)) - F(\gamma(0)) \tag{2.2.16}$$

Remark. If γ is piecewise C^1, this follows from the ordinary real variable fundamental theorem of calculus (see Problem 3). One can then prove the result in general by using Proposition 2.2.3.

Proof. The proof is essentially the same as for the real variable case. By Theorem 2.1.2 with $K = \gamma([0,1])$, for any ε, there is a δ so that if $|\gamma(t) - \gamma(s)| < \delta$, then

$$|F(\gamma(t)) - F(\gamma(s)) - f(\gamma(s))(\gamma(t) - \gamma(s))| \le \varepsilon|\gamma(t) - \gamma(s)| \tag{2.2.17}$$

It follows that if n is so large that $|\gamma(\frac{j+1}{n}) - \gamma(\frac{j}{n})| < \delta$ for all n, then (by summing over j, (2.2.17) for $s = \frac{j}{n}$, $t = \frac{j+1}{n}$)

$$\left|F(1) - F(0) - \sum_{j=0}^{n-1} f\left(\gamma\left(\frac{j}{n}\right)\right)\left(\gamma\left(\frac{j+1}{n}\right) - \gamma\left(\frac{j}{n}\right)\right)\right| \le \varepsilon L(\gamma)$$

Taking $n \to \infty$, we see the absolute value of the difference of the two sides of (2.2.16) is bounded by $\varepsilon L(\gamma)$. Since ε is arbitrary, (2.2.16) holds. □

The first corollary is our first version of the Cauchy integral theorem:

Theorem 2.2.5 (CIT for Derivatives). *If f is a continuous function on Ω which is the derivative of a function F holomorphic on Ω, then for any closed rectifiable curve in Ω,*

$$\oint f(z)\,dz = 0 \tag{2.2.18}$$

Remark. We'll eventually prove a converse (see Theorem 2.5.4), that is, if (2.2.18) holds for every curve, then f is the derivative of holomorphic functions.

Proof. $\gamma(0) = \gamma(1)$, so $F(\gamma(1)) - F(\gamma(0)) = 0$. □

Since $1 = \frac{dz}{dz}$, $z = \frac{d}{dz}(\frac{1}{2}z^2)$, we see that for any closed rectifiable curve, γ, in $\mathbb{C}$,

$$\int_\gamma dz = \int_\gamma z\,dz = 0 \tag{2.2.19}$$

Corollary 2.2.6. *If f is holomorphic on Ω and $f' = 0$ on all of Ω, then f is constant.*

Proof. Let $z_0 \in \Omega$ and $\delta > 0$ be chosen so $\mathbb{D}_\delta(z_0) \subset \Omega$. If $z \in \mathbb{D}_\delta(z_0)$, then with $\gamma(t) = (1-t)z_0 + tz$, we have by (2.2.16) that $f(z) = f(z_0)$, so f is constant on $\mathbb{D}_\delta(z_0)$.

Pick $w_0 \in \Omega$ and let $Q = \{z \in \Omega \mid f(z) = f(w_0)\}$. By the above, Q is open. By continuity of f, Q is closed. Since $w_0 \in Q$, Q is nonempty. By connectedness, $Q = \Omega$. □

Before leaving this subject, we want to note the relation between the scaling of arcs and lengths under holomorphic maps and to state one technical fact that we will often need later.

Let $f\colon \Omega \to \widetilde{\Omega}$ be a C^∞ conformal bijection. Let γ be a piecewise C^1 Jordan arc in Ω. Then $f \circ \gamma$ is a piecewise C^1 Jordan arc in $\widetilde{\Omega}$. We claim that

$$\ell(f \circ \gamma) = \int_0^1 |f'(\gamma(t))|\,|\gamma'(t)|\,dt \tag{2.2.20}$$

This follows from (2.2.3) and the real-variable chain rule result that $\frac{d}{dt}f(\gamma(t)) = \gamma'(t)f'(\gamma(t))$.

On the other hand, as we've seen, Df_{z_0} is multiplication by $f'(z_0)$, so a product of a rotation and scaling by $|f'(z_0)|$. Thus (see also the argument proving (2.1.22)),

$$\det(Df_{z_0}) = |f'(z_0)|^2 \tag{2.2.21}$$

The Jacobian formula for change of variables in two-dimensional volumes then implies that if $Q \subset \Omega$ is an open bounded set, then

$$\mathrm{vol}(f[Q]) = \int_{z\in Q} |f'(z_0)|^2 \, d^2z \tag{2.2.22}$$

These relations between length and volume will be needed in Section 5.6 and in Sections 12.5 and 16.1 of Part 2B.

As for the technical fact, we will need the following compactness result whose proof is left to the problems (see Problem 7); the name comes from the fact that the U_j pave the curve if we make a cover by disks:

Proposition 2.2.7 (The Paving Lemma). *Let γ be a curve in $\mathbb{C}$ and $\{U\}_{U\in\mathcal{U}}$, an open cover of* $\mathrm{Ran}(\gamma)$. *Then there exists a partition* $\mathbf{t} = \{t_j\}_{j=0}^n$ *of* $[0,1]$ *and $U_1,\dots,U_n \in \mathcal{U}$ so that $\gamma([t_{j-1},t_j]) \subset U_j$.*

Notes and Historical Remarks. Contour integrals appear explicitly in Cauchy's 1825 mémoire [**104**], where the Cauchy integral theorem appears. Cauchy considered mainly circles and rectangles and didn't discuss general curves. They were clearly known to Gauss. More generally, they are essentially line integrals which were explored in $\mathbb{R}^2$ and $\mathbb{R}^3$ in the eighteenth century.

That the Cauchy integral theorem is essentially equivalent to a holomorphic function on a simple connected region being the derivative of a holomorphic function is due to Weyl [**590**].

(2.2.22) is sometimes called the *Lusin area integral.*

Problems

1. (a) If $\Omega \subset \mathbb{C}$ is open and $z_0 \in \Omega$ and δ is such that $\mathbb{D}_\delta(z_0) \subset \Omega$, then any $z_1 \in \mathbb{D}_\delta(z_0)$ can be connected to z_0 by a straight line lying in $\mathbb{D}_\delta(z_0)$. Use this to prove that $\{z_1 \mid z_1$ can be connected to z_0 by a curve$\}$ is both open and closed, and so is all of Ω if Ω is a region. Conclude that any region is arcwise connected.

   (b) Prove that any arcwise connected set is connected.

   (c) By the same argument, show that any two points can be connected by a polygonal path.

   (d) By smoothly replacing the corners of the polygon, show that any two points can be connected by a C^∞ path.

2. Prove that reparametrization is an equivalence relation.

3. If $F' = f$ is continuous and γ is a smooth curve, use the chain rule and (2.1.10) to provide another proof of (2.2.16).

4. Let $\gamma(t) = e^{2\pi it}$ be the curve of Example 2.2.2. Let $n \in \mathbb{Z}$, $n \neq -1$. Prove that

$$\oint_\gamma z^n \, dz = 0$$

(a) by direct calculation, as in Example 2.2.2. (*Hint*: For $n = -2, -3, \dots$, you'll need $z^{-1} = \bar{z}$ for $z \in \operatorname{Ran}(\gamma)$.)

(b) by using Theorem 2.2.5.

(c) For the same curve, compute $\oint_\gamma x \, dz$.

5. Let

$$\gamma(t) = \begin{cases} 2t, & 0 \le t \le \frac{1}{2} \\ 1 + i(2t-1), & \frac{1}{2} \le t \le 1 \end{cases}$$

Find a reparametrization, $\tilde{\gamma}$, so that $\tilde{\gamma}$ is a C^1 function.

6. Prove (2.2.4).

7. Prove the paving lemma, Proposition 2.2.7. (*Hint*: Using the fact that any open subset of $\mathbb{R}$ is a union of open intervals, find an open cover of $[0,1]$ by intervals so that γ of each interval lies in a single U, pass to a finite cover, and use that.)

8. (a) If $\mathbf{t}, \mathbf{s}$ are subdivisions with $\mathbf{s} \rhd \mathbf{t}$ and $\mathbf{t} = \{t_j\}_{j=0}^n$ and if $I_{\mathbf{t}}(\gamma, f) = \sum_{j=0}^{n-1} f(\gamma(t_j))[\gamma(t_{j+1}) - \gamma(t_j)]$, prove that

$$|I_{\mathbf{t}}(\gamma, f) - I_{\mathbf{s}}(\gamma, f)| \le \ell(\gamma) \sup_{j=0,\dots,n-1} \sup_{u,w \in [t_j, t_{j+1}]} |f(u) - f(w)|$$

(b) Use this to prove the existence of a limit for the net $I_{\mathbf{t}}(\gamma, f)$.

(c) Use the ideas used in Problem 6 to show the limit in (b) is given by (2.2.4).

9. This problem will construct a Jordan curve (called the *Koch snowflake* after von Koch [**568**]) which is not rectifiable; remarkably, there is a regular tiling of the plane by Koch snowflakes of two different sizes (see [**82**]). Let $\gamma_1(t)$ be a triangle starting at the lower corner and going at constant length. It has three line segments. $\gamma_n(t)$ will have $3 \times 4^{n-1}$ segments. Each segment is split in three and the middle segment is replaced by the other two sides of an equilateral triangle pointing out. We don't change the parametrization on the old segments but use constant length on the new ones. Figure 2.2.2 shows the first four iterations. Show that $\gamma_\infty(t) = \lim_{n\to\infty} \gamma_n(t)$ exists for all t and defines a continuous Jordan curve which is not rectifiable.

Remark. It can be shown γ is nowhere differentiable.

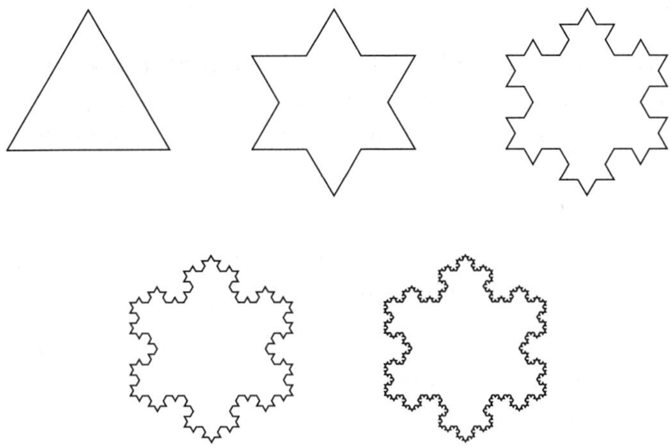

Figure 2.2.2. First five of the Koch Snowflake constructions.

2.3. Analytic Functions

Having seen the differential/integral calculus (Cauchy) and geometric (Riemann) aspects of complex functions, we turn to the third view—as power series (Weierstrass). Many of the later developments have power series proofs, but we'll often use other proofs which are less calculational. But, at least for this section, power series is king.

A *power series* is a formal sum

$$\sum_{n=0}^{\infty} a_n w^n \tag{2.3.1}$$

where $\{a_n\}_{n=1}^{\infty}$ and w are complex numbers. Sums and differences of formal series are obvious. We discuss products in Problem 21. We fix the coefficients $\{a_n\}_{n=0}^{\infty}$ and think of varying w. The basic fact is:

Theorem 2.3.1 (Cauchy Radius Formula). *Let*

$$R = (\limsup|a_n|^{1/n})^{-1} \tag{2.3.2}$$

Then for $|w| < R$, the series (2.3.1) *converges absolutely; indeed, uniformly and absolutely in $\{w \mid |w| \le R - \varepsilon\}$ for any $\varepsilon > 0$, and for $|w| > R$,* $\limsup_{n\to\infty} a_n|w|^n = \infty$ *and the series is divergent.*

Remarks. 1. R can be 0 or ∞, in which case one of the two assertions is empty (and if $R = \infty$, "$R - \varepsilon$" is replaced by "any finite K").

2. (2.3.2) is called the *Cauchy radius formula* or the *Cauchy–Hadamard radius formula.* It appeared first in Cauchy's 1821 *Cours d'Analysis* [**103**] but was forgotten; it became widely known only with Hadamard's rediscovery

in 1888 [**228**]. Recently, it has been found in lecture notes of Riemann from 1855–56.

3. On the circle $|w| = R$, called the *circle of convergence*, there are many kinds of allowed behavior—of course, absolute convergence can happen either for all or for no points on this circle. But if there is not absolute convergence, one can have conditional convergence at some, no, or all points on the circle; see Problem 5.

4. As we will discuss in the Notes and Problem 8, if the series converges conditionally at $Re^{i\theta}$, then this sum is $\lim_{\varepsilon\downarrow 0}\sum_{n=0}^\infty a_n[Re^{i\theta}(1-\varepsilon)]^n$.

5. Later, we'll discuss a different way of classifying points on the circle of convergence as regular or singular.

Proof. Given $\rho < R$, pick ρ' so that $\rho < \rho' < R$. Then for large n, $|a_n|^{1/n} \le (\rho')^{-1}$. So for some C and all n,

$$|a_n| \le C(\rho')^{-n} \tag{2.3.3}$$

Thus, if $|w| \le \rho$,

$$\sum_{n=0}^\infty |a_n w^n| \le C\sum_{n=0}^\infty \biggl(\frac{\rho}{\rho'}\biggr)^n < \infty \tag{2.3.4}$$

so the series is absolutely convergent. Since, by the same bound,

$$\sum_{n=N}^M |a_n w^n| \le C\biggl(\frac{\rho}{\rho'}\biggr)^N \biggl(1-\frac{\rho}{\rho'}\biggr)^{-1} \tag{2.3.5}$$

we see the convergence is uniform.

On the other hand, there is a subsequence $n(j)$, so

$$\lim_{j\to\infty} |a_{n(j)}|^{1/n(j)} = R^{-1}$$

Thus, if $|w| > R+\varepsilon$,

$$\lim_{j\to\infty} |a_{n(j)}w^{n(j)}| \ge \lim_{j\to\infty} (w(R+\varepsilon)^{-1})^{n(j)} = \infty \qquad \Box$$

Definition. Let f be a complex-valued function on a region, Ω. f is called *analytic at* z_0 if and only if there is a power series $\sum_{n=0}^\infty a_n w^n$ with a nonzero radius convergence and a δ so that $\mathbb{D}_\delta(z_0) \subset \Omega$ and so that δ is smaller than the radius of convergence of the series, and for $z \in \mathbb{D}_\delta(z_0)$,

$$f(z) = \sum_{n=0}^\infty a_n(z-z_0)^n \tag{2.3.6}$$

f is called *analytic in* Ω if it is analytic at each $z_0 \in \Omega$. $\mathfrak{A}(\Omega)$ will denote the family of all functions analytic in Ω.

A central theorem of complex analysis is that

$$f \text{ is analytic in } \Omega \Leftrightarrow f \text{ is holomorphic in } \Omega \tag{2.3.7}$$

We will actually prove more than just the equality of the local conditions near each $z_0 \in \Omega$. We actually prove that the radius of convergence of the series at $z_0 \in \Omega$ is at least $\operatorname{dist}(z_0, \partial\Omega)$ (see Theorem 3.1.2). The direction $\Rightarrow$ is straightforward and will be proven next. The direction $\Leftarrow$ is more subtle and will appear in Section 3.1. To prove $\Rightarrow$, we need two simple lemmas:

Lemma 2.3.2. *For each $z, w \in \mathbb{C}$ and $n = 1, 2, \dots,$ we have*

$$|(z+w)^n - z^n - nwz^{n-1}| \le \tfrac{1}{2}\, n(n-1)|w|^2(|z|+|w|)^{n-2} \tag{2.3.8}$$

Remark. We will provide an analytic proof; see Problem 11 for an algebraic proof.

Proof. Let $g(t) = (z+tw)^n$. Taylor's theorem with remainder (see Theorem 1.4.1) for g says that the left-hand side of (2.3.8) is equal to $\frac{1}{2}n(n-1)|z+tw|^{n-2}$ for some $t \in [0,1]$, from which (2.3.8) is immediate. □

Lemma 2.3.3. *For $0 < r < 1$, $\sum_{n=0}^\infty nr^n$ and $\sum_{n=0}^\infty n(n-1)r^n$ are absolutely convergent.*

Proof. Differentiating the finite geometric series (see Problem 12) implies that the partial sums are bounded by $r(1-r)^{-2}$ and $2r^2(1-r)^{-3}$, respectively. □

Theorem 2.3.4. *If $f(z)$ is given by a convergent power series, (2.3.6), in some $D_\delta(z_0)$ with $\delta > 0$, then in that disk, f is holomorphic with*

$$f'(z) = \sum_{n=0}^\infty a_n n(z-z_0)^{n-1} \tag{2.3.9}$$

Remark. The proof shows that the $o(|z-\zeta|)$ term in $f(\zeta)-f(z)-(\zeta-z)g(z)$ below is uniformly bounded in z as z runs through any $D_{\delta'}(z_0)$ with $\delta' < \delta$.

Proof. For notational simplicity, take $z_0 = 0$, which is no loss if we use $z - z_0$ everywhere that z appears below. By Lemma 2.3.3 and the argument in the proof of Theorem 2.3.1 (or by $\sqrt[n]{n} \to 1$), the series, (2.3.9), has the same radius of convergence as the series in (2.3.6), so the right side of (2.3.9) defines a function $g(z)$ in $D_\delta(0)$. Let $z, \zeta \in D_\delta(0)$ and let $w = \zeta - z$ in (2.3.8), which becomes

$$|\zeta^n - z^n - n(\zeta-z)z^{n-1}| \le \tfrac{1}{2}\, n(n-1)|\zeta-z|^2(|z|+|\zeta-z|)^{n-1} \tag{2.3.10}$$

By taking finite sums (from $n = 0$ to $n = N$) and then taking $N \to \infty$, this implies that

$$|f(\zeta)-f(z)-(\zeta-z)g(z)| \leq \tfrac{1}{2}\,|\zeta-z|^2 \sum_{n=0}^{\infty} n(n-1)|a_n|(|z|+|\zeta-z|)^{n-1} \quad (2.3.11)$$

Fix z. If $\zeta - z$ is so small that $|z| + |\zeta - z| \leq \frac{1}{2}(|z| + \delta) < \delta$, the sum on the right of (2.3.11) is uniformly bound for such ζ, so

$$|f(\zeta) - f(z) - (\zeta - z)g(z)| = O(|\zeta - z|^2) = o(|\zeta - z|) \quad (2.3.12)$$

proving that f is holomorphic at z and $f'(z) = g(z)$. □

In particular, this proof shows inductively that f has complex derivatives of all orders and that $a_1 = f'(z_0)$, and then by induction that $n!a_n = f^{(n)}(z_0)$, that is, the convergent power series is

$$f(z) = \sum_{n=0}^{\infty} \frac{f^{(n)}(z_0)}{n!}\,(z - z_0)^n \quad (2.3.13)$$

which is, of course, a *Taylor series*.

We also get, by applying this to $f^{(k)}$, that

$$f^{(k)}(z) = \sum_{n=0}^{\infty} \frac{f^{(n+k)}(z_0)}{n!}\,(z - z_0)^n \quad (2.3.14)$$

We can also prove that a function given in a disk by a convergent power series is analytic in the entire disk:

Theorem 2.3.5. *Let f be a function given by a convergent power series in some disk $D_\delta(z_0)$. Then f is analytic at each $z \in D_\delta(z_0)$ with a power series that is convergent in $D_{\delta-|z-z_0|}(z)$.*

Remark. If $R(z_0)$ is the radius of convergence of the Taylor series at z_0, this says

$$R(z) \geq R(z_0) - |z - z_0| \quad (2.3.15)$$

By interchanging z and z_0,

$$R(z) \leq R(z_0) + |z - z_0| \quad (2.3.16)$$

so $|R(z) - R(z_0)| \leq |z - z_0|$.

Proof. As above, without loss, take $z_0 = 0$. We'll give a proof up to resummation, leaving the justification to Problem 13 (essentially the double sum in j and m below is absolutely summable if $|z| + |\zeta - z| < \delta$ and so can be rearranged):

$$f(\zeta) = \sum_{n=0}^{\infty} \frac{f^{(n)}(0)}{n!}\,(\zeta - z + z)^n$$

$$= \sum_{n=0}^{\infty} \frac{f^{(n)}(0)}{n!} \sum_{j=0}^{n} \binom{n}{j} z^{n-j}(\zeta - z)^j \tag{2.3.17}$$

$$= \sum_{j=0}^{\infty} \frac{(\zeta - z)^j}{j!} \sum_{m=0}^{\infty} \frac{z^m}{m!} f^{(m+j)}(0) \tag{2.3.18}$$

$$= \sum_{j=0}^{\infty} \frac{f^{(j)}(z)}{j!} (\zeta - z)^j \tag{2.3.19}$$

We get (2.3.17) by using the binomial theorem, (2.3.18) by using $\binom{n}{j} = n!/j!(n-j)!$, and (2.3.19) from (2.3.14). $\square$

We are heading towards a proof that zeros of an analytic function (which is not identically zero) are isolated. We'll do this in two steps: first local and then local to global.

Proposition 2.3.6. *Let $f \in \mathfrak{A}(\Omega)$. Then zeros of f are isolated: If $f(z_0) = 0$, either there is $\delta > 0$ so f is nonvanishing in $\{z \mid 0 < |z - z_0| < \delta\}$ or there is $\delta > 0$ with $f(z) = 0$ for all z in $D_\delta(z_0)$.*

Proof. Let f have the form (2.3.6) in $D_{\delta_0}(z_0)$. By hypothesis, $a_0 = 0$. Suppose some $a_n \neq 0$ and let

$$m = \min\{n \mid a_n \neq 0\} \tag{2.3.20}$$

Let

$$g(z) = \sum_{n=0}^{\infty} a_{n+m}(z - z_0)^n \tag{2.3.21}$$

Then $g(z_0) \neq 0$ and

$$f(z) = (z - z_0)^m g(z) \tag{2.3.22}$$

g is analytic in $D_{\delta_0}(z_0)$, so continuous. Then for some $\delta < \delta_0$, g is nonvanishing in $D_\delta(z_0)$. Thus, by (2.3.22), f is nonvanishing in $D_\delta(z_0) \setminus \{z_0\}$.

If all a_n are zero, $f(z) = 0$ in $D_{\delta_0}(z_0)$. $\square$

Proposition 2.3.7. *Let $f \in \mathfrak{A}(\Omega)$. Suppose, for some $z_0 \in \Omega$, there is $\delta > 0$ so that f is identically zero in $D_\delta(z_0)$. Then f is identically zero in all of Ω.*

Proof. Let $A = \{z \in \Omega \mid \text{for all } n,\ f^{(n)}(z) = 0\}$. Then A is clearly closed since each $f^{(n)}(z)$ is continuous. A is open since $z_1 \in A$ implies the Taylor series at z_1 is identically zero, so $f(z) \equiv 0$ near z_1. A is nonempty since $z_0 \in A$. Thus, by connectedness, $A = \Omega$. $\square$

Putting this together, we obtain

Theorem 2.3.8 (Identity Theorem for Analytic Functions). *Let $f \in \mathfrak{A}(\Omega)$. Suppose that $f(z_n) = 0$ for a sequence $\{z_n\} \in \Omega$ where $z_n \to z_\infty \in \Omega$. Then f is identically zero in Ω. In particular, if $f, g \in \mathfrak{A}(\Omega)$ and $f(z_n) = g(z_n)$ for $z_n \to z_\infty \in \Omega$, then $f \equiv g$.*

Proof. The first part is immediate from the last two propositions. The second part follows by applying the first part to $f - g$. □

Corollary 2.3.9. *If $\Omega, \widetilde{\Omega}$ are two regions with $\Omega \subset \widetilde{\Omega}$ and $f \in \mathfrak{A}(\Omega)$, there is at most one $g \in \mathfrak{A}(\widetilde{\Omega})$ with $g \restriction \Omega = f$.*

This is, of course, immediate. When this occurs, we say g is the *analytic continuation* of f. We caution that it can happen that $\Omega \subset \Omega_1$, $\Omega \subset \Omega_2$, and $f \in \mathfrak{A}(\Omega)$ have continuations g_1 to Ω_1 and g_2 to Ω_2, but g_1 and g_2 are not equal on $\Omega_1 \cap \Omega_2$. The simplest example is to define

$$S_{\alpha,\beta} = \{re^{i\theta} \mid \alpha < \theta < \beta;\ 0 < r < \infty\} \tag{2.3.23}$$

with $\Omega = S_{-\frac{\pi}{2},\frac{\pi}{2}}$, $\Omega_1 = S_{-\frac{\pi}{2},\frac{3\pi}{2}}$, $\Omega_2 = S_{-\frac{3\pi}{2},\frac{\pi}{2}}$ and $f(z) = \sqrt{z}$ with $\sqrt{x} > 0$ for $x > 0$. Then $g_1(-1) = i$ and $g_2(-1) = -i$. Section 11.2 will deal with the issues of how to track all analytic continuations and the related notions of global analytic functions and their Riemann surfaces.

Related to this notion of continuation, we have the following which is a direct consequence of the locality of the definition of holomorphic.

Proposition 2.3.10. *Let $\Omega, \widetilde{\Omega}$ be two regions with $\Omega \cap \widetilde{\Omega} \neq \emptyset$. Let $f, \widetilde{f}$ be functions on Ω and $\widetilde{\Omega}$ respectively, each holomorphic. If $f(z) = \widetilde{f}(z)$ for all $z \in \Omega \cap \widetilde{\Omega}$, then the extension of f to $\Omega \cup \widetilde{\Omega}$, obtained by setting it equal to $\widetilde{f}$ on $\widetilde{\Omega}$, is holomorphic on $\Omega \cup \widetilde{\Omega}$.*

As the above example shows, it can happen that $\Omega \cap \widetilde{\Omega}$ has more than one component and $f = \widetilde{f}$ on one component, but not the other. However,

Proposition 2.3.11. *Under the hypotheses of Proposition* 2.3.10, *if f and $\widetilde{f}$ are assumed analytic rather than just holomorphic and if $\Omega \cap \widetilde{\Omega}$ is connected and $f = \widetilde{f}$ in some disk $\mathbb{D}_\delta(x_0)) \subset \Omega \cap \widetilde{\Omega}$, then the extension of f to $\Omega \cup \widetilde{\Omega}$, obtained by setting it equal to $\widetilde{f}$ on $\widetilde{\Omega}$, is holomorphic on $\Omega \cup \widetilde{\Omega}$.*

Proof. Given the connectedness of $\Omega \cap \widetilde{\Omega}$, Theorem 2.3.8 implies $f = \tilde{f}$ on all of $\Omega \cap \widetilde{\Omega}$, so we can apply the last proposition. □

If Ω is a region, $f \in \mathfrak{A}(\Omega)$, and $z_0 \in \partial\Omega$, we say that f is *regular* at z_0 if and only if there is $\delta > 0$, g analytic in $D_\delta(z_0)$ so that $f(z) = g(z)$ for all $z \in \Omega \cap D_\delta(z_0)$. If f is not regular at z_0, we say that f is *singular* at z_0.

Typically, singularities are associated with some $f^{(k)}(z)$ not having a finite limit for some $z_n \in \Omega$ with $z_n \to z_0$, but there are other possibilities (see Problem 9). The set of regular points is clearly open, so the singular points are closed.

If $f \in \mathfrak{A}(\Omega)$ and every point in $\partial\Omega$ is singular for f, we say $\partial\Omega$ is a *natural boundary* for f. Problems 17, 18, and 19 will give examples of $f \in \mathfrak{A}(\mathbb{D})$ for which $\partial\mathbb{D}$ is a natural boundary. In Section 9.5, we'll prove that every $\partial\Omega$ is a natural boundary for some $f \in \mathfrak{A}(\Omega)$.

In Section 3.1 (see Theorem 3.1.7), we'll prove that if a power series has a finite nonzero radius of convergence, the analytic function it defines always has a singularity at some point on the boundary of the disk of convergence. Problem 5 explores relations (or rather, lack of) between conditional convergence on this boundary and regularity or singularity.

Finally, we turn to some important power series. Since $\{m\}_{m=1}^n$ is larger than $n/2$ for $n/2$ factors, $n! \geq (n/2)^{n/2}$ so $(n!)^{1/n} \geq (n/2)^{1/2} \to \infty$. Thus, $\limsup(1/n!)^{1/n} = 0$ so $\sum_{n=0}^\infty w^n/n!$ has an infinite radius of convergence. Thus, for any $z \in \mathbb{C}$, we define

$$\exp(z) = \sum_{n=0}^\infty \frac{z^n}{n!} \tag{2.3.24}$$

Theorem 2.3.12. (a) $\exp(z)$ *is nonvanishing on all of* $\mathbb{C}$.

(b)
$$\exp(z+w) = \exp(z)\exp(w) \tag{2.3.25}$$

for all $z, w \in \mathbb{C}$.

(c) *For* θ *real,*

$$\exp(i\theta) = \cos\theta + i\sin\theta \tag{2.3.26}$$

(d) exp *is strictly monotone on* $\mathbb{R}$ *and maps* $\mathbb{R}$ *onto* $(0,\infty)$.

(e) exp *maps* $\mathbb{C}$ *onto* $\mathbb{C}^\times$.

(f)
$$\exp(z) = \exp(w) \Leftrightarrow z - w \in 2\pi i\mathbb{Z} \tag{2.3.27}$$

Remarks. 1. (2.3.26) is called *Euler's formula* (one of many results with that name!) and implies his famous result[3] that $e^{i\pi} + 1 = 0$. Once we write $\exp(z) = e^z$, it is consistent with the shorthand $e^{i\theta}$ we use for the right side of (2.3.26).

2. If we define $e = \exp(1)$, (2.3.25) implies for x real that $\exp(x) = e^x$ and leads to using e^z for $\exp(z)$.

[3] As stated by Bottazzini-Gray [**69**, p. 83], while Euler did have (2.3.26), he never seems to have explicitly used $e^{i\pi} + 1 = 0$!

Proof. (a), (b) We begin by noting that since $n/n! = 1/(n-1)!$, (2.3.9) implies that

$$\left(\frac{d}{dz}\exp\right)(z) = \exp(z) \tag{2.3.28}$$

Fix $u \in \mathbb{C}$ and define

$$g(z) = \exp(z)\exp(u-z) \tag{2.3.29}$$

Then, using (2.3.28), $g'(z) = 0$. So, by Corollary 2.2.6, $g(z) = \exp(u)$. Letting $w = u - z$ so $u = w + z$, we obtain (2.3.25).

Since $\exp(0) = 1$ and exp is continuous, for some $r > 0$, $\exp(z) \neq 0$ on $\overline{\mathbb{D}_r(0)}$. For any $z \in \mathbb{C}$, there is an integer $n \in \mathbb{Z}_+$, so $|z/n| \leq r$. Then, by (2.3.25), $\exp(z) = (\exp(z/n))^n$ is nonzero.

(c) For θ real, let

$$g(\theta) = \exp(-i\theta)[\cos\theta + i\sin\theta] \tag{2.3.30}$$

Then $g'(\theta) = 0$, so $g(\theta) = 1$ since $g(0) = 1$. Since $\exp(i\theta)\exp(-i\theta) = 1$ by (2.3.25), we obtain (2.3.26).

(d) For x real, $\exp(x)$ is real, so $\exp(x) = \exp(x/2)^2$ is positive. By (2.3.28), we see that exp is strictly monotone on $\mathbb{R}$. Thus, $e = \exp(1) > \exp(0) = 1$ so $e^n \to \infty$, implying $\lim_{x\to\infty}\exp(x) = \infty$. Since $\exp(-x) = (\exp(x))^{-1}$, we see that $\lim_{x\to-\infty}\exp(x) = 0$.

(e) By (2.3.25),

$$\exp(z) = \exp(\operatorname{Re} z)\exp(i\operatorname{Im} z) \tag{2.3.31}$$

By (2.3.26), this is the polar decomposition. If $w \in \mathbb{C}\setminus 0$, $w = |w|e^{i\theta}$, we pick $\operatorname{Im} z = \theta$ and, by (d), let $\exp(\operatorname{Re} z) = |w|$. Thus, there is a z with $\exp(z) = w$.

(f) If $\exp(z) = \rho e^{i\theta}$, the above shows that ρ determines $\operatorname{Re} z$ and $\operatorname{Im} z$ determines θ up to 2π ambiguity. □

Since $\cos(-\theta) = \cos\theta$ and $\sin(-\theta) = -\sin\theta$, (2.3.26) implies

$$\cos\theta = \tfrac{1}{2}\left(e^{i\theta} + e^{-i\theta}\right), \qquad \sin\theta = \tfrac{1}{2i}\left(e^{i\theta} - e^{-i\theta}\right) \tag{2.3.32}$$

so we define cos, sin for complex z by

$$\cos z = \tfrac{1}{2}\left(e^{iz} + e^{-iz}\right), \qquad \sin z = \tfrac{1}{2i}\left(e^{iz} - e^{-iz}\right) \tag{2.3.33}$$

leading to the familiar power series formulae for cos and sin:

$$\cos z = \sum_{n=0}^{\infty}\frac{(-z^2)^n}{(2n)!}, \qquad \sin z = \sum_{n=0}^{\infty}\frac{(-1)^n}{(2n+1)!}z^{2n+1} \tag{2.3.34}$$

Finally, we turn to the $\log(z)$. $z \mapsto \exp(z)$ is many-to-one, but it is locally one-one, so we can define a local inverse; indeed, one on $\mathbb{C}\setminus(-\infty,0]$. Given z is this set, let $z = \rho e^{i\theta}$ be its polar form where θ is chosen in

$(-\pi, \pi)$. Since $\exp(\cdot)$ is strictly monotone from $\mathbb{R}$ onto $(0, \infty)$, we can define an inverse which we'll call log. We then define

$$\log(\rho e^{i\theta}) = \log(\rho) + i\theta \tag{2.3.35}$$

Clearly, $\exp(\log(z)) = z$, so Theorem 2.1.1(d) shows that log is holomorphic and that

$$\left(\frac{d}{dz} \log\right)(z) = \frac{1}{z} \tag{2.3.36}$$

so

$$\left(\left(\frac{d}{dz}\right)^k \log\right)(z) = \frac{(-1)^{k+1}(k-1)!}{z^k} \tag{2.3.37}$$

This leads to the Taylor series for log at $z = 1$:

$$\log(1 - z) = -\sum_{n=1}^{\infty} \frac{z^n}{n} \tag{2.3.38}$$

which, by the Cauchy radius formula, has radius of convergence 1.

There are no difficulties in extending the notion of analytic function to vector-valued function, even on infinite-dimensional vector spaces, so long as there is a norm. If X is a Banach space (which could be $\mathbb{C}^n$ with the Euclidean norm) and $f\colon \Omega \subset \mathbb{C} \to X$ is a function, we say it is analytic at $z_0 \in \Omega$ if it has an expansion of the form (2.3.6), where now $a_n \in X$. One still has a Cauchy radius formula in the form

$$R = (\limsup \|a_n\|^{1/n})^{-1} \tag{2.3.39}$$

Notes and Historical Remarks. In a real-variable setting, Taylor series appeared first in a 1715 work of Brook Taylor (1685–1731) [**552**], although its significance wasn't widely appreciated until work of Lagrange in 1772 (talk about a slower pace of mathematical life!). In some older books, a Taylor series about $z = 0$ is called a Maclaurin series. We return to the history of power series in complex variables in the Notes to Section 3.1.

This book, like most modern ones on complex variables, emphasizes the geometric and calculus points of view of Riemann and Cauchy. While we will often exploit power series, they definitely play a secondary role. Some classic books that emphasize the power series point of view include Courant–Hurwitz [**272**], Dienes [**142**], Dinghas [**144**], and Whittaker–Watson [**592**]. More recent are Bourbaki [**70**], Cartan [**100**], and Henrici [**252**].

It should be emphasized that the power series approach extends most naturally to several complex variables and that, given Weierstrass' fascination with higher abelian functions (i.e., integrals, like elliptic integrals, but with polynomials of degree larger than 4) which require several complex variables, this may explain his preference for this approach.

Especially in such an approach, manipulations of series is significant, including if f and g are the formal series, their sum, product and numeric inverse (discussed in Problems 21 and 22), and composition and functional inverse (see Problem 1 of Section 3.4). There is a huge literature on when power series have natural boundaries on their circle of convergence. An admirable summary can be found in Chapter 11 of Remmert [**477**].

The first examples were found by Weierstrass (whose life is discussed in the Notes to Section 9.4) and Kronecker. The latter found that $\sum_{n=0}^{\infty} z^{n^2}$, relevant in the theory of theta functions (see Section 10.5), has a natural boundary on $\partial\mathbb{D}$. That this has a natural boundary will be proven by the reader in Problem 11 of Section 6.2.

The earliest systematic theory was found by Hadamard [**229**] (whose life is discussed in the Notes to Section 9.10), who considered power series of the form $\sum_{k=0}^{\infty} a_k z^{n_k}$ where the n_k grow very fast—in his case, $n_{k+1} \geq (1+\delta)n_k$ for some $\delta > 0$ (Hadamard gap theorem). In that case, Hadamard showed the circle of convergence is a natural boundary. Problem 19 leads you through a slick and simple proof of this, due to Mordell [**395**], although earlier Porter [**456**] essentially had the same ideas (but he published in a then obscure journal called the Annals of Mathematics!). Fabry [**183**] later improved the result to only require $n_k/k \to \infty$. In fact, Fabry only allowed $n_{k+1} - n_k \to \infty$, but Faber [**182**] later realized his proof worked if $n_k/k \to \infty$. Agmon [**5**] (see also Breuer–Simon [**72**]) has proven that if $\sup_n |a_n| < \infty$ and there exists n_k so that $\lim_{k\to\infty} a_{n_k+m} = 0$ for $m < 0$ and $= \alpha \neq 0$ for $m = 0$, then $\sum_{k=0}^{\infty} a_k z^k$ has a natural boundary on $\partial\mathbb{D}$; see Problem 11 in Section 6.2.

One tends to think of natural boundaries as pathological (which they are) and unusual, but in various ways they are generic and common. For example, there is a theorem of Szegő [**549**] that says if $\{a_n\}_{n=0}^{\infty}$ takes only finitely many different values (in which case, assuming $f(z) = \sum_{n=0}^{\infty} a_n z^n$ is not a polynomial, the radius of convergence is one), then either a_n is eventually periodic (in which case f is a rational function) or else $\partial\mathbb{D}$ is a natural boundary for f. Also, there is a theorem of Steinhaus [**537**] that says if $\sum_{n=0}^{\infty} a_n z^n$ is a power series with radius of convergence 1 and if $\omega = \{\omega_j\}_{j=0}^{\infty}$ are independent, identically distributed random variables uniformly distributed on $\partial\mathbb{D}$ (i.e., $\omega \in \times_{j=0}^{\infty} \partial\mathbb{D}$ with a product measure), then for a.e. ω, $\sum_{n=0}^{\infty} a_n \omega_n z^n$ has a natural boundary on $\partial\mathbb{D}$. Paley–Zygmund [**426**] extended this to allow $\omega \in \times_{j=0}^{\infty}\{\pm 1\}$ with equal distribution, and Kahane has a chapter in his book on random functions [**296**] on theorems of this genre. Breuer–Simon [**72**] has results on random a_n's forcing natural boundaries without requiring independence.

Finally, if $K \subset \mathbb{C}$ is compact and $Q = K^\infty = \{(a_n)_{n=0}^\infty \mid a_n \in K\}$ is given the weak (product) topology, then Q is a complete metric space and $\{(a_n) \in Q \mid \sum_{n=0}^\infty a_n z^n$ has a natural boundary on $\partial\mathbb{D}\}$ is a dense G_δ. This result of Breuer–Simon [**72**] will be proven in Problem 9 of Section 6.2.

Euler's formula, (2.3.26), first appeared in part in his 1748 book [**178**], but he was quoting early variants of it already in letters to Bernoulli and Goldbach in the early 1740s. A formula equivalent to $-i\phi = \log[\cos(\phi) - i\sin(\phi)]$ appears in passing in the 1714 paper of Roger Cotes (1682–1716) [**123**] but was little noted by his contemporaries, and perhaps by Cotes himself.

We presupposed that trigonometric functions, their derivatives and periodicities were known to the reader—and the derivative formula comes from the addition formula. However, one can use Euler's formula to define sin and cos and prove that $e^{i\theta}$ is periodic, and define π to be half its minimal period (and then that 2π is the circumference of the unit circle). The addition formulae for sin and cos then come from $e^{i\theta}e^{i\varphi} = e^{i(\theta+\varphi)}$.

We also note that Euler's formula implies de Moivre's formula:

$$\cos(n\theta) + i\sin(n\theta) = (\cos\theta + i\sin\theta)^n \tag{2.3.40}$$

which leads to explicit formulae in terms of binomial coefficients.

The fact (see Problem 8) that if $\sum_{n=0}^N a_n$ is convergent, then $\lim_{N\to\infty}\sum_{n=0}^N a_n = \lim_{r\uparrow 1}\sum_{n=0}^\infty a_n x^n$ is a discovery of Abel [**1**], whose life is discussed in the Notes to Section 10.3. In Chapter 13 of Part 2B, Section 6.9 of Part 3 and Sections 6.11/6.12 of Part 4, we'll return to (extensions of) Abel's theorem and its partial converses, known as Tauberian theorems.

Problems

1. Let a_n be the coefficients of a power series. Suppose $\lim_{n\to\infty}|a_n/a_{n+1}|$ exists. Prove that this is the radius of convergence of the series.

2. (a) For $z_0 \in \mathbb{D}$ and γ the curve $\gamma(t) = e^{2\pi i t}$, prove that

$$\frac{1}{2\pi i}\oint_\gamma (z-z_0)^{-1} z^n\, dz = \begin{cases} z_0^n, & n \ge 0 \\ 0, & n < 0 \end{cases}$$

(*Hint*: Expand $(z-z_0)^{-1} = \sum_{j=0}^\infty z_0^j z^{-j-1}$, prove you can interchange the sum and integral, and use Problem 4 of Section 2.2.)

(b) If $f(z) = \sum_{n=0}^\infty a_n z^n$ has radius of convergence greater than 1, prove that

$$f(z) - \sum_{n=0}^N a_n z^n = \frac{z^{n+1}}{2\pi i}\oint_\gamma \frac{f(w)}{w^{n+1}(w-z)}\, dw$$

3. This will lead to an alternate proof of Theorem 2.3.4. Define $g(z) = \text{rhs}$ of (2.3.9)

$$f_N(z) = \sum_{n=0}^{N} a_n(z - z_0)^n, \qquad g_N(z) = f_N'(z)$$

(a) For any $\delta' \in (0, \delta)$, prove that uniformly in $D_{\delta'}(z_0)$, $f_N \to f$, $g_N \to g$.
(b) By using the one-variable fundamental theorem of calculus and taking limits, prove for any $z \in D_\delta(z_0)$ and w with $|w| < \delta - |z - z_0|$, one has $f(z + w) = f(z) + \int_0^1 g(z + sw)w\, ds$.
(c) Prove that $f(z + w) - f(z) = g(z)w + o(|w|)$.

4. Let $\{a_n\}_{n=0}^\infty$ be a sequence of complex numbers with $a_n \to 0$ and

$$\sum_{n=0}^{\infty} |a_{n+1} - a_n| < \infty \tag{2.3.41}$$

Prove that $\sum_{n=0}^\infty a_n z^n$ is convergent (perhaps only conditionally) for all $z \in \partial\mathbb{D}$ with $z \neq 1$. (*Hint.* Compute $(1 - z)(\sum_{N+1}^M a_n z^n)$.) Note (2.3.41) holds, in particular, if $a_n \geq a_{n+1} \geq \cdots > 0$; this special case goes back to Picard.

5. Fix $\alpha \in \mathbb{R}$ and let $a_n = (n + 1)^\alpha$.

(a) Prove that the radius of convergence of (2.3.1) is 1 for all α.

(b) If $\alpha < -1$, prove the power series, (2.3.1), is absolutely convergent on $\partial\mathbb{D}$.

(c) If $\alpha \geq 0$, prove the series is divergent for any $z \in \partial\mathbb{D}$.

(d) If $-1 \leq \alpha < 0$, prove that the series is divergent at $z = 1$ and conditionally convergent if $z \in \partial\mathbb{D} \setminus \{1\}$. (*Hint.* Use Problem 4.)

(e) For $\alpha < 0$, let $c_\alpha = (\int_0^\infty x^{-\alpha-1} e^{-x}\, dx)^{-1}$. Note that $(n + 1)^\alpha = c_\alpha \int_0^\infty x^{-\alpha-1} e^{-(n+1)x}\, dx$ and conclude an integral formula for

$$\sum_{n=0}^{\infty} (n + 1)^\alpha z^n \tag{2.3.42}$$

that allows you to show this function has an analytic continuation to $\mathbb{C} \setminus [1, \infty)$. In particular, this function is regular at all points in $\partial\mathbb{D} \setminus \{1\}$.

Note. That the integral formula defines an analytic function requires some work. We prove a general theorem (Theorem 3.1.6) about the analyticity of such integrals of analytic functions. You may use this fact.

(f) If (2.3.42) defines the function $f_\alpha(z)$ for $z \in \mathbb{D}$, prove that

$$f_{\alpha+1}(z) = \left[\frac{d}{dz}(zf_\alpha)\right](z) \tag{2.3.43}$$

and conclude that f_α has an analytic continuation to $\mathbb{C} \setminus [1,\infty)$ also for all $\alpha > 0$. Thus, each f_α is regular at all points in $\partial\mathbb{D} \setminus \{1\}$.

(g) Prove that f_α is singular at $z = 1$ for any α. (*Hint.* Look at $\lim_{r\uparrow 1} f_\alpha^{(k)}(r)$.)

6. This will prove Appell's theorem that if $\alpha > 0$ and

$$\lim_{n\to\infty} n^{-\alpha} a_n = A \tag{2.3.44}$$

then

$$\lim_{x\uparrow 1}(1-x)^{\alpha+1} \sum_{n=0}^{\infty} a_n x^n = A\Gamma(\alpha+1) \tag{2.3.45}$$

where

$$\Gamma(\alpha+1) = \int_0^\infty e^{-t} t^\alpha \, dt \tag{2.3.46}$$

(a) If f is C^1 on $[0,\infty)$ and $\int_0^\infty (|f(y)| + |f'(y)|)\, dy < \infty$, prove that

$$\left| \sum_{n=0}^{\infty} f(n) - \int_0^\infty f(y)\, dy \right| \le \int_0^\infty |f'(y)|\, dy \tag{2.3.47}$$

(b) If $x = e^{-1}$, prove that as $t \downarrow 0$,

$$\left| \sum_{n=0}^{\infty} n^\alpha x^n - \int_0^\infty e^{-ty} y^\gamma \, dy \right| = O(t^{-\gamma}) \tag{2.3.48}$$

(c) Prove (2.3.46) for $a_n = n^\alpha$ and $A = 1$.

(d) Prove that (2.3.44) implies (2.3.45).

7. Let a_n be complex numbers so $|\frac{a_{n+1}}{a_n}| = 1 + \frac{\mu}{n} + o(\frac{1}{n})$.

(a) Prove the Taylor series $\sum_{n=0}^\infty a_n z^n$ has radius of convergence 1.

(b) If $\mu > 0$, prove $|a_n| \to \infty$, so on $\partial\mathbb{D}$, the series is divergent.

(c) If $\mu < -1$, prove the series is absolutely convergent on $\partial\mathbb{D}$.

(d) If $-1 < \mu < 0$, prove the series is not absolutely convergent on $\partial\mathbb{D}$ but can be conditionally convergent.

Remark. This is called Gauss' criteria.

8. (a) Prove Abel's convergence theorem [**1**]: If $\sum_{n=0}^N a_n \equiv S_N \to \alpha$, then $\lim_{r\uparrow 1} \sum_{n=0}^\infty a_n r^n \to \alpha$. (*Hint.* Prove first that $\alpha - \sum_{n=0}^\infty a_n r^n = (1-r)\sum_{n=0}^\infty (\alpha - S_n) r^n$.)

(b) Compute $1 - \frac{1}{2} + \frac{1}{3} - \frac{1}{4} + \dots$.

9. (a) Suppose f is analytic in $\mathbb{D}$, not identically zero, and $\lim_{r\uparrow 1} f^{(k)}(r) = 0$ for all k. Prove that $z = 1$ is a singularity of f.

(b) For $z \in \mathbb{D}$, let

$$f(z) = \exp\left(-\frac{1}{(1-z)^{1/2}}\right)$$

with the $\frac{1}{2}$ power chosen to be positive for $z \in (0,1)$ and continuous on $\mathbb{D}$. Prove that $z = 1$ is a singularity of f even though $\lim_{z\to 1,\, z\in\mathbb{D}} f(z) = 0$.

10. The binomial theorem says that $(z+w)^n = \sum \binom{n}{j} z^j w^{n-j}$, where $\binom{n}{j} = n!/[j!(n-j)!]$.

(a) Use the binomial theorem to provide another proof that $\exp(z+w) = \exp(z)\exp(w)$.

(b) Conversely, show that $\exp(z+w) = \exp(z)\exp(w)$ can be used to prove the binomial theorem.

11. Use the binomial theorem to prove (2.3.8). (*Hint*: Show first that for $j \geq 2$, $\binom{n}{j} \leq \frac{n(n-1)}{2}\binom{n-2}{j-2}$.)

12. By differentiating $\sum_{n=0}^{N} r^n = \frac{1-r^{N+1}}{1-r}$, find formulae for $\sum_{n=0}^{N} nr^{n-1}$ and $\sum_{n=0}^{N} n(n-1)r^{n-2}$ and prove the bounds claimed in the proof of Lemma 2.3.3.

13. Using $\sum_{j=0}^{n} \binom{n}{j} |z|^{n-j} |\zeta - z|^j \leq (|z| + |\zeta - z|)^n$, justify the resummation in the proof of Theorem 2.3.5.

14. (a) Prove $n\log(1+\frac{z}{n}) \to z$ and conclude (Euler, 1784) that $\lim_{n\to\infty}(1+\frac{z}{n})^n = e^z$.

(b) Let $f(x)$ be defined and C^2 on (a,b) with $f''(x) \leq 0$. Prove that for all $x, y \in (a,b)$, $f(\theta x + (1-\theta)y) \geq \theta f(x) + (1-\theta)f(y)$. (*Hint*: First show that one need only prove the result if $f(x) = f(y) = 0$, note this means there is a point in between $f'(z) = 0$, and see what this says about f' on $[x,z]$ and $[z,y]$.)

(c) If $f(x) = \log(x)$, show $f''(x) \leq 0$ and conclude for $x > -n$, $(n+1)\log(1+\frac{x}{n+1}) \geq n\log(1+\frac{x}{n})$.

(d) For $x > 0$, show $(1+\frac{x}{n})^n$ is monotone increasing in n. What does this say about compound interest?

(e) For $x > 0$, prove $(1-\frac{x}{n})^n$ is monotone decreasing in n for all $n > x$.

15. Find the various values of i^i. That this number is real was a popular curiosity in lectures given in the nineteenth century.

16. Let $\sum_{n=0}^\infty a_n(z-z_0)^n$ be a power series about z_0 with radius of convergence $r \in (0,\infty)$, so that $a_n \ge 0$ for all n. This problem will prove that z_0+r is a singularity, that is $f(z)=\sum_{n=0}^\infty a_n(z-z_0)^n$ cannot be analytically continued to any $\mathbb{D}_{\delta+r}(z_0)$ with $\delta>0$. Without loss, for notational convenience, you can suppose $z_0=0$, $r=1$. The proof will rely on the fact (see Theorem 1.4.4) that sets of positive numbers can be arbitrarily arranged without changing convergence vs. divergence. Assume analyticity in $\mathbb{D}_\delta(1)$ for $\delta>0$.

    (a) Let $a_n = f^{(n)}(0)/n!$, $b_n = f^{(n)}(1)/n!$. Prove that $b_n = \sum_{k=0}^\infty a_{k+m}\binom{k+n}{n}$. (*Hint:* $f^{(n)}(1)=\lim_{\varepsilon\uparrow 1} f^{(n)}(1-\varepsilon)$.)

    (b) Prove that $\sum_{n=0}^\infty b_n(\frac{\delta}{2})^n \pm \sum_{\ell=0}^\infty a_\ell(1+\frac{\delta}{2})^n$.

    (c) Conclude that the radius of convergence of $\sum_{n=0}^\infty a_n z^n$ is at least $1+\frac{\delta}{2}$, and so get a contradiction.

    Note. The theorem proven in this problem is due to Pringsheim [**459**] and Vivanti [**565**]. Its descendant in the hands of Landau (see Theorem 13.2.15 of Part 2B) will play a major role in Chapter 13 of Part 2B.

17. For $z\in\mathbb{D}$, let $f(z)=\sum_{n=1}^\infty z^{n!}$. Prove that $\lim_{r\uparrow 1}|f(re^{2\pi i p/q})|=\infty$ for any rational p/q. Conclude that $\partial\mathbb{D}$ is a natural boundary for f. Prove the same for $\sum_{n=1}^\infty z^{2^n}$.

18. (Lambert's Series) Let $d(n)$ be the number of divisors of n (i.e., integers among $1,\dots,n$ which divide n). For $z\in\mathbb{D}$, let

    $$f(z)=\sum_{n=1}^\infty d(n)\, z^n$$

    (a) Prove that the series has radius of convergence 1.

    (b) Show that

    $$f(z)=\sum_{n=1}^\infty\sum_{m=1}^\infty z^{nm}$$

    and conclude that

    $$f(z)=\sum_{n=1}^\infty \frac{z^n}{1-z^n}$$

    (c) Suppose that $n\not\equiv 0 \bmod q$. Prove that for $r<1$ and $z=re^{2\pi i p/q}$ that $|1-z^n|^2\ge 4r^n\sin^2\pi/q$ and conclude that

    $$\left|\sum_{n\not\equiv 0 \bmod q}\frac{z^n}{1-z^n}\right|\le (1-r)^{-1}\left(\sin\frac{\pi}{q}\right)^{-1}$$

(d) For $z = re^{2\pi ip/q}$, prove that

$$\left| \sum_{n \equiv 0 \bmod q} \frac{z^n}{1-z^n} \right| \geq (q(1-r))^{-1} \log(1-r^q)^{-1}$$

Conclude that for any p/q, $\lim_{r\uparrow 1} |f(re^{2\pi ip/q})| = \infty$ and therefore that f has a natural boundary on $\partial\mathbb{D}$.

Remarks. 1. We'll use the arguments in this problem again in Problem 7 of Section 15.4 of Part 2B.

2. Closely related to this example is the function

$$f(z) = \sum_{n=0}^{\infty} \frac{z^n}{1+z^{2n}} \tag{2.3.49}$$

of Weierstrass [**587**]. As in this problem, f is analytic on $\mathbb{D}$ and has a natural boundary on $|z| = 1$. But, by writing

$$f(z) = \sum_{n=0}^{\infty} \frac{1}{z^n + z^{-n}} \tag{2.3.50}$$

one sees that the sum also converges on $\mathbb{C} \setminus \overline{\mathbb{D}}$. Thus, the "same" expression represents two functions unrelated by continuation. Motivated by this example, Borel tried to develop a theory of continuing functions across barriers. This idea consumed him during most of his career. While his theory never attracted many followers, in its development he found major results on compact sets (Heine–Borel property), the notion of Borel sum, and his ideas in measure theory.

19. A power series of the form

$$f(z) = \sum_{k=0}^{\infty} a_{n_k} z^{n_k} \tag{2.3.51}$$

is called *Hadamard lacunary* if and only if for some $\delta > 0$ and all k,

$$n_{k+1} \geq (1+\delta)n_k \tag{2.3.52}$$

(i.e., most a_j are zero with longer and longer gaps). The *Hadamard gap theorem* says that any Hadamard lacunary series with radius of convergence 1 has $\partial\mathbb{D}$ as a natural boundary. Problem 17 is a simple example of this theorem. The present problem leads you through an elegant proof of Mordell of this result. Problem 11 of Section 6.2 has a closely related result with a different proof. You will need to assume two results mentioned in this section, but not proved until Section 3.1: holomorphic $\Leftrightarrow$ analytic and that a power series must have a singularity on its circle of convergence.

(a) Given any function f analytic in $\mathbb{D}$, for each integer m, let

$$g_m(z) = f(\tfrac{1}{2}\, z^m(1+z)) \tag{2.3.53}$$

Prove that if f is regular at $z=1$, then $g_m(z)$ is regular at all points of $\partial\mathbb{D}$. (*Hint*: If $z \in \partial\mathbb{D}$ with $z \neq 1$, $\frac{1}{2}z^m(1+z)$ is in $\mathbb{D}$.)

(b) If m is so large that $1/m < \delta$ and n_{k+1}, n_k obey (2.3.52), show that the power series in $[z^m(1+z)]^{n_k}$ are all less than the powers in $[z^m(1+z)]^{n_{k+1}}$. Use this and the Cauchy radius formula to conclude that for such m, the radius of convergence of the power series for g_m and for f are identical. (*Hint*: If $g_m(z) = \sum_{j=0} b_j z^j$, prove that $|b_{mn_k+[\frac{1}{2}n_k]}| \geq |a_n|/(n_k+1)$ to see $\limsup|b_j|^{1/j} \geq \limsup|a_j|^{1/j}$.)

(c) Combine (a) and (b) to show that $z=1$ is a singularity of f given by (2.3.51). By applying the same argument to $f_\theta(z) = \sum_{k=0}^\infty a_{n_k} e^{in_k\theta} z^{n_k}$, prove that f is singular at each point in $\partial\mathbb{D}$ and so conclude the Hadamard gap theorem.

20. Let f be defined in $\mathbb{D}$ by

$$f(z) = \sum_{n=0}^\infty \frac{z^{n!}}{\exp(\sqrt{n!})} \tag{2.3.54}$$

Prove that

(a) f has a natural boundary on $\partial\mathbb{D}$ using the Hadamard gap theorem.

(b) The power series for f and all its derivatives converge uniformly on $\overline{\mathbb{D}}$. In particular, $\lim_{r\uparrow 1} f(re^{i\theta})$ defines a C^∞ function of $e^{i\theta}$.

Remark. The function $\sum_{n=0}^\infty a^n z^{n^2}$ for $0 < |a| < 1$, first noted by Fredholm [**195**], has similar properties (if one uses the Fabry gap theorem). The reader might wish to check this.

21. Let $\sum_{n=0}^\infty a_n z^n$ and $\sum_{n=0}^\infty b_n z^n$ be two power series and define

$$c_n = \sum_{j=0}^n a_j b_{n-j} \tag{2.3.55}$$

(a) Prove that $\sum_{n=0}^\infty c_n z^n$ is the formal product in that for all N,

$$\left(\sum_{n=0}^N a_n z^n\right)\left(\sum_{n=0}^N b_n z^n\right) - \sum_{n=0}^{2N} c_n z^n$$

is a polynomial of degree $2N$ with powers $\{z^j\}_{j=N+1}^{2N}$ or a subset thereof.

(b) If R_a, R_b, R_c are the radius of convergence of the corresponding power series, prove that

$$R_c \geq \min(R_a, R_b) \tag{2.3.56}$$

and if $|z| < \min(R_a, R_b)$, then $(\sum_{n=0}^{\infty} a_n z^n)(\sum_{n=0}^{\infty} b_n z^n) = \sum_{n=0}^{\infty} c_n z^n$.

(c) Find an example where one does not have equality in (2.3.56).

22. A Toeplitz matrix is an infinite matrix $T = (t_{ij})_{1 \leq i,j < \infty}$ where $t_{ij} = \tau_{j-i}$ for a sequence $\{\tau_n\}_{n=-\infty}^{\infty}$. An upper-triangular matrix is one with $t_{ij} = 0$ if $i > j$. A UTT (for upper-triangular Toeplitz matrix) is the one with $\tau_n = 0$ for $n < 0$. Thus, $(\tau_0, \tau_1, \dots)$ corresponds to

$$U(\tau) = \begin{pmatrix} \tau_0 & \tau_1 & \tau_2 & \tau_3 & \cdots \\ 0 & \tau_0 & \tau_1 & \tau_2 & \cdots \\ 0 & 0 & \tau_0 & \tau_1 & \cdots \\ \vdots & \vdots & \vdots & \vdots & \ddots \end{pmatrix} \tag{2.3.57}$$

(a) If $\{a_n\}_{n=0}^{\infty}$, $\{b_n\}_{n=0}^{\infty}$ are two sequences of complex numbers, prove that the matrix product $U(a)U(b)$ is a $U(c)$, where c is given by (2.3.55) so that this is a matrix realization of formal power series multiplication.

(b) If $\{a_n\}_{n=0}^{\infty}$ is a formal power series with $a_0 \neq 0$, and $\{b_n\}_{n=0}^{\infty}$ is the series of $1/\sum_{n=0}^{\infty} a_n z^n$, prove $U(b) = U(a)^{-1}$.

(c) Use (b) and Cramer's rule to prove *Wronski's formula*

$$b_n = \frac{(-1)^n}{a_0^{n+1}} \det \begin{pmatrix} a_1 & a_2 & \cdots & a_n \\ a_0 & a_1 & \cdots & a_{n-1} \\ 0 & a_0 & \cdots & a_{n-2} \\ \vdots & \vdots & \vdots & \vdots \\ 0 & 0 & a_0 & a_1 \end{pmatrix}$$

Note. This formula was found by Wronski in 1811.

2.4. The Goursat Argument

In this section, we'll prove that if f is holomorphic in Ω, then $\oint_\gamma f(z)\,dz = 0$ for all triangles γ whose interior lies in Ω. The proof, following an argument of Goursat, is by contradiction: If the integral is $c \neq 0$, then by breaking the triangle in four, as in Figure 2.2.1, we get some triangle where the integral is at least $\frac{1}{4}c$ in magnitude. Iterating, we get triangles so small that the approximation guaranteed by being holomorphic gets a contradiction. Goursat's big technical advance is that this argument does not require a priori continuity of f'. And as we'll explore in the rest of this chapter, we can boost this result to a much more general one.

As already discussed in Section 2.2, a *triangle*, $T, [z_1, z_2, z_3]$, is the contour made up of three line segments joined together, where we'll suppose z_1, z_2, z_3 are not colinear to avoid the degenerate case. The *inside* of the triangle, $\mathrm{ins}(T)$, is the convex combinations of z_1, z_2, z_3, that is, $\{sz_1 + tz_2 + (1 - s - t)z_3 \mid s, t \geq 0,\, s + t \leq 1\}$. $\mathrm{ins}(T)$ includes T and is closed. Here's the implementation of the notion that (2.1.1) lets us estimate integrals over very small triangles:

Lemma 2.4.1. *Let f be holomorphic in a region Ω. Then for any $z_0, \varepsilon > 0$, there is a $\delta > 0$ so that $D_\delta(z_0) \subset \Omega$, and for all triangles $T \subset D_\delta(z_0)$, we have*

$$\left| \oint_T f(z)\, dz \right| \leq \varepsilon\, \ell(T) \sup_{z \in T} |z - z_0| \tag{2.4.1}$$

Proof. By (2.2.19),

$$\oint_T [f(z_0) + (z - z_0) f'(z_0)]\, dz = 0 \tag{2.4.2}$$

so

$$\text{LHS of (2.4.1)} = \left| \oint_T [f(z) - f(z_0) - (z - z_0) f'(z_0)]\, dz \right| \tag{2.4.3}$$

Thus, by (2.2.10),

$$\begin{aligned} \text{LHS of (2.4.1)} &\leq \ell(T) \sup_{z \in T} |f(z) - f(z_0) - f'(z_0)(z - z_0)| \\ &\leq \ell(T) o\big(\sup_{z \in T} |z - z_0|\big) \end{aligned} \tag{2.4.4}$$

by the assumption that f is holomorphic at z_0. This proves (2.4.1). $\square$

Theorem 2.4.2 (CIT for Triangles). *Suppose that f is holomorphic in a region, Ω. Let T_0 be a triangle with* $\mathrm{ins}(T_0) \subset \Omega$. *Then*

$$\oint_{T_0} f(z)\, dz = 0 \tag{2.4.5}$$

Proof. Suppose the integral is $c \neq 0$. Break T_0 into four triangles by using the midpoints of the edges as in Figure 2.2.1. The resulting four integrals sum to c, so at least one, call it T_1, must have an integral of magnitude at least $c/4$.

Iterating this process, we get triangles $T_0, T_1, T_2, \dots$ similar to T_0 with

$$\left| \oint_{T_n} f(z)\, dz \right| \geq \frac{1}{4^n} |c| \tag{2.4.6}$$

$$\mathrm{ins}(T_{n+1}) \subset \mathrm{ins}(T_n) \tag{2.4.7}$$

$$\ell(T_n) = \frac{1}{2^n} \ell(T_0) \tag{2.4.8}$$

$$\operatorname{diam}(T_n) = \frac{1}{2^n} \operatorname{diam}(T_0) \tag{2.4.9}$$

where

$$\operatorname{diam}(S) = \sup\{|x - y| \mid x, y \in S\} \tag{2.4.10}$$

By compactness and (2.4.7), $\bigcap_n \operatorname{int}(T_n) \neq \emptyset$ and, by (2.4.9), it is a single point, call it $z_0 \in \operatorname{int}(T_0) \subset \Omega$. By the lemma and (2.4.9), for any ε, there is an N so that for $n \geq N$,

$$\left| \oint_{T_n} f(z)\, dz \right| \leq \varepsilon\, \ell(T_n) \operatorname{diam}(T_n) \tag{2.4.11}$$

By (2.4.8) and (2.4.9), if we pick ε so that $\varepsilon\, \ell(T_0) \operatorname{diam}(T_0) < |c|$, this is inconsistent with (2.4.6). This contradiction proves that the integral is zero. □

Notes and Historical Remarks. In 1884, Goursat [**212**] gave a proof of the Cauchy integral theorem that involved subdivision into tiny rectangles and the fact that the $f(z_0) + (z - z_0)f'(z_0)$ has zero integral. Sixteen years later, in [**213**], he remarked: "I have known for some time that the proof of the theorem of Cauchy I gave in 1883 need not suppose the continuity of the derivative." Pringsheim [**460**], complaining about Goursat's direct use for general curves, gave the modern version using triangles to prove the existence of a global antiderivative for star-shaped regions. I have kept the common name "Goursat argument" although Goursat–Pringsheim argument or even Pringsheim argument might be more appropriate.

An even stronger result than Goursat's is the Looman–Menchoff theorem: In place of it being required that f have a complex derivative at each z_0, one demands that f is continuous, have directional derivatives $\frac{\partial f}{\partial x}$ and $\frac{\partial f}{\partial y}$ at each z_0, and obey the Cauchy–Riemann equations $\frac{\partial f}{\partial x} + i\frac{\partial f}{\partial y} \equiv 0$. This was first stated by Montel [**391**], a proof first appeared in Looman [**366**], and a gap filled in by Menchoff [**379**]. For a textbook presentation, see Narasimhan–Nievergelt [**404**]. For a simple proof and discussion of alternatives, see Gray–Morris [**218**]. Note the example in Problem 5 of Section 2.1 is not a counterexample since f is not continuous at 0.

We should note that there are two other proofs of the CIT in cases where at least C^1 f is assumed. One, essentially the proof of Cauchy, uses the Gauss–Green theorem, that is what we called Stokes' theorem in Section 1.5, for the case of one-dimensional curves surrounding two-dimensional regions: essentially if Q is a smooth image of a disk with simple smooth boundary, ∂Q, all in Ω, Stokes' theorem says that

$$\int_{\partial Q} f\, dz = \int_Q d(f\, dz) \tag{2.4.12}$$

and we have seen (see (2.1.37)) that $d(f\, dz) = 0$ if $\bar\partial f = 0$.

The other proof (which is essentially a variant of the one just given) shows $\oint_\gamma f\,dz = 0$ for rectangles by smoothly moving two opposite sides together. If the bottom is moved up by changing a parameter, t, then the t-derivatives of the integral has three terms: two come from the sides and, by the fundamental theorem of calculus, are the difference of f at the two bottom corners. The third term, which is the derivative of the bottom, is the integral along the bottom of $\partial f/\partial y$. But, by the CR equations, this is $-i\partial f/\partial x$ and this derivative can then be integrated to cancel the other two terms (the $-i$ is relevant since the sides have $dz = i\,dy$ and so also have i's relative to the $dz = dx$). This proof has the advantage of also showing directly the Cauchy integral formula for rectangles.

Problems

1. Let f be a C^∞ function of x and y in a neighborhood of $\overline{\mathbb{D}}$. Prove that

$$\oint_{|z|=1} f(z)\,dz = 2i\int_{|z|\le 1} (\bar\partial f)(z)\,d^2z \tag{2.4.13}$$

(*Hint.* Look at (2.4.12).)

2.5. The CIT for Star-Shaped Regions

Definition. A region, Ω, is called *star-shaped* if and only if there exists $z_0 \in \Omega$ so that for any $z_1 \in \Omega$, $[z_0, z_1] = \{(1-t)z_0 + tz_1 \mid 0 \le t \le 1\}$ lies in Ω. We'll call z_0 a *hub* for Ω.

Recall that a region is *convex* if for all $z_0, z_1 \in \Omega$, $[z_0, z_1] \subset \Omega$, so any convex set is star-shaped and any of its points is a hub.

Our main purpose in this section is to prove

Theorem 2.5.1 (CIT for Star-Shaped Regions). *Let Ω be a star-shaped region, f holomorphic in Ω, and γ a closed rectifiable curve. Then*

$$\oint_\gamma f(z)\,dz = 0 \tag{2.5.1}$$

Along the way, we'll also prove a result we'll dub the pre-Morera's theorem, since once we prove holomorphic functions are analytic, it will imply Morera's theorem (which is a converse to the CIT). We begin with

Theorem 2.5.2. *Let f be holomorphic in a star-shaped region, Ω. Then there is a holomorphic function, F, on Ω with $F' = f$.*

Proof. Pick z_0, a hub for Ω. Define, for $z \in \Omega$,

$$F(z) = \int_{[z_0,z]} f(w)\,dw \tag{2.5.2}$$

Given z_1, pick δ so $D_\delta(z_1) \subset \Omega$. Then for each $z_2 \in D_\delta(z_1)$, $[z_1, z_2] \subset \Omega$, so each line from z_0 to a point in $[z_1, z_2]$ is in Ω, so $\mathrm{int}([z_0, z_1, z_2]) \in \Omega$. Thus, by Theorem 2.4.2,

$$\int_{[z_0,z_1,z_2]} f(z)\,dz = 0 \tag{2.5.3}$$

which implies

$$\begin{aligned} F(z_2) - F(z_1) &= \int_{[z_1,z_2]} f(z)\,dz \\ &= f(z_1)(z_2 - z_1) + \int_{[z_1,z_2]} (f(z) - f(z_1))\,dz \end{aligned} \tag{2.5.4}$$

Thus, by (2.2.10),

$$|F(z_2) - F(z_1) - (z_2 - z_1)f(z_1)| \le |z_2 - z_1| \sup_{z\in[z_1,z_2]} |f(z) - f(z_1)| \tag{2.5.5}$$

By continuity of f, this last sup is $o(1)$, so the left-hand side of (2.5.5) is $o(|z_2 - z_1|)$, that is, F is holomorphic at z_1 and $F'(z_1) = f(z_1)$. $\square$

Proof of Theorem 2.5.1. Holomorphic functions are continuous, so F in the last theorem is holomorphic with a continuous derivative. Thus, (2.5.1) follows from Theorem 2.2.5. $\square$

The argument in the proof of Theorem 2.5.2 only used f holomorphic when it used CIT for triangles. Thus:

Theorem 2.5.3 (Pre-Morera Theorem). *If f is continuous in some $D_\delta(z_0)$ and obeys* (2.5.1) *for all triangles in $D_\delta(z_0)$, then there exists F holomorphic in $D_\delta(z_0)$ so that $F' = f$.*

We also have the following converse to Theorem 2.2.5:

Theorem 2.5.4. *Let f be a continuous function in a region, Ω, so that* (2.5.1) *holds for every closed rectifiable path. Then there exists a holomorphic function, F, on Ω so that $F' = f$.*

Proof. Fix $z_0 \in \Omega$ and let $z \in \Omega$. Then, if $\gamma, \tilde\gamma$ are two rectifiable curves from z_0 to z, $\tilde\gamma^{-1} * \gamma$ is a closed curve, so we can define $F(z) = \oint_\gamma f(w)\,dw$, which is the same as $\tilde\gamma$. Thus, $F(z)$ is well-defined. If $x \in D_\delta(z) \subset \Omega$, and then $F(x) - F(z) = \int_{[zx]} f(w)\,dw$ by the path independence, so $F' = f$ by the same argument used in the proof of Theorem 2.5.2. $\square$

2.6. Holomorphically Simply Connected Regions, Logs, and Fractional Powers

A good theorem deserves a definition.

Definition. A region, Ω, is called *holomorphically simply connected* (hsc) if and only if for all holomorphic functions, f, on Ω and all closed rectifiable curves, γ, we have (2.5.1).

We've seen star-shaped regions are hsc. We'll eventually prove (see Theorem 8.1.2) that Ω is hsc $\Leftrightarrow$ Ω is topologically simply connected (tsc). In this section, we'll first show that nonvanishing holomorphic functions on hsc regions have logs and fractional powers, then show that biholomorphic images of hsc regions are hsc, and use that to establish that annuli with a radius removed are hsc, a result we'll need in the next section. Then we'll prove tsc $\Rightarrow$ hsc and provide a second proof of the cut annulus result.

Theorem 2.6.1. *Let Ω be a holomorphically simply connected region and f a holomorphic function on Ω with f' holomorphic. Suppose f is nowhere vanishing on Ω. Then*

(a) *There exists h holomorphic on Ω with*

$$e^h = f \tag{2.6.1}$$

(b) *For each $n = 2, 3, \dots,$ there exists a holomorphic function g_n on Ω with*

$$(g_n)^n = f \tag{2.6.2}$$

Remarks. 1. Of course, we'll soon see that it is automatic that f' holomorphic if f is.

2. h is unique up to an overall additive $2\pi i n$. g_n is unique up to multiplication by an nth root of unity.

3. We'll sometimes write $\log f$ for h and $f^{1/n}$ for g_n.

Proof. Pick z_0 and then h_0 so $e^{h_0} = f(z_0)$. By hypothesis, f'/f is holomorphic, so since Ω is hsc, Theorem 2.5.4 says there is a holomorphic function F with $F' = f'/f$. Let $g = f\exp(-h_0 - F(z) + F(z_0))$. Then, by a direct calculation, $g' = 0$ and $g(z_0) = 1$. Therefore, by Corollary 2.2.6, $g(z) \equiv 1$, that is, if $h = F(z) + h_0 - F(z_0)$, then (2.6.1) holds.

If $g_n = \exp(h/n)$, then (2.6.2) holds. $\square$

Theorem 2.6.2 (Conformal Equivalence of hsc). *Let Ω be a holomorphically simply connected region. Let $\widetilde{\Omega}$ be another region for which there is a bijection, $g\colon \Omega \to \widetilde{\Omega}$ with g, g^{-1} holomorphic and g' holomorphic. Then $\widetilde{\Omega}$ is holomorphically simply connected.*

Remark. After Sections 3.1 and 3.4, we'll know that any holomorphic bijection, g, automatically has g^{-1} and g' holomorphic. Such maps are called *conformal equivalences* or *biholomorphic maps.*

Proof. Immediate from Theorem 2.2.1 (the change of variable in contour integrals) for given h holomorphic on $\widetilde{\Omega}$ and $\tilde{\gamma}$ a closed rectifiable curve in $\widetilde{\Omega}$, let $f = h \circ g$, $\gamma = g^{-1} \circ \tilde{\gamma}$ in (2.2.8). $\square$

Finally, we turn to a special class of regions. Given $0 \le r < R < \infty$, we define the *annulus*,

$$\mathbb{A}_{r,R} = \{z \mid r < |z| < R\} \tag{2.6.3}$$

and for $\alpha < \beta$ real with $|\beta - \alpha| \le 2\pi$, we define the sector,

$$\mathbb{S}_{\alpha,\beta} = \{re^{i\theta} \mid 0 < r, \alpha < \theta < \beta\} \tag{2.6.4}$$

For example, $\mathbb{S}_{-\pi,\pi} = \mathbb{C} \setminus (-\infty, 0]$. We are heading towards a proof of

Theorem 2.6.3. *For any $0 \le r < R < \infty$, and α, β with $\alpha < \beta$ and $|\beta - \alpha| \le 2\pi$, $\mathbb{A}_{r,R} \cap \mathbb{S}_{\alpha,\beta}$ is holomorphically simply connected.*

Lemma 2.6.4. *If $\alpha < \beta$, $|\beta - \alpha| \le 2\pi$, and*

$$\cos\left(\frac{\beta - \alpha}{2}\right) > \frac{r}{R} \tag{2.6.5}$$

then $\mathbb{A}_{r,R} \cap \mathbb{S}_{\alpha,\beta}$ is star-shaped.

Remark. (2.6.5) then implies $|\beta - \alpha| < \pi$.

Proof. If $r = 0$ and $|\beta - \alpha| < \pi$, the region is convex, so we need only consider the case $r > 0$. By rotation, we can suppose $\alpha = -\beta$. By some simple geometry (see Figure 2.6.1), the two tangents to $\mathbb{D}_0(r)$ at $re^{\pm i\beta}$ meet at the point $r/\cos\beta = z_0$. This is in $\mathbb{A}_{r,R} \cap \mathbb{S}_{-\beta,\beta}$ if (2.6.5) holds and the same geometry shows that every z in this region has $[z_0, z]$ in the region. So, the region is star-shaped. $\square$

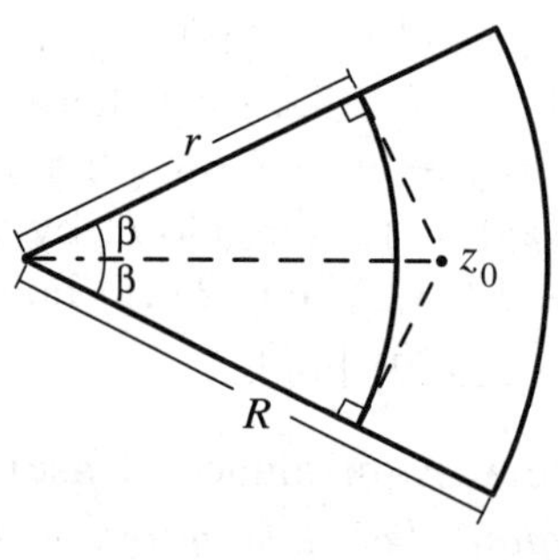

Figure 2.6.1. Showing an annular slice is star-shaped.

Proof of Theorem 2.6.3. Without loss, we suppose $\alpha = -\beta$ with $\beta \in (0,\pi]$. $\mathbb{C}\setminus(-\infty,0]$ is star-shaped about $z_0 = 1$. So, by Theorem 2.6.1, $g(z) = z^{1/n}$ is holomorphic on $\mathbb{C}\setminus(-\infty,0]$, where we take the branch with $g(1) = 1$. g maps $\Omega = \mathbb{A}_{r,R}\cap\mathbb{S}_{-\beta,\beta}$ to $\mathbb{A}_{r^{1/n},R^{1/n}}\cap\mathbb{S}_{-\frac{\beta}{n},\frac{\beta}{n}} \equiv \widetilde{\Omega}'_n$. Since $(\frac{r}{R})^{1/n} = 1 - \frac{1}{n}\log(\frac{R}{r}) + O(\frac{1}{n^2})$ with $\cos(\frac{\beta}{n}) = 1 - \frac{\beta^2}{6n^2} + O(\frac{1}{n^2})$, for large n, $(\frac{r}{R})^{1/n} < \cos(\frac{\beta}{n})$ which, by the lemma, says that for n large, $\widetilde{\Omega}_n$ is star-shaped. Thus, Ω, as the image of $\widetilde{\Omega}_n$ under $z \mapsto z^n$, is holomorphically simply connected. □

Finally, we turn to issues of homotopy and invariance of contour integrals under homotopic changes. Recall (see Section 1.7) that two closed curves, $\gamma,\tilde{\gamma}\colon [0,1]\to\Omega$, with $\gamma(0)=\gamma(1)=\tilde{\gamma}(0)=\tilde{\gamma}(1)=z_0$ are called *homotopic* (in Ω) if and only if there exists a continuous function $\Gamma\colon [0,1]\times[0,1]\to\Omega$ so that

$$\Gamma(t,0)=\gamma(t), \qquad \Gamma(t,1)=\tilde{\gamma}(t), \qquad \Gamma(0,s)=\Gamma(1,s)=z_0 \tag{2.6.6}$$

and that a region, Ω, is called *topologically simply connected* (tsc) if and only if every closed curve, γ, is homotopic to the trivial curve $\tilde{\gamma}(t)\equiv z_0$ (this is easily shown (see Problem 1) to be independent of z_0). We'll eventually prove that

$$\Omega \text{ tsc} \Leftrightarrow \Omega \text{ hsc} \tag{2.6.7}$$

Here we'll prove $\Rightarrow$ from

Theorem 2.6.5. *Let f be holomorphic on a region, Ω. Let $\gamma,\tilde{\gamma}$ be the rectifiable curves which are homotopic. Then*

$$\oint_\gamma f(z)\,dt = \oint_{\tilde{\gamma}} f(z)\,dt \tag{2.6.8}$$

In particular, tsc $\Rightarrow$ hsc.

Remark. We are dealing here with homotopy as maps from $[0,1]$ with fixed points. One can just as well deal with homotopies as maps from S^1 without fixed endpoints by noting that by splicing on a rectifiable curve from a fixed point to the image of S^1 at the start and end, one can reduce to the case of fixed endpoints in $[0,1]$.

Proof. We will need a technical fact whose proof we'll leave to Problem 2: If $\gamma,\tilde{\gamma}$ are rectifiable and homotopic, then they are homotopic under a homotopy Γ with

$$\gamma_s(\cdot)\equiv\Gamma(\,\cdot\,,s) \tag{2.6.9}$$

rectifiable for all s. We'll also need a simple fact (Problem 3): A locally constant function, f, on $[0,1]$ is constant (i.e., if for all $s_0\in[0,1]$, $\exists\varepsilon$ so $f(s)=f(s_0)$ for $s_0-\varepsilon<s<s_0+\varepsilon$, then f is constant).

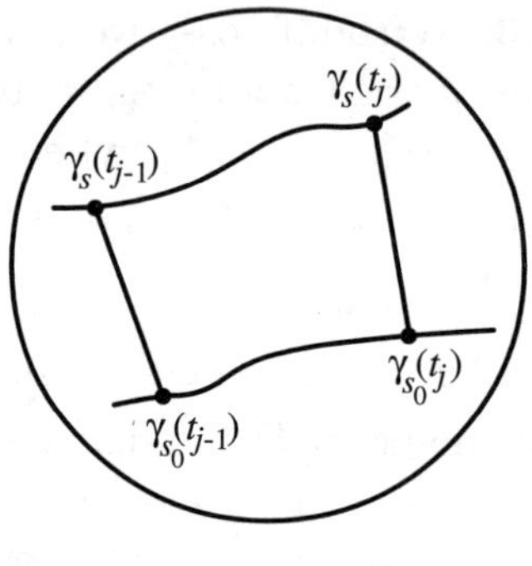

Figure 2.6.2. Subdividing a homotopy.

So suppose that Γ is a homotopy with each γ_s rectifiable and define for $s \in [0,1]$,

$$C(s) = \oint_{\gamma_s} f(z)\, dz \tag{2.6.10}$$

By Problem 3, we need only show C is locally constant, so pick $s_0 \in [0,1]$. Let $\delta = \frac{1}{2}\text{dist}(\text{Ran}(\gamma_{s_0}), \mathbb{C}\backslash\Omega)$ (or $\delta = 1$ if $\Omega = \mathbb{C}$), so $\{U_t \equiv \mathbb{D}_\delta(\gamma_{s_0}(t))\}$ is an open cover of $\text{Ran}(\gamma_{s_0})$ lying in Ω. By the paving lemma (Proposition 2.2.7), we can find $t_0 = 0 < t_1 < \cdots < t_n = 1$ and $U_1, \dots, U_n$ among these disks so that $\gamma_{s_0}([t_{j-1}, t_j]) \subset U_j$.

Pick ε, so if $|s - s_0| < \varepsilon$ and $t \in [t_{j-1}, t_j]$, then $\gamma_s(t) \subset U_j$, which we can do since γ is jointly uniformly continuous. Given $s \in (s_0 - \varepsilon, s_0 + \varepsilon)$, let η_j be the contour which goes from $\gamma_{s_0}(t_{j-1})$ to $\gamma_{s_0}(t_j)$ along γ_{s_0}, from $\gamma_{s_0}(t_j)$ to $\gamma_s(t_j)$ by a straight line, from $\gamma_s(t_j)$ to $\gamma_s(t_{j-1})$ by going backwards along γ_s, and then going from $\gamma_s(t_{j-1})$ to $\gamma_{s_0}(t_{j-1})$ by a straight line (see Figure 2.6.2).

Since U_j is a disk, so star-shaped, and η_j is a closed rectifiable curve in U_j,

$$\oint_{\eta_j} f(z)\, dz = 0$$

Because the added linear paths cancel in this sum

$$\sum_{j=1}^{n-1} \oint_{\eta_j} f(z)\, dz = \oint_{\gamma_{s_0}} f(z)\, dz - \oint_{\gamma_s} f(z)\, dz \tag{2.6.11}$$

proving the required local constancy. □

Here is another proof of Theorem 2.6.3: Notice first that if Ω is star-shaped with z_0 as a hub, then for any closed curve starting and ending at z_0,

$$\Gamma(t,s) = (1-s)\gamma(t) + sz_0 \tag{2.6.12}$$

is a homotopy to the trivial curve, so Ω is tsc. Moreover, if f is a homeomorphism of Ω to $\widetilde{\Omega}$ and Ω is tsc, then so is $\widetilde{\Omega}$, since f sets up bijections between closed curves and between homotopies.

In particular, $F\colon (0,1)\times(0,1)\to \mathbb{A}_{r,R}\cap \mathbb{S}_{\alpha,\beta}$ by

$$F(t,s) = \big[(1-t)r + tR\big]\,\exp\big(i\big[(1-s)\alpha + s\beta\big]\big)$$

is a homeomorphism, providing a second proof of Theorem 2.6.3.

Notes and Historical Remarks. The most complicated part of the homotopy argument we give is the fact proven in Problem 2. For the most important application we give, namely, that biholomorphic images of star-shaped regions are hsc, it can be avoided: for a star-shaped region, Ω, with hub z_0, the homotopy $\Gamma(t,s) = (1-s)\gamma(t) + sz_0$ clearly has rectifiable intermediate curves if γ is rectifiable. Applying the biholomorphic map to Γ shows any rectifiable curve in a biholomorphic image of a star-shaped region is homotopic to the trivial curve under a homotopy with intermediate rectifiable curves.

There is a proof of the homotopy equivalence result (Theorem 2.6.5), due to Výborný [**569**] and rediscovered by Hanche-Olsen [**235**], that mimics the Goursat strategy of subdivision and control of small rectangles.

Problems

1. Let Ω be an arcwise connected set. Suppose that any closed curve starting and ending at z_0 is homotopic to the constant curve. Prove that the same is true for any $z_1\in\Omega$. (*Hint*: Pick γ_0 with $\gamma_0(0)=z_1$, $\gamma_0(1)=z_0$. Given any closed curve with $\gamma(0)=\gamma(1)=z_1$, show it is homotopic to the curve $\gamma_0^{-1}*\gamma_0*\gamma*\gamma_0^{-1}*\gamma_0$, and then that this is homotopic to $\gamma_0^{-1}*\gamma_0$ which is homotopic to the trivial curve.)

2. This problem will lead you through a proof that two rectifiable homotopic curves are homotopic through a homotopy whose intermediate curves are all rectifiable.

   (a) Let γ_0,γ_1 be two curves with $\gamma_j(0)=z_0$, $\gamma_j(1)=z_1$ for $j=0,1$. Prove that $\gamma_t(s)\equiv(1-t)\gamma_0(s)+t\gamma_1(s)$ is rectifiable for each $t\in(0,1)$.

   (b) Let γ_0,γ_1 be homotopic rectifiable curves, homotopic under a homotopy $\Gamma\colon[0,1]\times[0,1]\to\Omega$ with $\gamma_j(s)=\Gamma(j,s)$, $\Gamma(t,0)=z$, $\Gamma(t,1)=z_1$. For each $N=1,2,\dots$, let $\Delta_{j,k}^N=\{(t,s)\mid \frac{j-1}{N}\le t\le\frac{j}{N}, \frac{k-1}{N}\le s\le\frac{k}{N}\}$ for $j,k=1,\dots,N$. Prove there is N so large that for each $j,k\in\{1,\dots,N\}$, there is a disk $\mathbb{D}_{jk}$ with $\Gamma([\Delta_{j,k}^N])\subset\mathbb{D}_{jk}\subset\Omega$.

   (c) On $\Delta_{j,k}^N$, define $\widetilde{\Gamma}$ so that at the four corners $\widetilde{\Gamma}(s,t)=\Gamma(s,t)$. If a top or bottom edge is part of γ_0 or γ_1, define $\widetilde{\Gamma}(j,s)=\gamma(j,s)$. On all other top and bottom edges, define $\widetilde{\Gamma}$ to be linear along the edge of $\Delta_{j,k}^N$. Fill in each square linearly, that is, for $j = 1, \dots, N$ and $\tau \in (0,1)$, for all s, define $\widetilde{\Gamma}$ by

$$\widetilde{\Gamma}\left((1-\tau)\frac{j-1}{N} + \tau\frac{j}{N}, s\right) = (1-\tau)\widetilde{\Gamma}\left(\frac{j-1}{N}, s\right) + \tau\widetilde{\Gamma}\left(\frac{j}{N}, s\right)$$

Prove that $\widetilde{\Gamma}$ is continuous with values in Ω.

(d) Prove that $\widetilde{\Gamma}(\,\cdot\,, s)$ is rectifiable.

3. Prove that a locally constant function is constant. (*Hint*: Show $\{s \mid f(s) = f(0)\}$ is open and closed.)

2.7. The Cauchy Integral Formula for Disks and Annuli

Our main goal in this section is to prove:

Theorem 2.7.1 (CIF for the Disk). *Let f be holomorphic in a neighborhood of $\overline{\mathbb{D}}$. Let $|z| = 1$ denote the counterclockwise contour around $\partial\mathbb{D}$ (used in Example* 2.2.2 *). Then for any $z_0 \in \mathbb{D}$,*

$$f(z_0) = \frac{1}{2\pi i}\int_{|z|=1} \frac{f(z)}{z - z_0}\, dz \tag{2.7.1}$$

Later, we'll also prove a similar formula for annuli.

Lemma 2.7.2. *Let $z_0 \in \mathbb{D}$. Let $\rho < 1 - |z_0|$. Let γ_1 be the contour indicated by $|z| = 1$ and γ_2 the analogous counterclockwise contour around the boundary of $\mathbb{D}_\rho(z_0)$. Then for any function, g, holomorphic in a neighborhood of $\overline{\mathbb{D}} \setminus \{z_0\}$, we have*

$$\int_{\gamma_1} g(z)\, dz = \int_{\gamma_2} g(z)\, dz \tag{2.7.2}$$

Proof. Let $z_0 = |z_0|e^{i\varphi}$. Let γ_ε be the boundary of $\mathbb{D} \setminus [\overline{\mathbb{D}_\rho(z_0)} \cup \{re^{i(\theta+\varphi)} \mid |\theta| < \varepsilon,\, r > 0\}]$ oriented so the part on $\partial\mathbb{D}$ is counterclockwise. γ_ε besides this arc on $\partial\mathbb{D}$ has, depending on ε and the relation of ρ to $|z_0|$, two or four linear pieces, and zero or two arcs of $\partial\mathbb{D}_\rho(z_0)$; see Figure 2.7.1. Notice that if $\delta > 0$ is chosen so that f is analytic in $\mathbb{D}_{1+\delta}(0)$, then f is analytic in $\mathbb{A}_{0,1+\delta} \cap \mathbb{S}_{\varphi+\frac{\varepsilon}{2}, 2\pi-\varphi-\frac{\varepsilon}{2}}$, a region that includes γ_ε. Thus, by Theorem 2.6.3, $\int_{\gamma_\varepsilon} g(z)\, d\varepsilon = 0$.

Letting $\varepsilon \downarrow 0$, the linear segments cancel and $\int_{\gamma_\varepsilon} \to \int_{\gamma_1} - \int_{\gamma_2}$, proving (2.7.2). $\square$

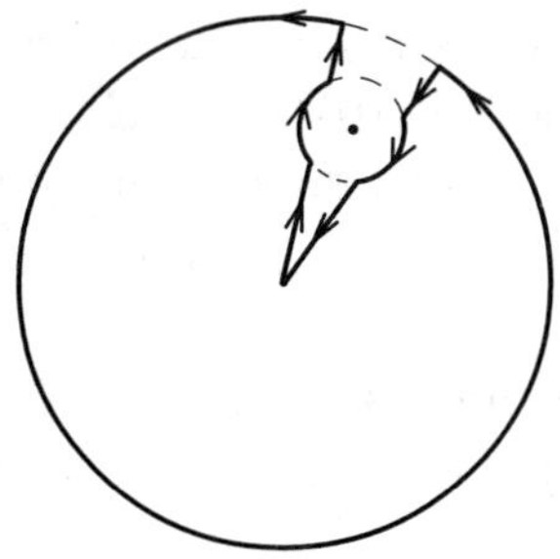

Figure 2.7.1. A keyhole contour.

The same proof shows

Theorem 2.7.3. *Let $z_1, \dots, z_k \in \mathbb{D}$ be distinct points. Let $\rho < \min\big(\min_j(1-|z_j|), \frac{1}{2}\min_{i,j}|z_i - z_j|\big)$. Let γ_0 be the curve we called $|z| = 1$ and γ_j the curve $|z - z_j| = \rho$ counterclockwise. Then*

$$\oint_{\gamma_0} g(z)\, dz = \sum_{j=1}^{k} \oint_{\gamma_j} g(z)\, dz \tag{2.7.3}$$

for any function analytic in a neighborhood of $\overline{\mathbb{D}} \setminus \{z_1, \dots, z_k\}$.

Lemma 2.7.4. *For any h continuous at $z = 0$,*

$$\lim_{\varepsilon \downarrow 0} \oint_{|z|=\varepsilon} \frac{h(z)}{z}\, dz = 2\pi i h(0) \tag{2.7.4}$$

Proof. By (2.2.7),

$$\oint \frac{h(0)}{z}\, dz = 2\pi i h(0)$$

so we need

$$\lim_{\varepsilon \downarrow 0} \oint_{|z|=\varepsilon} \frac{h(z) - h(0)}{z}\, dz = 0 \tag{2.7.5}$$

By (2.2.10), the integral is bounded by $(2\pi)\sup_{|z|=\varepsilon}|(h(z) - h(0))|$ which goes to zero since h is continuous. □

Proof of Theorem 2.7.1. By Lemma 2.7.2, the integral in (2.7.1) is equal to the integral over $|z - z_0| = \varepsilon$ for any ε. Taking $h(z) = f(z + z_0)$ and letting $\varepsilon \downarrow 0$, Lemma 2.7.2 proves this is $f(z_0)$. □

Theorem 2.7.5 (CIF for the Annulus). *Let f be holomorphic in a neighborhood of $\bar{\mathbb{A}}_{r,R}$ for some annulus with $r < R$. Then for any $z_0 \in \mathbb{A}_{r,R}$,*

$$f(z_0) = \frac{1}{2\pi i} \oint_{|z|=R} \frac{f(z)}{z - z_0}\, dz - \frac{1}{2\pi i} \oint_{|z|=r} \frac{f(z)}{z - z_0}\, dz \tag{2.7.6}$$

Proof. By essentially the same argument in Lemma 2.7.2 (but now γ_ε has arcs from both $\partial\mathbb{D}_R(0)$ and $\partial\mathbb{D}_r(0)$), one shows the integral on the right is $(2\pi i)^{-1}\oint_{|z-z_0|=\rho} f(z)(z-z_0)^{-1}\,dz$ for any $\rho < \min(R-|z_0|, |z_0|-r)$. Lemma 2.7.4 and the same limit argument as above completes the proof. $\square$

Notes and Historical Remarks. If f is a C^∞ function in a neighborhood of $\overline{\mathbb{D}}$, we have (see Problem 1) a "corrected" form of the CIF that reduces to the CIF if $\bar\partial f = 0$:

$$f(z_0) = \frac{1}{2\pi i}\oint \frac{f(z)}{z-z_0}\,dz - \frac{1}{\pi}\int \frac{(\bar\partial f)(z)}{z-z_0}\,d^2z \tag{2.7.7}$$

for $z_0 \in \mathbb{D}$. This is known as *Pompeiu's formula* after a work of Dimitrie Pompeiu (1873–1954) [**454**].

For another aspect of the CIF, see the proof of the Cauchy jump formula in Problem 1 of Section 4.8.

Problems

1. Prove Pompeiu's formula (2.7.7). (*Hint*: Prove first that away from $z = z_0$, $\bar\partial[f/(\cdot - z_0)] = \bar\partial f/(\cdot - z_0)$ and then apply Stoke's formula (see (2.4.13)) to $\overline{\mathbb{D}}$ with a small disk about z_0 removed.)

Chapter 3

Consequences of the Cauchy Integral Formula

It is to be noticed that this proof belongs to the most elementary class of proofs in that it calls for no explicit representation of the functions entering (e.g., by Cauchy's integral or by a power series).

—*Osgood* [**420**][1]

Big Notions and Theorems: Holomorphic ⇒ Analytic, Morera's Theorem, Weierstrass Convergence Theorem, Cauchy Estimates, Liouville's Theorem, Fundamental Theorem of Algebra, Weakly Analytic Function, Dunford's Theorem, Improved Cauchy Estimate, Borel–Carathéodory Theorem, Argument Principle, Rouché's Theorem, Winding Number, Inverse Function Theorem, Puiseux Series, Open Mapping Theorem, Maximum Principle, Schwarz Lemma, Laurent Series, Riemann Removable Singularity Theorem, Pole, Essential Singularity, Casorati–Weierstrass Theorem, Meromorphic Function, Rational Functions, Periodic Analytic Functions

Our last result, the Cauchy integral formula, implies a cornucopia of results. In Section 3.1, in rapid fire, we'll get holomorphic ⇒ analytic (power series), the Weierstrass convergence theorem, Cauchy estimates, Liouville's theorem, and the fundamental theorem of algebra. We'll also discuss Dunford's theorem on Banach space valued analytic functions. Section 3.2 will

[1] From Osgood's classic 1906 text on Complex Analysis discussing Morera's theorem.

address the following somewhat specialized result, but one that will be critical in Section 9.10. Cauchy estimates give information on Taylor coefficients in terms of $\max_{0\le\theta\le 2\pi}|f(re^{i\theta})|$. Suppose f is nonvanishing and we have control of $|f(z)|$. It will be natural to look at $\log(f(z)) = g(z)$, so control of $|f(z)|$ turns into upper bounds on $\operatorname{Re} g(z)$—not only isn't there $\operatorname{Im} g(z)$ but we don't have lower bounds. Remarkably, we'll still get information on Taylor coefficients. Section 3.3 will discuss the winding number: an integer associated to any closed rectifiable contour, γ, and point $z_0 \notin \operatorname{Ran}(\gamma)$ that captures the notion of how many times γ winds down around z_0. It will be the key to the ultimate CIF in the next chapter. Section 3.3 will also discuss the argument principle, the basic tool in counting and controlling zeros of analytic functions and use it to prove Rouché's theorem. Sections 3.4, 3.5, and 3.6 will discuss the local behavior of analytic functions, f. If $f(z)-f(z_0)$ has an nth order zero at z_0, then $f(z)$ will be locally n to 1 near z_0, and this will prove two global results, including the maximum principle: If f is nonconstant and analytic on Ω, $|f(z_0)|$ does not have any local maxima on Ω. Sections 3.7, 3.8, and 3.9 focus on isolated singularities, that is, what can happen if Ω is open, $z_0 \in \Omega$, and f is analytic on $\Omega \setminus \{z_0\}$. Finally, in Section 3.10, we'll discuss periodic functions, especially functions analytic on $\mathbb{C}$ with $f(z+\tau) = f(z)$ for some $\tau \neq 0$ and all z.

3.1. Analyticity and Cauchy Estimates

Here is a strong version of the fact that holomorphic implies analytic, thereby concluding the proof of (2.3.7).

Theorem 3.1.1 (Cauchy Power Series Theorem). *If f is holomorphic in $\mathbb{D}_\delta(z_0)$ for some $\delta > 0$, then f is analytic at z_0. Indeed,*

$$f(z) = \sum_{n=0}^{\infty} a_n (z-z_0)^n \tag{3.1.1}$$

converges on all of $\mathbb{D}_\delta(z_0)$ with

$$a_n = \frac{1}{2\pi i}\int_{|z-z_0|=\delta'} f(z)(z-z_0)^{-n-1}\,dz \tag{3.1.2}$$

for any $\delta' < \delta$.

Proof. By translation and scaling, we can suppose $z_0 = 0$, $\delta' = 1$, and $\delta > 1$. Then, by Theorem 2.7.1, we have

$$f(w) = \frac{1}{2\pi i}\int_{|z|=1} f(z)(z-w)^{-1}\,dz \tag{3.1.3}$$

For any $z \in \partial\mathbb{D}$, $w \in \mathbb{D}$, and N, we have

$$\begin{aligned}(z-w)^{-1} &= z^{-1}(1-wz^{-1}) \\ &= \sum_{n=0}^{N} w^n z^{-n-1} + w^{N+1} z^{-N-1}(z-w)^{-1}\end{aligned} \tag{3.1.4}$$

which, plugging into (3.1.3), implies

$$f(w) = \sum_{n=0}^{N} a_n w^n + R_N(w) \tag{3.1.5}$$

where a_n is given by (3.1.2) and

$$R_N(w) = w^{N+1}(2\pi i)^{-1} \int_{|z|=1} z^{-n-1} f(z)(z-w)^{-1}\, dz \tag{3.1.6}$$

Thus, by (2.2.10),

$$|R_N(w)| \le |w|^{N+1}(1-|w|)^{-1} \sup_{|z|=1} |f(z)| \tag{3.1.7}$$

goes to zero as $N \to \infty$ uniformly in $\{w \mid |w| < 1-\varepsilon\}$ for any $\varepsilon > 0$. $\square$

Putting together Theorems 2.3.4 and 3.1.1, we get

Theorem 3.1.2 (Fundamental Theorem of Complex Analysis). *Let f be a function on Ω, a region. If f is analytic on Ω, it is holomorphic there. Conversely, if f is holomorphic on Ω, it is analytic, and the radius of convergence of the series at $z_0 \in \Omega$ is at least* $\text{dist}(z_0, \partial\Omega)$.

Since f analytic $\Rightarrow$ f' analytic:

Corollary 3.1.3. *If f is holomorphic on Ω, so is f'.*

This corollary in turn implies:

Theorem 3.1.4 (Morera's Theorem). *Let f be a continuous function on a region, Ω, so that for all $z_0 \in \Omega$, there is $\delta > 0$, so $\oint_\gamma f\, dz = 0$ for all triangles, γ, in $\mathbb{D}_\delta(z_0)$. Then f is holomorphic on Ω.*

Proof. By Theorem 2.5.3, for every z_0, there is F holomorphic in $\mathbb{D}_\delta(z_0)$, so $F' = f$. By Corollary 3.1.3, f is holomorphic in $\mathbb{D}_\delta(z_0)$ and so at z_0. Since z_0 is arbitrary, f is holomorphic in Ω. $\square$

We want to emphasize one aspect of Theorem 3.1.4 that is important in some applications (see, e.g., Section 5.5). One does not need δ to be bounded away from zero as z_0 varies on compacts $K \subset \Omega$ (although, once one has the conclusion, $\delta = \text{dist}(K, \mathbb{C} \setminus \Omega)$ does work on all of K.)

As corollaries to this, we obtain two theorems:

Theorem 3.1.5 (Weierstrass Convergence Theorem). *Let Ω be a region and f_n a family of holomorphic functions on Ω. Let f be a function on Ω so for each compact $K \subset \Omega$, $\lim_{n\to\infty} \sup_{z\in K}|f(z) - f_n(z)| = 0$. Then f is holomorphic.*

Proof. For any $z \in \Omega$, pick δ so $\mathbb{D}_{2\delta}(z_0) \subset \Omega$. Then $f_n \to f$ uniformly on $\overline{\mathbb{D}_\delta(z_0)}$, so f is continuous on that set, so at z_0, so on Ω. By the Cauchy integral theorem for disks, for any triangle $\gamma \subset \mathbb{D}_\delta(z_0)$, $\int_\gamma f_n(z)\,dz = 0$. Since $\text{Ran}(\gamma)$ is compact, $f_n \to f$ uniformly on $\text{Ran}(\gamma)$, so by (2.2.11), $\int_\gamma f(z)\,dz = 0$. By Morera's theorem, f is holomorphic. □

Theorem 3.1.6. *Let Ω be a region, Q an abstract, σ-compact space, and $d\mu$ a complex Baire measure on Q. Suppose for each $\alpha \in Q$, we have an analytic function, $f(z;\alpha)$, so that $(z,\alpha) \mapsto f(z;\alpha)$ is jointly measurable and so that for any compact $K \subset \Omega$,*

$$\int_Q \sup_{z\in K}|f(z;\alpha)|\,d|\mu|(\alpha) < \infty \tag{3.1.8}$$

Then

$$z \mapsto \int_Q f(z;\alpha)\,d\mu(\alpha) \equiv g(z) \tag{3.1.9}$$

is an analytic function in Ω.

Proof. Let $\mathbb{D}_\delta(z_0)$ be an open disk in Ω and γ a triangle in $\mathbb{D}_\delta(z_0)$. By (3.1.8), Fubini's theorem implies that

$$\oint_\gamma g(z)\,dz = \int_Q \left[\oint_\gamma f(z;\alpha)\,dz\right] d\mu(\alpha) = 0 \tag{3.1.10}$$

By (3.1.8) and the dominated convergence theorem, g is continuous, so it is holomorphic by Morera's theorem. □

Remarks. 1. σ-compact spaces are defined in Section 2.3, measures on σ-compact spaces are discussed in Section 4.10, and Fubini's theorem in Section 4.11, all in Part 1.

2. Since we've noted that contour integrals are Stieltjes integrals, α can be another complex variable w and the integral over Q a contour integral.

Theorem 3.1.7. *Let $f(z) = \sum_{n=0}^\infty a_n(z-z_0)^n$ be analytic near z_0 with a finite, nonzero radius of convergence, R. Then f has a singularity on the circle of convergence.*

Proof. Without loss, by translation and scaling, we can take $z_0 = 0$ and $R = 1$. If $\partial\mathbb{D}$ has no singularities for every $z_1 \in \partial\mathbb{D}$, there is $\delta_1 > 0$, so

$f(z)$ can be continued to $\mathbb{D} \cup \mathbb{D}_{\delta_1}(z_1)$. By compactness, find $z_1, \dots, z_k$ and $\delta_1, \dots, \delta_k$ so $\bigcup_{j=1}^k \mathbb{D}_{\delta_j}(z_j)$ covers $\partial\mathbb{D}$.

Since, if $\mathbb{D}_{\delta_j}(z_j) \cap \mathbb{D}_{\delta_\ell}(z_\ell) \neq \emptyset$, this set also intersects $\mathbb{D}$, we can, by Corollary 2.3.9, continue to $\mathbb{D} \cup \bigcup_{j=1}^k \mathbb{D}_{\delta_j}(z_j)$, which is open, and contains $\overline{\mathbb{D}}$, and so some $\mathbb{D}_r(0)$ for $r > 1$. But then, by Theorem 3.1.2, the radius of convergence is $\geq r$, contradicting that $R = 1$. Thus, our assumption that $\partial\mathbb{D}$ has no singularities must be wrong! □

The explicit formula (3.1.2) immediately leads to bounds that will often be useful.

Theorem 3.1.8 (Cauchy Estimates). *If f is holomorphic in a neighborhood of $\overline{\mathbb{D}_R(0)}$ and $\sum_{n=0}^\infty a_n z^n$ is its power series expansion at $z_0 = 0$, then*

$$\text{(a) } |a_n| \leq R^{-n} \int_0^{2\pi} |f(Re^{i\theta})| \frac{d\theta}{2\pi} \tag{3.1.11}$$

$$\text{(b) } |a_n| \leq R^{-n} \sup_{0 \leq \theta \leq 2\pi} |f(Re^{i\theta})| \tag{3.1.12}$$

Remarks. 1. (3.1.12) is called a *Cauchy estimate.* It is sometimes written

$$|f^{(n)}(z_0)| \leq n! R^{-n} \sup_{0 \leq \theta \leq 2\pi} |f(z_0 + Re^{i\theta})| \tag{3.1.13}$$

if f is analytic in a neighborhood of $\overline{\mathbb{D}_{\mathrm{R}}(z_0)}$.

2. By a limit argument (from $R - \varepsilon$ to R), rather than assuming analyticity on a neighborhood of $\overline{\mathbb{D}_R(0)}$, one need only suppose f is continuous on $\overline{\mathbb{D}_R(0)}$ and analytic on $\mathbb{D}_R(0)$.

3. These estimates combine nicely to prove what is called the Weierstrass double series theorem; see Problem 3.

Proof. Since $\int_0^{2\pi} |g(e^{i\theta})| \frac{d\theta}{2\pi} \leq \|g\|_\infty$, (3.1.11) implies (3.1.12) and (3.1.11) is immediate from (3.1.2), (2.2.10), and $|z| = R$ on $\partial\mathbb{D}_R(0)$. □

A function, f is called *entire* if it is analytic on the entire complex plane, $\mathbb{C}$.

Theorem 3.1.9. *If f is an entire function and for some $\alpha \geq 0$ and $C < \infty$, we have that for all $z \in \mathbb{C}$,*

$$|f(z)| \leq C(|z| + 1)^\alpha \tag{3.1.14}$$

Then f is a polynomial of degree at most $[\alpha]$, the largest integer less than α. If $\alpha = m$, an integer, and

$$\lim_{|z| \to \infty} |z|^{-m} |f(z)| = 0 \tag{3.1.15}$$

then f is a polynomial of degree at most $m-1$ (where for $m=0$, this means $f\equiv 0$).

Proof. If $f(z)=\sum_{n=0}^{\infty} a_n z^n$, we know this converges for all $z\in\mathbb{C}$. The Cauchy estimate and (3.1.14) implies

$$|a_n| \le C(R+1)^\alpha R^{-n}$$

which implies, by taking $R\to\infty$, that $a_n=0$ if $n>\alpha$. That means $f(z)=\sum_{n=0}^{[\alpha]} a_n z^n$.

If $\alpha=m$ and (3.1.15) holds, the same argument shows that $a_m=0$. □

The special case $\alpha=0$ is called Liouville's theorem.

Theorem 3.1.10 (Liouville's Theorem). *A bounded entire function is a constant.*

As a consequence of this:

Theorem 3.1.11 (Fundamental Theorem of Algebra). *Every nonconstant polynomial, P, has a zero in $\mathbb{C}$.*

Proof. Since P is nonconstant, we can write

$$P(z)=\sum_{n=0}^{N} a_n z^n \tag{3.1.16}$$

with $a_N\ne 0$ and $N>0$. Suppose P is nonvanishing on all of $\mathbb{C}$. Then $f(z)=1/P(z)$ is an entire, nowhere vanishing function. Since $|P(z)|\,|z|^{-N}\to a_n$ as $z\to\infty$, we have in the same limit that

$$|f(z)|\to 0 \tag{3.1.17}$$

By Theorem 3.1.9 with $\alpha=0$, $f(z)\equiv 0$. This is a contradiction. □

Once one has this, standard algebra (see Problem 2) implies that if $\deg P=N$, there are N points $(z_1,\dots,z_N)$ (maybe not distinct) so that

$$P(z)=a_N\prod_{j=1}^{N}(z-z_j) \tag{3.1.18}$$

Later, following Theorem 3.3.4, we'll find a way to use complex variables to prove the existence of N zeros in one step.

Finally, we turn to analytic functions with values in a Banach space, X. This material will depend on some of the theory of Banach spaces discussed in Part 1. A function $f\colon\Omega\to X$ is called *weakly analytic* if and only if

$\ell \circ f : \Omega \to \mathbb{C}$ is analytic for all $\ell \in X^*$. One deep result about this notion is:

Theorem 3.1.12 (Dunford's Theorem). *If $f : \Omega \to X$, a complex Banach space, is weakly analytic, then it is analytic.*

Remarks. 1. Analyticity for Banach space-valued functions is defined at the end of Section 2.3.

2. While we have not stated it explicitly in the theorem, an important fact we'll prove below is a Cauchy integral formula for Banach space-valued functions, where the contour integral is a limit of Banach space-valued Riemann–Stieltjes sums. This in turn implies norm Cauchy estimates. The same method of proof shows that one can extend the CIF even in the general forms we'll discuss in Chapter 4. Gantmacher [**200**] discusses matrix-valued analytic functions. Basically, any result on bounds that holds in the scalar-valued case extends to the Banach space-valued case. What do not extend are results on zeros and factorization since one cannot take products of vector-valued functions. One can, of course, take products of matrix-valued functions and, as we'll explain in the Notes to Section 9.9, there are extensions of factorization results to the case of finite matrix-valued analytic functions. Because matrix products are noncommutative, they have new subtleties.

3. The same proof shows that if X and Y are complex Banach spaces and $f : \Omega \to \mathrm{BLT}(X, Y)$, the bounded linear transformation from X to Y, is such that $f(z)x$ is a strongly analytic Y-valued function, then $z \to f(z)$ is a strongly analytic $\mathrm{BLT}(X, Y)$-valued function. In particular, if $\ell(f(z)x)$ is analytic for all $x \in X$ and $\ell \in Y^*$, $z \to f(z)$ is strongly analytic.

4. Once one knows the equivalence, we can extend many of the scalar results we'll prove later, for example, the ultimate Cauchy formula, to the Banach space-valued case.

Proof. Fix $z_0 \in \Omega$. We'll start by proving continuity at z_0. Pick $\delta > 0$ so that $\overline{\mathbb{D}_\delta(z_0)} \subset \Omega$. Since $\ell \circ f$ is analytic in a neighborhood of $\overline{\mathbb{D}_\delta(z_0)}$, it is continuous there, and thus, for each $\ell \in X^*$,

$$\sup_{z \in \partial \mathbb{D}_\delta(z_0)} |\ell(f(z))| < \infty \tag{3.1.19}$$

Each $x \in X$ defines $\tilde{x}$ in X^{**} by $\tilde{x}(\ell) = \ell(x)$ with

$$\|\tilde{x}\| = \|x\| \tag{3.1.20}$$

by the Hahn–Banach theorem (Theorem 5.5.5 of Part 1). (3.1.19) says that for each $\ell \in X^*$,

$$\sup_{z\in\partial\mathbb{D}_\delta(z_0)} |\widetilde{f(z)}(\ell)| < \infty \tag{3.1.21}$$

so the uniform boundedness principle (Theorem 5.4.9 of Part 1) and (3.1.20) says that

$$\sup_{z\in\partial\mathbb{D}_\delta(z_0)} \|f(z)\| = F < \infty \tag{3.1.22}$$

Since $\ell \circ f$ obeys a CIF, we see, for $w \in \mathbb{D}_{\delta/2}(z_0)$, that

$$|\ell(f(w) - f(z_0))| \le \frac{1}{2\pi}\int_{|z-z_0|=\delta} \left|\frac{1}{z-w} - \frac{1}{z-z_0}\right| |\ell(f(z))|\,|dz| \tag{3.1.23}$$

$$\le \frac{2|z_0 - w|}{\delta} F\|\ell\| \tag{3.1.24}$$

so, by (3.1.20),

$$\|f(w) - f(z_0)\| \le 2|z_0 - w|\delta^{-1}F \tag{3.1.25}$$

proving norm continuity of f.

Once one has norm continuity, one can form the contour integral $(2\pi i)^{-1}\int f(z)(z-w)^{-1}\,dz$ for $w \in \mathbb{D}_\delta(z_0)$. Since this integral is a norm limit of Riemann–Stieltjes sums, one can interchange the application of $\ell \in X^*$ and integration, so

$$\ell\left(\frac{1}{2\pi i}\int_{|z-z_0|=\delta} f(z)(z-w)^{-1}\,dz\right) = \frac{1}{2\pi i}\int_{|z-z_0|=\delta} \ell(f(z))(z-w)^{-1}\,dz$$
$$= \ell(f(w)) \tag{3.1.26}$$

It follows that one has a Banach space-valued CIF:

$$f(w) = \frac{1}{2\pi i}\int_{|z-z_0|=\delta} f(z)(z-w)^{-1}\,dz \tag{3.1.27}$$

Given the CIF, one can mimic the proof of Theorem 3.1.1 to get norm analyticity and Cauchy estimates. □

Notes and Historical Remarks. Cauchy (who, you will recall, had his integral theorem by 1825 and integral formula by 1831) found the power series via geometric series and Cauchy estimates via the integral formula (3.1.2) by 1835. Without knowing of Cauchy's work, Weierstrass rediscovered Cauchy estimates in 1841, not using an integral formula but by series manipulations (see Problem 1). In the same work, he proved an earlier variant of the Weierstrass convergence theorem by power series manipulations.

Morera found the proof we use here. This earlier variant considered convergence of f's of the form $f_n(z) = \sum_{m=0}^{\infty}\left(\sum_{j=0}^{n} a_{jm}\right) z^m$ as $n \to \infty$ and so it is sometimes called the *Weierstrass double series theorem.*

Giacinto Morera (1856–1909) proved what we now call Morera's theorem in 1896 in [**396**]. In the same year, William Osgood (1864–1943) [**417**] found a similar result. Weierstrass, whose capsule biography appears in Section 9.4, found many facts about foundations of analysis (he invented $\varepsilon - \delta$ notation) and complex functions starting in the 1840s, but he published little. After he became a professor in Berlin in 1856, he attracted many students and his lectures were famous. Many of his results only became known through works of his students or when they pulled together his complete works in the 1890s.

What is called Liouville's theorem (a name going back at least to 1879) was first published in a note by Cauchy in 1844 [**107**].

The historical record [**370, 406, 519**] provides support for naming the theorem after Liouville. A month before Cauchy's note, Liouville did make a comment at an Academy meeting [**363**] that he had proven that every entire elliptic function is constant, a special case of what we now call Liouville's theorem (the special case appears later as Theorem 10.3.1). Liouville's notebooks have the full result. Liouville never published his results on these elliptic functions, but he explained them to two visiting German mathematicians, Borchardt and Joachimsthal in 1847 and he included them in his lectures at Collège de France in 1851. Through reports by Borchardt and Joachimsthal to Jacobi and Dirichlet and through the influential 1875 book of Briot and Bouquet [**75**] that codified much of complex analysis, this work became known widely enough that in 1880, Borchardt published what were essentially his lecture notes [**64**].

Remarkably, according to Remmert [**475**], the simple proof of Liouville's theorem from Cauchy estimates only appeared in 1883 in Jordan's *Cours d'Analyse.*

The notion of complex-valued functions of a complex variable was late in coming, and given the lack of closed formulae for solutions of polynomials of degree five or more, the notion we now call the fundamental theorem of algebra was only carefully considered around 1800, although there were several eighteenth century attempts, notably by Euler [**179**] in 1749 who handled degree 4 and 5. The first clear statement was in 1746 by Jean le Rond d'Alembert (1717–1783), whose proof was lacking. The first more-or-less full proof is due to Gauss around 1799. These initial works looked only at polynomials with real coefficients and, instead of talking about complex roots, stated that any real polynomial could be factored into real linear and real quadratic factors. Dunham [**153**] has a fascinating analysis of Euler's paper.

Dunford's theorem was proven in Dunford [**152**]. The proof we give is his proof.

Problems

1. This problem will lead you through Weierstrass' proof of Cauchy estimates.

(a) Let $\{b_j\}_{j=-m}^n$ be a finite set of complex numbers and let

$$G(z) \equiv \sum_{j=-m}^{n} b_j z^j \tag{3.1.28}$$

Prove that

$$|b_0| \le \max_{|z|=1} |G(z)| \equiv M \tag{3.1.29}$$

(*Hint*: Pick $\omega = \exp(2\pi i/(|n| + |m| + 1))$, so $\omega^j \neq 1$ for any $j \in \{-m, \dots, n\} \setminus \{0\}$ and note that

$$\sum_{\ell=0}^{k-1} G(\omega^\ell) = kb_0 + \sum_{j\neq 0} b_j \left[\frac{\omega^{jk} - 1}{\omega^j - 1}\right]$$

and conclude that

$$|b_0| \le M + \frac{1}{k}\left[\sup_{\substack{j\in\{-m,\dots,n\}\\ j\neq 0}} \frac{2}{|\omega^j - 1|} \sum_{j\neq 0} |b_j|\right]$$

and get (3.1.29) by taking $k \to \infty$.)

(b) Prove for any polynomial $P(z) = \sum_{n=0}^N a_n z^n$ and any k that

$$|a_k| R^k \le \sup_{|z|=R} |P(z)|$$

by reducing to $R = 1$ and then looking at $G(z) = z^{-k} P(z)$. From this, provide another proof of (3.1.12).

2. If P is a polynomial and $P(z_1) = 0$, prove that $P(z) = (z - z_1)Q(z)$, and so inductively, given the fundamental theorem of algebra, show that any polynomial of order N has the form (3.1.18).

3. Prove the *Weierstrass double series theorem*: If $f_n(z) = \sum_{m=0}^\infty a_{mn} z^m$ are analytic in $\mathbb{D}_r(0)$ and $\sum_{n=1}^N f_n \to F$ uniformly on each $\overline{\mathbb{D}_{r-\varepsilon}(0)}$, then $A_m = \sum_{n=1}^\infty a_{mn}$ converges for each m, and on $\mathbb{D}_r(0)$, $F(z) = \sum_{m=0}^\infty A_m z^m$.

4. If f is analytic on $\mathbb{D}$, say $f(z) = \sum_{n=0}^{\infty} a_n z^n$ and $|f(z)| < 1$ everywhere, then the Cauchy estimate for f in $\overline{\mathbb{D}}_r$ optimizing in r yields $|a_n| \le 1$. Now suppose instead that

$$|f(z)| < \frac{1}{1 - |z|}$$

What is the best bound you can get from Cauchy estimates for $|a_n|$? How does the bound behave as $n \to \infty$?

5. Let $f(z)$ be a bounded entire function. Compute for R large and $w, y \in \mathbb{C}$,

$$\frac{1}{2\pi i} \oint_{|z|=R} \frac{f(z)}{(z-w)(z-y)}\, dz$$

By taking $R \to \infty$, obtain a new proof of Liouville's theorem.

6. This will obtain yet another new proof of Liouville's theorem.

(a) For any function analytic in a neighborhood of $\overline{\mathbb{D}_r(z_0)}$, prove that

$$f(z_0) = \frac{1}{\pi r^2} \int_{\mathbb{D}_r(z_0)} f(w)\, d^2 w \tag{3.1.30}$$

(b) For any $z_0, w_0 \in \mathbb{C}$, prove that (with $|\cdot| = \text{area}$)

$$(\pi r^2)^{-1} |\mathbb{D}_r(z_0) \cap \mathbb{D}_r(w_0)| \to 0 \tag{3.1.31}$$

as $r \to \infty$ (see also Theorem 3.1.2 of Part 3).

(c) If f is bounded and entire, prove that for any z_0, w_0, we have $f(z_0) = f(w_0)$, that is, f is constant.

7. Let f be an injective holomorphic function in the unit disk, with $f(0) = 0$ and $f'(0) = 1$. If we write $f(z) = z + a_2 z^2 + a_3 z^3 \cdots$, then Bieberbach conjectured that $|a_n| \le n$ for all $n \ge 2$; this was proved by de Branges about 70 years after the conjecture was made! This problem outlines an argument to prove the conjecture under the additional assumption that the coefficients a_n are real.

(a) Let $z = re^{i\theta}$ with $0 < r < 1$, and show that if $v(r, \theta)$ denotes the imaginary part of $f(re^{i\theta})$, then

$$a_n r^n = \frac{2}{\pi} \int_0^{\pi} v(r, \theta) \sin n\theta \, d\theta$$

(b) Show that for $0 \le \theta \le \pi$ and $n = 1, 2, \dots$, we have $|\sin n\theta| \le n \sin \theta$.

(c) Use the fact that $a_n \in \mathbb{R}$ to show that $f(\mathbb{D})$ is symmetric with respect to the real axis, and use this fact to show that f maps the upper half-disk into either the upper or lower part of $\mathbb{C}$.

(d) Show that for r small, $v(r,\theta) = r\sin\theta[1+O(r)]$ and use the previous part to conclude that $v(r,\theta) \geq 0$ for all $0 < r < 1$ and $0 \leq \theta \leq \pi$.

(e) Prove that $|a_n r^n| \leq nr$, and let $r \to 1$ to conclude that $|a_n| \leq n$.

(f) Check that the function $f(z) = z/(1-z)^2$ satisfies all the hypotheses and that $|a_n| = n$ for all n.

8. (a) Recall the de Moivre formula, (2.3.40),

$$\cos n\theta + i\sin n\theta = (\cos\theta + i\sin\theta)^n$$

Using $\sin^2\theta = 1 - \cos^2\theta$, prove that there are polynomials $T_n(x)$ and $U_n(x)$ (called *Chebyshev polynomials* of the first and second kind) so that, for $n = 0, 1, \ldots$,

$$\cos n\theta = T_n(\cos\theta), \qquad \frac{\sin(n+1)\theta}{\sin\theta} = U_n(\cos\theta) \tag{3.1.32}$$

(b) Prove the recursion relations:

$$xT_n(x) = \tfrac{1}{2}\,(T_{n+1}(x) + T_{n-1}(x)), \qquad n \geq 1 \tag{3.1.33}$$

$$xT_0(x) = T_1(x), \qquad T_0(x) \equiv 1 \tag{3.1.34}$$

(c) Prove that $T_n(x) = \frac{1}{2^{n-1}}x^n +$ lower order for $n \geq 1$.

(d) Prove the recursion relations:

$$xU_n(x) = \tfrac{1}{2}\,U_{n+1}(x) + U_{n-1}(x), \qquad n \geq 1 \tag{3.1.35}$$

$$xU_0(x) = \tfrac{1}{2}\,U_1(x) \tag{3.1.36}$$

(e) Prove that $U_n(x) = \frac{1}{2^n}x^n +$ lower order.

(f) Prove the orthogonality relations:

$$\int_{-1}^{1} T_n(x)T_m(x)\,d\mu_1(x) = \begin{cases} \frac{1}{2}\,\delta_{nm}, & n \neq 0 \\ 1, & n = m = 0 \end{cases} \tag{3.1.37}$$

$$\int_{-1}^{1} U_n(x)U_m(x)\,d\mu_2(x) = \delta_{nm} \tag{3.1.38}$$

where $d\mu_1, d\mu_2$ are the probability measures

$$d\mu_1(x) = \frac{dx}{\pi(1-x^2)^{1/2}}, \qquad d\mu_2(x) = \frac{2}{\pi}\,(1-x^2)^{1/2}\,dx \tag{3.1.39}$$

Note. These polynomials are discussed further in Example 14.4.17 of Part 2B.

9. One often defines, or at least analyzes, a sequence of functions $\{f_n(z)\}_{n=0}^\infty$ by finding an explicit formula for

$$F(z,w) = \sum_{n=0}^\infty w^n f_n(z) \tag{3.1.40}$$

F is then called a *generating function* for f_n. One thing F determines is $\limsup_{n\to\infty}|f_n(z)|^{1/n}$ for fixed z as the nearest singularity of $F(z,w)$ in the w variable.

(a) The Chebyshev polynomials, $T_n(z)$, are defined by (3.1.32). Prove that for any z and $|w|$ small,

$$\sum_{n=0}^\infty \frac{w^n}{n} T_n(z) = -\tfrac{1}{2}\log[1-2zw+w^2] \tag{3.1.41}$$

(b) Prove that for all z,

$$\limsup_{n\to\infty} |T_n(z)|^{1/n} = \left|z+\sqrt{z^2-1}\,\right| \tag{3.1.42}$$

with a suitable choice of square root.

Notes. 1. With more effort, one can prove lim, not just lim sup in (3.1.42). Asymptotics of polynomials like T_n are discussed further in Section 4.4 of Part 4 and the references in its Notes.

2. See Problem 2 of Section 3.7 for a generating function definition of Bessel functions of integral order.

3. See Section 14.4 of Part 2B for more on Chebyshev polynomials.

10. Generating function methods can also be used for sequences of numbers. The *Fibonacci numbers* are defined by

$$F_{n+1} = F_n + F_{n-1}, \quad n \geq 2; \qquad F_1 = F_2 = 1 \tag{3.1.43}$$

Define

$$G(x) = \sum_{n=1}^\infty F_n x^n \tag{3.1.44}$$

(a) Prove that

$$G(x) = \frac{x}{1-x-x^2} \tag{3.1.45}$$

(b) Deduce a formula first found by de Moivre

$$F_n = \frac{[\varphi_+^n - (-\varphi_-)^n]}{\sqrt{5}} \tag{3.1.46}$$

where $\varphi_\pm = (\sqrt{5}\pm 1)/2$. φ_+ is called the *golden mean*.

(c) Conclude that

$$(F_n)^{1/n} \to \varphi_+ \quad \text{and} \quad \frac{F_{n+1}}{F_n} \to \varphi_+ \tag{3.1.47}$$

11. The *Bernoulli numbers*, B_n, are defined by the generating function

$$\frac{x}{e^x - 1} = \sum_{n=0}^{\infty} \frac{B_n x^n}{n!} \tag{3.1.48}$$

(a) Prove that the B_n obey

$$B_n = \sum_{k=0}^{n} \binom{n}{k} B_k \tag{3.1.49}$$

for $n \geq 2$ so that B_{n-1} is determined by $\{B_k\}_{k=0}^{n-2}$.

(b) Prove $B_0 = 1$ and deduce that B_n is rational.

(c) Compute B_k for $k = 1, 2, 3, 4$.

(d) Prove that

$$B_1 = -\tfrac{1}{2}, \qquad B_{2k+1} = 0 \quad \text{for } k \geq 1 \tag{3.1.50}$$

Hint:

$$x(e^x - 1)^{-1} + \frac{1}{2} x = \frac{1}{2} x \coth\left(\frac{x}{2}\right) \tag{3.1.51}$$

(e) Prove that the Taylor series for $\tan z$ is given by

$$\tan z = \sum_{k=1}^{\infty} \frac{B_{2k}}{(2k)!} (1 - 2^{2k}) 2^{2k} (-1)^k z^{2k-1} \tag{3.1.52}$$

(*Hint:* Prove and use

$$\frac{x}{e^x + 1} = \frac{x}{e^x - 1} - \frac{2x}{e^{2x} - 1} \tag{3.1.53}$$

an analog of (3.1.51) and $B_{2k+1} = 0$ for $k \geq 1$.)

Notes. 1. The rationality of B_n also follows from the rationality of the Taylor coefficients of e^x, together with the fact that the power series quotient, sum, and product preserve rationality.

2. The B's get complicated quite rapidly. For example, $B_{30} = 8615841276005/14322$ and the coefficients of z^{31} in (3.1.52) is 129848163681107301953/122529844256906551386796875. The huge denominator is a divisor of 31! and, in fact, the coefficient of z^{2k-1} multiplied by $(2k-1)!$ is an integer; see Problem 16 in Section 9.7.

3. The Bernoulli numbers and some associated polynomials will reappear in Problem 4 of Section 9.2, prominently in Section 9.7 and its Problems, and in Problem 3 of Section 13.3 of Part 2B.

3.2. An Improved Cauchy Estimate

In this section, we'll focus on an improved Cauchy estimate needed to prove the following result, related to the fact that Theorem 3.1.9 says if $|f(z)| \leq C(1+|z|)^n$, then f is a polynomial of degree at most n. The material in this section is needed in this volume only for Section 9.10.

Theorem 3.2.1. *Suppose f is an entire, nowhere vanishing function that obeys*

$$|f(z)| \leq Ae^{B|z|^\alpha} \tag{3.2.1}$$

for some constants A and B and some positive real α. Then

$$f(z) = e^{P(z)}$$

for some polynomial of degree at most $[\alpha]$.

Since f is nowhere vanishing, $g = \log(f)$ exists and is entire by Theorem 2.6.1. But (3.2.1) only gives a bound on $\operatorname{Re} f$—and only an upper bound at that. Thus, we need a strengthening of the Cauchy estimate to require only this information.

Theorem 3.2.2. *Let f be analytic in a neighborhood of $\overline{D_R(0)}$ and let* (2.3.1) *be the power series about* 0. *Then for any $k \geq 1$,*

$$|a_k|R^k \leq 4 \max_{0\leq\theta\leq 2\pi} \operatorname{Re} f(Re^{i\theta}) + 4|f(0)| \tag{3.2.2}$$

Proof. If $g(z) = f(z) - f(0)$, (3.2.2) for g implies the result for f, so, without loss, we can suppose $f(0) = 0$. By (3.1.2), we then have

$$a_k R^k = \int f(Re^{i\theta})e^{-ik\theta}\, \frac{d\theta}{2\pi} \tag{3.2.3}$$

$$0 = \int f(Re^{i\theta})\, \frac{d\theta}{2\pi} \tag{3.2.4}$$

Moreover, since $f(z)z^{k-1}$ is analytic, $\oint_{|z|=R} f(z)z^{k-1}\, dz = 0$, so

$$0 = \int f(Re^{i\theta})e^{ik\theta}\, \frac{d\theta}{2\pi}$$

If $a_k = |a_k|e^{i\varphi}$, we thus have

$$|a_k|R^k = 2\int f(Re^{i\theta})\cos(k\theta+\varphi)\, \frac{d\theta}{2\pi} \tag{3.2.5}$$

$$= 2\int [\operatorname{Re} f(Re^{i\theta})]\cos(k\theta+\varphi)\, \frac{d\theta}{2\pi} \tag{3.2.6}$$

$$\leq 2\int |\operatorname{Re} f(Re^{i\theta})|\, \frac{d\theta}{2\pi} \tag{3.2.7}$$

$$= 2\int \left[|\mathrm{Re}\, f(Re^{i\theta})| + \mathrm{Re}\, f(Re^{i\theta})\right] \frac{d\theta}{2\pi} \tag{3.2.8}$$

$$\leq 4 \max_{0\leq\theta\leq 2\pi} \left[\mathrm{Re}\, f(Re^{i\theta})\right] \tag{3.2.9}$$

where (3.2.6) comes from noting that the left side of (3.2.5) is real, as is $\cos(k\theta+\varphi)$, (3.2.8) comes from (3.2.7) and (3.2.4), and (3.2.9) comes from noting that

$$|x| + x = 2\max(x, 0) \tag{3.2.10}$$

and that, since (3.2.4) holds, $m \equiv \max_{0\leq\theta\leq 2\pi} \mathrm{Re}(f(Re^{i\theta})) \geq 0$. So for all θ, $0 \leq m$ and $\max(\mathrm{Re}(f(Re^{i\theta}), 0)) \leq m$. □

Proof of Theorem 3.2.1. Since f is nonvanishing, $f = e^g$ with g entire (by Theorem 2.6.1). (3.2.1) then says

$$\mathrm{Re}\, g(z) \leq \log(A) + B|z|^\alpha \tag{3.2.11}$$

So, by (3.2.2), for all $k \geq 1$ and $g(z) = \sum b_k z^k$,

$$\tfrac{1}{4}\,|b_k| \leq \big(\log(A) + |g(0)|\big) R^{-k} + BR^{\alpha-k} \tag{3.2.12}$$

Thus, taking $R\to\infty$, $b_k = 0$ for $k > \alpha$, which says g is a polynomial of degree at most $[\alpha]$. □

Notes and Historical Remarks. The main inequality, Theorem 3.2.2, in this section is closely related to a famous theorem of Borel [**66**] and Carathéodory [**85**] that if $0 < r < R$, then

$$\max_{|z|\leq r} |f(z)| \leq \frac{2r}{R-r} \max_{|z|=R} \mathrm{Re}\, f(z) + \frac{R+r}{R-r}\,|f(0)| \tag{3.2.13}$$

In Problem 1, the reader is asked to show that when $f(0) = 0$, (3.2.2) leads to (3.2.13) with an extra factor of two. In Section 3.6, (3.2.13) will be discussed further (see Problem 12 of that section).

Problems

1. Suppose $f(0) = 0$. Let $r < R$ and f analytic in a neighborhood of $\overline{D_R(0)}$. Use (3.2.2) to prove that

$$\max_{|z|=r} |f(z)| \leq \frac{4r}{R-r} \max_{|z|=R} \mathrm{Re}\, f(z)$$

(see the discussion in the Notes).

3.3. The Argument Principle and Winding Numbers

This is the first of several sections that analyze the local behavior of an analytic function. Here we'll introduce the key tools used in this analysis. First, we'll state a result about counting zeros in a disk in terms of a contour integral on the boundary, use this to prove Rouche's theorem and a local counting result that will be the key to the local analysis of the next section. Then we'll discuss why the principle is called the argument principle. Finally, we'll begin a discussion, concluded in Section 4.3, on the argument principle for general curves and use this to define the winding number of curves about a point, a critical notion for the discussion of the ultimate CIT in Section 4.2.

We've already seen (see the proof of Proposition 2.3.6) that if f is analytic, $f(z_0) = 0$, and f is not identically zero, then there is an integer m so that

$$f(z) = (z - z_0)^m g(z_0) \tag{3.3.1}$$

with $g(z_0) \neq 0$. m is determined by $f(z_0) = f'(z_0) = \cdots = f^{(m-1)}(z_0) = 0$, $f^{(m)}(z_0) \neq 0$, that is, the first nonzero term in the Taylor series at z_0 is at m. We call m the *order* or *multiplicity* of the zero at z_0.

Because zeros are isolated, if $\bar{S} \subset \Omega$ with $\bar{S}$ compact, f has only finitely many zeros in S. We'll systematically use the phrase *the number of zeros of f in S*, not to refer to the number of points, z_0, in S where $f(z_0) = 0$, but rather to the sum of the multiplicities of the zeros in S. We'll sometimes say "the number of zeros of f in S counting multiplicities" for emphasis, but we *always* mean "counting multiplicities" in the terminology. We'll use $N_S(f)$ for this number, sometimes dropping the S if it is clear, and for any $a \in \mathbb{C}$,

$$N_S(f, a) = N_S(f - a) \tag{3.3.2}$$

Lemma 3.3.1. *Let f be analytic in a region, Ω. Let $z_0 \in \Omega$ be a zero of order m. Define*

$$g(z) = \begin{cases} f(z)/(z - z_0)^m, & z \neq z_0 \\ f^{(m)}(z_0)/m!, & z = z_0 \end{cases}$$

Then g is analytic on Ω.

Proof. By Theorem 2.1.1(b), f is holomorphic at any $z \neq z_0$. By writing out the Taylor series of f at z_0, one sees that g is analytic at z_0. □

Theorem 3.3.2 (Argument Principle). *Let f be analytic in a neighborhood of some $\overline{\mathbb{D}_r(z_0)}$ and suppose f is nonvanishing on all of $\partial\mathbb{D}_r(z_0)$. Then*

$$N_{\mathbb{D}_r(z_0)}(f) = \frac{1}{2\pi i} \oint_{|z-z_0|=r} \frac{f'(z)}{f(z)}\, dz \tag{3.3.3}$$

More generally, if q is another function analytic in a neighborhood of $\overline{\mathbb{D}_r(z_0)}$ and $\{z_1, \dots, z_k\}$ are the distinct zeros of f in $\mathbb{D}_r(z_0)$ with orders $m_1, \dots, m_k$, then

$$\frac{1}{2\pi i}\oint_{|z-z_0|=r} \frac{q(z)f'(z)}{f(z)}\,dz = \sum_{j=1}^{k} m_j q(z_j) \tag{3.3.4}$$

Remark. (3.3.4) for $q \equiv 1$ is (3.3.3).

Proof. Let $\{z_1, \dots, z_k\}$ be the distinct zeros of f in $\mathbb{D}$ and $m_1, \dots, m_k$ their orders. Since f is holomorphic in a neighborhood of $\overline{\mathbb{D}_r(z_0)}$ and zeros of holomorphic functions are isolated, there are only finitely many zeros. By iterating the above lemma, we can write

$$f(z) = \prod_{j=1}^{k} (z - z_j)^{m_j} g(z) \tag{3.3.5}$$

with g holomorphic in a neighborhood of $\overline{\mathbb{D}_r(z_0)}$ and nonvanishing there.

By Leibniz's rule, if $h = f_1 \dots f_\ell$, then

$$\frac{h'}{h} = \sum_{j=1}^{\ell} \frac{f_j'}{f} \tag{3.3.6}$$

so, on $\partial\mathbb{D}_r(z_0)$,

$$\frac{qf'}{f} = \frac{qg'}{g} + \sum_{j=1}^{k} \frac{m_j q}{z - z_j} \tag{3.3.7}$$

Since g'/g is holomorphic in a neighborhood of $\partial\mathbb{D}_r(z_0)$, the integral of this function over $|z - z_0| = r$ is 0. By the Cauchy integral formula for the function q, for any $z_j \in \mathbb{D}_r(z_0)$,

$$\frac{1}{2\pi i}\oint_{|z-z_0|=r} \frac{q(z)dz}{z - z_j} = q(z_j) \tag{3.3.8}$$

Thus, (3.3.4) holds by (3.3.7). For $q \equiv 1$, (3.3.3) is (3.3.4). $\square$

Corollary 3.3.3. *Let $f_t(z)$, $0 \le t \le 1$, be a family of functions analytic in some $\mathbb{D}_{r+\varepsilon}(z_0)$, for $\varepsilon > 0$. Suppose $f_t(z)$ is continuous in t for all $z \in \mathbb{D}_{r+\varepsilon}(z_0)$, uniformly in z in compact subsets of this set. Suppose that, for all $t \in [0,1]$, $f_t(z)$ is nonvanishing on $\partial\mathbb{D}_{r+\varepsilon}(z_0)$. Then $N_{\mathbb{D}_r(z_0)}(f_t)$ is independent of t.*

Proof. By the Cauchy formula (3.1.2) for $f_t'(z_0)$, this function is also continuous in t, uniformly in z_0 on compact subsets of $\mathbb{D}_{r+\varepsilon}(z_0)$. Thus, f_t'/f_t is continuous in t uniformly on $\partial\mathbb{D}_r(z_0)$, so by (3.3.3), $N_{\mathbb{D}_r(z_0)}(f_t)$ is continuous. Since it is always an integer, it is a constant. $\square$

I want to emphasize how remarkable this theorem is. In the case of real-valued functions of a real variable, its analog is false, as can be seen by looking at $f_t(x) = x^2 - t$ for $x \in [-1, 1]$ and $t \in [-1, 1]$. In a sense, a circle on which f_t has no zero serves like a Hadrian's wall: keep zeros from outside from sneaking in and zeros inside from sneaking out as a parameter is varied. We've gotten used to this fact, but it's amazing nonetheless. The corollary has a consequence:

Theorem 3.3.4 (Rouché's Theorem for the Disk). *Let f, g be analytic in a neighborhood of $\overline{\mathbb{D}_r(z_0)}$ and suppose that on $\partial\mathbb{D}_r(z_0)$, we have that*

$$|f(z) - g(z)| < |f(z)| + |g(z)| \tag{3.3.9}$$

Then, f and g are nonvanishing on $\partial\mathbb{D}_r(z_0)$ and

$$N_{\mathbb{D}_r(z_0)}(f) = N_{\mathbb{D}_r(z_0)}(g) \tag{3.3.10}$$

Remarks. 1. By the triangle inequality, (3.3.9) always holds if $<$ is replaced by $\leq$.

2. This result is stronger than the traditional Rouché's theorem; see the discussion in the Notes.

3. We'll eventually prove a version of this, allowing more general curves than circles; see Section 4.3.

Proof. For $t \in [0, 1]$, let

$$f_t(z) = (1 - t)f(z) + tg(z) \tag{3.3.11}$$

If $f_t(z_0) = 0$, then $f(z_0) = t\big(f(z_0) - g(z_0)\big)$, and similarly, $g(z_0) = (1 - t)\big(g(z_0) - f(z_0)\big)$. Thus,

$$|f(z_0)| + |g(z_0)| = [t + (1 - t)]\,|f(z_0) - g(z_0)|$$

violating (3.3.9).

We conclude that $f_t(z_0)$ is nonvanishing on $\partial\mathbb{D}_r(z_0)$, so that (3.3.10) follows from Corollary 3.3.3. $\square$

Second Proof of the Fundamental Theorem of Algebra (Theorem 3.1.11). This proof will show more, namely, it will show a polynomial, P, of degree N has exactly N zeros, counting multiplicity. Suppose P has the form (3.1.16) with $a_N \neq 0$. Let

$$f(z) = a_N z^N, \qquad g(z) = P(z) \tag{3.3.12}$$

If $|z| > (|a_1| + \cdots + |a_{N-1}| + |a_N|)/|a_N|$, then (since $|z| > 1$)

$$\begin{aligned}|f(z) - g(z)| &\leq |z|^{N-1}(|a_1| + \cdots + |a_{N-1}|)\\ &< |a_N|\,|z|^N = |f| \leq |f| + |g|\end{aligned}$$

so, by Rouché's theorem, for every such z, $g(z)$ has exactly N zeros in $\{w \mid |w| < |z|\}$. $\square$

We want to note a sharpening of Corollary 3.3.3 and so of Rouché's theorem that can be useful (see Problem 5):

Theorem 3.3.5. *Let $\{f_t(z)\}_{0\le t\le 1,\, z\in\overline{\mathbb{D}}}$ be a family of functions jointly continuous in t and z and analytic in z for $z \in \mathbb{D}$ and t fixed. Suppose $f_t(z) \neq 0$ for all $t \in [0,1]$ and $z \in \partial\mathbb{D}$. Then $N_{\mathbb{D}}(f_t)$ is finite for all t and independent of f.*

Proof. We claim that there is $\varepsilon > 0$, so $f_t(z_0) \neq 0$ for all $t \in [0,1]$ and $|z| \geq 1-\varepsilon$. For if not, we can find $t_n \in [0,1]$ and z_n with $|z_n| \to 1$ so $f_{t_n}(z_n) = 0$. By compactness, we can pass to a subsequence so $t_{n(j)} \to t_\infty$, $z_{n(j)} \to z_\infty \in \partial\mathbb{D}$ and $f_{t_\infty}(z_\infty) = 0$, contradicting the hypothesis.

Once we have this, $N_{\mathbb{D}}(f_t) = N_{\mathbb{D}_{1-\varepsilon}(0)}(f_t)$ and the result follows from Corollary 3.3.3. $\square$

Once we have this, the proof of Theorem 3.3.5 immediately shows

Theorem 3.3.6 (Sharp Rouché Theorem for the Disk)**.** *Let f and g be continuous in $\overline{\mathbb{D}_r(z_0)}$. Suppose for all z in $\partial\mathbb{D}_r(z_0)$, we have* (3.3.9)*. Then f and g are nonvanishing on $\partial\mathbb{D}_r(z_0)$ and* (3.3.10) *holds.*

Here's what we'll need in the next two sections. Interestingly, we will only need the case $n = 1$ even though we'll analyze higher-order zeros.

Theorem 3.3.7. *Let f be holomorphic near z_0 and have an nth-order zero at z_0. Then there exist $\varepsilon > 0$ and $\delta > 0$, so for $|w| < \varepsilon$, there are exactly n zeros, counting multiplicity, of $f(z) - w$ in $\{z \mid |z - z_0| < \delta\}$.*

Remark. We'll see (Theorem 3.5.1) that there are actually n distinct zeros for $w \neq 0$.

Proof. Write $f(z) = (z - z_0)^n h(z)$ with $h(z_0) \neq 0$. By continuity, pick δ so on $\overline{\mathbb{D}_\delta(z_0)}$, we have $|h(z)| \geq \frac{1}{2}|h(z_0)|$. Let $\varepsilon = |h(z_0)|\delta^n/4$. For $z \in \partial\mathbb{D}_\delta(z_0)$, $|f(z)| \geq \delta^n \frac{1}{2}|h(z_0)| \geq 2\varepsilon$. Thus, for $|w| < \varepsilon$, $f(z) - w$ is nonvanishing on $\partial\mathbb{D}_\delta(z_0)$. For such w, let $f_t(z) = f(z) - tw$ and apply Corollary 3.3.3. $\square$

We should next explain why (3.3.2) is called the argument principle. If $f(z_0) \neq 0$, then f is nonvanishing in some disk centered at z_0, and by Theorem 2.6.1, $f(z) = e^{L(z)}$ in that disk. So for A real,

$$f(z) = |f(z)|\, e^{iA(z)} \tag{3.3.13}$$

that is, f has a locally defined, C^∞ argument $A(z)$. Since $L'(z) = f'(z)/f(z)$, we see that if γ is a smooth curve, then

$$\frac{d}{dt}A(\gamma(t)) = \operatorname{Im}\left(\frac{f'(\gamma(t))}{f(\gamma(t))}\gamma'(t)\right)$$

That means that the right side of (3.3.3) is $(2\pi)^{-1}$ times the change of the argument of f as one goes around $\partial\mathbb{D}_r(z_0)$. More geometrically, it counts the number of times the image of $\gamma(t)$ under f winds around the point 0. This is not only a useful intuition but suggests that for any curve and any f analytic and nonvanishing in a neighborhood of that curve, the corresponding integral should be an integer. That's the next thing we'll prove. As preparation:

Lemma 3.3.8. *Let $[z_0, z_1]$ be a straight line in $\mathbb{C}$ and f a function analytic in a neighborhood of $[z_0, z_1]$ and nonvanishing there. Then*

$$\exp\left(\oint_{[z_0,z_1]}\frac{f'(z)}{f(z)}\,dz\right) = \frac{f(z_1)}{f(z_0)} \tag{3.3.14}$$

Proof. We can find $\delta > 0$ so that $\{w \mid \operatorname{dist}(w, [z_0, z_1]) < \delta\} \equiv \widetilde{\Omega} \subset \Omega$ and f is nonzero on $\widetilde{\Omega}$. $\widetilde{\Omega}$ is convex, so by Theorem 2.6.1, there is an analytic function, h, on $\widetilde{\Omega}$ so that $e^h = f$. Thus, $f'/f = h'$ and so, by the fundamental theorem of calculus,

$$\oint_{[z_0,z_1]}\frac{f'(z)}{f(z)}\,dz = h(z_1) - h(z_0) \tag{3.3.15}$$

Therefore,

$$\text{LHS of (3.3.14)} = \frac{e^{h(z_1)}}{e^{h(z_0)}} = \text{RHS of (3.3.14)} \qquad \square$$

Theorem 3.3.9. *Let γ be a closed rectifiable curve in $\mathbb{C}$. Let f be a function analytic in a neighborhood of* $\operatorname{Ran}(\gamma)$ *and nonvanishing on* $\operatorname{Ran}(\gamma)$. *Then*

$$\frac{1}{2\pi i}\oint\frac{f'(z)}{f(z)}\,dz \in \mathbb{Z} \tag{3.3.16}$$

Proof. First suppose that $\gamma = [x_1x_2\dots x_nx_1]$ is a polygonal path. By the lemma, with $x_{n+1} \equiv x$,

$$\begin{aligned}\exp\left(\oint\frac{f'}{f}\,dz\right) &= \prod_{j=1}^{n}\exp\left(\oint_{[x_jx_{j+1}]}\frac{f'}{f}\right)\\ &= \prod_{j=1}^{n}\frac{f(x_{j+1})}{f(x_j)} = 1\end{aligned} \tag{3.3.17}$$

proving (3.3.16).

By Proposition 2.2.3, any closed rectifiable contour is a limit of closed polygons, so by the earlier case, the integral in (3.3.16) is a limit of integers, so an integer. □

A special case of great interest is where $z_0 \notin \operatorname{Ran}(\gamma)$ and $f(z) = 1/(z - z_0)$. If you go back to the intuition of the argument principle, the integral in (3.3.16) captures the notion of how many times γ goes around z_0, so we define:

Definition. Let γ be a closed rectifiable contour and $z_0 \notin \operatorname{Ran}(\gamma)$. The *winding number* of γ about z_0, $n(\gamma, z_0)$, is defined by

$$n(\gamma, z_0) = \frac{1}{2\pi i} \oint_\gamma \frac{dz}{z - z_0} \tag{3.3.18}$$

By the last theorem, this is an integer. Moreover, $n(\gamma, z)$ is continuous in z for $z \notin \operatorname{Ran}(\gamma)$ and thus, it is constant on each connected component of $\mathbb{C} \setminus \operatorname{Ran}(\gamma)$. Since $n(\gamma, z) \to 0$ as $z \to \infty$ (on account of $\ell(\gamma) < \infty$ and $|z - w| \geq \operatorname{dist}(z, \operatorname{Ran}(\gamma))$ if $w \in \operatorname{Ran}(\gamma)$), we see that $n(\gamma, z) = 0$ on unbounded components of $\mathbb{C} \setminus \operatorname{Ran}(\gamma)$. The Notes discuss a topological notion of winding number that this agrees with for rectifiable curves.

We will eventually prove (see Proposition 4.8.3) that if t_0 is a point in $[0, 1]$ so that there is an ε with $\gamma(t)$, C^1 on $(t_0 - \varepsilon, t_0 + \varepsilon)$, $\gamma'(t_0) \neq 0$, and for some δ, $\gamma^{-1}[\mathbb{D}_\delta(\gamma(t_0))] \subset (t_0 - \varepsilon, t_0 + \varepsilon)$ (so there are no self-intersections near t_0), then $n(\gamma, z)$ jumps by ± 1 as z crosses γ near z_0.

There is a simple and intuitive action on curves that does the right thing with regard to winding number. If γ is a closed rectifiable curve and $z_0 \notin \operatorname{Ran}(\gamma)$ and $g \colon \widehat{\mathbb{C}} \to \widehat{\mathbb{C}}$ by $g(z) = (z - z_0)^{-1}$, then

$$\tilde{\gamma}(t) = g(\gamma(t)) \tag{3.3.19}$$

is called the *inversion* of γ in z_0. By the change of variables in contour integrals (Theorem 2.2.1), it is easy to see (Problem 8) that

$$n(\tilde{\gamma}, w) = n(\gamma, z_0) - n(\gamma, g(w)) \tag{3.3.20}$$

In particular, when γ is a simple curve, so $\mathbb{C} \setminus \operatorname{Ran}(\gamma) = \Omega_1 \cup \Omega_2$ with $n(\gamma, z) = 1$ (respectively, 0) for $z \in \Omega_1$ (respectively, $z \in \Omega_2$) and $z_0 \in \Omega_1$, then g inverts the inside and outside of γ.

Notes and Historical Remarks. The argument principle and what is called Rouché's theorem are often attributed to Eugène Rouché (1832–1910) [**493**], but they appeared already in an 1831 mémoire of Cauchy and the argument principle explicitly in two papers of Cauchy in 1855 [**108, 109**]. The traditional form has a stronger hypothesis than (3.3.9), namely, that $|f(z) - g(z)| < |f(z)|$ on $\partial\mathbb{D}_r(z_0)$. The stronger form that requires a weaker hypothesis that we state and prove goes back at least to Estermann [**169**,

p. 156] in 1962. It was rediscovered in 1976 by Glicksberg [**210**], who is usually given credit for it.

This is the first version of the argument principle and Rouché's theorem that we'll present, but improvements will come later. In Section 3.9, we'll consider functions with zeros and poles and extend Theorem 3.3.2 to that case. And in Section 4.3 we replace $\partial\mathbb{D}_r(z_0)$ with a wide range of other contours ("chains homologous to 0").

There is a topological definition of winding number that has the advantage of working for any curve, not just rectifiable curves. If $0 \notin \text{Ran}(\gamma)$, $\tilde{\gamma}(t) = \gamma(t)/|\gamma(t)|$ defines a continuous function of $[0,1]$ to $\partial\mathbb{D}$. Since $\exp(i\,\cdot\,)\colon \mathbb{R} \to \partial\mathbb{D}$ is a covering map (see Section 1.7), there is a unique map, $\gamma^\sharp\colon [0,1] \to \mathbb{R}$, with $\exp(i\gamma^\sharp(z)) = \tilde{\gamma}(t)$ and $\gamma^\sharp(0) \in [0,2\pi)$. $n(\gamma,0)$ is then defined by

$$n(\gamma,0) = \frac{[\gamma^\sharp(1) - \gamma^\sharp(0)]}{2\pi} \tag{3.3.21}$$

It is not hard to see that this agrees with the definition of (3.3.18) for γ a C^1 curve and then for rectifiable curves by an approximation argument.

Alternatively, given a curve, $\gamma\colon [0,1] \to \mathbb{C}$, with $0 \in \text{Ran}(\gamma)$, one can find $t_0 = 0 < t_1 < \cdots < t_n = 1 = t_{-1}$ and disks, $D_j = \mathbb{D}_{\delta_j}(z_j)$, $j = 0,\dots,n$, so that $\gamma(t_{j-1}), \gamma(t_j) \in D_j$ and $0 \notin D_j$. Since $0 \notin D_j$, we can define $\arg_j$ on D_j a branch of $\arg(z)$ and then define

$$n(\gamma,0) = \frac{1}{2\pi}\sum_{j=0}^{n}[\arg_j(\gamma(t_j)) - \arg_j(\gamma(t_{j-1}))] \tag{3.3.22}$$

There is thus a winding number restatement of the argument principle: If f is analytic in a neighborhood of $\overline{\mathbb{D}_r(z_0)}$ and nonvanishing on $\partial\mathbb{D}_r(z_0)$ and $\gamma(t) = z_0 + re^{2\pi i t}$, then $N_{\mathbb{D}_r(z_0)}(f)$ is the winding number of $f(\gamma(\,\cdot\,))$ about 0, that is,

$$N_{\mathbb{D}_r(z_0)}(f) = n(f \circ \gamma, 0) \tag{3.3.23}$$

Problems

1. How many zeros does $f(z) = z^{12} - 4z^8 + 9z^5 - 2z + 1$ have in $\{z \mid |z| < 1\}$?

2. How many zeros does $f(z) = z^4 - 6z + 3$ have in the annulus $\{z \mid 1 < |z| < 2\}$?

3. (a) Show that $f(z) = 10z^6 + 3z^4 + 3z^3 + 5z^2 - 3z$ has two real roots and determine the quadrants each of the roots lies in.

   (b) Analyze which quadrant the zeros of $z^4 - 2z^3 - z^2 + 2z + 11$ lie in.

4. (a) How many zeros does $f(z) = z^{12} - 4z^8 + 9z^5 - 2z + 1$ have in $\{z \mid |z| < 1\}$?

(b) How many zeros does $f(z) = z^4 - 6z + 3$ have in the annulus $\{z \mid 1 < |z| < 2\}$? (*Hint*: Rouché.)

5. Let f be continuous on $\overline{\mathbb{D}}$ and analytic on $\mathbb{D}$ and suppose for all $z \in \overline{\mathbb{D}}$, we have $|f(z)| < 1$. Prove there is a unique $z_0 \in \overline{\mathbb{D}}$ with $f(z_0) = z_0$ and that $z_0 \in \mathbb{D}$. (See also Problem 9 of Section 12.2 of Part 2B.)

6. (a) Let $a_1, \dots, a_\ell \in \{z \mid \operatorname{Re} z \le 0\}$. Prove that if $\operatorname{Re} z > 0$, then

$$\operatorname{Re}\left[\sum_{j=1}^{\ell} \frac{1}{z - a_j}\right] > 0 \tag{3.3.24}$$

and if $\operatorname{Re} z = 0$, then (3.3.24) holds, unless $\operatorname{Re} a_j = 0$ for all j.

(b) Let $P(z)$ be a polynomial all of whose zeros lie in $\{z \mid \operatorname{Re} z \le 0\}$. Prove that $P'(z) \neq 0$ if $\operatorname{Re} z > 0$, and if $P'(z) = 0$ for z with $\operatorname{Re} z = 0$, then either z is a double zero of P or all zeros are on the imaginary axis. (*Hint*: Look at P'/P.)

(c) Prove *Lucas' theorem*: If P is a polynomial, then the zeros of P' lie in the convex hull, cvh(zeros), of the zeros of P and that a zero of P' can only lie on $\partial(\text{cvh(zeros)})$ if P has a double zero on that boundary or all the zeros lie on a line.

Note. This is also sometimes called the Gauss–Lucas theorem or the Grace–Lucas theorem.

7. (a) If $z, z_0 \in \mathbb{C}$, prove that

$$\operatorname{Im}\left[\frac{1}{z - z_0} + \frac{1}{z - \bar{z}_0}\right] < 0 \tag{3.3.25}$$

if z is not in the closed disk centered at $\operatorname{Re} z_0$ ($\in \mathbb{R}$) of radius $\operatorname{Im} z_0$, called the *Jensen disk* of z_0. Similarly, if $z_0 \in \mathbb{R}$, $z \in \mathbb{C}_+$, prove (3.3.25) holds.

(b) Conclude the *Jensen–Walsh theorem*: If P is a polynomial with real coefficients and $z \in \mathbb{C}_+$ has $P'(z) = 0$, then z lies inside one of the Jensen disks of the nonreal zeros of P.

Note. This was stated by Jensen [**285**] without proof and proven in Walsh [**573**].)

8. Prove (3.3.20).

9. Given a function f on a region, Ω, or an interval, $I \subset \mathbb{R}$, its *divided differences* $[x_1, \dots, x_n; f]$ are defined inductively on $\{x_j \in \Omega \text{ or } I \mid x_j \neq x_k \text{ for all } j \neq k\}$ by

$$[x_1; f] = f(x_1) \tag{3.3.26}$$

$$[x_1, x_2, \dots, x_n; f] = \frac{[x_1, \dots, x_{n-1}; f] - [x_2, \dots, x_n; f]}{x_1 - x_n} \tag{3.3.27}$$

This problem will explore divided differences.

(a) For any f and $x_1, \dots, x_n$, prove that there is an entire function g with $[x_1, \dots, x_n; f] = [x_1, \dots, x_n; g]$.

(b) If $f(x) = (z - x)^{-1}$, find $[x_1, \dots, x_n; f]$.

(c) If C is a rectifiable Jordan contour that surrounds $x_1, \dots, x_n$ and f is analytic in a neighborhood of the curve and the region it surrounds, prove inductively that

$$[x_1, \dots, x_n; f] = \frac{1}{2\pi i} \oint_C f(z) \prod_{j=1}^{n} \frac{1}{z - x_j}\, dz \tag{3.3.28}$$

(d) For any function, f, prove that

$$[x_1, \dots, x_n; f] = \sum_{j=1}^{n} \frac{f(x_j)}{\prod_{k \neq j}(x_j - x_k)} \tag{3.3.29}$$

(e) Show that $[x_1, \dots, x_n; f]$ is a symmetric function of $x_1, \dots, x_n$.

(f) If f is analytic inside a rectifiable Jordan curve, prove for $x_1, \dots, x_n$ in that region, Ω, $[x_1, \dots, x_n; f]$ extends to $\Omega \times \cdots \times \Omega$ with

$$[x_1, \dots, x_n; f] = \sum_{j=1}^{\ell} \frac{1}{(m_j - 1)!} \left(\frac{d}{dx}\right)^{(m_j - 1)} \left[f(x) \prod_{k \neq j} (x - y_k)^{-m_k}\right]\Bigg|_{x = y_j} \tag{3.3.30}$$

if m_j x's are equal to y_j (so $m_1 + \cdots + m_\ell = n$).

(g) If f is $C^{n-1}(a, b)$, prove that (3.3.30) extends $[x_1, \dots, x_n; f]$ continuously to $(a, b)^n$.

(h) If $z_1, \dots, z_n$ are distinct points, prove that

$$p(z) = \sum_{j=1}^{n} [z_1, \dots, z_{j-1}; f] \prod_{\ell=1}^{j-2} (z - z_j) \tag{3.3.31}$$

(where $\prod_{\ell=1}^{0}$ is interpreted as 1) is the unique polynomial of degree at most $n - 1$ with $p(z_j) = f(z_j)$.

(i) If f is a polynomial of degree m, prove that for $n \leq m + 1$, $[x_1, \dots, x_n; f]$ is a polynomial of degree $m + 1 - n$ in x_n, and that for $n \geq m + 2$, $[x_1, \dots, x_n; f] = 0$.

10. This problem will prove and exploit the Kakeya–Eneström theorem. Throughout,

$$a_n > a_{n-1} \geq a_{n-2} \geq \cdots \geq a_0 \geq 0 \tag{3.3.32}$$

Let

$$P(z) = \sum_{j=0}^{n} a_j z^j \tag{3.3.33}$$

(a) Prove that

$$\begin{aligned} &\mathrm{Re}[z^{-(n+1)}(z-1)P(z)]\Big|_{z=e^{i\theta}} \\ &\quad = a_n + \sum_{j=1}^{n} (a_{n-j} - a_{n-j+1})\cos(j\theta) - a_0 \cos[(n+1)\theta] \end{aligned} \tag{3.3.34}$$

(b) Show that if $z \in \partial\mathbb{D} \setminus \{1\}$, then $\mathrm{Re}[z^{-(n+1)}(z-1)P(z)] > 0$ and conclude $P(z)$ is nonvanishing on $\partial\mathbb{D}$.

(c) Show $P(z)$ has all its zeros in $\mathbb{D}$ (Kakeya–Enström theorem).

(d) Prove that $\sum_{j=0}^{n} a_j \cos(j\theta)$ has n zeros for $\theta \in [0, \pi)$.

3.4. Local Behavior at Noncritical Points

Let f be a nonconstant analytic function on a region, Ω. A point, $z_0 \in \Omega$, is called a *critical point* if and only if

$$f'(z_0) = 0 \tag{3.4.1}$$

Since f' is not identically zero, the set of critical points is discrete in Ω. In this section, we analyze the behavior of f near a point that is not critical, and in the next section, near critical points. The main theorem in this section shows there is an analytic inverse near noncritical points.

Theorem 3.4.1 (Local Behavior at Noncritical Points). *Let f be analytic on a region, Ω, and $z_0 \in \Omega$ with*

$$f'(z_0) \neq 0 \tag{3.4.2}$$

Let $w_0 = f(z_0)$. Then there exist $\varepsilon > 0$ and $\delta > 0$ so that for $w \in \mathbb{D}_\varepsilon(w_0)$, there is a unique $z \in \mathbb{D}_\delta(z_0)$ solving

$$f(z) = w \tag{3.4.3}$$

Moreover, there is an analytic function, g, from $\mathbb{D}_\varepsilon(w_0)$ into $\mathbb{D}_\delta(z_0)$ so that

$$f(g(w)) = w \tag{3.4.4}$$

g is a bijection of $\mathbb{D}_\varepsilon(w_0)$ and $N \equiv g[\mathbb{D}_\varepsilon(w_0)]$ and $f \restriction N$ is its inverse. Moreover,

$$g'(w) = \frac{1}{f'(g(w))} \tag{3.4.5}$$

Proof. By (3.4.2), $f(z) - w_0$ has a first-order zero at $z = z_0$, so

$$f(z) - w_0 = (z - z_0)h(z)$$

where $h(z_0) \neq 0$. Following our proof of Theorem 3.3.7, pick δ so that $\mathbb{D}_\delta(z_0) \subset \Omega$ and $|h(z)| \geq \frac{1}{2}|h(z_0)|$ on $\mathbb{D}_\delta(z_0)$. So, on $|z - z_0| = \delta$,

$$|f(z) - w_0| \geq 2\varepsilon \tag{3.4.6}$$

with

$$\varepsilon = \tfrac{1}{4}\,|h(z_0)|\delta \tag{3.4.7}$$

By the argument principle, as in that theorem, if $w \in \mathbb{D}_\varepsilon(w_0)$, (3.4.4) has a unique solution with $z \in \mathbb{D}_\delta(z_0)$. That is, we use the continuity in w of

$$g(w) = \frac{1}{2\pi i}\oint_{|z-z_0|=\delta} \frac{f'(z)}{f(z) - w}\,dz \tag{3.4.8}$$

Define, for $w \in \mathbb{D}_\varepsilon(w_0)$,

$$g(w) = \frac{1}{2\pi i}\oint_{|z-z_0|=\delta} \frac{zf'(z)}{f(z) - w}\,dz \tag{3.4.9}$$

By (3.4.6), $f(z) - w$ is nonvanishing on the circle, so by (3.3.4), $g(w)$ is that unique z which solves $f(z) = w$. Clearly, g is holomorphic in w (by approximating the contour integral by a Riemann sum and using Theorem 3.1.5). $f(g(w)) = w$ by the above, so g is a bijection of $\mathbb{D}_\varepsilon(w_0)$ and its image, and thus on this image, $g(f(z)) = z$. By the chain rule applied to $f(g(w)) = w$, $f'(g(w))g'(w) = 1$, which is (3.4.5). $\square$

(3.4.8)/(3.4.9) are a powerful combination that allow extensions of Theorem 3.4.1, thought of as an inverse function theorem, to an implicit function theorem and also from analytic to C^k coefficients. A function, $F(z, w)$, from a neighborhood of $(z_0, w_0) \in \mathbb{C}^2$ to $\mathbb{C}$ is called analytic at (z_0, w_0) if near (z_0, w_0), it is given by the uniformly convergent power series

$$F(z, w) = \sum_{n,m=0}^{\infty} a_{nm}(z - z_0)^n (w - w_0)^m \tag{3.4.10}$$

Theorem 3.4.2 (Analytic Implicit Function Theorem). *Let F be defined and analytic near $(z_0, w_0) \in \mathbb{C}$. Suppose that we have the pair of conditions*

$$F(z_0, w_0) = a_{00} = 0 \tag{3.4.11}$$

and

$$\left.\frac{\partial F}{\partial z}\right|_{z=z_0,\, w=w_0} = a_{10} \neq 0 \tag{3.4.12}$$

Then there exist $\varepsilon > 0$ and $\delta > 0$ so that $\mathbb{D}_\delta(z_0) \times \mathbb{D}_\varepsilon(w_0)$ is in the neighborhood where F is defined and $g\colon \mathbb{D}_\varepsilon(w_0)$ into $\mathbb{D}_\delta(z_0)$ so that

$$F(g(w), w) = 0 \tag{3.4.13}$$

and for each $w \in \mathbb{D}_\varepsilon(w_0)$, $g(w)$ is the unique solution of (3.4.13) with $g(w) \in \mathbb{D}_\delta(z_0)$. Moreover, g is analytic on $\mathbb{D}_\varepsilon(w_0)$ and

$$g'(w) = -\frac{\frac{\partial F}{\partial w}(g(w), w)}{\frac{\partial F}{\partial z}(g(w), w)} \tag{3.4.14}$$

Remark. Theorem 3.4.1 is the special case $F(z, w) = f(z) - w$.

Proof. Since $F(z, w_0)$ has a first-order zero at $z = z_0$, we can write

$$F(z, w_0) = (z - z_0)h_{w_0}(z) \tag{3.4.15}$$

where $h_{w_0}(z_0) \neq 0$. Pick $\delta > 0$ so $\overline{\mathbb{D}_\delta(z_0)} \times \{w_0\} \subset N^{\text{int}}$, the domain of definition of F, and so that $|h_w(z)| \geq \frac{2}{3}|h_{w_0}(z_0)|$ if $z \in \mathbb{D}_{\delta_0}(z_0)$. Thus,

$$|z - z_0| = \delta \Rightarrow |F(z, w_0)| \geq \tfrac{2}{3}\,\delta|h_w(z_0)| \tag{3.4.16}$$

Now pick ε so $\overline{\mathbb{D}_\delta(z_0)} \times D_{2\varepsilon}(w_0) \subset N$ and

$$|z - z_0| = \delta\,|w - w_0| \leq \varepsilon \Rightarrow |F(z, w)| \geq \tfrac{1}{2}\,\delta|h_w(z_0)|$$

By analyticity of $F(z, w)$ and Theorem 3.1.6,

$$N(w) = \frac{1}{2\pi i}\int_{|z-z_0|=\delta} \frac{\frac{\partial F}{\partial z}(z, w)}{F(z, w)}\,dz \tag{3.4.17}$$

$$g(w) = \frac{1}{2\pi i}\int_{|z-z_0|=\delta} \frac{z\frac{\partial F}{\partial z}}{F(z, w)}\,dz \tag{3.4.18}$$

are defined and analytic for $w \in \mathbb{D}_\delta(w_0)$. Since N is an integer and $N(w_0) = 1$ (since (3.4.15) implies $F(z, w_0) \neq 0$ if $z \in \mathbb{D}_\delta(z_0) \setminus \{z_0\}$), $N(w) = 1$ and then $g(w)$ is the value of the unique zero of F. (3.4.14) follows from applying the chain rule to

$$\frac{\partial}{\partial w} F(g(w), w) = 0 \tag{3.4.19}$$

$\square$

Analyticity of F in w is only used to get analyticity of g in w. We say $F(z, w)$ is analytic in z and continuous (respectively, C^k) in w near w_0 if $F(z) = \sum_{n=0}^\infty a_n(w)(z - z_0)$ converging uniformly for $|z - z_0| < \delta_0$ (for each z_0 in a compact subset of allowed z's) and if the $a_n(w)$ are uniformly continuous and bounded by $C\delta_0^{-n}$ (respectively, C^k with derivatives bounded by $C\delta_0^{-n}$).

The argument above proves:

Theorem 3.4.3. *Let $z_0 \in \mathbb{C}$, $w_0 \in \mathbb{R}^k$. Suppose F is defined by $\{(z, w_0) \mid |z - z_0| < \delta_0, |w - w_0| < \varepsilon\}$ and F is analytic in z, continuous (respectively, C^k). Suppose (3.4.11) and (3.4.12) hold. Then there exist $\delta > 0$, $\varepsilon > 0$, and g mapping $\{|w - w_0| < \varepsilon\}$ to $\mathbb{D}_\delta(z_0)$ so that (3.4.13) holds and $g(w)$ is the unique solution in $\mathbb{D}_\delta(z_0)$. Moreover, g is continuous (respectively, C^k).*

Notes and Historical Remarks. In essence, our proof here of the existence of an inverse relied on the argument principle. There are at least two other approaches. One uses the fact that $|f'(z_0)| \neq 0$ implies the real variable differential, DF, of (2.1.17) is an invertible 2×2 matrix and then applies the real variable inverse function theorem. Another approach manipulates power series; see Problem 1.

Problems

1. (a) Let $z + \sum_{n=2}^\infty a_n z^n$ be a power series. We seek a power series $w + \sum_{n=2}^\infty b_n w^n$ so that $w = z + \sum_{n=2}^\infty a_n z^n$ is equivalent to $z = w + \sum_{n=2}^\infty b_n w^n$. By plugging the w series into the z series and identifying coefficients of w^k, show that this is equivalent to

$$b_k = \sum_{\ell=2}^k a_\ell P_{k,\ell}(b_1, \dots, b_{k-1}) \tag{3.4.20}$$

for polynomials $P_{k,\ell}$ and so conclude inductively that at the level of formal power series, there is a unique solution

$$b_k = Q_k(a_2, \dots, a_k) \tag{3.4.21}$$

where the Q_k are polynomials.

(b) Prove inductively that the coefficients of $-P_{k,\ell}(-b_1, \dots, -b_{k-1})$ are all positive. Conclude that $-Q_k(-a_2, \dots, -a_k)$ has all positive coefficients.

(c) Use (b) to show that if $|a_k| \le a_k^*$, then

$$|b_k| \le -Q_k(-a_1^*, \dots, a_k^*)$$

Then verify that a proof that the b series has a nonzero radius of convergence for the special case $a_k = -A^k$ (any $A > 0$) implies, in general, that the b series has a finite radius convergence if the a series does.

(d) Analyze this special case to see that the b series has a finite radius of convergence.

(e) Use these considerations for another proof of Theorem 3.4.1.

3.5. Local Behavior at Critical Points

In this section, we consider nonconstant analytic functions near a point z_0 with $f'(z_0) = 0$. The main theme is that while solutions of $w = f(z)$ for w near $w_0 = f(z_0)$ are many-to-one, they can be given in terms of w by a convergent series, called a Puiseux series, in a fractional power of $w - w_0$. Here is the result:

Theorem 3.5.1 (Local Behavior at Critical Points). *Let f be a holomorphic, nonconstant function in a region, Ω. Let $z_0 \in \Omega$, $w_0 = f(z_0)$, and suppose $f(z) - w_0$ has a pth order zero, $p \geq 2$, at z_0 (i.e., $f'(z_0) = \cdots = f^{(p-1)}(z_0) = 0$, $f^{(p)}(z_0) \neq 0$). Then*

(a) *There are $\varepsilon > 0$ and $\delta > 0$ so that for every $w \in \mathbb{D}_\varepsilon(w_0) \setminus \{w_0\}$, there are exactly p distinct solutions of*

$$f(z) = w \tag{3.5.1}$$

with $z \in \mathbb{D}_\delta(z_0)$. Moreover, for these solutions, $f(z) - w$ has a simple zero.

(b) *There is an analytic function, g, on $\mathbb{D}_{\varepsilon^{1/p}}(0)$ with $g(0) = 0$, $g'(0) \neq 0$, so that if $w \in \mathbb{D}_\varepsilon(w_0)$ and*

$$w = w_0 + \rho e^{i\varphi}, \qquad 0 < \rho < \varepsilon, \, 0 \leq \varphi < 2\pi \tag{3.5.2}$$

then the p solutions of (3.5.1) are given by

$$z = z_0 + g(\rho^{1/p} e^{i(\varphi + 2\pi j)/p}), \qquad j = 0, 1, \dots, p-1 \tag{3.5.3}$$

(c) *There is a power series, $\sum_{n=1}^\infty b_n x^n$, with radius of convergence at least ε, so the solutions of (3.5.1) are given by*

$$z = z_0 + \sum_{n=1}^\infty b_n (w - w_0)^{n/p} \tag{3.5.4}$$

where $(w - w_0)^{1/p}$ is interpreted as the pth roots of $(w - w_0)$ (same root taken in all terms of the power series).

Remark. (3.5.4) is called a *Puiseux series.*

Proof. Since $f(z) - w_0$ has a pth order zero, we can write

$$f(z) - w_0 = (z - z_0)^p q(z) \tag{3.5.5}$$

where $q(z_0) \neq 0$. Pick δ_1 so q is nonvanishing in $\mathbb{D}_{\delta_1}(z_0)$. By Theorem 2.6.1, there exists a function, $q_1(z)$, with $q_1(z)^p = q(z)$.

Define $h(z) = (z - z_0) q_1(z)$. Then

$$f(z) - w_0 = h(z)^p, \qquad h(z_0) = 0, \, h'(z_0) \neq 0 \tag{3.5.6}$$

Then (3.5.1) is equivalent to

$$(w - w_0)^{1/p} = h(z) \tag{3.5.7}$$

for any of the pth roots of $w - w_0$, p in all.

Theorem 3.4.1 says that, for η small, $h(z) = \eta$ has exactly one solution. So given the p roots, (3.5.7) has p solutions, proving (a). If g is the inverse function to h, we get (b), and if $\sum_{n=1}^{\infty} b_n \eta^n$ is the power series for g about $\eta = 0$, we get (c). □

Functions defined implicitly when $\frac{\partial F}{\partial z}(z, w) = 0$ are, in general, complicated so we'll restrict the discussion to algebraic and, more generally, algebroidal function (as defined below). The rest of this section should be regarded as bonus material; while it belongs here, we'll need to refer to a result (the monodromy theorem) only proven later and to use some exterior algebra, as discussed in Part 1.

Definition. An *algebroidal function* is a solution $w = f(z)$ of an equation of the form

$$w^n + a_{n-1}(z)w^{n-1} + \cdots + a_0(z) = 0 \tag{3.5.8}$$

where $\{a_j\}_{j=0}^{n-1}$ are analytic functions. If the $a_j(z)$ are rational functions, that is, ratios of polynomials, we say that f is an *algebraic function.*

Remarks. 1. In (3.5.8), the a_j are analytic in some region Ω and we seek f's defined on all or part of Ω. It can be proven (Problem 2) that a locally defined solution can be analytically continued throughout Ω if a discrete set of points in Ω is avoided.

2. Often one considers

$$b_n(z)w^n + b_{n-1}(z)w^{n-1} + \cdots + b_0(z) = 0 \tag{3.5.9}$$

where b_n is not identically zero. If $a_j = b_j/b_n$ for $j = 0, 1, \dots, n-1$, then this is equivalent to (3.5.8) in $\Omega \setminus Z$, where Z is the discrete set of zeros of b_n. In particular, algebraic functions can be alternatively defined by (3.5.9) where the b's are all polynomials, that is, $F(w, z) = 0$, where F is a polynomial in two variables.

The main result in the remainder of this section is:

Theorem 3.5.2. *Let $a_0, a_1, \dots, a_{n-1}$ be analytic functions on a region, Ω. Then there exists a discrete set $D \subset \Omega$ so that all the roots of (3.5.8) are analytic near any given $z_0 \in \Omega \setminus D$. If $z_0 \in D$, the roots of (3.5.9) are given by all the values of one or more Puiseux series about z_0.*

Remarks. 1. We emphasize that the analyticity result on $\Omega \setminus D$ is local, that is, the roots are the values of n functions analytic in a neighborhood of

z_0. They may not be globally defined *single-valued* functions as the example $w^2 - z = 0$ shows (then $D = \{0\}$, $w = \pm\sqrt{z}$).

2. To compare with Theorem 3.5.1, we note that the roles of z and w are reversed, that is, Theorem 3.5.2 implies the case of Theorem 3.5.1 where f is a polynomial, $p(z)$, in z and we are solving $p(w) - z = 0$.

Our proof of Theorem 3.5.2 will rely on the following piece of algebra whose proof we defer:

Theorem 3.5.3. *Fix n and $r < n$. View the monic polynomials of degree n as $\mathbb{C}^n$ under*

$$(a_0, \dots, a_{n-1}) \mapsto w^n + a_{n-1}w^{n-1} + \dots + a_0 \equiv P_a(w) \tag{3.5.10}$$

Then the set of $a \in \mathbb{C}^n$ for which $P_a(w)$ has r or fewer distinct roots is a polynomial variety, that is, given by the vanishing of a set of polynomials in $(a_0, \dots, a_{n-1})$.

Remark. We'll find $\binom{n+r-2}{n-r-1}$ such polynomials. In particular, if $r = n-1$, we have one polynomial which can be taken to be the famous discriminant which measures multiple roots (for $n = 2$, $r = 1$, it is $a_1^2 - 4a_0$, that is, $b^2 - 4ac$ in the usual quadratic formula).

Proof of Theorem 3.5.2 given Theorem 3.5.3. Write $p_z(w)$ for the polynomial in (3.5.8). Let $r(z)$ be the number of distinct roots of p_z. Let $r_\infty = \max\{r(z) \mid z \in \Omega\}$. Let

$$D = \{z \mid r(z) < r_\infty\} \tag{3.5.11}$$

By Theorem 3.5.3 with $r = r_\infty - 1$, D is given by the vanishing of some number of polynomials in $a_0(z), \dots, a_{n-1}(z)$, that is, by some number of analytic functions. These functions cannot be all identically zero on Ω since then $r(z) < r_\infty$ for all $z \in \Omega$. Thus, D is discrete.

If $z_0 \notin D$, there are r_∞ roots $w_1^{(0)}, \dots w_{r_\infty}^{(0)}$. By looking at

$$N_j(z) = \frac{1}{2\pi i} \int_{|w-w_j^{(0)}|=\delta} \frac{\frac{\partial p_z(w)}{\partial w}}{p_z(w)}\, dw \tag{3.5.12}$$

for $\delta < \frac{1}{2}\min_{j\neq k}|w_j^{(0)} - w_k^{(0)}|$, we see that for z near z_0, $p_z(w)$ has $N_1(z_0)$ roots (counting multiplicity) near $w_1^{(0)}, \dots, N_{r_\infty}(z_0)$ roots near $w_{r_\infty}^{(0)}$. If any of these roots split, that is, there is more than one distinct root near $w_j^{(0)}$, then $p_z(w)$ has at least $r_\infty + 1$ roots, contrary to the fact that r_∞ is the maximum. Thus, the roots have the same multiplicity and are given by the analytic functions

$$w_j(z) = \frac{1}{2\pi i} \int_{|w-w_j^{(0)}|=\delta} \frac{w \frac{\partial p_z(w)}{\partial w}}{p_z(w)}\, dw \tag{3.5.13}$$

If we now take a point in $z_0 \in D$, the various roots can be continued indefinitely in $\mathbb{D}_\delta(z_0) \setminus \{z_0\}$ for δ small and remain roots. Thus, circling z_0 once, these r_∞ roots are permuted by a permutation, $\pi \in \Sigma_{r_\infty}$. By the monodromy theorem (Theorem 11.2.1), this permutation is the same for all single loops about z_0. Let $w_j(z_1)$ be a root for z_1 near z_0 and suppose j is contained in a p-cycle of π. Then w_j returns to itself after circling p times. It follows that if written in the variable $\zeta = (z - z_0)^{1/p}$ that w_j is analytic in a punctured disk. We claim w_j is bounded for z near z_0. Accepting this, we see z_0 is a removable singularity and w_j is given by a Puiseux series. All the branches are roots since they are analytic continuations of roots.

To see that w_j is bounded, we note that if $p_{z_0}(w)$ has a root of multiplicity k at w_0, then for z near z_0, $p_{z_0}(w)$ has k roots near w_0. Thus, there are n roots near the roots of $p_{z_0}(w)$, that is, all roots of $p_z(w)$ are near roots of $p_{z_0}(w)$, so bounded. □

Next, we turn to the proof of Theorem 3.5.3. A first preliminary extends the notion that a square matrix, A, has a nontrivial kernel if and only if $\det(A) = 0$.

Proposition 3.5.4. *Let $T\colon \mathbb{C}^s \to \mathbb{C}^{k+s}$ for $s \geq 1$. Let $\{t_{ij}\}_{i=1,\dots,k+s,\, j=1,\dots,s}$ be its matrix so*

$$T\delta_j = \sum_{i=1}^{k+s} t_{ij}\widetilde{\delta}_i \tag{3.5.14}$$

where δ_j and $\widetilde{\delta}_i$ are the canonical bases for $\mathbb{C}^s$ and $\mathbb{C}^{k+s}$. For each $i_1 < i_2 < \cdots < i_k$ in $(1,\dots,k+s)$, let $T^{(i_1,\dots,i_k)}$ be the $s \times s$ square matrix with rows in $i_1,\dots,i_k$ removed. Then

$$\operatorname{Ker}(T) \neq \{0\} \Leftrightarrow \forall i_1 < \cdots < i_k,\ \det(T^{(i_1,\dots,i_k)}) = 0 \tag{3.5.15}$$

Remark. There are $\binom{k+s}{k}$ determinants.

Proof. This relies on exterior algebra, that is, $\bigwedge^\ell(\mathbb{C}^{k+s})$, as discussed in Section 3.8 of Part 1.

$$\begin{aligned} \operatorname{Ker}(T) \neq \{0\} &\Leftrightarrow \{T\delta_j\}_{j=1}^s \text{ are not independent in } \mathbb{C}^{k+s} \\ &\Leftrightarrow T\delta_1 \wedge \cdots \wedge T\delta_s = 0 \end{aligned} \tag{3.5.16}$$

Given $i_1 < \cdots < i_k$ in $\{1,\dots,k+s\}$, let $j_1 < j_2 < \cdots < j_s$ be the complementary set of indices and

$$\eta_{i_1,\dots,i_k} = \widetilde{\delta}_{j_1} \wedge \widetilde{\delta}_{j_2} \wedge \cdots \wedge \widetilde{\delta}_{j_s} \tag{3.5.17}$$

Then (Problem 3)

$$T\delta_1 \wedge \cdots \wedge T\delta_s = \sum_{i_1<\cdots<i_k} \det(T^{(i_1,\dots,i_k)})\eta_{i_1,\dots,i_k} \tag{3.5.18}$$

Since the η's are independent, (3.5.16) $\Leftrightarrow$ (3.5.15). □

All polynomials below are over the complex numbers.

Proposition 3.5.5. *Let $p(z), q(z)$ be monic polynomials of degree n and m, respectively. Let $\ell \le \min(n,m)$, $\ell > 0$. Then p and q have at least ℓ common zeros $\Leftrightarrow$ $\exists u, v$ obeying* (3.5.19),

$$\deg(u) \le m-\ell, \qquad \deg(v) \le n-\ell, \qquad up+vq=0 \tag{3.5.19}$$

Remark. By ℓ common zeros, we mean counting multiplicity, that is, there are not necessarily distinct $z_1, \dots, z_\ell$ so that $\prod_{j=1}^{\ell}(z-z_j)$ divides both p and q.

Proof. If p, q have ℓ common zeros, $z_1, \dots, z_\ell$, then take

$$u = \frac{q}{\prod_{j=1}^{\ell}(z-z_j)}, \qquad v = -\frac{p}{\prod_{j=1}^{\ell}(z-z_j)} \tag{3.5.20}$$

and note that (3.5.19) holds.

Conversely, if there are u, v so that (3.5.19) holds, then

$$up = -vq \tag{3.5.21}$$

Since v only has at most $n-\ell$ factors, of the n factors in the complete factorization of p, at least ℓ must be factors of v. So u and v have at least ℓ common factors. □

Theorem 3.5.6. *Let p, q be monic polynomials of degree n and m, respectively:*

$$p(z) = z^n + a_{n-1}z^{n-1} + \cdots + a_0, \qquad q(z) = z^m + b_{m-1}z^{m-1} + \cdots + b_0 \tag{3.5.22}$$

Fix $\ell \le \min(n,m)$, $\ell > 0$. Then there are $\binom{m+n-\ell-1}{\ell-1}$ polynomials, $\{R_\alpha\}$, in $\{a_i, b_j\}_{i=0,\dots,n-1;\, j=0,\dots,m-1}$ so that p and q have ℓ common factors if and only if $R_\alpha(a,b) = 0$ for all $\binom{n+m-\ell-1}{\ell-1}$ polynomials.

Remark. If $\ell = 1$, we have a single polynomial called the *resultant*. For $\ell \ge 2$, the polynomials are *subresultants*. The explicit polynomials enter in algebraic complexity theory and in computational algebraic geometry. See, for example, the discussion in Bürgisser et al. [**81**].

Proof. Fix p and q. Consider pairs of polynomials, u, v, of degree at most $m-\ell$ and $n-\ell$, respectively, with $u = \sum_{j=0}^{m-\ell} x_j z^j$, $v = \sum_{j=0}^{n-\ell} y_z z^j$, but demand

$$x_{m-\ell} = -y_{n-\ell} \tag{3.5.23}$$

Consider the map

$$T\colon (u,v) \mapsto up + vq \tag{3.5.24}$$

By (3.5.23), $up + vq$ has degree $n + (m - \ell) - 1$, so T maps

$$\mathbb{C}^{m-\ell+1} \oplus \mathbb{C}^{n-\ell} \to \mathbb{C}^{n+m-\ell} \tag{3.5.25}$$

This is a set up of Proposition 3.5.4 where $s = m + n - 2\ell + 1$ and $k = \ell - 1$. Thus, $\text{Ker}(T) \neq \{0\}$ if and only if $\binom{n+m-\ell-1}{\ell-1}$ polynomials in the matrix elements of T vanish. Proposition 3.5.5 says p and q have ℓ common factors if and only if $\text{Ker}(T) \neq \{0\}$. By the structure of polynomial multiplication, all the matrix elements of T are 0's, a's, or b's. $\square$

Proposition 3.5.7. *Let p be a monic polynomial of degree n. Let $r < n$. Then p has r or fewer distinct roots if and only if p and p' have at least $n-r$ common zeros.*

Proof. Let

$$p(z) = \prod_{j=1}^{r} (z - z_j)^{t_j} \tag{3.5.26}$$

Then the common roots of p and p' are $(z - z_j)^{t_j - 1}$ which number $\sum_{j=1}^{r}(t_j - 1) = (\sum_{j=1}^{r} t_j) - r = n - r$ since $\sum_{j=1}^{n} t_j = \deg(p) = n$. $\square$

Proof of Theorem 3.5.3. P_a and $n^{-1}P_a'$ are monic polynomials. By the proposition, we are interested in $\ell = n - r$ common zeros. Thus, $m = n - 1$, $n + m - \ell - 1 = n + r - 2$ and Theorem 3.5.6 gives us $\binom{n+r-2}{n-r-1}$ polynomials. $\square$

Notes and Historical Remarks. Puiseux series were formalized by V. A. Puiseux (1820–83) in 1850 [**463**].

Problems

1. Does there exist a holomorphic surjection from the unit disk to $\mathbb{C}$? (*Hint*: Move the upper half-plane "down" and then square it to get $\mathbb{C}$.)

2. (a) If w solves (3.5.8), prove that

$$|w| \leq \max(1, |a_0(z)| + \cdots + |a_{n-1}(z)|) \tag{3.5.27}$$

   (b) Using (a), prove that any local root of (3.5.8) can be continued along any curve in $\Omega \setminus Z$.

3. Verify (3.5.18).

3.6. The Open Mapping and Maximum Principle

As a bonus for the analysis of the local behavior of analytic functions, we get two global consequences that will often be useful. After proving them, we'll turn to three variants: the minimum modulus principle, the Schwarz lemma (which we'll use, in particular, in Sections 7.4, 8.1, and Section 12.3 of Part 2B), and the Hadamard three-circle theorem.

To appreciate how unusual these results are, we note that the real-valued, real-analytic function $f(x) = 1 - x^2$ on $(-1, 1)$ has values filling $(0, 1]$—note that the range is not open and also note that $f(x)$ has a local (indeed, a global) maximum at a point $(x = 0)$ in the open set. Of course, these two facts are not unrelated—the existence of a global maximum implies the range is not open. But analytic functions are different in both regards.

The absence of local maxima for $|f(z)|$ and, as we'll see, minima, if $f(z_0) \neq 0$, implies that nondegenerate critical points, that is, $f'(z_0) = 0$, $f''(z_0) \neq 0$ are saddle points of $|f(z)|$.

Theorem 3.6.1 (Open Mapping Principle). *Let f be a nonconstant analytic function in a region, Ω. Then* $\operatorname{Ran}(f)$ *is an open set.*

Proof. This follows from the detailed analysis of local singularities in the last two sections or even from just the argument principle (see Theorem 3.3.7). □

Theorem 3.6.2 (Maximum Principle, First Form). *Let f be analytic and nonconstant in a region, Ω. Then $|f(z)|$ has no local maxima, that is, for any z_0 and any $\delta > 0$, there is $z \in \mathbb{D}_\delta(z_0)$ so $|f(z)| > |f(z_0)|$.*

Proof. Immediate from applying the open mapping principle to $f \restriction \mathbb{D}_\delta(z_0)$. □

An equivalent form involves functions on compact subsets, K, of $\mathbb{C}$. A function, f, on K is called *regular* if and only if f is continuous on K and analytic on K^{int}. We emphasize this is a use of the term "regular" distinct from that used first in Section 4.3 (f is regular at a point $z_0 \in \partial\Omega$ if for some $r > 0$, f agrees on $\mathbb{D}_r(z_0) \cap \Omega$ with a function analytic on $\mathbb{D}_r(z_0)$). Both uses are standard. They are distinguished by prepositions: the earlier notion is regular *at* z_0 while the notion here is regular *on* K. These are also distinct notions even if $K = \overline{\Omega}$: "regular at" is local and involves analyticity at z_0, while "regular on K" only involves continuity but at each $z_0 \in \partial\Omega$.

Theorem 3.6.3 (Maximum Principle, Second Form). *Let f be regular on a compact set $K \subset \mathbb{C}$. Then f takes its maximum value on ∂K and, if K^{int} is connected with $\overline{K^{\text{int}}} = K$ and f is nonconstant, only at points in ∂K.*

Proof. Since K is compact, f takes its maximum value somewhere on K. If it takes its maximum at $z_0 \in K^{\text{int}}$, then f is constant on the connected component Q of K^{int} containing z_0, so constant on $\overline{Q}$ which contains points in ∂Q. This argument also proves the last sentence. $\square$

$z \mapsto \text{Re}(z)$ is an open mapping of $\mathbb{C}$ to $\mathbb{R}$ so if $f\colon \Omega \to \mathbb{C}$ is analytic and nonconstant, then $\text{Re}\, f$ is an open map of Ω to $\mathbb{R}$. Thus, we have (using $\inf_z \text{Re}\, f(z) = \sup_z \text{Re}(-f(z))$ that

Theorem 3.6.4 (Maximum Principle for Harmonic Functions). *Let f be regular on a compact set, $K \subset \mathbb{C}$. Then $\text{Re}\, f$ takes its maximum and also its minimum values on ∂K.*

Remarks. 1. $\text{Re}\, f$ is a harmonic function and locally any harmonic functions are real parts of analytic functions. We'll discuss harmonic functions on $\mathbb{C}$ in Section 5.3 below and even more thoroughly (on $\mathbb{R}^\nu$) in Chapter 3 of Part 3.

2. The maximum principle only needs that u (=$\text{Re}\, f$) be subharmonic (see Section 3.2 of Part 3).

3. The argument doesn't require f to be continuous on ∂K, indeed the same argument proves

Theorem 3.6.5. *Let f be analytic on a bounded region, Ω, in $\mathbb{C}$. Suppose $u(z) = \text{Re}\, f(z)$ and*

$$\sup_{z\in\Omega} u(z) = M < \infty \tag{3.6.1}$$

Then there exists $z_\infty \in \partial\Omega$ and $z_n \in \Omega$ with $z_n \to z_\infty$ so that

$$\lim_{n\to\infty} u(z_n) = M \tag{3.6.2}$$

It is difficult to avoid overdoing superlatives in discussing complex variables. But even if one tries to minimize raves, one has to conclude that the open mapping and maximal principles are central and very powerful. For example, in a few lines, the open mapping theorem will prove that a nonconstant analytic function between compact Riemann surfaces is surjective (see Theorem 7.1.7), the maximum principle will prove $\sum_{n=0}^\infty z^{n^2}$ has a natural boundary (see Problems 10 and 11 of Section 6.2), and the Schwarz lemma (a refined maximum principle) will describe all analytic bijections of $\mathbb{D}$ to itself (see Lemma 7.4.1).

Having seen there are no points of maximum modulus in Ω, we can ask about minimum modulus:

Theorem 3.6.6 (Minimum Modulus Principle). *If f is analytic and nonconstant in a region, Ω, and z_0 is a point where for some $\delta > 0$,*

$$|f(z_0)| = \min\{|f(w)| \mid w \in \mathbb{D}_\delta(z_0)\} \tag{3.6.3}$$

then $f(z_0) = 0$.

Proof. If $f(z_0) \neq 0$, $g(z) = f(z)^{-1}$ is analytic near z_0, so the maximum modulus principle for g implies that (3.6.3) is not true. □

This is a useful tool to assure certain functions have zeros—it provides another proof of the fundamental theorem of algebra (see Problem 1) and other results on the existence of zeros (see Problem 4).

Our last two results improve the maximum principle where there is additional information:

Theorem 3.6.7 (Schwarz Lemma). *Let f be a bounded analytic function on $\mathbb{D}$ with $f(0) = 0$. Then*

$$|f(z)| \leq |z| \sup_{w\in\mathbb{D}} |f(w)| \tag{3.6.4}$$

Moreover, unless $f(z) = cz$, $\leq$ in (3.6.4) is $<$.

Remark. By Theorem 3.6.3, if f is regular on $\overline{\mathbb{D}}$, (3.6.4) becomes

$$|f(z)| \leq |z| \sup_{w\in\partial\mathbb{D}} |f(w)| \tag{3.6.5}$$

Proof. Let $g(z) = f(z)/z$ and

$$M = \sup_{w\in\mathbb{D}} |f(w)| \tag{3.6.6}$$

Then, for $0 < r < 1$,

$$\sup_{\theta\in[0,2\pi]} |g(re^{i\theta})| \leq Mr^{-1} \tag{3.6.7}$$

so, by the maximum principle for g,

$$\sup_{|z|\leq r} |g(z)| \leq Mr^{-1} \tag{3.6.8}$$

As $r \uparrow 1$, the left side is monotone increasing and the right side decreasing, so

$$\sup_{|z|<1} |g(z)| \leq M \tag{3.6.9}$$

and, by the maximum principle, unless g is constant, also $|g(z)| < M$ for all z. Since $|f(z)| = |z|\,|g(z)|$, we have the claimed results. □

In Problem 12, the reader will use the Schwarz lemma to prove the Borel–Carathéodory inequality (3.2.13). Our final result is for an annulus; in Problem 15, the reader will show it provides another proof of the Schwarz lemma.

Theorem 3.6.8 (Hadamard Three-Circle Theorem). *Let $0 < r < R$ and f be regular in the closed annulus $\bar{\mathbb{A}}_{r,R}$. Define for any $\rho \in [r, R]$,*

$$M(\rho) \equiv \sup_{\theta\in[0,2\pi]} |f(\rho e^{i\theta})| \tag{3.6.10}$$

Then for any $s \in [0,1]$,

$$M(r^s R^{1-s}) \le M(r)^s M(R)^{1-s} \tag{3.6.11}$$

Remarks. 1. The inequality can be rephrased as saying that $\log(M(\rho))$ is a convex function of $\log(\rho)$.

2. In Problem 13, the reader will show (3.6.11) is equivalent to

$$\det \begin{pmatrix} \log(M(\rho)) & \log(\rho) & 1 \\ \log(M(r)) & \log(r) & 1 \\ \log(M(R)) & \log(R) & 1 \end{pmatrix} \ge 0 \tag{3.6.12}$$

for $r \le \rho \le R$.

Proof. Define for each $\alpha \in \mathbb{R}$,

$$M_\alpha(\rho) = \rho^\alpha M(\rho) \tag{3.6.13}$$

Suppose we prove that for all α that

$$M_\alpha(\rho) \le \max(M_\alpha(r), M_\alpha(R)) \tag{3.6.14}$$

Then, optimizing in α (see Problem 14) proves (3.6.11). Thus, we need only prove (3.6.14) and, by continuity, it suffices to prove it for $\alpha = p/q$ rational, with $q > 0$ and $p, q \in \mathbb{Z}$.

Given such an α, define g on $\bar{\mathbb{A}}_{r^{1/q}, R^{1/q}}$ by

$$g(w) = w^p f(w^q) \tag{3.6.15}$$

Then

$$\sup_{\theta \in [0,2\pi]} g(\rho^{1/q}) = M_{p/q}(\rho) \tag{3.6.16}$$

and the ordinary maximum principle for g implies (3.6.14). □

Notes and Historical Remarks. The maximum principle for harmonic functions was stated explicitly by Riemann in 1851, but its use for holomorphic functions seems to have come into use only after 1890.

The Schwarz lemma in a more general form was stated in passing in 1870, but it was only Carathéodory in 1912 [**87**], who realized its importance and gave the now standard proof we use (mentioning that he learned it from Schmidt).

Hermann Schwarz (1843–1921) was a student of Weierstrass. He married Kummer's daughter. His name comes up most in the Cauchy–Schwarz inequality and Schwarz lemma, but his deepest work was on conformal mappings (especially of polygonal regions, see Example 8.4.6) and his wonderful analysis of which hypergeometric functions are algebraic and the related discovery of Schwarzian derivatives (described in Osgood [**416**]). For a discussion of Schwarz's work in historical context, see Gray [**217**].

Schwarz's work on conformal mapping was related to his attempts to make sense out of the Riemann mapping theorem—for a time, he had the best rigorous results on the subject; he invented the first notion of universal covering and his work was important in uniformization. Eventually, in 1892, he was appointed to Weierstrass' chair in Berlin but he had done most of his research eaarlier while at Halle, ETH, and then Göttingen. Under Klein, Göttingen took the lead over Berlin under Frobenius and Schwarz as seen by Berlin's inability to lure Hilbert away from Göttingen. Schwarz's students include Fejér, Koebe, and Zermelo.

The three-circle theorem is usually presented after more developments using either subharmonic functions or the three-line theorem, which requires a variant of the Phragmén–Lindelöf method. The three-circle theorem was attributed to Hadamard in Bohr–Landau [**59**] and their name has stuck.

The three-circle theorem is interesting even for functions on $\mathbb{D}$ if one restates it to say $\log(M(r))$ is convex in $\log(r)$. In Theorem 5.1.2 of Part 3, we'll prove an L^p variant of the three-circle theorem, due to Hardy, that says for any function, f, analytic in $\mathbb{D}$ and any $p \in (0, \infty)$, $r \to \log(\int_0^{2\pi} |f(re^{i\theta})|^p \frac{d\theta}{2\pi})^{1/p}$ is convex in $\log(r)$.

Problems

1. Let P be a polynomial. Use the minimum modulus principle to prove that P has a zero. (*Hint*: Prove first that $\lim_{|z|\to\infty} |P(z)| = \infty$.)

2. Let Ω be the region surrounded by a polygon in the plane. Prove that the maximum over $\overline{\Omega}$ of the product of distances to the vertices of Ω occurs on $\partial\Omega$.

3. (a) If f is analytic in a neighborhood of $\overline{\mathbb{D}_r(z_0)}$ and f is constant on $\partial\mathbb{D}_r(z_0)$, prove that f is constant.

   (b) Using the Cauchy formula $f(z_0) = \frac{1}{2\pi} \int f(z_0 + re^{i\theta}) \frac{d\theta}{2\pi}$, prove that for all $r > 0$, either $f(z_0 + re^{i\theta})$ is constant for all θ or $|f(z_0)| < \sup_{\theta\in[0,2\pi)} |f(z_0 + re^{i\theta})|$. Conclude a second proof of the maximum principle.

4. Let K be a compact set with K^{int} connected. Suppose f is regular on K and $|f(z)|$ is constant on ∂K. Prove that either f is constant on K^{int} or f has a zero in K^{int}.

5. The function

$$\beta(z, w) = \frac{z - w}{1 - \bar{w}z} \tag{3.6.17}$$

for $w \in \mathbb{D}$, $z \in \mathbb{D}$, helps one analyze functions f regular on $\overline{\mathbb{D}}$ where $|f(z)| = 1$ on $\partial\mathbb{D}$ (up to a phase, it is the Blaschke factor we'll look at in Sections 9.8 and 9.9).

(a) Prove for w fixed that β is regular on $\overline{\mathbb{D}}$ and $|\beta(e^{i\theta}, w)| = 1$.

(b) Prove that f has finitely many zeros in $\mathbb{D}$, $w_1, \dots, w_k$ (counting multiplicity).

(c) Prove that for some ω with $|\omega| = 1$, we have

$$f(z) = \omega \prod_{j=1}^{k} \beta(z, w_j)$$

6. Let g be analytic in a region Ω and f in a region $\widetilde{\Omega}$ containing $g[\Omega]$. Suppose $f(g(z))$ is constant. Prove either f is constant or g is constant.

7. Let $f_1, \dots, f_n$ be analytic in a region Ω. Suppose $\sum_{j=1}^{n} |f_j|^2$ is constant. Prove that each f_j is constant. (*Hint*: Look at when equality holds in the Cauchy–Schwarz inequality on $\mathbb{C}^n$.)

8. If f is analytic and nonconstant on $\mathbb{D}$ has zeros at $z_1, \dots, z_k \in \mathbb{D}$, prove that

$$|f(0)| < \Big[\sup_{|z|<1} |f(z)|\Big] \prod_{j=1}^{k} |z_k| \tag{3.6.18}$$

(*Hint*: If $\beta(\,\cdot\,, w)$ is given by (3.6.17), apply the maximum principle to $[f(z)/\prod_{j=1}^{k} \beta(z, z_k)]$.)

9. (a) Prove that if $f\colon \mathbb{D}_\alpha(0) \to \mathbb{D}_\beta(0)$ is analytic and $f(0) = 0$, then $|f(z)| \le \beta|z|/\alpha$.

(b) Prove that if f is entire and $|f(z)|/|z| \to 0$ as $|z| \to \infty$, then $f(z) = f(0)$ for all z. (*Hint*: Take $\beta = \sup_{|z|\le\alpha} |f(z) - f(0)|$, use the maximum principle on $f(z) - f(0)$, and take $\alpha \to \infty$.)

10. Let $f\colon \mathbb{D} \to \mathbb{D}$. Prove the Schwarz–Pick lemma: that then

$$\left| \frac{f(z) - f(w)}{1 - \overline{f(z)}\, f(w)} \right| \le \frac{|z - w|}{|1 - \bar{z} w|} \tag{3.6.19}$$

(*Hint*: If $\beta(\,\cdot\,, w)$ is the function in (3.6.17), apply the Schwarz lemma to $\beta(\,\cdot\,, f(w)) \circ f \circ \beta(\,\cdot\,, w)^{-1}$.)

11. Let $f\colon \mathbb{D} \to \mathbb{D}$. Use Problem 10 to prove that

$$|f'(z)| \le \frac{1 - |f(z)|^2}{1 - |z|^2} \tag{3.6.20}$$

Remark. This result will be important in Section 12.2 of Part 2B.

12. Prove the Borel–Caretheódory theorem—inequality (3.2.13). (*Hint*: First show that it suffices to prove it in case $f(0) = 0$ and in that case, apply the Schwarz lemma to $f(z)/(2A - f(z))$ where $A = \max_{|z|=R}[\operatorname{Re} f(z)]$. It will simplify matters to first see what the map $q(w) = w/(2A - w)$ does to the set $\{w \mid \operatorname{Re} w \le A\}$.)

13. Prove that (3.6.12) is equivalent to (3.6.11). (*Hint*: Use row operations to prove that if $\rho = r^s R^{1-s}$, then $[\det(\dots)] = (\log(r) - \log(R))(\log(M(\rho)) - s\log(M(r)) - 1 - s)\log(M(R))$.)

14. For any r, R, s and $\rho = r^s R^{1-s}$, pick α in (3.6.14) so $r^{-\alpha} M_\alpha(r) = R^{-\alpha} M_\alpha(R)$ and show this implies (3.6.11). Then show this is the optimal choice for α.

15. Use the three-circle theorem to prove the Schwarz lemma. (*Hint*: Use that $f(0) = 0$ implies $|f(z)| \le Cz$ for some C and look at $\overline{\mathbb{A}}_{\varepsilon,1-\varepsilon}$ as $\varepsilon \downarrow 0$.)

16. The Schwarz lemma implies that if $f\colon \mathbb{D} \to \mathbb{D}$ and $f(0) = 0$, then $|f'(0)| \le 1$. Prove the following vast generalization of this fact: If Ω is a bounded region and $f\colon \Omega \to \Omega$, and for some $a \in \Omega$, $f(a) = a$, then $|f'(a)| \le 1$. (*Hint*: First show, by a Cauchy estimate, that given a, there is an M so that for all $g\colon \Omega \to \Omega$, $|g'(a)| \le M$. Then define $f^{[n]}$ inductively by $f^{[1]} = f$ and $f^{[n+1]}(z) = f(f^{[n]}(z))$. Prove that $(f^{[n]})'(a) = (f'(a))^n$ and so deduce $|f'(a)| \le 1$.)

    Note. This theme is studied further in the problems to Section 7.4.

3.7. Laurent Series

The existence of convergent Taylor series in Theorem 3.1.1 followed easily from the CIF for disks and the geometric series with remainder formula. We have seen there is a CIF for annuli and it, too, leads to a convergent series, albeit more complicated than a Taylor series.

Theorem 3.7.1 (Laurent's Theorem). *Let f be holomorphic in an annulus, $\mathbb{A}_{r,R}$ for some $r < R$. Then there is a two-sided sequence $\{a_n\}_{n=-\infty}^{\infty}$ so that*

$$f(z) = \sum_{n=-\infty}^{\infty} a_n z^n \tag{3.7.1}$$

converging uniformly on $\overline{\mathbb{A}}_{r(1+\varepsilon),R(1-\varepsilon)}$ for any $0 < \varepsilon < \frac{1}{2}(R-r)$. Moreover,

$$a_n = \frac{1}{2\pi i} \oint_{|z|=\rho} z^{-n-1} f(z)\, dz \tag{3.7.2}$$

for any $\rho \in (r, R)$.

Proof. By Theorem 2.7.5 for $z_0 \in \bar{\mathbb{A}}_{r(1+\varepsilon),R(1-\varepsilon)}$, we have

$$f(z_0) = \frac{1}{2\pi i} \oint_{|z|=R(1-\varepsilon/2)} (z - z_0)^{-1} f(z)\, dz - \frac{1}{2\pi i} \oint_{|z|=r(1+\varepsilon/2)} (z - z_0)^{-1} f(z)\, dz \tag{3.7.3}$$

In the first integral, use

$$(z - z_0)^{-1} = \sum_{n=0}^{N} z^{-n-1} z_0^n + z_0^{N+1} z^{-N-1} (z - z_0)^{-1} \tag{3.7.4}$$

and in the second, interchange the role of z and z_0 to get

$$(z - z_0)^{-1} = -\sum_{n=0}^{N} z^n z_0^{-n-1} + z^{N+1} z_0^{-N-1} (z - z_0)^{-1} \tag{3.7.5}$$

$$= -\sum_{m=-N-1}^{-1} z^{-m-1} z_0^m + z^{N+1} z_0^{-N-1} (z - z_0)^{-1} \tag{3.7.6}$$

by changing from n to $m = -n - 1$. The result is, with a_n given by (3.7.2),

$$f(z) - \sum_{n=-N-1}^{N} a_n z^n = R_N(z_0) \tag{3.7.7}$$

where

$$|R_N(z_0)| \le \left| \frac{z_0}{R(1 - \frac{\varepsilon}{2})} \right|^{N+1} S_+(z_0) + \left| \frac{r(1 - \frac{\varepsilon}{2})}{z_0} \right|^{N+1} S_-(z_0) \tag{3.7.8}$$

with

$$\begin{aligned} S_+(z_0) &= R\left(1 - \frac{\varepsilon}{2}\right) \sup_{|z|=R(1-\varepsilon/2)} \left| \frac{f(z)}{z - z_0} \right| \\ S_-(z_0) &= r\left(1 + \frac{\varepsilon}{2}\right) \sup_{|z|=r(1-\varepsilon/2)} \left| \frac{f(z)}{z - z_0} \right| \end{aligned} \tag{3.7.9}$$

Since, for ε fixed, S_+, S_- are uniformly bounded in $z_0 \in \bar{\mathbb{A}}_{r(1+\varepsilon),R(1-\varepsilon)}$, $R_N(z_0) \to 0$ uniformly on the set, and thus, (3.7.1) and (3.7.2) hold. $\square$

Series like the right side of (3.7.1) are called *Laurent series*. If R_+ is the radius of convergence of $\sum_{n=0}^{\infty} a_n w^n$ and R_- of $\sum_{n=0}^{\infty} a_{-n} w^n$, then, if $R_-^{-1} < R_+$, such series converge in $\mathbb{A}_{R_-^{-1},R_+}$ uniformly on compacts.

(3.7.2) immediately implies an analog of estimates that we'll still call Cauchy estimates:

Theorem 3.7.2 (Cauchy Estimates for Laurent Series). *If f is holomorphic in a neighborhood of $\bar{\mathbb{A}}_{r,R}$, then the coefficients of the Laurent series obey*

$$|a_n| \leq \max\biggl(r^{-n}\int_0^{2\pi} |f(re^{i\theta})| \,\frac{d\theta}{2\pi},\ R^{-n}\int_0^{2\pi} |f(Re^{i\theta})|\,\frac{d\theta}{2\pi}\biggr) \tag{3.7.10}$$

Remark. For $|n|$ large, the R term is smaller if $n > 0$ and the r term if $n < 0$.

A finite Laurent series (i.e., one with $a_n = 0$ if $|n| \geq N_0$ for some N_0) is called a *Laurent polynomial.* If $a_n = 0$ for $n < N_1$ and $a_{N_1} \neq 0$, the *degree* of the Laurent polynomial, q, is the degree of the ordinary polynomial $z^{-N_1}g(z)$.

The essence of the proof of Laurent series for an annulus is the following:

Theorem 3.7.3 (Laurent Splitting). *Let f be analytic in an annulus $\mathbb{A}_{r,R}$ for $0 \leq r < R \leq \infty$. Then f can be written*

$$f(z) = f_+(z) + f_-(z) \tag{3.7.11}$$

where f_+ is analytic in $\{z \mid |z| > r\}$ and goes to zero at ∞ and f_- is analytic in $\{z \mid |z| < R\}$. Moreover, $f_\pm$ are uniquely determined by these conditions.

Remarks. 1. This immediately implies the Laurent theorem by writing out the Taylor series for $f_+(z^{-1})$ and $f_-(z)$. Conversely, one can derive this from Laurent's theorem.

2. $\mathbb{A}_{r,R} = \Omega \setminus K$ where $\Omega = \mathbb{D}_R$, $K = \overline{\mathbb{D}}_r$. In Section 4.4, we'll prove an analog for any domain Ω and compact subset $K \subset \Omega$.

Proof. In (3.7.3), let f_- be the first term and f_+ the second (with the minus sign). This defines f_+ analytic on $\mathbb{C} \setminus \overline{\mathbb{D}}_{r(1+\varepsilon/2)}$ and f_- on $\mathbb{D}_{R(1-\varepsilon/2)}$ so that (3.7.11) holds on $\overline{\mathbb{A}}_{r(1+\varepsilon),R(1-\varepsilon)}$.

By the Cauchy integral formula, for $z_0 \in \mathbb{D}_\rho$, $f_-(z_0)$ is independent of ε so long as $\rho < R(1-\varepsilon/2)$. Since ε is arbitrary, we get an analytic function on all of $\mathbb{D}_R$ and similarly for f_+. This proves existence.

If $f_+ + f_- = g_+ + g_-$ on $\mathbb{A}_{r,R}$, where f_+, g_+ (respectively, f_-, g_-) obey the stated properties of f_+ (respectively, f_-), then

$$h \equiv f_+ - g_+ = -f_- + g_- \tag{3.7.12}$$

on $\mathbb{A}_{r,R}$. But $f_+ - g_+$ is analytic on $\mathbb{C} \setminus \overline{\mathbb{D}_r}$ and $-f_- + g_-$ on $\mathbb{D}_R$, so h can be continued to an entire function which goes to zero at ∞ (since $f_+ - g_+ \to 0$). Thus, $f_+ - g_+ = 0 = f_- - g_-$ proving uniqueness. $\square$

Notes and Historical Remarks. Pierre Laurent (1813–54), a construction engineer in the French army (like Cauchy once was), found his series for functions in the annulus and submitted it to a prize competition of the French Academy in 1843. Alas, he missed the deadline, so the paper was not considered, but it attracted Cauchy's attention. He reported on it to the Academy, but his request that the Academy publish it was ignored.

In his 1841 lectures (only published in 1894!) [**582**], Weierstrass also had this theorem, so some of the older literature calls the series Laurent–Weierstrass series.

In the engineering literature, the map $\{a_n\}_{n=-\infty}^{\infty} \to \sum_{n=-\infty}^{\infty} a_n z^n$ is often called "the z transform." Restricted to $z = e^{i\theta}$, it is, of course, just a concrete Fourier series (see Section 3.5 of Part 1). Section 3.10 explores in more detail the connection between Fourier series and Laurent series.

Problems

1. Let $f(z) = (z-1)^{-1} + (z-2)^{-1}$ on $\mathbb{C}\setminus\{1,2\}$. How many different Laurent series are associated to f? Find them.

2. (a) Prove that for any $z \in \mathbb{C}$,
$$F(z,w) = \exp\left(\frac{z}{2}\,(w - w^{-1})\right) \tag{3.7.13}$$
is analytic for $w \in \mathbb{C}^\times$ and so defines a sequence of functions $\{J_n(z)\}_{n=-\infty}^{\infty}$ by
$$F(z,w) = \sum_{n=-\infty}^{\infty} J_n(z) w^n \tag{3.7.14}$$
J_n is the *Bessel function* (of the first kind) of order n. Bessel functions of general order are the subject of Sections 14.5 and 14.7 of Part 2B.

   (b) Prove that
$$J_n(z) = (-1)^n J_n(z) \tag{3.7.15}$$
and for $n \geq 0$,
$$J_n(z) = \left(\frac{z}{2}\right)^n \sum_{k=0}^{\infty} \frac{(-1)^k (\frac{z}{2})^{2k}}{k!(n+k)!} \tag{3.7.16}$$

   (c) By setting $w = e^{i\theta}$, prove that
$$J_n(z) = \frac{1}{\pi} \int_0^\pi \cos(z \sin\theta - n\theta)\, d\theta \tag{3.7.17}$$

   (d) Prove that (*Hint*: $w = ie^{i\theta}$)
$$e^{ikr\cos\theta} = J_0(kr) + \sum_{n=1}^{\infty} 2i^n J_n(kr) \cos(n\theta) \tag{3.7.18}$$

Note. This is sometimes called the *multipole expansion* of a plane wave.

(e) Prove that $J_n(z)$ solves the differential equation

$$z^2u''(z) + zu'(z) + (z^2 - n^2)u(z) = 0 \tag{3.7.19}$$

(f) This will relate two-dimensional Fourier transforms written in polar coordinates to Bessel functions (and assumes your knowledge of Fourier transforms; see Chapter 6 of Part 1). $\mathcal{S}(\mathbb{C})$ will denote functions in $\mathcal{S}(\mathbb{R}^2)$ with $\mathbb{C}$ thought of as $\mathbb{R}^2$. $\widehat{\ }$ maps $\mathcal{S}(\mathbb{C})$ to $\mathcal{S}(\mathbb{C})$ via

$$\hat{f}(w) = (2\pi)^{-1} \int f(z) \exp(-i\,\mathrm{Re}(\bar{w}z))\, d^2z \tag{3.7.20}$$

and is essentially the usual Fourier transform. For any function f in $\mathcal{S}(\mathbb{C})$, define $f_n^\sharp(r)$ by

$$f_n^\sharp(r) = \int_0^{2\pi} f(re^{i\theta})e^{-in\theta}\, \frac{d\theta}{2\pi} \tag{3.7.21}$$

Prove that

$$(\hat{f})_n^\sharp(\rho) = i^n \int_0^\infty f_n^\sharp(r) r J_n(r\rho)\, dr \tag{3.7.22}$$

(*Hint*: See (3.7.18).)

Remark. There are versions of (3.7.22) on $\mathbb{R}^\nu$ and not just $\mathbb{R}^2$. In that case, $e^{in\theta}$ is replaced by spherical harmonics (see Section 3.5 of Part 3) of degree ℓ, and J_n is replaced by $J_{\ell+\frac{1}{2}(\nu-2)}$ (and $r\,dr$ by $r^{n/2}\,dr$). For ν odd, this involves Bessel functions of odd half-integral order (or, up to a power of r, spherical Bessel functions.)

3.8. The Classification of Isolated Singularities; Casorati–Weierstrass Theorem

We say an analytic function has an *isolated singularity* at $z_0 \in \mathbb{C}$ if and only if f is analytic in $\mathbb{D}_\delta(z_0) \setminus \{z_0\}$ for some $\delta > 0$. In this section, we'll study such points, presenting first in terms of the associated Laurent series, a trichotomy into removable singularity, pole, and essential singularity. We'll then refine this in terms of the limiting behavior of $f(z)$ as $z \to z_0$ where a key is the Riemann removable singularity theorem that if $f(z)$ is bounded near z_0, one can assign a value to $f(z_0)$ so that f is analytic at z_0. Finally, we'll prove a theorem about asymptotic behavior near z_0 in the essential singularity case. We'll study poles further in the next section and essential singularities in Sections 11.3, 11.4, Section 12.4 of Part 2B, and Section 3.3 of Part 3.

If f is analytic in $\mathbb{D}_\delta(z_0) \setminus \{z_0\}$, there is a Laurent series

$$f(z) = \sum_{n=-\infty}^{\infty} a_n (z - z_0)^n \tag{3.8.1}$$

Let $N_{\min}(f) = \min\{n \mid a_n \neq 0\}$ with$N_{\min}(f) = -\infty$ if there are infinitely many negative points in the set and $N_{\min}(f) = \infty$ if all a_n are zero.

Definition. We say that f has a *removable singularity* if $N_{\min}(f) \geq 0$, an *essential singularity* if $N_{\min}(f) = -\infty$, and a *pole* if $-\infty < N_{\min}(f) < 0$. The *order* of the pole is $-N_{\min}(f)$. A pole of order 1 is called a *simple pole.*

The reason for the name "removable singularity" is that in that case, we can define $f(z_0) = a_0$ and extend f to a function analytic in $\mathbb{D}_\delta(z_0)$.

Theorem 3.8.1 (Riemann Removable Singularity Theorem). *If f is bounded on some $\mathbb{D}_{\delta'}(z_0) \setminus \{z_0\}$, then f has a removable singularity at z_0.*

Remarks. 1. Our proof shows that it suffices to have the weaker condition

$$\lim_{r \downarrow 0} r\Big[\sup_{\theta \in [0,2\pi]} |f(re^{i\theta})|\Big] = 0 \tag{3.8.2}$$

2. In contrast, if f has a pole of order k at z_0, $(z - z_0)^k f(z) \to a_k$ as $z \to z_0$, so $\lim_{|z-z_0| \to 0} |f(z)| = \infty$.

Proof. By a Cauchy estimate, (3.7.10), for any k, we have for any $r > 0$,

$$|a_k| \leq r^{-k} \sup_{\theta \in [0,2\pi]} |f(re^{i\theta})| \tag{3.8.3}$$

which goes to zero as $r \downarrow 0$ if $k < 0$. Thus, f bounded $\Rightarrow$ all $a_k = 0$ for $k < 0$. □

As an immediate corollary, we have

Theorem 3.8.2 (Casorati–Weierstrass Theorem). *If f has an essential singularity at z_0, then for any $\delta_1 > 0$, the set of values of f in $\mathbb{D}_{\delta_1}(z_0) \setminus \{z_0\}$ is dense in $\mathbb{C}$, that is, for any $\alpha \in \mathbb{C}$, we can find $z_n \to z_0$ so that $f(z_n) \to \alpha$.*

Proof. If not, there is an $\alpha, \varepsilon > 0$ and $\delta_1 > 0$ so

$$0 < |z - z_0| < \delta_1 \Rightarrow |f(z) - \alpha| > \varepsilon \tag{3.8.4}$$

Let

$$g(z) = \frac{1}{f(z) - \alpha} \tag{3.8.5}$$

on $\mathbb{D}_{\delta_1}(z_0) \setminus \{z_0\}$. By (3.8.4), $|g(z)| \leq \varepsilon^{-1}$, so by Theorem 3.8.1, g has a removable singularity, that is, we can give g a value $g(z_0)$ so g is analytic at z_0.

If $g(z_0) \neq 0$, then

$$f(z) = g(z)^{-1} + \alpha \tag{3.8.6}$$

is bounded near z_0, and thus, f has a removable singularity at z_0, not an essential singularity.

If $g(z_0) = 0$, since g is nonvanishing near z_0, $g(z)$ has a zero of some finite order $k > 0$, and thus, f, given by (3.8.6), has a pole of order k at z_0, not an essential singularity. □

In Section 11.3, we'll prove a considerable strengthening of this last theorem: what is called the great Picard theorem which says that in any neighborhood of an essential singularity, every value is taken with at most one exception.

To summarize: we see that

- f removable singularity at $z_0 \Leftrightarrow f$ bounded near $z_0 \Leftrightarrow \lim_{|z-z_0|\to 0} f(z)$ exists and is finite
- f pole at $z_0 \Leftrightarrow \lim_{|z-z_0|\to 0}|f(z)| = \infty$
- f essential singularity at $z_0 \Leftrightarrow$ limit points of $f(z)$ as $z \to z_0$ are all of $\widehat{\mathbb{C}}$.

If f is analytic in $\{z \mid |z| > R\}$—in particular, if f is an entire function—then

$$g(z) = f(z^{-1}) \tag{3.8.7}$$

has an isolated singularity at $z = 0$, and we can use the ideas here to analyze the behavior near infinity. The removable singularity theorem is then a local variant of Liouville's theorem and the Casorati–Weierstrass theorem says something about the behavior of entire functions which are not polynomials.

Finally, we want to note that f'/f can be a powerful indicator of whether an isolated singularity of f is essential. An extension of the following will play an important role in Chapter 14 of Part 2B (see Lemma 14.3.4 of Part 2B).

Theorem 3.8.3. *Let f be analytic and not identically zero on $\mathbb{D}_\delta(z_0) \setminus \{z_0\}$ for some $\delta > 0$. Then*

(a) *If f has zeros arbitrarily near z_0, then z_0 is an essential singularity of f.*

(b) *If f is nonvanishing near z_0, then either*

$$g(z) = \frac{f'(z)}{f(z)} \tag{3.8.8}$$

is singular at $z = z_0$ or f has a removable singularity with $f(z_0) \neq 0$.

For (c) *and* (d) *below, we suppose* f *is nonvanishing near* z_0.

(c) *The residue* (*i.e., the coefficient of* $(z-z_0)^{-1}$ *in the Laurent series of* g) *of* g *at* z_0 *is always an integer. If* f *has a pole at* z_0, *the residue is* $-m$, *where* m *is the order of the pole. If* f *has a removable singularity and a zero of order* $m > 0$, *then the residue is* m.

(d) *The following are equivalent:*
   (1) f *does not have an essential singularity at* z_0.
   (2) g *has a simple pole or removable singularity at* z_0 *and the integer in* (c) *is the order of the pole or zero of* f *at* z_0.
   (3) *For some* C *and* z *near* z_0,

$$|g(z)| \le C|z-z_0|^{-1} \tag{3.8.9}$$

Remark. Problem 1 explores what happens if g has a finite-order pole of order larger than 1.

Proof. For simplicity of notation, we can take $z_0 = 0$ and $\delta = 1$, and in (b)–(d), suppose f is nonvanishing in $\mathbb{D}^\times$.

(a) If f has a finite-order pole or removable singularity at zero, for some k, $z^k f(z) \equiv g(z)$ is analytic at $z = 0$. So $z = 0$ cannot be a limit point of zeros of g, and so not of f.

(c) This is immediate from the extended argument principle in the form of Theorem 3.3.9 and the formula (3.7.2) for a_{-1}.

(d) <u>(1) ⇒ (2).</u> If $f(z) = z^k h(z)$ with $h(0) \ne 0$ and $k \in \mathbb{Z}$, then $g(z) = kz^{-1} + h'(z)/h(z)$ has a simple pole (if $k \ne 0$) or removable singularity (if $k = 0$).

<u>(2) ⇒ (3).</u> is trivial.

<u>(3) ⇒ (1).</u> (3.8.9) implies for all r small,

$$\left| \int_{1/2}^{r} g(\rho e^{i\theta})\, dr \right| \le C \log(2r)^{-1} \tag{3.8.10}$$

Since $\frac{d}{dr} \log|f(re^{i\theta})| = \operatorname{Re} g(e^{i\theta})$, (3.8.10) implies that for a constant D and $|z| < \frac{1}{2}$,

$$|f(z)| \le D|z|^{-C} \tag{3.8.11}$$

which implies Laurent coefficients vanish for $n < -C$.

(b) follows from the proof of (1) ⇒ (2). □

Notes and Historical Remarks. In his 1851 dissertation, Riemann proved that if a function was analytic in a punctured disk and continuous in the whole disk, it was analytic in the whole disk. In 1905, Landau

[**339**] (whose biographical sketch is in the Notes to Section 9.10) noted that boundedness is enough (earlier, Osgood [**419**] discussed the same result).

There are vast generalizations of the removable singularities theorem of Riemann. For example, if Ω is a domain in $\mathbb{C}$, $E \subset \Omega$ a closed subset of linear Lebesgue measure zero, and $f\colon \Omega \setminus E \to \mathbb{C}$ a bounded analytic function, then f has an analytic extension to all of Ω. For a discussion of this result (including the definition of linear Lebesgue measure zero), see the discussion of "Painlevé's theorem" in Iwaniec–Martin [**279**] or see Pajot [**424**].

These two results show that there are really two meanings for a set E to be removable: we can require $f\colon \Omega \setminus E \to \mathbb{C}$ bounded and analytic implies f can be extended to all of Ω, or we can require that only for f's which we already know have continuous extensions to Ω. The function $f(z) = \int_{-1}^{1} \frac{dx}{x-z}$ with $\Omega = \mathbb{C}$, $E = [-1,1]$ has no analytic or even continuous extension to all of $\mathbb{C}$ so E does not have the stronger kind of removability but Theorem 5.5.2 implies it does have the weaker kind. We'll have a lot more to say about these questions in Section 8.8 and its Notes.

By the method used to prove Theorem 5.5.2, one can prove that a continuous function on Ω and analytic on $\Omega \setminus E$, where E is the image of a rectifiable curve in Ω, is analytic in all of Ω. See the Notes to Section 5.5 and the discussion in Section 8.8. See also Theorem 3.6.12 of Part 3.

The Casorati–Weierstrass theorem was proven by the Italian mathematician Felice Casorati (1835–90) in 1868 [**101**] and later in an 1876 work of Weierstrass [**585**]. It was also found in 1868 by Sokhotskii [**527**] in his master's thesis, the first Russian publication on complex variables. In the Russian literature, the result is often called Sokhotskii's theorem or the Casorati—Sokhotskii–Weierstrass theorem. Neuenschwander [**406**] discusses the history and concludes it is likely that Weierstrass lectured on this result as early as 1863.

Problems

1. Let f be analytic and nonvanishing on $\mathbb{D}^\times$ so that g given by (3.8.8) has a pole of finite order at $z = 0$. Prove that there is an integer k, an analytic function h on $\mathbb{D}$ with $h(0) \neq 0$, and a polynomial P so that $f(z) = z^k h(z) \exp(P(1/z))$.

3.9. Meromorphic Functions

Essential singularities are quite far from analytic functions, but if g has a pole at z_0, then $f(z) = g(z)^{-1}$ has a removable singularity at z_0, so poles can be thought of as points where g is "analytic" but just happens to have the value infinity. This is a view we'll make precise when we discuss Riemann surfaces in Section 7.1. This leads to:

Definition. Let Ω be a region. A *meromorphic* function, f, on Ω is a function from Ω to $\widehat{\mathbb{C}}$ where $\mathfrak{P} = \{z \mid f(z) = \infty\}$ is a discrete subset of Ω (i.e., no limit points in Ω), where f is analytic on $\Omega \setminus \mathfrak{P}$, and each $z_j \in \mathfrak{P}$ is a pole of some order. If $\Omega = \mathbb{C}$, we will speak of *entire meromorphic* functions (perhaps "meromorphic in all of $\mathbb{C}$" is more common but this term is shorter and shouldn't lead to confusion).

We see meromorphic functions arise most often via:

Theorem 3.9.1. *Let f, g be analytic functions on a region, Ω, with g not identically zero. Then there exists a meromorphic function, h, so for all $z \in \Omega$ with $g(z) \neq 0$,*

$$h(z) = \frac{f(z)}{g(z)} \tag{3.9.1}$$

Remark. As the proof shows, if $g(z)$ has a zero of order ℓ at z_0 and $f(z_0) \neq 0$, h has a pole of order ℓ at z_0, and if $f(z_0) = 0$ and k is the order of its zero, then h has a pole of order $\ell - k$ if $k < \ell$ and a removable singularity if $k \geq \ell$.

Proof. (3.9.1) defines a function analytic on $\Omega \setminus \{z \mid g(z) = 0\}$ and by looking at the zero factors at the zeros of g, one sees that these zeros are poles or removable singularities of h. □

The same argument shows that the result is true if f, g are meromorphic functions, so that the meromorphic functions are a field.

What is true, but we'll only see in Section 9.5, is that every meromorphic function is such a ratio, that is, the field of meromorphic functions is the field of fractions of the ring of analytic functions.

At any pole, z_0, f has an expansion

$$f(z) = \sum_{j=-\ell}^{\infty} b_j (z - z_0)^j \tag{3.9.2}$$

where $\ell > 0$ is the order of the pole and $b_{-\ell} \neq 0$. We define the *residue*, $\text{Res}(f, z_0)$, at the pole to be b_{-1} and

$$\sum_{j=-\ell}^{-1} b_j (z - z_0)^j$$

to be the *principal part* of the pole.

The basis of the use of complex analysis to evaluate integrals (discussed further in Section 5.7) comes from the following (which we'll extend to arbitrary contours in Theorem 4.3.1):

Theorem 3.9.2 (Residue Theorem for $\mathbb{D}$). *Let f be meromorphic in a neighborhood of the closed disk, $\overline{\mathbb{D}}$, with no poles on $\partial\mathbb{D}$. Then f has finitely many poles in $\mathbb{D}$ and*

$$\frac{1}{2\pi i}\oint f(z)\,dz = \sum_{\text{poles } z_j \text{ in } \mathbb{D}} \text{Res}(f, z_j) \tag{3.9.3}$$

Proof. Since the poles are discrete in the region of analyticity and $\overline{\mathbb{D}}$ is compact, there can only be finitely many poles, say $\{z_j\}_{j=1}^{\ell}$ with $\rho_j(z)$ their principal parts. Then

$$f(z) = \sum_{j=1}^{\ell} \rho_j(z) + g(z) \tag{3.9.4}$$

with g analytic in a neighborhood of $\overline{\mathbb{D}}$.

By the CIT, $\oint g(z)\,dz = 0$, so (3.9.3) follows from

$$\frac{1}{2\pi i}\oint \frac{dz}{(z-z_0)^k} = \delta_{k1} \tag{3.9.5}$$

for $k = 1, \dots$. (3.9.5) for $k = 1$ follows from the CIF and for $k > 1$ from the CIT for derivatives (Theorem 2.2.5) and the fact that $(z - z_0)^{-k} = -\frac{1}{k-1}\frac{d}{dz}(z - z_0)^{-(k-1)}$. □

If f is meromorphic in a neighborhood of $\overline{\mathbb{D}}$ (and not identically zero), so is f'/f with poles only at the zeros and poles of f with residues n_j at a zero of order n_j and $-m_j$ at a pole of order m_j. We thus have the following version of the argument principle directly from Theorem 3.9.2.

Theorem 3.9.3 (Argument Principle for Meromorphic Functions). *Let f be meromorphic in a neighborhood of $\overline{\mathbb{D}}$ with no zeros or poles on $\partial\mathbb{D}$. Then f has finitely many zeros and poles in $\mathbb{D}$, say zeros $\{z_j\}_{j=1}^{N_z}$ of order $\{n_j\}_{j=1}^{N_z}$, and poles $\{p_j\}_{j=1}^{N_p}$ of order $\{m_j\}_{j=1}^{N_p}$. Moreover,*

$$\sum_{j=1}^{N_z} n_j - \sum_{j=1}^{N_p} m_j = \frac{1}{2\pi i}\int_{|z|=1} \frac{f'(z)}{f(z)}\,dz \tag{3.9.6}$$

A *rational function* is a ratio of two polynomials $P(z)/Q(z)$. Problem 1 has a partial fraction result for rational functions derived from analytic function theory.

Notes and Historical Remarks. In 1859, Briot and Bouquet [**75**] produced the first systematic textbook of complex function theory. Their 1875 second edition introduced the terms holomorphic and meromorphic and popularized the term pole which appeared earlier in Neumann [**407**].

Problems

1. (a) Let $f(z) = P(z)/Q(z)$ be a rational function. Show it has finitely many poles.

   (b) Let $p_1, \dots, p_\ell$ be the poles and $\rho_1(z), \dots, \rho_\ell(z)$ the principal parts. Show that

$$f(z) = \sum_{j=1}^{\ell} \rho_j(z) + R(z)$$

   where R is a polynomial of degree exactly $\deg(P) - \deg(Q)$ (with the convention, $R = 0$, if this difference is negative). In particular, if $\deg(P) < \deg(Q)$, $f = \sum_{j=1}^{\ell} \rho_j$. (*Hint*: Use Theorem 3.1.9.)

   The next three problems have a complex variable approach to the Jordan normal form following Kato [**299**].

2. Let A be an $n \times n$ matrix. Let $\sigma(A) = \{z \mid \det(A - z\mathbb{1}) = 0\}$. Define $R(z) = (A - z)^{-1}$ on $\mathbb{C} \setminus \sigma(A)$. Prove that $R(z)$ is a matrix-valued analytic function by one of the following methods:

   (a) The method of minors (Cramer's rule).

   (b) $(A - z)^{-1} = \sum_{j=0}^{\infty} (z - z_0)^j (A - z_0)^{-j-1}$

3. Let V be a finite-dimensional vector space.

   (a) Let N be a linear map on V with $N^\ell = 0$ for some ℓ. Prove that there is a basis for V for which N is a direct sum of Jordan blocks of the form (1.3.1) with $\lambda = 0$. (*Hint*: Look at $\text{Ran}(N^{\ell-1}) \subset \text{Ran}(N^{\ell-2}) \subset \dots$.)

   (b) If N is an operator on V and if $\lim_{n\to\infty} \|N^n\|^{1/n} = 0$, prove that 0 is the only eigenvalue of N. Then prove $N^\ell = 0$ if $\ell = \dim(V)$. (*Hint*: You need to know that if $p(\lambda) = \det(\lambda - N)$, then $p(N) = 0$.)

4. Given a finite matrix, A, with $\sigma(A) = \{\lambda_j, \dots, \lambda_k\}$, this problem will show there exist operators $P_1, \dots, P_k$, $N_1, \dots, N_k$, so that

$$P_j P_k = \sigma_{jk} P_j\,, \quad N_j P_j = P_j N_j = N_j\,, \quad N_j^n = 0 \tag{3.9.7}$$

$$A = \sum_{j=1}^{k} \lambda_j P_j + N_j \tag{3.9.8}$$

   (a) Prove that the existence of these operators proves Theorem 1.3.1. (*Hint*: See Problem 3.)

(b) Prove $R(z) = (A - z)^{-1}$ is analytic on $\mathbb{C} \setminus \{\lambda_j, \dots, \lambda_k\}$ and write

$$R(z) = -\sum_{n-1}^{\infty} C_n^{(j)}(z - \lambda_j)^{-n} + \text{analytic at } \lambda_j \tag{3.9.9}$$

for the Laurent series at λ_j.

(c) Prove that for each j and all $n, m \geq 1$,

$$C_n^{(j)} C_m^{(j)} = C_{n+m-1}^{(j)} \tag{3.9.10}$$

(*Hint*: Write a contour integral for $C_n^{(j)}$ and use a slightly bigger contour for $C_m^{(j)}$. Then use $(A-z)^{-1}(A-w)^{-1} = [(A-z)^{-1}-(A-w)^{-1}](z-w)^{-1}$.) Conclude that if

$$P_j = C_1^{(j)}, \qquad N_j = C_2^{(j)} \tag{3.9.11}$$

then that $P_j^2 = P_j$ and $N_j P_j = P_j N_j = N_j$ and that for $m \geq 2$, $C_m^{(j)} = N_j^{m-1}$.

(d) Prove $\|N_j^m\|^{1/m} \to 0$ and conclude that $N_j^\ell = 0$ for $\ell = \dim(\operatorname{Ran}(P_j))$.

(e) Prove that $P_j P_k = 0$ if $j \neq k$.

(f) Prove (3.9.8).

Remark. The approach in this problem is further discussed in Section 2.3 of Part 4.

3.10. Periodic Analytic Functions

Recall that if $f(x)$ is a complex-valued function on $\mathbb{R}$ with $f(x+1) = f(x)$ and

$$\int_0^1 |f(x)|^2 \, dx < \infty \tag{3.10.1}$$

then one defines its *Fourier coefficients* for $n \in \mathbb{Z}$ by

$$a_n = \int_0^1 e^{-2\pi i n x} f(x) \, dx \tag{3.10.2}$$

and in some sense,

$$f(x) = \sum_{n=-\infty}^{\infty} a_n e^{2\pi i n x} \tag{3.10.3}$$

In this section, we'll see what happens to these formulae when f is analytic and see that (3.10.3) is essentially a Laurent series! Indeed, the results can be proven and can be understood without knowing about Fourier analysis.

Here is the main theorem:

Theorem 3.10.1. *For $a, b > 0$, let $\Omega_{a,b}$ be the strip*

$$\Omega_{a,b} = \{z \mid -a < \operatorname{Im} z < b\} \tag{3.10.4}$$

Any function f on $\Omega_{a,b}$ which is analytic and obeys

$$f(z+1) = f(z) \tag{3.10.5}$$

has an expansion

$$f(z) = \sum_{n=-\infty}^{\infty} a_n e^{2\pi i n z} \tag{3.10.6}$$

converging uniformly on compact subsets of $\Omega_{a,b}$ where, for any $y \in (-a, b)$,

$$a_n = \int_0^1 f(x+iy) e^{-2\pi i n(x+iy)}\, dx \tag{3.10.7}$$

Moreover, for any $\varepsilon > 0$, there is C_ε so that

$$|a_n| \le C_\varepsilon(\min(e^{-2\pi(a-\varepsilon)n}, e^{2\pi(b-\varepsilon)n}) \tag{3.10.8}$$

for all n. Conversely, if $\{a_n\}_{n=-\infty}^{\infty}$ is a sequence of numbers obeying (3.10.8) for all n and ε, then the series in (3.10.1) converges on compact subsets of $\Omega_{a,b}$ and defines an analytic function, f, obeying (3.10.5).

Remark. The proof shows more than uniform convergence on every compact set, K. It shows uniform convergence on each $\overline{\Omega}_{a-\varepsilon,b-\varepsilon}$.

Proof. Let

$$h(z) = e^{2\pi i z} \tag{3.10.9}$$

h is a many-to-one map of the strip, $\Omega_{a,b}$, to the annulus, $\mathbb{A}_{r,R}$, where

$$r = e^{-2\pi b}, \qquad R = e^{2\pi a} \tag{3.10.10}$$

Since $h(z) = h(w)$ if and only if $h(z) - h(w) \in \mathbb{Z}$, we see that f is constant on $\{z \mid h(z) = \zeta\}$ for all $\zeta \in \mathbb{A}_{r,R}$. So there is a well-defined function, g, on $\mathbb{A}_{r,R}$ with

$$f(z) = g(h(z)) = g(e^{2\pi i z}) \tag{3.10.11}$$

Since $h'(z) \neq 0$ for all z, by Theorem 3.4.1, h is locally one-one with a local analytic inverse. Thus, with h^{-1} a local analytic inverse near ζ_0, we have that $g(\zeta) = f(h^{-1}(\zeta))$ is analytic near ζ_0, so g is analytic in the annulus.

The series (3.10.6) is simply the Laurent series for g, (3.10.7) is just (3.7.2), (3.10.8) is just (3.7.10), and the uniform convergence follows from Theorem 3.7.1.

For the converse, given (3.10.9), it is easy to see the uniform convergence of the sum in (3.10.6), so f is analytic by the Weierstrass convergence theorem (Theorem 3.1.5). Given that

$$\int_0^1 e^{-2\pi i(x+iy)n} e^{2\pi i(x+iy)m} \, dx = \delta_{nm} \tag{3.10.12}$$

and the uniform convergence of the series (3.10.6), we see that the a_n's obey (3.10.7). □

Sometimes, we only need the following, which we just proved along the way:

Theorem 3.10.2. *Let f be analytic in $\Omega_{a,b}$ where a can be $-\infty$ or b can be ∞. Let f obey (3.10.5). Let r, R be given by (3.10.10). Then there is an analytic function g on $\mathbb{A}_{r,R}$ so that*

$$f(z) = g(e^{2\pi i z}) \tag{3.10.13}$$

Of interest is the case where f is analytic in $\bigcup_{a,b>0} \Omega_{a,b} = \mathbb{C}$, that is, entire periodic functions:

Definition. An entire function, f, is said to have *period* $\tau \in \mathbb{C}^\times$ if and only if for all integers n and all $z \in \mathbb{C}$,

$$f(z + n\tau) = f(z) \tag{3.10.14}$$

Theorem 3.10.3. *Every periodic entire function with period τ has an expansion,*

$$f(z) = \sum_{n=-\infty}^{\infty} a_n e^{2\pi i z \tau^{-1}} \tag{3.10.15}$$

where, for each $B > 0$, there is C_B so

$$|a_n| \le C_B e^{-B|n|} \tag{3.10.16}$$

Conversely, if $\{a_n\}_{n=-\infty}^{\infty}$ obeys (3.10.16) *for each $B > 0$, then the series in* (3.10.15) *defines an entire function obeying* (3.10.14).

Proof. If

$$g(z) = f(z\tau^{-1})$$

then g obeys the hypotheses of Theorem 3.10.1 on $\Omega_{a,b}$ for all $a, b > 0$, so this just becomes a consequence of that theorem. □

Notes and Historical Remarks. The theorems of this section can be summarized as: For periodic functions, analyticity conditions are equivalent to exponential decay hypotheses on its Fourier series coefficients. If one drops the periodicity requirement and replaces Fourier series by Fourier transform, there are analogous theorems associated with the work of Paley and Wiener. We return to them in Section 11.1.

For any entire function, f, the set of $\tau \in \mathbb{C}$ for which (3.10.14) holds is called the periods of f. If f is nonconstant, this set is a discrete, additive subgroup of $\mathbb{C}$. We analyze the possibilities for this in Section 10.2, where we see the set is either $\{n\tau_1\}_{n\in\mathbb{Z}}$ or $\{n\tau_1 + m\tau_2\}_{n,m}$ for suitable nonzero τ_1, τ_2 (and $\tau_2/\tau_1 \notin \mathbb{R}$). In the latter case, f is called elliptic or doubly periodic. In fact, one needs to allow entire meromorphic, not just entire analytic functions for nonconstant f to exist. Chapter 10 will study such functions.

Problems

1. The purpose of this problem is to show that if $f(z)$ obeys $f(z+1) = f(z)$ and $f(z) = f(-z)$, then $f(z)$ has a convergent expansion

$$f(z) = \sum_{n=0}^{\infty} a_n(\cos(2\pi z))^n \tag{3.10.17}$$

with

$$|a_n| \le C_K e^{-Kn} \tag{3.10.18}$$

for all $K > 0$.

(a) Prove that for $w \neq \pm 1$, $\cos(2\pi z) = w$ has exactly two solutions in $0 \le \operatorname{Re} z < 1$, and for $w = 1$ or $w = -1$, exactly one solution. (*Hint*: Look for η with $\eta + \eta^{-1} = w$.)

(b) Prove there is a well-defined function, g, on $\mathbb{C}$ so that $g(\cos(2\pi z)) = f(z)$.

(c) Prove that $w = \cos(2\pi z)$ is locally an analytic bijection if $w \neq \pm 1$ and conclude g is analytic on $\mathbb{C} \setminus \{\pm 1\}$.

(d) Prove that f only has even terms in its power series about $z = 0$ and about $z = \frac{1}{2}$ and conclude g is an entire function.

(e) Conclude (3.10.17) and (3.10.18) hold.

Chapter 4

Chains and the Ultimate Cauchy Integral Theorem

The year 1885 has a special significance in the history of approximation theory. It was then that Weierstrass published his famous result which says that a continuous function on a closed bounded interval can be uniformly approximated by polynomials. The same year saw the birth of holomorphic approximation in the celebrated paper of Runge.

—*Stephen Gardiner* [**201**]

Big Notions and Theorems: Chains, Homologous Chains, Ultimate CIT, Ultimate CIF, Ultimate Argument Principle, Mesh-Defined Chains, Simply Connected Regions, Multiply Connected Regions, Ultra CIT, Ultra CIF, Runge's Theorems, Jordan Curve Theorem

Having seen some applications of the CIT and CIF, we return to our study of the CIT, that is, for a rectifiable contour γ and analytic function f,

$$\oint_\gamma f(z)\, dz = 0 \tag{4.0.1}$$

We want to know for which f and for which γ this is valid. There is actually a third player here, namely, the region Ω on which f is assumed analytic. We already began the discussion in Section 2.6 of which Ω have (4.0.1) for *all* f, γ, and we'll continue that in this chapter (and complete it in Section 8.1).

We have already found in Theorems 2.2.5 and 2.5.4 a general answer to the question of which f's have (4.0.1) for all γ, and one of our main focuses in the start of this chapter will be determining for which γ's (4.0.1) holds for all f's analytic in Ω: we'll actually want more. In essence, in Section 2.7, we showed for the annulus $\mathbb{A}_{r,R}$ if $\rho, \rho' \in (r, R)$, then

$$\oint_{|z|=\rho} f(z)\, dz - \oint_{|z|=\rho'} f(z)\, dz = 0 \tag{4.0.2}$$

So we'll want to consider "sums of contours" and this notion of "chain" will be the focus of Section 4.1.

The canonical example of the failure of (4.0.1) is for $\Omega = \mathbb{A}_{0,1+\varepsilon}$ where

$$\frac{1}{2\pi i} \oint_{|z|=1} \frac{dz}{z} = 1 \tag{4.0.3}$$

The moral of this chapter is that, in some sense, this is the only example! (4.0.1) only fails for γ's that have nonzero winding numbers for some $z_0 \notin \Omega$. This will be the ultimate CIT, which we'll prove in Section 4.2 following a remarkably simple argument of Dixon. We'll then use this to prove an ultimate CIF, ultimate argument principle, ultimate Rouché theorem, and ultimate residue theorem in Section 4.3.

I know some mathematicians who, on philosophical grounds, object strongly to this approach pioneered by Artin and Ahlfors. Since this contrary view is illuminating, I want to describe it, even though, for pedagogical reasons, I will follow the Artin–Ahlfors approach.

The point is, that at a deep level, homology is really about the ability or inability to fill in chains with disks, often described in terms of singular homology as described in Section 1.8. It is a fact that for subsets, Ω, of $\mathbb{C}$, the first homology group can be described in terms of winding numbers. But it is winding about points in the complement, $\mathbb{C} \setminus \Omega$, of Ω. What if there is no complement, as in the study of Riemann surfaces. For a complex torus, an object we'll study in Sections 7.1 and in Chapter 10, homology is described in terms of wrapping around the basic cycles. So the reader needs to at least understand the approach in this chapter is not suitable for more general objects than subsets of $\mathbb{C}$.

We caution the reader that, despite the name, there are variants of the CIT not covered by the ultimate CIT. For example, if f is regular on $\overline{\mathbb{D}}$, that is, continuous there and analytic on $\mathbb{D}$, then $\oint_{|z|=1} f(z)\, dz = 0$ by taking limits of $\oint_{|z|=1-\varepsilon} f(z)\, dz$ as $\varepsilon \downarrow 0$. In fact, if γ is a closed Jordan curve oriented so $n(\gamma, z_0) = 1$ in the bounded component of $\mathbb{C} \setminus \text{Ran}(\gamma)$, if Ω is this bounded component, and if f is regular in $\overline{\Omega}$, then for z_0 in the bounded component,

$$f(z_0) = \frac{1}{2\pi i} \oint_\gamma \frac{f(z)}{z - z_0}\, dz$$

We'll prove this in Section 4.6 (see also the Notes to Section 4.7).

The moral of the ultimate CIT is that holes in Ω destroy holomorphic simple connectivity. We'll make this precise in Section 4.5 and prove that if $\mathbb{C} \setminus \Omega$ has a bounded component, then Ω is not hsc. To do this requires a useful technique for constructing polygonal contours, which we present in Section 4.4. Since that technique will easily provide a second proof of the ultimate CIT, we'll present that also. There is a third proof of the ultimate CIT sketched in the Notes to Section 4.7.

The technique will also provide a proof (in Section 4.7) of Runge's theorem that if K is compact, $\mathbb{C}\setminus K$ connected, and f analytic in a neighborhood of K, then f can be uniformly approximated on K by polynomials. Finally, in Section 4.8, we'll close this chapter on contours and winding numbers by proving the Jordan curve theorem for C^1 Jordan curves.

4.1. Homologous Chains

Here we define the notion of formal sums of contours, aka chains, and the key notion of homologous chains. Let $\mathcal{C}(\Omega)$ denote the set of all closed rectifiable contours γ with $\mathrm{Ran}(\gamma) \subset \Omega$. The chains in Ω, $\mathrm{Chain}(\Omega)$, is the free abelian group generated by $\mathcal{C}(\Omega)$. More specifically, a *chain*, Γ, is an assignment, $N_\Gamma(\gamma)$, of an integer to each $\gamma \in \mathcal{C}(\Omega)$ so that $\{\gamma \mid N_\Gamma(\gamma) \neq 0\}$ is a finite set. If Γ_1 and Γ_2 are two chains, their sum $\Gamma_1 + \Gamma_2$ is the chain with

$$N_{\Gamma_1+\Gamma_2}(\gamma) = N_{\Gamma_1}(\gamma) + N_{\Gamma_2}(\gamma) \tag{4.1.1}$$

Clearly, this makes $\mathrm{Chain}(\Omega)$ into an abelian group. We define $\mathrm{Ran}(\Gamma) = \cup\{\mathrm{Ran}(\gamma) \mid N_\Gamma(\gamma) \neq 0\}$.

If f is a continuous function on $\mathrm{Ran}(\Gamma)$, we define

$$\oint_\Gamma f(z)\, dz = \sum_{\gamma \mid N_\Gamma(\gamma)\neq 0} N_\Gamma(\gamma) \oint_\gamma f(z)\, dz \tag{4.1.2}$$

Winding numbers are defined as in (3.3.18) by

$$n(\Gamma, z_0) = \frac{1}{2\pi i} \oint_\Gamma \frac{dz}{z - z_0} = \sum_{\gamma \mid N_\Gamma(\gamma)\neq 0} N_\Gamma(\gamma) n(\gamma, z_0) \tag{4.1.3}$$

One can invert chains (see (3.3.19)) about $z_0 \notin \mathrm{Ran}(\Gamma)$ and get

$$n(\widetilde{\Gamma}, w) = n(\Gamma, z_0) - n(\Gamma, g(w)) \tag{4.1.4}$$

the analog of (3.3.20).

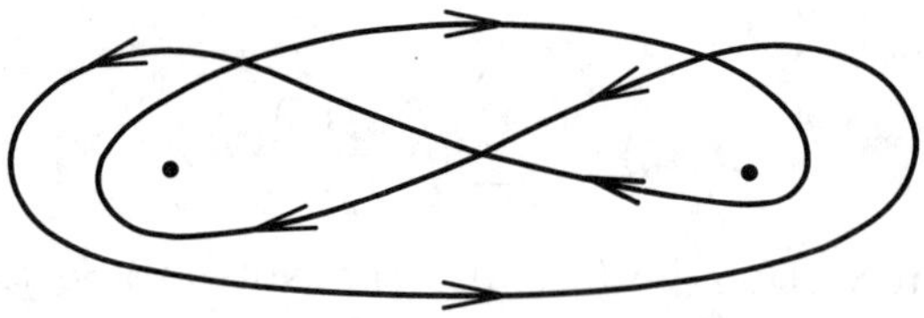

Figure 4.1.1. Homologous to, but not homotopic to, zero.

While we have taken $N_\Gamma(\gamma) \in \mathbb{Z}$, it is sometimes useful to allow coefficients in $\mathbb{Q}$, $\mathbb{R}$, or $\mathbb{C}$; for example, principal value integrals have natural coefficients of $\frac{1}{2}$.

Definition. Two chains, Γ_1 and Γ_2, are called *homologous* in Ω if and only if for all $z_0 \notin \Omega$,

$$n(\Gamma_1, z_0) = n(\Gamma_2, z_0) \tag{4.1.5}$$

A chain, Γ, is called *homologous to zero* in Ω if and only if it is homologous to the zero chain, that is, for all $z_0 \notin \Omega$,

$$n(\Gamma, z_0) = 0 \tag{4.1.6}$$

We associate a contour, γ, with the chain, Γ, with $N_\Gamma(\gamma') = \delta_{\gamma\gamma'}$ so that we can say a single contour is homologous to zero.

Since homotopy preserves contour integrals, it preserves winding numbers, and thus, homotopic contours are homologous by Theorem 2.6.5, but the converse is definitely not true, that is, homology is weaker than homotopy. Figure 4.1.1 shows a curve in $\mathbb{C} \setminus \{0, 1\}$ which is homologous to zero but not homotopic to zero—we will say more about this in the Notes.

We can now state the two main theorems of this chapter:

Theorem 4.1.1 (The Ultimate CIT). *Let Ω be a region, f an analytic function on Ω, and Γ a chain homologous to zero in Ω. Then*

$$\oint_\Gamma f(z)\, dz = 0 \tag{4.1.7}$$

Theorem 4.1.2 (The Ultimate CIF). *Let Ω be a region, f an analytic function on Ω, and Γ a chain homologous to zero in Ω. Then, for all $w \in \Omega \setminus \mathrm{Ran}(\Gamma)$,*

$$\frac{1}{2\pi i} \oint_\Gamma \frac{f(z)}{z-w}\, dz = n(\Gamma, w) f(w) \tag{4.1.8}$$

We want to note for now that either theorem implies the other:

Theorem 4.1.1 $\Leftrightarrow$ Theorem 4.1.2. Given $w \in \Omega \backslash \mathrm{Ran}(\Gamma)$, and $f \in \mathfrak{A}(\Omega)$, define g by

$$g(z) = \frac{f(z) - f(w)}{z - w} \tag{4.1.9}$$

Then $g \in \mathfrak{A}(\Omega)$ and every g arises via this relation (take $f(z) = g(z)(z-w)$). Since

$$(4.1.8) \Leftrightarrow \int g(z)\, dz = 0 \tag{4.1.10}$$

we see the equivalence of (4.1.8) for all f to (4.1.7) for all f. $\square$

In Section 4.4 we will prove another major result which is motivated by Theorem 3.7.3.

Theorem 4.1.3 (Ultimate Laurent Splitting). *Let $K \subset \Omega \subset \mathbb{C}$ with K compact and Ω open and connected. Let f be analytic on $\Omega \setminus K$. Then one can write*

$$f = f_+ + f_- \tag{4.1.11}$$

where f_+ is analytic on $\mathbb{C} \setminus K$ with $\lim_{z\to\infty} f_+(z) = 0$ and f_- analytic on Ω. This decomposition is unique. Moreover, if f is bounded on $\Omega \setminus K$, then f_+ (respectively, f_-) is bounded on $\mathbb{C} \setminus K$ (respectively, Ω).

As a final topic, we briefly consider when two chains, Γ_1, Γ_2, are *equivalent* in that

$$\oint_{\Gamma_1} f(z)\, dz = \oint_{\Gamma_2} f(z)\, dz \tag{4.1.12}$$

for all continuous functions on Ω (not just analytic functions). There are four types of equivalence we mention.

(1) *Running backwards*: One can replace γ by γ^{-1} and flip signs. Namely, in Γ_1, replace a γ by γ^{-1} and make $N_{\Gamma_2}(\gamma^{-1}) = -N_{\Gamma_1}(\gamma)$.

(2) *Combining γ's*: If Γ_1 contain $\gamma, \tilde{\gamma}$ with $\gamma(1) = \tilde{\gamma}(0)$ and $N_{\Gamma_1}(\gamma) = N_{\Gamma_1}(\tilde{\gamma})$, one can drop $\gamma, \tilde{\gamma}$ and take $N_{\Gamma_2}(\tilde{\gamma} * \gamma) = N_{\Gamma_1}(\gamma)$.

(3) *Cyclic reparametrization*: Replace γ by $\gamma \restriction [0,t] * \gamma \restriction [t,1]$.

(4) (the one we'll need later) *Cancellation*: Suppose $\gamma, \tilde{\gamma}$ are two closed curves and $\gamma \restriction [s_1, t_1]$ is a linear reparametrization of $\tilde{\gamma} \restriction [s_2, t_2]$ but run backwards (i.e., for $0 \le \theta \le 1$),

$$\gamma(\theta s_1 + (1-\theta)t_1) = \tilde{\gamma}(\theta t_2 + (1-\theta)s_2) \tag{4.1.13}$$

Thus, these cancel and one can drop $\gamma, \tilde{\gamma}$ (i.e., decrease their N_Γ's by one) and add $\gamma \restriction [t_1, 1] * \tilde{\gamma} \restriction [0, s_2] * \tilde{\gamma} \restriction [t_2, 1] * \gamma \restriction [0, s_1]$; see Figure 4.1.2. We'll call this *matched contour cancellation.*

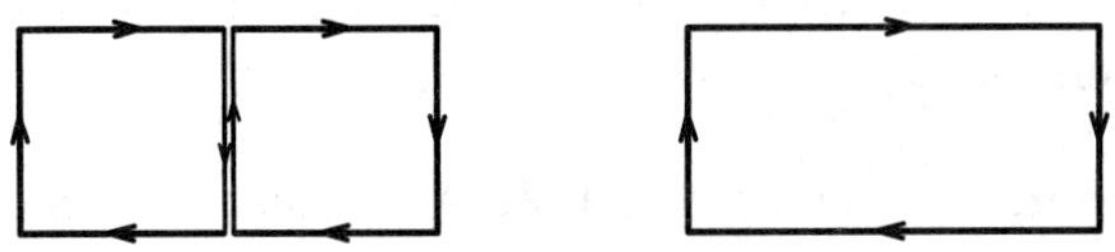

Figure 4.1.2. These two chains are equivalent.

Notes and Historical Remarks. The idea of stating a CIT for general regions and curves in terms of homology, as defined by winding numbers, was in the first edition of Ahlfors' book [**9**], following a suggestion of E. Artin.

Homology is usually defined (see standard books on algebraic topology such as Hatcher [**244**] or see Section 1.8) in terms of curves or formal sums of curves bounding areas. Intuitively, at least, this should be equivalent to winding number and that can be proven using singular homology for the open regions in $\mathbb{C}$ we are considering.

The family of chains is an abelian group and the set of chains homologous to zero is a subgroup, so the quotient of equivalence classes of homologous chains is an abelian group, called the first homology group. The group of homotopy classes of curves, aka fundamental group, is, in general, nonabelian. Hurewicz's theorem (see Theorem 1.8.1) says that, in fact, the homology group is the abelianization of the fundamental group for arcwise connected spaces. Figure 4.1.1 exactly shows a commutator.

4.2. Dixon's Proof of the Ultimate CIT

Quite remarkably, the proof of the ultimate CIT found by Dixon is elegant and really simple.

Lemma 4.2.1. *Let Ω be a region and $f \in \mathfrak{A}(\Omega)$. Define for $z, w \in \Omega$,*

$$G(z,w) = \begin{cases} \dfrac{f(z)-f(w)}{z-w}, & z \neq w \\ f'(z), & z = w \end{cases} \tag{4.2.1}$$

Then, G is jointly continuous, and for each fixed z, $w \to G(z,w)$ is analytic in $w \in \Omega$.

Proof. Continuity at points in $\{(z_0, w_0) \mid z_0 \neq w_0\}$ is trivial and at $z = w = z_0$ by writing $f(z) = \sum_{n=0}^\infty a_n (z-z_0)^n$, so

$$G(z,w) = \sum_{n=1}^\infty a_n \sum_{j=0}^{n-1} (z-z_0)^j (w-z_0)^{n-1-j}$$

we get continuity.

Analyticity in $w \neq z$ is trivial and $w = z$ is handled either by writing out the Taylor series or appealing to the Riemann removable singularity theorem (Theorem 3.8.1). □

Proof of Theorems 4.1.1 and 4.1.2. As noted, it suffices to prove the ultimate CIF. Let

$$\widetilde{\Omega} = \{z \notin \mathrm{Ran}(\Gamma) \mid n(\Gamma, z) = 0\}$$

By hypothesis, $\Omega \cup \widetilde{\Omega} = \mathbb{C}$, and clearly, both are open since n is continuous and integer-valued. Since $\mathbb{C}$ is connected, $\Omega \cap \widetilde{\Omega} \neq \emptyset$.

For $w \in \Omega$, define

$$F(w) = \frac{1}{2\pi i} \oint_\Gamma \frac{f(z) - f(w)}{z - w}\, dz \tag{4.2.2}$$

and for $w \in \widetilde{\Omega}$, define

$$\widetilde{F}(w) = \frac{1}{2\pi i} \oint_\Gamma \frac{f(z)}{z - w}\, dz \tag{4.2.3}$$

F and $\widetilde{F}$ are analytic in their regions of definition by Theorem 3.1.6.

In Ω, clearly

$$F(w) = \widetilde{F}(w) - f(w) n(\Gamma, w) \tag{4.2.4}$$

This implies that $F = \widetilde{F}$ on $\Omega \cap \widetilde{\Omega}$ so, by Proposition 2.3.10, F can be continued to $\mathbb{C}$ and so defines an entire function.

Clearly,

$$|\widetilde{F}(w)| \leq \sup_{z \in \mathrm{Ran}(\Gamma)} |f(z)| \Big(\inf_{z \in \mathrm{Ran}(\Gamma)} |z - w| \Big)^{-1} \to 0$$

as $w \to \infty$. So, by Liouville's theorem (Theorem 3.1.10), $F \equiv 0$. Thus, by (4.2.2), (4.1.8) holds. $\square$

Remark. This proof is due to Dixon in 1971 [**146**].

4.3. The Ultimate Argument Principle

Given the ultimate CIF and the steps we used to go from the ordinary CIF to the argument principle and residue theorem, their ultimate versions are easy.

Given Γ, a chain in Ω, we define $\mathrm{out}(\Gamma)$, the *outside* of Γ, to be those $z_0 \in \mathbb{C} \backslash \mathrm{Ran}(\Gamma)$ with $n(\Gamma, z_0) = 0$ and $\mathrm{ins}(\Gamma)$, the *inside* of Γ, to be $\mathbb{C} \backslash \mathrm{out}(\Gamma)$, so it is $\mathrm{Ran}(\Gamma)$ union those $z_0 \in \mathbb{C} \setminus \mathrm{Ran}(\Gamma)$ with $n(\Gamma, z_0) \neq 0$. If Γ is homologous to 0 in Ω, $\mathrm{ins}(\Gamma) \subset \Omega$ and, always, $\mathrm{out}(\Gamma)$ is open and $\mathrm{ins}(\Gamma)$ closed, indeed compact.

If f is a meromorphic function on Ω, then, since $\mathrm{ins}(\Gamma)$ is compact, f has only finitely many poles in $\mathrm{ins}(\Gamma)$ and we can find $\widetilde{\Omega} \subset \Omega$, a region, so that the only poles of f in $\widetilde{\Omega}$ lie in $\mathrm{ins}(\Gamma)$. Clearly, if Γ is Ω-homologous to zero, it is $\widetilde{\Omega}$-homologous to zero.

Theorem 4.3.1 (Ultimate Residue Theorem). *Let Ω be a region, Γ a chain homologous to zero, and f a meromorphic function on Ω with no poles on*

Ran(Γ). *Let* $\{z_j\}_{j=1}^N$ *be the positions of the poles of* f *in* ins(Γ). *Then*

$$\frac{1}{2\pi i}\oint_\Gamma f(z)\,dz = \sum_{j=1}^{N} n(\Gamma, z_j)\mathrm{Res}(f; z_j) \tag{4.3.1}$$

Proof. Let $\rho_j(z)$ be the principal part of f at z_j. Then (3.9.4) holds with g analytic on $\widetilde{\Omega}$, so (4.3.1) follows from

$$\frac{1}{2\pi i}\oint_\Gamma \rho_j(z)\,dz = n(\Gamma, z_j)\mathrm{Res}(f; z_j) \tag{4.3.2}$$

which in turn follows from

$$\frac{1}{2\pi i}\oint_\Gamma \frac{dz}{(z-z_j)^k} = n(\Gamma, z_j)\delta_{k1} \tag{4.3.3}$$

for $k = 1, 2, \ldots$. For $k = 1$, this is the definition of winding number, and for $k > 1$, it is zero by noting that $(z - z_j)^{-k}$ is a global derivative on Γ. □

If f has no zeros or poles on Ran(Γ), f'/f has poles exactly at the positions of the zeros (respectively, poles) of f with residues the order of the zero (respectively, the negative of the order of the pole). Thus,

Theorem 4.3.2 (Ultimate Argument Principle). *Let* Ω *be a region,* Γ *a chain homologous to zero, and* f *a meromorphic function on* Ω *with no zeros or poles on* Ran(Γ). *Let* $\{z_j\}_{j=1}^{N_z}$ *be the zeros of* f *in* ins(Γ) *with order* $\{n_j\}_{j=1}^{N_z}$ *and let* $\{p_j\}_{j=1}^{N_p}$ *be the poles of* f *with orders* $\{m_j\}_{j=1}^{N_p}$. *Then*

$$\frac{1}{2\pi i}\oint_\Gamma \frac{f'(z)}{f(z)}\,dz = \sum_{j=1}^{N_z} n_j n(\Gamma, z_j) - \sum_{j=1}^{N_p} m_j n(\Gamma, p_j) \tag{4.3.4}$$

As usual, we get an ultimate Rouché theorem:

Theorem 4.3.3 (Ultimate Rouché Theorem). *For* f *nonvanishing on* Ran(Γ), *let* $N_\Gamma(f)$ *denote either side in* (4.3.4). *If* f *and* g *are analytic functions in* Ω, Γ *is homologous to zero on* Ω, *and on* Ran(Γ),

$$|f(z) - g(z)| < |f(z)| + |g(z)| \tag{4.3.5}$$

then

$$N_\Gamma(f) = N_\Gamma(g) \tag{4.3.6}$$

Proof. Identical to the proof of Theorem 3.3.4. □

Notes and Historical Remarks. There is a version of (4.3.4) analogous to (3.3.23). If Γ is a chain with continuous paths, homologous to zeros on Ω in the sense that $n(\Gamma, w) = 0$ for $w \notin \Omega$, if f is meromorphic on Ω with no zeros or poles on $\operatorname{Ran}(\Gamma)$ and if $\{n_j\}_{j=1}^{N_z}$, $\{p_j\}_{j=1}^{N_p}$ are the zeros and poles at points in Ω with $n(\Gamma, z) \neq 0$, then

$$n(f \circ \Gamma, 0) = \sum_{j=1}^{N_z} n_j n(\Gamma, z_j) - \sum_{j=1}^{N_p} m_j n(\Gamma, p_j) \tag{4.3.7}$$

4.4. Mesh-Defined Chains

One of our goals in this section is to prove

Theorem 4.4.1. *Let Ω be a region and K a compact subset of Ω. Then there exists a chain, Γ, homologous to zero in Ω so that* $\operatorname{Ran}(\Gamma) \cap K = \emptyset$ *and*

$$n(\Gamma, z_0) = \begin{cases} 1, & z_0 \in K \\ 0, & z_0 \notin \Omega \\ 0 \text{ or } 1, & \text{for every } z_0 \notin \operatorname{Ran}(\Gamma) \end{cases} \tag{4.4.1}$$

This result will be the key to our proof in the next section that

$$\Omega \text{ hsc} \Leftrightarrow \text{every component of } \mathbb{C} \setminus \Omega \text{ is unbounded} \tag{4.4.2}$$

and of Runge's first theorem that for any compact K, any analytic function in a neighborhood of K can be uniformly approximated on K by rational functions.

By slightly modifying the construction, we'll also find a second proof, due to Beardon, of the ultimate CIT. Given the simplicity of the Dixon proof, we don't have a need for a second proof, but since it is just a few lines, given the machinery to prove Theorem 4.4.1, we give it.

All the chains in this section are not only consisting of curves but of polygons, indeed ones whose sides are parallel to the axes. Here is the basic construction: Given $\delta > 0$ and $z_0 = x_0 + iy_0 \in \mathbb{C}$, we define the associated *mesh* to be the family of open squares indexed by $\alpha \in \mathbb{Z}^2$ with

$$\Delta_\alpha = \left\{ z = x + iy \,\middle|\, |x - x_0 - \alpha_1 \delta| < \frac{\delta}{2},\ |y - y_0 - \alpha_2 \delta| < \frac{\delta}{2} \right\} \tag{4.4.3}$$

As we run through α, the different Δ_α are disjoint with $\overline{\Delta}_\alpha \cap \overline{\Delta}_\beta$ either empty or a single side or a single corner. Moreover,

$$\bigcup_\alpha \overline{\Delta}_\alpha = \mathbb{C} \tag{4.4.4}$$

Given δ, z_0, and a finite subset, $J \subset \mathbb{Z}^2$, we define a contour Γ_J as follows. We start with each $\partial\Delta_\alpha$, thought of as a closed contour going counterclockwise around the sides of the square, starting at the lower-left corner. $\widetilde{\Gamma}_J = \sum_{\alpha \in J} \partial\Delta_\alpha$. If J contains a nearest neighbor pair, (α, β) (i.e.,

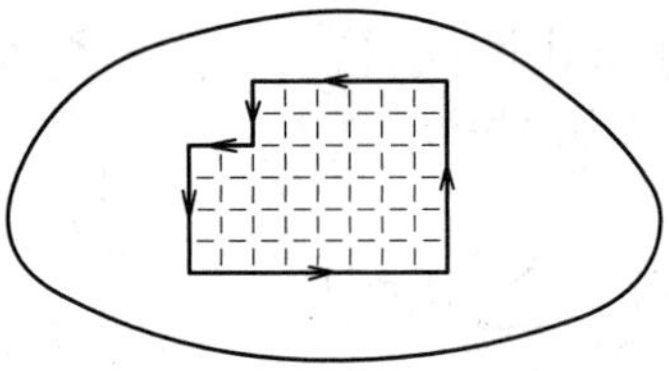

Figure 4.4.1. A mesh-defined contour.

with $|\alpha - \beta| = 1$), then $\partial\Delta_\alpha$ and $\partial\Delta_\beta$ have a single side in common, but the directions are opposite. So we can remove them by matched contour cancellation (see the end of Section 4.1). By doing this successively for each neighboring pair, we get Γ_J, a collection of closed polygons so that each segment bounds two squares in the mesh, one in J and one not in J (see Figure 4.4.1). We'll also define

$$\Delta_J = \bigcup_{\alpha \in J} \overline{\Delta}_\alpha \tag{4.4.5}$$

Γ_J is the boundary of Δ_J, wrapped around counterclockwise. Viewed this way, there is an ambiguity at points which are corners of exactly two squares touching in a point (see Figure 4.4.2(a)). The rule at such corners is to take the paths that bend (Figure 4.4.2(b)). This issue only arises if we think of Γ_J as $\partial\Delta_J$—if we think of it as $\widetilde{\Gamma}_J$ with removed edges, the orientations of $\widetilde{\Gamma}_J$ determine the behavior at such corners.

Proposition 4.4.2. *We have that*

(a) $\mathrm{Ran}(\Gamma_J) = \partial\Delta_J$ (4.4.6)

(b) $n(\Gamma_J, z_0) = 0$ *for all* $z_0 \in \mathbb{C} \setminus \Delta_J$ (4.4.7)

(c) $n(\Gamma_J, z_0) = 1$ *for all* $z_0 \in \Delta_J \setminus \partial\Delta_J$ (4.4.8)

Proof. (a) $\mathrm{Ran}(\widetilde{\Gamma}_J) = \bigcup_{\alpha \in J} \partial\Delta_\alpha$ and contour cancellation removes the borders of this union that are in $\mathrm{ins}(\Delta_J)$.

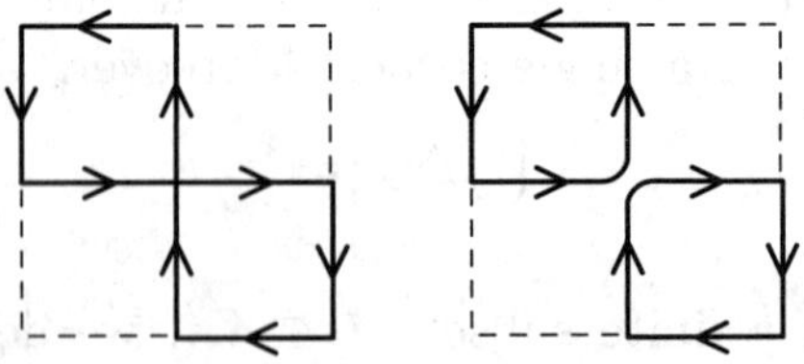

(a) (b)

Figure 4.4.2. At crossing boundary lines, choose the path that bends.

(b), (c) We note that

$$n(\partial\Delta_\alpha, z_0) = \begin{cases} 1, & z_0 \in \Delta_\alpha \\ 0, & z_0 \in \mathbb{C} \setminus \overline{\Delta}_\alpha \end{cases} \tag{4.4.9}$$

Thus, $n(\widetilde{\Gamma}_J, z_0) = 1$ if $z_0 \in \bigcup_{\alpha\in J} \Delta_\alpha$ and 0 if $z_0 \in \mathbb{C} \setminus \Delta_J$. Since Γ_J and $\widetilde{\Gamma}_J$ are equivalent, for $z_0 \notin \operatorname{Ran}(\Gamma_J) \cup \operatorname{Ran}(\widetilde{\Gamma}_J)$, $n(\widetilde{\Gamma}_J, z_0) = n(\Gamma_J, z_0)$, proving (b) and (c) for $z_0 \in \bigcup_{\alpha\in J} \Delta_\alpha$. Since $n(\Gamma_J, \cdot\,)$ is continuous on components of $\mathbb{C} \setminus \Gamma_J$, we get (c) in general. □

Proof of Theorem 4.4.1. Pick $\delta > 0$ so that

$$\delta\sqrt{2} < \operatorname{dist}(K, \mathbb{C} \setminus \Omega) \tag{4.4.10}$$

Pick $z_0 = 0$ and let

$$J = \{\alpha \mid K \cap \overline{\Delta}_\alpha \neq \emptyset\} \tag{4.4.11}$$

Let Γ_J be the chain constructed above and Δ_J given by (4.4.5). By (4.4.10), if $z \in \Delta_J$, then $z \in \Omega$ (for $z \in$ some $\overline{\Delta}_\alpha$, so that there is $w \in K$ with $|z - w| \le \delta\sqrt{2}$). Thus, since $\operatorname{Ran}(\Gamma_J) \subset \Delta_J$, we see that Γ_J is a chain in Ω.

Since we have proven that $\Delta_J \subset \Omega$, by (4.4.7), Γ is homologous to zero, and the middle of (4.4.1) holds.

$K \cap \operatorname{Ran}(\Gamma) = 0$, since any edge in $\operatorname{Ran}(\Gamma)$ is in some $\overline{\Delta}_\beta$ with $\beta \notin J$. Thus, since $K \subset \Delta_J$, (4.4.8) implies the first of (4.4.1). Finally, by (4.4.7) and (4.4.8), we have the final line of (4.4.1). □

For the second proof of the ultimate CIT, we need a strengthened form of Theorem 4.4.1.

Theorem 4.4.3. *In Theorem 4.4.1, suppose Ω is bounded. Then the contour can be picked so that for every $w \in \operatorname{Ran}(\Gamma)$, there is $z \in \mathbb{C} \setminus \Omega$ and a curve γ with $\gamma(0) = w$, $\gamma(1) = z$, and $\operatorname{Ran}(\gamma) \cap K = \emptyset$. In addition, we have for any $z_0 \in K$ and any f analytic in Ω, that*

$$f(z_0) = \frac{1}{2\pi i} \oint_\Gamma \frac{f(z)}{z - z_0}\, dz \tag{4.4.12}$$

Proof. Pick δ so that (4.4.10) holds, $z_0 = 0$, but now

$$J = \{\alpha \mid \overline{\Delta}_\alpha \subset \Omega\} \tag{4.4.13}$$

which is finite since Ω is bounded.

If $w \in \operatorname{Ran}(\Gamma_J)$, then the edge containing w must be contained in both a square, $\overline{\Delta}_\alpha$, in Ω and one, $\overline{\Delta}_\beta$, not entirely in Ω, that is, there is $z \in \overline{\Delta}_\beta \cap (\mathbb{C} \setminus \Omega)$. Thus, by (4.4.10), $\overline{\Delta}_\beta \cap K = \emptyset$, so the straight line from w to z is the required γ. Also, by (4.4.10), $K \subset \Delta_J$.

That leaves (4.4.12). For $z_0 \in \Delta_\alpha$ for some $\alpha \in J$, and all $\beta \in \mathbb{Z}^2$,

$$\frac{1}{2\pi i} \oint_{\partial\Delta_\beta} \frac{f(z) - f(z_0)}{z - z_0}\, dz = 0 \tag{4.4.14}$$

since Δ_β is convex and $g(z) = (f(z) - f(z_0))/(z - z_0)$ is analytic in a neighborhood of $\overline{\Delta}_\beta$. By (4.4.9), we see

$$\frac{1}{2\pi i} \oint_{\partial\Delta_\beta} \frac{f(z)}{z - z_0}\, dz = \delta_{\alpha\beta} f(z_0) \tag{4.4.15}$$

which proves (4.4.12) for such z_0 and Γ_J replaced by $\widetilde{\Gamma}_J$. But then by equivalence, we can replace $\widetilde{\Gamma}_J$ by Γ_J and use continuity to get the result for $\Delta_J \setminus \operatorname{Ran}(\Gamma_J) \supset K$. □

Second Proof of Theorems 4.1.1 and 4.1.2. Let Γ be the given chain and $K = \operatorname{Ran}(\Gamma)$. Pick R so large that $K \subset \{z \mid |z| < R\}$. For $|z| \geq R$, $n(\Gamma, z) = 0$ since $n(\Gamma, z) \to 0$ as $z \to \infty$, $n(\Gamma, \cdot\,)$ is continuous on $\{z \mid |z| \geq R\}$ integral. Then Γ is homologous to zero with respect to $\Omega \cap \{z \mid |z| < R\}$, so without loss, we can suppose Ω is bounded.

Therefore, Theorem 4.4.3 is applicable and we let $\widetilde{\Gamma}$ be the contour guaranteed by it. For $z \in \mathbb{C} \setminus \Omega$, by hypothesis, $n(\Gamma, z) = 0$, so for $w \in \operatorname{Ran}(\widetilde{\Gamma})$,

$$n(\Gamma, w) = 0 \tag{4.4.16}$$

because $n(\Gamma, z)$ is continuous on the curve γ guaranteed by the theorem. By (4.4.12) for $z \in \operatorname{Ran}(\Gamma)$,

$$f(z) = \frac{1}{2\pi i} \oint_{\widetilde{\Gamma}} \frac{f(w)}{w - z}\, dz \tag{4.4.17}$$

Thus, by Fubini's theorem and boundedness of $f(w)/(w-z)$ for $w \in \operatorname{Ran}(\widetilde{\Gamma})$, $z \in \operatorname{Ran}(\Gamma)$,

$$\begin{aligned}\oint_\Gamma f(z)\, dz &= \frac{1}{2\pi i} \oint_\Gamma dz \oint_{\widetilde{\Gamma}} \frac{f(w)}{w - z}\, dw \\ &= \oint_{\widetilde{\Gamma}} f(w)\, dw \left(\oint_\Gamma \frac{dz}{w - z} \right) \\ &= -\oint_{\widetilde{\Gamma}} f(w) n(\Gamma, w)\, dw \\ &= 0 \end{aligned} \tag{4.4.18}$$

by (4.4.16). □

Proof of Theorem 4.1.3. (Ultimate Laurent Splitting). Let $\Gamma_1^{(0)}$ be a mesh chain as given by Theorem 4.4.1. Define

$$\operatorname{ins}(\Gamma_1^{(0)}) = \{z \notin \operatorname{Ran}(\Gamma_1^{(0)}) \mid n(\Gamma_1^{(0)}, z) = 1\} \tag{4.4.19}$$

and

$$\text{out}(\Gamma_1^{(0)}) = \{z \notin \text{Ran}(\Gamma_1^{(0)}) \mid n(\Gamma_1^{(0)}, z) = 0\} \tag{4.4.20}$$

Since $\text{ins}(\Gamma_1^{(0)}) \cup \text{Ran}(\Gamma_1^{(0)})$ is compact, we can find a mesh chain, $\Gamma_2^{(0)}$, with $\text{ins}(\Gamma_1^{(0)}) \cup \text{Ran}(\Gamma_1^{(0)}) = \overline{\text{ins}(\Gamma_1^{(0)})} \subset \text{ins}(\Gamma_2^{(0)})$.

Define $f_-^{(0)}$ (respectively, $f_+^{(0)}$) on $\text{ins}(\Gamma_2^{(0)})$ (respectively, $\mathbb{C} \setminus \overline{\text{ins}(\Gamma_1^{(0)})}$) by

$$f_-^{(0)}(z) = \int_{\Gamma_2^{(0)}} \frac{f(w)dw}{w-z}, \quad f_+^{(0)}(z) = -\int_{\Gamma_1^{(0)}} \frac{f(w)dw}{w-z} \tag{4.4.21}$$

It is easy to see that $\Gamma_2^{(0)} - \Gamma_1^{(0)}$ is $\Omega \backslash K$ homologous to 0 and $n(\Gamma_2^{(0)} - \Gamma_1^{(0)}, z) = 1$ for all $z \in \text{ins}(\Gamma_2^{(0)}) \setminus \overline{\text{ins}(\Gamma_1^{(0)})}$, so by the ultimate Cauchy formula, for such z,

$$f(z) = f_+(z) + f_-(z) \tag{4.4.22}$$

By shrinking mesh size, it is not hard to find $\Gamma_1^{(n)}$, $\Gamma_2^{(n)}$ inductively so that

$$\overline{\text{ins}\big(\Gamma_1^{(n+1)}\big)} \subset \text{ins}\big(\Gamma_1^{(n)}\big), \quad \overline{\text{ins}\big(\Gamma_2^{(n)}\big)} \subset \text{ins}\big(\Gamma_2^{(n+1)}\big) \tag{4.4.23}$$

and

$$\bigcap_{n=1}^{\infty} \overline{\text{ins}(\Gamma_1^{(n)})} = K, \quad \bigcup_{n=1}^{\infty} \text{ins}(\Gamma_1^{(n)}) = \Omega \tag{4.4.24}$$

If we define $f_\pm^{(n)}$ by the analog (4.4.21), by the ultimate Cauchy formula again $f_\pm^{(n)}(z) = f_\pm^{(n+1)}(z)$ on the sets where both $f^{(n)}$ and $f^{(n+1)}$ are defined. Thus we get a single function f_+ analytic on $\mathbb{C} \setminus K$ and f_- analytic on Ω, so that (4.4.22) holds on $\Omega \setminus K$. Clearly $f_+^{(n)}(z) \to 0$ as $|z| \to \infty$. This shows existence.

We get uniqueness as in the proof of Theorem 3.7.3. If $f_+ + f_- = g_+ + g_-$, $f_+ - g_+ = -f_- + g_-$ on $\Omega \setminus K$, so we get an entire function h going to zero at infinity, so 0, i.e., $f_+ = g_+$, $f_- = g_-$.

Since $f_+ \to 0$ at infinity and is analytic in a neighborhood of $\text{Ran}(\Gamma_1^{(0)})$, it is bounded on $\mathbb{C} \setminus \text{ins}\big(\Gamma_1^{(0)}\big)$. Since f_- is analytic in a neighborhood of $\text{ins}\big(\Gamma_2^{(0)}\big)$, it is bounded on $\text{ins}\big(\Gamma_2^{(0)}\big)$. If f is bounded on $\Omega \setminus K$, $f_+ = f - f_-$ is bounded on $\text{ins}\big(\Gamma_1^{(0)}\big) \setminus K$ and, therefore, on $\mathbb{C} \setminus K$. A similar argument works for f_- on Ω. □

Notes and Historical Remarks. Mesh-defined contours go back at least to Saks–Zygmund [**500**] (they appear on page 155 of the third English edition (1971); the Polish first edition was published in 1938). They have been used in Ahlfors' book [**9**] where the Beardon proof was presented.

While the subtlety shown in Figure 4.4.2 is easy to overcome, Ash–Novinger [**25**] has an interesting way to avoid it: define contours based on hexagons rather than squares. In that case, three edges come together at each point and, after removing duplicate edges, only two edges can come into a vertex, and there is no ambiguity to resolve.

4.5. Simply Connected and Multiply Connected Regions

In Section 2.6, we defined Ω to be holomorphically simply connected if and only if $\oint_\gamma f(z)\,dz = 0$ for all $f \in \mathfrak{A}(\Omega)$ and all closed rectifiable curves, γ, in Ω. The ultimate CIT implies that this is equivalent to $n(\gamma, z) = 0$ for all such γ and all $z \notin \Omega$. Here we'll pursue this further by proving:

Theorem 4.5.1. *Let Ω be a region in $\mathbb{C}$. The following are equivalent:*

(1) *In $\widehat{\mathbb{C}}$, $\widehat{\mathbb{C}} \setminus \Omega$ is connected.*
(2) *Every component of $\mathbb{C} \setminus \Omega$ is unbounded.*
(3) *Ω is hsc.*

Remarks. 1. For now, $\widehat{\mathbb{C}}$ is just the one-point compactification of $\mathbb{C}$, that is, $\mathbb{C}$ together with an extra point called $\{\infty\}$, so that the open sets in $\widehat{\mathbb{C}}$ are the open sets in $\mathbb{C}$ plus the complements in $\widehat{\mathbb{C}}$ of compact subsets of $\mathbb{C}$. We'll put a lot more structure on $\widehat{\mathbb{C}}$ in Chapters 6 and 7.

2. If Ω is bounded, (1) is equivalent to $\mathbb{C} \setminus \Omega$ being connected, but as $\Omega = \{z \mid |\mathrm{Im}\, z| < 1\}$ shows, if Ω is unbounded, (1) can hold even though $\mathbb{C} \setminus \Omega$ is not connected.

3. Recall (see Section 2.1 of Part 1) that a component is a maximal connected subset, that such sets are always (relatively) closed, and that the components are a disjoint decomposition of the space.

Since $n(z, \gamma)$ is constant on each component of $\mathbb{C} \setminus \mathrm{Ran}(\gamma)$ and $\lim_{z\to\infty} n(z, \gamma) = 0$, $n(z, \gamma) = 0$ on each unbounded component of $\mathbb{C} \setminus \Omega$, so by the ultimate CIT, (2) $\Rightarrow$ (3). We will prove (3) $\Rightarrow$ (1) below. Thus, (2) $\Rightarrow$ (1). That (2) $\Leftrightarrow$ (1) is a general topological fact that only depends on $\widehat{\mathbb{C}}$ being a compact, connected metric space. We proved it as Theorem 5.4.24 of Part 1. That the result (1) $\Rightarrow$ (2) is subtle can be seen that it can fail if Ω is not assumed open (see Example 5.4.23 of Part 1).

Proof that (3) $\Rightarrow$ (1) in Theorem 4.5.1. We'll prove $\sim$ (1) $\Rightarrow$ $\sim$ (3) which is equivalent. If $\widehat{\mathbb{C}} \setminus \Omega$ is not connected, we can find A, B disjoint subsets of this space both relatively open and closed. Suppose $\infty \in B$. Then A is a compact subset of $\mathbb{C}$ since $B \cup \Omega$ is an open neighborhood of ∞. Thus, in $\mathbb{C}$, $\mathrm{dist}(A, B) > 0$, and thus, $\widetilde{\Omega} = \Omega \cup A$ is open in $\mathbb{C}$. By Theorem 4.4.1 with $K = A$ and $\Omega = \widetilde{\Omega}$, there is chain Γ with $\mathrm{Ran}(\Gamma) \subset \Omega$

$(= \widetilde{\Omega} \setminus A)$ so $n(\Gamma, z_0) = 1$ for any $z_0 \in A$. But then $1/(z - z_0) \in \mathfrak{A}(\Omega)$ and $\oint_\gamma f(z)\,dz \neq 0$, so Ω is not hsc. $\square$

A region Ω so $\widehat{\mathbb{C}} \setminus \Omega$ is connected is sometimes called *simply connected.* We already have used hsc and tsc (in Section 2.6). We've just seen this notion is equivalent to hsc and we'll see eventually (Theorem 8.1.2) that these are equivalent, so the name is reasonable.

If $\widehat{\mathbb{C}} \setminus \Omega$ has more than one component, then it is called *multiply connected*; if there are n components, it is called *n-connected.* One can speak of doubly connected if $n = 2$. An annulus is doubly connected. A disk with k disjoint closed disks removed is $(k+1)$-connected.

4.6. The Ultra Cauchy Integral Theorem and Formula

Our goal in this section is to prove the following, which says that if f is continuous on a rectifiable Jordan curve, γ, and if f has an analytic continuation to the inside of γ, then

$$\oint_\gamma f(z)\,dz = 0 \tag{4.6.1}$$

Theorem 4.6.1 (Ultra CIT/CIF). *Let γ be a rectifiable Jordan curve and Ω the bounded component of $\mathbb{C} \setminus \mathrm{Ran}(\gamma)$. Orient γ so that $n(\gamma, z) = 1$ for $z \in \Omega$. Let f be regular on $\overline{\Omega} = \Omega \cup \mathrm{Ran}(\gamma)$. Then* (4.6.1) *holds, and for any $z_0 \in \Omega$,*

$$f(z_0) = \frac{1}{2\pi i} \oint_\gamma \frac{f(z)}{z - z_0}\,dz \tag{4.6.2}$$

Our proof will exploit the Banach indicatrix theorem (see Theorem 4.15.7 of Part 1). The Notes to Section 4.7 will sketch another proof, given an approximation theorem (Mergelyan's Theorem) that we will prove in Section 6.10 of Part 4.

Proof of Theorem 4.6.1. Let $L =$ length of γ. For $n = 2, 3, \dots$, let $\delta_n = L/n$ and pick n points $z_1^{(n)}, \dots, z_n^{(n)}$ on $\mathrm{Ran}(\gamma)$ so that they lie consecutively in a counterclockwise direction, and the length of each segment, obtained by deleting these points, is L/n.

Consider lines $\mathrm{Re}\, z = \alpha$. Since the x-component of γ has bounded variation, Banach's indicatrix theorem (see Theorem 4.15.7 of Part 1) implies that for almost every α, these lines have finitely many intersections with γ. The same is true for the lines $\mathrm{Im}\, z = \beta$. It follows that if we consider grids with spacing, δ_n, for almost every choice of the center of squares Δ_0, all the grid lines have only finitely many intersections with γ, so since γ is bounded, the intersections with the grids are finite. Fix such a choice.

Break all $\{\Delta_\alpha\}_{\alpha\in\mathbb{Z}^2}$ into three collections: those with $\Delta_\alpha^{\text{int}} \subset \Omega$, the bounded component of $\mathbb{C}\setminus \operatorname{Ran}(\gamma)$; those with $\Delta_\alpha^{\text{int}} \subset \mathbb{C}\setminus\overline{\Omega}$, the unbounded component; and those with $\Delta_\alpha^{\text{int}} \cap \operatorname{Ran}(\gamma) \neq \emptyset$. Given that there are only finitely many intersections, we include all squares. We'll call them inner, outer, and boundary squares. Any collection of more than four squares must have two whose boundaries are disjoint, so a distance $\geq \delta_n$ from each other. It follows that each segment of γ intersects at most four squares, so there are at most $4n$ boundary squares.

For each inner and boundary square, Δ_α, let γ_α be the boundary of $\Delta_\alpha^{\text{int}} \cap \Omega$, that is, $\partial\Delta_\alpha$ if Δ_α is inner and pieces of $\partial\Delta_\alpha$, and of γ if Δ_α is a boundary square. The parts that are boundaries of Δ_α cancel, so

$$\oint_\gamma f(z)\,dz = \sum_{\{\alpha|\Delta_\alpha \text{ inner or boundary}\}} \oint_{\gamma_\alpha} f(z), dz \tag{4.6.3}$$

$$= \sum_{\{\alpha|\Delta_\alpha \text{ boundary}\}} \oint_{\gamma_\alpha} [f(z) - f(z_\alpha)]\,dz \tag{4.6.4}$$

since $\oint_{\gamma_\alpha} f(z)\,dz = 0$ for inner squares (by the usual CIT or a limit if γ_α contains single points of γ). In adding $-f(z_\alpha)$ with $z_\alpha \in \Delta_\alpha \cap \Omega$, we use the fact that if γ_α is a closed rectifiable curve, then $\oint_{\gamma_\alpha} dz = 0$.

It follows from (4.6.4) that

$$\left|\oint_\gamma f(z)\,dz\right| \leq \sup_{\substack{z,w\in\Omega \\ |z-w|\leq\delta_n\sqrt{2}}} |f(z)-f(w)| \sum_{\{\alpha|\Delta_\alpha \text{ boundary}\}} L(\gamma_\alpha) \tag{4.6.5}$$

$L(\gamma_\alpha)$ comes from parts of Δ_α of size at most $4\delta_n$ and a piece of γ—the sum of these pieces has length at most $L(\gamma)$, so

$$\begin{aligned}\text{RHS of (4.6.5)} &\leq \sup_{\substack{z,w\in\Omega \\ |z-w|\leq\delta_n\sqrt{2}}} |f(z)-f(w)|[L(\gamma)+(4\delta_n)(4n)] \\ &= [17L(\gamma)] \sup_{\substack{z,w \\ |z-w|\leq\delta_n\sqrt{2}}} |f(z)-f(w)|\end{aligned}$$

Since $\delta_n \to 0$ as $n\to\infty$ and f is uniformly continuous on $\overline{\Omega}$, this goes to zero as $n\to\infty$. Thus, $\oint f(z)\,dz = 0$, proving the ultra CIT. As usual, applying the CIT to $(f(z)-f(z_0))/(z-z_0)$, we get the ultimate CIF. □

Notes and Historical Remarks. Theorem 4.6.1 is a result of Denjoy [**137**], Heilbronn [**248**], and Walsh [**574**] (see also Beckenbach [**35**]). The proof we give, which I learned from P. Deift and X. Zhou, is due to Chen [**114**]. For other proofs, see Kunugi [**333**] and the Notes to Section 4.7.

4.7. Runge's Theorems

In this section, we discuss approximation of analytic functions by polynomial and rational functions. This seems far afield from the subject matter of this chapter—it is here because the key to the proof is the mesh-defined contours of Section 4.4. In addition, as we explain in the Notes, there is a third proof of the ultimate CIF that depends on Runge's theorem. One uses the CIF for rectangles and mesh-defined contours to get Runge's theorem, proves the ultimate CIF for rational functions, and uses density and Runge's theorem to get the full ultimate CIF. Finally, we mention that in the Notes, we'll discuss how a different approximation argument implies a CIF for boundaries of Jordan regions.

For background, recall the classical Weierstrass approximation theorem (proven in Section 2.4 of Part 1): Given any continuous function, f, on $[0,1]$ and every $\varepsilon > 0$, there is a polynomial, p, so $\|f-p\|_\infty \equiv \sup_{z\in[0,1]}|f(z)-p(x)| < \varepsilon$. Equivalently, the polynomials are $\|\cdot\|_\infty$-dense in $\mathbb{C}([0,1])$. The analog for general sets in $\mathbb{C}$ is false because of a different result of Weierstrass: the Weierstrass convergence theorem (Theorem 3.1.8) that a uniform limit of analytic functions on an open set is analytic. Thus, for example, a uniform limit on $\overline{\mathbb{D}}$ of polynomials is regular, that is, continuous on $\overline{\mathbb{D}}$ and analytic on $\mathbb{D}$. The reader is asked in Problem 1 to prove that every regular function on $\overline{\mathbb{D}}$ is a uniform limit of polynomials.

The best approximation theorem for $K \subset \mathbb{C}$ compact with $\mathbb{C}\setminus K$ connected (Mergelyan's theorem, stated formally in the Notes) says that any regular function on K can be uniformly approximated by polynomials. We will not prove this result here (but will in Section 6.10 of Part 4) but only the weaker:

Theorem 4.7.1 (Runge's Second Theorem). *Let K be a compact subset of $\mathbb{C}$ with $\mathbb{C}\setminus K$ connected. Then given any $\varepsilon > 0$ and any function, f, analytic in a neighborhood of K, there is a polynomial, p, so that* (4.7.5) *holds.*

There is also a theorem for general compact K:

Theorem 4.7.2 (Runge's First Theorem). *Let K be a compact subset of $\mathbb{C}$. Then given any $\varepsilon > 0$ and any function, f, analytic in a neighborhood of K, there is a rational function, p/q, with all poles in $\mathbb{C}\setminus K$ so that*

$$\left\| f - \frac{p}{q} \right\|_K < \varepsilon \tag{4.7.1}$$

In this section, we'll prove the two Runge theorems (and a somewhat stronger version of the first Runge theorem). We emphasize that K need not be connected in Runge's theorem, and this is the basis for some remarkable constructions (Problems 4, 6, and 7).

Proof of Theorem 4.7.2. Let $\Omega \supset K$ be the open set on which f is analytic. Let Γ be the mesh-defined chain assured by Theorem 4.4.1. (Note that K, and so Ω, may not be connected, so Ω may not be a region, but the proof of Theorem 4.4.1 did not use connectedness of Ω.)

By (4.4.1) and the ultimate CIT, for all $z \in K$,

$$f(z) = \frac{1}{2\pi i} \oint_\Gamma \frac{f(w)}{w - z}\, dw \tag{4.7.2}$$

By definition of contour integrals, the right side of (4.7.2) is a limit of Riemann sums, and by the uniform continuity of $(w - z)^{-1}$ for $w \in \Gamma$, $z \in K$, these sums converge uniformly on K. Thus, $f \restriction K$ is a uniform limit of $\sum c_n f(w_n)/(w_n - z)$, which are rational functions. $\square$

Proposition 4.7.3. *Let $K \subset \mathbb{C}$ be compact and let Q be a connected subset of $\mathbb{C} \setminus K$. Fix $z_0 \in Q$ and let $\mathcal{R}_{z_0}$ be the closure in $C(K)$ in $\|\cdot\|_K$ of the polynomials in $(z - z_0)^{-1}$. Then $\mathcal{R}_{z_0}$ contains $(z - z_1)^{-k}$ for every $z_1 \in Q$ and every $k = 1, 2, \dots$. If Q is unbounded, one can replace $\mathcal{R}_{z_0}$ by $\mathcal{R}_\infty$, the closure of the polynomials.*

Proof. Let $Q_1 = \{z_1 \in Q \mid (z - z_1)^{-1} \in \mathcal{R}_{z_0}\}$. Q_1 is nonempty since $z_0 \in Q_1$. Since $z_n \to z_\infty$ all in Q implies $\|(\cdot - z_n)^{-1} - (\cdot - z_\infty)^{-1}\|_K \to 0$, Q_1 is closed. If $z_1 \in Q_1$ and $R = \operatorname{dist}(z_1, K) > 0$ and $|z_2 - z_1| \le R/2$, then

$$(z - z_2)^{-1} = \sum_{n=0}^\infty (z_2 - z_1)^n (z - z_1)^{-n-1} \tag{4.7.3}$$

gives a $\|\cdot\|_K$ convergent expansion in polynomials in $|z - z_1|^{-1}$. Since $(z - z_1)^{-1}$ is a uniform limit of polynomials in $(z - z_0)^{-1}$, we see that $z_2 \in Q_1$. Thus, Q_1 is open, and thus, by connectedness, $Q_1 = Q$.

When Q is unbounded, let $R = \sup\{|z| \mid z \in K\} < \infty$ and let $Q_1 = \{z_1 \in Q \mid (z - z_1)^{-1} \in \mathcal{R}_\infty\}$. Since $|z_1| \ge 2R$ implies $(z - z_1)^{-1} = -\sum_{n=0}^\infty z^n z_1^{-n-1}$, we see $z_1 \in Q_1$, so Q_1 is nonempty. That it is open and closed is the same as above. $\square$

Proof of Theorem 4.7.1. By a partial fraction expansion, any p/q with poles in $\mathbb{C} \setminus K$ is a sum of $(z - z_1)^{-k}$, so by the proposition, approximable by polynomials. $\square$

Remark. Proposition 4.7.3 also implies that Theorem 4.7.2 can be improved to allow q to have all its zeros in a fixed set, S, with one point in each bounded component of $\mathbb{C} \setminus K$.

Finally, when $K \subset \Omega$, we'd like to know any function analytic in a neighborhood of K can be approximated by functions analytic in Ω. The example $\Omega = \mathbb{D}$, $K = \{z \mid \frac{1}{2} \le |z| \le \frac{3}{4}\}$ and $f(z) = z^{-1}$ shows this isn't true in general. Basically, one needs every component of $\mathbb{C} \setminus K$ to contain points of $\mathbb{C} \setminus \Omega$ (with a special rule for unbounded components). Here is a case where it's true and which we'll need in Section 9.5.

Theorem 4.7.4. *Let $\Omega \subset \mathbb{C}$ be a region, and for $\varepsilon > 0$ and R finite, define the compact set K by*

$$K = \{z \in \Omega \mid |z| \leq R, \operatorname{dist}(z, \mathbb{C} \setminus \Omega) \geq \varepsilon\} \tag{4.7.4}$$

Then for any function, f, analytic in a neighborhood of K and $\delta > 0$, there is a rational function, g, with poles in $\mathbb{C} \setminus \Omega$ so that $\|f - g\|_K \leq \delta$. In particular, $\mathfrak{A}(\Omega)$ is dense in $\|\cdot\|_K$ in the functions analytic in a neighborhood of K.

Proof. Let Q be a bounded component of $\mathbb{C} \setminus K$. We claim Q contains points of $\mathbb{C} \setminus \Omega$. For, since Q is bounded, $Q \subset \{z \mid |z| \leq R\}$, and so if $y_0 \in Q$, $\operatorname{dist}(y_0, \mathbb{C} \setminus \Omega) < \varepsilon$. Pick $x_0 \in \mathbb{C} \setminus \Omega$ so that $\operatorname{dist}(y_0, x_0) < \varepsilon$. Then, with $\gamma(s)$ the line segment, $\gamma(s) = sx_0 + (1 - \varepsilon)y_0$, $\operatorname{dist}(\gamma(s), \mathbb{C} \setminus \Omega) < \varepsilon$, so $\gamma(s) \in \mathbb{C} \setminus K$. Thus, y_0 is the same component of $\mathbb{C} \setminus K$ as x_0.

By Theorem 4.7.2, there is h a rational function with poles in $\mathbb{C} \setminus K$ so that $\|f - h\|_K \leq \delta/2$. Then, by Proposition 4.7.3, we find g with poles in $\mathbb{C} \setminus \Omega$ for bounded components and a polynomial part (for the unbounded component), so $\|h - g\|_K \leq \delta/2$. $\square$

Remarks. 1. The above proof shows that if Ω is simply connected (in the sense of (1)–(3) of Theorem 4.5.1), then K given by (4.7.4) has $\mathbb{C} \setminus K$ connected. For since K is compact, $\mathbb{C} \setminus K$ has at most one unbounded component. If $\mathbb{C} \setminus K$ had a bounded component, by the above proof, it would contain points of $\mathbb{C} \setminus \Omega$, so a component of $\mathbb{C} \setminus \Omega$, so it wouldn't be bounded after all.

2. In Problem 3, the reader will prove an improvement of Theorem 4.7.4 that shows if f is nonvanishing on K, then g can be picked rational with no zeros or poles on Ω.

Notes and Historical Remarks. Carl David Tolmé Runge (1856–1927) was a German mathematician, experimental physicist, and applied mathematician. Carl was born in Bremen but spent his early years in Havana, Cuba where his father traveled as Danish consul. His mother was British and his parents spoke English at home so that their children would be fluent in the language. When Carl was seven, his father died and his mother returned to Bremen to raise her eight children. After he graduated from secondary school, his mother took him on a six-month tour of the cultural centers of Italy.

He began studying mathematics and physics in Munich where he met and became lifelong friends with Max Planck. After two years, he and

Planck moved to Berlin where, attracted by the courses of Kummer and Weierstrass, he shifted to pure mathematics. His 1880 thesis on differential geometry was formally supervised by Weierstrass but it was on a topic he came up with in discussion with the other students. He then worked in the group around Kronecker in Berlin but did not publish much until he visited Mittag-Leffler in Stockholm who encouraged him to publish and accepted a number of papers in *Acta* (the journal Mittag–Leffler had founded) in 1884–86 including the one on the theorems of this section.

In Berlin, Runge became friendly with professor of physiology, Emile du Bois-Reymond, brother of the mathematician Paul, and became engaged to Aimée, Emile's daughter, but her father wouldn't allow the marriage until Runge obtained a professorship! As a result, Runge moved to Hanover in 1906 where he was made a professor. He was in Hanover for 18 years during which time he mainly worked on experimental and theoretical spectroscopy, including the determination of the spectrum of the recently discovered helium. He also worked on numerical analysis producing in 1895 a method for solving ordinary differential equations that, after a refinement by Martin Kutta (1887–1944), became known as the Runge–Kutta method.

Recognizing his talents, Klein arranged an appointment as a professor of applied mathematics in Göttingen where he joined Hilbert, Klein, and Minkowski starting in 1904 until his retirement in 1923 (the special nature of what Klein had done is seen by the fact that his successor, Gustav Herglotz (1881–1953) did not have "applied" in his title). Max Born was a student of Runge.

One can ask how small the error is in Theorem 4.7.1 for $\|f - P_n\|$ as a function of $n = \deg(P_n)$. In Problem 8, the reader will reconstruct a proof of Szegő that, in the simply connected case, it is $O(e^{-\alpha n})$ for some $\alpha > 0$.

For simply connected K, the best general approximation result is

Theorem 4.7.5 (Mergelyan's Theorem). *Let K be a compact subset of $\mathbb{C}$ with $\mathbb{C} \setminus K$ connected. Then given any $\varepsilon > 0$ and any function, f, regular on K, there is a polynomial, p, so that*

$$\|f - p\|_K = \sup_{z \in K} |f(z) - p(z)| < \varepsilon \tag{4.7.5}$$

Runge proved his theorem in 1885 [**498**] (there was related work by Appell [**19, 20**] slightly earlier). Remarkably, while Mergelyan's theorem (Theorem 4.7.5) uses related ideas to Runge, it was only proven 67 years later in 1952 [**380**]. For a proof, see, Theorem 6.10.4 of Part 4 or, for example, Greene and Krantz [**220**, Ch. 12].

Just as Mergelyan's theorem extends Runge's second theorem, one might hope that there is a similar extension of the first Runge theorem, that is, that for any compact K, any f regular on K can be approximated uniformly on K by rational functions. This hope is unfounded—a counterexample can be found in Example 6.10.12 of Part 4.

Another theorem in the circle of ideas is the following, which we prove in Part 4 as Theorem 6.10.1:

Theorem 4.7.6 (Hartogs–Rosenthal Theorem). *If $K \subset \mathbb{C}$ is compact and of zero planar Lebesgue measure, then any continuous function on K is a uniform limit in $\|\cdot\|_K$ of rational functions.*

For a proof as well as a discussion of other results on rational and polynomial approximation, see Section 6.10 of Part 4 or Gamelin [**198**]. There is an enormous literature on analogs of Runge's theorem in several complex variables; see Gamelin [**198**], Stout [**545**], Browder [**78**], and Leibowitz [**352**].

While we used the ultimate CIF in our proof of Theorem 4.7.2, we used it only for a mesh-defined contour which is a sum of squares for which the CIF is known from the star-shaped case. Once we have Theorem 4.7.2, we can use it to prove the general ultimate CIF, which is easy for rational functions. This approach of proving Runge's first theorem first and using it to prove the ultimate CIF is the one of Saks–Zygmund [**500**].

Similarly, Mergelyan's theorem (proven in Section 6.10 of Part 4) allows a quick proof of Theorem 4.6.1.

Proof of Theorem 4.6.1. Mergelyan's theorem for $\overline{\Omega}$ shows that it suffices to prove (4.6.2) when f is a polynomial P. Let

$$Q(z) = \frac{P(z) - P(z_0)}{z - z_0} \tag{4.7.6}$$

This is also a polynomial, so a global derivative, so $\oint_\gamma Q(z)\,dz = 0$. Thus,

$$\text{RHS of (4.6.2)} = P(z_0)n(\gamma, z_0) = P(z_0) \tag{4.7.7}$$

□

A different approach to polynomial and rational approximation to analytic functions using approximations by lemniscates is due to Hilbert [**258**]; see also Walsh [**575**] and Hille [**262**, Ch. 16 of Vol. 2].

Problems

1. Let f be analytic in $\mathbb{D}$, continuous in $\overline{\mathbb{D}}$.

   (a) Prove that if $f_r(z) = f(rz)$, then $\lim_{r\uparrow 1}\|f - f_r\|_{\overline{\mathbb{D}}} = 0$.

   (b) Prove that the Taylor series for f_r converges uniformly on $\overline{\mathbb{D}}$. Conclude f is a uniform limit on $\overline{\mathbb{D}}$ of polynomials.

2. Prove a converse to Runge's second theorem as follows.

(a) Given a compact $K \subset \mathbb{C}$, let Ω be the unbounded component of $\mathbb{C}\setminus K$ and let $\widetilde{K} = \mathbb{C}\setminus\Omega$. Prove that any function, f, analytic in a neighborhood of $\widetilde{K}$ has

$$\sup_{z\in\widetilde{K}} |f(z)| = \sup_{z\in K} |f(z)| \tag{4.7.8}$$

(b) If $\widetilde{K} \neq K$, $z_0 \in \widetilde{K} \setminus K$, and p is a polynomial such that

$$\sup_{z\in K} \left| \frac{1}{z-z_0} - p(z) \right| \leq \tfrac{1}{2} \sup_{z\in\widetilde{K}} |z - z_0|$$

prove that $f(z) = 1 - (z-z_0)p(z)$ violates (4.7.8).

(c) Prove that if $K \subset \mathbb{C}$ is compact and has the property that any function analytic on K can be approximated in $\|\cdot\|_K$ by polynomials, then $\mathbb{C} \setminus K$ is connected.

3. Under the hypothesis of Theorem 4.7.4, suppose also that f is nonvanishing on K. This problem will prove g can then be picked nonvanishing also.

(a) For any $a \in \mathbb{C} \setminus K$, prove that there is $b \in \mathbb{C} \setminus \Omega$ and a polygonal path, $\Gamma_{ab} \subset \mathbb{C} \setminus K$, connecting a to b.

(b) Prove that there is h_{ab} analytic in a neighborhood of K so that $\exp(h_{ab}(z)) = (z-b)/(z-a)$ on K. (*Hint*: Prove for any path γ in $\mathbb{C} \setminus \Gamma_{ab}$, with $H(z) = (z-b)/(z-a)$, we have $\frac{1}{2\pi i} \oint_\gamma \frac{H'(z)}{H(z)}\, dz = 0$ and define h_{ab} so $h'_{ab} = H'/H$.)

(c) Given any $a_1, \dots, a_m \in \mathbb{C}\setminus K$ and any ε, show there exist $b_1, \dots, b_m \in \mathbb{C} \setminus \Omega$ and h analytic on Ω so that

$$\sup_{z\in K} \left| \prod_{j=1}^{m} \frac{z-b_j}{z-a_j} - e^{h(z)} \right| < \varepsilon$$

(d) Prove the g in Theorem 4.7.6 can be picked nonvanishing on Ω.

4. The purpose of this problem is to lead the reader through a rather remarkable application of Runge's theorem, the existence of a sequence of polynomials, p_n (note that p_n is not claimed to have degree n) so that

$$\lim_{n\to\infty} p_n(0) = 1; \qquad \text{For any fixed } z \neq 0, \lim_{n\to\infty} p_n(z) = 0 \tag{4.7.9}$$

We'll comment on the significance of these sequences below.

(a) Let (see Figure 4.7.1)

$$K_n = \left[\{z \mid |z| \leq n\} \setminus \{z \mid \operatorname{dist}(z, [0,n]) < n^{-1}\}\right] \cup \{0\} \cup \left[\frac{1}{n}, n\right]$$

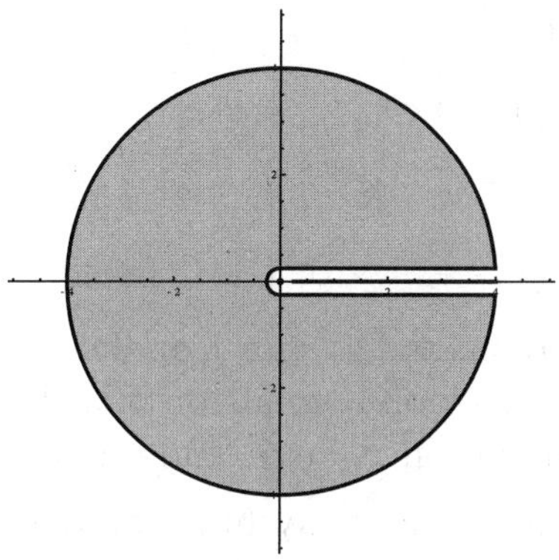

Figure 4.7.1. The set K_n for $n = 4$.

Prove that K_n is compact and $\mathbb{C} \setminus K_n$ is connected.

(b) Let f_n be the continuous function on K_n with $f_n(0) = 1$ and $f_n(z) = 0$ for $z \in K_n \setminus \{0\}$. Use Runge's second theorem to find a polynomial, p_n, with

$$\sup_{K_n} |p_n(z) - f_n(z)| \leq \frac{1}{n} \tag{4.7.10}$$

(c) Show that p_n obeys (4.7.9).

Remark. We've already seen in the Weierstrass convergence theorem (Theorem 3.1.5) that analyticity is preserved under uniform convergence on compact subsets of Ω, and we'll see more about the suitability of this topology in Chapter 6. This example dramatically shows that analyticity is not preserved under pointwise limits. At first blush, one might think the maximum principle (or the Vitali convergence theorem; see Theorem 6.2.8) forbids examples like this. They do not, but they do imply p_n obeys $\sup_{|z|\leq\varepsilon}|p_n(z)| \to \infty$ for any $\varepsilon > 0$! With a slightly more involved Runge theorem construction (see [**477**, Sect. 2.3.1]), one can arrange that also, for all $k \geq 1$, $f_n^{(k)}(z) \to 0$ for each fixed $z \in \mathbb{C}$. In spite of this example, if $g_n \to g_\infty$ pointwise for $g_n \in \mathfrak{A}(\Omega)$, there is a dense open subset, Ω', of Ω so that $g_\infty \restriction \Omega'$ is analytic; see Problem 13 of Section 6.2.

5. By mimicking the construction in Problem 4, find a sequence of polynomials so $P_n \to f_\infty$ pointwise and $f_\infty = \frac{1}{2}$ on $\cup_{n=1}^\infty \partial\mathbb{D}_n(0)$, $f_\infty = 0$ on $\bigcup_{n=1}^\infty \mathbb{D}_{2n}(0) \setminus \overline{\mathbb{D}_{2n-1}(0)}$, and $f_\infty = 1$ on $\mathbb{D} \cup \bigcup_{n=2}^\infty \mathbb{D}_{2n+1}(0) \setminus \overline{\mathbb{D}_{2n}(0)}$.

6. The purpose of this problem is to construct a function f analytic on $\mathbb{D}$ so that for any $\theta \in [0, 2\pi)$, the set of limit points of $f(re^{i\theta})$ as $r \uparrow 1$ is all of $\mathbb{C}$. Let $\{\omega_j\}_{j=1}^\infty$ be a subset of $\mathbb{C}$ so that $\{\omega_{2j}\}_{j=1}^\infty$ and $\{\omega_{2j+1}\}_{j=1}^\infty$ are

both dense in $\mathbb{C}$. Let

$$A_j = \begin{cases} \{(1-2^{-j})e^{i\theta} \mid \theta \le \frac{3\pi}{4}\}, & j \text{ odd} \\ \{(1-2^{-j})e^{i\theta} \mid |\theta - \frac{\pi}{2}| \le \frac{3\pi}{4}\}, & j \text{ even} \end{cases}$$

and $K_j = A_j \cup \{z \mid |z| \le 1 - 2^{-(j-1)}\}$.

(a) Prove $\mathbb{C} \setminus K_j$ is connected so any f analytic in a neighborhood of K_j can be approximated by polynomials on K_j. In particular, this can be done if f is a polynomial on $K_j \setminus A_j$ and any constant on A_j.

(b) Show that one can inductively find polynomials f_j so that $f_1 = \omega_1$ and

$$\sup_{|z| \le 1 - 2^{-(j-1)}} |f_j(z) - f_{j-1}(z)| \le 2^{-j}$$

$$\sup_{z \in A_j} |f_j(z) - \omega_j| \le 2^{-j}$$

(c) Prove that $\lim_{j\to\infty} f_j(z) \equiv f(z)$ exists uniformly on compact subsets of $\mathbb{D}$ and that $\sup_{z\in A_j} |f(z) - \omega_j| \le 2^{-j+1}$.

(d) Prove that for any θ, the set of limit points of $f(re^{i\theta})$ as $r \uparrow 1$ is all of $\mathbb{C}$.

7. The purpose of this problem is to find an entire function, f, in $\mathfrak{A}(\mathbb{C})$ whose translates are dense in $\mathfrak{A}(\mathbb{C})$!

(a) Let $\{Q_n\}_{n=1}^\infty$ be a counting of the set of all polynomials with rational coefficients. Prove this is dense in $\mathfrak{A}(\mathbb{C})$.

(b) Define $K_n = A_n \cup B_n$ where $A_n = \overline{\mathbb{D}_{n^2}(0)}$ and $B_n = \overline{\mathbb{D}_n(\frac{1}{2}n^2 + \frac{1}{2}(n+1)^2)}$. Note that A_n and B_n are disjoint. Define analytic functions, g_n, and polynomials, P_n, inductively as follows:

$$g_1 = P_1 = Q_1$$

$$g_{n+1} \restriction A_n = P_n \restriction A_n, \qquad g_{n+1} \restriction B_n = Q_{n+1}(\cdot + \tfrac{1}{2}\, n^2 + \tfrac{1}{2}\,(n+1)^2)$$

P_{n+1} is a polynomial so that

$$\sup_{z\in K_n} |g_{n+1}(z) - P_{n+1}(z)| \le 2^{-n} \tag{4.7.11}$$

(c) Prove that in the topology on $\mathfrak{A}(\mathbb{C})$ of uniform convergence on compacts, $P_n(z)$ converges to an entire function f.

(d) Prove that if $f_n(z) = f(z - \frac{1}{2}n^2 - \frac{1}{2}(n+1)^2)$, then $\sup_{|z|\le n} |f_n(z) - Q_{n+1}(z)| \le 2^{-n+1}$, and conclude that the translates of f are dense in $\mathfrak{A}(\mathbb{C})$.

Remark. The existence of functions with the property above is due to Birkhoff [**52**]; the construction using Runge's theorem is due to Seidel–Walsh [**515**].

8. The purpose of this problem is to lead the reader through a proof of Szegő's result that, in the context of Runge's second theorem, the error can be made exponentially small in the degree of the approximating polynomial. Fix K compact, connected, and simply connected, f analytic in Ω a simply connected neighborhood of f, and Γ a closed mesh contour homologous to 0 in Ω with $n(\Gamma, z) = 1$ for $z \in K$. Let $R = \max(|z-w| \mid z \in K,\, w \in \mathrm{Ran}(\Gamma))$.

   (a) Prove that Γ can be decomposed into segments $\Gamma_1, \dots, \Gamma_\ell$ (which can be smaller than the mesh size) so that for each Γ_j,
$$\sup\left\{\left|\frac{1}{w-z} - \frac{1}{\tilde{w}-z}\right| \;\middle|\; z \in K,\, w, \tilde{w} \in \Gamma_j\right\} \le (4R)^{-1}$$

   (b) Pick a point $w_j \in \Gamma_j$ and polynomial p_j so
$$\left\| p_j(\cdot) - \frac{1}{w_j - \cdot} \right\|_K \le (4R)^{-1}$$
Prove that $\sup_{w \in \Gamma_j,\, z \in K} |1 - (w-z)p_j(z)| \le \frac{1}{2}$.

   (c) Let
$$Q_m(z) = \sum_{j=1}^{\ell} (2\pi i)^{-1} \int_{\Gamma_j} \frac{1 - (1-(w-z)p_j(z))^m}{w-z} f(w)\, dw$$
Prove that for some constant d, Q_m is a polynomial of degree at most dm.

   (d) Prove that $\|f - Q_m\|_K \le C2^{-m}$ and conclude the error is $O(\exp(-\alpha \deg(Q_m)))$ for some $\alpha > 0$.

For further examples of the use of Runge's theorem, see Rubel [**494**].

4.8. The Jordan Curve Theorem for Smooth Jordan Curves

Recall that a closed C^1 curve is a C^1 function $\gamma = [0,1] \to \mathbb{C}$ so $\gamma'(z) \neq 0$ for all $t \in (0,1)$ and $\gamma'(0+) = \gamma'(1-) \neq 0$. This curve is instead called piecewise C^1 if there exist $t_0 = 0 < t_1 < \cdots < t_n \le 1$ so that γ' is C^1 on $\bigcup_{j=0}^n (t_j, t_{j+1})$ and at each t_j, $\gamma'(t_j \pm 0)$ exist but they may not be equal. We again require that all derivatives are nonzero. A *cusp* is a t_j (or 0) with
$$\gamma'(t_j + 0) = -\gamma_j'(t_j - 0) \tag{4.8.1}$$
(see Figure 4.8.1).

Figure 4.8.1. A cusp.

That is, at a cusp, the curve comes in and leaves in opposite directions. At noncusps, there is either smoothness (i.e., $\gamma'(t_j+0)=\gamma'(t_j-0)$) or a finite nonzero angle between the incoming and outgoing curves. In this section, we'll prove:

Theorem 4.8.1 (Jordan Curve Theorem). *Let γ be a closed, simple piecewise C^1 Jordan curve in $\mathbb{C}$ with no cusps. Then $\mathbb{C}\setminus\mathrm{Ran}(\gamma)$ is a disjoint union $S_+\cup S_-$ of open connected sets (i.e., regions) where S_+ is unbounded and S_- bounded. $n(\gamma,z_0)=0$ for all $z_0\in S_+$ and for all $z_0\in S_-$, either $n(\gamma,z_0)=+1$ or $n(\gamma,z_0)=-1$.*

Remark. The result is true for any continuous simple curve, but it is easier in the special case, and the proof we give uses some complex function techniques.

The sketch of the proof is as follows:

(1) Using the implicit function theorem, we'll prove a local version, namely, for all $t_0\in[0,1]$, there is a $\delta>0$ so that $\mathrm{Ran}(\gamma)$ divides $\{z\mid |z-\gamma(t_0)|<\delta\}$ into two connected components, $S_{+,t}$ and $S_{-,t}$.
(2) The winding number for γ jumps by one as we cross from S_{+,t_0} to S_{-,t_0}.
(3) By picking a finite number of points, $t_1,\dots,t_\ell$, in $[0,1]$, we get an open connected neighborhood, N, of $\mathrm{Ran}(\gamma)$ with at most two components for $N\setminus\mathrm{Ran}(\gamma)$.
(4) By a simple argument, $\mathbb{C}\setminus\mathrm{Ran}(\gamma)$ has at most two components.
(5) By a winding numbers argument, there are precisely two components.

Proposition 4.8.2. *For every $t_0\in[0,1]$, there is a δ so $\mathbb{D}_\delta(\gamma(t_0))$ is divided into two connected components, S_{+,t_0} and S_{-,t_0}, by removing $\mathrm{Ran}(\gamma)$ from that set.*

Proof. Suppose first that t_0 is a point where γ is C^1 and that $\gamma'(t_0)$ is real. Write, for γ near t_0,

$$\gamma(t)=x(t)+iy(t) \tag{4.8.2}$$

so by hypothesis,

$$x'(t_0)\neq 0, \qquad y'(t_0)=0 \tag{4.8.3}$$

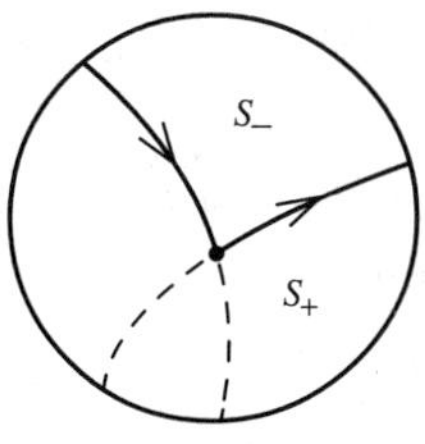

Figure 4.8.2. Corner of a piecewise C^1 curve.

By the real variable implicit function theorem (see Theorem 1.4.2), there is a $\delta > 0$ and function Y defined on $(x(t_0)-\delta, x(t_0)+\delta)$ so that $y(t) = Y(x(t))$ for $t \in \{s \mid |x(s) - x(t_0)| < \delta\}$. Define

$$S_{+,t_0} = \{(x,y) \in \mathbb{D}_\delta(\gamma(t_0)) \mid y > Y(x)\} \tag{4.8.4}$$

$$S_{-,t_0} = \{(x,y) \in \mathbb{D}_\delta(\gamma(t_0)) \mid y < Y(x)\} \tag{4.8.5}$$

Then these two sets are arcwise connected and cover $\mathbb{D}_\delta(\gamma(t_0)) \setminus \operatorname{Ran}(\gamma)$.

If γ is C^1 at t_0, the same construction works in a rotated coordinate system $(\tilde{x}, \tilde{y})$ where $\tilde{x}' \neq 0$, $\tilde{y}' = 0$. If t_0 is a noncusp singularity, each half-line can be extended and $\mathbb{D}_\delta(x_0)$ is broken in four pieces, three of which combine to give S_+ on one S_- (see Figure 4.8.2). $\square$

Proposition 4.8.3. *Under the above hypotheses, $n(\gamma, z_0)$ is constant on S_{+,t_0} and on S_{-,t_0} and the values differ by 1 in absolute value.*

Proof. By the above analysis, γ crosses $\overline{\mathbb{D}_\delta(\gamma(t_0))}$ in a curve from one side of the circle to the other, say from $\gamma(t_1)$ to $\gamma(t_2)$, $t_2 > t_1$. Let C^+, C^- be the closed curves obtained by going counterclockwise from $\gamma(t_1)$ to $\gamma(t_2)$ along $\partial\mathbb{D}_\delta(\gamma(t_0))$ or from $\gamma(t_2)$ to $\gamma(t_1)$ and closed by using $\gamma \restriction [t_1, t_2]$ or going counterclockwise from $\gamma(t_1)$ to $\gamma(t_2)$ and running $\gamma \restriction [t_1, t_2]$ backwards. After cancellation, $C^+ + C^-$ is $\partial\mathbb{D}_\delta(\gamma(t_0))$ counterclockwise. So C^+ (respectively, C^-) is the boundary of S_{+,t_0} (respectively, S_{-,t_0}). For points z_0 outside S_{+,t_0} (respectively, S_{-,t_0}), we have (see Figure 4.8.3)

$$n(C^+, z_0) = 0 \tag{4.8.6}$$

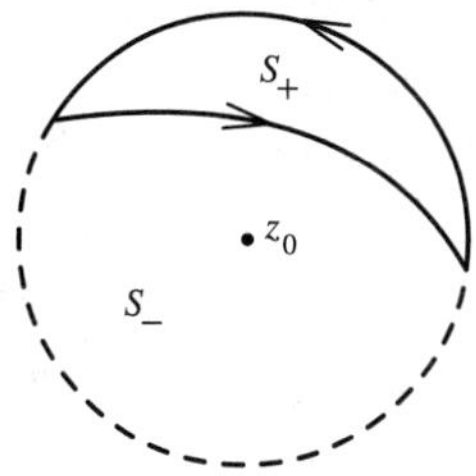

Figure 4.8.3. Closing a contour.

(respectively, $n(C^-, z_0) = 0$), since z_0 can be connected to infinity. On the other hand,

$$n(C^+, z_0) + n(C^-, z_0) = n(\partial\mathbb{D}_\delta(z_0)) = 1$$

by the Cauchy integral formula.

We conclude that the discontinuity of $n(C^+, z_0)$ is 1 as z_0 crosses γ inside $\mathbb{D}_\delta(\gamma(t_0))$. On the other hand, by cancellation, $n(C^+, z_0) + n(\gamma, z_0)$ is continuous across γ in $\mathbb{D}_\delta(\gamma(t_0))$, so the discontinuity of $n(\gamma, z_0)$ is one as z_0 crosses γ inside $\mathbb{D}_\delta(\gamma(t_0))$. □

In Problem 1, we'll extend this discontinuity of winding number argument to more general integrals of the form $\oint_\gamma \frac{f(z)}{z-z_0}\,dz$ where f is nice enough as a function on $\mathrm{Ran}(\gamma)$. These jump calculations of Cauchy integrals are often useful.

Proposition 4.8.4. *There exists $0 = t_0 < \cdots < t_k < 1$ so that $\cup(S_{+,t_j} \cup S_{-,t_j}) \cup \mathrm{Ran}(\gamma)$ is a connected neighborhood, N, of $\mathrm{Ran}(\gamma)$ so that $\mathrm{Ran}(\gamma)$ breaks N into the connected subsets, each of which $n(\gamma, z_0)$ is constant with values differing by 1.*

Proof. By the paving lemma, Proposition 2.2.7, we find $0 = t_0 < t_1 < \cdots < t_k < 1$ so that the corresponding $\mathbb{D}_\delta(\gamma(t_j))$ cover $\mathrm{Ran}(\gamma)$. With this choice, the neighborhoods $S_j^\pm$ fit together into two connected sets as claimed. □

Proof of Theorem 4.8.1. Given any $z_0 \in \mathbb{C}\backslash\mathrm{Ran}(\gamma)$, let $\tilde{\gamma}(s)$ be a straight line from $\tilde{\gamma}(0) = z_0$ to a point $\tilde{\gamma}(1) \in \mathrm{Ran}(\gamma)$. Let $s_0 = \inf\{s \mid \tilde{\gamma}(s) \in \mathrm{Ran}(\gamma)\}$. Then for ε small, $\tilde{\gamma}(s_0 - \varepsilon)$ is in N, and then, z_0 is in the same connected component of $\mathbb{C}\setminus\mathrm{Ran}(\gamma)$ as either $\cup S_j^+$ or $\cup S_j^-$. Thus, $\mathbb{C}\setminus\mathrm{Ran}(\gamma)$ is either one of two components. Since on the two halves of $N(\mathrm{Ran}(\gamma))$, $n(\gamma, \cdot)$ are 0 and ± 1, there must be two components and one must be bounded. The other is obviously unbounded. □

Notes and Historical Remarks. The Jordan curve theorem first appeared in the 1887 first edition of his *Cours d'Analyse* [**289**]. Early twentieth-century geometers complained that it was not rigorous so that they would quote Veblen [**561**] or Brouwer [**77**] for a complete proof, so much so that some call it the Jordan–Brouwer theorem. But many contemporary mathematicians think the proof is essentially correct and complete; see Hales [**234**], the Wikipedia page[1], or the page on Andrew Ranicki's homepage [**470**].

Camille Jordan (1838–1922), a student of Puiseux, was a French mathematician who lived long enough to see three of his six sons killed in the

[1] `http://en.wikipedia.org/wiki/Jordan_curve_theorem`

First World War. Like Cauchy, he was trained as an engineer at École Polytechnique where he also taught for much of his career. A key contribution was the development of the notion of a group (as a group of permutations) and the related notion of composition series for solvable groups. He studied finite subgroups of the group of rotations and crystal groups (discrete subgroups of the affine group), work that strongly influenced Lie and Klein who studied with him. He also discovered the Jordan normal form for matrices (Weierstrass had earlier not unrelated ideas) and was the first to define the fundamental group. He realized the need to actually prove the intuitively obvious result we now call the Jordan curve theorem and attempted it in his multivolume *Cours d'Analyse*.

For the record, here is a precise statement of the Jordan curve theorem. It relies on the notion of winding number for general closed (not necessarily rectifiable) curves described in the Notes to Section 3.3.

Theorem 4.8.5 (Jordan Curve Theorem for General Curves). *Let γ be a closed simple continuous curve. Then $\mathbb{C} \setminus \text{Ran}(\gamma)$ has two components, C_+ and C_-. On the unbounded component, C_+, $n(\gamma, \cdot\,) \equiv 0$, and on the bounded component, C_-, either $n(\gamma, \cdot\,) \equiv 1$ or $n(\gamma, \cdot\,) \equiv -1$. Moreover, every neighborhood of $z_0 \in \text{Ran}(\gamma)$ intersects both C_+ and C_-.*

A simple closed curve is called a Jordan curve. In the case $n(\gamma, \cdot\,) \equiv 1$ on C_-, we say γ is positively oriented. For proofs of the general Jordan curve theorem, see Burckel [**80**], Hatcher [**244**, p. 169], or Leoni [**354**]

While our desire to avoid mainly topological arguments has restricted us to the smooth case, there are costs in limitations. Consider, for example, the following theorem of Darboux:

Theorem 4.8.6. *Let γ be a Jordan curve. Let $\Omega = C_-$, the bounded component, and let f be a continuous function on $\Omega \cup \text{Ran}(\gamma)$ which is analytic on Ω. If $f \restriction \text{Ran}(\gamma)$ is one-one, then f is one-one on all of $\Omega \cup \text{Ran}(\gamma)$.*

Given the full Jordan curve theorem and full argument principle (4.3.7) (extended to regular functions), the proof is extremely simple: If f were constant, it would be constant on $\text{Ran}(\gamma)$, so we know f is an open map. Since f is one-one, $f \circ \gamma$ is a simple closed curve. Let $\widetilde{C}_\pm$ be the components of $\mathbb{C} \setminus \text{Ran}(f \circ \gamma)$. If $z_0 \in \Omega$, $f(z_0) \notin \widetilde{C}_+$ by the argument principle applied to $f(z) - f(z_0)$. Thus, by the last sentence in Theorem 4.8.5 and the open mapping theorem, $f(z_0) \notin \text{Ran}(f \circ \gamma)$ either. Thus, $f(z_0) \in \widetilde{C}_-$ and the argument principle applies to $f(z) - f(z_0)$ to show there is no other point, z_1, in Ω with $f(z_1) = f(z_0)$, so f is one-one on Ω. As we've seen, values there are disjoint from $f[\text{Ran}(\gamma)]$, so f is one-one on all of $\Omega \cup \text{Ran}(\gamma)$.

Clearly, for f merely continuous on $\text{Ran}(\gamma)$, this argument requires the full Jordan curve theorem. But even if f is analytic in a neighborhood of $\text{Ran}(\gamma)$ and γ obeys the hypotheses of Theorem 4.8.1, $f \circ \gamma$ may not, if $f'(z_0) = 0$ for some $z_0 \in \text{Ran}(\gamma)$.

We note that the converse of Theorem 4.8.6 is false. If γ is the curve which goes from -1 to 1 along $\mathbb{R}$ and returns along $\partial\mathbb{D}$ in the upper half-plane and $f(z) = z^2$, then f is one-one on $\Omega = C_-$ but not on $\text{Ran}(\gamma)$.

Problems

1. Let γ be a simple C^1 arc and f a C^1 function on $\text{Ran}(\gamma)$ supported on $\gamma([\varepsilon, 1-\varepsilon])$ for some ε. For $w \notin \text{Ran}(\gamma)$, define the contour integral (*Cauchy transform* of f),

$$C_f(w) = \frac{1}{2\pi i} \int \frac{f(z)}{w-z}\, dz$$

Fix $\varepsilon > 0$. For $z_0 \in \text{Ran}(\gamma)$, define $\Gamma^{\pm}(z_0) = \{w \mid w = z_0 \pm re^{i(\theta+\theta_0)} \mid 0 < r < \rho,\ |\theta| < \frac{\pi}{2} - \varepsilon\}$, where ρ is picked so small that $\Gamma^{\pm} \cap \text{Ran}(\gamma) = \emptyset$, and $\theta_0(z)$ is defined so that $e^{-i\theta_0(z)}\gamma'(z_0)$ is a positive imaginary (normal and to the left; see Figure 4.8.4).

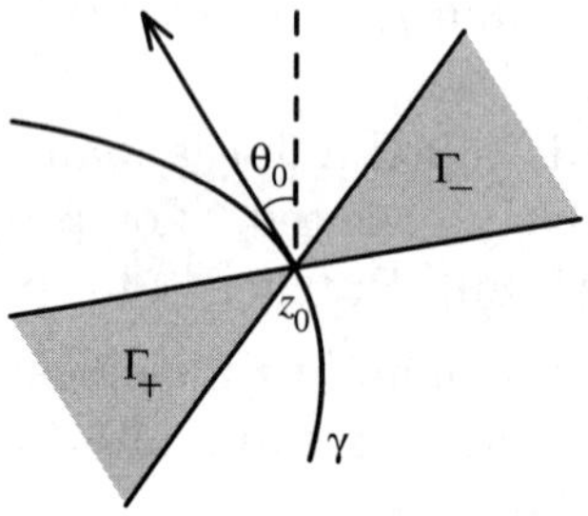

Figure 4.8.4. Nontangential limits.

(a) If $f(z_0) = 0$, prove that $C_f(w)$ can be extended to $\Gamma^+(z_0) \cup \Gamma^-(z_0) \cup \{z_0\}$ to be continuous at z_0 (i.e., $C_f(w)$ has equal nontangential limits at z_0).

(b) Using Proposition 4.8.3, prove the *Cauchy jump formula* that for any such f and all z_0, we have

$$\lim_{\substack{w \to z_0 \\ w \in \Gamma^+(z_0)}} C_f(z) - \lim_{\substack{w \to z_0 \\ w \in \Gamma^-(z_0)}} C_f(z) = f(z_0) \tag{4.8.7}$$

Chapter 5

More Consequences of the CIT

> Riesz was troubled by the fact that his proof did not extend convexity to the entire unit square and at the end of his paper he even conjectured that this was not possible. But ten years later the extension was done by Riesz's student Olof Thorin.
>
> —*Lars Gårding* [**202**]

Big Notions and Theorems: Phragmén–Lindelöf Method, Three-Line Theorem, Riesz–Thorin Theorem, Complex Interpolation, Complex Poisson Representation, Real Poisson Representation, Harmonic Functions Defined by Mean Value Property, Equivalence of Three Definitions of Harmonic Function, Maximum Principle for Harmonic Functions, Dirichlet Problems for the Disk, Painlevé's Theorem, Reflection Principle, Strong Reflection Principle, Analytic Arcs, Continuity in Corners Theorem, Contour Integral Calculational Methods, Gaussian Integrals, Fresnel Integrals

Having taken a break from studying consequences of the CIT and CIF to prove the ultimate CIT, we return to a diverse set of applications of the CIT. The first two sections extend our study of the maximum principle from Section 3.6. Section 5.2 includes an application to L^p space mappings (the Riesz–Thorin theorem), filling in something used in Part 1. This latter half of Section 5.2 can be skipped by readers focusing on complex analysis per se. The next two sections discuss some simple aspect of harmonic function theory in the plane—a subject, extended to $\mathbb{R}^n$, that we return to in Chapter 3 of Part 3. This is a preliminary for a strong version of the reflection principle, the subject of Sections 5.5 and 5.6. A final section is an exposé of using complex analysis to compute definite integrals, a subject usually treated in greater depth in the more elementary texts.

5.1. The Phragmén–Lindelöf Method

Recall that f is called regular on $\overline{\Omega}$ with Ω a region in $\mathbb{C}$ if f is continuous on $\overline{\Omega}$ and analytic in Ω. In one form, the maximum principle (see Theorem 3.6.3) says that if f is regular and $\overline{\Omega}$ is compact, then

$$\sup_{z\in\Omega}|f(z)| = \sup_{z\in\partial\Omega}|f(z)| \tag{5.1.1}$$

There are two further facts—namely, that the sup is taken at a point $z_0 \in \partial\Omega$, and if the sup is taken at a point $z_0 \in \Omega$, then f is constant. We will not always be explicit about similar situations below, although they sometimes, even often, hold. In general, (5.1.1) fails for unbounded Ω. Here is the canonical example:

Example 5.1.1. Let $\Omega = \{z \mid \operatorname{Re} z > 0\}$. Let

$$f(z) = e^z \tag{5.1.2}$$

Then f is regular on $\overline{\Omega}$,

$$\sup_{\Omega}|f(z)| = \infty \tag{5.1.3}$$

but on $\partial\Omega = \{iy \mid y \in \mathbb{R}\}$, $|f(z)| \equiv 1$. □

Our goal here is to prove in fairly great generality that for suitable Ω, rather weak a priori growth conditions on f, together with boundedness on $\partial\Omega$, imply boundedness of f and (5.1.1). For example, we'll prove below the following that shows (5.1.2) is essentially the slowest growth consistent with (5.1.3) even though f is bounded on $\partial\Omega$.

Theorem 5.1.2. *Suppose that f is continuous in $\overline{\Omega} = \{z \mid \operatorname{Re} z \geq 0\}$ and analytic in $\Omega = \{z \mid \operatorname{Re} z > 0\}$. Suppose also that*

(i) $$\sup_{y\in\mathbb{R}}|f(iy)| < \infty \tag{5.1.4}$$

(ii) *For some C_1, C_2 and some $\alpha \in (0,1)$,*

$$|f(z)| \leq C_1 \exp(C_2|z|^\alpha) \tag{5.1.5}$$

for all $z \in \Omega$. Then f is bounded and (5.1.1) *holds.*

By applying this to $f(z) = g(e^{i\theta}z^\gamma)$ for suitable real θ and γ, one gets

Corollary 5.1.3. *Suppose $\Omega = \{z|\alpha < \arg(z) < \beta\}$ and that g is analytic on Ω and continuous on $\overline{\Omega}$ with*

$$|g(z)| \leq C_1 \exp(C_2|z|^\nu) \tag{5.1.6}$$

where

$$(\beta - \alpha)\nu < \pi \tag{5.1.7}$$

and

$$\sup_{z \in \partial\Omega} |g(z)| < \infty \tag{5.1.8}$$

Then g is bounded and (5.1.1) *holds for g.*

Lest the reader think that the idea of this section has to do with unbounded regions, we note the following theorem that illustrates the essence of the Phragmén–Lindelöf method:

Theorem 5.1.4. *Let f be analytic on $\mathbb{D}$ with*

$$\sup_{z \in \mathbb{D}} |f(z)| < \infty \tag{5.1.9}$$

Fix $z_0 \in \partial\mathbb{D}$ and suppose that for all $z_1 \in \partial\mathbb{D}$, $z_1 \neq z_0$ we have

$$\limsup_{z \to z_1} |f(z)| \leq 1 \tag{5.1.10}$$

Then

$$\sup_{z \in \mathbb{D}} |f(z)| \leq 1 \tag{5.1.11}$$

Proof. Without loss, suppose $z_0 = 1$ and define on $\mathbb{D}$

$$g_\varepsilon(z) = f(z) + \varepsilon \log\bigl(\tfrac{1}{2}(1 - z)\bigr) \tag{5.1.12}$$

where the branch of log with $\log\bigl(\frac{1}{2}(1-z)\bigr)$ real on $(-1, 1)$ is picked. g is analytic on $\mathbb{D}$ and since

$$\operatorname{Re} \log\bigl(\tfrac{1}{2}(1 - z)\bigr) = \log\bigl(\tfrac{1}{2}\,|1 - z|\bigr) \leq 0 \tag{5.1.13}$$

on $\mathbb{D}$, (5.1.9) implies that

$$\sup_{z \in \mathbb{D}} \operatorname{Re} g_\varepsilon(z) \equiv M < 1 \tag{5.1.14}$$

By Theorem 3.6.5, there is $z_1 \in \mathbb{D}$ and $z_n \to z_1$ such that $\operatorname{Re} g_\varepsilon(z_n) \to M$. By (5.1.9) and $\log\bigl(\frac{1}{2}(1-z)\bigr) \to -\infty$ as $z \to 1$, we see that $z_1 \neq 1$. Thus by (5.1.10) and (5.1.13), $M < 1$, i.e., we have proven that

$$\operatorname{Re}\bigl[f(z) + \varepsilon \log\bigl(\tfrac{1}{2}(1 - z)\bigr)\bigr] \leq 1 \tag{5.1.15}$$

for all $z \in \mathbb{D}$ and $\varepsilon > 0$. Taking $\varepsilon \downarrow 0$, we see that for all $z \in \mathbb{D}$

$$\operatorname{Re}\bigl[f(z)\bigr] \leq 1 \tag{5.1.16}$$

Applying this argument to $e^{i\theta} f$ and using $|f(z)| = \sup_{e^{i\theta} \in \mathbb{D}} \operatorname{Re}\bigl[e^{i\theta} f(z)\bigr]$, we get (5.1.11). $\square$

Remarks. 1. The essence of this method is that for some boundary points (which may be ∞!) not to matter, we need a function h singular (in this case $\log(\frac{1}{2}|1-z|)$) at the boundary points and an apriori bound on $\operatorname{Re} f$ by εg.

2. The result is really something about harmonic (even subharmonic) functions, not analytic functions, so there are versions on $\mathbb{R}^\nu$ and this is a result in "real analysis," not complex analysis.

3. As a result on harmonic functions using that $\log(d^{-1}|z-z_0|)$ is harmonic, one can replace $\mathbb{D}$ by any bounded domain Ω.

4. Clearly $\{z_0\}$ can be replaced by an finite set. More can be said; the set on which one has no apriori information can be a set of logarithmic capacity zero; see Theorem 3.6.12 of Part 3.

5. As we'll see in Section 3.6 of Part 3, this result says something about uniqueness of solutions of the Dirichlet problem.

We begin the discussion of unbounded regions with the elementary:

Proposition 5.1.5. *Suppose f is regular on $\overline{\Omega}$ and*

$$\lim_{\substack{|z|\to\infty \\ z\in\overline{\Omega}}} |f(z)| = 0 \tag{5.1.17}$$

Then f is bounded on $\overline{\Omega}$ and (5.1.1) *holds.*

Proof. By the compactness of $\overline{\Omega}\cup\{\infty\}$ and (5.1.17) which says f is continuous on $\overline{\Omega}\cup\{\infty\}$ if we set $f(\infty)=0$, f is bounded.

For $R<\infty$, let $\Omega_R = \{z \mid z\in R,\ |z|<R\}$, so

$$\partial\Omega_R = [\partial\Omega \cap \{z \mid |z|<R\}] \cup \{z\in\overline{\Omega} \mid |z|=R\} \tag{5.1.18}$$

We define the two sets whose union is given in (5.1.18) as $\partial^{(1)}\Omega_R$ and $\partial^{(2)}\Omega_R$. Clearly, by (5.1.17),

$$\lim_{R\to\infty} \Big[\sup_{z\in\partial^{(2)}\Omega_R} |f(z)|\Big] = 0 \tag{5.1.19}$$

By Theorem 3.6.3,

$$\begin{aligned}
\sup_{z\in\Omega_R} |f(z)| &\le \sup_{z\in\partial\Omega_R} |f(z)| \\
&\le \max\Big(\sup_{\partial^{(1)}\Omega_R} |f(z)|,\ \sup_{\partial^{(2)}\Omega_R} |f(z_0)|\Big) \\
&\le \max\Big(\sup_{\partial\Omega} |f(z)|,\ \sup_{\partial^{(2)}\Omega_R} |f(z)|\Big)
\end{aligned}$$

Thus, by (5.1.19),

$$\sup_{z\in\Omega}|f(z)| = \lim_{R\to\infty}\sup_{z\in\Omega_R}|f(z)| \le \sup_{\partial\Omega}|f(z)| \tag{5.1.20}$$

□

Theorem 5.1.6 (Phragmén–Lindelöf Method). *Let Ω be a region and f a function regular on $\overline{\Omega}$. Suppose there exists a function g regular on $\overline{\Omega}$ so that*

(a) $\operatorname{Re} g \ge 0$ *on* $\overline{\Omega}$ (5.1.21)

(b) $\lim\limits_{\substack{|z|\to\infty \\ z\in\overline{\Omega}}} \operatorname{Re} g(z) = \infty$ (5.1.22)

If f is bounded on $\partial\Omega$ and, for all $\varepsilon > 0$, there exists C_ε so that for all $z \in \overline{\Omega}$,

$$|f(z)| \le C_\varepsilon e^{\varepsilon \operatorname{Re} g(z)} \tag{5.1.23}$$

Then f is bounded on $\overline{\Omega}$ and (5.1.1) holds.

Proof. Let

$$h_\varepsilon(z) = f(z)e^{-\varepsilon g(z)} \tag{5.1.24}$$

which is regular on $\overline{\Omega}$. Moreover, by (5.1.23) with ε replaced by $\varepsilon/2$,

$$|h_\varepsilon(z)| \le C_{\varepsilon/2} e^{-\varepsilon/2 \operatorname{Re} g(z)} \tag{5.1.25}$$

By (5.1.22), h_ε obeys (5.1.17), so h_ε obeys (5.1.1), that is, for all $z \in \overline{\Omega}$,

$$\begin{aligned} |h_\varepsilon(z)| &\le \sup_{w\in\partial\Omega} |h_\varepsilon(w)| \\ &\le \sup_{w\in\partial\Omega} |f(w)| \end{aligned} \tag{5.1.26}$$

by (5.1.21).

Fix $z \in \Omega$ and take $\varepsilon \downarrow 0$ in (5.1.26) to get (5.1.1) for f, and so boundedness of f. □

In applications, one often proves (5.1.23) only for $\varepsilon = 1$ but has two comparison functions to juggle:

Corollary 5.1.7. *Let Ω, f, g be as in the first sentence of Theorem 5.1.6 (i.e., g obeys (5.1.21) and (5.1.22)). Suppose there is a positive function, h, on $\overline{\Omega}$ so that for all ε, there is D_ε so that*

$$|h(z)| \le \varepsilon \operatorname{Re} g(z) + D_\varepsilon \tag{5.1.27}$$

and that f obeys

$$|f(z)| \le Ce^{|h(z)|} \tag{5.1.28}$$

for all $z \in \overline{\Omega}$. Then if f is bounded on $\partial\Omega$, it is bounded on $\overline{\Omega}$ and (5.1.1) holds.

Proof. (5.1.27) and (5.1.28) imply that

$$|f(z)| \le Ce^{D_\varepsilon} e^{\varepsilon \operatorname{Re} g(z)} \tag{5.1.29}$$

and so (5.1.23). □

Corollary 5.1.8. *Let Ω be a region, f regular on $\overline{\Omega}$ and bounded on Ω. Suppose a function, g, exists obeying* (5.1.21) *and* (5.1.22). *Then* (5.1.1) *holds.*

Proof. Let $h(z) \equiv 1$. Then, since $\operatorname{Re} g \ge 0$, (5.1.27) holds with $D_\varepsilon \equiv 1$. Since f is bounded, (5.1.27) holds. Now use the last corollary. □

Proof of Theorem 5.1.2. Pick β so $\alpha < \beta < 1$. Let $h(z) = |z|^\alpha$ and $g(z) = z^\beta$. Then

$$\operatorname{Re} g(z) = |z|^\beta \cos(\beta[\arg(z)]) \tag{5.1.30}$$

Since $|\arg(z)| \le \pi/2$ in the right half-plane,

$$\cos(\beta[\arg(z)]) \ge \cos\left(\frac{\beta\pi}{2}\right) > 0 \tag{5.1.31}$$

so (5.1.21) holds, and clearly, (5.1.22) holds since $|z|^\beta \to \infty$ as $|z| \to \infty$.

Since $\alpha < \beta$, for any ε, (5.1.27) holds. Thus, this theorem follows from Corollary 5.1.7. □

A similar argument proves:

Theorem 5.1.9. *Let f be regular on $\overline{\Omega}$ where $\Omega = \{z \mid 0 < \operatorname{Re} z < 1\}$. Suppose f is bounded on $\partial\Omega$ and obeys*

$$|f(z)| \le \exp(A \exp(\alpha|z|)) \tag{5.1.32}$$

for some $\alpha < \pi$ and some A. Then f is bounded and (5.1.1) *holds.*

Remarks. 1. $f(z) = \exp(i \exp(i\pi z))$ is unbounded on Ω but bounded on $\partial\Omega$, showing $\alpha < \pi$ is optimal.

2. To get the applicability of Corollary 5.1.8, one can use the simpler calculations with

$$g(z) = -z^2 + 1$$

rather than the g used in Problem 1.

Proof. Left to Problem 1. □

The Phragmén–Lindelöf idea of bounding f by studying fh for suitable h has other applications; see the next section and Problems 2 and 3.

Notes and Historical Remarks. The Phragmén–Lindelöf principle is named after Edvard Phragmén (1863–1937), a Swedish student of Mittag-Leffler, and Ernst Lindelöf (1870–1946), the founder of the Finnish school of complex analysis (that later included Nevanlinna and Ahlfors). A first version appeared in a 1904 paper of Phragmén alone [**429**]; its final form, including the method discussed here, is in a 1908 joint paper [**430**]. A detailed history can be found in Chapter 3 of Gårding [**202**].

Problems

1. Prove Theorem 5.1.9. (*Hint*: Translate so $\Omega = \{z \mid |\operatorname{Re} z| < \frac{1}{2}\}$ and let $g(z) = e^{i\beta z} + e^{-i\beta z}$ where $\alpha < \beta < \pi$.)

2. The purpose of this problem is to lead the reader through a proof of the following theorem: Let $S_{\alpha,\beta}$ be the sector, (2.6.4), with $\beta - \alpha < 2\pi$. Suppose f is regular and bounded in $\bar{S}_{\alpha,\beta}$ and that for some $a \in \mathbb{C}$,

$$\lim_{r\to\infty} f(re^{i\beta}) = \lim_{r\to\infty} f(re^{i\alpha}) = a \tag{5.1.33}$$

Then, uniformly in $\bar{S}_{\alpha,\beta}$,

$$\lim_{|z|\to\infty} f(z) = a \tag{5.1.34}$$

The key will be the use of the maximum principle on

$$F_n(z) = \frac{z}{z+n} f(z) \tag{5.1.35}$$

a Phragmén–Lindelöf-type of idea.

(a) Prove that it suffices to consider the case $a = 0$, $\beta = -\alpha = \pi/4$, which we henceforth assume.

(b) Prove that for all n,

$$|F_n(z)| \le |f(z)|, \qquad |F_n(z)| \le \frac{|z|}{|n|}\,|f(z)| \tag{5.1.36}$$

$$|f(z)| = \left(1 + \frac{n}{|z|}\right)|F_n(z)| \tag{5.1.37}$$

(c) Find (recall that $a = 0$), $R_1(\varepsilon)$ so that for all n,

$$\sup_{r\ge R_1} (|F_n(re^{i\pi/4})|, |F_n(re^{-i\pi/4})|) \le \frac{\varepsilon}{2} \tag{5.1.38}$$

and then $n(\varepsilon)$ so large that

$$\sup_{\substack{|z|=R_1(\varepsilon)\\ z\in\bar{S}_{-\frac{\pi}{4},\frac{\pi}{4}}}} |F_{n(\varepsilon)}(z)| \le \frac{\varepsilon}{2} \tag{5.1.39}$$

(d) Prove that if $|z| \ge \max(n(\varepsilon), R_1(\varepsilon))$ and in $\bar{S}_{-\frac{\pi}{4},\frac{\pi}{4}}$, then $|f(z)| < \varepsilon$, and so conclude that $f \to 0$ uniformly in the sector.

3. In the theorem in Problem 2, replace (5.1.33) by

$$\lim_{r\to\infty} f(re^{i\beta}) = a, \qquad \lim_{r\to\infty} f(re^{i\alpha}) = b \tag{5.1.40}$$

Prove that $a = b$ and (5.1.34) holds. (*Hint*: Consider $g(z) = (f(z) - a)(f(z) - b)$.)

4. Let $\Omega = \{z \mid a < \operatorname{Re} z < b\}$. Let f be analytic on Ω, continuous on $\overline{\Omega}$, and bounded on $\overline{\Omega}$. Suppose $\lim_{y\to+\infty} f(a+iy) = \alpha$ and $\lim_{y\to\infty} f(b+iy) = \beta$. Prove that $\alpha = \beta$ and, uniformly in $x \in [a, b,]$, $\lim_{y\to\infty} f(x+iy) = \alpha$. (*Hint*: See Problems 2 and 3, noting the continuity at 0 played no role in those problems.)

5.2. The Three-Line Theorem and the Riesz–Thorin Theorem

Here we'll prove the analog of the Hadamard three-circle theorem (Theorem 3.6.8) for a strip and apply it to interpolation theory for L^p spaces. The main result is:

Theorem 5.2.1 (Hadamard Three-Line Theorem). *Let f be regular and bounded on the closure of the strip $\{z \mid \operatorname{Re} z \in (0,1)\}$. Let*

$$M_t(f) \equiv \sup_{y\in\mathbb{R}} |f(t+iy)| \tag{5.2.1}$$

for $0 \le t \le 1$. Then

$$M_t \le M_0^{1-t} M_1^t \tag{5.2.2}$$

Remarks. 1. By Theorem 5.1.9, we can replace boundedness on the entire strip by boundedness on the boundary and (5.1.31) for some $\alpha < \pi$.

2. As in the circle case, this implies $t \to \log M_t$ is convex.

3. One can deduce the three-circle theorem from the result (see Problem 1).

Proof. Define for any $A \in \mathbb{R}$,

$$g_A(z) = f(z) e^{Az} \tag{5.2.3}$$

Then g_A is regular and bounded, so by the maximum principle for the strip,

$$M_t(g_A) \le \max(M_0(g_A), M_1(g_A)) \tag{5.2.4}$$

Since $|e^{A(t+iy)}| = e^{At}$, we have

$$M_t(g_A) = e^{At} M_t(f) \tag{5.2.5}$$

so (5.2.4) implies

$$M_t(f) \le \max(e^{-At} M_0(f), e^{(1-t)A} M_1(f)) \tag{5.2.6}$$

Now choose A so that

$$e^A = \frac{M_0}{M_1} \tag{5.2.7}$$

(if $M_1 = 0$ or $M_0 = 0$, use (5.2.7) with M_0 or M_1 replaced by $\varepsilon > 0$ and take $\varepsilon \downarrow 0$) and get (5.2.2). $\square$

The remainder of this section fills in a "gap" in Part 1 and may be skipped by the reader uninterested in L^p spaces.

Theorem 5.2.2 (The Riesz–Thorin Theorem). *Let (Ω, μ) be a measure space. Let $p_0, p_1, q_0, q_1 \in [1, \infty]$ and define, for $t \in (0,1)$,*

$$p_t^{-1} = (1-t)p_0^{-1} + tp_1^{-1}, \qquad q_t^{-1} = (1-t)q_0^{-1} + tq_1^{-1} \tag{5.2.8}$$

Let T be a linear transformation from $L^{p_0} \cap L^{p_1}$ to $L^{q_0} \cap L^{q_1}$ with

$$\|Tf\|_{q_j} \le C_j \|f\|_{p_j} \tag{5.2.9}$$

for $j = 0, 1$ and all $f \in L^{p_0} \cap L^{p_1}$. Then for all $t \in (0,1)$ and $f \in L^{p_0} \cap L^{p_1}$,

$$\|Tf\|_{q_t} \le C_t \|f\|_{p_t} \tag{5.2.10}$$

where

$$C_t = C_0^{1-t} C_1^t \tag{5.2.11}$$

Remark. Since $L^{p_0} \cap L^{p_1}$ is dense in each L^{p_t}, T extends to a bounded map of L^{p_t} to L^{q_t}.

Proof. Fix $t \in (0,1)$. Let r_t be the dual index to q_t (see Section 4.9 of Part 1 for a discussion of L^p duality), that is, $r_t^{-1} = 1 - q_t^{-1}$ so that

$$r_t^{-1} = (1-t)r_0^{-1} + tr_1^{-1} \tag{5.2.12}$$

It suffices, by a density argument, to prove (5.2.10) when f is a simple function (if $q_0 = q_1 = \infty$ and $\mu(X) = \infty$, this is not true, but the case $q_0 = q_1$ is easy to see directly; see Problem 3), that is,

$$f = \sum_{j=1}^N \alpha_j \chi_{E_j} \tag{5.2.13}$$

where $\mu(E_j) < \infty$ and the E_j are disjoint.

By duality, to get (5.2.10), we need only prove

$$\left| \int g(x)(Tf)(x)\, d\mu(x) \right| \le C_t \|f\|_{p_t} \|g\|_{r_t} \tag{5.2.14}$$

and again, by density, it suffices for g a simple function, that is,

$$g = \sum_{k=1}^M \beta_k \chi_{F_k} \tag{5.2.15}$$

with $\mu(F_k) < \infty$ and the F_k disjoint. By replacing f by $f/\|f\|_{p_t}$ and g by $g/\|g\|_{r_t}$, we can suppose

$$\sum_{j=1}^{N} |\alpha_j|^{p_t} \mu(E_j) = 1 = \sum_{k=1}^{M} |\beta_k|^{q_k} \mu(F_k) \tag{5.2.16}$$

Write

$$\alpha_j = |\alpha_j| e^{i\theta_j}, \qquad \beta_k = |\beta_k| e^{i\psi_k} \tag{5.2.17}$$

and define for $z \in \{z \mid 0 \le \operatorname{Re} z \le 1\}$,

$$F(z) = \sum_{\substack{j=1,\dots,N \\ k=1,\dots,M}} e^{i\theta_j} e^{i\psi_k} |\alpha_j|^{\ell(z)} |\beta_j|^{\lambda(z)} \int_{E_j} (T\chi_{F_k})(x)\, d\mu(x) \tag{5.2.18}$$

where

$$\ell(z) = p_t[(1-z)p_0^{-1} + zp_1^{-1}], \qquad \lambda(z) = r_t[(1-z)r_0^{-1} + zr_1^{-1}] \tag{5.2.19}$$

Clearly, $F(z)$, as a finite sum of exponentials, is analytic and bounded on the strip.

We claim that:

(1) $F(t) = \displaystyle\int g(x)(Tf)(x)\, d\mu(x)$ (5.2.20)

(2) $|F(iy)| \le C_0$ (5.2.21)

(3) $|F(1+iy)| \le C_1$ (5.2.22)

Assuming this, we get (5.2.14) since the three-line theorem says, given (2) and (3), that $|F(t)| \le C_0^{1-t} C_1^t$.

(5.2.20) is immediate from $\ell(t) = \lambda(t) = 1$. To get (5.2.21) and (5.2.22), define

$$f(x,z) = \sum_{j=1}^{N} e^{i\theta_j} |\alpha_j|^{\ell(z)} \chi_{E_j}$$

$$g(x,z) = \sum_{k=1}^{M} e^{i\psi_k} |\beta_k|^{\lambda(z)} \chi_{F_k}$$

and note that

$$F(z) = \int g(x,z)\big(Tf(\cdot,z)\big)(x)\, d\mu(x) \tag{5.2.23}$$

and that, by (5.2.16),

$$\|f(\,\cdot\,,t+iy)\|_{p_t} = 1 = \|g(\,\cdot\,,t+iy)\|_{r_t}$$

so (5.2.21) and (5.2.22) follow from Hölder's inequality and (5.2.9). □

Notes and Historical Remarks. The three-line theorem is a variant of the three-circle theorem, so it is often named after Hadamard. Since it first appeared explicitly in Doetsch [**147**], it is sometimes called the Doetsch three-line theorem. Between Hadamard and Doetsch, Lindelöf [**359**] had a closely related result.

M. Riesz in 1927 in his studies of conjugate harmonic functions on the disk [**487, 488**] initially proved that H, the map from a function to its harmonic conjugate (aka Hilbert transform), was bounded from L^p to L^p if $p = 2, 4, 6 \dots$. He then invented interpolation and proved (5.2.2) in case $q_j \geq p_j$. The general result was proven by Olof Thorin, a student of Riesz, in 1938 [**554**], with a detailed exposition in his final thesis after the war in 1948 [**555**]. By then, Thorin had left mathematics to go into the insurance business. We discuss Riesz's theorem on conjugate functions (proven by other means) in Section 5.7 of Part 3.

Abstract versions for suitable pairs of norms on vector spaces were developed by Calderón [**84**] and Lions [**360, 361, 362**]. E. Stein [**536**] proved a variant that allows the map T to depend on a complex parameter, see Problem 4. A typical application of his approach shows that if e^{-zH} is bounded from L^1 to L^1 for $z \in [0, \infty)$ and from $L^2 \to L^2$ for $\operatorname{Re} z \geq 0$, then for $p \in (1, 2)$, one has boundedness for z in a sector.

Problems

1. By mapping $\{z \mid 0 \leq \operatorname{Re} z \leq 1\}$ to an annulus via $z \to e^{cz}$, deduce the three-circle theorem from the three-line theorem.

2. Prove that the choice (5.2.7) optimizes A in (5.2.6).

3. Prove Theorem 5.2.2 directly (i.e., without complex analysis) in the special case $q_0 = q_1$.

4. Prove the Stein iterpolation theorem. If $p_0, q_0, p_1, q_1, p_z, q_z$ are as in Theorem 5.2.2 and $T(z)$ is a function from $\{z \mid 0 \leq \operatorname{Re} z < 1\}$ to the linear transformations from $L^{p_0} \cap L^{p_1}$ to $L^{q_0} \cap L^{q_1}$, so that for all $y \in \mathbb{R}$,
$$\|T(iy)f\|_{q_0} \leq C_0 \|f\|_{p_0}; \quad \|T(1+iy)f\|_{q_1} \leq C_1 \|f\|_{p_1} \tag{5.2.24}$$
Then, for all $t \in (0, 1)$, $T(t)$ map L^{p_z} to L^{q_z} and
$$\|T(t)f\|_{q_z} \leq C_0^{1-t} C_1^t \|f\|_{p_z} \tag{5.2.25}$$
(Hint: Go through the proof above but in (5.2.19) replace T by $T(z)$.)

5.3. Poisson Representations

The identity principle for analytic functions (see Theorem 2.3.8) implies the values of such a function, f, on a sequence of points $\{z_n\}_{n=1}^{\infty}$ with a limit point in the region, Ω, of analyticity determine f. But what about $\operatorname{Re} f$: do its values on a sequence determine $\operatorname{Re} f$ or even f? The answer is no. A function like $f(z) = iz$ is analytic on $\mathbb{C}$ has $\operatorname{Re} f(z) = 0$ on $\mathbb{R}$ but $\operatorname{Re} f$ is not identically zero.

Remarkably though, if f is analytic in a neighborhood of $\overline{\mathbb{D}}$, then the values of $\operatorname{Re} f$ on $\partial\mathbb{D}$ determine f up to a single imaginary constant, and even via an explicit formula. That's our main goal in this section. It will be critical in the next section and also in Chapter 3 of Part 3. In particular, that chapter will refine our results in this section, and, more importantly, will discuss extensions to functions on $\mathbb{R}^\nu$.

Lemma 5.3.1. *For all $z \in \mathbb{D}$, $e^{i\theta} \in \partial\mathbb{D}$, we have that*

$$1 + 2\sum_{n=1}^{\infty} e^{-in\theta} z^n = \frac{e^{i\theta} + z}{e^{i\theta} - z} \tag{5.3.1}$$

where the convergence is uniform in $(z, e^{i\theta}) \in \overline{\mathbb{D}}_\rho \times \partial\mathbb{D}$ for each $\rho < 1$.

Proof. By the geometric series with remainder, we deduce the claimed uniform convergence of the sum on the left side of (5.3.1) to

$$\frac{2}{1 - ze^{-i\theta}} - 1 = \frac{1 + ze^{-i\theta}}{1 - ze^{-i\theta}} = \text{RHS of (5.3.1)}$$ □

Theorem 5.3.2 (Complex Poisson Representation). *Let f be a regular function in $\overline{\mathbb{D}}$. Then for all $z \in \mathbb{D}$,*

$$f(z) = i \operatorname{Im} f(0) + \int_0^{2\pi} \frac{e^{i\theta} + z}{e^{i\theta} - z} \operatorname{Re} f(e^{i\theta}) \frac{d\theta}{2\pi} \tag{5.3.2}$$

Remarks. 1. The function

$$C(z, e^{i\theta}) = \frac{e^{i\theta} + z}{e^{i\theta} - z} \tag{5.3.3}$$

is called the *complex Poisson kernel* or *Schwarz kernel.*

2. For (partially) alternative proofs, see Problems 1, 2, and 3.

3. (5.3.2) is sometimes called the *Schwarz integral formula.*

Proof. Let

$$f(z) = \sum_{n=0}^{\infty} a_n z^n \tag{5.3.4}$$

be the Taylor series for f at $z = 0$. Since f is analytic in $\mathbb{D}$, Cauchy's formula (3.1.2) for f says that for any $\rho < 1$,

$$a_n = \int_0^{2\pi} f(\rho e^{i\theta}) \rho^{-n} e^{-in\theta} \frac{d\theta}{2\pi} \tag{5.3.5}$$

(since, for $\gamma(\theta) = \rho e^{i\theta}$, $dz/iz = d\theta$). Since f is uniformly continuous on $\overline{\mathbb{D}}$, we can take $\rho \uparrow 1$ and conclude

$$a_n = \int_0^{2\pi} f(e^{i\theta}) e^{-in\theta} \frac{d\theta}{2\pi} \tag{5.3.6}$$

Again, by the analyticity and the CIT for $n = 1, 2, \dots,$

$$\frac{1}{2\pi i} \oint_{|z|=\rho} f(z) z^{n-1}\, dz = 0 \tag{5.3.7}$$

which, as with the argument leading to (5.3.6), proves that

$$0 = \int_0^{2\pi} f(e^{i\theta}) e^{in\theta} \frac{d\theta}{2\pi} \tag{5.3.8}$$

Taking complex conjugates for $n = 1, 2, \dots,$

$$0 = \int \overline{f(e^{i\theta})}\, e^{-in\theta} \frac{d\theta}{2\pi} \tag{5.3.9}$$

so, by (5.3.6), for $n = 1, 2, \dots,$

$$a_n = 2 \int [\operatorname{Re} f(e^{i\theta})] e^{-in\theta} \frac{d\theta}{2\pi} \tag{5.3.10}$$

On the other hand, by (5.3.6), for $n = 0$,

$$a_0 = i \operatorname{Im} f(0) + \int_0^{2\pi} [\operatorname{Re} f(e^{i\theta})] \frac{d\theta}{2\pi} \tag{5.3.11}$$

By (5.3.4), for $z \in \mathbb{D}$,

$$f(z) = i \operatorname{Im} f(0) + \int_0^{2\pi} [\operatorname{Re} f(e^{i\theta})] \left[1 + 2 \sum_{n=1}^{\infty} z^n e^{-in\theta} \right] \frac{d\theta}{2\pi} \tag{5.3.12}$$

where the uniform convergence in $e^{i\theta}$ justifies the interchange of sum and integral. (5.3.1) yields (5.3.2). $\square$

The *Poisson kernel* (or real Poisson kernel) is defined for $r \in [0, 1]$, $\theta, \psi \in [0, 2\pi]$, by

$$P_r(\theta, \psi) = \operatorname{Re} C(re^{i\psi}, e^{i\theta}) \tag{5.3.13}$$

$$= \frac{1 - r^2}{1 + r^2 - 2r\cos(\theta - \psi)} \tag{5.3.14}$$

by a direct calculation (Problem 4). Clearly, by applying Re to both sides of (5.3.2), we get

Theorem 5.3.3 (Poisson Representation). *If f is regular on $\overline{\mathbb{D}}$, then for all $r \in [0, 1)$, $\psi \in [0, 2\pi)$,*

$$\operatorname{Re} f(re^{i\psi}) = \int_0^{2\pi} P_r(\psi, \theta) \operatorname{Re} f(e^{i\theta}) \frac{d\theta}{2\pi} \tag{5.3.15}$$

Here is a consequence of this:

Theorem 5.3.4. *Let u be a continuous real-valued function on $\overline{\mathbb{D}}$. Then u is the real part of a function, f, analytic on $\mathbb{D}$ if and only if u obeys*

$$u(re^{i\psi}) = \int_0^{2\pi} P_r(\psi,\theta)u(e^{i\theta})\,\frac{d\theta}{2\pi} \tag{5.3.16}$$

f is uniquely determined by u up to a single additive purely imaginary constant.

Remark. $u \restriction \mathbb{D}$ is thus only dependent on $f = u \restriction \partial\mathbb{D}$ so (5.3.16) allows one to solve the Dirichlet problem for $\partial\mathbb{D}$ (i.e., given $f \in C(\partial\mathbb{D})$, find u harmonic on $\mathbb{D}$, continuous on $\overline{\mathbb{D}}$ so $u \restriction \partial\mathbb{D} = f$). This is a theme we return to in Section 3.1 of Part 3.

Proof. If $u = \operatorname{Re} f$, then, by Theorem 5.3.3 applied to $f_\rho(z) = f(\rho z)$, for any $\rho < 1$,

$$u(\rho re^{i\psi}) = \int_0^{2\pi} P_r(\psi,\theta)u(\rho e^{i\theta})\,\frac{d\theta}{2\pi} \tag{5.3.17}$$

By the uniform continuity of u on $\overline{\mathbb{D}}$, we can take $\rho \uparrow 1$ and see that (5.3.16) holds.

Conversely, if (5.3.16) holds, define f by

$$f(z) = \int \frac{e^{i\theta}+z}{e^{i\theta}-z}\,u(e^{i\theta})\,\frac{d\theta}{2\pi} \tag{5.3.18}$$

It is easy to see (Problem 5) that f is analytic in $\mathbb{D}$, and clearly, by (5.3.16), $\operatorname{Re} f = u$ in $\mathbb{D}$.

Finally, to prove the uniqueness statement, suppose g and f are two analytic functions on $\mathbb{D}$ with $\operatorname{Re} f = \operatorname{Re} g$ $(= u)$. It is easy to see (Problem 6) that $f - g$ is a pure imaginary constant. $\square$

Notes and Historical Remarks. Poisson never discussed complex analysis, but he did solve the problem of expressing the value of a harmonic function inside a sphere in terms of the values on the sphere. The importance of this result to complex analysis and what we've called the complex Poisson representation is due to Schwarz and especially Fatou many years after Poisson's work.

Siméon Poisson (1781–1840) was a contemporary of Cauchy and, like him, a student at École Polytechnique. He made fundamental contributions to electrostatics, in which he developed the Poisson equation ($\Delta\varphi = -4\pi^2\rho$ in three dimensions, where ρ is charge density and φ electric potential), and the three-dimensional kernels, and to probability theory

(Poisson distribution, which is discussed in Section 7.4 of Part 1; but also early versions of the weak law of large numbers and the central limit theorem).

(5.3.2) appeared first in Schwarz [**511**], which is why it is sometimes called the Schwarz integral formula. The Poisson kernel arises in the discussion of Abelian summation of Fourier series; see Problem 12 of Section 3.5 of Part 1.

Notice that there is a subtle difference between Theorems 5.3.3 and 5.3.4. The function f in Theorem 5.3.3 is regular, that is, continuous on $\overline{\mathbb{D}}$. Clearly, in Theorem 5.3.4, if u is given, $\operatorname{Re} f$ is continuous in $\overline{\mathbb{D}}$, but we have not claimed that $\operatorname{Im} f$ is. Indeed, there are examples (see Problem 7) where $\operatorname{Im} f$ is not continuous. In Chapter 5 of Part 3, this continuity problem will be studied further.

If we make the choice (5.3.18), which fixes the arbitrary constant by setting $\operatorname{Im} f(0) = 0$, we get a natural map from continuous functions on $\overline{\mathbb{D}}$ obeying (5.3.16) to functions on $\mathbb{D}$ which are also the real part of an analytic function (since $\operatorname{Im} f = \operatorname{Re}(-if)$). This is called the map from harmonic functions to their *harmonic conjugate*. It will be studied heavily in Sections 5.7, 5.8 and 5.11 of Part 3.

Problems

1. (a) Prove that the real Poisson kernel obeys $P_r(\theta,\varphi) > 0$, $\int P_r(\theta,\varphi)\frac{d\varphi}{2\pi} = 1$, and for all $\varepsilon > 0$,

$$\lim_{r\uparrow 1} \int_{|\theta-\varphi|\geq\varepsilon} P_r(\theta,\varphi)\,\frac{d\varphi}{2\pi} = 0 \tag{5.3.19}$$

(b) Use (a) to prove that if $f \in C(\partial\mathbb{D})$ and if

$$P_r(f)(\theta) = \int P_r(\theta,\varphi)f(\varphi)\,\frac{d\varphi}{2\pi} \tag{5.3.20}$$

then $\|P_r f - f\|_\infty \to 0$ as $r \uparrow 1$.

(c) Prove $G(re^{i\theta}) = P_r(f)(\theta)$ is harmonic in $z = re^{i\theta}$.

(d) Assuming an independent proof of the fact that the harmonic functions are determined by their boundary values, prove that if F is regular on $\overline{\mathbb{D}}$ and $f(e^{i\theta}) = \operatorname{Re} F(e^{i\theta})$, then $\operatorname{Re} F(re^{i\theta}) = P_r(f)(\theta)$. Use this to provide a new proof of (5.3.2).

2. Show that the CIF can be rewritten as

$$f(z) = \tfrac{1}{2}\int C(z,e^{i\theta})f(e^{i\theta})\,\frac{d\theta}{2\pi} + \tfrac{1}{2}\,f(0)$$

and manipulate this and its complex conjugate to prove (5.3.2).

3. For f analytic in a neighborhood of $\overline{\mathbb{D}}$, we have, of course, that

$$f(0) = \int_0^{2\pi} f(e^{i\theta})\,\frac{d\theta}{2\pi} \tag{5.3.21}$$

Given $a \in \mathbb{D}$, let

$$h_a(z) = \frac{z+a}{1+\bar{a}z}$$

which maps $\mathbb{D}$ to $\mathbb{D}$, $\partial\mathbb{D}$ to $\partial\mathbb{D}$, and 0 to a. (Check these facts.) Let $g(z) = f(h_a(z))$. By applying (5.3.21) to g, deduce that if $a = |a|e^{i\psi}$, then

$$f(a) = \int_0^{2\pi} P_{|a|}(\psi,\theta) f(e^{i\theta})\,\frac{d\theta}{2\pi}$$

and deduce the complex Poisson representation from this.

4. Verify the calculation needed to get from (5.3.13) to (5.3.14).

5. Check that f given by (5.3.18) is analytic in $z \in \mathbb{D}$. (*Hint*: Morera.)

6. Suppose f and g are analytic in $\mathbb{D}$ and $\operatorname{Re} f = \operatorname{Re} g$.

   (a) Use the CR equations to prove that $\operatorname{grad}(\operatorname{Im} f - \operatorname{Im} g) = 0$.

   (b) Conclude that

$$f(z) - i\operatorname{Im} f(0) = g(z) - i\operatorname{Im} g(0)$$

7. Let $u(e^{i\theta}) = \operatorname{sgn}(\theta)(-\log(|\sin(\theta)|))^{-\alpha}$ and let f be given by (5.3.18). Show that for all $\alpha > 0$, $\operatorname{Im} f$ has continuous boundary values on $(0,\pi) \cup (\pi, 2\pi)$. For which α are these boundary values continuous at $\theta = 0$?

8. A *Carathéodory function* is an analytic function on $\mathbb{D}$ with $F(0) = 1$ and $\operatorname{Re} F(z) \geq 0$ on $\mathbb{D}$. Suppose

$$F(z) = 1 + \sum_{n=1}^{\infty} c_n z^n \tag{5.3.22}$$

Prove (a theorem of Borel [**66**]) that for such functions, $|c_n| \leq 2$. (*Hint*: Use the complex Poisson representation to prove for each $r \in (0,1)$,

$$c_n = 2r^{-n} \int e^{-in\theta} \operatorname{Re} F(re^{i\theta})\,\frac{d\theta}{2\pi} \tag{5.3.23}$$

and take $r \to 1$ after using $\int \operatorname{Re} F(re^{i\theta})\frac{d\theta}{2\pi} = 1$.)

9. Prove the following theorem of Rogosinski [**490**]: If $f\colon \mathbb{D} \to \mathbb{D}$ is analytic and everywhere nonvanishing, then $|f'(0)| \leq 2/e$ by the following steps:

   (a) Without loss, show one can suppose that $f(0) > 0$.

   (b) Let $L(z)$ be the analytic branch of $\log(1/f(z))$ with $L(0) > 0$. Prove that $\operatorname{Re} L(z) > 0$ and that $L(z)/L(0) = F(z)$ is a Carathéodory function.

(c) Prove $|L'(0)| \le 2|L(0)|$. (*Hint*: See Problem 8.)

(d) Prove that $|f'(0)| \le -2|f(0)|\,\log|f(0)|$.

(e) Prove that $\max_{0<y\le 1} -y\log(y) = e^{-1}$ and conclude $|f'(0)| \le 2/e$.

10. (a) Let f be bounded and analytic in $\mathbb{D}$ and suppose $\lim_{r\uparrow 1} \operatorname{Re} f(re^{i\theta}) \equiv \operatorname{Re} f(e^{i\theta})$ exists for a.e. θ. Prove that (5.3.15) is valid.

(b) Let f be analytic and bounded in $\mathbb{C}_+$ and continuous in $\overline{\mathbb{C}_+}$. For $y > 0$, prove that

$$\operatorname{Re} f(x_0 + iy_0) = \frac{1}{\pi}\int_{\infty}^{\infty} \frac{y_0 \operatorname{Re} f(x)}{(x-x_0)^2 + y_0^2}\,dx \tag{5.3.24}$$

(*Hint*: Apply (a) to $g(z) = f(i(1-z)/(1+z))$.)

Remarks. 1. (a) requires you to know the dominated convergence theorem (see Section 4.6 of Part 1).

2. $P_{x_0,y_0}(x) = y_0\pi^{-1}[(x-x_0)^2 + y_0^2]$ is called the *Poisson kernel* for the plane. (5.3.24) is the *Poisson formula* for the plane. We'll say a lot more about this in Sections 3.1 and 5.9 of Part 3.

5.4. Harmonic Functions

We will define a function to be harmonic via the following rather weak condition:

Definition. Let Ω be a region in $\mathbb{C}$. A function $u\colon \Omega \to \mathbb{R}$ is said to be *harmonic at* $z_0 \in \Omega$ if and only if there is $\delta > 0$ so that $\mathbb{D}_\delta(z_0) \subset \Omega$, u is continuous on $\mathbb{D}_\delta(z_0)$, and for any $\rho \in (0,\delta)$,

$$\int_0^{2\pi} u(z_0 + \rho e^{i\theta})\,\frac{d\theta}{2\pi} = u(z_0) \tag{5.4.1}$$

If u is harmonic at each $z_0 \in \Omega$, we say u is *harmonic in* Ω.

Remarks. 1. (5.4.1) is called the *mean value property* (MVP).

2. For our application in the next section, it is critical that in the definition of "harmonic in Ω," δ can be z_0-dependent and is not a priori assumed to be bounded from below on compact subsets, although a posteriori, it is (see Problem 1). In some ways, the fact that the a priori requirement is totally local is a harmonic function analog of Morera's theorem (Theorem 3.1.4), which is also totally local.

3. Chapter 3 of Part 3 is an extensive study of harmonic functions on $\mathbb{R}^\nu$.

The main result of this section, towards which we are heading, is:

Theorem 5.4.1. *Let $u\colon \Omega \to \mathbb{R}$ be a real-valued function on a region, Ω. The following are equivalent:*

(1) *u is harmonic in* Ω.
(2) *For every* $z_0 \in \Omega$, *there is a* $\delta > 0$ *and function,* f, *analytic in* $\mathbb{D}_\delta(z_0)$, *so that*

$$u(z) = \operatorname{Re} f(z) \tag{5.4.2}$$

for $z \in \mathbb{D}_\delta(z_0)$.
(3) *u is* C^2 *and* $\Delta u = 0$ *on* Ω.

One can show (Problem 3) that if Ω is hsc, we can find an analytic function, f, on all of Ω so that (5.4.2) holds. However,

Example 5.4.2. Let $\Omega = \mathbb{D}^\times$ and $u(z) = \log|z|$. Then for any $z_0 \in \Omega$, if $\delta = \min(|z_0|, 1 - |z_0|)$, then $\mathbb{D}_\delta(z_0) \subset \Omega$ and is hsc, and $g(z) = z$ is nonvanishing on $\mathbb{D}_\delta(z_0)$. Thus, by Theorem 2.6.1 there exists f analytic in $\mathbb{D}_\delta(z)$ so $e^f = z$ and $\operatorname{Re} f = \log|z|$. Thus, u obeys (2). But it is not hard to see (Problem 2) that there is no global f on Ω, so that (5.4.2) holds. □

Proposition 5.4.3 (Maximum Principle for Harmonic Functions). *Let u be harmonic on* Ω. *If there exists* $z_0 \in \Omega$ *so that for some* $\delta > 0$,

$$u(z_0) = \sup\{u(z) \mid |z - z_0| < \delta\} \tag{5.4.3}$$

then u is constant on $\mathbb{D}_\delta(z_0)$. *If*

$$u(z_0) = \sup\{u(z) \mid z \in \Omega\} \tag{5.4.4}$$

then u is constant on Ω.

Remark. Once one knows (1) ⇒ (2) in Theorem 5.4.1, u constant on $\mathbb{D}_\delta(z_0)$ implies u constant on all of Ω.

Proof. If (5.4.3) holds, then for any $\rho \in (0, \delta)$, the MVP and continuity of $u(z_0 + \rho e^{i\theta})$ in θ imply

$$u(z_0 + \rho e^{i\theta}) = u(z_0) \tag{5.4.5}$$

for all $\theta \in [0, 2\pi)$, so u is constant in $\mathbb{D}_\delta(z_0)$. This proves the first statement.

If (5.4.4) holds, let $Q = \{z_1 \mid u(z_1) = u(z_0)\}$. Since $z_0 \in Q$, it is not empty. Since u is continuous, Q is closed. By the first part, Q is open. By connectedness of Ω, $Q = \Omega$. □

Given $g \in C(\partial\mathbb{D})$, we say u solves the *Dirichlet problem of* $\mathbb{D}$ for g if and only if u is a function continuous on $\overline{\mathbb{D}}$, harmonic on $\mathbb{D}$, and $u \restriction \partial\mathbb{D} = g$.

Corollary 5.4.4. *The solution of the Dirichlet problem, if it exists, is unique.*

Proof. If u_1, u_2 solve the Dirichlet problem for g, then $u_3 \equiv u_1 - u_2$ solves it for $g \equiv 0$. Thus, we need only show that if u_3 is continuous on $\overline{\mathbb{D}}$, harmonic in $\mathbb{D}$, and $u_3 \restriction \partial\mathbb{D} = 0$, then $u_3 \equiv 0$.

Let

$$M_+ = \max_{z\in\overline{\mathbb{D}}} u_3(z) = \sup_{z\in\mathbb{D}} u_3(z) \tag{5.4.6}$$

If $u_3(z_0) = M_+$ for $z_0 \in \mathbb{D}$, then by Proposition 5.4.2, u is constant on $\mathbb{D}$, so on $\overline{\mathbb{D}}$, so $M_+ = 0$.

If $u_3(z_0) < M_+$ for all $z_0 \in \mathbb{D}$, then $\max u_3(z)$ occurs on $\partial\mathbb{D}$ (since, by continuity and compactness, it occurs somewhere). Thus, again $M_+ = 0$. We thus see that

$$u_3(z) \le 0 \tag{5.4.7}$$

for all $z \in \overline{\mathbb{D}}$.

Applying the same argument to $-u_3$, we see $u_3(z) \ge 0$ for all $z \in \overline{\mathbb{D}}$, that is, $u_3 \equiv 0$. □

Theorem 5.4.5. *For any $g \in C(\partial\mathbb{D})$, there is a unique solution of the Dirichlet problem, and on $\mathbb{D}$, it is the real part of an analytic function, f.*

Proof. Uniqueness has just been proven. For existence, define f by (5.3.18) where $u(e^{i\theta})$ is $g(e^{i\theta})$. By Problem 5 of Section 5.3, f is analytic in $\mathbb{D}$. $u \equiv \operatorname{Re} f$ is given by (5.3.16) with $u(e^{i\theta}) = g(e^{i\theta})$. So by Problem 1 of Section 5.3, u is continuous on $\overline{\mathbb{D}}$ if we set $u \restriction \partial\mathbb{D} = g$. By the Cauchy formula (3.1.2) for any z_0 and $\rho < 1 - |z_0|$,

$$f(z_0) = \int_0^{2\pi} f(z_0 + \rho e^{i\theta})\, \frac{d\theta}{2\pi} \tag{5.4.8}$$

so $\operatorname{Re} f$ is harmonic. □

Proof of Theorem 5.4.1. <u>(2) ⇒ (3).</u> As noted already in (2.1.39), if f is holomorphic and C^∞ (we now know C^∞ is implied by being holomorphic), then f obeys $\Delta f = 0$, so $\operatorname{Re}(\Delta f) = \Delta(\operatorname{Re} f) = 0$ and, of course, $\operatorname{Re} f$ is C^∞, and so C^2.

<u>(3) ⇒ (1).</u> Obviously, if u is C^2, it is continuous, so we only need to check the MVP. In polar coordinates (r,θ), Δ has the form (Problem 4)

$$\Delta = \frac{1}{r}\frac{\partial}{\partial r}\left(r\frac{\partial}{\partial r}\right) + \frac{1}{r^2}\frac{\partial^2}{\partial\theta^2} \tag{5.4.9}$$

away from $r = 0$.

For $z_0 \in \Omega$ and $r \in [0, \operatorname{dist}(z_0, \mathbb{C}\setminus\Omega))$, let

$$g(r,\theta) = u(z_0 + re^{i\theta}) \tag{5.4.10}$$

$$\begin{aligned} A(r) &\equiv \int_0^{2\pi} u(z_0 + re^{i\theta})\, \frac{d\theta}{2\pi} \\ &= \int_0^{2\pi} g(r,\theta)\, \frac{d\theta}{2\pi} \end{aligned} \tag{5.4.11}$$

By integrating the derivatives and using that $\partial g/\partial\theta$ is periodic in θ of period 2π, for $r \neq 0$,

$$\frac{1}{r^2}\int_0^{2\pi} \frac{\partial^2 g}{\partial^2\theta^2}\,\frac{d\theta}{2\pi} = 0 \tag{5.4.12}$$

Since g is C^2 in r, A is C^2 away from $r = 0$ and

$$\begin{aligned}\frac{1}{r}\frac{d}{dr}\,r\,\frac{dA}{dr} &= \int_0^{2\pi}\left(\frac{1}{r}\frac{\partial}{\partial r}\,r\,\frac{\partial}{\partial r}\right)g(r,\theta)\,\frac{d\theta}{2\pi} \\ &= \int_0^{2\pi}\left[(\Delta g)(r,\theta) - \frac{1}{r^2}\frac{\partial}{\partial\theta^2}\,g(r,\theta)\right]\frac{d\theta}{2\pi} \\ &= 0\end{aligned} \tag{5.4.13}$$

by (5.4.12) and $\Delta u = 0$.

Thus,

$$r\,\frac{dA}{dr} = c_1 \tag{5.4.14}$$

If $c_1 \neq 0$, $A = c_2 + c_1\log(r)$ diverges as $r \downarrow 0$. But

$$\lim_{r\downarrow 0} A(r) = u(z_0) \tag{5.4.15}$$

since u is continuous. It follows that $c_1 = 0$, so A is constant and (5.4.15) implies $A \equiv u(z_0)$, that is, the MVP holds.

(1) $\Rightarrow$ (2). Pick δ so $\mathbb{D}_{2\delta}(z_0) \subset \Omega$. let $g = u \restriction \partial\mathbb{D}_\delta(z_0)$. Then u solves the Dirichlet problem for g on $\overline{\mathbb{D}_\delta(z_0)}$, so by Theorem 5.4.5, u is the real part of an analytic function on $\mathbb{D}_\delta(z_0)$. □

The above proof shows the MVP holds for any disk in Ω; see Problem 1.

Note. Many of the results of this section are extended to $\mathbb{R}^\nu$ in Chapter 3 of Part 3.

Notes and Historical Remarks. For a discussion of the history of the Dirichlet problem, see the Notes to Section 3.6 of Part 3.

Harmonic functions and the Dirichlet problem arise in many physical situations: first, via fluid flow, as explained in the Notes to Section 2.1; second, in electrostatics since the electric field in a region void of sources is given by $\vec{E} = \text{grad}(u)$ where u is harmonic, and also in acoustics. For example, if one has a circle in the plane made of two arcs, A_1 and A_2, which are perfect conductors joined by insulators (so $\overline{A_1 \cup A_2} = \partial\mathbb{D}$ and $A_1 \cap A_2 = \emptyset$), the potential inside the circle solves a Dirichlet problem with a discontinuous g equal to different constants on A_1 and A_2.

Problems

1. If u is harmonic in Ω, show that (5.4.1) holds for all $\rho < \text{dist}(z_0, \partial\Omega)$.

2. Let $u = \log|z|$ on $\mathbb{D}^\times$. Show there exists no analytic function, f, on all of $\mathbb{D}^\times$ so that $\text{Re}\, f = u$.

3. (a) Let u be harmonic on a region, Ω. Let f_1 and f_2 be two analytic functions on some $\mathbb{D}_\delta(z_0) \subset \Omega$, so $\text{Re}\, f_1 = \text{Re}\, f_2 = u \restriction \mathbb{D}_\delta(z_0)$. Prove that $f_1' = f_2'$.

   (b) Conclude there is an analytic function, g, on all of Ω so that if f is analytic on some $\mathbb{D}_\delta(z_0)$ with $\text{Re}\, f = u \restriction \mathbb{D}_\delta(z_0)$, then $f' = g$.

   (c) If Ω is hsc, prove that there is an analytic function, f, on all of Ω so $\text{Re}\, f = u$. (*Hint*: Theorem 2.5.4.)

4. (a) Using $r^2 = x^2 + y^2$, $\theta = \arctan(x/y)$, prove that

$$\frac{\partial}{\partial x} = \frac{x}{r}\frac{\partial}{\partial r} + \frac{y}{r^2}\frac{\partial}{\partial \theta}, \qquad \frac{\partial}{\partial y} = \frac{y}{r}\frac{\partial}{\partial r} - \frac{x}{r^2}\frac{\partial}{\partial \theta}$$

   (b) Prove that

$$\Delta u = \frac{\partial^2 u}{\partial r^2} + \frac{1}{r}\frac{\partial u}{\partial r} + \frac{1}{r^2}\frac{\partial^2 u}{\partial \theta^2}$$

   (c) Deduce (5.4.9).

5. Let u be nonnegative and harmonic in a neighborhood of $\overline{\mathbb{D}}$.

   (a) Prove that the Poisson kernel obeys

$$\sup_{\theta,\varphi} P_r(\theta,\varphi) = \frac{1+r}{1-r}, \qquad \inf_{\theta,\varphi} P_r(\theta,\varphi) = \frac{1-r}{1+r} \tag{5.4.16}$$

   (b) Prove that for any $z \in \mathbb{D}$ (Harnack's inequality)

$$\frac{1-|z|}{1+|z|}\, u(0) \le u(z) \le \frac{1+|z|}{1-|z|}\, u(0) \tag{5.4.17}$$

   (c) Let $0 < u_n \le u_{n+1} \le \dots$ for functions harmonic in a neighborhood of $\overline{\mathbb{D}}$. Prove Harnack's principle:

$$\sup_n u_n(z) < \infty \text{ for one } z \in \mathbb{D} \Leftrightarrow \sup_n u_n(z) < \infty \text{ for all } z \in \mathbb{D}$$

   Note. Problems 6–11 assume a knowledge of distribution theory; see Chapter 6 of Part 1.

6. Let f be a C^∞ function of compact support on $\mathbb{C}$ and let

$$g(z) = \frac{1}{2\pi}\log(|z|) \tag{5.4.18}$$

   (a) Prove g is C^∞ on $\mathbb{C}^\times$ and that on that set, $\Delta g = 0$.

(b) Prove that

$$\int g(z)(\Delta f)(z)\, d^2z = f(0)$$

so that, in the distributional sense,

$$\Delta g = \delta(z) \tag{5.4.19}$$

(*Hint*: Use $\int_{\varepsilon<|z|<R}(g\Delta f - f\Delta g)\, d^2z = \int_{\varepsilon<|z|<R} \operatorname{grad}(g \operatorname{div} f - f \operatorname{div} g)\, d^2z$, followed by Gauss' theorem, and then take $\varepsilon \downarrow 0$.)

7. Prove that, in the distributional sense,

$$\bar\partial\left(\frac{1}{\pi z}\right) = \delta(z) \tag{5.4.20}$$

(*Hint*: Use (5.4.19), (2.1.38), and Problem 8 of Section 2.1.)

8. Given a complex measure μ of compact support on $\mathbb{C}$, prove that

(a) $$\int d^2z \int \frac{d|\mu(w)|}{|w-z|} < \infty \tag{5.4.21}$$

(b) $$C_{d\mu}(z) = \int \frac{d\mu(w)}{w-z} \tag{5.4.22}$$

is defined for a.e. z and defines an $L^1(\mathbb{C}, d^2z)$ function. It is called the *Cauchy transform* of μ.

(c) Prove that, in the distributional sense,

$$\bar\partial C_{d\mu} = -\pi\, d\mu \tag{5.4.23}$$

Remarks. 1. There are differing conventions on whether $C_{d\mu}$ is given by (5.4.22) or has a $(2\pi)^{-1}$, π^{-1}, or $(2\pi i)^{-1}$. Our definition is the most common.

2. There are whole books on the Cauchy transform [**38, 117**].

9. Let γ be a rectifiable Jordan curve, Ω its interior, and orient γ so the winding number of γ in Ω is 1. Compute $\bar\partial\chi_\Omega$ and $\partial\chi_\Omega$ where χ_Ω is the characteristic function of Ω and the derivatives are distributional.

10. Let f be regular on $\overline{\mathbb{D}}$. Let $d\mu = f\, dz$ where dz is the line integral on $\partial\mathbb{D}$, counterclockwise. Compute $C_{d\mu}$. Comment on the relation of the Cauchy integral theorem and (5.4.23).

11. Let $g \in C_0^\infty(\mathbb{C})$ and let

$$f(z) = -\frac{1}{\pi}\int \frac{g(w)}{w-z}\, d^2w \tag{5.4.24}$$

Prove that f is C^∞ and solves $\bar\partial f = g$. (*Hint*: Use Problem 7 to first check that it is a distributional solution in that, for all $h \in C_0^\infty$,

$-\int f(z)(\bar\partial h)(z)\,d^2z = \int g(z)h(z)\,d^2z$, and then knowing that f is C^∞, it obeys $\bar\partial f = g$.)

Note. There is a connection between this result and Pompeiu's formula; see Problem 1 of Section 2.7 and (2.7.7).

5.5. The Reflection Principle

The main result of this section is:

Theorem 5.5.1 (Reflection Principle). *Let $\Omega \subset \mathbb{C}_+ = \{z \mid \operatorname{Im} z > 0\}$ be a region and $I \subset \mathbb{R}$, an open set (in the topology of $\mathbb{R}$), so that for any $x \in I$, there is δ so that $\mathbb{D}_\delta(x_0) \cap \mathbb{C}_+ \subset \Omega$* (see Figure 5.5.1). *Suppose f, given on $\Omega \cup I$, is analytic in Ω and continuous on $\Omega \cup I$. Suppose also that f is real-valued on I. Then the function, F, on $\widetilde{\Omega} \equiv \Omega \cup \bar\Omega$ by*

$$F(z) = \begin{cases} f(z), & z \in \Omega \cup I \\ \overline{f(\bar z)}, & z \in \bar\Omega \end{cases} \tag{5.5.1}$$

is analytic on $\widetilde{\Omega}$.

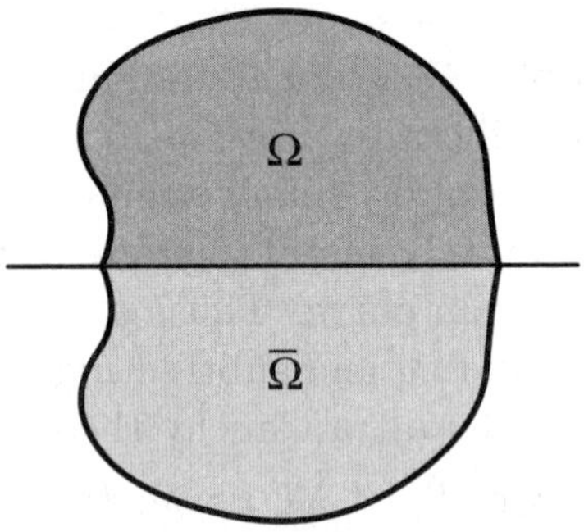

Figure 5.5.1. A reflected region.

Remark. We emphasize that here $\bar\Omega$ means $\{z \mid \bar z \in \Omega\}$, not the closure of Ω!

We will deduce this from:

Theorem 5.5.2 (Painlevé's Theorem). *Let $\widetilde{\Omega}$ be a region in $\mathbb{C}$. Suppose γ is a simple rectifiable curve with $\gamma[(0,1)] \subset \widetilde{\Omega}$, $\gamma(0), \gamma(1) \in \partial\widetilde{\Omega}$, and so that $\widetilde{\Omega} \setminus \operatorname{Ran}(\gamma)$ is the union of two connected components Ω^+ and Ω^-* (see Figure 5.5.2). *Suppose that $f^\pm$ are two functions defined and continuous on $\Omega^\pm \cup \operatorname{Ran}(\gamma)$ and analytic on $\Omega^\pm$, and that*

$$f^+ \restriction \gamma[(0,1)] = f^- \restriction \gamma[(0,1)] \tag{5.5.2}$$

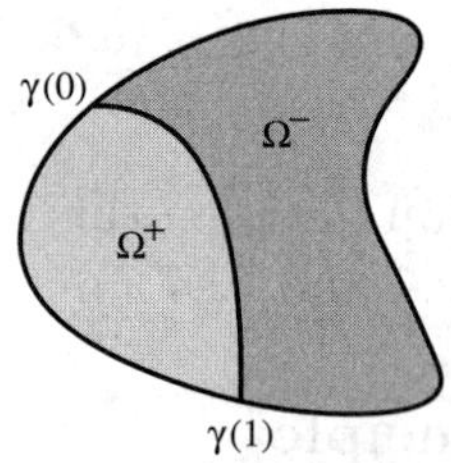

Figure 5.5.2. A divided region.

Then the function F on $\widetilde{\Omega}$ defined by

$$F(z) = \begin{cases} f^+(z), & z \in \Omega^+ \cup \gamma[(0,1)] \\ f^-(z), & z \in \Omega^- \cup \gamma[(0,1)] \end{cases} \tag{5.5.3}$$

is analytic on $\widetilde{\Omega}$.

Remarks. 1. In other words, if f is continuous in Ω and analytic on $\Omega^+ \cup \Omega^-$, it is analytic on Ω.

2. This result may fail if γ is not rectifiable; see the Notes.

Proof. We use a variant of the argument used in the proof of Theorem 4.6.1. By Morera's theorem, it suffices to prove $\oint_T f(z)\,dz = 0$ for every triangle, T in Ω. If T intersects γ in finitely many points, $\text{ins}(T) \setminus \text{Ran}(\gamma)$, the inside of T with γ removed, is finitely many regions, each entirely in Ω^+ or Ω^- and each bounded by a Jordan curve. Thus, $\oint_T f(z)\,dz$ is equal to a sum of finitely many integrals by adding and subtracting integrals of f along pieces of γ, each of which is 0 by the ultra Cauchy theorem (Theorem 4.6.1).

For an arbitrary triangle, T, let λT be the triangle with the same orthocenter as T but sides scaled by a factor of λ. For $|\lambda - 1|$ small, $\lambda T \subset \Omega$ and, by the rectifiability of γ and the same Banach indicatrix theorem argument used in the proof of Theorem 4.6.1, λT intersects γ finitely many times for a.e. λ. Since $\lambda \mapsto \oint_{\lambda T} f(z)\,dz$ is continuous in λ by the continuity of f, we see $\oint_T f(z)\,dz = 0$. □

Proof of Theorem 5.5.1. Let I_0 be an interval in I. Let $\Omega^+ = \Omega$, $\Omega^- = \bar{\Omega}$, and $F^\pm$ be defined on $\Omega^\pm \cup I$ by

$$F_+(z) = \begin{cases} f(z), & z \in \Omega \\ f(z), & z \in I_0 \end{cases} \tag{5.5.4}$$

$$F_-(z) = \begin{cases} \overline{f(\bar{z})}, & z \in \bar{\Omega} \\ f(z), & z \in I_0 \end{cases} \tag{5.5.5}$$

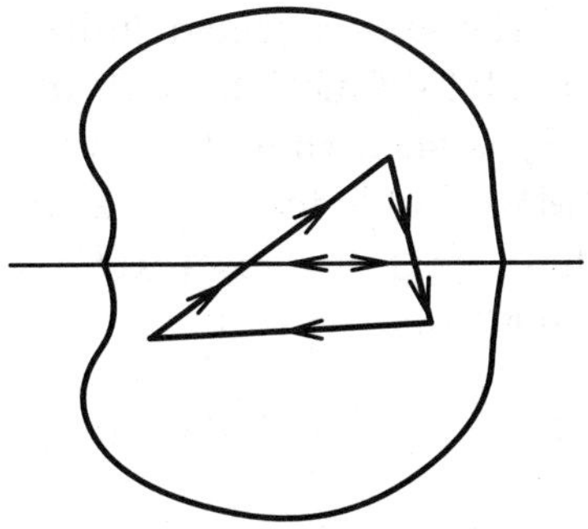

Figure 5.5.3. Slicing a triangle.

Since $\partial\bar{g} = \overline{\partial g}$ and if $h(z) = g(\bar{z})$, then $(\bar\partial h)(z) = (\partial g)(\bar{z})$, we see that if $g(z) = \overline{f(\bar{z})}$ and $h(z) = f(\bar{z})$, then

$$\bar\partial g = \overline{\partial h} = \overline{\bar\partial f} = 0 \qquad (5.5.6)$$

so F is analytic on Ω^-. Moreover, since f is real on I_0, $F^+ \restriction I_0 = F^- \restriction I_0$. Now use Theorem 5.5.2. □

Remark. The Morera theorem argument used in the proof of Theorem 5.5.2 can be specialized to this case. There are only two regions (see Figure 5.5.3) where the Banach indicatrix argument isn't needed. Moreover, the ultra Cauchy theorem is trivial at a flat edge by a simple continuity argument.

We are interested in extending Theorem 5.5.1 in several ways. The most significant for later applications concerns the following: In the hypotheses of Theorem 5.5.1, we need to suppose that $\operatorname{Im} f(z) \to 0$ as $z \to I$ and that $\operatorname{Re} f(z)$ has a continuous limit as $z \to I$. Remarkably, it is actually true that if $\operatorname{Im} f(z) \to 0$, then $\operatorname{Re} f$ automatically has continuous boundary values on I.

Theorem 5.5.3 (Strong Reflection Principle). *Let Ω, I, and $\widetilde{\Omega}$ be as in Theorem* 5.5.1. *Suppose f is analytic in Ω and the function v on $\Omega \cup I$ given by*

$$v(z) = \begin{cases} \operatorname{Im} f(z), & z \in \Omega \\ 0, & z \in I \end{cases} \qquad (5.5.7)$$

is continuous. Then there is an analytic function F on Ω so $F \restriction \Omega = f$ and $(F \restriction \bar\Omega)(z) = \overline{f(\bar{z})}$. In particular, f has continuous boundary values on I.

Proof. Define v on $\widetilde{\Omega}$ by

$$v(z) = \begin{cases} \operatorname{Im} f(z), & z \in \Omega \\ 0, & z \in I \\ -\operatorname{Im} f(\bar{z}), & z \in \bar\Omega \end{cases} \qquad (5.5.8)$$

We claim v is harmonic in the sense of the definition involving (5.4.1). Since v is the real part of an analytic function in Ω, for $z_0 \in \Omega$, the MVP holds so long as the ρ in (5.5.8) is such that it obeys $\rho < \min(\operatorname{Im} z_0, \operatorname{dist}(\partial\Omega))$. By reflection in $\mathbb{R}$, the MVP holds for $z_0 \in \tilde{\Omega}$. If $z_0 \in I$ and $\rho < \delta$ where $\mathbb{D}_\delta(z_0) \cap \mathbb{C}_+ \subset \Omega$, we have $\{z \mid z = z_0 + \rho e^{i\theta}\} \subset \widetilde{\Omega}$ and $v(z_0 + \rho e^{i\theta}) = -v_0(z_0 + \rho e^{-i\theta})$. Averaging from $-\pi$ to π, we see

$$\frac{1}{2\pi}\int_0^{2\pi} v(z_0 + \rho e^{i\theta})\, d\theta = 0 = v(z_0) \tag{5.5.9}$$

so the MVP holds on I. Thus, v is harmonic as claimed.

By Theorem 5.4.1, for any $z_0 \in I$ and with δ as above, there is a function g analytic in $\mathbb{D}_\delta(z_0)$, so $v = \operatorname{Re} g$. Thus, on $\mathbb{D}_\delta(z_0) \cap \mathbb{C}_+$, $\operatorname{Re}(g + if) = 0$. So, by the Cauchy–Riemann equations, $(g + if)' = 0$ there, so for a constant c,

$$f = ig + c \tag{5.5.10}$$

on $\mathbb{D}_\delta(z_0) \cap \mathbb{C}_+$. Thus, f has an analytic continuation to $\mathbb{D}_\delta(z_0)$, and so f is continuous on I. Thus, by Theorem 5.5.1, there is the required analytic function F on Ω. $\square$

Next, we want to note an extension to $\partial\mathbb{D}$ in place of $\mathbb{R}$:

Theorem 5.5.4. *Let $\Omega \subset \mathbb{D}$ and $I \subset \partial\mathbb{D}$ an open set (in the topology of $\partial\mathbb{D}$) so that for $z_0 \in I$, there is $\delta > 0$ so that $\mathbb{D}_\delta(z_0) \cap \mathbb{D} \subset \Omega$. Suppose that f is analytic in Ω and the function v on $\Omega \cup I$ given by*

$$v(z) = \begin{cases} \operatorname{Im} f(z), & z \in \Omega \\ 0, & z \in \Omega \end{cases} \tag{5.5.11}$$

is continuous. Let

$$\Omega^\sharp = \left\{ \frac{1}{\bar{z}} \;\middle|\; z \in \Omega \right\} \tag{5.5.12}$$

and $\widetilde{\Omega} = \Omega \cup I \cup \Omega^\sharp$. Then $\operatorname{Re} f$ has a continuous extension to $\Omega \cup I$ and there exists F analytic in $\widetilde{\Omega}$ so that $F \restriction \Omega = f$, and for $z \in \Omega^\sharp$,

$$F(z) = \overline{F(1/\bar{z})} \tag{5.5.13}$$

Sketch. The map

$$u(z) = \frac{i(1+z)}{1-z} \tag{5.5.14}$$

maps $\mathbb{D}$ to $\mathbb{C}_+$ (since $U(e^{i\theta}) = -\cot(\theta/2)$ and $U(0) = i$). Moreover, U maps $\widehat{\mathbb{C}} \setminus \{1\}$ to $\mathbb{C}$ and

$$\overline{U(z)} = U(1/\bar{z}) \tag{5.5.15}$$

This theorem then follows by applying Theorem 5.5.3 to $f \circ U^{-1}$ on $U(\Omega)$ with boundary values on $U(I)$. The details are left to the reader (Problem 1). $\square$

The same idea, but applied to the values rather than the domain, leads to (the reader is asked to prove this in Problem 2):

Theorem 5.5.5. *Let* $\Omega, I, \widetilde{\Omega}$ *be as in Theorem* 5.5.1. *Suppose* f *is analytic on* Ω, $|f(z)|$ *has a continuous extension to* $\Omega \cup I$, *and for* $z_0 \in I$, *we have* $|f(z_0)| = 1$. *Then there is a meromorphic function,* F, *on* $\widetilde{\Omega}$, *so for* $z \in \Omega$, $F(z) = f(z)$, *and for* $z \in \bar{\Omega}$ *with* $f(1/\bar{z}) \neq 0$, *we have that*

$$F(z) = \frac{1}{\overline{f(\bar{z})}} \tag{5.5.16}$$

There is an additional facet of continuations guaranteed by the reflection principle that we should mention:

Theorem 5.5.6. *Let* f *be analytic in* $\Omega \subset \mathbb{C}_+$ *with* $I \subset \mathbb{R}$ *an interval in* $\partial\Omega$. *Let* $z_0 \in I$ *so that for some* $\delta > 0$, $N_0 = \mathbb{D}_\delta(z_0) \cap \mathbb{C}_+ \subset \Omega$ *and suppose that on* N_0, *either* $\operatorname{Im} f > 0$ *and* $\lim_{z \to I,\, z \in \Omega} (\operatorname{Im} f)(z) = 0$ *or* $|f(z_0)| < 1$ *and* $\lim_{z \to I,\, z \in \Omega} |f(z)| = 1$. *Then* f *has a meromorphic continuation to* $\mathbb{D}_\delta(z_0)$ *and* $f'(z_0) \neq 0$.

Proof. We consider the case $\operatorname{Im} f > 0$. By the strong reflection principle, f has an analytic continuation to $\mathbb{D}_\delta(z_0)$ with f real on $I \cap \mathbb{D}_\delta(z_0)$. If $f'(z_0) = 0$, f has a zero of order $k \geq 2$ at z_0, so $\arg(f)$ changes by πk in N_0 near z_0, inconsistent with $\operatorname{Im} f > 0$ there. □

Finally, we mention a somewhat special situation that one can analyze via a special reflection, and then leave to the problems a generalization that uses the mapping idea of the last two theorems.

Theorem 5.5.7. *Let* f *be analytic in* $\Omega = \{z \mid 0 < |z| < \rho,\ \arg(z) \in (0, \frac{\pi}{2})\}$ *and so that* $\operatorname{Im} f$ *has a continuous extension to* $\bar{\Omega}$ *with* $\operatorname{Im} f(z) = 0$ *if* $\arg(z)$ *is* 0 *or* $\frac{\pi}{2}$. *Then* f *has an analytic continuation to* $\mathbb{D}_\rho(0)$ *and the Taylor series at* $z = 0$ *has the form*

$$f(z) = \sum_{n=0}^{\infty} a_{2n} z^{2n} \tag{5.5.17}$$

with a_{2n} *real.*

Proof. By the reflection principle applied to $\arg(z) = 0$, f has a continuation to $\{z \mid 0 < z < \rho,\ \arg(z) \in (-\frac{\pi}{2}, \frac{\pi}{2})\}$. But then applying the reflection principle to $\arg(z) = \pm\frac{\pi}{2}$, we get an analytic continuation to $\mathbb{D}_\rho(0) \setminus \{0\}$. By reflection, we have $\operatorname{Im} f(z) \to 0$ as $|z| \to 0$.

This continuation, call it f also, cannot have a finite-order pole at 0, since if it is real on $\arg(z) = 0, \frac{\pi}{2}$, it must be imaginary at some sector in between, and then $\operatorname{Im} f(re^{i\phi}) \to \infty$ or $-\infty$ as $r \to 0$, contradicting the assumption $\lim_{r\to 0} \operatorname{Im} f(re^{i\phi}) = 0$. By the Casorati–Weierstrass theorem, it cannot be an essential singularity and have this behavior of $\operatorname{Im} f$ as $|z| \to 0$. Thus, it has a removable singularity.

We have $f(\bar{z}) = \overline{f(z)}$ and $f(\overline{iz}) = \overline{f(iz)}$, which implies $f(-z) = f(z)$. So there can only be even nonzero Taylor coefficients. Reality of f on $\mathbb{R}$ proves that all a_n are real. □

By a different method (see Problem 4), one proves:

Theorem 5.5.8. *Let f be analytic in $\Omega = \{z \mid 0 < |z| < \rho, \arg(z) \in (0, \frac{\pi}{\alpha})$ for some $\alpha \in [\frac{1}{2}, \infty)$ and suppose $\operatorname{Im} f$ has a continuation to $\bar{\Omega}$ with $\operatorname{Im} f(z) = 0$ if $\arg(z)$ is 0 or $\frac{\pi}{\alpha}$. Then there is a function g analytic at $w = 0$ with real Taylor coefficients so that*

$$f(z) = g(z^\alpha) \tag{5.5.18}$$

Moreover, if $\operatorname{Im} f > 0$ in Ω, then $g'(0) \neq 0$.

Notes and Historical Remarks. The reflection principle was discovered by Schwarz in 1869–70 in the same series of papers that had the Schwarz lemma. It is often called the Schwarz reflection principle.

Painlevé's theorem, in the form we refer to involving arbitrary rectifiable curves, can be found in Beckenbach [**35**] or Rudin [**495**]. It is false for nonrectifiable curves; see below. There are several different theorems called "Painlevé's theorem"; see the Notes to Section 3.8 and to Section 12.6 of Part 2B.

The issue of Painlevé's theorem and its failure for certain nonrectifiable curves is intimately connected to the discussion of extensions of the Riemann removable singularity theorem to sets more general than points. For example, the same argument that proved Theorem 5.5.2 shows if $E \subset \Omega$, E is the image of a simple rectifiable curve and $\Omega \setminus E$ is connected, then f continuous in Ω and analytic in $\Omega \setminus E$ implies analyticity in E. This question is discussed further in the Notes to Section 8.8.

Some of the counterexamples related to the Painlevé theorem in the nonrectifiable case are in this $\Omega \setminus E$ context where E may not be an image of a curve or even connected. The original constructions go back to Pompeiu [**453**] and Denjoy [**136**] and were raised to high art in Carleson [**93, 94**]. In particular, it follows from Carleson's results that if γ is the Koch snowflake, a nonrectifiable curve constructed in Problem 9 of Section 2.2, then there is a function f continuous on $\mathbb{C}$ and analytic on $\mathbb{C} \setminus \gamma$, but analytic at no point of γ. This uses the fact that the Koch snowflake has dimension $\log 4/\log 3 > 1$.

There are, of course, nonrectifiable curves for which Painlevé's theorem holds. For example, if γ is a parametrization of $\{x+iy \mid -1 \le x \le 1,\, y = x\sin(\frac{1}{x})\}$ and $\Omega = \{x+iy \mid -1 < x, y < 1\}$, then γ is not rectifiable, but Painlevé's theorem holds since γ is rectifiable except at $x+iy=0$, and so a combination of Theorem 5.5.2 and the Riemann removable singularity result proves that Painlevé's theorem holds in this case.

For the special case where $\mathrm{Ran}(\gamma)$ is a subset of $\mathbb{R}$ (used for Theorem 5.5.1), the generalization of Painlevé's theorem to higher complex dimension is called *the edge of the wedge theorem*—it is discussed in Rudin [**495**], Jost [**293**], and Streater–Wightman [**546**]. The name comes from the form of the theorem. The one-variable theorem can be rephrased as taking I open in $\mathbb{R}$ so that $\Omega^\pm = \{x+iy \mid x \in I,\, \pm y \in (0,\varepsilon)\}$ with analytic functions, $f^\pm$, on $\Omega^\pm$ and equal boundary values. In higher dimensions, I is open in $\mathbb{R}^n$, V is an open convex cone in $\mathbb{R}^n$, and $\Omega^\pm = \{x+iy \mid x \in I,\, \pm y \in V\}$, which are wedges.

There are distributional versions of Painlevé's theorem (and the edge of the wedge theorem). $f^\pm(x \pm i\varepsilon)$ are only required to converge to equal distributions in the sense of distributional convergence. Problem 5 has an L^1 version of this.

Problems

1. Fill in the details of the proof of Theorem 5.5.4.

2. Prove Theorem 5.5.5. (*Hint*: Prove $|f(z)|$ is harmonic in any disk free of zeros of f since $|f(z)| = \mathrm{Re}(\log(f(z)))$.)

3. Prove a version of Theorem 5.5.5 when $\Omega, \widetilde{\Omega}$ are like the sets in Theorem 5.5.4.

4. Prove Theorem 5.5.8. (*Hint*: Apply Theorems 5.5.3 and 5.5.6 to $g(z) = f(z^{1/\alpha})$ with $\arg(z) \in (0,\pi)$, $0 < |z| < \rho^\alpha$.)

   Note. The next problem requires the reader to know about L^1, convolutions, and approximate identities (see Theorem 3.5.11 of Part 1). It also uses the Vitali theorem of Section 6.2.

5. Let $f_\pm$ be analytic in $\Omega^\pm = \{x+iy \mid x \in (a,b),\, \pm y \in (0,\delta)\}$ and suppose for a function $g \in L^1((a,b),dx)$, we have that
$$\lim_{\varepsilon \downarrow 0} \int_a^b |f(x \pm i\varepsilon) - g(x)|\, dx = 0 \tag{5.5.19}$$
   This problem will lead you through a proof that g is continuous and there is an analytic function F on $\widetilde{\Omega} = \{x+iy \mid x \in (a,b),\, y \in (-\delta,\delta)\}$ so that $F \restriction \Omega^\pm = f^\pm$.

(a) For $\tau \in (0, \frac{1}{2}(b-a))$, let $\Omega_\tau^\pm = \{x+iy \mid x \in (a+\tau, b-\tau),\ \pm y \in (0,\delta)\}$. Let h_τ be a C^∞ approximate identity on $\mathbb{R}$ with support $h_\tau \in (-\tau,\tau)$. Define $f_\tau^\pm$ on $\Omega_\tau^\pm$ by

$$f_\tau^\pm(x+iy) = \int h_\tau(x-w) f^\pm(w+iy)\, dw \qquad (5.5.20)$$

Prove that $f_\tau^\pm$ are analytic on $\Omega_\tau^\pm$ and have equal continuous boundary values so that Theorem 5.5.2 implies there is an analytic F_τ on $\widetilde{\Omega}_\tau$ with $F_\tau \restriction \Omega_\tau^\pm = f_\tau^\pm$.

(b) For any δ, show there exists a C^∞, q supported in $\mathbb{D}_\delta(0)$, so if f is any function analytic in a neighborhood of $\overline{\mathbb{D}_\delta(0)}$, then

$$f(z) = \int q(w) f(z-w)\, d^2w \qquad (5.5.21)$$

(*Hint*: For $\rho \le \delta$, $f(z) = \int_0^{2\pi} f(z - i\rho e^{i\theta}) \frac{d\theta}{2\pi}$. Average this in ρ with a smooth function.)

(c) Using (b), prove that for any $\tau_0 > 0$, there is a uniform bound on $\{F_\tau\}_{\tau \le \tau_0}$ on any compact subset of $\widetilde{\Omega}_{\tau_0}$.

(d) Prove that in $\Omega^\pm$, $f_\tau(z) \to f(z)$ pointwise, and use (c) and Vitali's theorem to prove that the required F exists.

5.6. Reflection in Analytic Arcs; Continuity at Analytic Corners

We've seen two cases where continuation across a boundary is automatic if boundary values are real—when the piece of boundary is in $\mathbb{R}$ and when it is in $\partial\mathbb{D}$. The two have in common that they are real analytic curves in a sense we make precise below. After proving that there is a reflection principle for all such curves, we prove a continuity at a corner where two analytic arcs come together, a result we'll need in Sections 8.2 and 8.3. Given a simply connected region, Ω, there is a biholomorphic bijection $f\colon \Omega \to \mathbb{D}$. Its existence is asserted by the Riemann mapping theorem discussed in Section 8.1. In this section, we'll refer to this map, f, and call it a Riemann map.

Definition. A *real analytic curve* (also called an *analytic arc*) is a curve $\gamma\colon [0,1] \to \mathbb{C}$ which is simple and real analytic on $[0,1]$ with $\gamma'(t) \ne 0$ for all $t \in [0,1]$.

By real analytic, we mean for each $t_0 \in [0,1]$, there is a convergent Taylor series at t_0 with $\gamma(t) = \sum_{n=0}^\infty a_n(t_0)(t-t_0)^n$ for all t near t_0 and in $[0,1]$. Here's the general reflection principle:

Theorem 5.6.1. *Let Ω be a region and $\{\gamma(t)\}_{0\le t\le 1}$ a real analytic curve which is an edge of $\partial\Omega$ in the sense that for every $z_0=\gamma(t_0)$, $t_0\in(0,1)$, there is a disk $\mathbb{D}_\delta(z_0)$ so that $\mathrm{Ran}(\gamma)\cap\mathbb{D}_\delta(z)$ divides $\mathbb{D}_\delta(z_0)$ in two: $\mathbb{D}^+$ and $\mathbb{D}^-$, so that $\mathbb{D}_\delta(z_0)\cap\Omega=\mathbb{D}^+$ (i.e., $(\mathrm{Ran}(\gamma)\cap\mathbb{D}_\delta(z_0))\cup\mathbb{D}^-\subset\mathbb{C}\setminus\Omega$). Suppose that $f\in\mathfrak{A}(\Omega)$ and for every $z_n\in\Omega$, $z_n\to z_0=\gamma(t_0)$ with $t_0\in(0,1)$, we have $\mathrm{Im}\,f(z_n)\to 0$. Then f can be analytically continued from Ω to a neighborhood of $\gamma[(0,1)]$.*

Proof. This is a local result. We need only prove analyticity in a neighborhood of any $z_0=\gamma(t_0)$. Pick δ_0 so that the power series representation for γ converges in $\mathbb{D}_{\delta_0}(t_0)$, and let $\{\Gamma(z)\mid z\in\mathbb{D}_{\delta_0}(t_0)\}$ be the value of the power series. By shrinking δ_0, we can suppose $\Gamma[\mathbb{D}_\delta(t_0)]$ lies in the disk guaranteed by the conditions of the theorem and that Γ is one-one on $\mathbb{D}_\delta(t_0)$ (since $\gamma'(t_0)\neq 0$).

Then $\Gamma[(t_0-\delta,t_0+\delta)]=\gamma[(t_0-\delta,t_0+\delta)]\subset\mathrm{Ran}(\gamma)$. So $\Omega\cap\mathrm{Ran}(\Gamma)$ is either $\Gamma[\mathbb{C}_+\cap\mathbb{D}_\delta(z_0)]$ or $\Gamma[\mathbb{C}_-\cap\mathbb{D}_\delta(z_0)]$. For simplicity, suppose the former.

Define g on $\mathbb{C}_+\cap\mathbb{D}_\delta(z_0)$ by $g=f\circ\Gamma$. Then $\mathrm{Im}\,g(z_n)\to 0$ if $z_n\to t_\infty\in(t_0-\delta,t_0+\delta)$, so by the strong reflection principle (Theorem 5.5.3), g has an analytic continuation, $\tilde g$, through $\mathbb{D}_\delta(z_0)$. Thus, $\tilde f=\tilde g\circ\Gamma^{-1}$ extends f through $\Gamma[\mathbb{D}_\delta(z_0)]$. □

Next, we want to note two further aspects of the situation. First, instead of $\mathrm{Im}\,f(z_n)\to 0$, we could have used $|f(z_n)|\to 1$ and Theorem 5.5.4. Second, as with Theorem 5.5.6, if $|f(z)|<1$ for $z\in\mathbb{D}^+$ and $|f(z_n)|\to 1$ as $z_n\to\mathrm{Ran}(\gamma)\cap\mathbb{D}_\delta(z_0)$, then $f'(z_0)\neq 0$. Indeed, the sign of $if(\gamma(t))\frac{d}{dt}f(\gamma(t))$ is determined by the placement of $\mathrm{Ran}(\gamma)$ relative to $\mathbb{D}^+$ (if, as t increases, $\mathbb{D}^+$ is to the left of $\gamma(t)$, then this derivative is positive).

As a final result on reflections, we want to prove a continuity result at corners under special circumstances:

Theorem 5.6.2 (Continuity at analytic corners). *Let Ω be a region, $z_0\in\partial\Omega$, so that there are two curves starting at z_0, γ and $\tilde\gamma$ both real analytic at $t=0$ with $\gamma'(0)\neq 0\neq\tilde\gamma'(0)$ and with $\mathbb{D}_\delta(z_0)\cap\partial\Omega=(\mathrm{Ran}(\gamma)\cup\mathrm{Ran}(\tilde\gamma))\cap\mathbb{D}_\delta(z_0)$ for suitable small δ* (see Figure 5.6.1). *Let f be analytic in Ω. Suppose, for some $\delta>0$, $f\colon\Omega\cap\mathbb{D}_\delta(z_0)$ to $\mathbb{D}$ and is one-one there, and that*

$$\lim_{\substack{z_n\to z_\infty\\ z_n\in\mathbb{D}_\delta(z_0)\cap\Omega\\ z_\infty\in\partial\Omega}}|f(z)|=1 \tag{5.6.1}$$

For $z\in(\gamma[(0,1)]\cup\tilde\gamma[(0,1)])\cap\mathbb{D}_\delta(z_0)\equiv\Gamma$, define f to be limits of $f(z)$ guaranteed by Theorem 5.6.1. *Then*

(a) *f is one-one on Γ.*

(b) *$f(z_0)$ can be defined so that $f\restriction\Gamma\cup\{z_0\}$ is continuous.*

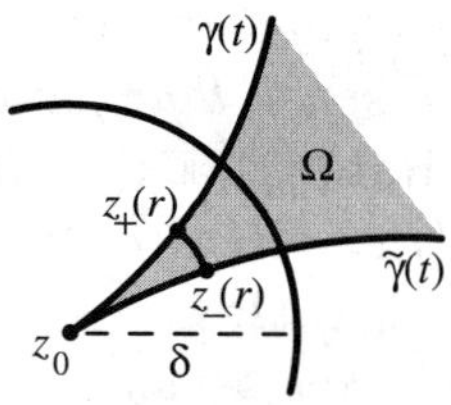

Figure 5.6.1. An analytic corner.

Remark. We emphasize that this theorem is local—it is not required that all of $\partial\Omega$ be analytic curves with C^1 corners. That said, if $\partial\Omega$ has this property, a Riemann map of Ω to $\mathbb{D}$ extends to a homeomorphism of $\overline{\Omega}$ and $\overline{\mathbb{D}}$.

Proof. By Theorem 5.6.1, we have continuous extensions of f to Γ, so for t small, $f(\gamma(t)) = e^{i\theta(t)}$ and $f(\tilde\gamma(t)) = e^{i\widetilde{\theta}(t)}$. By the remark about the sign of the derivative, either $\theta' > 0$ and $\widetilde{\theta}' < 0$ or vice-versa. Let us suppose the former.

Since $f'(z_1) \neq 0$ for $z_1 \in \Gamma$, f is a bijection near z_1, and since $|f(z)| > 1$ for $z \notin \Omega$ near z_1, we see f takes all values in $\mathbb{D}$ near $f(z_1)$ for z's near z_1. Since f is supposed one-one on Ω near z_0, we conclude that for t, s small, it cannot be that $f(\gamma(t)) = f(\tilde\gamma(s))$, that is, (a) holds.

By the monotonicity, $\lim_{t\downarrow 0} f(\gamma(t)) \equiv e^{i\theta(0)}$ and $\lim_{s\downarrow 0} f(\tilde\gamma(s)) = e^{i\tilde\theta(0)}$ exist, and by the monotonicity and one-one property, $\widetilde{\theta}(0) \le \theta(0)$ (we may need to adjust θ or $\widetilde{\theta}$ by 2π). The continuity result (b) that we need to prove is that

$$\tilde\theta(0) = \theta(0) \tag{5.6.2}$$

So suppose (5.6.2) is false.

By the geometric assumptions, for r small, $\partial\mathbb{D}_r(z_0)\cap\Omega$ is an arc running from $z_-(r) \in \operatorname{Ran}(\tilde\gamma)$ to $z_+(r) \in \operatorname{Ran}(\gamma)$ (see Figure 5.6.1). Write $z_-(r) = z_0 + re^{i\psi_-(r)}$, $z_+(r) = z_0 + re^{i\psi_+(r)}$ and define the arc $\eta_r\colon [\psi_-(r), \psi_+(r)] \to \Omega$ by

$$\eta_r(\psi) = z_0 + re^{i\psi} \tag{5.6.3}$$

By definition of $\theta, \widetilde{\theta}$, we have

$$f(\eta_r(\psi_-(r)) = e^{i\tilde\theta(t_-(r))}, \qquad f(\eta_r(\psi_+(r)) = e^{i\theta(t^+(r))} \tag{5.6.4}$$

for suitable $t^{\pm}(r)$.

By the monotonicity of $\theta, \widetilde{\theta}$, for r small,

$$|f(\eta_r(\psi_+(r))) - f(\eta_r(\psi_-(r)))| \ge |e^{i\tilde\theta(0)} - e^{i\theta(0)}| \equiv \beta > 0 \tag{5.6.5}$$

so, by (2.2.20), the length of the image of $\eta_r([\psi_-(r),\psi_+(r)])$ under f obeys

$$\int_{\psi_-(r)}^{\psi_+(r)} |f'(z_0+re^{i\psi})|r\,d\psi \ge \beta \tag{5.6.6}$$

By the Cauchy–Schwarz inequality,

$$\biggl(\int_{\psi_-(r)}^{\psi_+(r)} |f'(z_0+re^{i\psi})|^2\,d\psi\biggr)\biggl(\int_{\psi_-}^{\psi_+} d\psi\biggr) \ge \biggl(\int_{\psi_-(r)}^{\psi_+(r)} |f'(z_0+re^{i\psi})|\,d\psi\biggr)^2 \tag{5.6.7}$$

so, by (5.6.6),

$$\int_{\psi_-(r)}^{\psi_+(r)} |f'(z_0+re^{i\psi})|^2\,d\psi \ge \frac{1}{2\pi}\,\frac{\beta^2}{r^2} \tag{5.6.8}$$

On the other hand, by the conformality of f as expressed in (2.2.22),

$$\int_0^R \int_{\psi_-(r)}^{\psi_+(r)} |f'(z_0+re^{i\psi})|^2 r\,dr \le \pi R^2 \tag{5.6.9}$$

(5.6.9) and (5.6.8) imply

$$\beta^2 \int_0^R \frac{dr}{r} < \infty \tag{5.6.10}$$

This is a contradiction with $\beta \neq 0$, that is, (5.6.2) must hold. $\square$

If $\Omega = \widetilde{\Omega}\setminus \operatorname{Ran}(\gamma)$, where $\gamma\colon [0,1]$ is a Jordan curve from $z_0 \in \partial\Omega$ to $z_1 \in \widetilde{\Omega}$ with $\gamma(s)\in\widetilde{\Omega}$ for $s\in(0,1]$, we say Ω is $\widetilde{\Omega}$ with a slit removed. If γ is real analytic, we can apply the reflection principle on either side and see that if $f\colon\Omega\to\mathbb{D}$ is a Riemann map, f^{-1} can be continuously extended onto part of $\partial\mathbb{D}$, so that f^{-1} maps to each side of $\operatorname{Ran}(\gamma)$. If γ is real analytic at $s=1$ also, f^{-1} can be continued to part of $\partial\mathbb{D}$, so that a single arc gets mapped two-one onto $\gamma[(1-\varepsilon,1)]$ and is continuous at the tip. To see things locally,

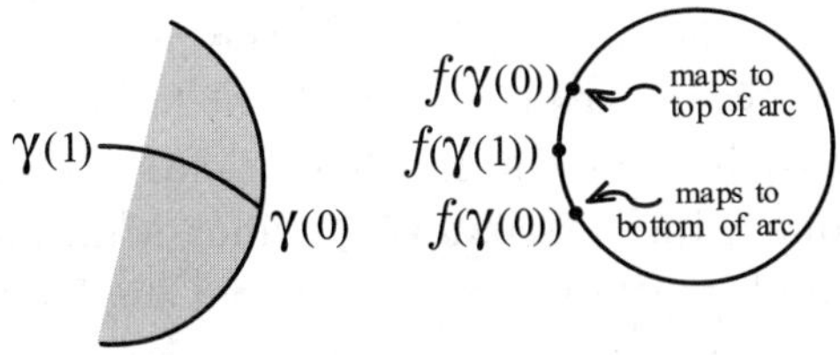

Figure 5.6.2. Boundary behavior of a Riemann map to a region with analytic slit.

map $\Omega_1 \equiv \mathbb{D}_\varepsilon(\gamma(1)) \setminus \text{Ran}(\gamma)$ to a "half-disk" by taking $G\colon z \to (z-\gamma(1))^{1/2}$. Under the map, Ω_1 has a boundary which is two analytic curves, real analytic at their ends coming together at angle π (see Figure 5.6.2). A Riemann map, f, of Ω defines a map of Ω_1, $w \mapsto f(\gamma(1)+w^2)$ to which Theorem 5.6.2 can be applied. We summarize in:

Theorem 5.6.3 (Continuity of Analytic Slit Maps). *Let $\Omega = \widetilde{\Omega} \setminus \text{Ran}(\gamma)$ be a slit domain whose slit is piecewise analytic with real analytic corners and real analytic ends. Let $f\colon \Omega \to \mathbb{D}$ be a map obeying* (5.6.1) *area points in* Ran(γ). *Then f^{-1} can be continued to an arc $(\alpha, \beta) \in \partial\mathbb{D}$, so that for $\kappa \in (\alpha, \beta)$, $f^{-1}(\kappa)$ is the tip of the slit and f^{-1} maps (α, κ) onto* $\gamma[(0,1)]$ *and also (κ, β). f is continuous as one approaches* Ran(γ) *from one side or the other as one approaches the tip.*

Notes and Historical Remarks. The lovely geometric argument used in the proof of Theorem 5.6.2 is due to Carathéodory [**88**] in his proof of continuity of Riemann mappings up to the boundary for interiors of Jordan curves. Earlier, Schwarz in his work on the reflection principle proved continuity up to the boundary for Jordan curves which are a finite union of analytic arcs coming together at nonzero angles. In our application in Sections 8.2 and 8.3, we'll need the result at cusps, that is, points where the angle of intersection is 0.

Constantin Carathéodory (1873–1950) was the son of a Greek diplomat (his father was of Greek descent but served the Ottoman Empire which controlled much of Greece at the time). Carathéodory was born in Berlin, spent much of his youth in Brussels, and was a student in Berlin and Göttingen where he got his degree from Minkowski in 1904. He married his aunt, who was eleven years younger than him, in 1909. He was at a remarkable number of different universities, in part because of the economic dislocations from the First World War and because of a period in the 1920s when he helped set up mathematics in several Greek universities. Most notably, he was Klein's successor at Göttingen and a professor in Munich from 1924 until his retirement in 1938, and again after the Second World War when he helped reestablish German mathematics.

Carathéodory was a key player in both the proof of the Riemann mapping theorem and, as we've seen, the study of boundary behavior of such maps. In this regard, he was the mathematician who realized the power of the Schwarz lemma which he named. Under Minkowski's influence, his early work was in the theory of convex sets and his focus on Carathéodory functions came in that context. He also did fundamental work in the foundations of the calculus of variations, in thermodynamics, and, as we've seen in Part 1 (Section 8.1), in measure theory.

5.7. Calculation of Definite Integrals

Cauchy first caught the imagination of the mathematical world with his ability to compute some definite integrals where an explicit antiderivative is not available, and this use of complex variables is still central to many applied areas. Since it it not central to the theoretical concerns that dominate this book, we'll not spend as long as some texts, but we'd be remiss if we didn't say something. The choice of contour is an art as much as a science, so we'll mainly study by example, which we'll break into six parts. All use the residue theorem, Theorem 4.3.1.

5.7.1. Periodic Functions Over a Period. These are the simplest since they essentially come as an integral over a closed contour.

Example 5.7.1. For $a > |b|$ both real, compute

$$\int_0^{2\pi} \frac{1}{a + b\cos\theta}\,\frac{d\theta}{2\pi} \tag{5.7.1}$$

Let $z = e^{i\theta}$ so $dz = iz\,d\theta$ and the integral in (5.7.1) becomes (with the usual counterclockwise contour on $\partial\mathbb{D}$)

$$\frac{1}{2\pi i}\oint \frac{1}{a + \frac{1}{2}b(z + z^{-1})}\,\frac{dz}{z} = \frac{1}{2\pi i}\oint \frac{2}{bz^2 + 2az + b}\,dz$$

The poles are at $(\pm\sqrt{a^2 - b^2} - a)/b = z_\pm$. Both are in $(-\infty, 0)$, one inside the circle, one outside. The polynomial is thus $b(z - z_+)(z - z_-)$ and the residue is $1/b(z_+ - z_-) = 1/2(\sqrt{a^2 - b^2})$, that is,

$$\int_0^{2\pi} \frac{1}{a + b\cos\theta}\,\frac{d\theta}{2\pi} = \frac{1}{\sqrt{a^2 - b^2}} \tag{5.7.2}$$

Taking $a = 1$ and expanding both sides as a power series in b (using the binomial theorem for $(1 - b^2)^{-1/2}$) one gets (one can also get this from the binomial theorem and go backwards to (5.7.2))

$$\int_0^{2\pi} \cos^{2n}(\theta)\,\frac{d\theta}{2\pi} = \frac{(2n)!}{2^{2n}(n!)^2} \tag{5.7.3}$$

□

5.7.2. Integrals of Rational Functions. Most of the remaining examples are $\int_{-\infty}^{\infty}$ or $\int_0^{\infty}$ and so are "improper," that is, over infinite intervals. There are three possible meanings of such an integral:

(a) $\int_0^\infty |f(x)|\,dx < \infty$ or $\int_{-\infty}^\infty |f(x)|\,dx < \infty$. Such integrals are said to be *absolutely convergent*.

(b) $\lim_{R\to\infty} \int_0^R f(x)\,dx$ exists but $\int_0^\infty |f(x)|\,dx = \infty$. Such integrals are said to be *conditionally convergent*.

(c) $\lim_{R\to\infty} \int_{-R}^{R} f(x)\,dx$ exists but $\lim_{R\to\infty} \int_0^R f(x)\,dx$ does not. Such integrals are called *principal values at infinity* (written pv).

Examples of the three types are $\int_{-\infty}^{\infty} \frac{dx}{1+x^2}$ (see Example 5.7.3 below), $\int_{-\infty}^{\infty} \frac{\sin x}{x}\,dx$ (see Example 5.7.5 below), and $\text{pv} \int_{-\infty}^{\infty} \frac{x\,dx}{x^2+1} = 0$. While on the subject of principal values, if $\int_{|x|>\varepsilon,\, x\in(a,b)} |f(x)|\,dx$ is finite for all $\varepsilon > 0$ but $\int_0^{\varepsilon} |f(x)|\,dx = \int_{-\varepsilon}^{0} |f(x)|\,dx \;(=\infty)$, we define for $a < 0 < b$,

$$\text{pv} \int_a^b f(x) \equiv \lim_{\varepsilon\downarrow 0} \left[\int_a^{-\varepsilon} f(x)\,dx + \int_{\varepsilon}^{b} f(x)\,dx \right] \tag{5.7.4}$$

if the limit exists. If the singular point is at x_0 rather than 0, we make a similar definition. If γ is a simple contour, C^1, near $\gamma(t_0)$, we can similarly define a principal value as integrating without $(t_0 - \varepsilon, t_0 + \varepsilon)$. Here is the key fact:

Theorem 5.7.2. *Let γ be a simple closed contour, f meromorphic in a neighborhood of* $\text{ins}(\gamma)$ *so that f has only simple poles on* $\text{Ran}(\gamma)$ *and with γ C^1 near these poles. If γ is oriented so that $n(\gamma, z) = 1$ on* $\text{ins}(\gamma)^{\text{int}}$, *then*

$$\frac{1}{2\pi i}\,\text{pv} \oint f(x)\,dz = \sum_{z_0\in \text{ins}(\gamma)^{\text{int}}} \text{Res}(f; z_0) + \tfrac{1}{2} \sum_{z_0\in\text{Ran}(\gamma)} \text{Res}(f; z_0) \tag{5.7.5}$$

Proof. Letting $\tilde{\gamma}_r$ be the contour

$$\tilde{\gamma}_r(t) = re^{\pi i t}, \qquad 0 \le t \le 1 \tag{5.7.6}$$

the standard calculation (see Example 2.2.2) shows the integral

$$\int_{\tilde{\gamma}_r} z^{-1}\,dz = \pi i \tag{5.7.7}$$

Thus, if $\gamma^{(\varepsilon)}$ is the contour γ with symmetric gaps of size 2ε about each point on γ and if $\gamma^{(\varepsilon)\sharp}$ is $\gamma^{(\varepsilon)}$ with semicircles running outside (see Figure 5.7.1)

$$\int_{\gamma^{(\varepsilon)}} f(z)\,dz = \int_{\gamma^{(\varepsilon)\sharp}} f(z)\,dz - \pi i \sum_{z_0\in\text{Ran}(\gamma)} \text{Res}(f; z_0) + O(\varepsilon) \tag{5.7.8}$$

From the residue theorem, we get (5.7.5). ☐

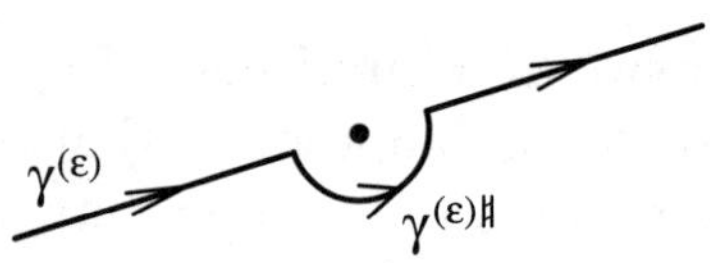

Figure 5.7.1. Resolving a principal value.

If P/Q is a rational function with no poles on $\mathbb{R}$, $\int_{-\infty}^{\infty} (P(x)/Q(x))\,dx$ is absolutely convergent if $\deg(Q) \ge 2 + \deg(P)$. If $\deg(Q) = 1 + \deg(P)$

so $P(x)/Q(x) = c/x + O(1/x^2)$ at infinity, then only the principal value at infinity exists and is easy to accommodate. Thus, we will limit ourselves here to $\deg(Q) \geq 2 + \deg(P)$. Using Theorem 5.7.2, it is easy to accommodate simple poles on $\mathbb{R}$.

Example 5.7.3. Compute for $a > 0$,

$$\int_{-\infty}^{\infty} \frac{dx}{x^2 + a^2} \tag{5.7.9}$$

This is the limit as $R \to \infty$ of $\int_{-R}^{R}$. We now "close the contour," that is, add and subtract the integral counterclockwise along $|z| = R$ from R to $-R$. If $R > a$, the amount subtracted is bounded by $\pi R(R^2 - a^2)^{-1}$, since $\sup_{|z|=R}|1/(z^2 + a^2)| \leq 1/(R^2 - a^2)$. This goes to zero as $R \to \infty$. Thus, if C_R is the closed contour (see Figure 5.7.2), the integral in (5.7.9) is the limit as $R \to \infty$ of the integral over the closed contour (which we'll see is R-independent if $R > a$). Inside the contour, $z^2 + a^2$ has one pole at $z = ia$ and the residue is $1/2ia$. Thus,

$$\int_{-\infty}^{\infty} \frac{dx}{x^2 + a^2} = \frac{\pi}{a} \tag{5.7.10}$$

The reader should do the computation if the contour is closed in the lower half-plane.

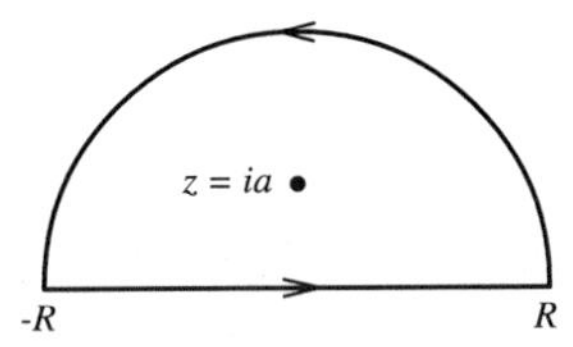

Figure 5.7.2. Contour for Example 5.7.3.

Since $(x^2 + a^2)^{-1} = a^{-1}\frac{d}{dx}\arctan(\frac{x}{a})$, this can also be computed by standard real variable methods. Indeed, by the method of partial fractions, every P/Q which is real on $\mathbb{R}$ with all poles in $\mathbb{C}\setminus\mathbb{R}$ can be reduced to sums of the form (5.7.9) and powers of that integrand. $\square$

5.7.3. Trigonometric Times Rational and Exponential Functions.

This class illustrates the power of complex variable methods since the integrand in (5.7.11) is not the derivative of any elementary function.

Example 5.7.4. Compute for $a > 0$ and b real with $b \neq 0$,

$$\int_{-\infty}^{\infty} \frac{\cos bx}{x^2 + a^2}\, dx \tag{5.7.11}$$

Without loss, suppose $b > 0$. Since $\mathrm{Re}(e^{ibx}) = \cos bx$, we need only compute the integral with e^{ibx} replacing $\cos bx$. We can close the contour in the upper but not lower half-plane (since $|e^{ib(x+iy)}| = e^{-by}$). Computing the residue at $x = ia$ yields

$$\int_{-\infty}^{\infty} \frac{\cos bx}{x^2 + a^2} = \frac{\pi e^{-|b|a}}{a} \tag{5.7.12}$$

□

Example 5.7.5. Compute for $a > 0$,

$$\int_{-\infty}^{\infty} \frac{\sin ax}{x}\, dx \tag{5.7.13}$$

This has four subtleties compared to the last example. First, it is not absolutely convergent, but by general principles (see Problem 3), it is conditionally convergent.

Second, we cannot close the contour naively in either half-plane because the function $\sin az/z$ grows badly in each half-plane. The obvious solution is to write $\sin az/z = (e^{iaz} - e^{-iaz})/2iz$, and for each term close in different half-planes. The problem is that while $\sin az/z$ is regular at $z = 0$, $e^{\pm iaz}/z$ are not. Here, what we do is replace $\int_{-\infty}^{\infty}$ by $\int_{\varepsilon}^{\infty} + \int_{-\infty}^{-\varepsilon} + \int_{C_\varepsilon}$ where C_ε lies in the lower half-plane as the semicircle from $-\varepsilon$ to ε. Since $\sin az/z$ is bounded on that circle uniformly in ε and the contour is $O(\varepsilon)$, we are sure to recover the given integral as $\varepsilon \downarrow 0$ (and, indeed, by the CIT, this sum is ε-independent!).

Third, we cannot close trivially with a large semicircle (although with extra work, one can (see Problem 4(b)). Instead, we close the contour for $e^{iaz}/2iz$ by going from $-R$ to R along $\mathbb{R}$ (with the ε-change above), go upwards to $R + i\sqrt{R}$, horizontally and backward to $-R + i\sqrt{R}$, and then down (see Figure 5.7.3). A simple estimate (Problem 4(a)) shows that as $R \to \infty$, this extra contour contributes zero.

There results two integrals, one involving $e^{iaz}/2iz$ in the upper half-plane and the other $e^{-iaz}/2iz$ in the lower half-plane. The second encloses no poles (with our choice of deformation near zero) and the other only a

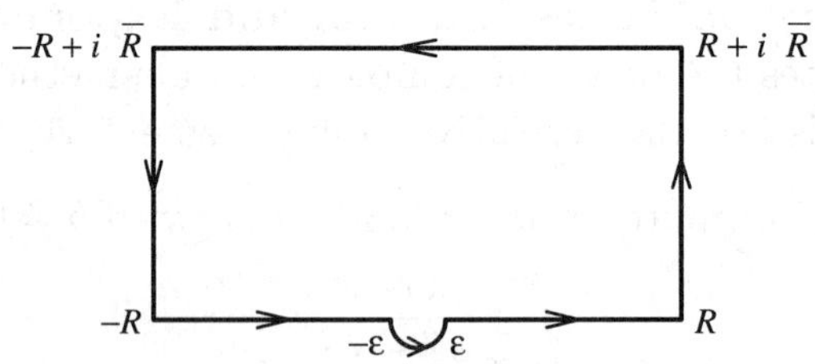

Figure 5.7.3. Contour for Example 5.7.5.

pole at zero. The result is $2\pi i/2i$, so

$$\int_{-\infty}^{\infty} \frac{\sin ax}{x}\, dx = \pi \tag{5.7.14}$$

The last subtlety concerns the fact that while (5.7.14) is independent of a, for $a = 0$, the integral is zero! This lack of continuity was discovered by Cauchy and concerned him. It shows the difficulty of interchanging limits and conditionally convergent integrals (or sums). □

Example 5.7.6. For $0 < a < 1$, compute

$$\int_{-\infty}^{\infty} \frac{e^{ax}}{1+e^x}\, dx \tag{5.7.15}$$

Exponentials and trigonometric functions are brothers. Since for $x > 0$, this function vanishes as $e^{-(1-a)x}$, and for $x < 0$ as $e^{-a|x|}$, the integral is absolutely convergent. Illustrating that contour integrals are an art, we close this contour by letting

$$F(z) = \frac{e^{az}}{1+e^z} \tag{5.7.16}$$

and note that

$$F(z+2\pi i) = e^{2\pi i a} F(z) \tag{5.7.17}$$

Thus, we take the rectangle in Figure 5.7.4 with corners at $\pm R$ and $\pm R+2\pi i$. The contributions of the vertical edges go to zero as $R \to \infty$. Thus, the contour integral as $R \to \infty$ goes to

$$(1 - e^{2\pi i a}) \int_{-\infty}^{\infty} \frac{e^{ax}}{1+e^x}\, dx \tag{5.7.18}$$

On the other hand, there is one pole within the contour at $z = i\pi$ where the residue is

$$e^{i\pi a} \left[\frac{d}{dz}(1+e^z) \Big|_{z=i\pi} \right]^{-1} = -e^{i\pi a}$$

Thus,

$$(5.7.18) = -2\pi i\, e^{i\pi a} \tag{5.7.19}$$

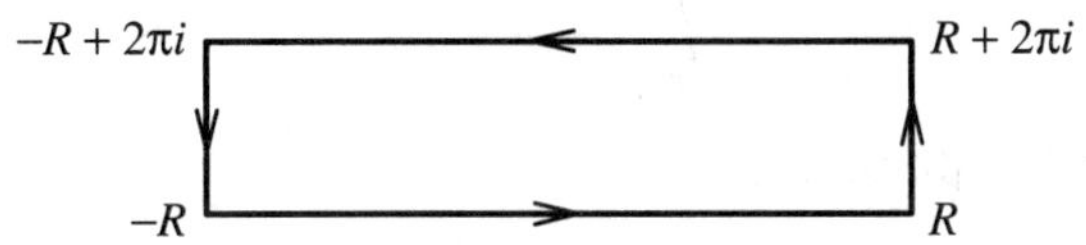

Figure 5.7.4. Contour for Example 5.7.6.

Since $-2\pi i e^{i\pi a}/(1-e^{2\pi i a}) = \pi/\sin(\pi a)$, we see that (see also (5.7.30) and Problem 10)

$$\int_{-\infty}^{\infty} \frac{e^{ax}}{1+e^x}\, dx = \frac{\pi}{\sin(\pi a)} \tag{5.7.20}$$

□

5.7.4. Examples with Branch Cuts.

Example 5.7.7. Compute for $a > 0$,

$$\int_0^{\infty} \frac{\log(x)}{x^2+a^2}\, dx \tag{5.7.21}$$

For $-\pi/2 < \arg(z) < 3\pi/2$, define

$$F(z) = \frac{\log(z)}{z^2+a^2} \tag{5.7.22}$$

and integrate $F(z)$ over the contour from $-R$ to R along $\mathbb{R}$ and then along the semicircle in the upper half-plane. The contribution of the semicircle is $O\big((R\log(R))/R^2\big)$ as $R \to \infty$, and so goes to zero.

Since $\log(-x) = \log(x) + i\pi$ for this branch of $\log(z)$, the integral over the real axis piece as $R \to \infty$ goes to

$$2\int_0^{\infty} \frac{\log(x)}{x^2+a^2} + i\pi \int_0^{\infty} \frac{dx}{x^2+a^2} \tag{5.7.23}$$

On the other hand, for large R, the contour has one pole at $z = ia$ with residue $\log(ia)/2ia = (\log(a) + i\pi/2)/2ia$. Thus, if I is the integral in (5.7.21), we get

$$2I + i\pi \int_0^{\infty} \frac{dx}{x^2+a^2} = \frac{\pi}{a}\left[\log(a) + \frac{i\pi}{2}\right] \tag{5.7.24}$$

We could just take real parts (but, in fact, the imaginary parts are equal by (5.7.10)). Thus,

$$\int_0^{\infty} \frac{\log(x)\, dx}{x^2+a^2} = \frac{\pi}{2a}\log(a) \tag{5.7.25}$$

In fact, this can be computed by conformal invariance alone (see Problem 6).

□

Example 5.7.8. This integral is also evaluated in Problem 10. For $0 < a < 1$, compute

$$\int_0^{\infty} \frac{x^{a-1}}{1+x}\, dx \tag{5.7.26}$$

We define on $\mathbb{C} \setminus [0, \infty)$,

$$f(z) = \frac{(-z)^{a-1}}{1+z} \tag{5.7.27}$$

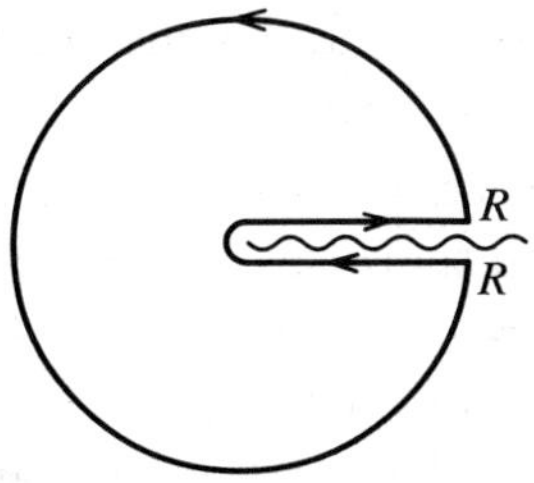

Figure 5.7.5. Contour for Example 5.7.8.

where we take the branch of $(-z)^{a-1}$ which is positive for $z < 0$, that is, for $z = re^{i\theta}$, $0 < \theta < 2\pi$,

$$(1+z)f(z) = r^{a-1}e^{i(\theta-\pi)(a-1)} \tag{5.7.28}$$

Thus, for $x > 0$,

$$f(x \pm i0) = \frac{e^{\pm i\pi(1-a)}}{1+x}\, x^{a-1} \tag{5.7.29}$$

Consider the contour that goes above the cut from 0 to R, circles to around just below R, and back to 0 (see Figure 5.7.5). The contribution of the circle to $\oint f(z)\,dz$ is bounded by $R^{a-1}(2\pi R)/(R-1)$ goes to zero as $R \to \infty$. Thus, as $R \to \infty$, the integral is equal to $[e^{i\pi(1-a)} - e^{-i\pi(1-a)}]$ (integral in (5.7.26)).

f has one pole at $z = -1$ and the residue is 1. Thus, since $e^{i\pi} = -1$, the integral is $(2\pi i)/(e^{i\pi a} - e^{-i\pi a}) = \pi/\sin(\pi a)$ and

$$\int_0^\infty \frac{x^{a-1}}{1+x}\, dx = \frac{\pi}{\sin(\pi a)} \tag{5.7.30}$$

□

The reader may have noticed that the right sides of (5.7.30) and (5.7.20) agree. This is no coincidence—the change of variable $x = e^y$ in (5.7.30) turns one integral into the other! (See also Problem 10.)

Example 5.7.9. In Section 5.7.2, we saw how to evaluate $\int_{-\infty}^\infty \frac{P(x)}{Q(x)} dx$ if

$$\deg Q \geq \deg P + 2 \tag{5.7.31}$$

where $Q(x)$ has no zeros on $\mathbb{R}$. One need only close the contour in the upper half-plane. Here, we'll consider $\int_0^\infty \frac{P(x)}{Q(x)}\, dx$ if Q has no zeros on $[0,\infty)$ and (5.7.31) holds. A simple example (which can be computed by other means!) is

$$\int_0^\infty \frac{dx}{(x+1)^2} \tag{5.7.32}$$

Let $f(z) = \log(-z)/(z+1)^2$ and consider the integral of f along the contour shown in Figure 5.7.5. Since $|f(z)| = o(\frac{1}{|z|})$ at infinity, the circle makes a vanishing contribution as $R \to \infty$. Since $\log(-x+i0) - \log(-x - i0) = -2\pi i$, the combined integral along the cut is $-2\pi i I$, where I is the integral in (5.7.32). Near $z=-1$, $\log(-z) = -(z+1) + O((z+1)^2)$, so the residue of f at -1 is -1. Therefore, $-2\pi i I = -2\pi i$, that is, $I = 1$ (which can be obtained from the antiderivatives). Clearly, the same method works for any $\int_0^\infty \frac{P(x)}{Q(x)}\,dx$ of the type discussed (see Problem 11). □

Example 5.7.10. Compute

$$\int_{-1}^{1} \frac{dx}{\sqrt{1-x^2}} \tag{5.7.33}$$

Of course, this can be done via trigonometric substitution $x = \sin\theta$ to get $\int_{-\pi/2}^{\pi/2} d\theta = \pi$. But we can do it and more complicated examples (see Problem 8) using contour integration. Put a branch cut for $(1-z^2)^{-1/2}$ from -1 to 1 and take the branch which is positive for $z = x+i0$, $-1<x<1$, so the branch which is positive for $z = iy$, $y>0$. The integrand at $z = x - i0$ is $-(1-x^2)^{-1/2}$, so a contour around the top returning on the bottom (which is clockwise) is $-2\times$ (5.7.33). This can be deformed to a contour around a large circle, so large that the Laurent series at infinity converges.

Near $z = \infty$,

$$f(z) = \frac{i}{\sqrt{z^2-1}} = \frac{i}{z}\left(1 - \frac{1}{z^2}\right)^{-1/2} \tag{5.7.34}$$

where we pick $\sqrt{-1} = i$ rather than $-i$ to have $f(iy) > 0$ for $y > 0$. Since $\oint \frac{dz}{z^\ell} = 0$ for $\ell > 1$, we see

$$(5.7.33) = -\frac{1}{2}(2\pi i)(i) = \pi \tag{5.7.35}$$

□

5.7.5. Gaussian and Related Integrals.

Example 5.7.11 (Gaussian Integral). Evaluate the Gaussian integral

$$\int_{-\infty}^{\infty} e^{-x^2/2}\,dx \tag{5.7.36}$$

using residue calculus.

Remark. The simplest method of evaluating this integral uses polar coordinates; see Proposition 4.11.10 of Part 1 or Theorem 9.6.6. Part 1 also had a third proof using the Fourier inversion formula; see Problem 11 of Section 6.2. We will also find a method using Euler's gamma function in Section 9.6 (see especially the remark following Corollary 9.6.5). Problems 26 and 28 sketch two other proofs.

This is a tricky example since there are no poles and no obvious closed contours to use.

The arithmetic will be easier for e^{-x^2} rather than $e^{-x^2/2}$; we'll then scale to get (5.7.36). A special role will be played by the complex number

$$\beta = \sqrt{\pi}\, e^{i\pi/4} \tag{5.7.37}$$

which obeys

$$\beta^2 = \pi i \tag{5.7.38}$$

Thus,

$$e^{-(z+\beta)^2} = e^{-z^2-\beta^2-2\beta z} = -e^{-z^2} e^{-2\beta z} \tag{5.7.39}$$

The minus sign is promising, but $e^{-2\beta z}$ looks problematic. However, since $e^{-2\beta^2} = e^{-2\pi i} = 1$, $h(z) = e^{-2\beta z}$ obeys $h(z+\beta) = h(z)$.

This leads us to define

$$f(z) = \frac{e^{-z^2}}{1+e^{-2\beta z}} \tag{5.7.40}$$

As noted, the denominator is invariant under $z \to z+\beta$, so by (5.7.39),

$$f(z) - f(z+\beta) = e^{-z^2} \tag{5.7.41}$$

We therefore consider the contour, Γ_R, which is a parallelogram with corners, $-R, R, R+\beta, -R+\beta$ (Figure 5.7.6). By (5.7.41),

$$\int_{-R}^{R} e^{-z^2}\, dz = \int_{\text{bottom–top of } \Gamma_R} f(z)\, dz \tag{5.7.42}$$

It is an easy exercise (Problem 12) to see the contribution of the sides goes to zero as $R \to \infty$, so by the residue theorem,

$$\int_{-\infty}^{\infty} e^{-z^2}\, dz = 2\pi i \sum_{\substack{\text{poles} \\ z_j \text{ of } f(z) \text{ in } \{z|0<\operatorname{Im} z<\operatorname{Im}\beta\}}} \operatorname{Res}(f; z_j) \tag{5.7.43}$$

The poles of $f(z)$ are solutions of

$$e^{-2\beta z} = -1 \tag{5.7.44}$$

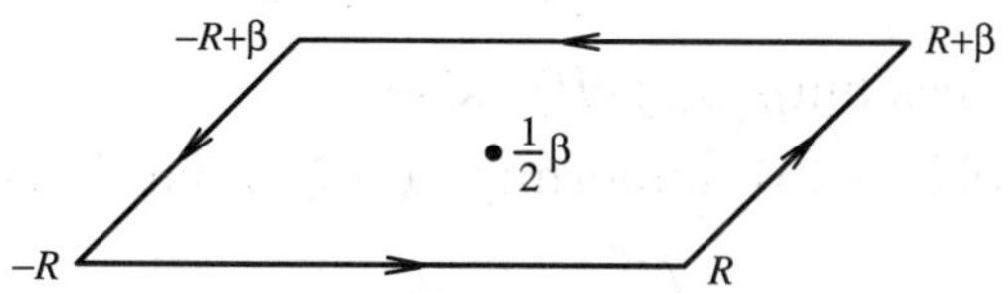

Figure 5.7.6. Contour for Example 5.7.11.

that is,

$$2\beta z = (2n+1)\pi i = (2n+1)\beta^2 \tag{5.7.45}$$

or

$$z_n = (n+\tfrac{1}{2})\beta, \qquad n = 0, \pm 1, \dots \tag{5.7.46}$$

Only z_0 is in the strip $\{z \mid 0 < \operatorname{Im} z < \operatorname{Im} \beta\}$ (see Figure 5.7.6) and the residue there is

$$\frac{e^{-\frac{1}{4}\beta^2}}{\frac{d}{dz}(1+e^{-2z\beta})\big|_{z=\beta/2}} = \frac{e^{-\beta^2/4}}{-2\beta e^{-\beta^2}} = \frac{e^{-i\pi/4}}{2\beta} = \frac{1}{2i\sqrt{\pi}}$$

Thus, $2\pi i \sum_{z \text{ in strip}} \operatorname{Res}(f; z) = \sqrt{\pi}$, and we see that

$$\int_{-\infty}^{\infty} e^{-x^2}\, dx = \sqrt{\pi} \tag{5.7.47}$$

By scaling, for $a > 0$,

$$\int_{-\infty}^{\infty} e^{-ax^2}\, dx = \sqrt{\pi/a} \tag{5.7.48}$$

and, in particular,

$$\int_{-\infty}^{\infty} e^{-\frac{1}{2}x^2}\, dx = \sqrt{2\pi} \tag{5.7.49}$$

From this, one can easily see (Problem 13) that for all $w \in \mathbb{C}$,

$$\int_{-\infty}^{\infty} e^{-\frac{1}{2}x^2} e^{iwx}\, \frac{dz}{\sqrt{2\pi}} = e^{-\frac{1}{2}w^2} \tag{5.7.50}$$

□

Example 5.7.12 (Fresnel Integrals). Evaluate

$$\int_0^{\infty} \sin(x^2)\, dx \quad \text{and} \quad \int_0^{\infty} \cos(x^2)\, dx \tag{5.7.51}$$

Obviously, these integrals are not absolutely convergent. They are conditionally convergent; indeed, if

$$f(R) = \int_0^R e^{ix^2}\, dx \tag{5.7.52}$$

then, changing variables to $y = x^2$, we have

$$f(R) = \int_0^{R^2} e^{iy}\, \frac{dy}{2\sqrt{y}} \tag{5.7.53}$$

and Problem 3 implies $\lim_{R\to\infty} f(R)$ exists.

In fact (see Problem 14), uniformly in $\alpha \geq 0$, for R fixed,

$$\sup_{R' > R \geq 1} \left| \int_R^{R'} e^{(i-\alpha)x^2}\, dx \right| \leq cR^{-1} \tag{5.7.54}$$

That implies $\lim_{R\to\infty} \int_0^R e^{(i-\alpha)x^2}\,dx$ converges uniformly in $\alpha \ge 0$, which justifies interchanging limits to conclude

$$\lim_{R\to\infty} \int_0^R e^{ix^2}\,dx = \lim_{\alpha\downarrow 0} \int_0^\infty e^{(i-\alpha)x^2}\,dx \qquad (5.7.55)$$

In the region $\operatorname{Re} a > 0$, both sides of (5.7.48) are analytic (the left by Theorem 3.1.6). Since they are equal for a real, they are equal for all a with $\operatorname{Re} a > 0$ and, in particular, for $a = \alpha - i$ with $\alpha > 0$.

Thus, by (5.7.55),

$$\int_0^\infty e^{ix^2}\,dx = \frac{1}{2}\sqrt{\frac{\pi}{-i}} = \frac{1}{2\sqrt{2}}\,(1+i)\sqrt{\pi} \qquad (5.7.56)$$

Taking real and imaginary parts,

$$\int_0^\infty \cos(x^2)\,dx = \int_0^\infty \sin(x^2)\,dx = \frac{\sqrt{\pi}}{2\sqrt{2}} \qquad (5.7.57)$$

These are called *Fresnel integrals*; more generally, the *Fresnel functions* are defined by

$$C(x) = \int_0^x \cos(y^2)\,dy, \qquad S(x) = \int_0^x \sin(y^2)\,dy \qquad (5.7.58)$$

The curve

$$E(t) = C(t) + iS(t) \qquad (5.7.59)$$

(see Figure 5.7.7) is called the *Euler* or *Cornu spiral*. It enters in optics and in civil engineering. There is more about these functions in the Notes and in Problems 15 and 16. □

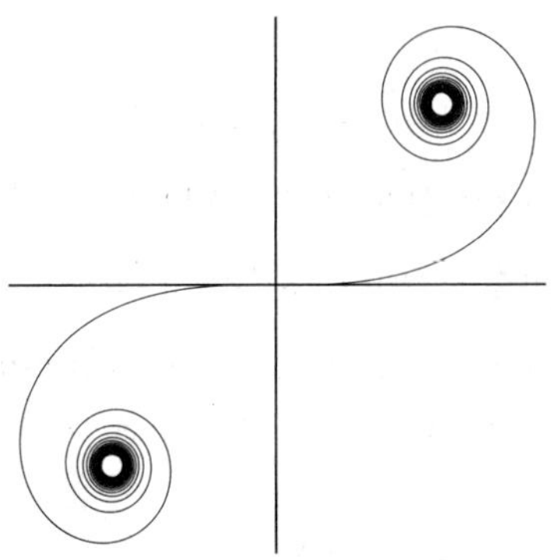

Figure 5.7.7. The Euler spiral, $x = C(t)$, $y = S(t)$.

5.7.6. Infinite Sums Via the Residue Calculus. Remarkably, some infinite sums can be calculated as finite sums of residues. Besides the theorem below, one can use contour integration on sums to prove the Poisson summation formula (Theorem 6.6.10 of Part 1), see Problem 29, and to prove a suitable version of the Nyquist–Shannon sampling theorem (Theorem 6.6.16 of Part 1), see Problem 20.

Theorem 5.7.13. *Let f be a rational function $f(z) = P(z)/Q(z)$, where*

(i) $\deg(Q) \geq 2 + \deg(P)$

(ii) *P and Q have no common zeros and Q is nonvanishing at every $n \in \mathbb{Z}$. Then*

$$\sum_{n=-\infty}^{\infty} f(n) = - \sum_{\text{zeros } z_k \text{ of } Q} \text{Res}(fg; z_k) \tag{5.7.60}$$

where

$$g(z) = \pi \cot(\pi z) \tag{5.7.61}$$

Remarks. 1. Since $|f(n)| \leq C|n|^{-2}$ for n large, the sum is absolutely convergent.

2. If f has a simple pole at z_k, then

$$\text{Res}(fg; z_k) = g(z_k)\text{Res}(f; z_k) \tag{5.7.62}$$

Proof. Let

$$F(z) = f(z)g(z) \tag{5.7.63}$$

which is meromorphic on all of $\mathbb{C}$. Let Γ_n be the rectangle with corners at $(\pm(n+\frac{1}{2}), \pm n)$ oriented counterclockwise. We claim that

$$\lim_{n\to\infty} \frac{1}{2\pi i} \oint_{\Gamma_n} F(z)\, dz = 0 \tag{5.7.64}$$

We note first that

$$g(z+1) = g(z) \tag{5.7.65}$$

since $\cos(\pi z)$ and $\sin(\pi z)$ are periodic. When $z = \frac{1}{2} + iy\, e^{\pm i\pi z} = \pm i e^{\pm \pi y}$, so

$$|g(\tfrac{1}{2} + iy)| = \pi|\tanh(\pi y)| \leq \pi \tag{5.7.66}$$

Therefore, on the vertical sides of Γ_n, $|g(z)|$ is bounded by π.

On the horizontal sides, where $|e^{i\pi z}|$ and $|e^{-i\pi z}|$ are $e^{\pi n}$ and $e^{-\pi n}$ or vice-versa,

$$|g(x \pm in)| \leq \pi\left(\frac{e^{2n\pi}+1}{e^{2n\pi}-1}\right) = \pi(1 + O(e^{-2n\pi})) \tag{5.7.67}$$

which, for n large, is certainly bounded by 2π.

Thus, for n large,

$$\sup_{z\in\Gamma_n} |F(z)| \le 2\pi \sup_{z\in\Gamma_n} |f(z)| \le Cn^{-2} \tag{5.7.68}$$

Since $|\Gamma_n| = O(n)$, we obtain (5.7.64).

By the residue theorem,

$$\lim_{n\to\infty} \left[\sum_{\substack{z_k \text{ poles of } f \\ w_\ell \text{ poles of } g \text{ in } \Gamma_n}} \text{Res}(fg; z_k \text{ or } w_\ell) \right] = 0 \tag{5.7.69}$$

The poles of fg are precisely the poles of f and the poles of g (for which $f(w_\ell) \neq 0$). g has a pole in Γ_n at $0, \pm 1, \dots, \pm n$ with residue 1 (by (5.7.65) and $\cos(0) = 1$, $\lim_{|z|\to 0,\, z\neq 0} \frac{\pi z}{\sin(\pi z)} = 1$). Thus, the sum over the w's is $\sum_{m=-n}^{n} f(m)$, which means that (5.7.69) implies (5.7.60). □

Example 5.7.14. For a real and $a \neq 0$, compute

$$\sum_{n=-\infty}^{\infty} \frac{1}{n^2 + a^2} \tag{5.7.70}$$

In Theorem 5.7.13, take $f(z) = (z^2 + a^2)^{-1}$. This has two poles, both simple, at $z_\pm = \pm ia$ with residues $\pm 1/2ia$. Taking into account that cot is odd, the two terms in (5.7.60) are equal, and by (5.7.62), the sum is $-2\pi \cot(\pi i a)/2ia = \pi \coth(\pi a)/a$, so

$$\sum_{n=-\infty}^{\infty} \frac{1}{n^2 + a^2} = \frac{\pi}{a} \coth(\pi a) \tag{5.7.71}$$

□

Remark. This can also be proven via a Poisson summation formula; see Example 6.6.12 in Section 6.6 of Part 1.

Example 5.7.15. Compute

$$\sum_{n=1}^{\infty} \frac{1}{n^2} \tag{5.7.72}$$

By dominated convergence for sums, this sum is $\lim_{a\downarrow 0} \sum_{n=1}^{\infty} (n^2 + a^2)^{-1}$, and by (5.7.71), this is

$$\lim_{a\downarrow 0} \frac{1}{2} \left[\frac{\pi}{a} \coth(\pi a) - \frac{1}{a^2} \right] \tag{5.7.73}$$

Since

$$\coth(x) = \frac{\cosh(x)}{\sin(x)} = \frac{1 + \frac{x^2}{2} + O(x^4)}{x(1 + \frac{x^2}{6} + O(x^4))} = x^{-1} + \frac{x}{3} + O(x^3) \tag{5.7.74}$$

we see the celebrated result of Euler (discussed further in Section 9.2),

$$\sum_{n=1}^{\infty} \frac{1}{n^2} = \frac{\pi^2}{6} \tag{5.7.75}$$

□

Notes and Historical Remarks. While the general theory was developed by Cauchy, Euler, using a variety of tricks, had earlier found the values of several definite integrals which couldn't be evaluated via antiderivatives. In some of his earliest work, Cauchy realized the need for and usefulness of principal values.

For the argument in Problem 3, monotonicity of f is not needed. It suffices that $\int_1^\infty |f'(s)|\,dx < \infty$. In fact, that f is C^1 is not needed. By integrating by parts in a Stieltjes integral, it suffices that f be of bounded variation.

While Cauchy's work was 200 years ago and Euler computed the Gaussian integral almost 300 years ago, it is less than 70 years ago that the calculation of the Gaussian integral via contour methods was found. Indeed, Copson's [**122**] and Watson's [**579**] complex variable books state explicitly that one cannot compute the Gaussian integral using the residue calculus. The first evaluation using $f(z) = e^{\pi i z^2} \tan(\pi z)$ along an oblique parallelogram was found by Pólya and presented in his talk in 1945 [**448**]; Mitrinović and Kečkić [**383**] have the details of Pólya's calculation and some more history. Their book also has lots of subtle examples. The proof here is a variant of that of Srinivasa Rao [**534**] (and follows Karunakaran [**297**]).

Some classic books with lots of examples of the use of the calculus of residues are Lindelöf [**356**], Copson [**122**], and Titchmarsh [**556**]. Ingham [**276**] has examples relevant to number theory. In particular, Lindelöf [**356**] was important historically.

The Fresnel functions come up in describing the diffraction pattern from a half-plane. Indeed, Augustin Fresnel (1788–1827) invented his functions for that purpose and placed them in an 1819 paper that won a French Academy prize competition.

There are differing normalization conditions on the Fresnel functions, so that others define a function that, in our notation, is $\alpha S(\beta x)$ for suitable α, β. Mathematica, for example, uses $S(u) = \int_0^u (\sin(\frac{1}{2}\pi x^2))\,dx$. *An Atlas of Functions* [**414**] has a whole chapter on $C(x)$ and $S(x)$. NIST has a chapter on the Fresnel and closely related error functions [**553**].

The Euler spiral (5.7.59) is also called the Cornu spiral or clothoid. Cornu (1841–1902), a French physicist, specialized in optics. Because of its property that the curvature is proportional to the speed (see Problem 16), the Euler spiral is used in railroad and highway construction.

The problem of finding $\sum_{n=1}^{\infty} 1/n^2$ was a celebrated one when Euler solved it in 1734 [**172**]. It established his fame. It was proposed by Pietro Mengoli in 1644 and popularized by Jacob Bernoulli in 1689. Jacob was the brother of Euler's teacher, Johann Bernoulli. It was called the Basel problem because the Bernoullis lived in Basel.

In this series, we give many proofs of Euler's formula: the one in this section, two using Fourier series (see Problem 16 of Section 3.5 and Example 6.6.12 in Section 6.6 of Part 1), Euler's original proof from his product formula (see Section 9.2), and three other proofs of Euler in Problems 10, 12, and 13 of Section 9.2. Chapman [**111**] has fourteen proofs, not including the one in this section (!) or the elementary one found in Wikipedia[1].

Problems

1. Evaluate the following integrals:

(a) $\displaystyle\int_0^{2\pi} \frac{dx}{(a+b\cos x)^2}$ for $|b| < a$

(b) $\displaystyle\mathrm{pv}\int_0^{2\pi} \frac{dx}{a+b\cos x}$ for $b > a > 0$

(c) $\displaystyle\int_0^{2\pi} \frac{\cos\theta}{a+\cos\theta}\,dx$ for $a > 1$

(d) $\displaystyle\int_0^{2\pi} \frac{dx}{1-2a\cos x+a^2}$ for $0 < a < 1$

(e) $\displaystyle\int_{-\pi}^{\pi} \frac{x\sin x\,dx}{1-2a\cos x+a^2}$ for $0 < a < 1$

(f) $\displaystyle\int_0^{2\pi} \frac{\cos mx\,dx}{1-2a\cos x+a^2}$ for $0 < a < 1$

(g) $\displaystyle\int_0^{2\pi} \exp(\cos(x))\,dx$

(h) $\displaystyle\int_0^{\pi} e^{\cos(x)}\cos(\sin x - nx)\,dx$

(i) $\displaystyle\int_0^{\pi} \frac{\sin^2 x}{a+b\cos x}\,dx$ for $0 < b < a$

[1] `http://en.wikipedia.org/wiki/Basel_problem`

(j) $\displaystyle\int_0^{\pi} \frac{dx}{a+\sin^2 x}$ for $a > 0$

(k) $\displaystyle\int_0^{2\pi} \frac{dx}{a^2\cos^2 x + b^2\sin^2 x}$ for $a \neq 0 \neq b$

(l) $\displaystyle\int_0^{\pi} \left(\frac{\sin nx}{\sin x}\right)^2 dx$

(*Hints*: (e) requires Section 5.7.4. (f) is simplified if you use $\cos mx = \mathrm{Re}(e^{imx})$. For (g), the answer will be an infinite series. For (h), look at the Taylor series for e^z.)

2. Evaluate the following integrals ($a > 0$):

(a) $\displaystyle\int_{-\infty}^{\infty} \frac{dx}{(x^2+a^2)^n}$

(b) $\displaystyle\int_{-\infty}^{\infty} \frac{x^2 dx}{(x^2+1)^2}$

(c) $\displaystyle\int_{-\infty}^{\infty} \frac{dx}{x^{2n}+a^{2n}}$

(d) $\displaystyle\mathrm{pv}\int_{-\infty}^{\infty} \frac{dx}{x^2-1}$

(e) $\displaystyle\int_{-\infty}^{\infty} \frac{x^2}{x^4+a^4}\,dx$

(f) $\displaystyle\int_{-\infty}^{\infty} \frac{dx}{x^2+2\cos\theta x+1}$, $\theta \in (0,\pi)$

(g) $\displaystyle\int_0^{\infty} \frac{dx}{x^3+a^3}$

(h) $\displaystyle\int_0^{\infty} \frac{dx}{(x^2+a^2)(x^2+b^2)}$

(i) $\displaystyle\int_0^{\infty} \frac{x^m}{x^n+a^n}\,dx$, $m < n-1$ integers

(*Hint*: For (g), consider symmetries of $f(z) = (z^3+a^3)^{-1}$.)

3. Let $f(x)$ be monotone decreasing and C^1 on $[1,\infty)$ and $\lim_{x\to\infty} f(x) = 0$. Prove for α real and nonzero $\lim_{R\to\infty}\int_1^R e^{i\alpha x} f(x)\,dx$ exists. (*Hint*: Use $e^{i\alpha x} = (i\alpha)^{-1}\frac{d}{dx}e^{i\alpha x}$ and integrate by parts.)

4. (a) For the contour used in Example 5.7.5, estimate the parts of the contour off $\mathbb{R}$ and show they go to zero.

(b) Close the contour in Example 5.7.5 by a semicircle and show it is bounded by $\int_0^\pi e^{-aR\sin\theta}\,d\theta$ and that this goes to zero as $R\to\infty$, indeed as $O((aR)^{-1})$.

(c) More generally, prove that if f is a continuous function in $\mathbb{C}_+$ with $\sup_{|z|=R,\,0<\arg z<\pi}|f(z)| = 0$, then the integral along a counterclockwise semicircle

$$\lim_{R\to\infty}\int_{\substack{|z|=R\\0<\arg z<\pi}} f(z)e^{iz}\,dz = 0$$

Remark. This is known as Jordan's lemma.

5. Evaluate the following integrals (in (a) for $0<\alpha<\beta$, $0<\alpha=\beta$, and $0<\beta<\alpha$):

(a) $\displaystyle\int_{-\infty}^{\infty}\frac{\sin\alpha x\cos\beta x}{x}\,dx$

(b) $\displaystyle\int_{-\infty}^{\infty}\frac{\sin^2 x}{x^2}\,dx$

(c) $\displaystyle\int_{-\infty}^{\infty}\frac{1-\cos x}{x^2}\,dx$

(d) $\displaystyle\int_{-\infty}^{\infty}\frac{\sin(x)}{x(x^2-\pi^2)}\,dx$

(e) $\displaystyle\int_{-\infty}^{\infty}\frac{\sin^3 x}{x^3}\,dx$

(f) $\displaystyle\int_{-\infty}^{\infty}\frac{\cos(ax)}{\cosh(bx)}\,dx$

(g) $\displaystyle\int_{-\infty}^{\infty}\frac{\sin(ax)}{\sinh(bx)}\,dx$

(h) $\displaystyle\int_{-\infty}^{\infty}\frac{x\,dx}{\sinh(ax)}$

(i) $\displaystyle\int_{-\infty}^{\infty}\frac{\cos x-\cos a}{x^2-a^2}\,dx$, $a\neq 0$ real

(j) $\displaystyle\int_0^{\infty}\frac{\cos x\,dx}{(x^2+a^2)(x^2+b^2)}$

(k) $\displaystyle\int_0^{\infty}\frac{x\sin bx}{x^2+a^2}\,dx$

(l) $\displaystyle\int_0^{\infty}\frac{\sin bx}{x(x^2+a^2)}\,dx$

6. For $a > 0$, let

$$I(a) = \int_0^\infty \frac{\log(x)}{x^2 + a^2}\, dx$$

(a) Prove that

$$I(a) = \frac{1}{a}\, I(1) + \frac{\log(a)}{a}\, \frac{\pi}{2}$$

(*Hint*: Change variables from x to xa.)

(b) Prove that $I(1) = -I(1)$. (*Hint*: Change variables from x to x^{-1}.)

(c) Conclude that (5.7.25) holds.

7. Compute the following integrals:

(a) $\displaystyle\int_0^\infty \frac{x^{a-1}}{x+b}\, dx$ for $b > 0$ and $0 < a < 1$

(b) $\displaystyle\int_0^\infty \frac{x^{a-1}\log(x)}{1+x}\, dx$ for $0 < a < 1$

(c) $\displaystyle\int_0^\infty \frac{x^{a-1}}{x + e^{i\eta}}\, dx$ for η real, $|\eta| < \pi$, and $0 < a < 1$

(d) $\displaystyle\int_0^\infty \frac{x^{m-1}}{x^n + e^{i\eta}}\, dx$ for η real, $|\eta| < \pi$, and $0 < m < n$

(e) $\displaystyle\int_0^\infty \frac{x^{m-1}}{x^{2n} + 2x^n\cos(\eta) + 1}\, dx$ for η real, $|\eta| < \pi$, $0 < m < n$

(*Hints*: Scaling for (a), analyticity (plus $\frac{d}{da}$) in a and b for (b) and (c), change of variables to $y = x^n$ for (d) and $x^{2n} + 2x^n\cos(\eta) + 1 = (x + e^{ik})(x + e^{-ik})$ for (e).)

8. Compute the following integrals ((b) is two integrals; the one for $x^2 - 1$ is intended as a PV integral)

(a) $\displaystyle\int_0^\infty \frac{(\log(x))^2}{x^2+1}\, dx$

(b) $\displaystyle\int_0^\infty \frac{\log(x)}{x^2 \pm 1}\, dx$

(c) $\displaystyle\int_0^\infty \frac{dx}{\sqrt{x}\,(x^2+1)}$

(d) $\displaystyle\int_0^\infty \frac{\sqrt{x}\, dx}{x^2+5x+6}$

(e) $\displaystyle\int_0^\infty \frac{\log x}{x^3+1}\, dx$

(f) $\displaystyle\int_0^\infty \frac{1}{x^3+1}\, dx$

(g) $\displaystyle\int_0^1 \frac{x^4}{\sqrt{x(1-x)}}\, dx$

(h) $\displaystyle\int_{-1}^1 \frac{\sqrt{1-x^2}}{1+x^2}\, dx$

9. For $\operatorname{Im} z > 0$, define $z^i = \exp(i \log z)$, where the branch of log is picked with $\operatorname{Im} \log z \in (0, \pi)$. Compute $\int_{-1}^{1} \lim_{\varepsilon \downarrow 0} f(x + i\varepsilon)\, dx$.

10. Compute the following integrals for $0 < a < 1$:

(a) $\displaystyle\int_0^\infty \frac{x^{a-1}}{1+x}\, dx$ (b) $\displaystyle\text{pv} \int_0^\infty \frac{x^{a-1}}{1-x}\, dx$

(*Hint*: You can evaluate (a) and (b) together by integrating $z^{a-1}(1-z)^{-1}$ around a suitable contour and taking the real and imaginary parts.)

Notes. 1. The integral in (a) and the integral $\int_0^\infty \frac{x^{a-1}-x^{b-1}}{1-x}\, dx$ for $0 < a, b < 1$ go back to Euler.

2. See also Example 5.7.8 for (a).

11. Compute the following integrals:

(a) $\displaystyle\int_0^\infty \frac{dx}{1+x^3}$

(b) $\displaystyle\int_0^\infty \frac{x^2\, dx}{(x^2+1)(x+4)^2}$

(c) $\displaystyle\int_0^\infty \frac{dx}{(x+1)^4}$

(d) $\displaystyle\int_0^\infty \frac{dx}{1+x^n}, \qquad n = 2, 3, 4, \dots$

(e) $\displaystyle\int_0^\infty \frac{x^{2m}}{x^{2n}+1}\, dx, \qquad n, m \in \mathbb{Z},\ 0 < m < n$

12. Prove that the contributions of the oblique edges of the parallelogram Γ_R in Figure 5.7.6 for $\oint_{\Gamma_R} f(z)\, dz$ with f given by (5.7.40) go to zero as $R \to \infty$.

13. (a) Prove (5.7.50) by proving it first for $w = iy$, $y \in \mathbb{R}$, and then using analyticity in w.

(b) Prove (5.7.50) by choosing contours for general $w \in \mathbb{C}$.

14. Prove (5.7.54). (*Hint*: Change variables from x to $y = x^2$ and use $\frac{1}{i-\beta} \frac{d}{dx} e^{(i-\beta)y} = e^{(i-\beta)y}$ and/or integration by parts.)

15. Let $C(x), S(x)$ be given by (5.7.58).

(a) Prove that as $x \to \infty$ along $[0, \infty)$,

$$C(x) = \frac{\sqrt{\pi}}{2\sqrt{2}} + \frac{\gamma}{x} \sin(x^2) + O\left(\frac{1}{x^2}\right)$$

for an explicit γ that you should compute. (*Hint*: See Problem 14.)

Note. We'll return to the asymptotics of $C(x), S(x)$ as $x \to \infty$ in Problem 4 of Section 15.1 of Part 2B.

(b) Write out the power series for $C(x), S(x)$ at 0. (They are related to confluent hypergeometric functions, it turns out; see (14.4.72) of Part 2B.) Conclude $C(z)$ and $S(z)$ are entire functions.

16. Let $E(x) = C(x) + iS(x)$ for $x \in (0, \infty)$. This is the Euler spiral.

(a) Prove $|E'(x)| = 1$ and conclude the length of the curve described by E grows linearly in x.

(b) If γ is a simple C^2 curve and it is reparametrized by arclength (i.e., so that $|\gamma'(s)| = 1$), then the curvature at $\gamma(s)$ is $x = \det\begin{pmatrix} \gamma_1' & \gamma_2' \\ \gamma_1'' & \gamma_2'' \end{pmatrix}$. Prove that the curve $E(x)$ at the point $E(x)$ has curvature $= 2x$ so it is a curve with curvature proportional to arclength.

(c) Prove that "essentially," the only curve with curvature proportional to arclength is the Euler spiral.

17. Evaluate the following sums for $a > 0$,

(a) $\displaystyle\sum_{n=-\infty}^{\infty} (n^2 + a^2)^{-2}$

(b) $\displaystyle\sum_{n=-\infty}^{\infty} (n^4 + a^4)^{-1}$

(c) $\displaystyle\sum_{n=1}^{\infty} n^{-4}$

(d) $\displaystyle\sum_{n=-\infty}^{\infty} (n^2 + n - \tfrac{3}{16})^{-1}$

(e) $\displaystyle\sum_{n=1}^{\infty} \frac{\coth n\pi}{n^2}$

(f) $\displaystyle\sum_{n=1}^{\infty} \frac{1}{n^4 + a^4}$

18. This problem uses results on convergence of Fourier series proven in Section 3.5 of Part 1.

(a) Let $a \in \mathbb{R}$. Consider the function $f(x) = \cosh(ax)$ on $[-\pi, \pi]$. Compute its Fourier coefficients in $L^2([-\pi, \pi], \frac{dx}{2\pi})$ in e^{inx} basis.

(b) Using Theorem 3.5.4 of Part 1, prove that the Fourier series at $x = \pi$ converges to $\cosh(a\pi)$.

(c) Provide a proof of (5.7.71) without contour integration.

19. (a) Under the hypotheses of Theorem 5.7.13, prove that

$$\sum_{n=-\infty}^{\infty} (-1)^n f(n) = - \sum_{\text{zeros } z_k \text{ of } Q} \operatorname{Res}(fg; z_k)$$

where $g(z) = \pi \csc(\pi z)$.

(b) Under the hypothesis of Theorem 5.7.13 modified for Q nonvanishing at $\{n+\frac{1}{2}\}_{n\in\mathbb{Z}}$, prove that

$$\sum_{n=-\infty}^{\infty} f\left(\frac{2n+1}{2}\right) = \sum_{\text{zeros } z_k \text{ of } Q} \operatorname{Res}(fg_1; z_k)$$

$$\sum_{n=-\infty}^{\infty} (-1)^n f\left(\frac{2n+1}{2}\right) = \sum_{\text{zeros of } z_k \text{ of } Q} \operatorname{Res}(fg_2; z_k)$$

where $g_1(z) = \pi\tan(\pi z)$, $g_2(z) = \pi\sec(\pi z)$. (*Hint*: There are two proofs: one uses a $z \to z+\frac{1}{2}$ change of variables and the other uses the method of proof of Theorem 5.7.13.)

20. This problem will provide an alternate proof of the strong form of the Nyquist–Shannon sampling theorem (Theorem 6.6.16 of Part 1), following closely the 1920 proof of Ogura [**412**]. We'll suppose $f(z)$ is an entire function obeying

$$|f(z)| \le C\exp((\pi-\varepsilon)|\operatorname{Im} z|) \tag{5.7.76}$$

Fix $z_0 \in \mathbb{C}\setminus\mathbb{Z}$. Let

$$g(z) = \frac{f(z)}{(z-z_0)\sin(\pi z)} \tag{5.7.77}$$

and R_k be the rectangle which is the boundary of $\{x+iy \mid |x| \le k+\frac{1}{2}, |y| \le k+\frac{1}{2}\}$ oriented clockwise.

(a) Prove that

$$\lim_{k\to\infty} \int_{R_k} |g(z)|\, d|z| = 0 \tag{5.7.78}$$

(b) Prove that

$$f(z_0) = \frac{\sin \pi z}{\pi} \lim_{k\to\infty} \sum_{n=-k}^{k} \frac{(-1)^n f(n)}{z_0 - n} \tag{5.7.79}$$

which is (6.6.76) of Part 1.

21. Evaluate $\sum_{n=-\infty}^{\infty}(n-\frac{1}{2})^{-2}$ and use that to evaluate $\sum_{n=0}^{\infty}(2n+1)^{-2}$. If the sum is S and T is $\sum_{n=1}^{\infty} n^{-2}$, prove that $T = S + \frac{1}{4}T$ and deduce Euler's formula.

22. Evaluate the following sums:

(a) $\displaystyle\sum_{n=-\infty}^{\infty} \frac{(-1)^n}{(n+a)^2}$, $a \notin \mathbb{Z}$

(b) $\displaystyle\sum_{n=0}^{\infty} \frac{(-1)^n}{(2n+1)^2}$

(c) $\displaystyle\sum_{n=1}^{\infty} \frac{(-1)^{n+1}}{n^2+a^2}$

(d) $\displaystyle\sum_{n=1}^{\infty} \frac{(-1)^{n-1} n \sin n\theta}{n^2+a^2}, \qquad \theta \in (-\pi, \pi)$

(e) $\displaystyle\sum_{n=0}^{\infty} (-1)^n \frac{2n+1}{(2n+1)^2+a^2}$

Remark. (e) can also be evaluated with a Poisson summation argument; see Example 6.6.13 in Section 6.6 of Part 1.

23. Compute $\sum_{n=0}^{\infty} \binom{2n}{n} \frac{1}{3^n}$.
(*Hint*: For any $r > 0$, $\binom{2n}{n} = \frac{1}{2\pi i} \oint_{|z|<r} \frac{(1+z)^{2n}}{z^{n+1}}\, dz$.)

24. The Dedekind sum is defined for relatively prime integers by

$$s(p,q) = \frac{1}{4q} \sum_{k=1}^{q-1} \cot\left(\frac{\pi k p}{q}\right) \cot\left(\frac{\pi k}{q}\right)$$

By considering integrals of $f(z) = \cot(\pi p z)\cot(\pi q z)\cot(\pi z)$ over contours over rectangles with corners $1 - \varepsilon \pm iR$ and $-\varepsilon \pm iR$ as $R \to \infty$, prove Dedekind reciprocity

$$s(p,q) + s(q,p) = -\frac{1}{4} + \frac{1}{12}\left(\frac{p}{q} + \frac{q}{p} + \frac{1}{pq}\right)$$

Rademacher–Grosswald [**465**] have six proofs of this result and a discussion of Dedekind sums.

25. (a) Extend Theorem 5.7.13 to allow Q to have zeros at $n \in \mathbb{Z}$, say, for simplicity at $n = 0$ and no other point in $\mathbb{Z}$.

(b) Use this extended theorem and $f(n) = n^{-2}$ to directly compute $\sum_{n=1}^{\infty} n^{-2}$.

26. This problem will have a real variable evaluation of the Gauss integral that is due to Stieltjes [**541**]. It is related to formal considerations of Euler.

(a) For $n = 0, 1, 2, \ldots$, let $S_n = \int_0^{\pi} \sin^n x\, dx$. Prove that $\lim_{n\to\infty} \frac{S_n}{S_{n+1}} = 1$. (*Hint*: Prove first, using the Cauchy–Schwarz inequality, that $1 \le S_n S_{n+1}^{-1} \le S_{n-1} S_n^{-1}$, and then that $(S_n)^{1/n} \to 1$.)

(b) Prove that for $n \geq 2$, $S_n = (\frac{n-1}{n})S_{n-2}$.
(*Hint*: Use $\int_0^\pi \frac{d}{dx}[(\sin^{n-1} x)\cos x] = 0$.)

(c) Prove that $S_0 = \pi$, $S_1 = 2$.

(d) Prove *Wallis' formula* that

$$\lim_{n\to\infty} \prod_{j=1}^{n} \frac{(2j)^2}{(2j+1)(2j-1)} = \frac{\pi}{2} \tag{5.7.80}$$

(e) Let W_n be the product in (5.7.80) and $\widetilde{W}_n = W_n \frac{2n+1}{2n}$ (i.e., the same products with the largest factor removed from the numerator and denominator). Prove $W_n \leq W_{n+1}$ and $\widetilde{W}_n \geq \widetilde{W}_{n+1}$ and deduce

$$W_n \leq W_{n+1} \leq \frac{\pi}{2} \leq \widetilde{W}_{n+1} \leq \widetilde{W}_n \tag{5.7.81}$$

and conclude $|W_n - \frac{\pi}{2}| \leq \frac{\pi}{4n}$.

(f) Let $G_n = \int_0^\infty x^n e^{-x^2}\, dx$ for $n = 0, 1, 2, \dots$. Prove that for $n \geq 1$, $G_n^2 \leq G_{n+1}G_{n-1}$.

(g) Prove that for $n \geq 2$, $G_n = (\frac{n-1}{2})G_{n-2}$. (*Hint*: $\frac{d}{dx}(e^{-x^2}) = -2xe^{-x^2}$.)

(h) Prove that $G_{2n+1}^2 \leq \frac{2n+1}{2} G_{2n}^2 \leq \frac{2n+1}{2n} G_{2n+1}^2$ and conclude that

$$\lim_{n\to\infty} \frac{2n+1}{2} \frac{G_{2n}^2}{G_{2n+1}^2} = 1$$

(i) Prove that $G_1 = \frac{1}{2}$ and then that $G_{2n+1} = \frac{1}{2}n!$.

(j) Using Wallis' formula, conclude $G_0 = \frac{1}{2}\sqrt{\pi}$ and so a new proof of (5.7.47).

27. Evaluate the integral $\int_0^\pi \sin^{2n}(\theta)\, d\theta$ by noting it is $\frac{1}{2}\int_0^{2\pi} \sin^{2n}(\theta)\, d\theta$ and using the method of Section 5.7.1 to relate this to $\oint (z - z^{-1})^{2n} \frac{dz}{z}$.

28. Gaussian sums are defined as

$$g_n = \sum_{j=0}^{n-1} e^{2\pi i j^2/n} \tag{5.7.82}$$

In this problem, use contour integrals to compute g_n and, as a bonus, Fresnel integrals and therefore Gaussian integrals!

(a) Let $f(z) = e^{2\pi i z^2/n}/(e^{2\pi i z} - 1)$. Let Γ_R be the rectangle with corners, $\pm iR$, $\frac{1}{2}n \pm iR$ oriented counterclockwise. There is a pole of f at 0, and if n is even at $n/2$, a principal part is needed. Prove that $\text{Res}(f; k) = (2\pi i)^{-1} e^{2\pi i k^2/n}$ and then that, for any R,

$$\tfrac{1}{2} g_n = \text{pv} \oint_{\Gamma_R} f(z)\, dz$$

(b) Let H_R and V_R be the horizontal and vertical sides of the contour. Prove that as $R \to \infty$, $H_R = O(1/R)$ by first proving that for R large,

$$|f(x+iR)| \le 2e^{-4\pi x R/n}, \qquad |f(x-iR)| \le 2e^{-4\pi(\frac{n}{2}-x)/n}$$

(c) Prove that $f(iy) + f(-iy) = -e^{-2\pi i y^2/n}$ and that $f(\frac{n}{2}+iy) + f(\frac{n}{2} - iy) = i^{3n} e^{-2\pi i y^2/n}$, and then that

$$V_R = i(1+i^{3n}) \int_0^R e^{-2\pi i y^2/n}\, dy$$

(d) Conclude that

$$g_n = 2i(1+i^{3n})\sqrt{n} \lim_{R\to\infty} \int_0^R e^{-2\pi i y^2}\, dy$$

(e) Compute g_4 and thereby that

$$\lim_{R\to\infty} \int_0^R e^{-2\pi i y^2}\, dy = \tfrac{1}{4}(1-i)$$

From this, compute the basic Gaussian integral.

(f) Prove that $g_n = \frac{1}{2}(1+i)(1+i^{3n})\sqrt{n}$.

29. This will prove the Poisson summation formula by a method different from that used in Part 1 (see Theorem 6.6.10 of Part 1 or Problem 7 of Section 9.7). In parts (a)–(c), we'll suppose f is analytic in $\{z \mid |\mathrm{Im}\, z| < a\}$ for some $a > 0$ with f and $z^2 f$ bounded on that set.

(a) Show that $g(z) \equiv f(z)/(e^{iz} - 1)$ has poles at $z = 2\pi n$ with residues $f(2\pi n)/i$.

(b) Prove that for $0 < b < a$,

$$\sum_{n=-\infty}^{\infty} f(2\pi n) = \int_{-\infty}^{\infty} [g(x-ib) - g(x+ib)]\, \frac{dx}{2\pi}$$

(c) Prove that for any $0 < b < a$ and $k \in \mathbb{R}$ with $\pm k \ge 0$, we have

$$\int_{-\infty}^{\infty} f(x \pm ib) e^{ik(x\pm ib)}\, dx = \int_{-\infty}^{\infty} f(x) e^{ikx}\, dx$$

(d) Prove for $0 < b < a$ fixed, uniformly in $x \in \mathbb{R}$,

$$\frac{1}{e^{i(x-ib)} - 1} = \sum_{n=0}^{\infty} e^{-in(x-ib)}$$

and

$$\frac{1}{e^{i(x+ib)} - 1} = -\sum_{n=1}^{\infty} e^{in(x+ib)}$$

(e) Prove the Poisson summation formula for f,

$$\sum_{n=-\infty}^{\infty} f(2\pi n) = (2\pi)^{-1/2} \sum_{n=-\infty}^{\infty} \widehat{f}(n)$$

where $\widehat{f}(k) = (2\pi)^{-1/2} \int_{-\infty}^{\infty} e^{-ikx} f(x)\, dx$.

(f) By an approximation argument, prove Theorem 6.6.10 of Part 1.

Chapter 6

Spaces of Analytic Functions

...having thought at length about this proof, I am sure that it is good, solid, and valid. I had to examine it all the more carefully because a priori it seemed to me that the theorem stated could not exist and had to be false.

—*Stieltjes* [**477**][1]

Big Notions and Theorems: Weierstrass Convergence Theorem, $\mathfrak{A}(\Omega)$ as a Fréchet Space, Montel's Theorem, Vitali's Theorem, Density Results (Runge Restatements), Hurwitz's Theorem, Normal Families, Spherical Metric, Marty's Theorem

Thus far, we we have focused on individual analytic functions and not so much on families of functions. The key exception was Theorem 3.1.5, the Weierstrass convergence theorem, which will play a central role in this chapter. Here we'll discuss $\mathfrak{A}(\Omega)$, the analytic functions as a topological vector space with the topology given by uniform convergence on compact subsets of $K \subset \Omega$. This topology is not given by a single norm, but by a countable family of norms, which means that it is a metric topology. The key result in Section 6.1 is that it is a complete metric space, that is, $\mathfrak{A}(\Omega)$ is a Fréchet space. See Section 6.1 of Part 1 for a review of the language of countably normed spaces. Section 6.2 has the remarkable fact, called Montel's theorem, that closed bounded subsets of $\mathfrak{A}(\Omega)$ are compact. It will

[1]Letter dated February 14, 1894 from Stieltjes to Hermite, about his discovery of his result (a precursor of Vitali's theorem) that uniform convergence of a sequence of bounded analytic functions on a subdomain implies convergence, uniformly on compacts, on the larger domain.

also provide a simple consequence of this, called Vitali's theorem, a sufficient condition for a sequence $f_n \in \mathfrak{A}(\Omega)$ to converge. We'll also define normal families and explain why they occur. In Section 6.3, we'll restate Runge's theorem as a density result for $\mathfrak{A}(\Omega)$, and in Section 6.4, an important fact, Hurwitz's theorem, on zeros under convergence. Section 6.5, a bonus section, has a discussion of an extension of compactness to meromorphic functions that include an extension of the notion of normal families, and a related variant of Montel's theorem, called Marty's theorem. The key will be the spherical metric.

6.1. Analytic Functions as a Fréchet Space

Let Ω be a region and $\mathfrak{A}(\Omega)$ the space of all analytic functions on Ω. For any compact set $K \subset \Omega$, we define

$$\|f\|_K = \sup_{z \in K} |f(z)| \tag{6.1.1}$$

On $C(\Omega)$, all continuous functions, this is only a seminorm (for $\|f\|_K$ can be zero even though f is not zero), but on $\mathfrak{A}(\Omega)$, this is a norm so long as K is not a finite set of points.

Definition. A *compact exhaustion* of Ω is a family $\{K_n\}_{n=1}^{\infty}$ of countably many nonfinite compact subsets of Ω so that

$$K_n \subset K_{n+1}^{\text{int}} \tag{6.1.2}$$

$$\bigcup_{n=1}^{\infty} K_n = \Omega \tag{6.1.3}$$

Exhaustions exist; for example, one can take $K_n = \{z \in \Omega \mid |z| \leq n, \operatorname{dist}(z, \partial\Omega) \geq n^{-1}\}$.

Proposition 6.1.1. *Let $\{K_n\}_{n=1}^{\infty}$ and $\{L_n\}_{n=1}^{\infty}$ be two compact exhaustions of Ω. Then $\{\|\cdot\|_{K_n}\}_{n=1}^{\infty}$ and $\{\|\cdot\|_{L_n}\}_{n=1}^{\infty}$ are equivalent sets of norms on $\mathfrak{A}(\Omega)$ in the following strong sense: for every n, there is an ℓ with*

$$\|f\|_{K_n} \leq \|f\|_{L_\ell} \tag{6.1.4}$$

and for every ℓ, there is an m with

$$\|f\|_{L_\ell} \leq \|f\|_{K_m} \tag{6.1.5}$$

Proof. Since $\cup L_\ell^{\text{int}} = \Omega$, K_n is contained in a finite union $L_{\ell_1}, \dots, L_{\ell_j}$. Let $\ell = \max(\ell_j)$. Then $K_n \subset L_\ell$, and so (6.1.4) holds. By symmetry, so does (6.1.5). □

Thus, the two countable families of norms generate the same topology. Given an exhaustion, $\{K_n\}_{n=1}^\infty$, we get a metric on $\mathfrak{A}(\Omega)$ by

$$\rho(f,g) = \sum_{n=1}^\infty \min(\|f-g\|_{K_n}, 2^{-n}) \tag{6.1.6}$$

Then, $\rho(f_m,f) \to 0$ if and only if, for all n, $\|f_m - f\|_{K_n} \to 0$ as $m \to \infty$ for each fixed n. Moreover, a sequence is Cauchy in ρ if and only if for each fixed n, it is Cauchy in $\| \ \|_{K_n}$. Here is the main theorem of this section:

Theorem 6.1.2. *$\mathfrak{A}(\Omega)$ with the metric* (6.1.6) *is complete, that is, $\mathfrak{A}(\Omega)$ is a Fréchet space.*

Proof. Let $\{f_m\}_{m=1}^\infty$ be a Cauchy sequence in $\mathfrak{A}(\Omega)$. Since $C(K_n)$ is complete in $\|\cdot\|_{K_n}$, there exists $g^{(n)} \in C(K_n)$ so that $\|f_m - g^{(n)}\|_{K_n} \to 0$. Clearly, since $K_n \subset K_{n+1}$, $g^{(n+1)} \restriction K_n = g^{(n)}$. Thus, there is a function $g \in C(\Omega)$ so

$$\|f_m - g\|_{K_n} \to 0 \tag{6.1.7}$$

By the Weierstrass convergence theorem (Theorem 3.1.5), $g \in \mathfrak{A}(\Omega)$, and by (6.1.7), $\rho(f_m, g) \to 0$. □

Theorem 6.1.3. *For any region, Ω, the map $f \mapsto f'$ from $\mathfrak{A}(\Omega)$ to itself is continuous in the metric* (6.1.6).

Proof. Since $K_n \subset K_{n+1}^{\text{int}}$, $\delta_n = \text{dist}(K_n, \Omega \setminus K_{n+1}) > 0$. By a Cauchy estimate, if $z \in K_n$,

$$|f'(z)| \le \delta_n^{-1} \sup_{|w-z|=\delta_n} |f(w)| \tag{6.1.8}$$

so

$$\|f'\|_{K_n} \le \delta_n^{-1} \|f\|_{K_{n+1}} \tag{6.1.9}$$

which implies continuity. □

Notes and Historical Remarks. Many books call convergence in $\mathfrak{A}(\Omega)$ normal convergence. Following Montel, we'll reserve that for a slightly more general notion.

While we will not focus on it, one can study $\mathfrak{A}(\Omega)$ as one would a general topological vector space and ask about its dual. It is also a topological algebra and there are questions related to that. There is a book by Luecking–Rubel [**369**] on these issues, and we'll explore them in problems below and again in the Problems to Sections 9.3 and 9.4. With regard to the algebra structure, the main theorem (Problem 9) is:

Theorem 6.1.4. *Let $T\colon \mathfrak{A}(\Omega) \to \mathfrak{A}(\Omega')$ be a topological algebra isomorphism, that is, a continuous linear bijection with $T(fg) = T(f)T(g)$. Then there is an analytic bijection $F\colon \Omega' \to \Omega$ so that $T(f) = f \circ F$.*

Bers [**44**] (see also [**369**, Ch. 16]) has analyzed ring isomorphisms (i.e., $T(f+g) = T(f) + T(g)$ and $T(fg) = T(f)T(g)$ are kept but linearity, i.e., $T(\lambda f) = \lambda T(f)$ for $\lambda \in \mathbb{C}$, is dropped) and found they are induced by conformal or anticonformal equivalences.

Theorem 6.1.4 implies the algebraic structure of $\mathfrak{A}(\Omega)$ encodes the topology of Ω. This is explored in Problem 10. We will further study the algebraic structure of $\mathfrak{A}(\Omega)$ in Problems 3 of Section 9.3 and 8 of Section 9.4.

As for the dual space structure, the main result is due independently to Grothendieck [**223**], Köthe [**328**], and da Silva Dias [**130**] (see also [**369**]). Given a closed set, G, in $\widehat{\mathbb{C}}$, the *germ* of an analytic function is an equivalence class of functions analytic in a neighborhood of G with $f \sim g$ if they agree in some neighborhood of G. If G^{int} is dense in G, then equivalence classes are determined by the function on G. In general, one needs all derivatives on G. For example, if $G = \{0\}$, a germ is just a Taylor series with a nonzero radius of convergence. These kinds of germs are a key element of our discussion of global analytic functions in Section 11.2. Here is the duality theorem:

Theorem 6.1.5 (Duality for $\mathfrak{A}(\Omega)$)**.** *Let Ω be a region in $\mathbb{C}$ and $\mathfrak{A}(\Omega)$ the Fréchet space of analytic functions in Ω with the topology of uniform convergence on compacts. Then $\mathfrak{A}(\Omega)^*$, its dual space, is naturally isomorphic to the germs of analytic functions, F, on $\widehat{\mathbb{C}} \setminus \Omega$ with $F(\infty) = 0$. Explicitly, if $L \in \mathfrak{A}(\Omega)^*$ and g is the function $g(w) = w$, then F is given by*

$$F(z) = \ell\left(\frac{1}{z-g}\right) \tag{6.1.10}$$

$$F^{(k)}(z) = \ell\left(\frac{(-1)^k k!}{(z-g)^{-k-1}}\right), \qquad z \neq \infty \tag{6.1.11}$$

$$\text{“}F^{(k)}(\infty)\text{”} = k!\ell(g^{k-1}), \qquad z = \infty,\ k \geq 1 \tag{6.1.12}$$

Conversely, L is associated to F via

$$L(f) = \frac{1}{2\pi i}\oint_\Gamma F(z) f(z)\, dz \tag{6.1.13}$$

where Γ is any mesh-defined contour in Ω that surrounds $K = \widehat{\mathbb{C}} \setminus \widetilde{\Omega}$ with $\widetilde{\Omega}$ an open set containing $\widehat{\mathbb{C}} \setminus \Omega$ on which F is analytic.

Remark. by “$F^{(k)}(\infty)$,” we mean the Taylor coefficients of F near infinity in z^{-1} coordinates, that is, “$F^{(k)}(\infty)$”$= \frac{d^k}{dw^k} F(w^{-1})\Big|_{w=0}$.

The reader will prove this in Problem 7 after exploring the case $\Omega = \mathbb{D}$ in Problems 5 and 6 and a preliminary determination in Problem 2.

Problems

1. (Weierstrass M-test) Let $\{f_j\}_{j=1}^{\infty} \subset \mathfrak{A}(\Omega)$. Suppose for every compact $K \subset \Omega$, $\sum_{j=1}^{\infty} \|f_j\|_K < \infty$. Prove there exists $g \in \mathfrak{A}(\Omega)$ so that $\sum_{j=1}^{n} f_j \to g$ in the natural topology on $\mathfrak{A}(\Omega)$.

   The remaining problems in this section require the reader to know various basics of real analysis from Part 1, including the Hahn–Banach theorem (Theorem 5.5.5 of Part 1) and the Riesz–Markov theorem (Section 4.5 of Part 1).

2. (a) Let ℓ be a continuous linear functional on $\mathfrak{A}(\Omega)$. Prove that there is a constant c and compact $K \subset \Omega$ so that

$$\ell(f) \le c\|f\|_K \tag{6.1.14}$$

   (*Hint*: Prove that, with ρ given by (6.1.6), if $\varepsilon > 0$ is given, and n is such that $2^{-n} < \varepsilon/2$, and $\|f\|_{K_n} < \varepsilon/2$, then $\rho(f,0) < \varepsilon$.)

   (b) If ℓ is a continuous linear function on $\mathfrak{A}(\Omega)$, prove there is a complex Baire measure μ of compact support $K \subset \Omega$ so that

$$\ell(f) = \int f(z)\, d\mu(z) \tag{6.1.15}$$

   (*Hint*: Use the Hahn–Banach theorem.)

   (c) Conversely, every $\mu \in \mathcal{M}(K)$ defines a continuous linear functional on $\mathfrak{A}(\Omega)$ via (6.1.15).

   (d) Is the μ of (6.1.15) unique?

   Remark. The nonuniqueness issue is resolved in Problem 7.

3. A set, $S \subset \mathfrak{A}(\Omega)$, is called bounded if and only if for every open set $U \ni 0$, there is a λ with $S \subset \lambda U \equiv \{\lambda f \mid f \in U\}$. Prove that S is bounded if and only if, for any compact K, $\sup_{f\in S}\|f\|_K < \infty$. (*Hint*: Use the hint for Problem 2(a).)

4. Prove that a set, $S \subset \mathfrak{A}(\Omega)$, is bounded in the sense of Problem 3 if and only if for every bounded linear functional, ℓ, $\sup_{f\in S}|\ell(f)| < \infty$.
   Hint: If $\sup_{f\in S}\|f\|_K = \infty$ for some K, construct ℓ as follows: Find $f_n \in S$ inductively so that $\|f_1\|_K \ge 1$ and $\|f_{n+1}\|_K \ge 8\|f_n\|_K$ and $z_n \in K$ so $|f_n(z_n)| = \|f_n\|_K$. Let

$$\ell(g) = \sum_{n=1}^{\infty} e^{i\psi_n} 2^n \|f_n\|^{-1} g(z_n) \tag{6.1.16}$$

   where ψ_n is picked inductively so that

$$\arg[e^{i\psi_n} g(z_n)] = \arg\left[\sum_{m=1}^{n-1} e^{i\psi_m} 2^m \|f_m\|^{-1} f_n(z_n)\right] \tag{6.1.17}$$

5. Let $L \in \mathfrak{A}(\mathbb{D})^*$, the set of continuous linear functionals on the Fréchet space, $\mathfrak{A}(\mathbb{D})$.

(a) Let $L_n \equiv L(z^n)$. Prove that for some $A > 0$, $r < 1$,

$$|L_n| \le Ar^n \tag{6.1.18}$$

and that

$$L\biggl(\sum_{n=0}^{\infty} a_n z^n\biggr) = \sum_{n=0}^{\infty} a_n L_n \tag{6.1.19}$$

(b) Prove that, given any sequence $\{L_n\}_{n=0}^{\infty}$ obeying (6.1.18), there is a unique $L \in \mathfrak{A}(\mathbb{D})^*$ with $L(z^n) = L_n$. Thus, $\mathfrak{A}(\mathbb{D})^*$ can be naturally realized as the space of sequences obeying (6.1.18).

6. Let $L \in \mathfrak{A}(\mathbb{D})^*$ and $L_n = L(z^n)$ as in Problem 5. Suppose L_n obeys (6.1.18).

(a) Prove that

$$F(z) = \sum_{n=0}^{\infty} L_n z^{-n-1} \tag{6.1.20}$$

is analytic in a neighborhood of $\widehat{\mathbb{C}} \setminus \mathbb{D}$ with $F(\infty) = 0$.

(b) If γ_ε is the circle of radius $1 - \varepsilon$ about 0, prove that for all ε small and all $f \in \mathfrak{A}(\mathbb{D})$,

$$\frac{1}{2\pi i} \oint_{\gamma_\varepsilon} f(z) F(z) = L(f) \tag{6.1.21}$$

so this sets up a one-one correspondence between $\mathfrak{A}(\mathbb{D})^*$ and functions, F, on $\widehat{\mathbb{C}} \setminus \mathbb{D}$ with $F(\infty) = 0$ such that F has an analytic continuation to a neighborhood of $\widehat{\mathbb{C}} \setminus \mathbb{D}$.

7. Let $\Omega \subset \mathbb{C}$ be a region and $L \in \mathfrak{A}(\Omega)^*$. Let K be a compact so that L has a representation of the form (6.1.11). For $z \in \widehat{\mathbb{C}} \setminus K$, define

$$F(z) = \int \frac{d\mu(w)}{z - w} \tag{6.1.22}$$

(a) Prove that F is analytic in a neighborhood of $\widehat{\mathbb{C}} \setminus \Omega$.

(b) If Γ is a mesh-defined contour in Ω with $n(\Gamma, z) = 1$ (respectively, 0) if $z \in K$ (respectively, $z \in \widehat{\mathbb{C}} \setminus \Omega$), prove that for all $f \in \mathfrak{A}(\Omega)$,

$$L(f) = \frac{1}{2\pi i} \oint_{\Gamma} f(z) F(z)\, dz \tag{6.1.23}$$

(c) Conclude Theorem 6.1.5.

8. An element, L, of $\mathfrak{A}(\Omega)^*$ is called a *multiplicative linear functional* if and only if $L(fg) = L(f)L(g)$ and $L(\mathbb{1}_\Omega) = 1$ where $\mathbb{1}_\Omega$ is the function identically 1 on Ω.

   (a) Prove that $L(z) = z_0 \in \Omega$. (*Hint*: If $z_0 \notin \Omega$, there is g analytic in Ω with $(\cdot - z_0)g = \mathbb{1}_\Omega$.)

   (b) Prove that for any $f \in \mathfrak{A}(\Omega)$, $L(f) = f(z_0)$. (*Hint*: There is a $g \in \mathfrak{A}(\Omega)$ with $(\cdot - z_0)g = f(\cdot) - f(z_0)$.) Thus, every multiplicative linear functional on $\mathfrak{A}(\Omega)$ is of the form $L_{z_0}(f) = f(z_0)$ for some $z_0 \in \Omega$.

   Remark. In Chapter 6 of Part 4, we study multiplicative linear functionals on Banach algebras (note that $\mathfrak{A}(\Omega)$ is a Fréchet space, *not* a Banach space).

9. Let T be a topological algebra isomorphism from $\mathfrak{A}(\Omega)$ to $\mathfrak{A}(\Omega')$ as in Theorem 6.1.4.

   (a) Prove $T(\mathbb{1}_\Omega) = T(\mathbb{1}_{\Omega'})$.

   (b) Prove there is a bijection $F\colon \Omega' \to \Omega$ so $T(f) = f \circ F$. (*Hint*: Use Problem 8 to define F via $T^*(L_w) = L_{F(w)}$.)

   (c) Prove F is holomorphic. (*Hint*: Let $f \in \mathfrak{A}(\Omega)$ by $f(z) = z$.)

10. A *unit* in $\mathfrak{A}(\Omega)$ is an element $f \in \mathfrak{A}(\Omega)$ so that there is $g \in \mathfrak{A}(\Omega)$ with $fg = \mathbb{1}_\Omega$. We say f has a square root if there is g with $g^2 = f$.

    (a) Prove f is a unit if and only if f is everywhere nonvanishing. f, a unit, has a square root if and only if for any closed curve γ in Ω,

    $$N_\gamma(f) = \frac{1}{2\pi i}\oint_\gamma \frac{f'(z)}{f(z)}\,dz \tag{6.1.24}$$

    obeys $N_\gamma(f) \in 2\mathbb{Z}$. (*Hint*: Try to define g by

    $$g(z) = \sqrt{f(z_0)}\,\exp\biggl(\int_{z_0}^{z} \frac{f'(z)}{2f(z)}\,dz\biggr) \tag{6.1.25}$$

    Conversely, if $f = g^2$, then $N_\gamma(f) = 2N_\gamma(g)$.)

    (b) Prove that every unit in $\mathfrak{A}(\Omega)$ has a square root if and only if Ω is simply connected.

    (c) Prove Ω is doubly connected if and only if there is a unit $f \in \mathfrak{A}(\Omega)$ with no square root, but for any two units, f_1 and f_1, at least one of f_1, f_2 and $f_1 f_2$ has a square root.

    See Luecking–Rubel [**369**, Ch. 20].

6.2. Montel's and Vitali's Theorems

The main result in this section (Montel's theorem) is that a subset, S, of $\mathfrak{A}(\Omega)$ is compact if and only if it is closed and bounded. Of course, this is true for $\mathbb{R}^n$, but it is never true for an infinite-dimensional Banach space (see Theorem 5.1.7 of Part 1). It is of general interest but will also be a critical part of some of our proofs of the Riemann mapping and the great Picard theorems (see Sections 8.1, 11.3, and 11.4). It will also allow us to state a convergence criterion called the Vitali convergence theorem.

We begin with two variants of Montel's theorem:

Theorem 6.2.1 (Montel's Theorem, First Form). *Let K_n be a compact exhaustion of a region Ω. Let $\{c_j\}_{j=1}^\infty$ be a sequence of positive numbers. Then the set*

$$S_\Omega(\{c_j\}_{j=1}^\infty) = \{f \in \mathfrak{A}(\Omega) \mid \|f\|_{K_j} \le c_j\} \tag{6.2.1}$$

is compact in the natural topology of $\mathfrak{A}(\Omega)$.

Proof. We'll give a proof here using the Ascoli–Arzelà theorem (see Theorem 2.3.14 of Part 1). A more bare-hands proof is sketched in Problem 1. We use S for the set in (6.2.1). Clearly, S is closed, so we need only show every sequence in S has a convergent subsequence.

We slightly refine the proof in Theorem 6.1.3. For each n, let

$$\delta_n = \text{dist}(K_n, \Omega \setminus K_{n+1}) > 0 \tag{6.2.2}$$

since $K_n \subset K_{n+1}^{\text{int}}$. Then, by a Cauchy estimate (see (3.1.12)),

$$\begin{aligned} \sup_{\{z \mid \text{dist}(z,K_n) \le \delta_n/3\}} |f'(z)| &\le \frac{3}{\delta_n} \sup_{K_{n+1}} |f(w)| \\ &\le \frac{3c_{n+1}}{\delta_n} \end{aligned} \tag{6.2.3}$$

Thus, for any $\zeta \in K_n$ and $w, z \in \overline{\mathbb{D}_{\delta_n/3}(\zeta)}$, we have, for all $f \in S$, that

$$|f(w) - f(z)| \le 3c_{n+1}\delta_n^{-1}|w - z| \tag{6.2.4}$$

Therefore, the Ascoli–Arzelà theorem (see Theorem 2.3.14 of Part 1) applies, and given any sequence, f_n, in S, we can find a subsequence so that f_{n_j} converges uniformly on $\overline{\mathbb{D}_{\delta_n/3}(\zeta)}$. Since K_n can be covered by finitely many $\{\mathbb{D}_{\delta_n/3}(\zeta)\}_{\zeta \in K_n}$, we can find a subsequence converging on K_n.

By the diagonalization trick, we can find a subsequence f_{n_j} converging uniformly on each K_m, and so in $\mathfrak{A}(\Omega)$. $\square$

An equivalent form is immediate:

Theorem 6.2.2 (Montel's Theorem, Second Form). *Let* $\{f_n\}_{n=1}^{\infty}$ *be a sequence in* $\mathfrak{A}(\Omega)$ *so that for any compact* $K \subset \Omega$,

$$\sup_n \|f_n\|_K < \infty \tag{6.2.5}$$

Then, there is a subsequence converging in $\mathfrak{A}(\Omega)$.

The last form uses the notion of bounded subsets of $\mathfrak{A}(\Omega)$ discussed in Problems 3 and 4 of Section 6.1. S is bounded if and only if for any $K \subset \Omega$ compact,

$$\sup_{f \in S} \|f\|_K < \infty \tag{6.2.6}$$

Theorem 6.2.3 (Montel's Theorem, Third Form). $S \subset \mathfrak{A}(\Omega)$ *is compact if and only if* S *is closed and bounded.*

Proof. If S is closed and bounded, by comparing (6.2.6) and (6.2.5), we see any sequence in S has a convergent subsequence. Thus, S is closed and precompact, and so compact (see Section 2.3 of Part 1).

Conversely, if S is compact, it is automatically closed, and since $f \to \|f\|_K$ is continuous for any K, (6.2.6) is automatic since a continuous real-valued function on a compact set is bounded. $\square$

Example 6.2.4 (Schur Functions). A *Schur function* is an analytic function $f\colon \mathbb{D} \to \mathbb{D}$. Let

$$S = \{f\colon \mathbb{D} \to \mathbb{C} \mid \sup_{z \in \mathbb{D}} |f(z)| \le 1\} \tag{6.2.7}$$

By Montel's theorem, this set is compact in the $\mathfrak{A}(\mathbb{D})$ topology. If $f \in S$ and $|f(z)| = 1$ for some $z_0 \in \mathbb{D}$, then by the maximum principle, $f(z)$ is a constant, c, and obviously with $|c| = 1$. Thus, the set S of (6.2.7) is the disjoint union of the Schur functions and $\{f(z) \equiv \zeta_0 \mid \zeta_0 \in \partial\mathbb{D}\}$.

Note that S is the unit ball in the Banach space, $H^\infty(\mathbb{D})$, of bounded analytic functions on $\mathbb{D}$ with $\|\cdot\|_\infty$-norm. It is not compact in this topology! One has to be careful to note the topology for compactness results. See Problem 14. $\square$

Example 6.2.5. Let S_0 consist of the Schur functions with $f(0) = 0$. By the Schwarz lemma, this is precisely the set of analytic functions obeying

$$|f(z)| \le |z| \tag{6.2.8}$$

for each $z \in \mathbb{D}$. This is a compact set where, for any f, $\|f\|_{\mathbb{D}} < 1$ and never equal one, but despite this strict inequality, the set is closed, indeed compact. It is critical, of course, that $\|\cdot\|_{\mathbb{D}}$ is not continuous in the topology in which S_0 is compact! $\square$

Example 6.2.6 (Carathéodory Functions). A *Carathéodory function* (defined already in Problem 8 of Section 5.3) is an analytic function f with $\operatorname{Re} f(z) > 0$ and $f(0) = 1$. We claim the set of such functions is compact. One way of seeing this is to note that the associations,

$$F(z) = \frac{1+f(z)}{1-f(z)} \tag{6.2.9}$$

$$f(z) = \frac{1-F(z)}{1+F(z)} \tag{6.2.10}$$

which are inverse to each other, set up a one-one correspondence between Carathéodory functions and Schur functions with $f(0) = 0$. Moreover (see Problem 2), $f \mapsto F$ is continuous in the topology of $\mathfrak{A}(\mathbb{D})$. Thus, the image is compact.

Alternatively, the Schwarz lemma and (6.2.9) imply that (see Problem 3) if F is a Carathéodory function, then

$$\frac{1-|z|}{1+|z|} \le |F(z)| \le \frac{1+|z|}{1-|z|} \tag{6.2.11}$$

The upper bounds and Montel's theorem show the set of Carathéodory functions are precompact and the maximum principle for the harmonic function $-\operatorname{Re} F(z)$ shows limits continue to be Carathéodory functions. $\square$

Example 6.2.7 (Herglotz Functions). A *Herglotz function* is an analytic map of $\mathbb{C}_+$ to $\mathbb{C}_+$. The map

$$G(z) = \frac{i-z}{z+i} \tag{6.2.12}$$

is a one-one map of $\mathbb{C}_+$ to $\mathbb{D}$ (we'll say more about this in Section 7.3, but it is easy to see directly that $|G(x+iy)| < 1$) and that

$$G^{-1}(z) = \frac{i(1-z)}{1+z} \tag{6.2.13}$$

has $\operatorname{Im} G(z) = (1-|z|^2)/(1+z)^2 > 0$ if $z \in \mathbb{D}$. Thus, F is Herglotz if and only if $G(F(G^{-1}(z)))$ is a Schur function. Thus, there is a one-one correspondence between Herglotz functions and Schur functions, but despite that, the Herglotz functions are not compact in $\mathfrak{A}(\mathbb{C}_+)$. Indeed, $f_n(z) = in$ has no limit point. This limitation will be addressed later in this section. $\square$

One simple consequence of Montel's theorem is:

Theorem 6.2.8 (Vitali Convergence Theorem). *Let Ω be a region and $f_n \in \mathfrak{A}(\Omega)$ a sequence obeying (6.2.5) for all compact $K \subset \Omega$. Let $z_m \in \Omega$ be a sequence with $z_m \to z_\infty \in \Omega$. Suppose that for each $m = 1, 2, \dots$, $\lim_{n\to\infty} f_n(z_m) = w_m$ exists. Then there exists $f \in \mathfrak{A}(\Omega)$ so $f_n \to f$ in the natural topology on $\mathfrak{A}(\Omega)$.*

Proof. By a standard argument in metric space theory (Problem 4), it suffices to show all limit points are equal, that is, if $f_{n_j} \to f$ and $f_{m_\ell} \to g$, then $f = g$.

By hypothesis, $f(z_m) = w_m = g(z_m)$, so $(f-g)(z_m) = 0$. It follows from Theorem 2.3.8 that $f - g = 0$, that is, $f = g$. $\square$

For typical applications, f_n converges on $\mathbb{R} \cap \Omega$, which one supposes contains an interval; see also Problem 5.

Finally, we want to return to the example of Herglotz functions. As we saw in Example 6.2.7, the set of functions taking $\mathbb{C}_+$ to $\mathbb{C}_+$ is not compact—the example we saw of functions with no limit point was $f_n(z) = in$, which converged to infinity uniformly on compacts. In fact, as we'll see, this is the only counterexample. This leads to:

Definition. A family, $S \subset \mathfrak{A}(\Omega)$, is called *normal* if and only if for every sequence f_n in S, there is a subsequence f_{n_j} which is either convergent uniformly on compacts $K \subset \Omega$ to an analytic function f, or else f_{n_j} converges uniformly on compacts to ∞, in the sense that for any compact $K \subset \Omega$, $\inf_{z\in K}|f_{n_j}(z)| \to \infty$; equivalently, $\|f_{n_j}^{-1}\|_K \to \infty$.

Example 6.2.7, revisited. Let f_n be a sequence of Herglotz functions and $g_n = G \circ f_n$ where G is given by (6.2.12). Then, g_n is a sequence of functions with values in $\mathbb{D}$. By the same analysis as in Example 6.2.4, there is g_{n_j} and g_∞ with $g_{n_j} \to g_\infty$ and g_∞ analytic, and either

$$\operatorname{Ran}(g_\infty) \subset \mathbb{D} \tag{6.2.14}$$

or

$$g_\infty(z) = e^{i\theta} \qquad \text{for some } \theta \in [0, 2\pi) \tag{6.2.15}$$

In the latter case, we separate out $\theta = \pi$ (i.e., $e^{i\theta} = -1$) since $G^{-1}(-1) = \infty$ and $\theta \neq \pi$. Apply G^{-1} to $g_n \to g_\infty$ using the fact that G^{-1} is continuous on $\overline{\mathbb{D}} \setminus \{-1\}$, we get three possibilities, uniformly on compacts $K \subset \mathbb{C}_+$:

(1) $f_{n_j} \to f$, a Herglotz function
(2) $f_{n_j} \to x \in \mathbb{R}$
(3) $f_{n_j} \to \infty$

This shows the Herglotz functions are a normal family. $\square$

The same argument yields the following (details are left to Problem 6):

Theorem 6.2.9. *Let Ω be a region and let $\{K_n\}_{n=1}^\infty$ be a compact exhaustion. Fix $w_0 \in \mathbb{C}$ and $\delta_n > 0$. Then*

$$\{f \in \mathfrak{A}(\Omega) \mid \min_{K_n}|f(z) - w_0| \geq \delta_n\} \tag{6.2.16}$$

is a normal family.

Notes and Historical Remarks. The Vitali and Montel theorems are intimately related. We've seen how to derive Vitali's theorem from Montel's, but one can do the opposite—since, by use of the diagonalization trick, one easily gets convergence of a subsequence at any countable subset.

Vitali's theorem was actually proven first—by Vitali [**564**] in 1903 and independently by Porter [**455**] in 1904. Earlier, Stieltjes [**542**] and Osgood [**419**] had precursors. In 1894, Stieltjes noted that boundedness on compacts plus convergence on a single open subset implies convergence on all compacts. In 1901, Osgood noted that boundedness on compacts, plus convergence on a dense subset of Ω suffices for convergence on all compacts. Of course, convergence on a dense set is all that is needed for the Montel via Vitali argument sketched in the last paragraph.

Montel [**389**] proved his theorem in 1907 in his thesis, using the Ascoli–Arzelà method we use here. Koebe [**325**] independently found the result in 1908. Koebe [**325**] was motivated by an 1899 work of Hilbert [**260**], who was interested in convergent subsequences to solve the Dirichlet problem. Both Hilbert and Koebe did their work in connection with proving the Riemann mapping theorem (see the Notes to Section 8.1). Montel didn't mention Hilbert or Vitali, but did quote Stieltjes and Osgood. Some of the older literature, for example, [**500**], calls what we call Montel's theorem the Stieltjes–Osgood theorem.

In 1912, Montel [**390**] introduced the term "normal family," eventually writing a whole monograph [**392**] on it.

Montel proved a strengthened result that extends his signature theorem:

Theorem 6.2.10 (Montel's Three-Value Theorem). *Let $\alpha \neq \beta$ lie in $\mathbb{C}$. The set of functions $f \in \mathfrak{A}(\Omega)$ with $\mathrm{Ran}(f) \subset \mathbb{C} \setminus \{\alpha, \beta\}$ is a normal family.*

We call it a three-value theorem because it implies fairly immediately (once we define normality for meromorphic functions)—see the Notes to Section 6.5—the normality of the set of functions meromorphic on Ω with $\mathrm{Ran}(f) \subset \widehat{\mathbb{C}} \setminus \{\alpha, \beta, \gamma\}$ for distinct α, β, γ in $\widehat{\mathbb{C}}$. Theorem 6.2.10 is then just $\gamma = \infty$.

The result is notable because we'll see it implies Picard's theorem rather directly (see Section 11.3) and we'll provide a proof of it as a road to Picard's theorem also in Section 11.4 and Section 12.4 of Part 2B. Theorem 6.2.10 implies Montel's theorem, since $\sup_n \|f_n\|_K < \infty$ then implies normality on K^{int}, and so, by a use of the diagonalization trick, normality when $\sup_n \|f_n\|_K < \infty$ for all compact K.

Theorem 6.2.10 is from Montel [**390**]. Somewhat earlier, Carthéodory and Landau [**92**] had proven a related result (i.e., a corollary of Theorem 6.2.10) that if $\{f_n\} \in \mathfrak{A}(\Omega)$ have values in $\mathbb{C} \setminus \{\alpha, \beta\}$ and $f_n(w_m)$

converges for all w_m in a sequence with a limit point in Ω, then f_n converges in $\mathfrak{A}(\Omega)$.

Paul Antoine Aristede Montel (1876–1975) and Guiseppe Vitali (1875–1932) both spent part of their careers as high school teachers, although the French Montel eventually rose to Dean of Sciences at the Sorbonne. His students include H. Cartan, Dieudonné, and Marty (whose theorem on normal families will be discussed in Section 6.5). In particular, Montel was Dean of Science during the Nazi occupation of Paris and managed to steer the Sorbonne through a difficult period.

Montel was born in Nice and got his degree from the École Normale Supérieure in Paris supervised by Lebesgue and Borel. The 1918 Academy of Sciences Grand Prix was for global features of iteration of maps and was won by Julia but a small monetary prize was awarded to Montel (who hadn't even entered the competition) because all the entries used normal families.

Vitali was born in Ravenna, Italy, studied initially at the nearby University of Bologna under Arzelà and Enriques and then at Pisa under Dini and Bianchi. As noted, Vitali did his early work (from 1899 until 1923) as a secondary school teacher—not only the convergence result mentioned here but seminal work on Lebesgue integration—including the Vitali set (see Section 4.3 of Part 1), convergence theorems (see the Notes to Section 4.6 of Part 1) and the Vitali covering lemma (see Theorem 2.3.2 of Part 3). In the later part of his career, he made important discoveries in the asymptotics of ordinary differential equations.

Despite many unsuccessful applications to chairs earlier, it was only in 1923 that he got an initial university appointment in Moderna. This followed with chairs in Padua and Bologna. He was only 56 when he died after fewer than ten years in higher academics.

There is an approach to proving the Vitali, and thereby the Montel, theorem that relies on the Schwarz lemma. It is due to Lindelöf [**357**] and Jentzsch [**286**] and is presented in Problem 8.

Carathéodory functions were introduced by him in [**85, 86**]. Schur functions are named after Schur [**507**], and Herglotz functions after [**254**]. For extensive discussion, see Simon [**524**, Sect. 1.3]. We discuss Schur functions further in Section 7.5.

A particularly interesting application of Vitali's theorem to physics is due to Lee and Yang [**347, 348**] and is discussed further after Problem 5.

Problems

1. The purpose of this problem is to lead the reader through an alternate proof of Montel's theorem. $\{f_n\}_{n=1}^\infty$ is a sequence in $\mathfrak{A}(\Omega)$ where Ω is a

region and where (6.2.5) holds. The goal is to find a convergent subsequence.

(a) Prove for any $z_0 \in \Omega$ and $\delta < \mathrm{dist}(z_0, \mathbb{C} \setminus \Omega)$, we have an $M(z_0, \delta)$ so that, for all n and all $k = 0, 1, 2, \ldots$,

$$\left| \frac{f_n^{(k)}(z_0)}{k!} \right| \le M\delta^{-k} \tag{6.2.17}$$

(b) Prove for any $z_0 \in \Omega$ and $\delta < \mathrm{dist}(z_0, \mathbb{C} \setminus \Omega)$, there is a subsequence f_{n_j} so that f_{n_j} converges uniformly on $\overline{\mathbb{D}_{\delta/2}(z_0)}$ to a limit $f_{z_0,\delta}$.

(c) Prove that Ω can be covered by countably many $\mathbb{D}_{\delta/2}(z_0)$ and then, by a use of the diagonalization trick, we can find $f_{n_j} \to f$ uniformly on each of these $\overline{\mathbb{D}_{\delta/2}(z_0)}$.

(d) Using the fact that any compact K can be covered by finitely many of the $\mathbb{D}_{\delta/2}(z_0)$, prove that f_{n_j} converges uniformly on every compact.

2. Prove that in the topology of $\mathfrak{A}(\mathbb{D})$, the map (6.2.7) from f to F is continuous on $\{f \mid \|f\|_{\mathbb{D}} \le 1,\ f(0) = 0\}$.

3. Prove (6.2.11) for Carathéodory functions.

4. Let Q be a compact subset of a metric space, X. Let $\{x_n\}_{n=1}^\infty$ be a sequence of points in X which has a unique limit point x_∞. Prove that $x_n \to x_\infty$. (*Hint*: If $x_n \to x_\infty$ is false, there exists $\varepsilon > 0$ and a subsequence x_{n_j} with $\rho(x_{n_j}, x_\infty) \ge \varepsilon$. Use this to show x_n has a limit point different from x_∞.)

5. Let $f_n(z)$ be a sequence of functions on $\mathbb{D}$ obeying for all n,
   (i) $f_n(x) > 0$ for $x \in (\alpha, \beta) \subset (0, 1)$.
   (ii) For any $\theta \in [0, 2\pi)$ and $x \in (\alpha, \beta)$, $|f_n(xe^{i\theta})| \le |f_n(x)|$.
   (iii) f_n is nonvanishing in $\{z \mid \alpha < |z| < \beta\}$.
   (iv) Uniformly for $x \in (\alpha, \beta)$, $\lim_{n\to\infty} \frac{1}{n} \log(f_n(x)) = p(x)$ exists and is nonzero.

   Prove that $p(x)$ has an analytic continuation to $\{z \mid \alpha < |z| < \beta\}$ and, in particular, is real analytic on (α, β).

   Remark. The problem has to do with an approach to phase transitions due to Lee and Yang [**347, 348**]. One puzzle of physics is how mechanics, which is analytic in all parameters, can produce the kinds of discontinuities seen in real matter. The idea is that true nonanalyticities occur in the idealized limit of infinite volume. f_n here is a finite volume partition function and p a pressure. The Lee–Yang idea is that nonanalyticity comes from zeros of f_n pinching the real axis. As this problem shows, the key to that view is the Vitali convergence theorem.

6. Prove Theorem 6.2.9 by considering $g(z) = |f(z) - w_0|^{-1}$.

7. Prove the following: If f_n is a sequence of functions in $\mathfrak{A}(\Omega)$ obeying (6.2.8), and for some $z_0 \in \Omega$ and all k, $f_n^{(k)}(z_0)$ has a limit, then f_n converges in $\mathfrak{A}(\Omega)$ to a function $f \in \mathfrak{A}(\Omega)$.

8. This problem will lead you through the Jentzsch–Lindelöf proof of the Vitali theorem using the Schwarz lemma instead of direct compactness arguments. So suppose f_n is a sequence of functions on $\mathbb{D}$, with $\sup_n \|f_n\|_{\mathbb{D}} < \infty$ and so that, as $n \to \infty$, $f_n(z_m) \to w_m$ for each z_m in a sequence converging to 0. Suppose also $z_m \neq 0$ for all m.

   (a) Prove that for some C and all m and n,
$$|f_n(z_m) - f_n(0)| \leq C|z_m| \tag{6.2.18}$$

   (b) Prove that $f_n(0)$ is Cauchy, and so has a limit a_0.

   (c) Let g_n be defined by
$$g_n(z) = \begin{cases} (f_n(z) - f_n(0))/z, & z \neq 0 \\ f_n'(0), & z = 0 \end{cases}$$
   Prove $\sup_n \|g_n\|_{\mathbb{D}} < \infty$ and conclude $f_n'(0)$ has a limit a_1. By induction, conclude for all k that $\lim_{n\to\infty} f_n^{(k)}(0) = a_k$ exists.

   (d) Using a Cauchy estimate, show $\sum_{k=0}^{\infty} a_k z^k/k!$ converges for $z \in \mathbb{D}$ and defines a function f which is the limit of f_n in $\mathfrak{A}(\mathbb{D})$.

   (e) Derive the full Vitali theorem from the special case above.

9. This problem assumes some familiarity with Baire category; see Section 5.4 of Part 1. Let $K \subset \mathbb{C}$ be compact with at least two points. Let $Q = K^\infty = \{\{a_n\}_{n=0}^\infty \mid a_n \in K\}$, thought of as the coefficients of a power series, and put the metric $\rho(a,b) = \sum_{n=0}^\infty 2^{-n}|a_n - b_n|$ on Q which turns it into a metric space (with the product topology).

   (a) Let $N, m \in \mathbb{Z}$ be positive and p, q rational with $p < q$. Let $S_{N,m,p,q} = \{\{a_n\} \in Q \mid \sum_{n=0}^\infty a_n z^n$ has an analytic continuation, f, to $T = \{z \mid |z| < 1 + m^{-1}, \frac{p}{2\pi} < \arg(z) < \frac{q}{2\pi}\}$ with $\sup_{z\in T}|f(z)| \leq N\}$. Prove that S is closed in Q.

   (b) Conclude $Q_1 = \{\{a_n\} \in Q \mid \sum_{n=0}^\infty a_n z^n$ has a natural boundary on $\partial\mathbb{D}\}$ is a G_δ.

   (c) Using the Weierstrass $\sum_{n=0}^\infty z^{n!}$ example, prove that Q_1 is nonempty.

   (d) Prove that any $\{a_n\}_{n=0}^\infty \in Q$ is a limit of points in Q_1 so that Q_1 is a dense G_δ, that is, Baire generic. (*Hint*: Replace the tail of $\{a_n\}_{n=0}^\infty$ by the example from (c).)

   Remark. This is a result from Breuer–Simon [**72**].

10. This problem involves the maximum principle only—it is here as preparation for the next problem which uses the Vitali theorem. This problem proves a lemma of M. Riesz [**486**]. Let $\{a_n\}_{n=0}^\infty$ be a set of power series coefficients with

$$\sup_n |a_n| < \infty \tag{6.2.19}$$

Suppose $f(z) = \sum_{n=0}^\infty a_n z^n$ has an analytic continuation into a neighborhood, Q, of $\{z \mid 0 \le |z| \le R,\ \alpha \le \arg(z) \le \beta\} \equiv S$ for some $\alpha, \beta \in [0, 2\pi]$ with $\alpha < \beta$ and $R > 1$. For $N = 1, \dots,$

$$f_+^{(N)}(z) = \sum_{n=0}^\infty a_{n+N} z^n \tag{6.2.20}$$

$$f_-^{(N)}(z) = \sum_{n=-N}^{-1} a_{n+N} z^n \tag{6.2.21}$$

(a) Prove $f_+^{(N)}$ is analytic on $\mathbb{D} \cup Q$ and $f_-^{(N)}$ on $\widehat{\mathbb{C}} \setminus \{0\}$.

(b) Prove that on $Q \setminus \{0\}$,

$$f_+^{(N)}(z) + f_-^{(N)}(z) = z^{-N} f(z) \tag{6.2.22}$$

(c) Prove that (the lemma of Riesz)

$$\sup_{z \in S, N} |f_+^{(N)}(z)| < \infty \tag{6.2.23}$$

(*Hint*: Pick α_0, β_0 with $\alpha_0 < \alpha < \beta < \beta_0$ so that $\widetilde{S} = \{z \mid 0 \le |z| \le R,\ \alpha_0 \le |\arg(z)| \le \beta_0\} \subset Q$. Apply the maximum principle on $\widetilde{S}$ to $g^N(z) = (z - e^{i\alpha_0})(z - e^{i\beta_0}) f_+^{(N)}(z)$. (6.2.22) will be useful.)

11. This problem will provide an approach to the study of power series with natural boundaries on $\partial\mathbb{D}$. It follows a paper of Breuer–Simon [**72**], who have lots of further applications. The actual result you'll prove is a special case of a theorem of Agmon [**5**]. Let $\{a_n\}_{n=0}^\infty$ be a set of power series coefficients obeying (6.2.19). Let $Q, S, f_\pm^{(N)}$ be as in Problem 10. We say $\{b_n\}_{n=-\infty}^\infty$ is a *right limit* of $\{a_n\}_{n=0}^\infty$ if and only if for some $N_j \to \infty$ and all $m \in \mathbb{Z}$,

$$a_{N_j+m} \to b_m \tag{6.2.24}$$

as $j \to \infty$.

(a) Under the existence of Q, S, and a right limit, define f_+ on $\mathbb{D}$ and f_- on $\widehat{\mathbb{C}} \setminus \overline{\mathbb{D}}$ by

$$f_+(z) = \sum_{n=0}^\infty b_n z^n, \qquad f_-(z) = \sum_{n=-\infty}^{-1} b_n z^n$$

Prove that f_+ has an analytic continuation to a neighborhood of S and on $S \cap \{z \mid |z| > 1\}$ obeys

$$f_+(z) + f_-(z) = 0 \tag{6.2.25}$$

(b) Prove that if $\{a_n\}_{n=0}^\infty$ is a sequence obeying (6.2.19), which has a right limit for which f_+ has a continuation outside $\mathbb{D}$ for which (6.2.25) fails, then $\sum_{n=0}^\infty a_n z^n$ has a natural boundary on $\mathbb{D}$.

(c) Let $\{a_n\}_{n=0}^\infty$ be a sequence so that for some $N_j \to \infty$, $a_{N_j-1} \to \alpha \neq 0$, and for all $m \geq 0$, $a_{N_j+m} \to 0$. Prove that $\sum_{n=0}^\infty a_n z^n$ has a natural boundary on $\partial\mathbb{D}$. (*Hint*: Use compactness to show that for a subsequence, there is a right limit with $f_+ \equiv 0$ but $f_- \not\equiv 0$.)

(d) Prove the special case of the Fabry gap theorem with $\sup_n |a_n| < \infty$. That is, if $f(z) = \sum_{n=0}^\infty a_j z^{n_j}$ with $\lim_{j\to\infty} n_j/j = \infty$, $\sup_j |a_j| < \infty$, and $\liminf |a_j| > 0$, then f has a natural boundary on $\partial\mathbb{D}$.

(e) Prove that $\sum_{n=0}^\infty z^{n^2}$ has a natural boundary on $\partial\mathbb{D}$ (Kronecker example).

Remarks. 1. While this method has the limitation that $\sup_n |a_n| < \infty$, it doesn't require that there only be gaps.

2. $\sum_{n=0}^\infty z^{n^2}$ will play a starring role in Section 10.5 and Sections 13.1 and 13.3 of Part 2B.

12. Normal families were used by Fatou and Julia to study iterates of analytic maps (and rational maps on all of $\widehat{\mathbb{C}}$). Let $f\colon \Omega \to \Omega$. Define $f^{[n]}$ by $f^{[1]} = f$, $f^{[n+1]} = f \circ f^{[n]}$. The *Fatou set* of $f = \{z_0 \in \Omega \mid \exists\delta$ so that $\{f^{[n]} \restriction \mathbb{D}_\delta(z_0)\}$ is normal$\}$. The complement is called the *Julia set*.

(a) Let $f(z) = z + 1$. Find $f^{[n]}$ and show that the Julia set is empty.

(b) Let $f(z) = z^2$. Find the Julia set.

Remark. The celebrated Mandelbrot set is $\{c \in \mathbb{C} \mid$ the Julia set of $f(z) = z^2 + c$ is connected$\}$.

13. This problem uses the Baire category theorem (see Theorem 5.4.1 of Part 1). Let Ω be a region in $\mathbb{C}$ and $g_n \in \mathfrak{A}(\Omega)$ so that for all $z \in \Omega$, $g_n(z) \to g_\infty(z)$ for some function g_∞ (finite everywhere). In this problem, you'll prove a theorem of Osgood that there is a dense open subset $\Omega' \subset \Omega$ so that $g_\infty \restriction \Omega'$ is analytic (but note that Ω' may not be connected; see Problem 5 of Section 4.7).

(a) Let $\Omega_m = \{\sup_n |g_n(z)| \leq m\}$. Prove that $\cup\Omega_m = \Omega$.

(b) Let $\Omega' = \cup_m \Omega_m^{\text{int}}$. Prove $g_\infty \restriction \Omega'$ is analytic.

(c) Prove some Ω_m has nonempty interior (use the Baire category theorem).

(d) Let $\widetilde{\Omega} = \Omega \setminus \overline{\Omega'}$. Apply (a)–(c) to $\widetilde{\Omega}$ to see that if $\widetilde{\Omega} = \emptyset$ for some m, $\Omega_m^{\text{int}} \cap \widetilde{\Omega} \neq \emptyset$, which is a contradiction. Conclude that Ω' is dense.

14. Let $f(z) = \exp(-[\frac{1+z}{1-z}])$ on $\mathbb{C} \setminus \{1\}$.

(a) Prove that $|f(z)| < 1$ if $|z| < 1$ and $|f(z)| \geq 1$ if $|z| \geq 1$.

(b) Prove that the limit points of $\{f(z)\}_{z\in\mathbb{D}}$ as $z \to 1$ are $\{w \mid |w| \geq 1\}$. (*Hint*: What kind of singularity does f have at $z = 1$?)

(c) Prove that on $\mathbb{D}$, f is not a limit in $\|\cdot\|_\infty$ of polynomials.

(d) Prove that in $\mathfrak{A}(\mathbb{D})$, f is a limit of polynomials, p_n, with $\|p_n\|_\infty \leq 1$ (on $\mathbb{D}$). (*Hint*: First show if $f_r(z) = f(rz)$, then in $\mathfrak{A}(\mathbb{D})$, $f_r \to f$ as $r \uparrow 1$, and then approximate f_r by truncated, normalized Taylor series.)

(e) Find a sequence of polynomials in the unit ball, $H^\infty(\mathbb{D})$, with no limit point.

6.3. Restatement of Runge's Theorems

Given our focus on $\mathfrak{A}(\Omega)$ and its topology, we want to note the following consequence of Runge's theorems:

Theorem 6.3.1. *For any region Ω, the rational functions with poles outside Ω are dense in $\mathfrak{A}(\Omega)$. If Ω is simply connected, then the polynomials are dense in $\mathfrak{A}(\Omega)$.*

Proof. Let K_n be given by

$$K_n = \{z \in \Omega \mid |z| \leq n,\ \text{dist}(z, \partial\Omega) \geq n^{-1}\}$$

which is a compact exhaustion (perhaps after dropping finitely many n for which K_n is empty or finite). By Theorem 4.7.6, given $f \in \mathfrak{A}(\Omega)$, there is a rational function, R_n, with poles in $\mathbb{C} \setminus \Omega$ so that

$$\|f - R_n\|_{K_n} \leq \frac{1}{n} \tag{6.3.1}$$

Then $R_n \to f$ uniformly on each K_n since for $m \geq n$, $K_n \subset K_m$, so by (6.3.1), $\|f - R_m\|_{K_n} \leq 1/m$.

If Ω is simply connected, by the remark after Theorem 4.7.6, $\mathbb{C} \setminus K_n$ is connected, so by Theorem 4.7.1, the function R_n in (6.3.1) can be taken to be a polynomial. □

Problems

1. (a) Use Theorem 6.1.5 to prove the following: If $\ell \in \mathfrak{A}(\Omega)^*$ and $\ell(f) = 0$ for every $f(z) = (z-w)^{-k}$ for $w \in \mathbb{C} \setminus \Omega$, $k = 0, 1, \dots$, and also for every $f(z) = z^k$, then $\ell = 0$. Also, if $\mathbb{C} \setminus \Omega$ is connected and $\ell(f) = 0$ for all polynomials f, then $f \equiv 0$.

   (b) Using (a) and the Hahn–Banach theorem, provide another proof of Theorem 3.6.1.

6.4. Hurwitz's Theorem

In this section, we'll prove the following result about preservation of zeros under convergence in $\mathfrak{A}(\Omega)$:

Theorem 6.4.1 (Hurwitz's Theorem). *Let $f_n \to f$ in $\mathfrak{A}(\Omega)$ where Ω is a region in $\mathbb{C}$. Suppose f is not the zero function. Then the zeros of f are the limits of zeros of f_n in the following precise sense:*

(a) *If $f(\zeta_0) \neq 0$, there is a $\delta > 0$ and integer, N, so that $f_n(z)$ has no zeros in $\mathbb{D}_\delta(\zeta_0)$ for $n > N$.*
(b) *If ζ_0 is a zero of order k of f, then for suitable $\delta > 0$ and N, for $n > N$, $f_n(z)$ has exactly k zeros in $\mathbb{D}_\delta(\zeta_0)$ counting multiplicity.*
(c) *If $K \subset \Omega$ is compact and f has no zeros on ∂K, then for some N large and $n > N$, the sum of orders of the zeros of f_n in K is the same as for f.*

Remark. This is, of course, closely related to the phenomenon of preservation of zeros under continuous deformation that we discussed in Corollary 3.3.3. As in that case, the analog is false for even very nice real-valued functions of a real variable: $f_n(x) = x^2 + 1/n$ has no zeros in $[-1, 1]$ but $\lim f_n(x) = f_\infty(x)$ vanishes at $x = 0$. Of course, f_n has zeros in the complex plane that converge to zero.

Proof. We'll prove (a) and (b) and leave (c) to the reader. (a) is (b) when $k = 0$, so we'll prove them both together. Since f is not the zero function, either $f(\zeta_0) \neq 0$ or f has an isolated zero at ζ_0. Thus, for some $\delta < \text{dist}(\zeta_0, \partial\Omega)$, we can arrange that f is nonvanishing on $\overline{\mathbb{D}_\delta(\zeta_0)} \setminus \{\zeta_0\}$.

Since f is nonvanishing on the compact set $\partial\mathbb{D}_\delta(\zeta_0)$ for n large, f_n is nonvanishing there, so $f_n'/f_n \to f'/f$ uniformly on $\partial\mathbb{D}_\delta(\zeta_0)$. Thus, with N given by (3.3.3),

$$N_{\mathbb{D}_\delta(\zeta_0)}(f_n) \to N_{\mathbb{D}_\delta(\zeta_0)}(f) \tag{6.4.1}$$

Since they are integers, for some M, they are equal for $n > M$. By the argument principle (Theorem 3.3.2), the zeros result holds. □

An immediate consequence is:

Corollary 6.4.2. *If $f_n \to f$ in $\mathfrak{A}(\Omega)$ and each f_n is nonvanishing on all of Ω, then either $f \equiv 0$ or f is nonvanishing on all of Ω.*

Definition. A *univalent function* (also called *schlicht*) on Ω is a one-one analytic function on Ω.

Corollary 6.4.3. *Let $f_n \in \mathfrak{A}(\Omega)$ be a family of univalent functions and let $f_n \to f$ in $\mathfrak{A}(\Omega)$. Then either f is constant or f is univalent.*

Proof. Fix $\zeta_0 \in \Omega$. Let

$$g_n(z) = f_n(z) - f_n(\zeta_0) \tag{6.4.2}$$

on $\widetilde{\Omega} \equiv \Omega \setminus \{\zeta_0\}$. Then g_n is nonvanishing on $\widetilde{\Omega}$, so $g(z) \equiv f(z) - f(\zeta_0)$ is either identically zero or nonvanishing.

If g is identically zero, then $f(z) \equiv f(\zeta_0)$ constant, so if f is nonconstant, $f(z) \neq f(\zeta_0)$ for $z \neq \zeta_0$. Since ζ_0 is arbitrary, either f is constant or schlicht. $\square$

Notes and Historical Remarks. Hurwitz proved his theorem in 1889 [**270**]. Adolf Hurwitz (1859–1919) was a German Jewish mathematician heavily influenced by Klein and, in turn, he was a significant influence and friend of Hilbert and Minkowski. Hilbert and Hurwitz took daily walks discussing mathematics while both were at Königsberg. Hurwitz spent the last twenty-seven years of his life, many in ill health (he had a diseased kidney removed in 1905), in Zurich where he was a professor at the ETH.

Problems

1. Prove the third conclusion in Theorem 6.4.1. (*Hint*: Use (a), (b), and compactness.)

   The next problems use the notion of iterates $f^{[n]}$ defined in Problem 12 of Section 6.2.

2. Let $f \in \mathfrak{A}(\Omega)$ for a region Ω. Suppose $\operatorname{Ran}(f) \subset \Omega$ and that for some $n_j \to \infty$, uniformly on compacts, $f^{[n_j]} \to \mathbb{1}_\Omega$, the identity function on Ω.

   (a) Prove that f is schlicht.

   (b) Prove that f is onto. (*Hint*: If $w \notin \operatorname{Ran}(f)$, look at $g_j(z) = f^{[n_j]}(z) - w$ and apply Corollary 6.4.2.)

3. Let Ω be a region and $\{g_n\}_{n=1}^\infty \subset \mathfrak{A}(\Omega)$ all take Ω to Ω. Suppose $g_n \to g$ in the topology of $\mathfrak{A}(\Omega)$. Prove that either $\operatorname{Ran}(g) \subset \Omega$ also, or else g takes a constant value in $\partial\Omega$. (*Hint*: If $g(z_0) = w_0$ for $z_0 \in \Omega$, $w_0 \in \partial\Omega$, apply Corollary 6.4.2 to $g(z) - w_0$.)

4. Let Ω be a bounded region and let $f\colon \Omega \to \Omega$ be an element of $\mathfrak{A}(\Omega)$ so that for some $n_j \to \infty$, $f^{[n_j]} \to g$ in $\mathfrak{A}(\Omega)$ and g is nonconstant.

   (a) Prove that $\operatorname{Ran}(g) \subset \Omega$ and is open. (*Hint*: Use Problem 3.)

   (b) By passing to a subsequence, suppose $n_{j+1} > n_j$ and let $m_j = n_{j+1} - n_j$ and let h be a limit point of $f^{[m_j]}$. Prove that $h \circ g = g$. (*Hint*: $f^{[n_{j+1}]} = f^{[m_j]} \circ f^{[n_j]}$.)

   (c) Prove that $h(z) = z$ for all z and conclude that f is a bijection. (*Hint*: Use (a) and then Problem 2.)

5. Let Ω be a bounded region. Let $f\colon \Omega \to \Omega$ be analytic and suppose that f has two distinct fixed points. Prove that f is a bijection. (*Hint*: Use Problem 4.)

   Note. The arguments in Problems 4 and 5 are from H. Cartan [**97**].

6. Prove that the conclusion of Problem 5 holds if, instead of assuming Ω is bounded, we suppose $\mathbb{C} \setminus \Omega$ has a nonempty interior.

6.5. Bonus Section: Normal Convergence of Meromorphic Functions and Marty's Theorem

In this section, we discuss the notion of normal convergence for meromorphic functions and prove an analog of Montel's theorem called Marty's theorem. The key will be to find a replacement for $\|\cdot\|_K = \sup_{z\in K}|f(z)|$. The problem, of course, is that $|\cdot|$ is not a good notion of size for functions taking the value infinity. Instead, we use:

Definition. Let $w, z \in \widehat{\mathbb{C}}$. Define the *spherical metric* $\sigma(w,z)$ as follows:

$$\sigma(w,z) = \begin{cases} \frac{|w-z|}{\sqrt{1+|z|^2}\,\sqrt{1+|w|^2}}, & w,z \in \mathbb{C} \\ \frac{1}{\sqrt{1+|w|^2}}, & w \in \mathbb{C},\ z = \infty \\ 0, & w = z = \infty \end{cases} \tag{6.5.1}$$

In order to understand where this comes from and to prove the triangle inequality, we introduce a map $Q\colon \widehat{\mathbb{C}} \to \mathbb{R}^3$ by

$$Q(z) = \begin{cases} \left(\frac{\operatorname{Re} z}{|z|^2+1}, \frac{\operatorname{Im} z}{|z|^2+1}, \frac{|z|^2}{|z|^2+1}\right), & z \in \mathbb{C} \\ (0,0,1), & z = \infty \end{cases} \tag{6.5.2}$$

For now, we take this out of a hat, but in Section 7.2, we'll see its geometric significance:

Proposition 6.5.1. (a) $|Q(z) - (0,0,\frac{1}{2})|^2 = \frac{1}{4}$ (6.5.3)

(b) *Q is a one-one map of $\widehat{\mathbb{C}}$ onto the sphere of radius $\frac{1}{2}$ centered at $(0,0,\frac{1}{2})$ with (for $(x_1,x_2,x_3)\neq(0,0,1)$)*

$$Q^{-1}(x_1,x_2,x_3)=\frac{x_1+ix_2}{1-x_3} \tag{6.5.4}$$

(c) $|Q(z)-Q(w)|=\sigma(z,w)$ (6.5.5)

(d) *σ obeys the triangle inequality, that is, for $z_1,z_2,z_3\in\mathbb{C}$,*

$$\sigma(z_1,z_3)\le\sigma(z_1,z_2)+\sigma(z_2,z_3) \tag{6.5.6}$$

(e) *Let*

$$\pi_\theta(x_1,x_2,x_3)=(x_1\cos\theta-x_2\sin\theta,x_2\cos\theta+x_1\sin\theta,x_3) \tag{6.5.7}$$

and

$$\tilde{\pi}(x_1,x_2,x_3)=(x_1,-x_2,1-x_3) \tag{6.5.8}$$

Then

$$Q(e^{i\theta}z)=\pi_\theta(Q(z)) \tag{6.5.9}$$

$$Q(1/z)=\tilde{\pi}(Q(z)) \tag{6.5.10}$$

(f) $\sigma(z^{-1},w^{-1})=\sigma(z,w)$ (6.5.11)

(g) $|z|,|w|\le R\Rightarrow\dfrac{|w-z|}{1+R^2}\le\sigma(w,z)\le|w-z|$ (6.5.12)

$$|z|,|w|\ge R^{-1}\Rightarrow\frac{|w^{-1}-z^{-1}|}{1+R^2}\le\sigma(w,z)\le|w^{-1}-z^{-1}| \tag{6.5.13}$$

Remark. (6.5.5) shows $\sigma(w,z)$ is a chordal distance on the three-sphere, that is, the length of the chord in a three-ball. The arc-distance, $\theta(w,z)$, that is, the length of the great circle in the sphere is given by

$$\theta(w,z)=\arctan\left(\frac{|w-z|}{|1+\bar{w}z|}\right) \tag{6.5.14}$$

Problem 1 has a proof of this. Of course, θ and σ determine equivalent metrics since they come from equivalent metrics on S^2. Indeed, since $2\pi^{-1}\theta\le2\sin(\theta/2)\le\theta$, we have that

$$\frac{2}{\pi}\theta(w,z)\le\sigma(w,z)\le\theta(w,z) \tag{6.5.15}$$

Proof. (a) The sphere of radius $\frac{1}{2}$ about $(0,0,\frac{1}{2})$ is given by

$$x_1^2+x_2^2+(x_3-\tfrac{1}{2})^2=\tfrac{1}{4}\Leftrightarrow x_1^2+x_2^2+x_3^2=x_3 \tag{6.5.16}$$

so (6.5.3) is

$$\frac{|z|^2}{(|z|^2+1)^2}+\frac{|z|^4}{(|z|^2+1)^2}=\frac{|z|^2}{|z|^2+1} \tag{6.5.17}$$

which is immediate.

(b) It is easy to check $Q^{-1} \circ Q(z) = z$, and that if (6.5.16) holds, then $Q \circ Q^{-1}(x) = x$, so Q is a bijection.

(c) By (6.5.16),

$$|Q(z)|^2 + |Q(w)|^2 = \frac{|z|^2}{1+|z|^2} + \frac{|w|^2}{1+|w|^2} = \frac{|z|^2 + |w|^2 + 2|w|^2|w|^2}{(1+|z|^2)(1+|w|^2)} \tag{6.5.18}$$

and, by (6.5.2),

$$Q(z) \cdot Q(w) = \frac{\operatorname{Re}(\bar{z}w) + |z|^2|w|^2}{(|z|^2+1)(|w|^2+1)} \tag{6.5.19}$$

so

$$|Q(w) - Q(z)|^2 = \frac{|z-w|^2}{(1+|z|^2)(1+|w|^2)} \tag{6.5.20}$$

which is (6.5.5).

(d) is immediate from (6.5.5); indeed, if z_1, z_2, z_3 are distinct, the inequality is strict.

(e) is a straightforward calculation.

(f) follows from (6.5.10) or from a direct calculation.

(g) (6.5.12) is elementary and (6.5.13) then follows from (6.5.11). □

Definition. We let $\mathcal{M}(\Omega)$ be the functions meromorphic in Ω together with the function $f \equiv \infty$.

In Sections 7.1 and 7.2, we give $\widehat{\mathbb{C}}$ the structure of a complex analytic manifold. In particular, we can define "analytic" functions from Ω to $\widehat{\mathbb{C}}$ as functions analytic in local coordinates. $\mathcal{M}(\Omega)$ will precisely be the "analytic" functions of Ω to $\widehat{\mathbb{C}}$.

Definition. We define for $K \subset \Omega$ compact,

$$\sigma_K(f,g) = \sup_{z \in K} \sigma(f(z), g(z)) \tag{6.5.21}$$

If $\{K_n\}_{n=1}^\infty$ is a compact exhaustion of Ω, we set

$$\rho(f,g) = \sum_{n=1}^\infty \min(\sigma_{K_n}(f,g), 2^{-n}) \tag{6.5.22}$$

If $\rho(f_n, f_\infty) \to 0$, we say f_n converges to f_∞ *normally*. A set $S \subset \mathcal{M}(\Omega)$ is called a *normal family* if and only if any sequence $\{f_n\}_{n=1}^\infty \subset S$ has a normally convergent subsequence.

Remark. We'll see below that this is consistent with our previous definition of normal family.

Theorem 6.5.2. (a) *Let* $\{f_n\}_{n=1}^\infty \subset \mathcal{M}(\Omega)$ *and let* f *be a function from* Ω *to* $\widehat{\mathbb{C}}$ *so that*

$$\sigma_{K_m}(f_n, f) \to 0 \tag{6.5.23}$$

for each $\mathcal{M}$. *Then* $f \in \mathcal{M}(\Omega)$.

(b) $\mathcal{M}(\Omega)$ *is a complete metric space in the metric,* ρ.

Proof. (a) Suppose $f(z_0) \neq \infty$. Since $Q(f_n(z)) \to Q(f(z))$ uniformly on compacts, $Q(f(z))$ is continuous, so for some $\delta > 0$, $|f(z)| \leq 2|f(z_0)|$ on $\overline{\mathbb{D}_\delta(z_0)}$. Thus, $|f_n(z)| \leq 4|f(z_0)|$ for n large and $z \in \overline{\mathbb{D}_\delta(z_0)}$. By (6.5.12), $f_n \to f$ uniformly on $\overline{\mathbb{D}_\delta(z_0)}$, so f is analytic in $\mathbb{D}_\delta(z_0)$.

If $f(z_0) = \infty$, we use $\sigma_{K_m}(f_n^{-1}, f^{-1}) = \sigma_{K_m}(f_n, f) \to 0$, and so f^{-1} is analytic in $\mathbb{D}_\delta(z_0)$, so either $f(z) \equiv \infty$ on $\mathbb{D}_\delta(z_0)$ or else f is meromorphic at z with a pole at z_0.

By a connectedness argument, using continuity of $Q(f(z))$, either $f \equiv \infty$ on Ω or f is meromorphic on Ω, that is, $f \in \mathcal{M}(\Omega)$.

(b) If $\sigma_{K_m}(f_n, f_m) \to 0$ as $n, m \to \infty$, then $\sup_K |Q(f_n(z)) - Q(f_m(z))| \to 0$. So, by completeness of continuous functions from Ω to S^2, there is an f so $\sigma_{K_m}(f_n, f) \to 0$. By (a), $f \in \mathcal{M}(\Omega)$. $\square$

Theorem 6.5.3. *Let* $\{f_n\}_{n=1}^\infty$ *and* f *be in* $\mathcal{M}(\Omega)$. *Suppose* (6.5.23) *holds and each* $f_n \in \mathfrak{A}(\Omega)$. *Then either* $f \in \mathfrak{A}(\Omega)$ *or* $f \equiv \infty$. *In the former case, the convergence is in* $\mathfrak{A}(\Omega)$. *In particular, normal family, as defined in Section* 6.2, *agrees with normal family as defined above.*

Proof. As a simple argument, the corollary of Hurwitz's theorem (Corollary 6.4.2) extends to $\mathcal{M}(\Omega)$. If f_n converges normally to f and f_n is nonvanishing, for all n, then either f is nonvanishing or $f \equiv 0$.

Let $g_n = f_n^{-1}$, $g = f^{-1}$ (with $\infty^{-1} = 0$). Then f_n analytic implies g_n is nonvanishing. Moreover, since $\sigma_{K_m}(f_n, f) = \sigma_{K_m}(g_n, g)$, we have g_n converging to g normally. Thus, by the above, either g is nonvanishing, that is, f is analytic, or $g \equiv 0$, that is, $f \equiv \infty$.

If f is analytic, for any m, f is bounded on K_m, so $\inf_{z \in K_m} \sigma(f(z), \infty) = \delta > 0$. Picking N so that for $n > N$, $\sigma_{K_m}(f, f_n) \leq \delta/2$, we can find $\inf_{z \in K_m} \sigma(f_n(z), \infty) \geq \delta/2$, so $\sup_{n \geq N} \|f_n\|_{K_M} < \infty$, and by (6.5.12), σ-convergence on K_m implies uniform convergence on K_m. $\square$

We define the *spherical derivative* of a function in $\mathcal{M}(\Omega)$ by

$$f^\sharp(z_0) = \lim_{\substack{z \to z_0 \\ z \neq z_0}} \frac{\sigma(f(z), f(z_0))}{|z - z_0|} \tag{6.5.24}$$

Proposition 6.5.4. (a) *If $f(z_0) < \infty$, we have that*

$$f^\sharp(z_0) = \frac{|f'(z_0)|}{1+|f(z_0)|^2} \tag{6.5.25}$$

(b) $(1/f)^\sharp = f^\sharp$ (6.5.26)

(c) *If $f \in \mathcal{M}(\Omega)$ and γ is a curve in Ω, then*

$$\sigma(f(\gamma(0)), f(\gamma(1))) \le \int_0^1 f^\sharp(\gamma(s))|d\gamma(s)| \tag{6.5.27}$$

Proof. (a) Since $(f(z)-f(z_0))/(z-z_0) \to f'(z_0)$, we see

$$\frac{\sigma(f(z), f(z_0))}{|z-z_0|} = \frac{1}{\sqrt{1+|f(z)|^2}\sqrt{1+|f(z_0)|^2}} \frac{|f(z)-f(z_0)|}{|z-z_0|} \tag{6.5.28}$$

converges to the right side of (6.5.25).

(b) is immediate from $\sigma(\frac{1}{w}, \frac{1}{z}) = \sigma(w,z)$ and (6.5.24).

(c) Suppose first that γ is piecewise C^1 and that $f(\gamma(z))$ is finite for all t. Then uniformly in s and $j = 1, 2, \dots, n$ as $n \to \infty$ and $j/n \to s$,

$$\frac{\sigma(f(\gamma(\frac{j}{n})), f(\gamma(\frac{j-1}{n})))}{|\gamma(\frac{j}{n}) - \gamma(\frac{j-1}{n})|} \to f^\sharp(\gamma(s)) \tag{6.5.29}$$

Thus, by the triangle inequality,

$$\sigma(f(\gamma(0)), f(\gamma(1))) \le \sum_{j=1}^n f^\sharp\left(\gamma\left(\frac{j}{n}\right)\right)\left|\gamma\left(\frac{j}{n}\right) - \gamma\left(\frac{j-1}{n}\right)\right| + O\left(\frac{1}{n}\right) \tag{6.5.30}$$

so, taking $n \to \infty$, we get (6.5.27).

By a polygonal path approximation, we get (6.5.27) for all rectifiable paths.

If $1/f(\gamma(s))$ is finite for all s, we do the same calculation for $1/f(\gamma(0))$, using the invariance of σ and $f^\sharp$ under $f \mapsto 1/f$. For general paths, we cover by finitely many intervals with f or $1/f$ finite on each interval and sum up using the triangle inequality. □

Lemma 6.5.5. *Let $\{f_n\}_{n=1}^\infty, f$ lie in $\mathcal{M}(\Omega)$. If f_n converges to f normally, then $f_n^\sharp \to f^\sharp$ uniformly on compacts.*

Proof. By a compactness argument, it suffices to show that for each $z_0 \in \Omega$, there is a $\delta > 0$ so that $f_n^\sharp \to f^\sharp$ uniformly in $\overline{\mathbb{D}_\delta(z_0)}$. If $f(z_0)$ is not infinite, $f_n \to f$ uniformly in some $\overline{\mathbb{D}_{2\delta}(z_0)}$, so $|f_n'|$ and $|f_n|$ converge uniformly, so

$f_n^\sharp$ does. If $f(z_0)$ is ∞, f_n^{-1} converges in f^{-1} uniformly in some $\overline{\mathbb{D}_{2\delta}(z_0)}$, so the invariance of $f^\sharp$ proves the needed convergence. $\square$

Theorem 6.5.6 (Marty's Theorem). *Let $S \subset \mathcal{M}(\Omega)$. Then S is a normal family if and only if for all compact K,*

$$\sup_{\substack{z\in K\\ f\in S}} f^\sharp(z) < \infty \tag{6.5.31}$$

Proof. If (6.5.31) fails, find K and $f_n \in S$ so

$$\lim_{n\to\infty} \sup_{z\in K} f_n^\sharp(z) = \infty \tag{6.5.32}$$

If f_{n_j} converges to f normally, by the lemma, for some J,

$$\lim_{j\to\infty} \sup_{z\in K} f_{n_j}^\sharp = \sup_{z\in K} f^\sharp < \infty$$

violating (6.5.32). Thus, S is not normal.

Conversely, suppose (6.5.31) holds. For any $z_0 \in \Omega$, if $\overline{\mathbb{D}_\delta(z_0)} \subset \Omega$, by (6.5.31) and (6.5.27), for all $f \in S$, $w \in \overline{\mathbb{D}_\delta(z_0)}$,

$$|(Q\circ f)(w) - (Q\circ f)(z)| \le \Big[\sup_{\zeta\in\mathbb{D}_\delta(z_0)} |f^\sharp(\zeta)|\Big]|w-z|$$

so the $\{Q\circ f \mid f\in S\}$ are locally uniformly equicontinuous. By the Ascoli–Arzelà theorem, we get the existence of convergent subsequences for $Q\circ f$, so normally convergent subsequences. Thus, S is normal. $\square$

Notes and Historical Remarks. The spherical metric is our first glimpse of two themes that will recur: $\widehat{\mathbb{C}}$ as a sphere, something we turn to in Sections 7.1 and 7.2, and the role of geometry and conformal metrics. In particular, the spherical metric $d^2z/(1+|z|^2)^2$ has constant positive curvature.

Many references use a projection of $\{(x_1,x_2,x_3) \mid |x|=1\}$ onto $\mathbb{C}\cup\{\infty\}$ (see the discussion in Section 7.2), so their σ is twice ours.

Marty's theorem and the use of the spherical derivative go back to Marty [**375**].

As we've defined it, the spherical derivative is asymmetric. A perhaps more natural definition would be

$$\lim_{\substack{z\to z_0\\ z\neq z_0}} \frac{\sigma(f(z),f(z_0))}{\sigma(z,z_0)} = f^\sharp(z_0)(1+|z_0|^2)$$

which is the definition used by Marty. It has the advantage of also working at $z_0=\infty$. The use of $f^\sharp$, as we use, has however become standard.

The strong Montel theorem, stated as Theorem 6.2.10 (see also Section 11.3 and 11.4), has an extension to meromorphic functions—one needs

α, β, γ all distinct in $\widehat{\mathbb{C}}$ and shows the set of functions $f \in \mathcal{M}(\Omega)$ with values in $\mathbb{C} \setminus \{\alpha, \beta, \gamma\}$ is normal. By replacing $\{f\}$ by $\{(f - \gamma)^{-1}\}$, this is reduced to Theorem 6.2.10.

Problems

1. (a) If x, y lie in a sphere of radius $\frac{1}{2}$ and make angle $\widetilde{\theta}(x, y)$, prove that $|x-y|^2 = \frac{1}{4}(1-\cos(\widetilde{\theta}(x, y)))$ so that the θ of (6.5.14) is given by $\sigma(w, z) = \frac{1}{4}(1 - \cos(\theta(w, z)))$.

   (b) From (a) and the formula for σ, deduce (6.5.14).

2. Prove by direct calculation from the definitions that $\sigma(z^{-1}, w^{-1}) = \sigma(z, w)$ and that $(1/f)^\sharp = f^\sharp$.

Chapter 7

Fractional Linear Transformations

> In addition to the two kinds of series presented above, we consider a third whose terms are connected by continuing division, so that it will be convenient to call these series continued fractions. Although this construction is less used than the other two, not only does it exhibit its value just as clearly but it is also very well suited to approximate computation.
>
> —L. Euler, *An Essay on Continued Fractions* [**175**][1]

Big Notions and Theorems: Riemann Surface, $\mathbb{A}\mathrm{ut}(\mathcal{S})$, Riemann Sphere, Complex Tori, Analytic Curve, Algebraic Curve, Elliptic Curve, Degree of a Meromorphic Function, Stereographic Projection, Complex Projective Space, Projective Curve, Resolution of Singularities, Normalization of a Curve, Fractional Linear Transformations, Cross-Ratio, Elliptic-Hyperbolic-Parabolic-Loxodromic Classification, Preservation of Circles/Lines, Groups of Conformal Automorphism for $\widehat{\mathbb{C}}, \mathbb{C}, \mathbb{D}$, Continued Fractions, Best Rational Approximation, Schur Algorithm

A fractional linear transformation (aka FLT) is a map defined by

$$f(z) = \frac{az+b}{cz+d} \tag{7.0.1}$$

where $ad - bc \neq 0$. FLTs appear frequently in both theory and applications; indeed, special cases mapping $\mathbb{D}$ to $\mathbb{C}_+$ or $\mathbb{H}_+$ to $\mathbb{D}$ have occurred earlier in this book. We'll especially need the family of all conformal bijections

[1]The other kinds mentioned by Euler are infinite series and infinite products which he illustrated by writing infinite expressions converging to $\pi/4$, one of them Wallis' formula.

of $\mathbb{D}$ to $\mathbb{D}$ in the next chapter. One of the things that we'll see in this chapter is that every conformal bijection of $\mathbb{D}$ to itself is an FLT. Because of their importance in applications, we'll make an extensive—rather than a quick—presentation.

Notice that $f(-d/c) = \infty$ so that we'll want to define f with values in $\widehat{\mathbb{C}}$. But also $f(z)$ has a limit as $|z| \to \infty$, so f is most naturally a function from $\widehat{\mathbb{C}}$ to $\widehat{\mathbb{C}}$. For this reason, we want to study what it means for a function to be analytic at ∞ and we do this in that context. We did this indirectly in Section 3.9, but we define this precisely not only for $\widehat{\mathbb{C}}$ but also for more general objects called Riemann surfaces, which we define and begin the discussion of in Section 7.1. In this book, Riemann surfaces are primarily a convenient language. Our study of them will be limited. Section 7.2 will then focus on $\widehat{\mathbb{C}}$ as a Riemann surface, most notably its realization as $\mathbb{CP}[1]$, the complex projective line. We'll also discuss stereographic projections. Section 7.3, the central one of this chapter, will discuss FLTs, primarily as maps induced by linear maps on $\mathbb{C}^2$. Section 7.4 will focus on conformal self-maps of $\mathbb{D}$ and bonus Section 7.5 will discuss continued fraction expansions, which are essentially a branch of the study of FLTs.

One of the surprises of this chapter is the rigidity of complex structures. The space of C^∞ diffeomorphisms of the two-dimensional sphere is an infinite-dimensional space. We'll see that it is much harder to be an analytic bijection of a Riemann surface. A major set of results of this chapter is that the conformal equivalences of $\widehat{\mathbb{C}}$ are a six-(real) parameter manifold; of $\mathbb{C}$, a four-parameter manifold; and of $\mathbb{D}$, a three-parameter manifold. As the set gets "smaller," its automorphism group "shrinks." In fact, if Ω is n-connected with $2 \le n < \infty$, then its automorphism group is finite. We'll not prove this, but the Problems and Notes will discuss special cases.

It must be emphasized that while FLTs seem innocent, they and their discrete subgroups (called Fuchsian and Kleinian groups) are incredibly rich and significant. Large swathes of modern geometry and number theory can be viewed as the study of them and their higher-dimensional analogs.

7.1. The Concept of a Riemann Surface

If you've seen the definition of a manifold, the definition of a Riemann surface is "the same" except C^∞ conditions are replaced by analyticity demands:

Definition. A *Riemann surface* is a connected Hausdorff topological space, $\mathcal{S}$, with a countable family $\{U_n\}_{n=1}^N$ of open sets in $\mathcal{S}$ and continuous open functions $f_n\colon U_n \to \mathbb{C}$, called *coordinate maps*, so that

(i) Each f_n is one-one, so a homeomorphism of U_n and $\mathrm{Ran}(f_n)$.
(ii) $\bigcup_{n=1}^N U_n = \mathcal{S}$.

(iii) If $U_n \cap U_m \neq \emptyset$, the *transition map*, $f_n \circ f_m^{-1}\colon f_m[U_n \cap U_m] \to f_n[U_n \cap U_m]$, is analytic.

In the above, N is a finite positive integer or ∞. Recall a map is open if and only if $f_n[V]$ is open for any open $V \subset U_n$. Thus, $f_n^{-1}\colon \operatorname{Ran}(f_n) \to U_n$ is a homeomorphism.

Definition. A family $\{U_n; f_n\}_{n=1}^{\infty}$ of maps making a Hausdorff space into a Riemann surface is called an *analytic structure*. Two analytic structures, $\{U_n; f_n\}_{n=1}^{\infty}$ and $\{V_m; g_m\}_{m=1}^{\infty}$, are called *compatible* if and only if each $f_n \circ g_m^{-1}\colon g_m[U_n \cap V_m] \to f_n[U_n \cap V_m]$ is analytic.

Example 7.1.1 (Regions). Let $\Omega \subset \mathbb{C}$ be a region. With $U_1 = \Omega$ and $g\colon U_1 \to \mathbb{C}$ by $g(z) = z$, it is a Riemann surface. □

Example 7.1.2 (Riemann Sphere). $\widehat{\mathbb{C}}$ is $\mathbb{C} \cup \{\infty\}$, the one-point compactification, topologized so that $\widehat{\mathbb{C}} \setminus \overline{\mathbb{D}_r(0)}$ are a neighborhood base for ∞. It has a natural Riemann surface realization with the following analytic structure: Take $U_1 = \mathbb{C}$ and $U_2 = \mathbb{C}^\times \cup \{\infty\}$. f_1 and f_2 map U_1 and U_2 bijectively to all of $\mathbb{C}$ by

$$f_1(z) = z \qquad f_2(z) = \begin{cases} z^{-1}, & z \in \mathbb{C}^\times \\ 0, & z = \infty \end{cases}$$

The transition map $g \equiv f_2 \circ f_1^{-1}\colon \mathbb{C}^\times \to \mathbb{C}^\times$ by $g(z) = z^{-1}$. □

Before turning to more examples, we consider five important definitions:

Definition. Let $\mathcal{S}_1, \mathcal{S}_2$ be Riemann surfaces with analytic structures $\{U_n, f_n\}_{n=1}^{\infty}$ and $\{V_m, g_m\}_{m=1}^{\infty}$, respectively. An *analytic function* on $\mathcal{S}_1$ is a map $f\colon \mathcal{S}_1 \to \mathbb{C}$ so that each $f \circ f_n^{-1}\colon [f_n[U_n]] \to \mathbb{C}$ is analytic. An *analytic function between* $\mathcal{S}_1$ *and* $\mathcal{S}_2$ is $f\colon \mathcal{S}_1 \to \mathcal{S}_2$ so that each $g_m \circ f \circ f_n^{-1}\colon f_n[U_n \cap f^{-1}[V_m]] \to g_m[f[U_n] \cap V_m]$ is analytic. A *conformal equivalence* of $\mathcal{S}_1$ and $\mathcal{S}_2$ is an analytic bijection of $\mathcal{S}_1$ onto $\mathcal{S}_2$. A *meromorphic function* on $\mathcal{S}_1$ is an analytic map of $\mathcal{S}_1$ to $\widehat{\mathbb{C}}$ with the analytic structure of Example 7.1.2. A function $u\colon U \subset \mathcal{S} \to \mathbb{R}$ with U open is called *harmonic* if for every $z \in U$, there is an open neighborhood N of z and $f\colon N \to \mathbb{C}$ analytic so that $u = \operatorname{Re} f$.

We use the symbol $\mathbb{A}\mathrm{ut}(\mathcal{S})$ for the set of all analytic bijections of $\mathcal{S}$ to itself. Because harmonic is defined locally, all the equivalences of Section 5.4 apply where Δ and circles can be taken in any compatible local coordinate system. The reader should check two analytic structures on $\mathcal{S}$ are compatible if and only if the identity map on $\mathcal{S}$ with the different structures on domain and range is a conformal equivalence, and that the new notation of meromorphic agrees with the old if $\mathcal{S}$ is given by Example 7.1.1.

One of the more intriguing parts of complex analysis is the determination of the set of all automorphisms of some region or Riemann surface. These results are spread out in the rest of this book; here is a summary: $\mathbb{A}\mathrm{ut}(\mathbb{C})$ is determined in Theorem 7.3.4, $\mathbb{A}\mathrm{ut}(\widehat{\mathbb{C}})$ in Theorem 7.3.5, and $\mathbb{A}\mathrm{ut}(\mathbb{D})$ in Theorem 7.4.2. The Notes to Section 7.4 include a theorem that if $2 \le n < \infty$ and Ω is n-connected, then $\mathbb{A}\mathrm{ut}(\Omega)$ is finite. Section 7.4 has problems that determine some $\mathbb{A}\mathrm{ut}(\Omega)$: $\mathbb{A}\mathrm{ut}(\mathbb{D}^\times)$ in Problem 4, $\mathbb{A}\mathrm{ut}(\mathbb{D}\setminus\{z_0,z_1\})$ in Problem 5, $\mathbb{A}\mathrm{ut}(\Omega\setminus\{z_0,z_1\})$ in Problem 6, and $\mathbb{A}\mathrm{ut}(\mathbb{A}_{r,R})$ in Problem 7. Problem 2 of Section 10.7 determines $\mathbb{A}\mathrm{ut}(\mathcal{J}_{\tau_1,\tau_2})$ for complex tori and Section 11.5 discusses some automorphism groups in higher complex dimension.

Example 7.1.3 (Complex Tori). Let $\tau_1,\tau_2 \in \mathbb{C}^\times$ with $\tau_2/\tau_1 \notin \mathbb{R}$. Let

$$\mathcal{L}_{\tau_1,\tau_2} = \{n_1\tau_1 + n_2\tau_2 \mid n_j \in \mathbb{Z}\} \tag{7.1.1}$$

the lattice generated by τ_1,τ_2; see Figure 7.1.1.

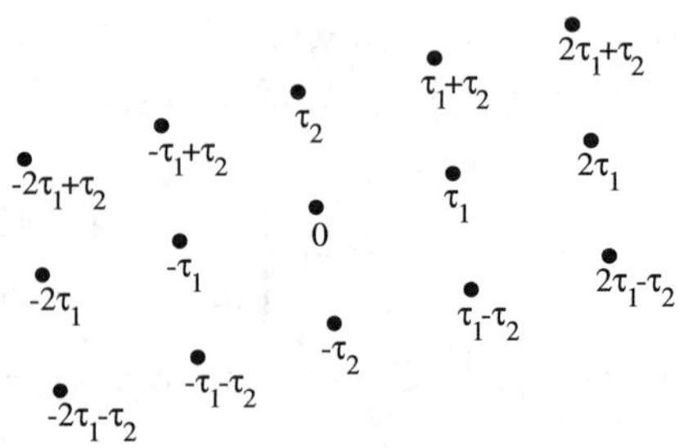

Figure 7.1.1. A portion of a lattice.

$\mathbb{C}$ is a topological group under addition and $\mathcal{L}_{\tau_1,\tau_2}$ a closed subgroup, so the quotient group is naturally a topological group. It is topologically a torus, as can be realized as follows. If we think of $\mathbb{C}$ as a two-dimensional real vector space, $\tau_2/\tau_1 \notin \mathbb{R}$ says τ_1 and τ_2 are linearly independent, so a basis. Thus, any z can be written uniquely as

$$z = x_1(z)\tau_1 + x_2(z)\tau_2, \qquad x_j(z) \in \mathbb{R}$$

Then, $\pi\colon \mathbb{C} \to \partial\mathbb{D}\times\partial\mathbb{D}$ with $\pi(z) = (e^{2\pi i x_1(z)}, e^{2\pi i x_2(z)})$ is a group homomorphism of the additive group $\mathbb{C}$ to the multiplicative group $\partial\mathbb{D}\times\partial\mathbb{D}$. The kernel of π is $\mathcal{L}_{\tau_1,\tau_2}$ and realizes an isomorphism and homeomorphism of the quotient and $\partial\mathbb{D}\times\partial\mathbb{D}$. π is (τ_1,τ_2)-dependent and we'll sometimes use π_{τ_1,τ_2}.

Let $V_1,\dots,V_5 \subset \mathbb{C}$ be defined by (see Figure 7.1.2)

$$V_1 = \{z \mid |x_1(z)| < \tfrac12,\ |x_2(z)| < \tfrac12\}, \qquad V_2 = \tfrac12(\tau_1+\tau_2) + V_1$$

$$V_3 = \tfrac12(\tau_1-\tau_2) + V_1, \qquad V_4 = \tfrac12(-\tau_1+\tau_2) + V_1$$

$$V_5 = \tfrac12(-\tau_1-\tau_2) + V_1$$

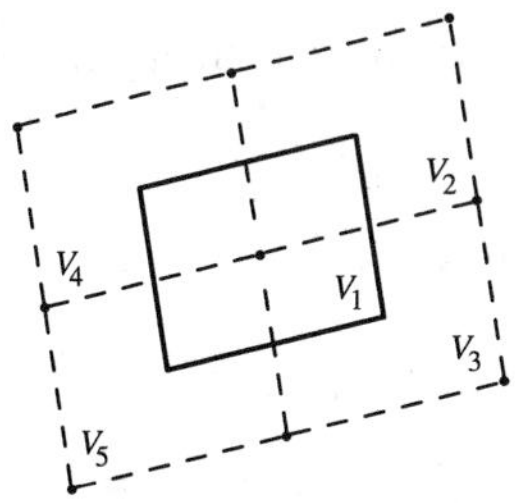

Figure 7.1.2. Coordinate cover of a torus.

Let $U_j = \pi[V_j]$. π is one-one on V_j so there are maps $f_j \colon U_j \to V_j \subset \mathbb{C}$ which are the inverses of π. It is easy to see (Problem 1) that $\bigcup_{j=1}^{5} U_j = \partial\mathbb{D}\times\partial\mathbb{D}$. $f_j \circ f_1^{-1} \colon V_1 \cap V_j \to V_1 \cap V_j$ is just the identity map and each $f_j \circ f_k^{-1}$ is $z + c_{jk}$, where c_{jk} is one of $\pm\tau_1$, $\pm\tau_2$, or $\pm\tau_1 \pm \tau_2$. Each map is clearly analytic. This defines an analytic structure on $\partial\mathbb{D}\times\partial\mathbb{D}$, and we'll call the resulting Riemann surface $\mathcal{J}_{\tau_1,\tau_2}$. As Figure 7.1.3 shows geometrically, the square with opposite sides identified is a torus.

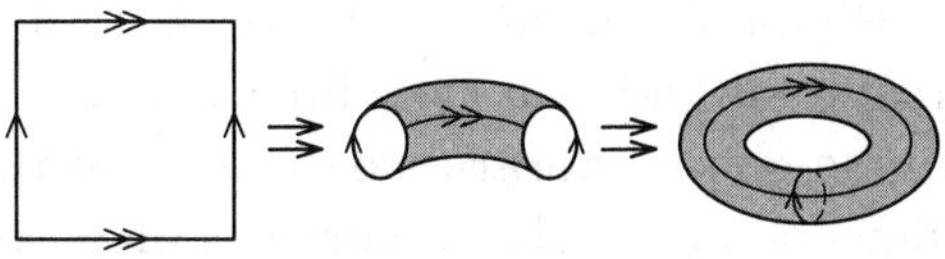

Figure 7.1.3. A scissors-and-paste construction shows that a square with opposite sides identified is a torus.

All $\mathcal{J}_{\tau_1,\tau_2}$ are isomorphic and homomorphic as groups and topological spaces, but it is not (yet) clear which are analytically equivalent, that is, which complex structures on $\partial\mathbb{D}\times\partial\mathbb{D}$ are equivalent. We'll answer this in Theorem 10.7.1, but for now, we'll note that we'll prove that there are uncountably many nonequivalent analytic structures among the $\mathcal{J}_{\tau_1,\tau_2}$. In the Notes to Section 10.7, we'll also show the $\mathcal{J}_{\tau_1,\tau_2}$ are the only analytic structures on the torus. We'll especially consider the special case $\tau_1 = 1$ and $\tau_2 = \tau \in \mathbb{C}_+$. In that case, we write $\mathcal{L}_\tau$ and $\mathcal{J}_\tau$. □

In some ways, the following is an abstraction of the initial construction of Riemann:

Example 7.1.4 (Analytic Curves). Let $F \colon \mathbb{C}^2 \to \mathbb{C}$ be an analytic function (convergent two-variable power series at each point), so that $\mathcal{S} = \{(z,w) \mid F(z,w) = 0\}$ is connected, and so that for each $(z_0,w_0) \in \mathcal{S}$, either $\frac{\partial F}{\partial z}(z_0,w_0) \neq 0$ or $\frac{\partial F}{\partial w}(z_0,w_0) \neq 0$.

For any such point, if $\frac{\partial F}{\partial w} \neq 0$, by the implicit function theorem (see Theorem 3.4.2 of Section 3.4), there are neighborhoods $N_{(z_0,w_0)}$ of z_0 and

$M_{(z_0,w_0)}$ of w_0 and an analytic function, $g(z_0, w_0)$, on $N_{(z_0,w_0)}$ so that

$$\begin{aligned} U_{(z_0,w_0)} &\equiv \{(z,w) \in N_{(z_0,w_0)} \times M_{(z_0,w_0)} \mid F(z,w) = 0\} \\ &= \{(z, g(z_0,w_0)) \mid z \in N_{(z_0,w_0)}\} \end{aligned} \tag{7.1.2}$$

Define $f_{(z_0,w_0)} \colon U_{(z_0,w_0)} \to N_{(z_0,w_0)}$ by

$$f_{(z_0,w_0)}(z,w) = z \tag{7.1.3}$$

If $\frac{\partial F}{\partial w}(z_0, w_0) = 0$, then $\frac{\partial F}{\partial z} \neq 0$, and we can interchange the roles of z and w. Since $\mathcal{S}$ has a countable base of open sets, we can find countably many $U_{(z_0,w_0)}$ that cover $\mathcal{S}$ (see Proposition 2.2.2 of Part 1). Every transition function is either the map $f(z) = z$ or a $g_{(z_0,w_0)}$. Thus, $\mathcal{S}$ is a Riemann surface (whose analytic structure is independent of the choice of which U's to take). This is the *Riemann surface of* F.

If, in the above example, F is a polynomial in z and w and the condition that $\nabla_{(z,w)}F$ is nonvanishing is imposed, $\mathcal{S}$ is called a *nonsingular algebraic curve*. Notice the somewhat confusing terminology: a surface is normally a two-dimensional object, and a curve one-dimensional, yet an analytic or algebraic curve is a Riemann surface! The resolution of this terminology puzzle is that, as a real manifold, it is two-dimensional—and it was in in that context that Riemann's contemporaries named his objects. As a complex object, it is one-dimensional. Algebraic curves started out as real curves in $\mathbb{R}^2$ given by polynomial equations. When complexified, it was natural to still call them curves.

More generally, an *algebraic curve*, $\mathcal{S}$, is the set of zeros of a polynomial $P(z,w)$ which is irreducible, that is, cannot be factored. The irreducibility implies that the set of points, (z_0, w_0), where $\operatorname{grad}(P(z_0, w_0)) = 0$, is finite. These points are the *singularities*. As in the last example, $\mathcal{S}$ is a Riemann surface if the singularities are removed. It is not at the singularities but there is always a Riemann surface, $\widetilde{\mathcal{S}}$, and a continuous map, $f \colon \widetilde{\mathcal{S}} \to \mathcal{S}$, which is one-one on f^{-1} [nonsingular points] and finite to 1 on f^{-1} [singularities]. $\widetilde{\mathcal{S}}$ is known as the *resolution of singularities* surface.

We can see this in detail in a special case of a simple double point. A *double point* is a singularity, (z_0, w_0), where all first derivatives are zero but some second derivative is nonzero. For simplicity of notation, suppose $(z_0, w_0) = (0,0)$. Then

$$P(z,w) = az^2 + 2bzw + cw^2 + O((|z|^2 + |w|^2)^{3/2}) \tag{7.1.4}$$

(z_0, w_0) is called a *simple double point* if $\left(\begin{smallmatrix} a & b \\ c & d \end{smallmatrix}\right)$ has empty kernel. If there is a kernel, we have a *cusp*. In the case of a simple double point, $az^2 + 2bzw + cw^2$ is a product of two distinct linear factors, so $\mathcal{S} \setminus \{0,0\}$ near $(0,0)$ is a union of two pieces, each of which is asymptotic to a punctured plane. If we add back $(0,0)$ to each piece, we get the intersection point that implies $\mathcal{S}$ is not a Riemann surface near $(0,0)$. If we add two separate points, we get a Riemann surface so far as the neighborhood of $(0,0)$ is concerned.

For example, if $F(z,w) = z^2 + w^2 + z^3$, then only $(0,0)$ is singular, the factorization is $(w+iz)(w-iz)$.

We note that for some algebraic curves and elliptic curves (see below), we will add some points at infinity (informally, later in this section for elliptic curves, and in general and formally, in the next section) to compactify the curve. We'll generally mean this compact Riemann surface when we speak of elliptic or algebraic curves. □

Example 7.1.5 (Riemann Surface of $\sqrt{z}$ and $\sqrt{P(z)}$)**.** The simplest example of an analytic curve is to take $F(z,w) = z - w^2$, so formally, $w = \sqrt{z}$. Instead of thinking of $\sqrt{z}$ with a branch cut at, say, $(-\infty, 0)$, we look at pairs (w,z) with $z = w^2$. $\frac{\partial F}{\partial w} = -2w = 0$ exactly at $w = z = 0$ so, except at this point, w is locally an analytic function z. Where $\frac{\partial F}{\partial w} = 0$, we have $\frac{\partial F}{\partial z} \neq 0$ (of course, in this case $\frac{\partial F}{\partial z}$ is globally nonzero, but see below). Thus, z is a local coordinate except where $z = 0$ and everywhere w is a possible local coordinate.

More generally, let $P(z)$ be a polynomial with all its zeros simple, that is,

$$P(z_0) = 0 \Rightarrow P'(z_0) \neq 0 \tag{7.1.5}$$

Let

$$F(z,w) = P(z) - w^2 \tag{7.1.6}$$

If $\frac{\partial F}{\partial w} = 0$, then $w = 0$, so by (7.1.5), $\frac{\partial F}{\partial z} \neq 0$. Thus, we have a curve since it is not hard to show that $\{(z,w) \mid w^2 = P(z)\}$ is connected (Problem 1). Now there are points with $\frac{\partial F}{\partial z} = 0$, but not also where $F = 0$ and $\frac{\partial F}{\partial w} = 0$.

In case $\deg(P)$ is 3 or 4, this is called an *elliptic curve*. We'll say a lot more about them in Chapter 10 and also some in the next section. □

Example 7.1.6 (Riemann Surface of $\log(z)$)**.** Now instead of solutions of $z = w^2$, we want solutions of $z = e^w$, so we take

$$F(z,w) = z - e^w \tag{7.1.7}$$

Here $\frac{\partial F}{\partial w} = -e^w$ is never 0, so z is a local coordinate everywhere.

Of course, w can only be given locally as a function of z. For any $z_0 \neq 0$ and any w_0 solving $z_0 = e^{w_0}$, we can initially define w as a function z, say, in $\mathbb{D}_{|z_0|}(z_0)$. We can then place branch cuts in various places but, for example, the neighborhood cannot include all of $\partial\mathbb{D}_{z_0}(0)$.

This example is subtly different from the square root. In both cases, $(z,w) \xrightarrow{f} z$ is an analytic function on $\mathcal{S}$. For the log case, f is locally one-one everywhere, but $\text{Ran}(f) \subset \mathbb{C}$ is only $\mathbb{C}^\times$. f is a covering map. For the square-root case, f is not locally one-one near $(z,w) = (0,0)$. In the proper local coordinate near there, which is w, $f(z,w) = w^2$ which has zero derivative. $\text{Ran}(f)$ is all of $\mathbb{C}$, and if we restrict to $f^{-1}(\mathbb{C}^\times)$, we again have a covering map. But the square-root case can be "filled in" at $z = 0$—the covering map is lost but there is still a local coordinate. □

While we have emphasized the view of the Riemann surface of $\sqrt{z}$ or $\log(z)$ as a subset of $\mathbb{C}^2$, the original view which captured the imaginations of Riemann's contemporaries was more geometric. They thought of the surface of $\sqrt{z}$ as taking two cut planes $\mathbb{C} \setminus (-\infty, 0]$, adding $(-\infty, 0)$ to each, but with the rule that passing from $\mathbb{C}_+$ in one plane through the cut took one to $\mathbb{C}_-$ on the other plane; see Figure 7.1.4.

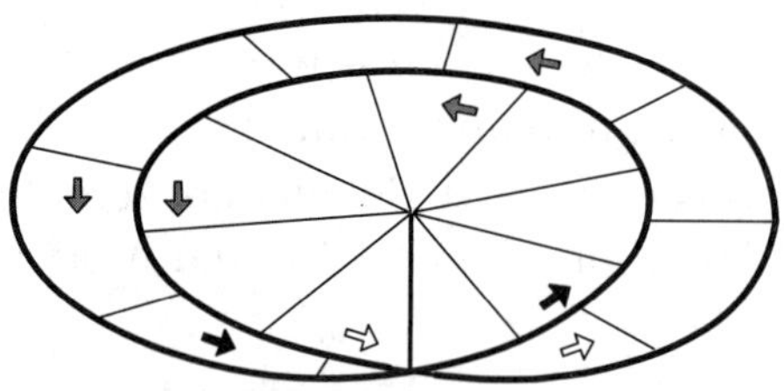

Figure 7.1.4. Idealization of the Riemann surface of $\sqrt{z}$.

For $\log(z)$, one needs infinitely many "sheets." For $w = \sqrt{P(z)}$, if $\deg(P)$ is even, we place cuts between pairs of zeros $\frac{1}{2}\deg(P)$ cuts in all (e.g., if all roots are real, we can write them as $x_1 < x_2 < \cdots < x_{2\ell}$ and place cuts in $[x_1, x_2], [x_3, x_4], \dots, [x_{2\ell-1}, x_{2\ell}]$); see Figure 7.1.5. If $\deg(P)$ is odd, we pair $\deg(P) - 1$ zeros and pair the remaining one with ∞.

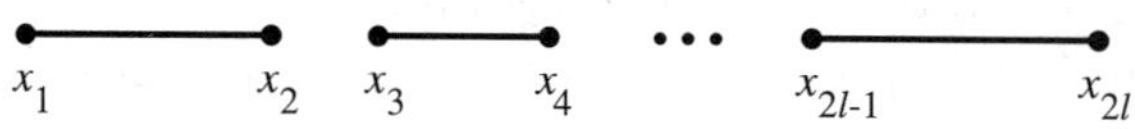

Figure 7.1.5. Cuts of the square root of a polynomial of degree 2ℓ.

In the polynomial case, we add the branch points to the surface. We can also understand how to compactify, at least for $w = \sqrt{P(z)}$. On each sheet we add a point at infinity, so we have two Riemann spheres with $\ell = \frac{1}{2}[\deg(P) + 1]$ cuts. When we glue them together, we get a compact orientable surface which is a sphere with $g = \ell - 1$ handles. For $g = 1$, this is illustrated in Figure 7.1.6. g is called the *genus* of the surface.

In particular, for the case of elliptic curves, $g = 1$ and this compactified surface is a torus. Its connection to some $\mathcal{L}_\tau$ will be clarified by the theory of elliptic functions; see Sections 10.4 and 10.7.

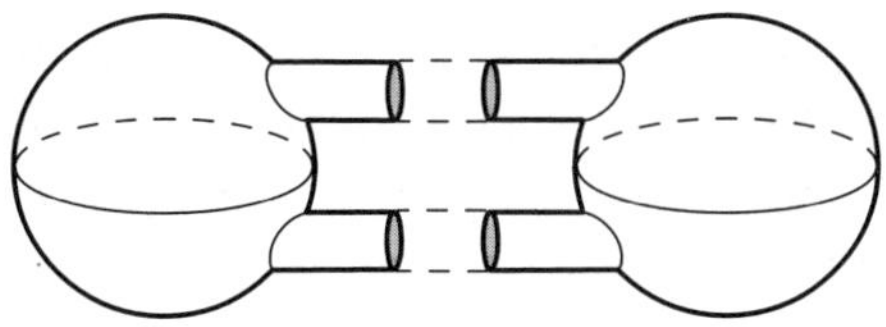

Figure 7.1.6. Two spheres with two cuts glued is a torus.

We end this section with a few simple observations about analytic and meromorphic functions on compact Riemann surfaces, our only results on general Riemann surfaces.

Definition. A Riemann surface, $\mathcal{S}$, is called *open* if it is not compact.

For this reason, compact surfaces are sometimes called "closed," but we avoid this terminology since "compact" is more informative. The following generalizes Liouville's theorem (part (a)) and the fundamental theorem of algebra (part (b)).

Theorem 7.1.7. *Let $\mathcal{S}$ be a compact Riemann surface and $\mathcal{J}$ a second Riemann surface. Let $f\colon \mathcal{S} \to \mathcal{J}$ be analytic. Then*

(a) *If $\mathcal{J}$ is open, f is constant.*
(b) *If $\mathcal{J}$ is compact and f is not constant, then f is surjective.*

Remarks. 1. If $f\colon \mathbb{C} \to \mathbb{C}$ is bounded, the singularity at ∞ is removable, so f extends to $\tilde{f}\colon \widehat{\mathbb{C}} \to \mathbb{C}$. By (a), this is constant, so (a) indeed implies Liouville's theorem. If f is a polynomial and $\widetilde{f}(\infty) = \infty$, f extends to an analytic map $\widetilde{f}$ of $\widehat{\mathbb{C}}$ to $\widehat{\mathbb{C}}$, so (b) indeed extends the fundamental theorem of algebra.

2. In particular, there are no nonconstant analytic functions on compact Riemann surfaces.

Proof. Any nonconstant analytic map is, by the open mapping theorem, locally open, so, since being open is a local property, $\operatorname{Ran}(f)$ is open. If $\mathcal{S}$ is compact, $\operatorname{Ran}(\mathcal{S})$ is compact (see Theorem 2.3.11 of Part 1), so closed. Thus, if f is nonconstant, $\operatorname{Ran}(f)$ is open and closed, so all of $\mathcal{J}$, since $\mathcal{J}$ is connected. This proves (b).

It also proves (a), since $\operatorname{Ran}(f)$ is compact and $\mathcal{J}$ is not, so f cannot be nonconstant. □

The second general result will show that if $f\colon \mathcal{S} \to \mathcal{J}$ for compact Riemann surfaces, every value gets taken the same number of times if we properly count multiplicity. We need to begin by describing multiplicity, for which the term "local degree" is used rather than "order."

Proposition 7.1.8. (a) *For any $f\colon \mathcal{S} \to \mathcal{J}$, a nonconstant analytic map between arbitrary Riemann surfaces, and any $P \in \mathcal{S}$ with $f(P) = Q \in \mathcal{J}$, there is a positive integer,* $\deg(f;P)$*, so that in any local coordinates g at P and h at Q, $h \circ f \circ g^{-1} - h(Q)$ has a zero of order* $\deg(f;P)$ *at $g(P)$.*

(b) *If $\mathcal{S}$ and $\mathcal{J}$ are compact, there are only finitely $P \in \mathcal{S}$ with* $\deg(f;P) > 1$.

Remarks. 1. $\deg(f;P)$ is called the *local degree* of f at P. Sometimes $\deg(f;P) - 1$ is called the *ramification index* of f at P. A point where $\deg(f;P) > 1$ is called a *critical point* of f. A point Q with $Q = f(P)$ for some critical point, P, is called a *critical value* or *ramification point* of f.

2. A similar argument shows that if $\mathcal{S}$ is compact for any $Q \in \mathcal{J}$, $\{P \mid f(P) = Q\}$ is finite.

3. Normally, critical points are defined by requiring $f'(z_0) = 0$. But f' is not a function since it is coordinate-dependent. There is, however, a differential form meaning; see the Notes.

Proof. (a) An easy calculation shows this is coordinate-independent. Essentially, $|h \circ f \circ g^{-1}(z) - h(Q)| = 0((z - g(P))^d)$ is coordinate-independent.

(b) If f is nonconstant and $h \circ f \circ g^{-1}(z) \equiv q(z)$, then if $q'(g(P)) = 0$, the derivative is nonzero nearby, that is, critical points are isolated. Compactness then implies they are finite in number. □

We want to use this notion plus the fact that the meromorphic functions are a field to state the following consequences of the nonexistence of analytic functions on a compact surface. We define the principal part of a meromorphic function, f, at a pole, z_0, to be the principal part, g, in a local coordinate system. Since g is determined by the requirements that $f - g$ is analytic near z_0 and $(1/g)(z_0) = 0$, this is coordinate-independent.

Theorem 7.1.9. (a) *If f, g are meromorphic functions on a compact Riemann surface and they have the same zeros and poles with the same local degrees, then $f = cg$ for a constant c.*

(b) *If f, g are meromorphic functions on a compact Riemann surface and they have the same poles and principal parts, then $f = g + c$ for a constant c.*

Proof. f/g in case (a) and $f - g$ in case (b) are functions without poles, so constant by Theorem 7.1.7(a). □

Theorem 7.1.10. *Let $f\colon \mathcal{S} \to \mathcal{J}$ be an analytic map between compact Riemann surfaces. Then*

$$\sum_{\{P \mid f(P)=Q\}} \deg(f;P) \equiv \deg(f) \tag{7.1.8}$$

is independent of Q.

Remark. $\deg(f)$ is called the *degree* of f.

Proof. Let $d(Q)$ be the left side of (7.1.8). By Remark 2 after Proposition 7.1.8, $d(Q) < \infty$ everywhere. Let $\mathcal{J}_m = \{Q \mid d(Q) \geq m\}$. We claim each $\mathcal{J}_m$ is open and closed, so either empty or all of $\mathcal{J}$. Since $\mathcal{J}_1$ is nonempty, there is a maximal m nonempty $\mathcal{J}_m$, that is, $\mathcal{J}_m = \mathcal{J}$, $\mathcal{J}_{m+1} = \emptyset$, so $d(Q) \equiv m$.

If $\deg(f;P_0) = \ell$, by shifting to local coordinates and using Theorems 3.3.7, 3.4.1, and 3.5.1, for Q near $Q_0 \equiv f(P_0)$, there are exactly ℓ points near P_0 with $f(P) = Q$, and for each of them, $\deg(f;P) = 1$. Thus, if $d(Q_0) = m$, there are, for all Q near Q_0, m points near those P with $f(P) = Q$. There might be additional solutions of $f(P) = Q$ but certainly $d(Q) \geq m$. Thus, $\mathcal{J}_m$ is open.

Let $Q_1, \dots, Q_n, \dots$ lie in $\mathcal{J}_m$ and let $Q_n \to Q_\infty$. By passing to a subsequence, we can suppose that the Q's are distinct and that no Q_j is a critical value (since we can throw out the finitely many critical values; of course, Q_∞ can be a critical value).

Thus, we can find $P_n^{(j)}$, $j = 1, \dots, m$, all distinct (for n fixed) with $f(P_n^{(j)}) = Q_n$. By passing to successive subsequences, we can suppose $P_n^{(j)} \to P_\infty^{(j)}$ for $j = 1, \dots, m$. By continuity of f, $f(P_\infty^{(j)}) = Q_\infty$. If $\ell, P_\infty^{(j)}$ agree with $P_\infty^{(1)}$, there are at least ℓ distinct points near $P_\infty^{(1)}$ with values near Q_∞. By Theorem 3.5.1, $\deg(f;P_\infty^{(1)}) \geq \ell$. Summing up over the distinct points among the $P_\infty^{(j)}$ shows that $d(Q_\infty) \geq m$. That is, $\mathcal{J}_m$ is closed. □

In Section 3.8 of Part 3, we'll show that every Riemann surface has nonconstant meromorphic functions.

Notes and Historical Remarks. The notion of Riemann surface appeared in what is almost surely the most significant work in all of complex analysis: Riemann's inaugural dissertation [**480**] of 1851. We've already seen that he had the Cauchy–Riemann equation and notion of conformality here but he also had the Riemann mapping theorem (we'll discuss Riemann's life in the Notes to Section 8.1 when we discuss that theorem). The work also had a profound impact on topology and algebraic geometry. Initially, the 1851 dissertation was not widely available and it was through the 1857 paper on abelian functions [**484**], where much of the discussion was reported and expanded, that the notion came into wide use.

He didn't define Riemann surfaces abstractly, but in terms of sheets pasted together. Puiseux [**463**] had partial results in this direction a year earlier, but nothing close to Riemann's deep vision. The modern notion—going beyond a scissors-and-paste construction—was developed by Hermann Weyl in his 1913 book, *The Concept of a Riemann Surface* [**590**]. In his preface, Weyl acknowledged the influence of Klein's book [**313**] which had emphasized the importance of Riemann surfaces as an object worthy of study. Over 45 years later, in 1955, Weyl published a substantially revised version of this classic.

Like us, Weyl assumed that the surface had a countable cover of coordinate neighborhoods. In 1935, Radó [**467**] showed this wasn't necessary and that any cover by coordinate neighborhoods had a countable subcover—the analogous result is false for general real manifolds and also for higher-dimensional complex manifolds. Since, in examples, the countability hypothesis is easy to check, we included it in our definition. A modern exposition of Radó's theorem can be found in Forster [**194**].

Hermann Weyl (1885–1955) was a German-born mathematician, one of the greatest of the twentieth century. He entered university in Munich in 1904, transferred to Göttingen to work with Hilbert, and remained for a time in Göttingen as an assistant. His book on Riemann surfaces was based on lectures he gave there. He left for Zurich in 1913 where he was a professor at the ETH until 1930. While in Zurich, he worked briefly with Einstein and worked in general relativity as a result, and later became friendly with Schrödinger (and his wife with whom he had a passionate affair) and wrote an influential book on the mathematics behind quantum theory. In 1930, Weyl returned to Göttingen to fill Hilbert's chair, but he left in 1933, concerned about the rise of the Nazis (in particular because his wife was Jewish), and spent the rest of his career at the Institute for Advanced Study in Princeton.

One can see the enormous breadth of Weyl's work and influence if I note that for me as a spectral theorist, Weyl's most important contributions are three enduring ones: his result on asymptotics of the number of eigenvalues of the Laplacian of a bounded region in $\mathbb{R}^n$ (see Theorem 7.5.29 in Part 4), his result on invariance of the essential spectrum (see Theorem 3.14.1 in Part 4), and his work on limit point/limit circle methods in ODEs (see Theorem 7.4.12 in Part 4). But none of these three are even mentioned in his Wikipedia biography which discusses many of his other contributions:

his classification theory of representations of compact Lie groups, the Peter–Weyl theorem, the equidistribution of reals (see Section 2.7 of Part 3), his development of the connections between geometry and gauge invariance in physics (which made him the grandfather of ideas that have dominated the last forty years of research in particle physics), and, of course, his work on Riemann surfaces.

We want to note that our definition of algebraic curve disagrees with the standard one. In the standard one, an algebraic curve (perhaps with singularities) is the zero set, $\mathcal{S}$, of a polynomial, $P(z,w)$, in two variables. If P is an irreducible polynomial, that is, cannot be factored into a product of nontrivial polynomials, $\mathcal{S}$ is connected. Conversely, if $\mathcal{S}$ is connected, P is a power of an irreducible polynomial, $P = Q^m$, and Q is then called the minimal polynomial. A point of $\mathcal{S}$ is called singular if $\text{grad}(Q) = 0$ at that point. Examples are double points, e.g., $y^2 = x^2 + x^3$ at $(x,y) = (0,0)$. Thus, our "nonsingular algebraic curves" are actually "irreducible algebraic curves with no singular points in F, the minimal polynomial." Since these algebraic curves are all Riemann surfaces, we use the shorter name as a slight abuse of terminology.

While we'll use the language of Riemann surfaces to organize many subjects later in this book, we'll hardly scratch the surface—the general results that appear at the end of this section capture one general theme, the use of local theory, but they don't touch the two most common themes of the general theory: the use of differential forms, needed for consideration of derivatives (if f is an analytic or meromorphic function, df is a one-form) but much more, and the use of harmonic functions/potential theory (but see Section 3.8 of Part 3). In particular, potential theory methods show any Riemann surface has meromorphic functions.

Forster [**194**], Griffiths–Harris [**222**], Miranda [**382**], and Springer [**533**]. For textbook presentations of the subject of Riemann surfaces, see Farkas–Kra [**186**],

Fischer [**189**], Griffiths [**221**], For textbooks on algebraic curves, see Brieskorn–Knörrer [**74**], and Ueno [**558**].

Problems

1. Prove $\{(w,z) \in \mathbb{C}^2 \mid w^2 - P(z) = 0\}$ is connected, where P is a polynomial with simple zeros. (*Hint*: Show if you place branch cuts, there are two connected sheets and going around a branch point moves between them.)

7.2. The Riemann Sphere as a Complex Projective Space

In the last section, we gave $\widehat{\mathbb{C}}$ the structure of a Riemann surface. Here we want to discuss two more concrete realizations as a sphere, one via stereographic projection and the other, in more detail as $\mathbb{CP}[1]$, one-dimensional complex projective line.

Stereographic projection concerns placing a two-dimensional sphere in $\mathbb{R}^3$ and mapping it to a plane. We'll place the north pole of the sphere at $\vec{n} = (0,0,1)$ and the plane as $\mathbb{P} = (x_1, x_2, 0)$, which we write as $x_1 + ix_2$). So we care about the following simple fact (proven in Problem 1):

Proposition 7.2.1. *Let $\vec{x} = (x_1, x_2, x_3)$ have $x_3 \neq 1$. The point in the plane $x_3 = 0$ and the straight line that contains $(0,0,1)$ and $\vec{x}$ is*

$$L(x_1, x_2, x_3) = \frac{x_1 + ix_2}{1 - x_3} \tag{7.2.1}$$

Let S_r be the sphere of radius r with $\vec{n} = (0,0,1)$ as north pole, that is,

$$S_r = \{\vec{x} \mid x_1^2 + x_2^2 + (x_3 - (1-r))^2 = r^2\} \tag{7.2.2}$$

We are mainly interested in the cases $r = \frac{1}{2}$, which we used in Section 6.5, and $r = 1$ used in some books, but the statements are true for all $r > 0$.

Every straight line through $\vec{n}$ not in the tangent plane to S_r at $\vec{n}$ intersects $\mathbb{C}$ in one point and S_r in one point and sets up a one-one correspondence, given by $L_r \equiv L \restriction S_r$, between $\mathbb{C}$ and $S_r \setminus \{n\}$. The inverse (Problem 1) is $Q_r = L_r^{-1}$

$$Q_r(z) = \left(\frac{2r}{|z|^2+1} \operatorname{Re} z, \frac{2r}{|z|^2+1} \operatorname{Im} z, \frac{|z|^2 + (1-2r)}{|z|^2+1} \right) \tag{7.2.3}$$

$r = \frac{1}{2}$ is (6.5.2) and $r = 1$ a commonly appearing case. The other point we'd make about this map $S_r \to \mathbb{R}$, called *stereographic projection*, is that it is conformal (Problem 2). See Figure 7.2.1 for the case $r = \frac{1}{2}$.

A rather different point of view concerns projective space. $\mathbb{CP}[n]$, *complex dimension-n projective space*, is the space of lines through 0 in $\mathbb{C}^{n+1}$. Explicitly, let $\vec{v}, \vec{w} \in \mathbb{C}^{n+1} \setminus \{0\}$. We say $\vec{v} \sim \vec{w}$ ($\vec{v}$ is equivalent to $\vec{w}$) if and only if $\vec{v} = \lambda \vec{w}$ for some $\lambda \in \mathbb{C}^\times$. It it easy to see that this is an equivalent relation. $\mathbb{CP}[n]$ is just the set of equivalence classes. $\pi\colon \mathbb{C}^{n+1} \setminus \{0\} \to \mathbb{CP}[n]$

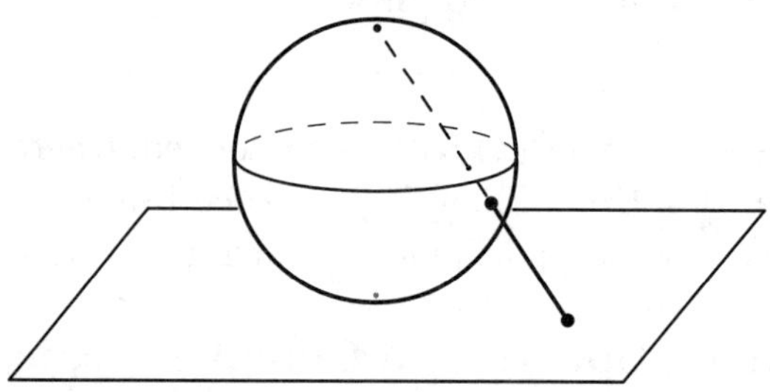

Figure 7.2.1. Stereographic projection (for $r = \frac{1}{2}$ where the sphere is tangent to the plane).

maps a vector to its equivalence class. We topologize $\mathbb{CP}[n]$ by saying $U \subset \mathbb{CP}[n]$ is open if and only if $\pi^{-1}[U]$ is open in $\mathbb{C}^{n+1} \setminus \{0\}$. This is the quotient topology (see Section 2.8 of Part 1) and is actually a metric topology; indeed (Problem 3), if $\langle \cdot\,, \cdot \rangle$ is the usual Euclidean inner product $(\sum_{j=1}^{n+1} \bar{u}_j v_j)$

$$d([u],[v]) = \sqrt{1 - \frac{|\langle u,v\rangle|}{\|u\|\|v\|}} \tag{7.2.4}$$

is a metric that induces this topology. Here we use $[u]$ for $\pi(u)$ when it is convenient.

We now specialize to $n = 1$, that is, consider $\mathbb{CP}[1]$. Let ∞ denote the point $[\binom{1}{0}]$. Every other point has a unique element of the form $\binom{z}{1}$, so

$$\mathbb{CP}[1] \setminus \{\infty\} = \left\{ \left[\begin{pmatrix} z \\ 1 \end{pmatrix}\right] \,\middle|\, z \in \mathbb{C} \right\}$$

and $f_\infty \colon \mathbb{CP}[1] \setminus \{\infty\} \to \mathbb{C}$ by

$$f_\infty\left(\begin{bmatrix} z \\ 1 \end{bmatrix}\right) = z \tag{7.2.5}$$

is easily seen to be a continuous map.

Similarly, if $0 = [\binom{0}{1}]$, then we can define $f_0 \colon \mathbb{CP}[1] \setminus \{0\} \to \mathbb{C}$ by

$$f_0\left(\begin{bmatrix} 1 \\ z \end{bmatrix}\right) = z \tag{7.2.6}$$

We easily see that $f_0 f_1^{-1} \colon \mathbb{C}^\times \to \mathbb{C}^\times$ is given by

$$f_0 f_1^{-1}(z) = z^{-1} \tag{7.2.7}$$

since $\left[\begin{smallmatrix} z \\ 1 \end{smallmatrix}\right] \sim \left[\begin{smallmatrix} 1 \\ z^{-1} \end{smallmatrix}\right]$. We conclude:

Proposition 7.2.2. $\mathbb{CP}[1]$ *has a natural Riemann surface structure making it conformally equivalent to* $\widehat{\mathbb{C}}$.

Indeed, the equivalence is given by $F([\begin{smallmatrix} z \\ 1 \end{smallmatrix}]) = z$, $F(\{\infty\}) = \infty$. The reader can check (Problem 5) that $\mathbb{CP}[n]$ has a similar coordinate system making it into an n-dimensional complex analytic manifold.

Next, we want to consider circles and lines in $\widehat{\mathbb{C}}$. For S^2 in $\mathbb{R}^3$, circles are exactly the intersection of circular cones and S^2. Such cones are described by the vanishing of a quadratic form, for example, $x_3^2 - x_1^2 - x_2^2$. This suggests that we want to take quadratic forms on $\mathbb{C}^2$ and look at where they vanish.

Thus, A will be a 2×2 Hermitian matrix and

$$q_A(u) = \langle u, Au \rangle \tag{7.2.8}$$

where $\langle\,\cdot\,,\,\cdot\,\rangle$ is the Euclidean inner product

$$\langle u,v\rangle=\sum_{j=1}^{2}\bar u_j v_j \tag{7.2.9}$$

q_A is not constant on a line since

$$q_A(\lambda u)=|\lambda|^2 q_A(u) \tag{7.2.10}$$

but whether q_A is 0 or positive is. Since we want q_A to vanish at some points, it better not be either strictly positive or strictly negative. Also, for the zero set to be more than a single point, we need $\ker(A)=0$. Thus,

Definition. An *indefinite quadratic form* is a q_A of the form (7.2.8) with $A=A^*$ and $\det(A)<0$. For such an A, we define $C_A, C_A^\pm$ in $\mathbb{CP}[1]$ by

$$C_A=\{[u]\mid q_A(u)=0\},\qquad C_A^\pm=\{[u]\mid \pm q_A(u)>0\}$$

C_A is the "circle" defined by A and C_A^+ the "disk."

Theorem 7.2.3. *The intersection of every C_A and $\mathbb{C}$ is either a circle or straight line and every circle and straight line is a $C_A\cap\mathbb{C}$. C_A^+ can be either the interior of the circle or it can be the exterior $\cup\{\infty\}$. Both possibilities occur.*

Proof. Write $A=\left(\begin{smallmatrix}\alpha&\gamma\\ \bar\gamma&\beta\end{smallmatrix}\right)$ for $\alpha,\beta\in\mathbb{R}$, $\gamma\in\mathbb{C}$.

Case 1 ($\alpha=0$) Then

$$q_A\left(\begin{pmatrix}z\\1\end{pmatrix}\right)=\bar\gamma z+\gamma\bar z+\beta=2(\operatorname{Re}\gamma)x+2(\operatorname{Im}\gamma)y+\beta \tag{7.2.11}$$

where $z=x+iy$, so $q_A(\left(\begin{smallmatrix}z\\1\end{smallmatrix}\right))=0$ is the equation of a straight line in $\mathbb{C}$.

Every straight line has the form $2ax+2by+\beta=0$ for $\beta\in\mathbb{R}$. This is C_A for $A=\left(\begin{smallmatrix}0&a+ib\\ a-ib&\beta\end{smallmatrix}\right)$ which has $\det(A)=-a^2-b^2<0$.

Case 2 ($\alpha\neq0$) If $\alpha\neq0$,

$$\begin{aligned}q_A\left(\begin{pmatrix}z\\1\end{pmatrix}\right)&=\alpha|z|^2+\bar\gamma z+\gamma\bar z+\beta\\ &=\alpha\left(\left|z+\frac{\gamma}{\alpha}\right|^2+\frac{\det(A)}{\alpha^2}\right)\end{aligned} \tag{7.2.12}$$

so C_A is $\partial\mathbb{D}_r(z_0)$, where

$$z_0=-\frac{\gamma}{\alpha}\qquad r=\sqrt{\frac{-\det(A)}{\alpha^2}} \tag{7.2.13}$$

In the other direction, $\partial\mathbb{D}_r(z_0)$ is C_A for $A=\left(\begin{smallmatrix}1&-z_0\\ -\bar z_0&-r^2+|z_0|^2\end{smallmatrix}\right)$.

By continuity of q_A, $C_A^{\pm}$ are two sides of the set obtained from removing C_A from $\widehat{\mathbb{C}}$. Since

$$C_{-A}^{\pm} = C_A^{\mp} \tag{7.2.14}$$

we can get either side by flipping the signs. $\square$

Remark. $q_A([\begin{smallmatrix}1\\0\end{smallmatrix}]) = \alpha$, so $\infty \in C_A \Leftrightarrow \alpha = 0$. Thus, straight lines contain ∞ and one can summarize this by saying "a straight line in $\mathbb{C}$ is precisely a 'circle' containing ∞."

The above argument actually proves more:

Proposition 7.2.4. *If A, B are indefinite quadratic forms, then*

$$C_A = C_B \Leftrightarrow A = \lambda B \qquad \text{for } \lambda \in \mathbb{C} \tag{7.2.15}$$

If also $C_A^+ = C_B^+$, then $\lambda > 0$.

Proof. (7.2.11) or (7.2.12) shows that the set of zeros determines β, γ (in the case of a line) and α, γ and $\sqrt{-\det(A)}$ (in the case of a circle) up to an overall common constant. That proves (7.2.15).

For $\lambda > 0$, $C_{\lambda A}^+ = C_A^+$, so this and (7.2.15) prove the final statement. $\square$

Algebraic curves can be naturally embedded into $\mathbb{CP}[2]$ to get a compact *projective curve* by adding points at infinity. Given an irreducible polynomial, $P(z, w)$, in two variables of degree d, define $\widetilde{P}(z, w, u)$ in three variables by

$$\widetilde{P}(z, w, u) = u^d P\left(\frac{z}{u}, \frac{w}{u}\right) \tag{7.2.16}$$

Then for $\lambda \in \mathbb{C}^\times$, $\widetilde{P}(z, w, u) = 0 \Leftrightarrow \widetilde{P}(\lambda z, \lambda w, \lambda u) = 0$, so the zero set is a subset of $\mathbb{C}^3$ which is zero, plus a union of elements of $\mathbb{CP}[2]$. The resulting subset of $\mathbb{CP}[2]$ is the projectivization, $\mathcal{S}_{\mathbb{P}}$, of $\mathcal{S} = \{(z, w) \mid P(z, w) = 0\}$. Under $(z, w) \mapsto [(z, w, 1)]$, $\mathcal{S}$ is a subset of $\mathcal{S}_{\mathbb{P}}$. The remainder is precisely those points where $u \equiv 0$.

For example, if $P(z, w) = w^2 - P_0(z)$, then

$$\widetilde{P}(z, w, u) = w^2 u^{d-2} - u^d P_0\left(\frac{z}{d}\right) \tag{7.2.17}$$

so, if $P_0(z) = \prod_{j=1}^d (z - z_j)$, then

$$\widetilde{P}(z, w, u) = w^2 u^{d-2} - \prod_{j=1}^{d} (z - z_j u)$$

For quadratics, $d = 2$, and the points at infinity ($u = 0$) are $w = \pm z$, that is $[(\pm 1, 1, 0)]$—two nonsingular points at infinity.

For cubics, $d = 3$, the only point at infinity is $z = 0$ (i.e., $[(0,1,0)]$, and $\frac{\partial \widetilde{P}}{\partial u}(0,1,0) = 1 \neq 0$, so this point is again nonsingular.

For quartics, $d = 4$, the points at infinity have $z = 0$ (i.e., $[(0,1,0)]$, and $\frac{\partial^2 \widetilde{P}}{\partial u^2}(0,1,0) \neq 0$, so we have a double point at infinity, but a cusp. Using u and z as local coordinates with $w \equiv 1$, $u = \pm z^2 + O(z^5)$, we again see that one needs to add two points to resolve the singularity. This is a projective interpretation of our naive addition of two points at infinity, as shown in Figure 7.1.5.

Notes and Historical Remarks. Stereographic projection goes back at least to Ptolemy in about 150 AD. That it is conformal was noted by Thomas Harriot (1560–1621) about 1590.

Projective geometry had its roots in understanding perspective in painting and issues of technical drawing. The more modern approach dates back to the work of Gaspard Monge (1746–1818) and Jean-Victor Poncelet (1788–1867) at the start of the nineteenth century. The special structure of complex projective space was studied by August Möbius (1790–1868) mid-century. The competition between the French and German schools continued with the work of Felix Klein (1849–1925) and Henri Poincaré (1854–1912), which will concern us further in the next section and in Section 8.7.

Problems

1. Prove that the line through $(0,0,1)$ and (x_1, x_2, x_3) is given by $(\theta x_1, \theta x_2, (1-\theta)+\theta x_3)$ and that the last coordinate is 0 at $\theta = (1-x_3)^{-1}$, and thereby verify (7.2.1). Verify also that (7.2.3) is the inverse of (7.2.1) restricted to the set (7.2.2).

2. This problem requires the reader to know about Riemann metrics on manifolds.

   (a) With Q_r given by (7.2.3) and σ by (6.5.1), prove that
$$|Q_r(z) - Q_r(w)| = r\sigma(z,w) \tag{7.2.18}$$
   (*Hint*: Prove that the square of the left-hand side of (7.2.18) is quadratic in r and compute for $r = 0, \frac{1}{2}, 1$. The case $r = \frac{1}{2}$ is Proposition 6.5.1(c).)

   (b) Prove the usual Riemann metric on S_r induces the Riemann metric $r^2 d^2 z/(1+|z|^2)^4$ on $\mathbb{C}$ and so stereographic projection is conformal (i.e., angle-preserving).

3. In this problem, let S^3 denote the unit sphere in $\mathbb{C}^2$, that is, $\{\vec{z} \mid |z| = 1\}$.

   (a) Given lines $\mathfrak{m}, \mathfrak{p} \in \mathbb{CP}[1]$, let
$$\rho(\mathfrak{m}, \mathfrak{p}) = \min\{|u - v| \mid u \in \mathfrak{m} \cap S^3,\ v \in \mathfrak{p} \cap S^3\} \tag{7.2.19}$$

Prove that ρ is a metric and that its topology is the one described in the text. (*Hint*: You will want to first prove that for any $u \in \mathfrak{m} \cap S^3$, $\rho(\mathfrak{m},\mathfrak{p}) = \min\{|u-v| \mid v \in \mathfrak{p} \cap S^3\}$.)

(b) For any $u \in \mathfrak{m} \cap S^3$, $v \in \mathfrak{p} \cap S^3$, prove that

$$\rho(\mathfrak{m},\mathfrak{p})^2 = \min_{e^{i\theta} \in \partial\mathbb{D}} |u - e^{i\theta} v|^2 = 2(1 - |\langle u, v\rangle|)$$

Conclude that d in (7.2.4) is $(\sqrt{2})^{-1}$ times ρ so that d is a metric inducing the topology on $\mathbb{CP}[1]$.

4. Let $\sigma(z,w)$ be given by (6.5.1) and d by (7.2.4).

(a) Prove

$$d\left(\left[\begin{pmatrix} z \\ 1 \end{pmatrix}\right], \left[\begin{pmatrix} w \\ 1 \end{pmatrix}\right]\right)^2 \left(1 + \frac{|\langle u, v\rangle|}{\|u\|\,\|v\|}\right) = \sigma(z,w)^2$$

if $u = \binom{z}{1}$, $v = \binom{w}{1}$.

(b) Conclude that $d([\binom{z}{1}], [\binom{w}{1}]) \le \sigma(z,w) \le \sqrt{2}\, d([\binom{z}{1}], [\binom{w}{1}])$.

5. In $\mathbb{CP}[n]$, let $U_1, \dots, U_{n+1}$ be the sets given by $U_j = \{[u] \mid u_j \ne 0\}$ and map $f_j \colon U_j \to \mathbb{C}^n$ by

$$f_j([u_1, \dots, u_{n+1}]) = \left(\frac{u_1}{u_j}, \dots, \frac{u_{j-1}}{u_j}, \frac{u_{j+1}}{u_j}, \dots, \frac{u_{n+1}}{u_j}\right)$$

Prove that the U_j's cover $\mathbb{CP}[n]$ and the f_j's give $\mathbb{CP}[n]$ the n-dimensional analog of a Riemann surface structure.

6. (a) Show that circles and lines in $\mathbb{C}$ are exactly those nonempty sets of the form ($z = x + iy$)

$$\alpha|z|^2 + \beta x + \gamma y + \delta = 0$$

with $\alpha, \beta, \gamma, \delta \in \mathbb{R}$ not all zero.

(b) Show that hyperplanes in $\mathbb{R}^3$ are given by $Ax_1 + Bx_2 + Cx_3 + D = 0$, with $A, B, C, D \in \mathbb{R}$ not all zero.

(c) Using the formula for Q_r, (7.2.3), prove that circles and lines in $\mathbb{C}$ are exactly the images under stereographic projection of the intersections of hyperplanes with S_r.

7.3. $\mathbb{PSL}(2,\mathbb{C})$

This is the central section of this chapter where we explore fractional linear transformations. The natural self-maps on a vector space are the linear transformations, so it is reasonable to ask what they induce on projective space. Since $\mathbb{CP}[1]$ is equivalence classes in $\mathbb{C}^2 \setminus \{0\}$, we want maps, T, of that set to itself, so $\ker(T) = \{0\}$, that is, T is invertible. $\mathbb{GL}(2,\mathbb{C})$ is the general linear group of 2×2 invertible matrices. Later we'll also look at $\mathbb{SL}(2,\mathbb{C})$, those $T \in \mathbb{GL}(2,\mathbb{C})$ with $\det(T) = 1$.

We'll begin by proving FLTs are characterized by what happens to 0, 1, and ∞ and that any three distinct image points occur. In particular, this will imply the FLTs are the quotient group $\mathbb{SL}(2,\mathbb{C})/\{\pm\mathbb{1}\}$. Then we'll classify all analytic bijections of $\mathbb{C}$ and of $\widehat{\mathbb{C}}$. Next will come the study of conjugacy classes in the group of FLTs and the classification into elliptic, parabolic, hyperbolic, and loxodromic, and the related issue of fixed points and their stability. We'll next prove that FLTs leave the set of all circles and lines fixed and then find which FLTs map $\mathbb{C}_+$ bijectively to itself, and similarly for $\mathbb{D}$.

If $u, v \in \mathbb{C}^2 \setminus \{0\}$ and $u \sim v$, that is, $u = \lambda v$ for $\lambda \in \mathbb{C}^\times$, then $Tu \sim Tv$ since $Tu = \lambda Tv$. Thus, there is map f_T from $\mathbb{CP}[1]$ to itself induced by T via

$$f_T([u]) = [Tu] \tag{7.3.1}$$

Clearly,

$$f_{TS} = f_T \circ f_S \tag{7.3.2}$$

and, in particular,

$$f_T^{-1} = f_{T^{-1}} \tag{7.3.3}$$

so each f_T is a bijection.

If $T = \left(\begin{smallmatrix} a & b \\ c & d \end{smallmatrix}\right)$ and we use the $[\left(\begin{smallmatrix} z \\ 1 \end{smallmatrix}\right)]$ representatives, (7.3.1) is equivalent to

$$f_T(z) = \left[\begin{pmatrix} az+b \\ cz+d \end{pmatrix}\right] = \frac{az+b}{cz+d} \tag{7.3.4}$$

and this is the more usual way to think of the f_T's. These maps are called *fractional linear transformations* or sometimes linear fractional transformations, bilinear transformations (the name comes from the fact that $w = f_T(z)$ is equivalent to $czw + dw - az - b = 0$, a biaffine relationship), or Möbius transformations.

We interpret $(az+b)/(cz+d)$ as a/c if $z = \infty$ or as ∞ if $z = -d/c$. This is consistent with ∞ as $[\left(\begin{smallmatrix} 1 \\ 0 \end{smallmatrix}\right)]$ and (7.3.1). One big advantage of this projective way of viewing FLTs is that it makes clear, via (7.3.2), why composition of FLTs is given by matrix multiplication. It also makes it easy to understand fixed points of an f_T. By (7.3.1) and the meaning of equivalence,

$$f_T(z) = z \Leftrightarrow \begin{pmatrix} z \\ 1 \end{pmatrix} \text{ is an eigenvector of } T \tag{7.3.5}$$

which is the key to

Proposition 7.3.1. (a) *f_T leaves three distinct points fixed if (and only if) $T = \lambda\mathbb{1}$.*

(b) $f_T = f_S \Leftrightarrow T = \lambda S$ $\quad$ (7.3.6)

Proof. (a) If T has three distinct eigenspaces, then T is a multiple of $\mathbb{1}$ (see Problem 2).

(b) By (7.3.2) and (7.3.3),

$$f_T = f_S \Leftrightarrow f_{TS^{-1}} = \mathbb{1} \Leftrightarrow TS^{-1} = \lambda\mathbb{1} \tag{7.3.7}$$

for some λ. □

As a consequence, we can replace $T \in \mathbb{GL}(2,\mathbb{C})$ by $T\det(T)^{-1/2} \in \mathbb{SL}(2,\mathbb{C})$, that is, every f_T is equal to an f_T with $\det(T) = \mathbb{1}$ so that, henceforth, we will (almost) always take $T \in \mathbb{SL}(2,\mathbb{C})$. The map $T \to f_T$ of $\mathbb{SL}(2,\mathbb{C})$ to FLTs has kernel (in the group theoretic sense, i.e., $\{T \mid f_T = \mathbb{1}\}$) exactly $\{\lambda\mathbb{1} \mid \det(\lambda\mathbb{1}) = \lambda^2 = 1\}$, that is,

Proposition 7.3.2. *The set of FLTs is*

$$\mathbb{PSL}(2,\mathbb{C}) \equiv \mathbb{SL}(2,\mathbb{C})/\{\pm\mathbb{1}\} \tag{7.3.8}$$

$\mathbb{P}$ stands for "projective." We are interested in the structure of this group of all FLTs. Proposition 7.3.1 has another consequence:

Theorem 7.3.3. *For any triples (z_1,z_2,z_3) and (w_1,w_2,w_3) of distinct points in $\widehat{\mathbb{C}}$ (i.e., $z_1 \neq z_2 \neq z_3 \neq z_1$ and $w_1 \neq w_2 \neq w_3 \neq w_1$ but w's can be z's), there is a unique f_T with*

$$f_T(z_j) = w_j \tag{7.3.9}$$

Proof. If (7.3.9) holds for T_1 and T_2, then by (7.3.2), $f_{T_1^{-1}T_2}$ leaves z_1, z_2, z_3 fixed. So, by Proposition 7.3.1, $T_1^{-1}T_2 = \lambda\mathbb{1}$, so $f_{T_1} = f_{T_2}$. This proves uniqueness.

Suppose for any distinct z_1, z_2, z_3, we find $f_{(z_1,z_2,z_3)} \in \mathbb{PSL}(2,\mathbb{C})$ with (7.3.9) for $(w_1,w_2,w_3) = (0,1,\infty)$. Then $f^{-1}_{(w_1,w_2,w_3)} f_{(z_1,z_2,z_3)}$ solves (7.3.9) in general. So it suffices to handle the case $(w_1,w_2,w_3) = (0,1,\infty)$. Let

$$f_{(z_1,z_2,z_3)}(z) = \frac{(z-z_1)(z_2-z_3)}{(z-z_3)(z_2-z_1)} \tag{7.3.10}$$

It is easy to see that $f(z_1) = 0$, $f(z_3) = \infty$, and $f(z_2) = 1$. □

The quantity (7.3.10) is called the *cross-ratio*, often written

$$[z,z_2,z_1,z_3] = \frac{(z-z_1)(z_2-z_3)}{(z-z_3)(z_2-z_1)} \tag{7.3.11}$$

As you might expect, there are different conventions, and some authors use $[z,z_1,z_2,z_3]$ for the right side of (7.3.11), but our convention is the most common.

A key fact (Problems 11 and 12) is that for any FLT, f,

$$[f(z), f(z_1), f(z_2), f(z_3)] = [z, z_1, z_2, z_3] \tag{7.3.12}$$

Theorem 7.3.3 implies $\mathcal{F}\colon \mathbb{PSL}(2,\mathbb{C}) \to \{(w_0, w_1, w_2) \in \widehat{\mathbb{C}}^3 \mid w_0 \neq w_1 \neq w_2 \neq w_0\}$ by $\mathcal{F}(f) = (f(0), f(1), f(\infty))$ is a bijection. This shows $\mathbb{PSL}(2,\mathbb{C})$ is a three-dimensional complex manifold, so six real dimensions. Of course, this also follows by noting a, b, d are free parameters if $a \neq 0$ since c can be adjusted to assure $ac - bd = 1$.

Many presentations of FLTs use cross-ratios as a fundamental object; for example, an important fact (Problems 15 and 16) is that $[z, z_1, z_2, z_3] \in \mathbb{R}$ if and only if z lies on the circle or line determined by z_1, z_2, z_3. We'll leave the study of cross-ratios to the Problems (see Problems 11–16).

Cross-ratios are an invariant of four points. Since any three points can be mapped into any other, there are no invariants of three points and cross-ratios are the simplest projective invariants.

One special reason for interest in FLTs is that they include all biholomorphic self-maps of $\widehat{\mathbb{C}}$ and $\mathbb{C}$. We begin with $\mathbb{C}$.

Theorem 7.3.4. *Let $f \in \mathbb{A}\mathrm{ut}(\mathbb{C})$. Then for some $a \in \mathbb{C}^\times$ and $b \in \mathbb{C}$,*

$$f(z) = az + b \tag{7.3.13}$$

Every such f is an analytic bijection.

Proof. The last sentence is immediate, so we can focus on characterizing analytic bijections. We first claim that

$$\lim_{|z|\to\infty} |f(z)| = \infty \tag{7.3.14}$$

If not, by compactness of $\widehat{\mathbb{C}}$, there exist $z_n \to \infty$ so $w_n \equiv f(z_n) \to w_\infty \in \mathbb{C}$. Since f is onto, there is $z_\infty \in \mathbb{C}$ with $f(z_\infty) = w_\infty$. Since f is analytic, the open mapping theorem implies that for n large, $w_n \in f[\mathbb{D}_1(z_\infty)]$. Since f is one-one, $|z_n - z_\infty| < 1$ for n large, contradicting that $z_n \to \infty$. Thus, (7.3.14) holds.

It follows that

$$|f(z)^{-1}| \to 0 \quad \text{as} \quad |z| \to \infty \tag{7.3.15}$$

By hypothesis and Theorem 3.5.1, f has a single simple zero, say at z_0. Thus, $h(z) \equiv (z - z_0)/f(z)$ is entire, and by (7.3.15), $|h(z)|/|z| = o(1)$, as $|z| \to \infty$. It follows, by Cauchy estimates, that h is constant, that is, for some α, $f(z) = \alpha(z - z_0)$.

Since f is one-one, $\alpha \neq 0$, that is, (7.3.13) holds. $\square$

Theorem 7.3.5. *Any FLT is an analytic bijection of $\widehat{\mathbb{C}}$. Conversely, every analytic bijection is an FLT. That is,*

$$\mathbb{A}\mathrm{ut}(\widehat{\mathbb{C}}) = \mathbb{PSL}(2,\mathbb{C}) \tag{7.3.16}$$

Proof. Since $f_{T^{-1}} = f_T^{-1}$, every FLT is a bijection and it is clearly analytic.

Given an analytic bijection, f, by Theorem 7.3.3, find an FLT, g, so that

$$g(0) = f(0), \qquad g(1) = f(1), \qquad g(\infty) = f(\infty) \tag{7.3.17}$$

Then $h = g^{-1}f$ leaves $0,1,\infty$ fixed. Since $h(\infty) = \infty$, h is an analytic bijection of $\mathbb{C}$, so $h(z) = az + b$ by Theorem 7.3.4. $h(0) = 0$ implies $b = 0$, and then $h(1) = 1$ implies $a = 1$, that is, $h(z) = z$, so $f = g \in \mathbb{PSL}(2,\mathbb{C})$. $\square$

Thus, we've proven that $\mathbb{A}\mathrm{ut}(\mathbb{C})$ is a four-(real) dimensional manifold and $\mathbb{A}\mathrm{ut}(\widehat{\mathbb{C}})$ is six-dimensional.

Next, we turn to studying conjugacy classes in $\mathbb{PSL}(2,\mathbb{C})$. Recall that x,y in a group, G, are called *conjugate* if and only if there is $z \in G$ so $x = zyz^{-1}$. The *conjugacy class* of $x \in G$ is the set of all y conjugate to it, that is, $\{zxz^{-1} \mid z \in G\}$. Elements, f, of $\mathbb{PSL}(2,\mathbb{C})$ are $f_{\pm T}$ for a unique $T \in \mathbb{SL}(2,\mathbb{C})$. Obviously, $\mathrm{Tr}(-T) = -\mathrm{Tr}(T)$, so we can define $\mathrm{tr}(f) \in \mathbb{C}/\{\pm 1\}$, that is, complex numbers modulo sign. Here is the basic result:

Theorem 7.3.6. (a) *If $T,S \in \mathbb{SL}(2,\mathbb{C})$, both different from $\pm\mathbb{1}$, then T is conjugate to S in $\mathbb{SL}(2,\mathbb{C})$ if and only if*

$$\mathrm{Tr}(T) = \mathrm{Tr}(S) \tag{7.3.18}$$

(b) *If $f,g \in \mathbb{PSL}(2,\mathbb{C})$, both different from the identity, then f is conjugate to g in $\mathbb{PSL}(2,\mathbb{C})$ if and only if*

$$\mathrm{tr}(f) = \mathrm{tr}(g) \tag{7.3.19}$$

Proof. (a) If $A \in \mathbb{SL}(2,\mathbb{C})$, its eigenvalues are λ and λ^{-1}. If λ is 1 or -1, then $\mathrm{Tr}(A) = 2$. Thus, $\mathrm{Tr}(A) \neq 2$ means A has distinct eigenvalues, and since they are the roots of $\lambda^2 - \mathrm{Tr}(A)\lambda + 1 = 0$, they are determined by $\mathrm{Tr}(A)$. By the diagonalizability of matrices with simple eigenvalues (see Theorem 1.3.1), there is $U \in \mathbb{GL}(n,\mathbb{C})$, so $UAU^{-1} = \left(\begin{smallmatrix} \lambda & 0 \\ 0 & \lambda^{-1} \end{smallmatrix}\right)$, and it is easy to see that U can be taken in $\mathbb{SL}(2,\mathbb{C})$. Thus, if $\mathrm{Tr}(T) = \mathrm{Tr}(S) \neq \pm 2$, we have conjugacy.

If $\mathrm{Tr}(A) = 2$, both eigenvalues are 1, so the Jordan normal form (see Theorem 1.3.1) is either $\left(\begin{smallmatrix} 1 & 0 \\ 0 & 1 \end{smallmatrix}\right)$ or $\left(\begin{smallmatrix} 1 & 1 \\ 0 & 1 \end{smallmatrix}\right)$. Thus, $\mathrm{Tr}(A) = 2$ and $A \neq \mathbb{1}$ implies A is conjugate to $\left(\begin{smallmatrix} 1 & 1 \\ 0 & 1 \end{smallmatrix}\right)$, which proves $\mathrm{Tr}(T) = \mathrm{Tr}(S) = \pm 2$ and $T \neq \mathbb{1} \neq S$, then T and S are conjugate.

Since $\operatorname{Tr}(AB) = \operatorname{Tr}(BA)$, we have that $\operatorname{Tr}(UTU^{-1}) = \operatorname{Tr}(T)$ and conjugate matrices have the same trace.

(b) Let $f = f_A$, $g = f_B$. Part (a) and $\operatorname{tr}(f) = \operatorname{tr}(g)$ implies $A = \pm TBT^{-1}$ for some $T \in \mathbb{SL}(2,\mathbb{C})$, so $f = f_T g f_T^{-1}$ and f and g are conjugate.

Conversely, if f and g are conjugate, $A = \pm TBT^{-1}$ and $\operatorname{tr}(f) = \operatorname{tr}(g)$. □

Definition. Let $f \in \mathbb{SL}(2,\mathbb{C})$. Then

(i) If $\operatorname{tr}(f) \in \pm(2,\infty)$, we say f is *hyperbolic*.
(ii) If $\operatorname{tr}(f) \in (-2,2)$, we say f is *elliptic*.
(iii) If $\operatorname{tr}(f) = \pm 2$ and $f \neq \mathbb{1}$, we say f is *parabolic*.
(iv) If $\operatorname{tr}(f) \notin \mathbb{R}$, we say f is *loxodromic*.

The following theorem shows why it makes sense to use this four-fold classification.

Theorem 7.3.7. (a) *If f is hyperbolic and $\operatorname{tr}(f) = \pm 2\cosh(\beta)$ with $\beta > 0$, then f is conjugate to*

$$f_\beta(z) = e^{-2\beta} z \tag{7.3.20}$$

f has two fixed points $z_\pm$, and if $w \neq z_\mp$, then $\lim_{n\to\infty} f^{[n]}(w) = z_+$ and $\lim_{n\to-\infty} f^{[n]}(w) = z_-$.

(b) *If f is parabolic, f is conjugate to*

$$f(z) = z + 1 \tag{7.3.21}$$

f has a single fixed point z_0, and for any w, $\lim_{n\to\pm\infty} f^{[n]}(w) = z_0$.

(c) *If f is elliptic and $\operatorname{tr}(f) = \pm 2\cos(\alpha)$ with $\alpha \in (0,\pi/2]$, then f is conjugate to*

$$f_\alpha(z) = e^{2i\alpha} z \tag{7.3.22}$$

f has two fixed points. If α/π is rational, the orbit $\{f^{[n]}(w) \mid n \in \mathbb{Z}\}$ for any fixed w is a finite set. If α/π is irrational, then $\overline{\{f^{[n]}(w) \mid n \in \mathbb{Z}\}}$ is a circle or straight line for any w distinct from the fixed points.

(d) *If f is loxodromic and $\operatorname{tr}(f) = \pm 2\cosh(\beta + i\alpha)$, with $\beta > 0$ and $0 < \alpha < \pi$, then f is conjugate to*

$$f_{\alpha,\beta}(z) = e^{-2(\beta+i\alpha)} z \tag{7.3.23}$$

f has two fixed points $z_\pm$ with the same attracting properties as in the hyperbolic case.

Remarks. 1. For $n > 0$, $f^{[n]}$ is defined as in Chapter 6 (see, e.g., Problem 12 of Section 6.2) as $f \circ \cdots \circ f$, n times. For $n < 0$, $f^{[n]} = (f^{[-n]})^{-1}$.

2. If $f^{[n]}(z) \to z_\infty$ as $n \to \infty$ for all points, z, which are not fixed points, we say z_∞ is a (globally) *attracting fixed point.* If that is true as $n \to -\infty$, we say it is globally *repelling.*

3. If $z \in \mathbb{C} \setminus \mathbb{R}$, either z or $-z$ is uniquely of the form $\lambda + \lambda^{-1}$ with $\lambda \in \mathbb{C}_+ \cap (\mathbb{C} \setminus \overline{\mathbb{D}})$, so $2\cosh(\beta + i\alpha)$ with $\beta > 0$ and $\alpha \in (0, \pi)$.

Proof. If $f = ghg^{-1}$ and $h(z_0) = z_0$, then $f(g(z_0)) = g(z_0)$ and $h^{[n]}(w) - z_0 \to 0$ if and only if $f^{[n]}(g(w)) - g(z_0) \to 0$. So we need only prove results for the canonical examples.

If $\pm\text{tr}(f) \neq 2$, $f = f_T$ with $T = \left(\begin{smallmatrix} \lambda & 0 \\ 0 & \lambda^{-1} \end{smallmatrix}\right)$ with $\lambda \neq 0, 1$, so $f(z) = \lambda^2 z$. This gives the canonical models (7.3.20), (7.3.22), and (7.3.23). In case f is parabolic, then the matrix for f is conjugate to $T = \left(\begin{smallmatrix} 1 & 1 \\ 0 & 1 \end{smallmatrix}\right)$ which has $f_T(z)$ given by (7.3.21).

Thus, all that remains is checking the fixed points ($z_+ = 0$, $z_- = \infty$ for case (a), (c), (d) and $z_0 = \infty$ in case (b)) and asymptotic behaviors, which we leave to the Problems (see Problem 5). $\square$

In the elliptic, nonperiodic case, the orbits are dense in a circle. In the hyperbolic and parabolic case, because there are limits at $\pm\infty$, there are at most two limit points, so that the orbits are not dense in a curve, but (Problem 25) there are invariant circles. The parts of these needed to contain an orbit are shown for typical hyperbolic and parabolic examples in Figure 7.3.1.

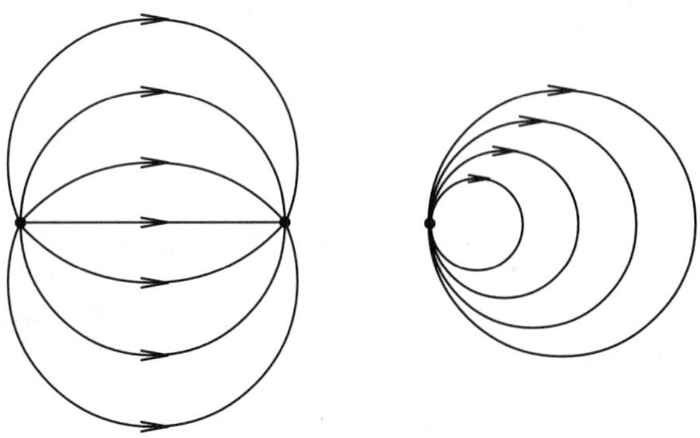

Figure 7.3.1. Flow lines for FLTs.

Next, we prove that

Theorem 7.3.8. *Let C be the family of all straight lines and circles in $\widehat{\mathbb{C}}$. Then*

(a) *The image of any element of C under an FLT is another element of C.*
(b) *For any two elements of C, there is an FLT that maps one setwise onto the other.*

Remark. Because the use of indefinite quadratic forms is nonstandard, most books prove this theorem by other means. See Problems 10 and 16 for alternate proofs of Theorem 7.3.8.

Proof. For any $T \in \mathbb{SL}(2,\mathbb{C})$,

$$\lambda_A(Tu) = \langle Tu, ATu\rangle = \lambda_{T^*AT}(u) \tag{7.3.24}$$

Thus, f_T maps C_A to $C_{(T^{-1})^*AT^{-1}}$. By Theorem 7.2.3, f_T thus maps $\mathcal{C}$ to $\mathcal{C}$, proving (a).

One can prove (b) by analyzing what T^*AT can occur (Problem 6), but instead we'll use some simple geometry. Clearly, any line can be mapped to any other by translation and rotation, and any circle can be mapped to any other circle by translation of centers and scaling. So, affine maps, $z \to az+b$, suffice to map any circle to any circle and any line to any line.

So it suffices to map one circle to one line with some FLT. The circle of radius $\frac{1}{2}$ and center $\frac{1}{2}$ is given by

$$|z|^2 = \operatorname{Re} z \tag{7.3.25}$$

The FLT, $w = 1/z$, takes this to

$$\frac{1}{|w|^2} = \operatorname{Re}\left(\frac{1}{w}\right) = \frac{\operatorname{Re} w}{|w|^2}$$

that is, the circle (7.3.25) goes to the straight line

$$\operatorname{Re} w = 1 \tag{7.3.26}$$

□

As a final topic, we want to see which FLTs map $\mathbb{C}_+$ to $\mathbb{C}_+$ or $\mathbb{D}$ to $\mathbb{D}$.

Theorem 7.3.9. *Let $T \in \mathbb{SL}(2,\mathbb{C})$. f_T maps $\mathbb{C}_+$ to $\mathbb{C}_+$ and $\mathbb{R}$ to $\mathbb{R}$ if and only if T is a real matrix.*

Proof.

$$\mathbb{C}_+ = \{z \mid \operatorname{Im} z > 0\} = \{z \mid q_A(z) > 0\} \tag{7.3.27}$$

where

$$A = \begin{pmatrix} 0 & i \\ -i & 0 \end{pmatrix} \tag{7.3.28}$$

by (7.2.11).

By (7.3.24), f_T maps $\mathbb{C}_+$ to $\mathbb{C}_+$ if and only if

$$C_{T^*AT} = C_T, \qquad C^+_{T^*AT} = C^+_A \tag{7.3.29}$$

By Proposition 7.2.4, (7.3.29) holds if and only if $T^*AT = \lambda A$ for $\lambda > 0$. But then $\det(T^*AT) = \lambda^2 \det(A)$, so, since $\det(T) = 1$, $\lambda = 1$ (since $\lambda > 0$). Thus, in general,

$$(7.3.29) \Leftrightarrow T^*AT = A \Leftrightarrow T^* = AT^{-1}A^{-1} \tag{7.3.30}$$

For A of the form (7.3.28), where $A^2 = \mathbb{1}$, a simple calculation shows that if $\det(T) = 1$, then

$$\begin{pmatrix} 0 & i \\ -i & 0 \end{pmatrix} T^{-1} \begin{pmatrix} 0 & i \\ -i & 0 \end{pmatrix}^{-1} = T^t \tag{7.3.31}$$

Thus, for A of the form (7.3.28),

$$(7.3.29) \Leftrightarrow T^* = T^t \Leftrightarrow \bar{T} = T \tag{7.3.32}$$

if and only if T is real. □

Thus,

$$\begin{aligned} \{f_T \mid f_T \text{ map } \mathbb{C}_+ \text{ onto } \mathbb{C}_+\} &= \{f_T \mid T \in \mathbb{SL}(2,\mathbb{R})\} \\ &= \mathbb{PSL}(2,\mathbb{R}) \end{aligned}$$

To map $\mathbb{D}$ to $\mathbb{D}$, we note that if $J = \left(\begin{smallmatrix} 1 & 0 \\ 0 & -1 \end{smallmatrix}\right)$, then $q_J(\left(\begin{smallmatrix} z \\ 1 \end{smallmatrix}\right)) = |z|^2 - 1$, so

$$J = \begin{pmatrix} 1 & 0 \\ 0 & -1 \end{pmatrix} \Rightarrow C_J^- = \mathbb{D} \tag{7.3.33}$$

We have thus proven that if

$$\mathbb{SU}(1,1) = \{T \in \mathbb{SL}(2,\mathbb{C}) \mid T^*JT = J\} \tag{7.3.34}$$

then

Theorem 7.3.10. *Let $T \in \mathbb{SL}(2,\mathbb{C})$. f_T map $\mathbb{D}$ to $\mathbb{D}$ if and only if $T \in \mathbb{SU}(1,1)$.*

It can be shown (Problem 7) that

$$T \in \mathbb{SU}(1,1) \Leftrightarrow T = \begin{pmatrix} \alpha & \gamma \\ \bar{\gamma} & \bar{\alpha} \end{pmatrix} \qquad \text{with } |\alpha|^2 - |\gamma|^2 = 1 \tag{7.3.35}$$

Notes and Historical Remarks. Some textbooks that emphasize complex geometry and FLTs in complex variable are Jones–Singerman [**287**] and Schwerdtfeger [**513**]. For a delightful excursion through the remarkable world of FLTs and their discrete subgroups, see Mumford et al. [**397**].

Just as the translations and rotations of Euclidean geometry can be supplemented by reflections, FLTs can be supplemented by reflections, that is, anticonformal maps of the form $z \to f_T(\bar{z})$, where $f_T \in \mathbb{PSL}(2,\mathbb{C})$. The simplest example is, of course, $z \to \bar{z}$. As fundamental are the conjugates of this map which are reflections in circles—the reflection in $\mathbb{D}_r(z_0)$ is

$$z = z_0 + \frac{r^2}{\bar{z} - \bar{z}_0} \tag{7.3.36}$$

Ford [**191, 193**], in particular, has emphasized reflections, proving that every elliptic, parabolic, or hyperbolic (but not loxodromic) FLT is a product of two reflections in circles or lines. This result and the related notion of isometric circles will appear in Problems 26 and 27.

The history of fractional linear transformations combines several different mathematical threads. On the analytic side, continued fractions are intimately connected to FLTs and one thread, which we'll say more about in the Notes to Section 7.5, involves Brouncker, Wallis, and Euler.

The main thread, however, is geometric, especially elements of projective geometry, some of it discussed in the Notes to the last section. Cross-ratios, in particular, have roots in projective geometry, predating the early nineteenth-century work. They were invented by Girard Desargues (1591–1661) [**139**] as a geometric object. Möbius [**387**] introduced the algebraic formulae for complex numbers.

Fractional linear transformations in the complex plane appeared first in passing in Euler [**181**] and were systematically developed by Möbius [**388**], after whom they are sometimes named.

A final thread is due to Poincaré and Klein who studied discrete subgroups of FLTs. In a famous lecture given in Erlangen in 1872, Felix Klein (1849–1925) introduced what came to be called the "Erlangen program" based on the notion that different geometries were associated with different groups and, in particular, projective geometry was associated with the group of FLTs. The abstract group concept was late in coming and, prior to Klein, one didn't think of the family of FLTs as a group. Many of the standard views of FLTs are in Poincaré [**438, 439**] who, in particular, discussed $\mathbb{PSL}(2, \mathbb{C})$.

Klein's father was a high-ranking Prussian official. Klein went to Bonn intending to be a physicist but the professor of experimental physics, Julius Plücker (1801–1867), was also a mathematician and he convinced Felix to become a geometer. Klein was only 19 when he got his degree and by age 23 was well enough regarded to be appointed a Professor in Erlangen—the 1872 lecture was his inaugural talk as a professor.

He was next a professor in Munich and then in Leipzig during the key years 1881–83 where he drove himself in competition with Poincaré leading to a nervous breakdown in 1883 and deep depression in 1884–85. He accepted a chair at Göttingen in 1886 turning his energies more to administration than research. Under his influence, Göttingen became a major center, most notably attracting Hilbert. He turned *Mathematische Annalen* into an important journal and became a major force in mathematics education. His students include Bieberbach, Bochner, Harnack, Hurwitz, Lindemann (my own scientific ancestor, five generation-mainly physicists-removed), Kasner, Ostrowski, and Grace Chisolm Young.

August Ferdinand Möbius (1790–1868) was a German astronomer whose training was partly under Gauss, who served as the director of the Göttingen Observatory (Gauss' reputation in the nonmathematical world relied on his computing the orbit of the asteroid Ceres which was "lost" (by going behind the sun) shortly after its discovery!). While Möbius did study mathematics under Pfaff, he spent most of his career as a professor of astronomy. Besides his work on projective geometry and FLTs, he is known for his work in number theory (Möbius function and Möbius inversion; see Section 13.2 of Part 2B) and, of course, for the Möbius strip.

An interesting subtopic in the study of fractional linear transformations concerns finite subgroups of $\mathbb{PSL}(2,\mathbb{C})$. It was started by Schwarz [**512**] who asked which hypergeometric functions were algebraic, which led to finite monodromy groups (we'll discuss monodromy groups in Section 11.2). Klein [**308**], in extending this work, realized one needed finite subgroups of $\mathbb{PSL}(2,\mathbb{C})$ and he related that to finite groups of rotations, which he analyzed in [**308, 312**]. The purely group theoretic analysis was streamlined by Weyl [**591**]. Problems 19 and 20 discuss this further.

Problems

1. If $f_T(z)$ is given by (7.3.4) and $(c,d) \neq (0,0)$ but $ad-bc=0$, prove that $f(z)$ is a constant function.

2. Prove that if a 2×2 matrix has three eigenvectors, each not a multiple of the others, then it is a multiple of the identity.

3. Prove directly that $f_T(z)=z$ has at most two solutions unless $f_T(z)\equiv z$ for all z.

4. Once one has (7.3.14), prove you can conclude that f has the form (7.3.13) by analyzing $f(z^{-1})^{-1}$ near $z^{-1}=0$. This provides an alternate to the proof of Theorem 7.3.4.

5. (a) If $f=hgh^{-1}$, prove $f^{[n]}=hg^{[n]}h^{-1}$ and that $g(w)=w$ if and only if $f(h(w))=h(w)$. Conclude that the notion of globally attracting or repelling fixed points is conjugacy class invariant.

   (b) If $f(z)=e^{-2\beta}z$ or $e^{-2(\beta+i\alpha)}z$ with $\beta>0$, prove that 0 is a globally attracting and ∞ a globally repelling fixed point. Show that the approach to the limit is exponentially fast in the σ-metric of Section 6.5.

(c) If $f(z) = z + 1$, prove ∞ is both a globally attracting and globally repelling fixed point. Show that the approach in σ-metric (defined in Section 6.5) is $O(n^{-1})$.

(d) If $f(z) = e^{i\theta}z$, prove that if $\theta/2\pi$ is rational, then f has a finite period (i.e., $f^{[p]} = \mathbb{1}$ for some p), and if $\theta/2\pi$ is irrational, then $\{f^{[n]}(z)\}_{n=-\infty}^{\infty}$ is dense in a circle if $z \neq 0, \infty$.

6. Prove that if A and B are any two indefinite quadratic forms, there is a T and $\lambda \in (0, \infty)$ so that $A = \lambda T^* B T$. (*Hint*: Do this first for the diagonal case with a positive number in the upper-left corner and then use the spectral theorem for matrices (Theorem 1.3.3).)

7. If $A \in \mathbb{SU}(1,1)$, prove that $A = \left(\begin{smallmatrix} \alpha & \gamma \\ \bar{\gamma} & \bar{\alpha} \end{smallmatrix}\right)$ for $|\alpha|^2 - |\gamma|^2 = 1$.

8. By manipulating the FLTs directly, prove that any FLT is a finite product of affine maps (i.e., $z \to az + b$) and the inversion $z \to z^{-1}$. (*Hint*: If $c \neq 0$, $\frac{az+b}{cz+d} = \frac{a}{c} - \frac{\det(A)}{c(cz+d)}$.)

9. As in Problem 22 of Section 2.3, $\left(\begin{smallmatrix} a & b \\ c & d \end{smallmatrix}\right)$ is *upper triangular* if $c = 0$ and *lower triangular* if $b = 0$.

(a) Prove any invertible $A \in \mathbb{GL}(2, \mathbb{C})$ is the product of an upper-triangular and a lower-triangular matrix.

(b) If $K = \left(\begin{smallmatrix} 0 & 1 \\ 1 & 0 \end{smallmatrix}\right)$, prove L is lower triangular if and only if KLK^{-1} is upper triangular.

(c) Prove any invertible matrix is a product of at most two upper-triangular matrices and K at most two times.

(d) Prove any $f \in \mathbb{PSL}(2, \mathbb{C})$ is a finite product of affine maps and the map $z \mapsto 1/z$.

10. (a) Prove in $\mathbb{C}$ that affine maps take circles to circles and lines to lines.

(b) Prove $z \mapsto 1/z$ takes circles and lines to circles and lines by considering what the map does to $\alpha|z|^2 + \beta x + \gamma y$.

(c) If L is stereographic projection for S_r, Q_r its inverse, and $I(z) = z^{-1}$, prove that $Q_r I L$ is rotation by 180° about the axis of S_r parallel to the x-axis.

(d) Given Problem 6 of Section 7.2, prove again that $z \mapsto 1/z$ maps circles and lines to circles and lines.

(e) Using Problem 8 or 9 above, prove that any FLT takes lines and circles to lines and circles.

11. (a) Let $f(z, z_1, z_2, z_3)$ be the FLT that maps z_1 to 0, z_2 to 1, and z_3 to infinity. For any FLT, g, prove that

$$f(\,\cdot\,, g^{-1}(z_1), g^{-1}(z_2), g^{-1}(z_3)) = f(\,\cdot\,, z_1, z_2, z_3) \circ g \tag{7.3.37}$$

(b) Prove that any FLT, g, preserves cross-ratios.

12. (This problem supposes that the reader knows exterior algebra, aka wedge products; see Section 3.8 of Part 1 or Section 1.3 of Part 4.) Given nonzero $v, w \in \mathbb{C}^2$ and a choice of nonzero $\omega \in \wedge^2(\mathbb{C}^2)$, define $v \times w \in \mathbb{C}$ by

$$v \wedge w = (v \times w)\omega \tag{7.3.38}$$

(a) Given v_1, v_2, v_3, v_4 nonzero vectors in $\mathbb{C}^2$, show that $(v_1 \times v_3)(v_2 \times v_4)/(v_1 \times v_4)(v_2 \times v_3)$ is invariant under multiplying the v_j by perhaps different scalars, and if $v_j = \binom{z_j}{1}$ is exactly the cross-ratio, $[z_1, z_2, z_3, z_4]$.

(b) Using the fact that $Tv \wedge Tw = \det(T)(v \wedge w)$, prove that the cross-ratio, $[z_1, z_2, z_3, z_4]$, is invariant under FLTs.

13. (a) Show that there are exactly four permutations of four distinct points z_0, z_1, z_2, z_3 that leave the cross-ratio, $[z_0, z_1, z_2, z_3]$, invariant, so under permutation, there are six values of the cross-ratio so long as the cross-ratio is not 0, 1, or ∞.

(b) If the cross-ratio is λ, show that the six values are λ, $1-\lambda$, $1/\lambda$, $\lambda/(\lambda-1)$, $1/(1-\lambda)$, and $1-1/\lambda$.

14. Show that any four disjoint points (z_1, z_2, z_3, z_4) can be taken to $1, -1, k, -k$ (i.e., z_1 to 1, z_2 to -1, etc.) by a suitable FLT (k will depend on the points). How many solutions are there?

15. Using the known invariance of the cross-ratio and transitivity of the action of $\mathbb{PSL}(2,\mathbb{C})$ on circles and lines, prove that

$$z \in \text{ circle or line generated by } z_0, z_1, z_2 \Leftrightarrow [z, z_0, z_1, z_2] \in \mathbb{R}$$

16. (a) Let θ be the angle at vertex z in the triangle $z_3 z z_1$ and ϕ at vertex z_2 of $z_3 z_2 z_1$. Prove that the argument of the cross-ratio, $[z, z_1, z_2, z_3]$, is $\theta - \phi$.

(b) Prove the geometric fact that z lies on the circle generated by z_1, z_2, z_3 in the same arc as z_2 if and only if $\theta = \phi$ and in the opposite arc if and only if $\theta = \pi + \phi$.

(c) Prove directly that z lies in the circle generated by z_1, z_2, z_3 if and only if $[z, z_1, z_2, z_3]$ is real.

(d) Provide another proof, using cross-ratios, that FLTs leave circles and lines invariant.

17. Let $\mathbb{C}(X)$ be the family of rational functions of a formal variable X. Given an FLT, f, prove that

$$g \mapsto g \circ f \tag{7.3.39}$$

is an algebra homomorphism of $\mathbb{C}(X)$ leaving $\mathbb{C}$ fixed (where $\lambda \in \mathbb{C}$ corresponds to the constant function) and that, conversely, every such automorphism has the form. Thus, $\mathbb{PSL}(2,\mathbb{C})$ is the Galois group $[\mathbb{C}(X)\colon \mathbb{C}]$.

18. Prove that the FLT that takes $z_j \to w_j$ for $j = 1, 2, 3$ is given by $w = f(z)$, where

$$\det \begin{vmatrix} wz & z & w & 1 \\ w_1z_1 & z_1 & w_1 & 1 \\ w_2z_2 & z_2 & w_2 & 1 \\ w_3z_3 & z_3 & w_3 & 1 \end{vmatrix} = 0 \tag{7.3.40}$$

The purpose of the next two problems is two-fold. $\mathbb{SU}(2)$ is the set of 2×2 unitary matrices of determinant 1 and $\mathbb{PSU}(2)$ is the set of induced FLTs. $\mathbb{SO}(3)$ is the set of 3×3 orthogonal matrices of determinant 1, aka the rotations in 3-sphere. We want to show that the maps on the 2-sphere obtained by dragging elements of $\mathbb{PSU}(2)$ to S^2 under stereographic projection are precisely $\mathbb{SO}(3)$ so that

$$\mathbb{PSU}(2) \cong \mathbb{SO}(3) \tag{7.3.41}$$

For definiteness, we'll take our S^2 to be the sphere of radius 1 centered at 0. Secondly, we'll show that any finite subgroup of $\mathbb{PSL}(2,\mathbb{C})$ is conjugate to a finite subgroup of $\mathbb{PSU}(2)$, and so reduce the analysis of such finite subgroups to the analysis of finite subgroups of $\mathbb{SO}(3)$. We will not complete the analysis of finite subgroups of $\mathbb{SO}(3)$ but refer the reader to, for example, Simon [**523**, Sect. I.4]. We note these groups are the rotations about a fixed axis by angle $2\pi/n$ (the cyclic group, C_n), the rotational symmetries of a cylinder with base a regular n-gon (the dihedral group, D_{2n}, of order $2n$), and the symmetry groups, T, O, I of the tetrahedron (of order 12, isomorphic to the alternating group, A_4), octahedron (or cube, of order 24, isomorphic to the permutation group, S_4), and icosahedron (or dodecahedron, of order 60, isomorphic to the alternating group, A_5).

19. (a) Prove that any 2×2 unitary matrix has the form $\left(\begin{smallmatrix} e^{i\theta}\alpha & e^{i\theta}\beta \\ -e^{i\theta}\bar\beta & e^{i\theta}\bar\alpha \end{smallmatrix}\right)$ for α and β with $|\alpha|^2 + |\beta|^2 = 1$. It has determinant 1 if and only if $e^{i\theta} = 1$, so

$$\mathbb{SU}(2) \cong \left\{ \begin{pmatrix} \alpha & \beta \\ -\bar\beta & \bar\alpha \end{pmatrix} \,\middle|\, |\alpha|^2 + |\beta|^2 = 1 \right\}$$

In particular, topologically $\mathbb{SU}(2)$ is the same as S^3, the three-dimensional sphere.

(b) Let R be a rotation on the sphere and Q_1 the map from $\widehat{\mathbb{C}}$ to S^2 realized as S_1, that is, Q_1 is given by (7.2.3). Prove that $Q_1^{-1}RQ_1 \equiv \widetilde{Q}_1[R]$ is an FLT. (*Hint*: R is conformal and every conformal bijection of $\widehat{\mathbb{C}}$ is an FLT.)

(c) Prove $Q_1(-1/\bar{z}) = -Q_1(z)$ and conclude that if $f = Q_1^{-1}RQ_1$ for some $f \in \mathbb{PSL}(2,\mathbb{C})$, then

$$f\left(-\frac{1}{\bar{z}}\right) = -\frac{1}{\overline{f(z)}} \tag{7.3.42}$$

(d) On $\mathbb{C}^2$, let K be the map, $K\binom{z_1}{z_2} = \binom{\bar{z}_2}{-\bar{z}_1}$. Prove that for any $A \in \mathbb{SL}(2,\mathbb{C})$, $KAK^{-1} = (A^*)^{-1}$, and that K induces the map on $\mathbb{CP}[1]$, $z \to -1/\bar{z}$.

(e) Prove that for $A \in \mathbb{SL}(2,\mathbb{C})$, f_A obeys (7.3.42) if and only if $A^*A = \mathbb{1}$, that is, $A \in \mathbb{SU}(2)$. Thus, $\widetilde{Q}_1$ maps $\mathbb{SO}(3)$ into $\mathbb{PSU}(2)$, and because Q_1 is a bijection, this map is one-one.

(f) If R_θ is rotation by angle θ about the z-axis, prove that $\widetilde{Q}_1(R_\theta)(z) = e^{i\theta}z$ and that this is the image under $A \to f_A$ of $\begin{pmatrix} e^{i\theta/2} & 0 \\ 0 & e^{-i\theta/2} \end{pmatrix}$.

(g) For any $z \in \mathbb{C}$, show there is an $R^{(z)}$ with $\widetilde{Q}_1(R^{(z)})(0) = z$. (*Hint*: $\mathbb{SO}(3)$ acts transitively on S^2.)

(h) Show $\{\widetilde{Q}_1(R^{(z)}R_\theta(R^{(z)})^{-1})\} = \{f_A \mid A \in \mathbb{SU}(2),\ \binom{z}{1} \text{ is an eigenvector of } A\}$ and so, $\widetilde{Q}_1$ is onto all of $\mathbb{PSU}(2)$. Thus, $\widetilde{Q}_1$ is a bijection of $\mathbb{SO}(3)$ and $\mathbb{PSU}(2)$.

Remarks. 1. The map $\mathbb{SU}(2) \to \mathbb{SO}(3)$, which is a two-fold cover, is sometimes called the *Cayley–Klein parametrization.*

2. The map of $\mathbb{SU}(2)$ to $\widehat{\mathbb{C}}$ by $A \mapsto f_A(0)$ is thus a map of S^3 to S^2 with inverse image circles that are linked. This is an implementation of the celebrated *Hopf fibration.*

20. (a) Let $\mathbb{G}$ be a finite subgroup of $\mathbb{SL}(2,\mathbb{C})$. Define an inner product, $\langle\ ,\ \rangle^\sim$ on $\mathbb{C}^2$ by

$$\langle\varphi,\psi\rangle^\sim = \sum_{g\in\mathbb{G}} \langle g\varphi, g\psi\rangle \tag{7.3.43}$$

where $\langle\ ,\ \rangle$ is the usual inner product. Prove that for all $g \in \mathbb{G}$ and $\varphi,\psi \in \mathbb{C}^2$,

$$\langle g\varphi, g\psi\rangle^\sim = \langle\varphi,\psi\rangle^\sim \tag{7.3.44}$$

(b) Show there is an invertible map, T on $\mathbb{C}^2$ so that

$$\langle T\varphi, T\psi\rangle = \langle \varphi, \psi\rangle^{\sim} \tag{7.3.45}$$

(*Hint*: Look at orthonormal bases in both inner products.)

(c) Pick a constant c so that $S \in \mathbb{SL}(2,\mathbb{C})$ if $S = cT$ with T defined in (b). Prove that for all $g \in \mathbb{G}$, $\langle SgS^{-1}\varphi, SgS^{-1}\psi\rangle = \langle \varphi, \psi\rangle$. Thus, there is S with $S\mathbb{G}S^{-1} \subset \mathbb{SU}(2,\mathbb{C})$.

(d) Prove any finite subgroup of $\mathbb{PSL}(2,\mathbb{C})$ is conjugate to a finite subgroup of $\mathbb{PSU}(2)$.

21. If $T \in \mathbb{SL}(2,\mathbb{R})$, prove that $f_T(i) = i$ if and only if $\pm T \in \mathbb{SO}(2) = \{\left(\begin{smallmatrix} \cos\theta & \sin\theta \\ -\sin\theta & \cos\theta \end{smallmatrix}\right)\}$, the two-dimensional orthogonal group. Thus, the isotropy group of i for the action of $\mathbb{PSL}(2,\mathbb{R})$ on $\mathbb{C}_+$ is $\mathbb{SO}(2)$.

22. (a) By looking at $\mathrm{Tr}[\left(\begin{smallmatrix} 1 & 1 \\ 0 & 1 \end{smallmatrix}\right)\left(\begin{smallmatrix} 1 & 0 \\ c & 1 \end{smallmatrix}\right)]$, prove that every element of $\mathbb{PSL}(2,\mathbb{C})$ is a product of two parabolic elements.

(b) Prove that the product of two affine transformations $az+b$ and $cz+d$ is of the form $acz + x$ for some x. Conclude from this that if A and B are affine, then $ABA^{-1}B^{-1}$ is either the identity or parabolic.

(c) Prove that if A, B are two elements of $\mathbb{PSL}(2,\mathbb{C})$ with exactly one fixed point in common, at least one of A, B or $ABA^{-1}B^{-1}$ is parabolic.

(d) Prove that every normal subgroup of $\mathbb{PSL}(2,\mathbb{C})$ is either $\{\mathbb{1}\}$ or all of $\mathbb{PSL}(2,\mathbb{C})$, that is, $\mathbb{PSL}(2,\mathbb{C})$ is a simple group.

23. Let T be a hyperbolic map and S some element of $\mathbb{PSL}(2,\mathbb{C})$ that maps one of the fixed points of T to the other.

(a) Prove that $TST^{-1}S^{-1}$ is hyperbolic.

(b) Prove that $TSTS^{-1}$ is parabolic.

24. (a) Prove that if f in $\mathbb{PSL}(2,\mathbb{C})$ has two finite fixed points, z_1, z_2, then there is $\alpha \in \mathbb{C}^\times$, so that

$$\frac{f(z) - z_1}{f(z) - z_2} = \alpha\,\frac{(z - z_1)}{(z - z_2)}$$

(b) Given two points, $z_1, z_2 \in \mathbb{C}$, find $f \in \mathbb{PSL}(2,\mathbb{C})$ so $f \circ f = \mathbb{1}$ and z_1, z_2 are both fixed points of f.

25. A *fixed circle* of an FLT, f, is a circle or straight line left setwise fixed by f, that is, $f[C] = C$.

(a) Prove loxodromic f have no fixed circles.

(b) Prove the fixed circles of a hyperbolic f are exactly all those passing through the two fixed points. See Figure 7.3.1.

(c) Prove the fixed circles of a parabolic f are all circles through the fixed point tangent to a particular line through the fixed point. See Figure 7.3.1.

(d) Prove the fixed circles of an elliptic transformation are those orthogonal to all circles through the fixed points. (*Hint*: Consider first the case where one fixed point is ∞ and the other is 0, if there is a second.)

26. Show that in the group of all conformal or anticonformal maps of $\widehat{\mathbb{C}}$ to itself, $\mathbb{PSL}(2,\mathbb{C})$ is a subgroup of index 2 and that the other coset includes all reflections in circles given by (7.3.36).

27. (a) Let $f \in \mathbb{PSL}(2,\mathbb{C})$ be nonaffine (i.e., ∞ is not a fixed point of f). Show there is a circle $I_f = \{z \mid |f'(z)| = 1\}$ whose center is $f^{-1}(\infty)$, and that inside I_f, $|f'(z)| > 1$, and outside, $|f'(z)| < 1$. I_f is called an *isometric circle* for f.

(b) Show that f maps the outside of I_f to the inside of $I_{f^{-1}}$ and the inside of I_f to the outside of $I_{f^{-1}}$.

(c) In the hyperbolic case, prove that I_f and $I_{f^{-1}}$ are disjoint; in the parabolic case, they intersect in one point; and in the elliptic case where $f^2(\infty) \neq \infty$ also holds, the intersection is two points.

(d) Suppose $f^2(\infty) \neq \infty$ also holds so the centers of I_f and $I_{f^{-1}}$ are distinct. Let R_f be reflection in I_f (given by (7.3.36)) and K reflection in the perpendicular bisection of the line between the centers of I_f and $I_{f^{-1}}$. Prove that if E_θ is Euclidean rotation by angle θ centered at $f^{-1}(\infty)$, then there is a θ so that $f = KR_fE_\theta$.

(e) Prove $E_\theta \neq \mathbb{1}$ if and only if f is loxodromic. In particular, prove that every nonloxodromic FLT is a product of two reflections (i.e., two anticonformal maps, each of whose square is $\mathbb{1}$).

Remarks. 1. (d)+(e) is a theorem of Ford [**191, 193**]; see also [**525**] for an exposition.

2. The isometric circle is the points where infinitesimally f preserves Euclidean distance.

3. We emphasize the notions here are Euclidean and not invariant under nonaffine FLTs.

7.4. Self-Maps of the Disk

In this section, our main result will be to find $\mathbb{A}\mathrm{ut}(\mathbb{D})$, all analytic bijections of $\mathbb{D}$. We'll see it is identical to all FLTs that map $\mathbb{D}$ to itself. While we parametrized such maps in Theorem 7.3.10, we'll find another parametrization and not use (7.3.35) in this new proof.

Lemma 7.4.1. *Let $f\colon \mathbb{D} \to \mathbb{D}$ be an analytic bijection with $f(0) = 0$. Then*

$$f(z) = e^{i\theta} z \tag{7.4.1}$$

for some $e^{i\theta} \in \partial\mathbb{D}$.

Proof. By the Schwarz lemma (Theorem 3.6.7),

$$|f(z)| \le |z| \tag{7.4.2}$$

Let g be the functional inverse of f. Since g also maps $\mathbb{D}$ to $\mathbb{D}$, is analytic, and has $g(0) = 0$, for any $w \in \mathbb{D}$,

$$|g(w)| \le |w| \tag{7.4.3}$$

Applying this to $w = f(z)$ yields

$$|z| \le |f(z)| \tag{7.4.4}$$

so

$$|f(z)| = |z| \tag{7.4.5}$$

Let $h(z) = f(z)/z$. Then $|h(z)| = 1$, so by the maximum modulus principle, $h(z) = e^{i\theta}$ for some $e^{i\theta} \in \partial\mathbb{D}$. □

The other input we need is the function, β, of (3.6.17), that is, for $z_0 \in \mathbb{D}$,

$$f_{z_0}(z) = \frac{z - z_0}{1 - \bar{z}_0 z} \tag{7.4.6}$$

which is f_T for

$$T_{z_0} = \begin{pmatrix} 1 & -z_0 \\ -\bar{z}_0 & 1 \end{pmatrix} \tag{7.4.7}$$

(*Warning*: T_{z_0} does not have determinant 1!) By an easy calculation (see Problem 5 of Section 3.6),

$$|f_{z_0}(e^{i\theta})| = 1 \tag{7.4.8}$$

so $f_{z_0}\colon \mathbb{D} \to \mathbb{D}$ by the maximum principle and the fact that f_{z_0} is analytic on $\mathbb{D}$ since $|z_0| < 1$.

Since $T_{-z_0} T_{z_0} = (1 - |z_0|^2)\mathbb{1}$,

$$(f_{-z_0} \circ f_{z_0})(z) = z \tag{7.4.9}$$

so $f_{z_0} \in \mathbb{A}\mathrm{ut}(\mathbb{D})$.

Theorem 7.4.2. *Every $f \in \mathbb{A}\mathrm{ut}(\mathbb{D})$ is of the form*

$$f(z) = e^{i\theta} f_{z_0}(z) \tag{7.4.10}$$

for some $e^{i\theta} \in \partial\mathbb{D}$ and some $z_0 \in \mathbb{D}$ and every such map is in $\mathbb{A}\mathrm{ut}(\mathbb{D})$.

Proof. Since f is a bijection, there is $z_0 \in \mathbb{D}$ with $f(z_0) = 0$. Since $f_{-z_0}(0) = z_0$, we see $f \circ f_{-z_0}(0) = 0$. Thus, by Lemma 7.4.1, $(f \circ f_{-z_0})(w) = e^{i\theta} w$. Apply to $w = f_{z_0}(z)$ and use $f_{-z_0} \circ f_{z_0} = \mathbb{1}$ to conclude that (7.4.10) holds. $\square$

Thus, $\mathbb{A}\mathrm{ut}(\mathbb{D})$ is topologically $\mathbb{D} \times \partial\mathbb{D}$ and so is a three-(real) dimensional manifold. One can also see this by counting parameters in $\mathbb{SL}(2,\mathbb{R})$ (or $\mathbb{SU}(1,1)$).

Having proven the main result, we turn to a few further ones.

Theorem 7.4.3. *Every nonidentity element of $\mathbb{A}\mathrm{ut}(\mathbb{D})$ is hyperbolic, parabolic, or elliptic. Each of these possibilities occurs. Indeed, any nonloxodromic conjugacy class in $\mathbb{PSL}(2,\mathbb{C})$ intersects $\mathbb{A}\mathrm{ut}(\mathbb{D})$.*

Proof. Since there is an analytic bijection of $\mathbb{C}_+$ to $\mathbb{D}$, any $f \in \mathbb{A}\mathrm{ut}(\mathbb{D})$ is conjugate to an element of $\mathbb{PSL}(2,\mathbb{R})$. But the trace of any real matrix is real, so $\mathrm{tr}(f) \in \mathbb{R}$, and thus, f cannot be loxodromic. (Alternatively, by (7.3.37), $\mathrm{Tr}(A) = 2\,\mathrm{Re}(\alpha)$ is real.)

(7.4.1) is elliptic. The T of determinant one for f_{z_0} is $(1-|z_0|^2)^{-1/2} T_{z_0}$, which has trace $2(1-|z_0|^2)^{-1/2} > 2$, so $\{f_{z_0}\}$ are all hyperbolic.

The matrix $\left(\begin{smallmatrix} 1+i & 1 \\ 1 & 1-i \end{smallmatrix}\right)$ is in $\mathbb{SU}(1,1)$ by (7.3.35) and has trace 2, and so the map,

$$f(z) = \frac{(1+i)z+1}{z+(1-i)} \tag{7.4.11}$$

and so is a parabolic element of $\mathbb{A}\mathrm{ut}(\mathbb{D})$.

As z_0 runs through $(0,1)$, $2(1-z_0^2)^{-1/2}$ runs through $(2,\infty)$, so every hyperbolic class occurs. As θ runs through $(0,\pi)$, $2\cos(\theta/2)$ runs through $(0,2)$, so every elliptic class occurs. $\square$

Problem 2 will prove the following:

Theorem 7.4.4. *Elliptic elements of $\mathbb{A}\mathrm{ut}(\mathbb{D})$ have one fixed point in $\mathbb{D}$ and one in $\mathbb{C} \setminus \mathbb{D}$. Hyperbolic elements have both fixed points in $\partial\mathbb{D}$. Parabolic elements have their fixed point in $\partial\mathbb{D}$.*

Problem 3 will prove the following, which will be used in Section 8.6:

Theorem 7.4.5. (a) *Each hyperbolic conjugacy class in $\mathbb{A}\mathrm{ut}(\widehat{\mathbb{C}})$ intersects a single conjugacy class in $\mathbb{A}\mathrm{ut}(\mathbb{D})$. The hyperbolic elements, f, g, in $\mathbb{A}\mathrm{ut}(\mathbb{D})$ are conjugate if and only if $f = f_T$, $g = f_S$ with $\mathrm{Tr}(T) = \pm\mathrm{Tr}(S)$ and $|\mathrm{Tr}(T)| > 2$. In $\mathbb{A}\mathrm{ut}(\mathbb{C}_+)$, the hyperbolic conjugacy classes are labeled by $\lambda > 1$ with $|\mathrm{Tr}(T)| = \lambda + \lambda^{-1}$. One element of the conjugacy class is $z \mapsto \lambda^2 z$. If f in $\mathbb{A}\mathrm{ut}(\mathbb{D})$ is hyperbolic, f and f^{-1} are conjugate.*

(b) *The single parabolic conjugacy class in* $\mathbb{A}\mathrm{ut}(\widehat{\mathbb{C}})$ *intersects* $\mathbb{A}\mathrm{ut}(\mathbb{D})$ *in two conjugacy classes with* f *and* f^{-1} *in distinct classes. In* $\mathbb{A}\mathrm{ut}(\mathbb{C}_+)$*, representatives of the two classes are* $z \mapsto z \pm 1$.

(c) *Each elliptic conjugacy class in* $\mathbb{A}\mathrm{ut}(\widehat{\mathbb{C}})$ *(with one exception) intersects* $\mathbb{A}\mathrm{ut}(\mathbb{D})$ *in two conjugacy classes. One representative in each elliptic class is obtained by looking at* $z \mapsto e^{2i\theta}z$, $-\pi/2 < \theta < \pi/2$, *and* $z \mapsto -z$ *(*f_T *has* $\mathrm{Tr}(T) = 2\cos\theta$*). For* $0 < \theta < \pi/2$, $z \mapsto e^{2i\theta}z$ *and* $z \mapsto e^{-2i\theta}z$ *are conjugate in* $\mathbb{A}\mathrm{ut}(\widehat{\mathbb{C}})$ *(under* $z \to 1/z$*) but not in* $\mathbb{A}\mathrm{ut}(\mathbb{D})$.

The other important aspect of $\mathbb{A}\mathrm{ut}(\mathbb{D})$ is that they are precisely orientation-preserving isometries of $\mathbb{D}$ in the Poincaré metric. This will be studied in Section 12.2 of Part 2B.

Notes and Historical Remarks. The special role of $\mathbb{A}\mathrm{ut}(\mathbb{D})$ and its subgroups was emphasized by Klein and Poincaré in work discussed in the Notes of the preceding section. In particular, Poincaré in [**438**] introduced and realized the importance of the hyperbolic metric.

We saw in this section that $\mathbb{A}\mathrm{ut}(\mathbb{D})$ is a rich set, rich enough that it is *transitive*, that is, for any $z, w \in \mathbb{D}$, there is $f \in \mathbb{A}\mathrm{ut}(\mathbb{D})$ so $f(z) = w$. In Problems 4–7, we'll explore $\mathbb{A}\mathrm{ut}(\Omega)$ for other regions, including all annuli $\mathbb{A}_{r,R}$ for $0 \le r < R < \infty$ and see explicitly that they are not transitive. Indeed, in the Notes to Section 8.7, we'll explain that if $\mathbb{A}\mathrm{ut}(\Omega)$ is transitive for $\Omega \subset \mathbb{C}$, then Ω is $\mathbb{C}$ or is biholomorphically equivalent to $\mathbb{C} \setminus \{0\}$ or to $\mathbb{D}$.

In addition, if Ω is n-connected, with n finite, $n \ge 2$, then $\mathbb{A}\mathrm{ut}(\Omega)$ is finite—Heins [**249**] has proven optimal bounds on the order of $\mathbb{A}\mathrm{ut}(\Omega)$ as $2n$ if $n \ne 4, 6, 8, 12, 20$ with orders $12, 24, 24, 60, 60$ in these five special cases. Heins relies on showing $\mathbb{A}\mathrm{ut}(\Omega)$ is isomorphic to a finite subgroup of $\mathbb{SL}(2, \mathbb{C})$ (which we analyzed in Problem 19 and 20 of Section 7.3). The special cases are connected to the group of symmetries of the five regular solids! Note that the special values of n are the numbers of vertices of these solids. If $\mathbb{C} \setminus \Omega$ has infinitely many components, $\mathbb{A}\mathrm{ut}(\Omega)$ can be infinite, for example, $\Omega = \mathbb{C} \setminus \mathbb{Z}$ has $z \to z + k$ in $\mathbb{A}\mathrm{ut}(\Omega)$ for all $k \in \mathbb{Z}$.

Problems

1. Provide another proof of Theorem 7.4.2 as follows:

   (a) Prove if $f \in \mathbb{A}\mathrm{ut}(\mathbb{D})$, then $\lim_{|z|\uparrow 1} |f(z)| = 1$. (*Hint*: Look at the proof of (7.3.14).)

   (b) Prove any $f \in \mathbb{A}\mathrm{ut}(\mathbb{D})$ has a continuation in a neighborhood of $\overline{\mathbb{D}}$. (*Hint*: Theorem 5.5.5.)

   (c) Use the fact that f has a single zero in $\mathbb{D}$ to show that f has the form (7.4.10). (*Hint*: Problem 5 of Section 3.6.)

2. (a) Let $f \in \mathbb{A}\text{ut}(\mathbb{D})$ be extended to $\widehat{\mathbb{C}}$. Let $f(z) = z$ for some $z \in \widehat{\mathbb{C}}$. Prove $\bar{z}^{-1}$ is also a fixed point of f.

   (b) Prove that if f has a fixed point in $\mathbb{D}$, then f is elliptic. (*Hint*: If z_0 is the fixed point, look at $f_{z_0} f f_{-z_0}$.)

   (c) Prove that if f is hyperbolic or parabolic, its fixed points lie in $\partial\mathbb{D}$.

   (d) Let $g \in \mathbb{A}\text{ut}(\mathbb{C}_+)$. Show that if z_0 is a fixed point, so is $\bar{z}_0$. If g is elliptic and has a fixed point at $z_0 \in \mathbb{R}$, prove that there is an $h \in \mathbb{A}\text{ut}(\mathbb{C}_+)$ so that hgh^{-1} has z_0 and ∞ as its fixed points, and obtain a contradiction. Conclude g must have one fixed point in $\mathbb{C}_+$ and one in $\mathbb{C}_-$.

   (e) If f is elliptic, prove that f has a fixed point in $\mathbb{D}$ and one in $\mathbb{C} \setminus \overline{\mathbb{D}}$.

3. (a) Let $f \in \mathbb{A}\text{ut}(\mathbb{C}_+)$ be hyperbolic. Prove there is $g \in \mathbb{A}\text{ut}(\mathbb{C}_+)$ so that $h = gfg^{-1}$ has 0 and ∞ as its fixed points and so that, for some $\lambda > 0$, $h_\lambda(z) = \lambda z$. Prove that $\tilde{g}(z) = -z^{-1}$ lies in $\mathbb{A}\text{ut}(\mathbb{C}_+)$ and that $\tilde{g} h_\lambda \tilde{g}^{-1} = h_{\lambda^{-1}}$. Conclude that any hyperbolic $f \in \mathbb{A}\text{ut}(\mathbb{C}_+)$ is conjugate in $\mathbb{A}\text{ut}(\mathbb{C}_+)$ to an h_λ with $\lambda > 1$. Prove that for $1 < \lambda < \lambda'$, h_λ and $h_{\lambda'}$ are not conjugate. Prove that every hyperbolic conjugacy class in $\mathbb{A}\text{ut}(\widehat{\mathbb{C}})$ intersects exactly one conjugacy class in $\mathbb{A}\text{ut}(\mathbb{C}_+)$.

   (b) Let $f \in \mathbb{A}\text{ut}(\mathbb{C}_+)$ be parabolic. Prove there is $g \in \mathbb{A}\text{ut}(\mathbb{C}_+)$ and $y \in \mathbb{R}^\times$ so that $h = gfg^{-1}$ has the form for $h_y(z) = z + y$. Prove that h_y and $h_{y'}$ are conjugate in $\mathbb{A}\text{ut}(\mathbb{C}_+)$ if and only if $y/y' > 0$. Conclude that there are two parabolic conjugacy classes in $\mathbb{A}\text{ut}(\mathbb{C}_+)$ and that $h_y^{-1} = h_{-y}$ are in distinct classes.

   (c) Let $f \in \mathbb{A}\text{ut}(\mathbb{D})$ be elliptic, Prove there is $g \in \mathbb{A}\text{ut}(\mathbb{D})$ and $\theta \in [-\pi/2, \pi/2]$ so that $h_\theta = gfg^{-1}$ has $h_\theta(z) = e^{2i\theta} h(z)$. Prove that h_θ and $h_{-\theta}$ are conjugate in $\mathbb{A}\text{ut}(\widehat{\mathbb{C}})$ but not in $\mathbb{A}\text{ut}(\mathbb{D})$ except for $\theta = \pi/2$, and that except for the class with $\text{Tr}(T) = 0$, each elliptic class in $\mathbb{A}\text{ut}(\widehat{\mathbb{C}})$ intersects $\mathbb{A}\text{ut}(\mathbb{D})$ in exactly two conjugacy classes.

4. Prove that any element of $\mathbb{A}\text{ut}(\mathbb{D}^\times)$ is a rotation. (*Hint*: Think removable singularity.)

5. Let $z_0, z_1 \in \mathbb{D}$ be distinct points. Let $f \in \mathbb{A}\text{ut}(\mathbb{D} \setminus \{z_0, z_1\})$.

   (a) Prove that f has an extension to $\mathbb{A}\text{ut}(\mathbb{D})$ with either $f(z_0) = z_0$, $f(z_1) = z_1$, or $f(z_0) = z_1$, $f(z_1) = z_0$.

   (b) Prove $\mathbb{A}\text{ut}(\mathbb{D} \setminus \{z_0, z_1\})$ has exactly two elements.

   (c) Prove the same result if $\mathbb{D}$ is replaced by $\mathbb{C}$.

6. (a) Find an example of a bounded Ω, z_0, z_1 where $\mathbb{A}\mathrm{ut}(\Omega \setminus \{z_0, z_1\})$ has more than two elements. (*Hint*: $z \to z^{-1}$ is an automorphism of $\mathbb{A}_{r,r^{-1}}$ for $r < 1$ and so are rotations.)

   (b) Find an example of a bounded Ω, z_0, z_1 where $\mathbb{A}\mathrm{ut}(\Omega \setminus \{z_0, z_1\})$ is only the identity. (*Hint*: You may use the fact that $\mathbb{D}$ has a Riemann metric in which all elements of $\mathbb{A}\mathrm{ut}(\mathbb{D})$ are isometries.)

   Remark. This theme is explored further in Section 12.2 of Part 2B.

7. This problem will determine $\mathbb{A}\mathrm{ut}(\mathbb{A}_{r,R})$ for the annulus $\mathbb{A}_{r,R}$ with $0 < r < R < \infty$. Let $\rho = \sqrt{rR}$. Let g be given by $g(z) = \rho^2/z$. Let $f \in \mathbb{A}\mathrm{ut}(\mathbb{A}_{r,R})$.
   (a) Let γ be the curve $\gamma(s) = f(\rho e^{2\pi i s})$, $0 \le s \le 1$. γ is an analytic Jordan curve, so it divides $\mathbb{C}$ in two. Show that either f maps $\mathbb{A}_{r,\rho}$ into the bounded component or into the unbounded component, and then $\mathbb{A}_{\rho,R}$ goes into the other component. Thus, either f or $f \circ g$ maps $\mathbb{A}_{r,\rho}$ into the bounded component.

   (b) Suppose that f maps $\mathbb{A}_{r,\rho}$ to the bounded component. Prove that $\lim_{|z|\to r,\, z\in\mathbb{A}_{r,R}} |f(z)| = r$. (*Hint*: Look at the proof of (7.3.14).)

   (c) Prove that if f maps $\mathbb{A}_{r,\rho}$ to the bounded component, then f has an analytic continuation to $\mathbb{A}_{r^2/R,R}$ with $\lim_{|z|\to r^2/R} |f(z)| = r^2/R$.

   (d) Prove that if f maps $\mathbb{A}_{r,\rho}$ to the bounded component, then f has an analytic continuation to $\mathbb{D}_R(0) \setminus \{0\}$.

   (e) Prove every f in $\mathbb{A}_{r,\rho}$ is of one of the form $f(z) = e^{i\theta} z$ or $f(z) = e^{-i\theta}\rho^2/z$. (*Hint*: See Problem 4.)

8. If Ω is a region and $z_0 \in \Omega$, $\mathbb{A}\mathrm{ut}_{z_0}(\Omega)$ is the subgroup of $\mathbb{A}\mathrm{ut}(\Omega)$ of those f with $f(z_0) = z_0$. You'll prove the following (see Bieberbach [**49**]): If Ω is bounded, then $\mathbb{A}\mathrm{ut}_{z_0}(\Omega)$ is either a finite cyclic group or isomorphic to $\partial\mathbb{D}$. Henceforth, Ω is bounded.

   (a) Prove that $\mathbb{A}\mathrm{ut}_{z_0}(\Omega)$ is compact. (*Hint*: See Problem 3 of Section 6.4.)

   (b) Prove that if $f \in \mathbb{A}\mathrm{ut}_{z_0}(\Omega)$, then $|f'(z_0)| = 1$. (*Hint*: See Problem 16 of Section 3.6.)

   (c) If $f(z) = z + a_m(z - z_0)^m + O((z - z_0))^{m+1}$, prove by induction that $f^{[n]}(z) = z + a_m n(z - z_0)^m + O((z - z_0))^{m+1}$.

   (d) Prove that if $f \in \mathbb{A}\mathrm{ut}_{z_0}(\Omega)$ and $f'(z_0) = 1$, then $f(z) = z$.

   (e) Prove $f \in \mathbb{A}\mathrm{ut}_{z_0}(\Omega) \to f'(z_0)$ is a bijection of $\mathbb{A}\mathrm{ut}_{z_0}(\Omega)$ and a closed subgroup of $\partial\mathbb{D}$.

   (f) Prove that every closed subgroup of $\partial\mathbb{D}$ is either all of $\partial\mathbb{D}$ or $\{e^{2i\pi j/n}\}_{j=0}^{n-1}$.

7.5. Bonus Section: Introduction to Continued Fractions and the Schur Algorithm

We would be remiss if we left the subject of FLTs without mentioning the closely related subject of continued fractions which, as we'll see, is essentially a study of composition of certain FLTs. We begin with the continued fraction expansion of a real number—a subject seemingly disconnected to complex analysis—and then turn to using similar ideas to analyze Schur functions, that is, maps of $\mathbb{D}$ to $\mathbb{D}$ (which are not necessarily either one-one or onto).

Let $x \in (1, \infty)$. We can write

$$x = [x] + \{x\} \tag{7.5.1}$$

as the sum of the integral and fractional parts of x. This doesn't seem to bode well for iteration until one notes that $0 < \{x\} < 1 \Rightarrow \{x\}^{-1} \in (1, \infty)$. So if x is not an integer, we can define $a_0(x) = [x]$, $y_1(x) = 1/\{x\}$, and $x = a_0(x) + 1/y_1(x)$. This can now be iterated, so we define $a_n(x), y_n(x)$ inductively by

$$y_0(x) = x, \qquad a_n(x) = [y_n(x)], \qquad y_{n+1}(x) = \{y_n(x)\}^{-1} \tag{7.5.2}$$

so that

$$y_n(x) = a_n(x) + \frac{1}{y_{n+1}(x)} \tag{7.5.3}$$

or

$$x = a_0 + \cfrac{1}{a_1 + \cfrac{1}{a_2 + \cfrac{1}{\cdots + \cfrac{1}{y_{n+1}(x)}}}} \tag{7.5.4}$$

Notation like (7.5.4), called a *continued fraction*, is unwieldy so there have been various attempts at alternates; for example, (7.5.4) is written as

$$x = a_0 + \mathop{\mathbf{K}}_{j=1}^{n+1} \frac{1}{\tilde{a}_j}, \qquad \tilde{a}_j = a_j, \quad j \le n, \qquad \tilde{a}_{n+1} = y_{n+1} \tag{7.5.5}$$

More generally, if the 1's in (7.5.4) are replaced by $b_1, b_2, \dots, b_{n+1}$ and y_{n+1} by a_{n+1}, we write $a_0 + \mathbf{K}_{j=1}^{n+1} b_j/a_j$ where, of course, the object makes sense for any positive real a_j, b_j and even for arbitrary complex a_j, b_j so long as we don't need to divide by 0 (and, we'll see later, even that is no problem!). As with iterated sums and products, one can try to make sense of $\mathbf{K}_{j=1}^{\infty} b_j/a_j$, but there is a convergence issue that, for the case $b_j \equiv 1$, $a_j \in \{1, 2, \dots\}$, we'll see is always solvable.

The potential fly in the ointment is if some $y_n(x) \in \mathbb{Z}_+$, then $y_{n+1}(x)$ is undefined (or ∞ if you prefer). In this case (7.5.4) with $1/y_{n+1}$ replaced by 0 shows x is rational, so if x is irrational, the process goes on indefinitely.

The converse is also true: If x is rational, then eventually $y_n \in \mathbb{Z}_+$ (Problem 1). We want to note that in this case, there are actually two representations as (7.5.4) with $a_j > 0$ possible. By our algorithm, $y_{n+1}(x) = \{y_n(x)\}^{-1}$, and since $\{\cdot\} \in [0,1)$, if $y_{n+1}(x) \in \mathbb{Z}$, then $a_{n+1} \geq 2$. In that case, in addition to a finite continued fraction $x = a_0 + \mathbf{K}_{j=1}^{n+1} 1/a_j$, we can define $\tilde{a}_j$ by

$$\tilde{a}_j = \begin{cases} a_j, & j \leq n \\ a_{n+1} - 1, & j = n+1 \\ 1, & j = n+2 \end{cases} \tag{7.5.6}$$

so with $m = n+1$,

$$x = a_0 + \mathop{\mathbf{K}}_{j=1}^{m} \frac{1}{a_j} = \tilde{a}_0 + \mathop{\mathbf{K}}_{j=1}^{m+1} \frac{1}{\tilde{a}_j} \tag{7.5.7}$$

Thus, any irrational in $(1, \infty)$ is associated to an infinite sequence $(a_0(x), a_1(x), \dots)$ in $\mathbb{Z}_+^\infty$, and if we allow $a_0 \in \mathbb{Z}$ any irrational real number. We'll see shortly this map is onto all of $\mathbb{Z}_+^\infty$.

Looking at (7.5.4), the natural temptation is to replace $1/y_{n+1}(x)$ by 0 and get a rational, $x_n(x)$, with

$$x_n(x) = a_0 + \mathop{\mathbf{K}}_{j=1}^{n} \frac{1}{a_j} \tag{7.5.8}$$

This is called the nth *continued fraction approximant* and we'll see they are, in a precise sense, best possible rational approximations.

One quickly gets a headache if one tries to naively simplify the fractions in (7.5.8). The key is to realize that

$$y \to a + \frac{1}{y} = \frac{ay+1}{y} \tag{7.5.9}$$

is an FLT, and so x_n is obtained by iteration of FLTs, which we know is done via multiplication of matrices! The matrix in (7.5.9) is $\left(\begin{smallmatrix} a & 1 \\ 1 & 0 \end{smallmatrix}\right)$ (this doesn't have determinant 1 but rather -1, but one can define f_T for any T in $\mathbb{GL}(2, \mathbb{C})$ after all). We'll focus on $x \in (0,1)$, that is, $a_0 = 0$.

Thus, given a sequence of positive integers $a_1, a_2, \dots$, we define integers p_n, q_n inductively, for $n \geq 0$, by

$$T_n \equiv \begin{pmatrix} q_n & q_{n-1} \\ p_n & p_{n-1} \end{pmatrix} = \begin{pmatrix} q_{n-1} & q_{n-2} \\ p_{n-1} & p_{n-2} \end{pmatrix} \begin{pmatrix} a_n & 1 \\ 1 & 0 \end{pmatrix} \tag{7.5.10}$$

or

$$q_n = a_n q_{n-1} + q_{n-2}, \qquad p_n = a_n p_{n-1} + p_{n-2} \tag{7.5.11}$$

We want $T_0 = \mathbb{1}$, so we have the initial conditions

$$q_0 = p_{-1} = 1, \qquad q_{-1} = p_0 = 0 \tag{7.5.12}$$

(7.5.11) are called the *Euler–Wallis equations.* (7.5.10) has built into it the fact that since the second column of $\left(\begin{smallmatrix} a & 1 \\ 1 & 0 \end{smallmatrix}\right)$ is $\left(\begin{smallmatrix} 1 \\ 0 \end{smallmatrix}\right)$, the second column of T_n will be the first column of T_{n-1}. That this summarizes simplifying denominators is encapsulated in:

Theorem 7.5.1. *Given $a_1, a_2, \dots \in \mathbb{Z}_+$, define p_n, q_n by the Euler–Wallis equations with initial conditions* (7.5.12). *Then*

$$\mathop{\mathbf{K}}_{j=1}^{n} \frac{1}{a_j} = \frac{p_n}{q_n} \tag{7.5.13}$$

and for any real $x \in (0,1)$, with $a_n = a_n(x)$, we have that (7.5.4) *becomes*

$$x = \frac{p_n y_{n+1} + p_{n-1}}{q_n y_{n+1} + q_{n-1}} \tag{7.5.14}$$

Proof. Let $f_n(z; a_1, \dots, a_n)^{-1}$ be expressions in (7.5.4) when $a_0 = 0$, where $1/y_{n+1}$ is replaced by z. Then (7.5.4) says

$$\begin{pmatrix} f_n(z; a_1, \dots, a_n) \\ 1 \end{pmatrix} = c \begin{pmatrix} a_1 & 1 \\ 1 & 0 \end{pmatrix} \begin{pmatrix} f_{n-1}(z; a_2, a_3, \dots, a_n) \\ 1 \end{pmatrix} \tag{7.5.15}$$

where $c = (f_{n-1}(z; a_1, \dots, a_n))^{-1}$. The initial condition is determined by

$$f_0(z; a_n) = a_n + z = \begin{pmatrix} a_n & 1 \\ 1 & 0 \end{pmatrix} \begin{pmatrix} 1 \\ z \end{pmatrix} \tag{7.5.16}$$

so $\left(\begin{smallmatrix} f_n \\ 1 \end{smallmatrix}\right) = cT_n\left(\begin{smallmatrix} 1 \\ z \end{smallmatrix}\right)$, which is solved by

$$f_n = \frac{p_n + p_{n-1}z}{q_n + q_{n-1}z} \tag{7.5.17}$$

$z = 0$ gives (7.5.13) and $z = 1/y_{n+1} \equiv z_{n+1}$ gives (7.5.14). $\square$

Remark. These results work just as well for $\mathbf{K}_{j=1}^{n} b_j/a_j$. In (7.5.10), $\left(\begin{smallmatrix} a_n & 1 \\ 1 & 0 \end{smallmatrix}\right)$ becomes $\left(\begin{smallmatrix} a_n & 1 \\ b_n & 0 \end{smallmatrix}\right)$ and the Euler–Wallis equations are now

$$q_n = a_n q_{n-1} + b_n q_{n-2} \tag{7.5.18}$$

and similarly for p. However, the convergence result we prove below relies on $b_n \equiv 1$. In general, see the discussion in the books in the Notes.

We can now state all we plan to prove about the continued fraction approximants of real numbers in this section (we return to the subject in Section 2.8 of Part 3):

Theorem 7.5.2. *Let $\{a_n\}_{n=0}^{\infty}$ be a sequence in $\mathbb{Z}_+$. Let p_n, q_n be the solution of the Euler–Wallis recursion. If the a_n's come from the continued fraction of an irrational real x, we'll write it as $a_n(x)$.*

(a) *For* $n \geq 0$ *(with* $q_{2n+1} > q_{2n}$ *if* $n \geq 1$*),*

$$q_{2n+1} \geq q_{2n} \geq 2^n \tag{7.5.19}$$

so (using a similar argument for p_n*)*

$$q_n \geq 2^{n/2}, \qquad p_n \geq 2^{(n-1)/2} \tag{7.5.20}$$

(b) $$\left|\frac{p_{n+1}}{q_{n+1}} - \frac{p_n}{q_n}\right| = \frac{1}{q_{n+1}q_n} \tag{7.5.21}$$

(c) $$\left|x - \frac{p_n}{q_n}\right| < \left|\frac{p_n}{q_n} - \frac{p_{n+1}}{q_{n+1}}\right| < \left|\frac{p_n}{q_n} - \frac{p_{n-1}}{q_{n-1}}\right| \tag{7.5.22}$$

(d) $$\frac{p_{2n}}{q_{2n}} < \frac{p_{2n+2}}{q_{2n+2}} < x < \frac{p_{2n+1}}{q_{2n+1}} < \frac{p_{2n-1}}{q_{2n-1}} \tag{7.5.23}$$

(e) $\lim_{n\to\infty} p_n/q_n$ *exists, lies in* $(1,\infty)$*, and is equal to* x *if the* a_n *come from an* x*.*

(f) *The* $a_n(y)$ *for the limit* y *of* p_n/q_n *are the given* a_n*.*

(g) $$\left|x - \frac{p_n}{q_n}\right| \leq \frac{1}{q_n^2} \tag{7.5.24}$$

(h) *Either* $|x - p_n/q_n| \leq 1/2q_n^2$ *or* $|x - p_{n+1}/q_{n+1}| \leq 1/2q_{n+1}^2$*.*

(i) *If* p/q *is a rational such that* $|x - p/q| \leq 1/2q^2$*, then* p/q *is among the* p_n/q_n*.*

Remark. One can improve the consequence of the above that for any x, there are rational a_n/b_n with $b_n \to \infty$ so $|x - a_n/b_n| \leq 1/2b_n^2$ to $1/\sqrt{5}\,b_n^2$; see the Notes.

Proof. (a) $q_{2n+1} = a_{2n+1}q_{2n} + q_{2n-1} \geq q_{2n}$ since $q_j \geq 0$ and $a_j \geq 1$. Since $q_j > 0$ for $j \geq 0$ (by induction), we get strict inequality if $n \geq 1$.

Clearly, $q_0 = 1 \geq 2^0$ and, by the first inequality and $a_{2n+2} \geq 1$, $q_{2n+2} \geq q_{2n+1} + q_{2n} \geq 2q_{2n}$, so inductively, $q_{2n} \geq 2^n$.

Since q has initial conditions $q_{-1} = 0$, $q_0 = 1$, and p has $p_0 = 0$, $p_1 = 1$, $p_n(a_1, a_2, \dots, a_n) = q_{n-1}(a_2, a_3, \dots, a_n)$, $q_n \geq 2^{n/2} \Rightarrow p_n \geq 2^{(n-1)/2}$.

(b) $\det\left(\begin{smallmatrix} a_n & 1 \\ 1 & 0 \end{smallmatrix}\right) = -1$, so $\det(T_n) = (-1)^n$ (since $\det(T_0) = 1$). Thus, $p_nq_{n+1} - p_{n+1}q_n = \det(T_{n+1}) = (-1)^{n+1}$. So

$$\frac{p_{n+1}}{q_{n+1}} - \frac{p_n}{q_n} = \frac{(-1)^n}{q_nq_{n+1}} \tag{7.5.25}$$

from which (7.5.21) is immediate.

(c), (d) Let

$$g_n(y) = \frac{p_ny + p_{n-1}}{q_ny + q_{n-1}} \tag{7.5.26}$$

as FLTs on $\mathbb{C}$. Then g_n maps $\mathbb{R}\cup\{\infty\}$ to $\mathbb{R}\cup\{\infty\}$ bijectively with $g_n^{-1}(\infty) < 0$, $g_n(0) = p_{n-1}/q_{n-1}$, $g_n(\infty) = p_n/q_n$, so by (7.5.25),

$$(-1)^n g_n(0) > (-1)^n g_n(\infty) \tag{7.5.27}$$

Since g_n is a bijection, it is monotone on $(0,\infty)$, increasing if n is odd and decreasing if n is even. Since $x = g_n(y_{n+1})$ by (7.5.14) and $g_n(y) = \infty$ for some $y < 0$, we see x lies between $g_n(0)$ and $g_n(\infty)$. This and (7.5.27) imply (7.5.23) and the first inequality in (7.5.22). The second inequality in (7.5.22) is immediate from (7.5.21) and $q_{n-1} < q_{n+1}$.

(e),(f) By (7.5.21) and (7.5.19),

$$\left|\frac{p_{n+1}}{q_{n+1}} - \frac{p_n}{q_n}\right| \le 2^{-n} \tag{7.5.28}$$

By (7.5.23), all p_m/q_m for $m \ge N$ lie in an interval of size the left-hand side of (7.5.28), so

$$\sup_{m,k\ge N}\left|\frac{p_m}{q_m} - \frac{p_k}{q_k}\right| \le 2^{-N} \tag{7.5.29}$$

so p_m/q_m is Cauchy, establishing the existence of the limit. By (7.5.23), in the case $a_n = a_n(x)$, x lies in all these shrinking intervals, and so is the limit.

The limit, y, of p_n/q_n is in each interval, so by Proposition 7.5.3 below, $a_j(y) = a_j$ for all j.

(g) By (7.5.23) and (7.5.21),

$$\left|x - \frac{p_n}{q_n}\right| \le \frac{1}{q_n q_{n+1}} \le \frac{1}{q_n^2}$$

since $q_{n+1} \ge q_n$.

(h) By (7.5.23) and then (7.5.21),

$$\begin{aligned}\left|x - \frac{p_n}{q_n}\right| + \left|x - \frac{p_{n+1}}{q_{n+1}}\right| &= \left|\frac{p_n}{q_n} - \frac{p_{n+1}}{q_{n+1}}\right| \\ &= \frac{1}{q_n q_{n+1}} &\quad (7.5.30) \\ &\le \frac{1}{2q_n^2} + \frac{1}{2q_{n+1}^2} &\quad (7.5.31)\end{aligned}$$

since $xy = \frac{1}{2}x^2 + \frac{1}{2}y^2 - \frac{1}{2}(x-y)^2 \le \frac{1}{2}x^2 + \frac{1}{2}y^2$, and we can take $x = q_n^{-1}$, $y_n = q_{n+1}^{-1}$. This proves (h).

(i) p/q is rational, so by (7.5.7), we can write p/q as a finite continued fraction with our choice of an even or odd number of terms. If $x < p/q$, use an odd number of terms, and if $x > p/q$, an even number. Use this finite sequence to define $\{q_j\}_{j=1}^n$ and $\{p_j\}_{j=1}^n$. Of course, $q_n = q$, $p_n = p$.

Since

$$\left|x - \frac{p_n}{q_n}\right| \leq \frac{1}{2q_n^2} \leq \frac{1}{2q_nq_{n-1}} \leq \frac{1}{q_nq_{n-1}} = \left|\frac{p_{n-1}}{q_{n-1}} - \frac{p_n}{q_n}\right| \tag{7.5.32}$$

and x and p_{n-1}/q_{n-1} are on the same side of p_n/q_n by our odd-even choice, we know for some $y \in (0, \infty)$,

$$x = g_n(y) \tag{7.5.33}$$

where g_n is given by (7.5.26). (We emphasize for now that the a_n's are for p/q, not for x.)

We claim $y > 1$, postponing the proof for a moment. Thus, y has an infinite continued fraction expansion $\{a_j(y)\}_{j=1}^\infty$ with $a_j(y) \geq 1$. If $\tilde{a}$ is defined by

$$\tilde{a}_j = \begin{cases} a_j(\frac{p}{q}), & j = 1, \dots, n \\ a_{j-n-1}(y), & j = n+1, \dots \end{cases} \tag{7.5.34}$$

then, by (7.5.33), $x = \mathbf{K}_{j=1}^\infty 1/\tilde{a}_j$, which implies $a_j(x) = \tilde{a}_j$ by (f) above. Thus, p/q is the nth approximant of x, as was to be proven.

Thus, we need only prove $y > 1$. This is implied by

$$\left|x - \frac{p_n}{q_n}\right| \leq \left|g_n(1) - \frac{p_n}{q_n}\right| \tag{7.5.35}$$

since g_n is monotone and $p_n/q_n = g(\infty)$. To prove (7.5.35), we note that, since $|p_{n-1}q_n - p_nq_{n-1}| = 1$,

$$\begin{aligned} \left|g_n(1) - \frac{p_n}{q_n}\right| &= \left|\frac{p_{n-1}+p_n}{q_{n-1}+q_n} - \frac{p_n}{q_n}\right| \\ &= \frac{1}{q_n(q_{n-1}+q_n)} > \frac{1}{2q_n^2} \\ &\geq \left|x - \frac{q_n}{p_n}\right| \end{aligned} \tag{7.5.36}$$

since $2q_n > q_n + q_{n-1}$. □

Proposition 7.5.3. *Let $\{a_n\}_{n=0}^\infty$ be an infinite sequence in $\mathbb{Z}_+$, let $\{p_n, q_n\}_{n=1}^\infty$ be given by the Euler–Wallis relations, and let $\{g_n\}_{n=1}^\infty$ be given by (7.5.26). Then for each fixed n, the following are equivalent for $x \in (1, \infty)$:*

(1) *x lies between p_{n-1}/q_{n-1} and p_n/q_n.*
(2) *$x = g_n(y)$ for some $y \in (0, \infty)$.*
(3) *$a_j(x) = a_j$ for $j = 1, \dots, n-1$* (7.5.37)

Proof. (1)⇔(2) is immediate from the monotonicity of g_n and $g_n(0) = p_{n-1}/q_{n-1}$, $g_n(\infty) = p_n/q_n$, $g_n^{-1}(\infty) \in (-\infty, 0)$.

(3)⇒(2) follows from (7.5.14).

(1)+(2)⇒(3). Since $a_2 \geq 1$, we have

$$\frac{q_0}{p_0} = a_1 \leq a_1 + \frac{1}{a_2} = \frac{q_1}{p_1} \leq a_1 + 1 \tag{7.5.38}$$

so by (7.5.23), for all j,

$$a_0 \leq \frac{q_j}{p_j} \leq a_0 + 1 \tag{7.5.39}$$

and so $a_1 \leq x^{-1} \leq a_1 + 1$ if (1) holds. Thus,

$$a_1(x) = a_1 \tag{7.5.40}$$

(2) can be rewritten using the $\mathbb{CP}[1]$ equivalence relation as

$$\left[\begin{pmatrix} x \\ 1 \end{pmatrix}\right] = \left[T_n \begin{pmatrix} y \\ 1 \end{pmatrix}\right] \tag{7.5.41}$$

for $y \in (0, \infty)$.

$x^{-1} = a_1 + z_1(x)$ can be rewritten as

$$\left[\begin{pmatrix} 1 \\ x \end{pmatrix}\right] = \left[\begin{pmatrix} a_1 & 1 \\ 1 & 0 \end{pmatrix} \begin{pmatrix} 1 \\ z_1(x) \end{pmatrix}\right] \tag{7.5.42}$$

which implies

$$\left[\begin{pmatrix} 1 \\ z_1(x) \end{pmatrix}\right] = \left[T_{n-1}(a_2, \dots, a_n) \begin{pmatrix} 1 \\ z_n(x) \end{pmatrix}\right] \tag{7.5.43}$$

which is (2) for the objects with one a stripped off the start. This implies (1) from $(a_1, a_2, \dots, a_n)$, which implies, by (7.5.40),

$$a_1(z_1(x)) = a_2(x) = a_2 \tag{7.5.44}$$

Iterating yields (7.5.37). □

This completes what we wanted to say about the continued fraction expansion of a real. We want to turn now to the use of FLT iteration in the study of Schur functions, leaving most detailed proofs to Problems 4–6.

Recall that a Schur function is an analytic map of $\mathbb{D}$ to $\mathbb{D}$. We'll use H_1^∞ for the set of Schur functions which are most of the unit ball in the set of bounded analytic functions on $\mathbb{D}$. The true unit ball, which we'll denote as $\overline{H_1^\infty}$, adds the constant functions with a value in $\partial\mathbb{D}$.

Given $f \in H_1^\infty$, what's a good analog of the integral part of a real? If you think of the Taylor series as an analog of a decimal expansion, you'll realize that

$$\gamma_0(f) = f(0) \tag{7.5.45}$$

is a good analog. For the analog of $\{x\}$, we can't just subtract for $f - f(0)$ is no longer a Schur function. Instead we use an FLT mapping $\mathbb{D}$ to $\mathbb{D}$! Thus, we look at $(f-\gamma_0)/(1-\bar{\gamma}_0 f)$. This is still a Schur function but it vanishes at 0. To get a next function, an analog of $y_1(x)$, we need to divide by z using the Schwarz lemma (Theorem 3.6.7) to assure we are still a Schur function (or at least an element of $\overline{H_1^\infty}$). Therefore, we define

$$f_1(x) = \frac{1}{z}\,\frac{f(z)-\gamma_0}{1-\bar{\gamma}_0 f(z)} \tag{7.5.46}$$

so inverting

$$f(z) = \frac{\gamma_0 + z f_1(z)}{1+\bar{\gamma}_0 z f_1(z)} \tag{7.5.47}$$

As with numeric continued fractions, we iterate and so define $\gamma_1(f), f_2, \gamma_2(f), \dots$ inductively by

$$\gamma_n(f) = \gamma_0(f_n) = f_n(0) \tag{7.5.48}$$

$$f_{n+1}(z) = \frac{1}{z}\,\frac{f_n(z)-\gamma_n}{1-\bar{\gamma}_n f_n(z)} \tag{7.5.49}$$

so

$$f_n(z) = S_{\gamma_n,z}(f_{n+1}(z)) \tag{7.5.50}$$

where $S_{\alpha,z}$ is the z-dependent FLT

$$S_{\alpha,z}(w) = \frac{1+\alpha z w}{1+\bar{\alpha} z w} \tag{7.5.51}$$

Thus,

$$f(z) = S_{\gamma_0,z}(S_{\gamma_1,z}(\dots S_{\gamma_n,z}(f_{n+1}(z)))) \tag{7.5.52}$$

The $\{\gamma_n(f)\}_{n=0}^\infty$ are the *Schur parameters*, the $\{f_n(z)\}_{n=0}^\infty$ the *Schur iterates*, and the association of $f \in H_1^\infty$ to a sequence $\{\gamma_n(f)\}_{n=0}^\infty \in \mathbb{D}^\infty$ is called the *Schur algorithm.*

There is a special case, however: If at some some point, $f_n(z)$ is a constant function in $\partial\mathbb{D}$, we have to stop, although we still define $\gamma_n(f)$ by (7.5.48). As in the real case, where the finite strings are associated with rational numbers, one can show (Problem 4) that the functions with finite strings of Schur parameters that end at $\gamma_n(f) \in \partial\mathbb{D}$ are precisely those $f \in H_1^\infty$ which are products of exactly n factors of the form $(z-z_j)/(1-\bar{z}_j z)$ (called Blaschke factors; see Section 9.9) and an $e^{i\theta}$.

The analogs of the continued fraction approximants of a real are the Schur approximants (not to be confused with the other meaning we've given to $f^{[n]}$ as iterated composition)

$$f^{[n]}(z) = S_{\gamma_0,z}(S_{\gamma_1,z}\cdots(S_{\gamma_n,z}(0))) \tag{7.5.53}$$

where f_{n+1} in (7.5.52) is replaced by 0. Thus, $f^{[n]}$ is a Schur function with

$$\gamma_j(f^{[n]}) = \begin{cases} \gamma_j(f), & j=0,\dots,n \\ 0, & j \ge n+1 \end{cases} \tag{7.5.54}$$

We already see a subtle distinction from the real-number case. There, the approximants were rational numbers whose $a_n(x)$-strings were finite. Here, the approximants are rational functions but not the special ones with finite strings.

The analog of $\left(\begin{smallmatrix} a & 1 \\ 1 & 0 \end{smallmatrix}\right)$ is the matrix

$$U_{\alpha,z} = \begin{pmatrix} z & \alpha \\ \bar{\alpha} z & 1 \end{pmatrix} \tag{7.5.55}$$

The analog of T_n of (7.5.10) is

$$T_n(z) = U_{\gamma_0,z} \dots U_{\gamma_j,z} \tag{7.5.56}$$

As with the numeric case where the four elements of T_n defined only two sequences, that's true here. One sees (Problem 5) that there are polynomials, $A_n(z), B_n(z)$, of degree at most n, called *Wall polynomials*, so that T_n has the form

$$T_n(z) = \begin{pmatrix} zB_n^*(z) & A_n(z) \\ zA_n^*(z) & B_n(z) \end{pmatrix} \tag{7.5.57}$$

where * is defined by

$$B_n^*(z) = z^n \, \overline{B_n(1/\bar{z})}, \qquad A_n^*(z) = z^n \, \overline{A_n(1/\bar{z})} \tag{7.5.58}$$

which are the polynomials of degree at most n obtained by reversing the order of the coefficients and conjugating them.

The analog of the Euler–Wallis equations are:

$$A_n(z) = A_{n-1}(z) + \gamma_n z B_{n-1}^*(z) \tag{7.5.59}$$

$$B_n(z) = B_{n-1}(z) + \gamma_n z A_{n-1}^*(z) \tag{7.5.60}$$

$$A_n^*(z) = z A_{n-1}^*(z) + \bar{\gamma}_n B_{n-1}(z) \tag{7.5.61}$$

$$B_n^*(z) = z B_{n-1}^*(z) + \bar{\gamma}_n A_{n-1}(z) \tag{7.5.62}$$

$$A_0(z) = \gamma_0 \tag{7.5.63}$$

$$B_0(z) = 1 \tag{7.5.64}$$

By (7.5.57), (7.5.52) becomes

$$f(z) = \frac{A_n(z) + zB_n^*(z) f_{n+1}(z)}{B_n(z) + zA_n^*(z) f_{n+1}(z)} \tag{7.5.65}$$

so, in particular,

$$f^{[n]}(z) = \frac{A_n(z)}{B_n(z)} \tag{7.5.66}$$

Since $\det(U_{\alpha,z}) = z(1 - |\alpha|^2)$, we see that

$$B_n^*(z)B_n(z) - A_n^*(z)A_n(z) = z^n \prod_{j=0}^{n} (1 - |\gamma_j|^2) \tag{7.5.67}$$

As in the real-number case (Problem 6), $f^{[n]}(z) \to f(z)$. Indeed,

$$\left| f(z) - \frac{A_n(z)}{B_n(z)} \right| \le 2|z|^{n+1} \tag{7.5.68}$$

and the image of the Schur algorithm on those $f \in H_1^\infty$ which do not have finite strings is all of D^∞ (Problem 6).

Notes and Historical Remarks. Continued fractions have a mathematical fascination that has spawned numerous books, of which we mention [**73, 253, 277, 301, 393, 415, 571**]. In particular, Khinchin [**301**] and Wall [**571**], still readable and informative, were historically significant. For a book presentation of the relevance of continued fractions to rational (and algebraic-number) approximations, see Bugeaud [**79**, Ch. 1].

The subject of continued fractions can be viewed as of Greek origin since (see Problem 1) the Euclidean algorithm can be thought of as a continued fraction expansion, and some date them from work of Rafael Bombelli (1526–72) and Pietro Cataldi (1548–1626). But it is probably best to say that their systematic theory dates to the 1650's work of the British/Irish mathematicians, William Brouncker (1620–84) and John Wallis (1616–1703). Brouncker, the first president of the Royal Society of London and an Irish lord, is less known because Euler mistakenly named an equation that Brouncker studied after Pell!

The Euler–Wallis equations appeared first in Wallis' 1656 *Arithmetica Infinitorum* [**572**]. Euler's work is in [**175**]. Nineteenth-century results on continued fraction expansions of real numbers include work of Dirichlet, Jacobi, and Legendre. In particular, part (i) of Theorem 7.5.2 is a discovery of Adrien-Marie Legendre (1752–1833) [**349**].

In Section 2.8 of Part 3, we'll prove some remarkable facts about the continued fraction approximants, for example, that for Lebesgue a.e. $x \in [0,1]$, $n^{-1} \log q_n(x) \to \pi^2/(12 \log 2)$.

Theorem 7.5.2 implies that for any rational x, there exist a_n/b_n rational with a_n relatively prime to b_n and $b_n \to \infty$, so

$$\left| x - \frac{a_n}{b_n} \right| \le \frac{\alpha}{b_n^2} \tag{7.5.69}$$

for $\alpha = \frac{1}{2}$. In fact, Hurwitz [**271**] has shown there are infinitely many a_n/b_n with $\alpha = 1/\sqrt{5}$, and the case where x has continued fraction $[1, 1, 1 \dots]$ shows this is best possible (see Problem 2). Ford [**192**] found a wonderful geometric proof described in the delightful book of Rademacher [**464**]. The Ford circles are not unrelated to our construction in Section 8.3 and discussed further there.

An early success in the use of continued fractions was the proof of the irrationality of π by J. H. Lambert (1728–77) in 1761 [**337**] (see [**233, 335**] for modern variants). Lambert proved that $\tan(x)$ has a continued fraction expansion

$$\tan(x) = \mathop{\mathbf{K}}_{n=0}^{\infty} \frac{a_n(x)}{2n+1}$$

where $a_0(x) = x$, and for $n \geq 1$, $a_n(x) = -x^2$ and that this implies the tangent of a rational is irrational (Problem 3 has a relative of such an irrationality proof.). If π were rational, $1 = \tan(\pi/4)$ would be irrational, so Lambert concluded that π was irrational.

The Schur algorithm, while involving iterated FLTs as do continued fractions, may not seem to be precisely a continued fraction, but since (7.5.47) can be rewritten as

$$f(z) = \gamma_0 + \cfrac{1 - |\gamma_0|^2}{\bar{\gamma}_0 + \cfrac{1}{zf_1(z)}} \tag{7.5.70}$$

it is a continued fraction expansion although, because of the two inversions, the $f^{[n]}$ are only half the continued fraction approximants.

Functional continued fractions like those that arise in the Schur algorithm appeared already in Wallis, but came into their own with work of Jacobi, Chebyshev, Markov, Hermite, and Stieltjes in the nineteenth century.

The Schur algorithm and virtually all the developments discussed here are from Schur [**507**]. The name "Wall polynomials," given by Khrushchev, is taken from Wall [**570**], but they already appeared twenty-six years earlier in Schur's papers. For books on the Schur algorithm, see [**12, 30, 119**].

Issai Schur (1875–1941) was born in Belarus, but he was a student of Frobenius in Germany and regarded himself as German and not a Russian or a Jew. He became a professor in Berlin in 1916 and was celebrated as a teacher—his students include Brauer, Prüfer, and Wielandt. As clouds darkened in the early thirties, he turned down several offers outside Germany because he couldn't foresee problems. He was dismissed from his professorship in 1935 and forced to resign from the Prussian Academy in 1938. Pressure came especially from Bieberbach who wrote: "I find it surprising that Jews are still members of academic commissions." Schur left Germany in 1938 for Israel, but with his resources depleted by the departure tax he had to pay and an inability to find a position in Israel, he was destitute and in ill health, passing away in 1941.

Schur's greatest contributions were in the theory of group representations, where he founded the theory of representations of continuous groups. His paper on the Schur algorithm included a formula, called the method of

Schur complements, for the inverse of an operator written as $\left(\begin{smallmatrix} A & B \\ C & D \end{smallmatrix}\right)$ in terms of blocks. He also made major contributions to algebra. For more on his life and work, see [**290**].

The map $f \mapsto \frac{1+zf(z)}{1-zf(z)} = F(z)$ sets up a one–one correspondence between Schur functions and functions, F, on $\mathbb{D}$, called Carathéodory functions, with $F(0) \equiv 1$ and $\operatorname{Re} F(z) > 0$. We'll say more about this subject in Section 5.4 of Part 3 and in Section 5.5 of Part 4.

Continued fractions are intimately connected with the history and development of the theory of orthogonal polynomials; for a discussion of this, see Khrushchev [**304, 305**]. In particular, Geronimus [**207**] proved the Schur parameters are essentially the recursion coefficients (called Verblunsky coefficients and discussed in Section 4.4 of Part 4) for orthogonal polynomials on the unit circle (see [**524**] for several proofs and Section 4.4 of Part 4 for a discussion of OPUC), and the Schur algorithm is central to some work of Khrushchev [**302, 303**].

Problems

1. The *Euclidean algorithm*, which appears in *Euclid's Elements* [**170**], finds the greatest common divisor (GCD) of two numbers in $\mathbb{Z}_+$, $p > q$, as follows:

$$\begin{aligned} p &= a_0 q + r_1, && 0 \le r_1 < q \equiv r_0 \\ r_{j-1} &= a_j r_j + r_{j+1}, && 0 \le r_{j+1} < r_j\,; \quad j = 1, 2, \dots \end{aligned} \tag{7.5.71}$$

repeated until $r_{j+1} = 0$ and then r_j is the GCD.

(a) Prove this stops after at most q steps, say J steps.

(b) Prove that $a_0, \dots, a_J$ are continued fraction integral parts for $x = p/q$ so that any rational has a representation as a finite continued fraction.

(c) Prove $a_J \ge 2$ and reinterpret the odd/even wiggle in (7.5.7).

2. (a) If $a_{n+p}(x) = a_n(x)$ for some p and all n, prove that x obeys a quadratic equation $\alpha x^2 + \beta x + \gamma = 0$ with $\alpha, \beta, \gamma \in \mathbb{Z}$ (a result of Euler).

(b) Find x with $a_n = 1$ for all n (golden mean). Do you recognize the p_n's and q_n's?

The rest of this problem deals with this x, p_n, q_n.

(c) Show $q_{n+1} = p_n$.

(d) Prove that $\lim_{n\to\infty} p_n^2 |x - q_n/p_n| = 1/\sqrt{5}$ and conclude that (7.5.69) cannot hold for this x with any $\alpha < 1/\sqrt{5}$.

3. This problem will prove that if $0 < b_n < a_n$ are integers, then $\mathbf{K}_{n=1}^N\, b_n/a_n$ converges and the limit is irrational (the text did this for $b_n = 1$).

(a) Let p_n, q_n be defined by (7.5.18) and the analog for q_n. Prove $p_n q_{n+1} - p_{n+1} q_n = (-1)^{n+1} b_1 \dots b_{n+1}$ and then that $p_{n+1}/q_{n+1} - p_n/q_n = \Delta_n$ has alternating signs, so it suffices to prove $|\Delta_n| \to 0$ to get convergence.

(b) Prove by induction that

$$p_{2n+1} \geq \prod_{j=1}^{n} (a_{2j+1} + b_{2j+1}), \qquad p_{2n} \geq \prod_{j=1}^{n} (a_{2j} + b_{2j})$$

and then that $|\Delta_n| \to 0$, proving convergence.

(c) If $1 > \alpha = A/B > 0$, $B > 0$, $p < q$, and $\alpha = p/(q + \beta)$, prove that $\beta = C/D$ with $D < B$ so that, as a rational, the denominator of β is less than B.

(d) If $\oplus_{n=1}^{\infty} b_n/a_n$ is rational, prove $\oplus_{n=2}^{\infty} b_n/a_n$ is rational with strictly smaller denominator, and so get a contradiction. Conclude (an argument of Legendre) that $\oplus_{n=1}^{\infty} b_n/a_n$ is irrational.

4. (a) Let f and f_1 be replaced by (7.5.47). Prove that if f is analytic in a neighborhood of $\partial\mathbb{D}$ with $|f(e^{i\theta})| = 1$, then so is f_1 with $|f_1(e^{i\theta})| = 1$, and f_1 has one fewer zero than f.

(b) Conversely, if f_1 is analytic in a neighborhood of $\partial\mathbb{D}$ with $|f_1(e^{i\theta})| = 1$, then so is f with $|f(e^{i\theta})| = 1$, and that f has one more zero than f_1.

(c) Show that if after n steps, the Schur algorithm has to stop because $f_n(z) \equiv e^{i\theta}$, then f has an analytic continuation into a neighborhood of $\partial\mathbb{D}$, $|f(e^{i\theta})| = 1$, and f has n zeros in $\mathbb{D}$. Find all f's with terminating Schur algorithm using Problem 5 of Section 3.6.

5. Prove the matrix (7.5.56) has the form (7.5.57). (*Hint*: Use induction showing the recursion relations (7.5.60)–(7.5.63).)

6. (a) Let f, f_1, g, g_1 be related by (7.5.47) (same γ_0). Let $k \in \{0, 1, 2, \dots\}$. If $f_1 - g_1 = O(z^k)$ at $z = 0$, prove that $f - g = O(z^{k+1})$ at $z = 0$.

(b) If f and g are two Schur functions with

$$\gamma_j(f) = \gamma_j(g), \qquad j = 0, 1, \dots, n \tag{7.5.72}$$

prove that $f - g = O(z^{n+1})$ at $z = 0$.

(c) If f and g are Schur functions with (7.5.72), prove that $|f(z) - g(z)| \leq 2|z|^{n+1}$. (*Hint*: $\frac{1}{2}(f - g)$ is a Schur function.)

(d) Prove (7.5.13).

(e) Given $\{\gamma_j\}_{j=0}^{\infty} \subset \mathbb{D}^{\infty}$, let $f^{[n]}$ be the Schur approximants. Prove that $|f^{[n]}(z) - f^{[m]}(z)| \leq 2|z|^{\min(n,m)+1}$ and deduce $f^{[n]}$ has a limit f whose Schur parameters are $\gamma_n(f) = \gamma_n$.

7. (a) Prove $B_n(z)$ is nonvanishing on $\mathbb{D}$. (*Hint*: (7.5.66) and (7.5.67).)
   (b) On $\partial\mathbb{D}$, prove $|B_n(z)|^2 - |A_n(z)|^2 = \prod_{j=0}^{n}(1 - |\gamma_j|^2)$.
   (c) Prove B_n is nonvanishing on $\partial\mathbb{D}$.

Chapter 8

Conformal Maps

> Very few mathematical papers have exercised an influence on the later development of mathematics which is comparable to the stimulus received from Riemann's dissertation. It contains the germ to a major part of the modern theory of analytic functions, it initiated the systematic study of topology, it revolutionized algebraic geometry, and it paved the way for Riemann's own approach to differential geometry.
>
> —*Lars Ahlfors* [**7**]

Big Notions and Theorems: Riemann Mapping Theorem, hsc $\Leftrightarrow$ tsc, Carathéodory–Osgood–Taylor Theorem, Painlevé Theorem, Accessible Points, Elliptic Modular Function, Hyperbolic Triangles, Farey Series, Farey Tesselation, Stern–Brocot Tree, Elliptic Integrals, Joukowski Map, Schwarz–Christoffel Maps, Classification of Complex Annuli, Moduli, Uniformization Theorem, Green's Function, Bipolar Green's Function, Ahlfors Function, Analytic Capacity, Uniqueness of Ahlfors Function, $\boldsymbol{A}(\mathfrak{e}) \leq \boldsymbol{C}(\mathfrak{e})$, Removable Set, Painlevé Problem, Denjoy Domain, Pommerenke's Theorem, Ahlfors Function For Finite Gap Set

As we've seen, a conformal map is an analytic map $f\colon \Omega \to \Omega'$ which is locally a bijection; equivalently, f' is everywhere nonvanishing. In this chapter, we study such maps, focusing especially on global analytic bijections. Of course, we've already discussed this subject—indeed, Chapter 7 can be viewed as a part of this chapter!

The central theorem in the subject is the Riemann mapping theorem that if $\Omega \subset \mathbb{C}$ but not equal to it and is simply connected, then there exists an analytic bijection of Ω to $\mathbb{D}$, the unit disk. One might ask if one means topologically or holomorphically simply connected but, in fact, by the time the smoke has cleared, we'll have completed the proof that these notions are equivalent. In fact, our proof in Section 8.1 will not even require holomorphic simple connectivity, but the a priori weaker property that any $f \in \mathfrak{A}(\Omega)$, which is everywhere nonvanishing, has a holomorphic square root.

Section 8.2 will turn to the issue of when the conformal bijection $f\colon \Omega \to \mathbb{D}$ extends to the boundary, that is, there exists a continuous bijection $\widetilde{f}\colon \overline{\Omega} \to \overline{\mathbb{D}}$ so $\widetilde{f} \restriction \Omega = f$. We've already laid the groundwork for this in Sections 5.5 and 5.6, so the section will be brief.

Section 8.3 will have an interesting application of the Riemann mapping theorem. We'll construct a function $\lambda\colon \mathbb{C}_+ \to \mathbb{C} \setminus \{0,1\}$, which is onto, a local bijection, and so that there is a group, Γ, of automorphisms of $\mathbb{C}_+$ so $\lambda(w) = \lambda(z) \Leftrightarrow z = \gamma(w)$ for some $\gamma \in \Gamma$. The existence of this function will be 80% of one of the proofs of Picard's theorems (see Section 11.3). Another proof of the existence of this function will appear in Section 10.6, and that construction gives λ its name: elliptic modular function, but elliptic functions, per se, make no appearance in Section 8.3 nor in the application to Picard's theorems. Yet another way to understand the existence of this map is to prove that the universal cover of $\mathbb{C} \setminus \{0,1\}$ is $\mathbb{D}$, which follows from the uniformization theorem discussed in Section 8.7.

Conformal maps are important in applications in part because they allow the transfer of things like solutions of the Dirichlet problem from $\mathbb{D}$ to some Ω. The exact form of the maps is important in practical applications, so Section 8.4 will give lots of examples, including the interiors of polygons, where the maps are known as Schwarz–Christoffel maps.

The last three sections turn to what happens if Ω is not simply connected. Section 8.6 has a complete analysis of the doubly connected case, that is, $\Omega \subset \mathbb{C}$ is such that $\mathbb{C} \setminus \Omega$ has a single bounded component and at least one unbounded component. It will turn out that every such region is conformally equivalent to some annulus $\mathbb{A}_{r,R}$, and $\mathbb{A}_{r,R}$ is equivalent to $\mathbb{A}_{r',R'}$ if and only if $r/R = r'/R'$.

Sections 8.5 and 8.7 look beyond global bijections and study the universal cover of Riemann surfaces. Section 8.5 looks at $\Omega \subset \mathbb{C}$, using essentially the same argument used in Section 8.1, to show the universal cover is $\mathbb{D}$ if $\mathbb{C} \setminus \Omega$ has at least two points. Section 8.7, relying on harmonic function results that we'll only prove in Part 3, proves the Poincaré–Koebe theorem that the universal cover of any Riemann surface is one of $\widehat{\mathbb{C}}$, $\mathbb{C}$, or $\mathbb{D}$.

8.1. The Riemann Mapping Theorem

Definition. We say Ω, a region in $\mathbb{C}$, has the *square root property* if $f \in \mathfrak{A}(\Omega)$ with $f(z) \neq 0$ for all z in Ω implies there exists $g \in \mathfrak{A}(\Omega)$ with $g^2 = f$.

Our main result in this section is:

Theorem 8.1.1 (Riemann Mapping Theorem). *If Ω has the square root property and $\Omega \neq \mathbb{C}$, then for any $z_0 \in \Omega$, there exists a unique analytic bijection, $f\colon \Omega \to \mathbb{D}$ with $f(z_0) = 0$ and $f'(z_0) > 0$.*

Uniqueness is immediate from Lemma 7.4.1: If f and g are two such maps, then $h \equiv f \circ g^{-1}\colon \mathbb{D} \to \mathbb{D}$ is an analytic bijection and $h(0) = 0$, $h'(0) > 0$, implying $h(z) \equiv z$ by that lemma. The maps of Section 7.4 also show that if there is any analytic bijection Ω to $\mathbb{D}$, there is one with $f(z_0) = 0$, $f'(z_0) > 0$. Thus, we will focus on existence only.

Before doing that, we note one consequence of the Riemann mapping theorem that completes the discussion started in Sections 2.6 and 4.5.

Theorem 8.1.2. *Let $\Omega \subset \mathbb{C}$ be a region with $\Omega \neq \mathbb{C}$. The following are equivalent:*

(1) *Ω has the square root property.*
(2) *Ω is hsc.*
(3) *Ω is tsc.*
(4) *In $\widehat{\mathbb{C}}$, $\widehat{\mathbb{C}} \setminus \Omega$ is connected.*
(5) *Every component of $\mathbb{C} \setminus \Omega$ is unbounded.*

Proof. Theorem 4.5.1 proved (2), (4), and (5) are equivalent and Theorem 2.6.5 proved (3) $\Rightarrow$ (2). Thus, it suffices to prove (2) $\Rightarrow$ (1) $\Rightarrow$ (3).

<u>(2) $\Rightarrow$ (1)</u> is part of Theorem 2.6.1.

<u>(1) $\Rightarrow$ (3)</u>. By Theorem 8.1.1, if (1) holds, there is a biholomorphic bijection of Ω and $\mathbb{D}$. A fortiori, the two are homeomorphic. Since tsc is invariant under homeomorphism and $\mathbb{D}$ is tsc, so is Ω. $\square$

The proof depends on looking at

$$\begin{aligned} \mathcal{R} \equiv \{f \in \mathfrak{A}(\Omega) \mid \sup_{z\in\Omega}|f(z)| \le 1,\ f(z_0) = 0,& \\ f'(z_0) > 0,\ f(w) = f(z) \Rightarrow z = w\}& \end{aligned} \tag{8.1.1}$$

The idea will be to prove $\mathcal{R}$ is nonempty and there is $f \in \mathcal{R}$ which maximizes $f'(z_0)$ and to show that this f is onto $\mathbb{D}$.

Lemma 8.1.3. *$\mathcal{R}$ is nonempty.*

Proof. This step is where we use $\Omega \neq \mathbb{C}$ (see Problem 1). Pick $w_0 \in \mathbb{C} \setminus \Omega$. Then $h(z) = (z - w_0)/(z_0 - w_0)$ is nonvanishing on $\mathbb{C}$. By the assumption that Ω has the square root property, there exists $g \in \mathfrak{A}(\Omega)$ so that $g(z_0) = 1$ and $g(z)^2 = h(z)$.

Since h is not constant, neither is g, so by the open mapping theorem, there is $\delta \in (0,1)$ so $\mathbb{D}_\delta(1) \subset \operatorname{Ran}(g)$. If $w \in \operatorname{Ran}(g)$, there is z_1 with $g(z_1) = w$. There cannot be z_2 with $g(z_2) = -w$, for if there were, $h(z_1) = h(z_2) \Rightarrow z_1 = z_2 \Rightarrow w = 0$. Thus, $\mathbb{D}_\delta(-1) \cap \operatorname{Ran}(g) = \emptyset$, that is, for all $z \in \Omega$, $|g(z)+1| \geq \delta$. Thus,

$$G(z) = \frac{\delta}{g(z)+1} \tag{8.1.2}$$

maps Ω to $\mathbb{D}$.

Composing G with a suitable element of $\mathbb{A}\mathrm{ut}(\mathbb{D})$ can arrange $f\colon \Omega \to \mathbb{D}$ with $f(z_0) = 0$ and $f'(z_0) > 0$. Finally, $f(z) = f(w) \Rightarrow G(z) = G(w) \Rightarrow g(z) = g(w) \Rightarrow h(z) = h(w) \Rightarrow z = w$. □

Lemma 8.1.4. *Let 0 denote the zero function. Then $\mathcal{R} \cup \{0\}$ is compact in the topology of $\mathfrak{A}(\Omega)$.*

Proof. By Montel's theorem (Theorem 6.2.1), $\overline{\mathcal{R} \cup \{0\}}$ is compact, so it suffices to show $\mathcal{R} \cup \{0\}$ is closed. Let $f_n \in \mathcal{R}$ and $f_n \to f_\infty$ uniformly on compacts. Then $\|f_\infty\|_\infty \leq 1$ and $f_\infty(z_\infty) = 0$, so by the maximum modulus principle, $\operatorname{Ran}(f_\infty) \subset \mathbb{D}$. Clearly, $f'_\infty(z_0) \geq 0$.

By the corollary to Hurwitz's theorem (Corollary 6.4.3), either f_∞ is one-one or f_∞ is constant, so since $f_\infty(z_\infty) = 0$, $f_\infty \equiv 0$. Thus, either $f_\infty \in \mathcal{R}$ or $f_\infty \equiv 0$ and $\mathcal{R} \cup \{0\}$ is closed. □

Lemma 8.1.5. *If φ is analytic on $\mathbb{D}$, $\operatorname{Ran}(\varphi) \subset \mathbb{D}$, $\varphi(0) = 0$, and $\varphi'(0) > 0$, then either $\varphi(z) \equiv z$ or $\varphi'(0) < 1$.*

Proof. Let $\psi(z) = \varphi(z)/z$. By the Schwarz lemma (Theorem 3.6.7), $\sup_{z\in\mathbb{D}}|\psi(z)| \leq 1$, so by the maximum principle, either $|\psi(0)| < 1$ or $\psi(z) \equiv e^{i\theta}$. Since $\psi(0) = \varphi'(0) > 0$, in the latter case, $\psi(z) \equiv 1$, so $\varphi(z) \equiv z$. If $|\psi(0)| < 1$, since $\varphi'(0) > 0$, we see $\varphi'(0) < 1$. □

Lemma 8.1.6. *Suppose $f \in \mathcal{R}$ and $\operatorname{Ran}(f) \neq \mathbb{D}$. Then there is $g \in \mathcal{R}$ and $\varphi\colon \mathbb{D} \to \mathbb{D}$ with $\varphi(z) \not\equiv z$ so that*

$$f(z) = \varphi(g(z)) \tag{8.1.3}$$

In particular,

$$f'(z_0) < g'(z_0) \tag{8.1.4}$$

Proof. Suppose $w_0 \in \mathbb{D} \setminus \operatorname{Ran}(f)$. Define

$$h(z) = f_{w_0}(f(z)) \tag{8.1.5}$$

where f_{w_0} is given by (7.4.6). Since $f_{w_0}(\zeta) = 0$ only if $\zeta = w_0$, h is nonvanishing, so we can define

$$r(z) = \sqrt{f_{w_0}(f(z_0))} \tag{8.1.6}$$

where we take either possible square root. Notice that

$$f(z) = f_{-w_0}(r(z)^2) \tag{8.1.7}$$

Finally, let

$$g(z) = \frac{|r'(z_0)|}{r'(z_0)} f_{r(z_0)}(r(z)) \tag{8.1.8}$$

(we'll see $r'(z_0) \neq 0$ in a moment). Then, with

$$\varphi(w) = f_{-w_0}\left(\left[f_{-r(z_0)}\left(\frac{r'(z_0)}{|r'(z_0)|}\, w\right)\right]^2\right) \tag{8.1.9}$$

we have (8.1.3) on account of (8.1.8) and (8.1.7).

It remains to see that $g \in \mathcal{R}$, $\varphi(0) = 0$, and $\varphi(z) \not\equiv z$, for (8.1.3) then implies that $f'(z_0) = \varphi'(0)g'(z_0)$ so that $\varphi'(0) > 0$, and so (8.1.4) is implied by Lemma 8.1.5.

That $g(z_0) = 0$ follows from $f_{r(z_0)}(r(z_0)) = 0$. Noticing (Problem 2) that for any $w_1 \in \mathbb{D}$, $f'_{w_1}(w_1) > 0$, we see

$$g'(z_0) = \frac{|r'(z_0)|}{r'(z_0)} f'_{r(z_0)}(r(z_0))r'(z_0)$$

so $g'(z_0) > 0$.

Notice that, since f_w is a bijection, h is one-one, so r is one-one (and $r'(z_0) \neq 0$), so g is one-one. Thus, $g \in \mathcal{R}$. Since (8.1.3) holds, taking $z = z_0$ implies $\varphi(0) = 0$. Since $f_{-r(z_0)}$ is one-one, pick $w_\pm$ so $f_{-r(z_0)}(\frac{r'(z_0)}{|r'(z_0)|}w_\pm) = \pm\frac{1}{2}$. By (8.1.9), since $(\frac{1}{2})^2 = (-\frac{1}{2})^2$, $\varphi(w_+) = \varphi(w_-)$, so φ is not one-one, and thus, $\varphi(z_0) \not\equiv z$. $\square$

Remark. Rather than see that $\varphi'(0) < 1$ by using the Schwarz lemma, one can compute $\varphi'(0)$ directly to see that

$$\varphi'(0) = \frac{2\sqrt{|w_0|}}{1 + |w_0|} \tag{8.1.10}$$

and conclude $\varphi'(0) < 1$ (see Problem 3).

Proof of Theorem 8.1.1. We already proved uniqueness, so we only need to prove some $f \in \mathcal{R}$ is onto $\mathbb{D}$. Let

$$Q(f) = f'(z_0) \tag{8.1.11}$$

which is a continuous function on $\mathcal{R} \cup \{0\}$, not identically zero since $\mathcal{R}$ is nonempty by Lemma 8.1.3. Since $\mathcal{R} \cup \{0\}$ is compact (by Lemma 8.1.4), there exists $\widetilde{f}$ maximizing Q, that is,

$$\widetilde{f}'(z_0) \geq g'(z_0) \tag{8.1.12}$$

for all $g \in \mathcal{R}$. Since $\mathcal{R}$ is nonempty, $\widetilde{f} \in \mathcal{R}$. If $\widetilde{f}$ is not onto $\mathbb{D}$, then by Lemma 8.1.6, there is $g \in \mathcal{R}$, violating (8.1.12). We conclude $\text{Ran}(\widetilde{f}) = \mathbb{D}$, so $\widetilde{f}$ is the required bijection. □

Notes and Historical Remarks.

> Two given simply connected plane surfaces can always be mapped onto one another in such a way that each point of the one corresponds to a unique point of the other in a continuous way and the correspondence is conformal; moreover, the correspondence between an arbitrary interior point and an arbitrary boundary point of the one and the other may be given arbitrarily, but when this is done the correspondence is determined completely.[1]
>
> —*B. Riemann*, Doctoral Thesis, 1851, [**480**] as translated in [**216**]

> Riemann's writings are full of almost cryptic messages to the future. For instance, Riemann's mapping theorem is ultimately formulated in terms which would defy any attempt of proof, even with modern methods.
>
> —*L. Ahlfors*, 1953 [**7**]

In his 1851 inaugural dissertation, Riemann [**480**] stated and "proved" a version of the Riemann mapping theorem—namely, for bounded, simply connected regions with smooth boundaries, he stated the existence of an analytic bijection. To Riemann, a region, Ω, is simply connected if every curve γ with $\gamma\bigl[(0,1)\bigr] \subset \Omega$ and $\gamma(0)$, $\gamma(1) \in \partial\Omega$ divides Ω in two. The reader should consider this notion where $\Omega = \mathbb{D}$ or $\Omega = \{z \mid 0 < \text{Re}\, z < 1\}$ or $\Omega = \mathbb{A}_{\frac{1}{2},1}$, the annulus.

His proof used harmonic function ideas assuming that one could solve the Dirichlet problem (find a harmonic function with given continuous boundary values) via Dirichlet's minimum principle (see the discussion later in these Notes.). Weierstrass, in particular, in 1870 [**584**] complained about the error of not proving that a minimum occurred. He encouraged Schwarz, his student, to fix the proof, and for many years, the best results were Schwarz's, who required a piecewise smooth boundary without cusps. It was in connection with this effort that he did his work on the reflection principle.

In 1900, Osgood [**418**], using in part ideas of Schwarz and Poincaré, found the first proof of the full result. But he was isolated in the U.S. and published in an American journal, so his work was then, and since, little noted. It was a series of papers by Koebe [**324, 325, 326, 327**] and Carathéodory [**87**] that established the result and introduced many of the techniques that have dominated the field, including many of the results on the boundary behavior. Koebe first used the square–root-type construction

[1] Many years later Carathéodory emphasized that rather than use boundary points (because maps may not extend nicely to the boundary), one should use $f(z_0)$ and $\arg(f'(z_0))$.

central to Lemma 8.1.6, but it was Carathédory [**87**] who made it a central point.

Carathéodory and Koebe had explicit iterative constructions—the wonderful idea of using a variational principle, but for $f'(z_0)$, is from Radó [**466**], who says he got the idea from Fejér and F. Riesz. Remmert [**477**] and Veech [**562**] present the more constructive Carathéodory–Koebe proof. All authors who used the square-root idea before 1924 computed the derivative directly (see Problem 3). Ostrowski [**422**] had the idea of using the Schwarz lemma, as we do.

The proof relied on the remarkable fact that maximizing $\operatorname{Re} f'(z_0)$ (which maximizes $|f'(z_0)|$ and removes the $f \mapsto e^{i\theta} f$ nonuniqueness) among all functions $f\colon \Omega \to \mathbb{D}$ with $f(z_0) = 0$ and f one-one yields a unique function and one that is onto all of $\mathbb{D}$, so a Riemann mapping. But much more is true even if we drop the requirements that $f(z_0) = 0$ and that f is one-one, the function f that maximizes $\operatorname{Re} f'(z_0)$ among all $f\colon \Omega \to \mathbb{D}$ is unique and obeys f is a bijection with $f(z_0) = 0$. The key to this is the Schwarz lemma and the reader will prove this in Problems 7 and 8. The proof uses the existence of a Riemann mapping so we still need to consider the restricted problem. These notion for general Ω will be discussed further in Section 8.8.

Georg Friedrich Bernard Riemann (1826–66) is arguably the most influential mathematician of the nineteenth century. His publications are few, but each has multiple important contributions. For example, the same work discussed both the basis of integration theory (Riemann integral) and results on Fourier transforms (including the Riemann–Lebesgue lemma). He began as a student at Göttingen but Gauss was not much concerned with students, so in 1847 he moved to Berlin where Jacobi, Eisenstein, Steiner, and especially Dirichlet influenced him. (As unintended payback, Dirichlet got credit for work on harmonic functions such as the Dirichlet problem, Dirichlet principle, and Dirichlet boundary conditions because Riemann learned of them from Dirichlet's lectures and named them after him—Gauss in 1839 [**206**], Green in 1828 [**219**], and Thomson had preceded Dirichlet whose work was published in 1850 [**145**].)

Riemann returned to Göttingen in 1849 where he spent the rest of his career. Among his students was Dedekind. Riemann always had a sickly constitution and died of complications of pleurisy at the age of only 39. Two of his students, Hankel and Roch, also died at early ages (34 and 26, respectively).

Riemann had six monumental works: the inaugural dissertation [**480**] (where the Riemann mapping theorem, Cauchy–Riemann equations, Riemann surfaces appeared), his work on abelian integrals [**484**] that we'll discuss in the Notes to Section 10.5, his work on integration and trigonometric series mentioned above [**481**], his work on number theory [**485**] (Riemann zeta function, Riemann hypothesis discussed in Chapter 13 of Part 2B), his work on differential geometry [**482**] (Riemann metric, Riemann curvature, Riemann tensor), and his 1857 paper on hypergeometric functions [**483**] (that included the precursor of the monodromy group and the analysis of what have come to be called Riemann's P-functions; see Section 14.4 of Part 2B).

We've presented the most common textbook presentation of the Riemann mapping theorem, but we should mention two other approaches. One involves the Bergman kernel, which will be discussed further in Chapter 12 of Part 2B.

The other is a Green's function approach, which is a direct descendant of Riemann's original idea. This approach is especially significant for some ways of attacking regularity at the boundary, and a variant will be central to the discussion of uniformization in Section 8.7. Here we'll discuss the idea when a *classical Green's function* exists (we add "classical" here because Section 8.7 will have a weaker notion of Green's function that agrees with this one in case a classical Green's function exists; see Problem 5 of Section 8.7). Green's functions are discussed further in Chapter 3 of Part 3.

Let Ω be a bounded open region in $\mathbb{C}$. We say a classical Green's function exists at $z_0 \in \Omega$ if there is a real-valued function $G(z, z_0)$ for $z \in \Omega \setminus \{z_0\}$ with

$$\text{(i) } G(z, z_0) + \log|z - z_0| = v(z) \tag{8.1.13}$$

can be given a value at $z = z_0$ so that v is harmonic on all of Ω.

$$\text{(ii) } \lim_{\substack{z\to\partial\Omega \\ z\in\Omega}} G(z, z_0) = 0 \tag{8.1.14}$$

Notice that since $G \to \infty$ at $z = z_0$ and $G \to 0$ at $\partial\Omega$, the maximum principle applied to $-G$ implies

$$G(z, z_0) > 0 \qquad \text{on } \Omega \setminus \{z_0\} \tag{8.1.15}$$

We also note that if the Dirichlet problem can be solved, that is, if given f on $\partial\Omega$, we can find u continuous on $\overline{\Omega}$, harmonic on Ω so $u \restriction \partial\Omega = f$, then a Green's function exists—just take $f(z) = \log|z - z_0|$ for $z \in \partial\Omega$, and $G(z, z_0) = u(z) + \log|z - z_0|^{-1}$.

What Riemann called the Dirichlet principle noted that if $\Delta u = 0$ on Ω and u has a continuus extension to $\partial\Omega$ with $u \restriction \partial\Omega = f$ and if h is continuous on $\overline{\Omega}$, $h \restriction \partial\Omega = 0$ and $h \in C^2$, then

$$\int_\Omega |\nabla(u+h)|^2\, d^2x = \int_\Omega (|\nabla u|^2 + |\nabla h|^2)\, dx + 2\int_\Omega \nabla u \cdot \nabla h\, d^2x \tag{8.1.16}$$

Since $\Delta u = 0$, $\nabla u \cdot \nabla h = \nabla \cdot \bigl(h \nabla u\bigr)$ so by Gauss' theorem

$$\int_\Omega \nabla u \cdot \nabla h \, d^2x = \int_{\partial\Omega} h \frac{\partial u}{\partial n} = 0$$

since $h \restriction \partial\Omega = 0$. Thus, formally, the solution to the Dirichlet problem for f obeys the *Dirichlet principle*:

$$u \text{ minimizes } \int_\Omega |\nabla u|^2 \, d^2x \text{ among } u\text{'s with } u \restriction \partial\Omega = f$$

and one can go backwards. Thus, Riemann argued that solutions to the Dirichlet problem existed by minimizing the Dirichlet integral $\int_\Omega |\nabla u|^2 \, dx^2$ and it was this step that Weierstrass objected to.

The point is if a classical Green's function exists if Ω is simply connected, it is easy to map Ω conformally to $\mathbb{D}$. If v is the harmonic function given by (8.1.13), there is an analytic function, g, on Ω so $v = \operatorname{Re} g$ (see Proposition 8.7.4 later). Let $f(z) = (z - z_0)e^{-g(z)}$. Then $|f(z)| = \exp(-v(z) + \log|z - z_0|)$, so

$$|f(z)| = e^{-G(z,z_0)} \tag{8.1.17}$$

Since $G > 0$, f is a map of Ω to $\mathbb{D}$ and

$$f(z) = 0 \Leftrightarrow z = z_0 \tag{8.1.18}$$

We claim f is a bijection. For given $\delta < \operatorname{dist}(z_0, \partial\Omega)/\sqrt{2}$, let J be the family of squares in (4.4.13). Let $\widetilde{J}$ be the subset of those squares that are in the connected component of $\cup_{\alpha \in J} \Delta_\alpha \setminus \{\delta n \mid n \in \mathbb{Z}^2\}$. Then $\Gamma_{\widetilde{J}}$ is a connected contour whose interior contains z_0 and within δ of the boundary (Problem 5). Moreover, if $\Gamma^{(m)}$ is this contour for $\delta = 2^{-m}$ and $\Omega^{(m)}$ its interior, then $\Omega^{(m-1)} \subset \Omega^{(m)}$ and $\cup_m \Omega^{(m)} = \Omega$ (Problem 5).

By (8.1.17), $|f(z)| \to 1$, so

$$\rho_m = \min_{z \in \Gamma^{(m)}} |f(z)| \to 1 \tag{8.1.19}$$

For $\zeta \in \mathbb{D}_{\rho_m}(0)$, define

$$w_m(\zeta) = \frac{1}{2\pi i} \oint_{\Gamma_m} \frac{f'(z)}{f(z) - \zeta} \, dz \tag{8.1.20}$$

Then w_m is an integer and analytic on $\mathbb{D}_{\rho_m}(0)$ and $w_m(0) = 1$ by the argument principle, since f has exactly one 0 on Ω and it is in $\Omega^{(m)}$. By (8.1.19), f is onto $\mathbb{D}$, and since $\cup \Omega^{(m)} = \Omega$, f is one-one. This shows that f is a bijection as claimed.

Problem 6 has an alternate proof that f is a bijection.

In Section 3.6 of Part 3, we'll turn this argument around and show that the Riemann mapping theorem yields the existence of a classical Green's function for any simply connected region of $\widehat{\mathbb{C}}$ missing at least two points.

Problems

1. Prove that there are no nontrivial analytic maps of $\mathbb{C}$ to $\mathbb{D}$.

2. Let $f_w(z) = (z - w)/(1 - \bar{w}z)$. Prove that $f'_w(w) > 0$.

3. Verify (8.1.10).

4. Suppose $f\colon \Omega \to \mathbb{D}$ extends to a continuous bijection $\tilde{f}\colon \bar{\Omega} \to \overline{\mathbb{D}}$. Let g be a continuous function on $\partial\Omega$ and let u solve the Dirichlet problem for $\mathbb{D}$ with boundary data $u(e^{i\varphi}) = g(\tilde{f}(e^{i\varphi}))$. Prove that $G(z) = u(\widetilde{f}(z))$ solves the Dirichlet problem for Ω.

   Remark. In the Problems of Section 8.2, we'll explore this issue further.

5. Provide the details of the argument in the Notes about $\Gamma^{(m)}$ and $\Omega^{(m)}$.

6. This problem will lead the reader through another proof that the f constructed in the Notes (obeying (8.1.17)) is a bijection. So we suppose that Ω is a bounded region, $f : \Omega \to \mathbb{D}$ is analytic and obeys: (i) $\lim_{z\to\partial\Omega}|f(z)| = 1$; (ii) For a fixed $z_0 \in \Omega$, we have $f(z) = 0 \Leftrightarrow z = z_0$ and $f'(z_0) \neq 0$. This problem will prove that such an f is a bijection.

   (a) If $w \in \mathbb{D}$, $f(z) = w$ has only finitely many solutions (counting multiplicity) since $|f(z)| \to 1$ as $z \to \partial\Omega$. Let $N_w(f)$ be the number of solutions and $B(k) = \{w \in \mathbb{D} \mid N_w(f) = k\}$. Using the analysis found in Sections 3.4 and 3.5, prove that $\bigcup_{k\geq 2} B(k)$ is open.

   (b) If $w_\ell \to w_\infty \in \mathbb{D}$ and $f(z_\ell) = w_\ell$, prove that there is $z_\infty \in \Omega$ with $f(z_\infty) = w_\infty$ and conclude that $\cup_{k\geq 1} B(k)$ is closed so that $B(1)$ is closed.

   (c) Suppose $w_1 \in B(1)$ and let $g(z) = \frac{f(z)-w_1}{1-\bar{w}_1 f(z)}$. Prove that $g(z) = 0$ has a unique solution z_1.

   (d) Find $\delta > 0$ and an open neighborhood of z_1, $N \subset \Omega$, with $\operatorname{dist}(N, \partial\Omega) > 0$ and $|g(z)| = \delta$ on ∂N, ∂N an analytic Jordan curve and g one-one on N with $g[N] = \{w \mid |w| < \delta\}$.

   (e) Prove that $h(z) = g(z)^{-1}$ is analytic on $\Omega \setminus N$ and $\sup_{z\in\Omega\setminus N} |h(z)| = \delta^{-1}$ using the maximum principle.

   (f) Conclude that $\{w \mid B_w(f) = 1\}$ contains a disk about w_1 so $\{w \mid B_w(f) = 1\}$ is open and closed. Prove that it is also nonempty.

   (g) Conclude that f is a bijection.

7. (a) Let $\Omega \subset \mathbb{C}$ be a domain and $f\colon \Omega \to \mathbb{D}$. Define

$$g(z) = \frac{f(z) - f(z_0)}{1 - \overline{f(z_0)}f} \tag{8.1.21}$$

for some z_0. Prove that

$$g'(z_0) = f'(z_0)\big(1 - |f(z_0)|^2\big)^{-1} \tag{8.1.22}$$

and conclude that

$$\sup\{|f'(z_0)| \mid f\colon \Omega \to \mathbb{D}, f(z_0) = 0\} = \sup\{|f'(z_0)| \mid f\colon \Omega \to \mathbb{D}\} \tag{8.1.23}$$

(b) If there is a function $h\colon \Omega \to \mathbb{D}$ so that

$$h'(z_0) = \sup\{|f'(z_0)| \mid f\colon \Omega \to \mathbb{D}\} \tag{8.1.24}$$

prove that $h(z_0) = 0$.

8. (a) Prove that

$$\sup\{\operatorname{Re} f'(0) \mid f\colon\ \colon \mathbb{D} \to \mathbb{D}\} = \sup\{|f'(0)| \mid f\colon \mathbb{D} \to \mathbb{D}\} = 1 \tag{8.1.25}$$

and that the unique maximizer is the function $f(z) = z$. (*Hint*: Use Problem 7 and the Schwarz lemma.)

(b) If $\Omega \subset \mathbb{C}$ is a simply connected proper domain, prove there is for each $z_0 \in \Omega$ a unique $f\colon \Omega \to \mathbb{D}$ maximizing $\operatorname{Re} f(z_0)$ and that is a bijection and so the (unique) maximizer. (*Hint*: Use the Riemann mapping theorem.)

8.2. Boundary Behavior of Riemann Maps

In this section, Ω will be a simply connected region different from $\mathbb{C}$ and $\varphi\colon \Omega \to \mathbb{D}$ a Riemann mapping that is an analytic bijection. A natural issue is to what extent we can extend φ to $\overline{\Omega}$, or if Ω is unbounded, $\overline{\Omega} \cup \{\infty\}$ as a map to $\overline{\mathbb{D}}$, and whether the extension remains a bijection. One cares also about the similar question of φ^{-1} extending to $\overline{\mathbb{D}}$. Of course, if φ extends to a bijection of $\overline{\Omega}$ and $\overline{\mathbb{D}}$, φ^{-1} extends to a bijection, but it can happen that φ^{-1} extends but the extension is no longer one-one (see Example 8.2.5, below).

This is a huge subject on which there are whole books or multiple chapters in others, so we'll only try to hit some of the main points here. We'll begin with a general result about limits, indicate the two main results about Jordan regions, and prove one result about Jordan regions with piecewise smooth boundaries, a result that provides an alternate to an argument in the next section. Finally, we'll say something, mainly through examples, about the variety of boundary behavior possible in the general case.

There is one general and also simple result:

Theorem 8.2.1. *If φ is an analytic bijection between a region, Ω, and $\mathbb{D}$, for every $\varepsilon > 0$, there is a compact K_ε in Ω so that $|\varphi(z)| > 1 - \varepsilon$ if $z \notin K_\varepsilon$. In particular, $\lim_{z \to \partial\Omega \cup \{\infty\}} |\varphi(z)| = 1$, and if $z_n \in \Omega$ obeys $z_n \to z_\infty \in \partial\Omega \cup \{\infty\}$ and $w_n \equiv \varphi(z_n) \to w_\infty$, then $w_\infty \in \partial\mathbb{D}$. Similarly, if $w_n \in \mathbb{D}$ has $|w_n| \to 1$, then $\varphi^{-1}(w_n)$ approaches $\partial\Omega \cup \{\infty\}$.*

Proof. Since φ^{-1} is continuous, $K_\varepsilon = \varphi^{-1}(\{z \mid |z| \le 1 - \varepsilon\})$ is compact. That implies the first result, which easily implies the others. □

This is tailor-made for using a reflection principle, so

Theorem 8.2.2. *Let φ be an analytic bijection of Ω, a region, and $\mathbb{D}$. Suppose γ is an analytic arc which is an edge of $\partial\Omega$ as defined in Theorem 5.6.1. Then φ has an analytic continuation to an entire neighborhood of $\gamma[(0,1)]$.*

Remark. In particular, if Ω has a real analytic boundary, then φ has an analytic continuation to a neighborhood of $\overline{\Omega}$.

Proof. This is immediate from Theorem 8.2.1 and the $\partial\mathbb{D}$ analog of Theorem 3.1.1. □

If $\psi \equiv \varphi^{-1}$ extends to a bijection, $\tilde{\psi}$, of $\overline{\mathbb{D}}$ and $\overline{\Omega}$, then $\tilde{\psi} \restriction \partial\mathbb{D}$ is a simple closed curve, so assuming Ω is bounded, it is a Jordan region, that is, the interior of a Jordan curve. Thus, a necessary condition for a bijective extension to exist is that Ω be the interior of a Jordan curve (or the exterior if we include ∞ in Ω). Remarkably, this necessary condition is also sufficient—a theorem of Osgood–Taylor and Carathéodory. There is also a theorem of Painlevé that if the boundary curve is C^∞, then φ is C^∞ up to the boundary as well. Since we will not prove them in this section, we'll only state the formal theorems of Carathéodory–Osgood–Taylor (COT), and Painlevé in the Notes. We'll prove Painlevé's theorem in Section 12.6 of Part 2B. We do, however, state and prove the following special case of the COT theorem that follows easily from Theorem 5.6.2:

Theorem 8.2.3. *Let Ω be a bounded, simply connected region whose boundary is a piecewise analytic curve which is also piecewise C^1, i.e., each segment is analytic with γ' having limits at the endpoints. Then any Riemann mapping $\varphi\colon \Omega \to \mathbb{D}$ extends to a continuous bijection $\tilde{\varphi}$ of $\bar{\Omega}$ to $\overline{\mathbb{D}}$.*

Remark. We emphasize that the corners can be cusps.

Proof. By Theorem 8.2.2, φ has an analytic continuation, $\tilde{\varphi}$, up to each C^1 piece of $\partial\Omega$, and as one goes around these arcs with Ω to the left, $\arg(\tilde{\varphi})$

increases. By Theorem 5.6.2, $\tilde{\varphi}$ is continuous at each corner, so $\tilde{\varphi}$ is continuous on all of $\partial\Omega$, and by the monotonicity (and the one-one nature of φ) one-one on $\partial\mathbb{D}$. For topological reasons, $\tilde{\varphi}$ must be onto $\partial\mathbb{D}$. $\square$

Both Theorem 8.2.3 and the more general COT theorem extend to slit domains. Let $\widetilde{\Omega}$ be a region. As noted in Section 5.6, $\Omega = \widetilde{\Omega}\setminus\operatorname{Ran}(\gamma)$, where γ is a Jordan curve with $\gamma(0) \in \partial\widetilde{\Omega}$ and $\gamma[(0,1]] \subset \widetilde{\Omega}$ is called $\widetilde{\Omega}$ with a slit removed. By using Theorem 5.6.3, one gets

Theorem 8.2.4. *Let $\widetilde{\Omega}$ be a region with a piecewise analytic, C^1, Jordan boundary as in Theorem 8.2.3. Let $\gamma_1, \dots, \gamma_\ell$ be piecewise analytic Jordan curves with disjoint ranges so that γ_j is analytic at each endpoint, $\gamma_j(0) \in \partial\widetilde{\Omega}$ are disjoint, and $\gamma_j[(0,1]) \in \widetilde{\Omega}$. Let $\Omega = \widetilde{\Omega} \setminus \bigcup_{j=1}^{\ell} \operatorname{Ran}(\gamma_j)$, $\widetilde{\Omega}$ with ℓ analytic slits removed. Let $\varphi\colon \Omega \to \mathbb{D}$ be a Riemann map and $\psi\colon \mathbb{D} \to \Omega$ its inverse. Then ψ has a continuous extension from $\overline{\mathbb{D}}$ to $\overline{\Omega}$ so ψ is two-one at points in $\gamma_j([0,1))$ and otherwise one-one. φ is continuous as one approaches $\partial\Omega \setminus \{\gamma_j(0)\}$ and at $\{\gamma_j(1)\}$ and continuous on $\gamma_j([0,1))$ if one approaches from one side of* $\operatorname{Ran}(\gamma_j)$ (see Figure 5.6.2).

Remark. Using the ideas in the proof of the COT theorem, the result extends to the case where $\widetilde{\Omega}$ is the interior of an arbitrary Jordan curve and $\gamma_1, \dots, \gamma_\ell$ are only required to be continuous.

Example 8.2.5. Let Ω be $\mathbb{C} \setminus (-\infty, 0]$. The map $z \to z^{1/2}$ (with $x^{1/2} > 0$ if $x \in \mathbb{R}$) maps Ω bijectively to $\mathbb{H}$ and then $z \to (z-1)/(z+1)$ maps $\mathbb{H}$ to $\mathbb{D}$, that is,

$$\varphi(z) = \frac{\sqrt{z}-1}{\sqrt{z}+1}, \qquad \varphi^{-1}(z) = \left(\frac{1+w}{1-w}\right)^2 \tag{8.2.1}$$

Clearly, φ^{-1} has a continuous extension, $\widetilde{\varphi^{-1}}$, to $\overline{\mathbb{D}}$ as a map to $\overline{\Omega} \cup \{\infty\}$. But $\widetilde{\varphi^{-1}}$ is not one-one, $\widetilde{\varphi^{-1}}(e^{i\theta}) = \widetilde{\varphi^{-1}}(e^{-i\theta})$. Put differently, φ has limits $\varphi(-x \pm i0) = \lim_{\varepsilon\downarrow 0} \varphi(-x \pm i\varepsilon)$ for $x > 0$, but the limits are different, so φ itself does not have a continuous extension. While we presented this map in a form that is especially simple, one could just as well have taken $\mathbb{D}\setminus(-1,0]$ to have a bounded region. $\square$

This example suggests that one focus on extending $\psi = \varphi^{-1}$ defined on $\mathbb{D}$ to Ω to a map of $\overline{\mathbb{D}}$ to $\overline{\Omega} \cup \{\infty\}$. That this is sensible is seen by the following that at least works if Ω is bounded. It is a theorem of Fatou (which we'll prove in Section 5.2 of Part 3) that for any bounded analytic function, ψ, on $\mathbb{D}$, $\lim_{r\uparrow 1} \psi(re^{i\theta})$ exists for Lebesgue a.e. $\theta \in [0, 2\pi)$. Here is an example that shows that this ψ may not always be onto $\overline{\Omega}$. Let Ω be a region. An *end-cut* in Ω is a Jordan arc, $\gamma\colon [0,1] \to \overline{\Omega}$ with $\gamma([0,1)) \subset \Omega$. A point, $z_0 \in \partial\Omega$, is called *accessible* if there is an end-cut with $\gamma(1) = z_0$.

Clearly, if ψ is a bijection of $\mathbb{D}$ to Ω and $\lim_{r\uparrow 1}\psi(re^{i\theta})$ exists, then $\psi(e^{i\theta})$ is an accessible point of Ω.

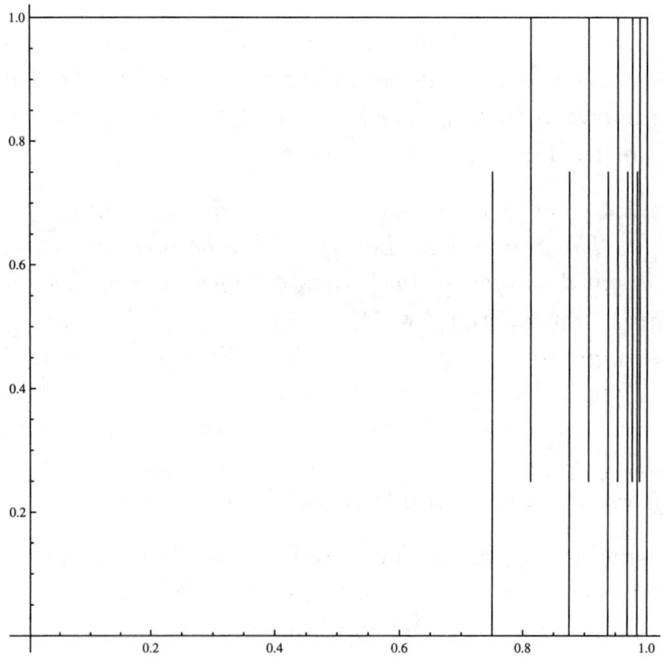

Figure 8.2.1. A region with inaccessible points.

Example 8.2.6. Let $\Omega=(0,1)\times(0,1)\setminus\bigcup_{n=1}^{\infty}\{(1-\frac{1}{2n},y)\cup\{1-\frac{1}{2n+1},1-y)\mid y\in(0,\frac{3}{4})\}$ (see Figure 8.2.1). Then it is not hard to see (Problem 1) that points of the form $(1,y)$ with $y\in[0,1]$ are in $\overline{\Omega}$ but not accessible. $\square$

Here is a final pathology where a point is accessible but in uncountably many ways!

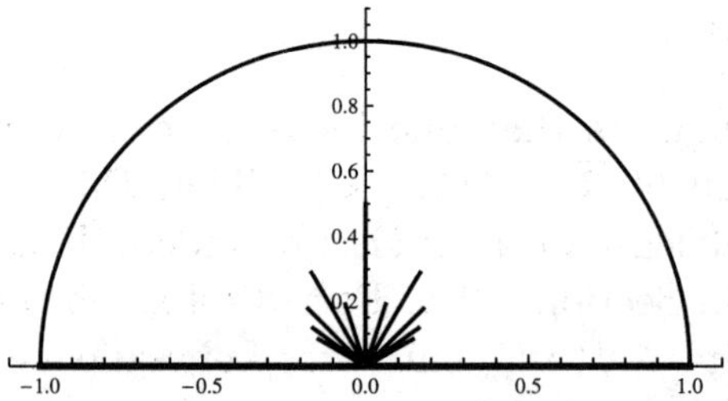

Figure 8.2.2. A region where a single boundary point corresponds to infinitely many points on $\partial\mathbb{D}$.

Example 8.2.7. Let $\Omega = (\mathbb{D} \cap \mathbb{C}_+) \setminus \bigcup_{q=2}^{\infty} \bigcup_{p=1}^{q-1} \{re^{i\pi p/q} \mid 0 < r < \frac{1}{q}\}$; see Figure 8.2.2. The reader will prove (see Problem 2) that if θ_0, θ_1 are distinct irrational multiples of π (in $(0, \pi)$) and φ is a Riemann map, then $\lim_{r\downarrow 0} |\varphi(re^{i\theta_0}) - \varphi(re^{i\theta_1})| \neq 0$. $\square$

There is a key notion for going further with extensions, called *prime ends*, but describing it in detail would take more space than we can invest here.

Notes and Historical Remarks. Because Riemann's original construction used regularity on the boundary, it has long been a focus of work. Already in 1887 in his thesis [**423**], Paul Painlevé (1863–1933), who served twice as prime minister of France and was minister of war during part of World War I, proved:

Theorem 8.2.8 (Painlevé's Smoothness Theorem). *Let γ be a C^∞ Jordan curve and Ω the bounded component of $\mathbb{C} \setminus \mathrm{Ran}(\gamma)$. Then for any $z_0 \in \Omega$, the unique Riemann map, f, with $f(z_0) = 0$, $f'(z_0) > 0$ extends to a bijection, $\widetilde{f}(z)$, of $\overline{\Omega}$ and $\overline{\mathbb{D}}$ and all derivatives of f extend continuously to $\overline{\Omega}$ and all derivatives of f^{-1} extend continuously to $\overline{\mathbb{D}}$.*

We'll prove this theorem in Section 12.6 of Part 2B.

The general Jordan curve result is due to Osgood–Taylor [**421**] and Carathéodory [**88**] about the same time, with alternate approaches by Courant [**124**] and Lindelöf [**358, 359**]. The theorem says that

Theorem 8.2.9 (Carathéodory–Osgood–Taylor Theorem). *Let γ be a Jordan curve and Ω the bounded component of $\mathbb{C} \setminus \mathrm{Ran}(\gamma)$. Then for any $z_0 \in \Omega$, the Riemann mapping $f\colon \Omega \to \mathbb{D}$, which is an analytic bijection with $f(z_0) = 0$, $f'(z_0) > 0$, extends to a continuous bijection $\widetilde{f}\colon \overline{\Omega} \to \overline{\mathbb{D}}$.*

For proofs, see Burckel [**80**] or Krantz [**331**, Ch. 5].

An interesting approach uses the notion of simple points. $z_\infty \in \partial\Omega$ is called a *simple point*, if whenever $\{z_n\}_{n=1}^{\infty}$ is a sequence of points in Ω with $z_n \to z_\infty$, there is a continuous curve γ with $\gamma([0,1)) \subset \Omega$, $\gamma(1 - 1/n) = z_n$ and $\gamma(1) = z_\infty$. It is a theorem (proven, e.g., in Ash–Novinger [**25**]) that Ω is bounded and simply connected, and if every point in $\partial\Omega$ is simple, then any Riemann map $f\colon \Omega \to \mathbb{D}$ has a unique continuous extension from $\overline{\Omega}$ to $\overline{\mathbb{D}}$ which is a bijection with continuous inverse.

For going beyond Jordan regions, there is extensive discussion in Markushevich [**372**] and Conway [**120**] as well as several books on the subject (see Bell [**38**], Ohtsuka [**413**], and Pommerenke [**451, 452**]).

The theory of prime ends goes back to Carathédory [89] in 1913. For modern expositions, see Ahlfors [8], Conway [121], Epstein [165], Mather [376], or Pommerenke [452].

Problems 4 and 5 indicate the applications of the Riemann mapping theorem to potential theory. Further developments concern the situation where Ω is a bounded and simply connected region so that $\{z \mid z = 0$ or $z^{-1} \in \mathbb{C} \setminus \Omega\}$ is connected and simply connected. One then defines an *exterior Green's function* by mapping $(\mathbb{C} \setminus \Omega) \cup \{\infty\}$ to $\mathbb{D}$. In particular, if $F\colon (\mathbb{C} \setminus \Omega) \cup \{\infty\} \to \mathbb{D}$ with $F(\infty) = 0$, $\text{Residue}(F, \infty) > 0$, then $\varphi(z) = \frac{1}{2\pi} \log(F(z)^{-1})$ is the *exterior potential* of Ω, and if φ is set to zero on Ω, the distributional derivative $\Delta\varphi$ is the equilibrium measure of Ω. It follows that if $F(z) = cz + O(1)$ near infinity, then c is the logarithmic capacity of Ω. For more on potential theory, see Ransford [471] or Landkof [343] or the discussion of Sections 3.6 and 3.8 of Part 3. Chapter 3 of Part 3 has several sections on potential theory.

The Green's function approach to constructing the Riemann mapping discussed in the Notes to Section 8.1 is ideal for local regularity results. If Ω is a Jordan region and f is a Riemann mapping with $f(z_0) = 0$, then Carathéodory's theorem implies that $G(z, z_0) = -\log|f(z)|$ is a classical Green's function. If now the boundary is smooth, say C^∞, near some point, ζ_0, on $\partial\Omega$, PDE methods can be used to prove that G is C^∞ near ζ_0. This shows the inverse map $g = f^{-1}$ has $|g|$, C^∞ near $f(\zeta_0)$, and then conjugate harmonic function arguments prove g (and so f) is C^∞ near $f(\zeta_0)$ (respectively, ζ_0).

Problems

1. In Example 8.2.6, prove that points of the form $(1, y)$, $0 \le y \le 1$, are not accessible.

2. Prove the assertion at the end of Example 8.2.7.

3. Prove that the set of accessible points is dense in $\partial\Omega$.

4. Let Ω be a bounded, simply connected region, and given $z_0 \in \Omega$, let f_{z_0} be the Riemann map to $\mathbb{D}$ with $f_{z_0}(z_0) = 0$, $f'_{z_0}(z_0) > 0$. Define $G_\Omega(z, z_0) = -\frac{1}{2\pi} \log|f_{z_0}(z)|$. Prove that $G_\Omega(\,\cdot\,, z_0)$ is harmonic in $\Omega \setminus \{z_0\}$, obeys $\lim_{|z| \to \partial\Omega} G_\Omega(z, z_0) = 0$, and that in the distributional sense, $\Delta_z G_\Omega(z, z_0) = -\delta(z - z_0)$.

   Remark. This is the *classical Green's function* for Ω.

5. Let Ω be a bounded, simply connected region. Let $f_{z_0}(z)$ and $G_\Omega(z, z_0)$ be as in Problem 4. Suppose $I \subset \partial\Omega$ is an open arc on which $f_{z_0}(z)$ is C^1 up to $\partial\Omega$. Let g be a continuous function on I whose support lies a

finite distance from ∂I. Prove that

$$h(z_0) = \int_I g(z) \frac{\partial G}{\partial n} (z, z_0) \, d|z| \tag{8.2.2}$$

is harmonic in Ω and obeys $\lim_{z_0 \to z_\infty} h(z_0) = g(z_0)$ for any $z_0 \in \partial\Omega$.

8.3. First Construction of the Elliptic Modular Function

Our main goal in this section will be to use the Riemann mapping theorem to construct a function with miraculous properties:

Theorem 8.3.1. *There exists a function λ on $\mathbb{C}_+$ so that*

(i) *λ is onto $\mathbb{C} \setminus \{0, 1\}$.*
(ii) *$\lambda'(z) \neq 0$ for all z.*
(iii) *There exists a group, $\Gamma(2)$, of elements of $\mathbb{A}\mathrm{ut}(\mathbb{C}_+)$ so that*

$$\lambda(z) = \lambda(w) \Leftrightarrow \exists \gamma \in \Gamma(2) \text{ so that } \gamma(z) = w \tag{8.3.1}$$

λ is called the *elliptic modular function.* The reasons for the name and calling the group $\Gamma(2)$ will be discussed in the Notes and in Section 10.6, where we will give another construction of λ. The structure of the group as a Fuchsian group (aka discontinuous group) will be discussed at the end of this section and in the Notes.

Calling this function "the" elliptic modular function may seem strange because it is clearly not unique: If γ_0 is any element of $\mathbb{A}\mathrm{ut}(\mathbb{C}_+)$, then $\lambda \circ \gamma_0$ also has the properties of λ (the group is just $\gamma_0^{-1}\Gamma(2)\gamma_0$). In fact, as we'll see (see Problem 5 of Section 8.5), every such function is related to λ in this way, so λ is "essentially" unique. Our choice is normalized by the requirement that $\lim_{\mathrm{Im}\, z \to \infty} \lambda(z) = 0$, $\lim_{\mathrm{Im}(-z^{-1}) \to \infty} \lambda(z) = 1$, $\lim_{\mathrm{Im}(-(1-z)^{-1}) \to \infty} \lambda(z) = \infty$.

λ may seem to be very specialized, but we'll see that it is the key to one of the proofs of Picard's theorems (see Section 11.3) and we'll also use it in Section 8.5.

We begin with the region

$$\mathcal{F}^\sharp = \{\tau \mid \mathrm{Im}\, \tau > 0, \, 0 < \mathrm{Re}\, \tau < 1, \, |\tau - \tfrac{1}{2}| > \tfrac{1}{2}\} \tag{8.3.2}$$

the shaded region on the right side of Figure 8.3.1. Bearing in mind that under FLTs, circles and lines are equivalent, this is like a triangle. Indeed, for reasons we discuss in the Notes, we'll call a region in $\mathbb{C}_+$ which is bounded by three parts of lines or circles, each of which, when extended, is orthogonal to $\mathbb{R}$, a *hyperbolic triangle.* The ones we'll discuss here, like $\mathcal{F}^\sharp$, are unusual in that all vertices are cusps, that is, lie on $\partial(\mathbb{C}_+ \cup \{\infty\})$, and have zero internal angle.

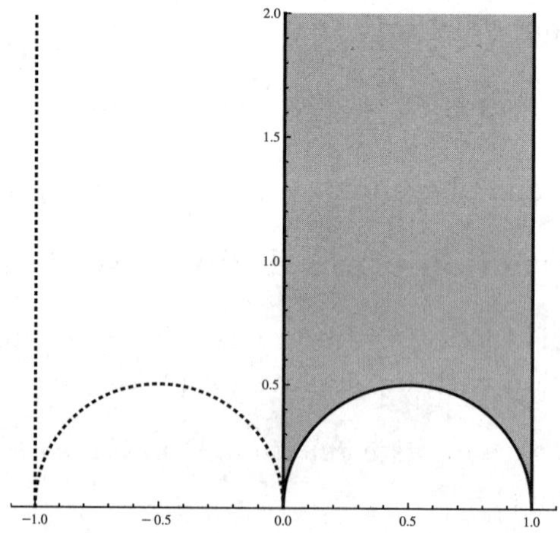

Figure 8.3.1. The regions $\mathcal{F}^\sharp$ and $\widetilde{\mathcal{F}}$.

$\mathcal{F}^\sharp$ is clearly simply connected (e.g., it is an image under an FLT of a symmetric triangle in $\mathbb{D}$ which is star-shaped; see the Notes and Figure 8.3.5), so there are Riemann maps from $\mathcal{F}^\sharp$ to $\mathbb{D}$ and then biholomorphically to $\mathbb{C}_+$ (since $z \mapsto i(1+z)/(1-z)$ maps $\mathbb{D}$ to $\mathbb{C}_+$). By the reflection principle (see Theorems 5.5.1 and 5.5.4), these Riemann maps have analytic continuations through the three edges of $\mathcal{F}^\sharp$, monotone along each edge, so if we can show these maps are continuous at each of the three corners, we get a bijection of $\overline{\mathcal{F}^\sharp} \cup \{\infty\} \to \overline{\mathbb{C}}_+ \cup \{\infty\}$ analytic on $\overline{\mathcal{F}^\sharp} \setminus \{0,1\}$ (and so the derivative is nonvanishing there). Because each pair of these Riemann maps, $\{\lambda_j\}_{j=1}^2$, are related by $\lambda_1 = g \circ \lambda_2$, where g is an FLT of $\overline{\mathbb{D}}$ to $\overline{\mathbb{D}}$, it suffices to prove one is continuous at each corner.

Moreover, if

$$f(z) = \frac{1}{1-z} \tag{8.3.3}$$

then f maps $\{0,1,\infty\}$ to $\{1,\infty,0\}$. Since it maps $\mathbb{R}\cup\{\infty\}$ to itself, it maps lines and circles orthogonal to $\mathbb{R}$ to other lines and circles orthogonal to $\mathbb{R}$. Thus, f maps $\mathcal{F}^\sharp$ to itself and ∞ to 0 and f^2 also maps $\mathcal{F}^\sharp$ to itself and ∞ to 1. Therefore, it suffices to show one Riemann mapping λ of $\mathcal{F}^\sharp$ to $\mathbb{C}_+$ is continuous at infinity to get continuity at all three corners. (For the image of $\mathcal{F}^\sharp$ in $\mathbb{D}$ shown in Figure 8.3.5, f is just rotation about 0 by angle $\frac{2\pi}{3}$.)

Let

$$F(z) = -\cos(\pi z) \tag{8.3.4}$$

Then F maps $\Omega \equiv \{z \mid \operatorname{Im} z \ge 0,\, 0 \le \operatorname{Re} z \le 1\}$ to $\overline{\mathbb{C}}_+$ and is an analytic bijection on Ω^{int}. For F maps $[0,1]$ monotonically to $[-1,1]$, $F(iy) = -\cosh(y)$ and $F(1+iy) = -F(iy) = \cosh(y)$, so F maps $[i\infty, 0]$ to

$[-\infty, -1]$, and $[1, 1+i\infty]$ to $[1, \infty]$. By the argument principle, it maps Ω^{int} to $\mathbb{C}_+$. Moreover, it is easy to see $\lim_{|z|\to\infty, z\in\Omega} F(z) = \infty$.

F is not conformal at the corners of Ω. At 0 and 1, F has a double zero so angles double and tangent curves remain tangent. But F has an essential singularity at ∞, and so it can open the cusp out. Indeed, if $G(w) = 1/F(w)$, $G[\mathcal{F}^\sharp]$ has a continuous extension to closures and ∞ goes to $0 \in [-1, 1]$.

λ is a composition of G with a Riemann map of $G[\mathcal{F}^\sharp]$ to $\mathbb{C}_+$ which, by the reflection principle, is continuous at 0, showing λ is continuous at ∞. Note we could have used Theorem 8.2.3 modified via fractional linear transformation to handle infinity to get continuity at the cusp, but we can be explicit here, so we have been.

We claim we can pick the Riemann map extended to closures $\widetilde{\lambda}\colon \overline{\mathcal{F}^\sharp} \cup \{\infty\} \to \overline{\mathbb{C}}_+ \cup \{\infty\}$ so that $\widetilde{\lambda}(\infty) = 0$, $\widetilde{\lambda}(0) = 1$, and $\widetilde{\lambda}(1) = \infty$. For pick any Riemann map, f, extended and follow it with the FLT that takes $f(\infty)$ to 0, $f(0)$ to 1, and $f(1)$ to ∞. Since $0, 1, \infty$ are on $\mathbb{R}$, this FLT takes $\mathbb{R}$ to $\mathbb{R}$, and by tracking orientation (Problem 2), it takes $\mathbb{C}_+$ to $\mathbb{C}_+$. Thus, by restricting $\widetilde{\lambda}$ to $\overline{\mathcal{F}^\sharp} \cap \mathbb{C}_+$, we have proven:

Theorem 8.3.2. *There exists a map $\lambda\colon \overline{\mathcal{F}^\sharp} \setminus \{0,1\}$ to $\overline{\mathbb{C}}_+ \setminus \{0,1\}$ analytic on $\mathcal{F}^\sharp$ and continuous on $\overline{\mathcal{F}^\sharp} \setminus \{0,1\}$ so that λ is a bijection from $\overline{\mathcal{F}^\sharp} \setminus \{0,1\}$ to $\overline{\mathbb{C}}_+ \setminus \{0,1\}$. Moreover, λ is real-valued on $\partial\mathcal{F}^\sharp \setminus \{0,1\}$, with an analytic continuation to a neighborhood of this boundary with $\lambda'(z) \neq 0$ at any point of $\overline{\mathcal{F}^\sharp} \setminus \{0,1\}$ including the boundary.*

Now consider

$$\widetilde{\mathcal{F}} = \{\tau \mid \operatorname{Im}\tau > 0,\ -1 < \operatorname{Re}\tau \le 1,\ |\tau - \tfrac{1}{2}| \ge \tfrac{1}{2},\ |\tau + \tfrac{1}{2}| > \tfrac{1}{2}\} \tag{8.3.5}$$

shown in Figure 8.3.1—the shaded region plus the region within the dotted line. It is $\mathcal{F}^\sharp \cup -\overline{\mathcal{F}^\sharp} \cup \{\tau \mid |\tau - \frac{1}{2}| = \frac{1}{2}, \operatorname{Im}\tau > 0\} \cup \{\tau \mid \operatorname{Im}\tau > 0, \operatorname{Re}\tau = 0, 1\}$. (Here the bar in $\overline{\mathcal{F}^\sharp}$ is complex conjugate, not closure.) Since λ is real on $\operatorname{Re}\tau = 0$, by the reflection principle, it can be continued to all of $\widetilde{\mathcal{F}}$. Moreover, because $\lambda \restriction \mathcal{F}^\sharp$ is a bijection to $\mathbb{C}_+$ and

$$\lambda(-\bar{z}) = \overline{\lambda(z)} \tag{8.3.6}$$

$\lambda \restriction -\overline{\mathcal{F}^\sharp}$ is a bijection to $\mathbb{C}_-$. Since λ is a bijection on

$$i(0,\infty) \cup \{\tau \mid |\tau - \tfrac{1}{2}| = \tfrac{1}{2}\} \cup 1 + i(0,\infty) \tag{8.3.7}$$

to $(0,1) \cup (1,\infty) \cup (-\infty, 0)$, we have

Theorem 8.3.3. *λ extends to a bijection of $\widetilde{\mathcal{F}}$ to $\mathbb{C} \setminus \{0,1\}$ analytic on $\widetilde{\mathcal{F}}^{\text{int}}$ and in a neighborhood of $\widetilde{\mathcal{F}}$ and $\lambda'(z) \neq 0$ for all $z \in \widetilde{\mathcal{F}}$.*

We want to use the fact that λ is real on the set in (8.3.7) to reflect in these three curves repeatedly and show this allows continuation of λ to all of $\mathbb{C}_+$. Let $\mathcal{R}_0, \mathcal{R}_{1/2}, \mathcal{R}_1$ be reflection in each of these curves, that is,

$$\mathcal{R}_0(z) = -\bar{z}, \qquad \mathcal{R}_{1/2}(z) = \tfrac{1}{2} + \frac{\frac{1}{4}}{\bar{z} - \frac{1}{2}}, \qquad \mathcal{R}_1(z) = 2 - \bar{z} \tag{8.3.8}$$

Let

$$S_n = \{\tau \mid \operatorname{Im}\tau > 0,\ n \le \operatorname{Re}\tau \le n+1\} \tag{8.3.9}$$

Notice that $\mathcal{R}_0$ maps S_0 to S_{-1} and that

$$\mathcal{R}_1(\mathcal{R}_0(z)) = 2 + z \tag{8.3.10}$$

so $(\mathcal{R}_1 \circ \mathcal{R}_0)^n \colon S_0 \cup S_{-1} \to S_{2n} \cup S_{2n-1}$, so we need to show we can analytically continue to all of S_0.

Define C_n and R_n inductively as follows: C_0 is the shaded region in Figure 8.3.1, that is, $\overline{\mathcal{F}^\sharp} \cap \mathbb{C}_+$. $\mathcal{R}_0 \equiv S_0 \setminus C_0$, that is, one semidisk. C_0 is a single hyperbolic triangle. Reflect each of its linear edges in the semicircular edge to get a single hyperbolic triangle with edges $\{\tau \mid \operatorname{Im}\tau > 0, |\tau - \frac{1}{2}| = \frac{1}{2}\}$, $\{\tau \mid \operatorname{Im}\tau > 0, |\tau - \frac{1}{4}| = \frac{1}{4}\}$, $\{\tau \mid \operatorname{Im}\tau > 0, |\tau - \frac{3}{4}| = \frac{1}{4}\}$. The largest shaded region in Figure 8.3.2 is the interior of this triangle plus the two smaller semicircles, and we'll call that C_1,

$$R_n = R_{n-1} \setminus C_n \tag{8.3.11}$$

Now take the triangle C_1 and reflect in each of the smaller edges. We get two hyperbolic triangles, $C_2^{(1)}$, $C_2^{(2)}$, shown as the two largest bounded

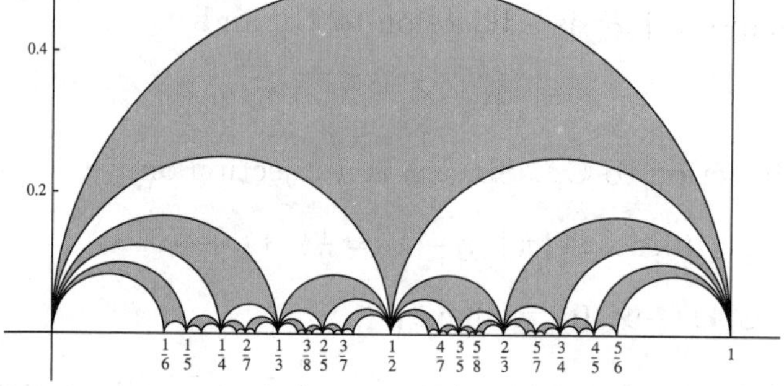

Figure 8.3.2. Six generations of reflected hyperbolic triangles.

unshaded regions in Figure 8.3.2. Repeat this process to get at level n, 2^{n-1} regions $C_n^{(j)}$, $j = 1, \dots, 2^{n-1}$. Figure 8.3.2 shows these regions for $n = 1, 2, \dots, 6$. Our goal is to prove that $\bigcup_{n=0}^{\infty} R_n = S_0$. In this regard, we need

Lemma 8.3.4. (a) *Let $d_{n-1}, d_n, d_{n+1}^{\pm}$ be the diameters of three successive generations: d_n, one of two triangles in d_{n-1} and $d_{n+1}^{\pm}$, the two inside the semicircle of diameter d_n* (see Figure 8.3.3), chosen so that $d_{n+1}^+ > d_{n+1}^-$. Then

$$d_{n+1}^{\pm} = \tfrac{1}{2}\, d_n \pm \frac{(\frac{1}{2}d_n)^2}{d_{n-1} - \frac{1}{2}d_n} \tag{8.3.12}$$

(b) If $x_n \equiv d_n/d_{n-1}$ and $x_{n+1}^{\pm} \equiv d_{n+1}^{\pm}/d_{n-1}$, then

$$x_{n+1}^+ = \frac{1}{2 - x_n} \qquad x_{n+1}^- = \frac{1 - x_n}{2 - x_n} \tag{8.3.13}$$

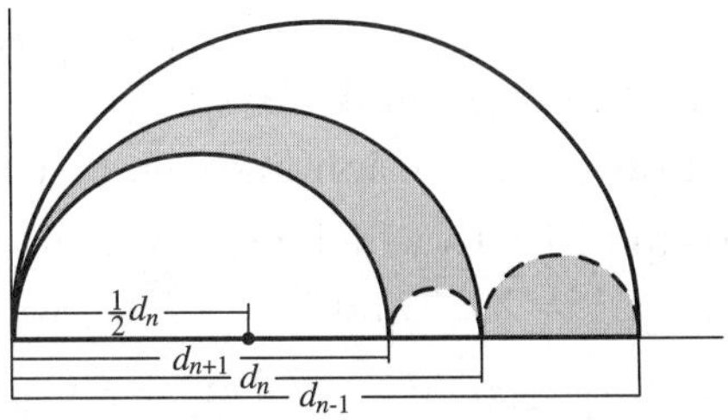

Figure 8.3.3. Three generations of semicircles.

Proof. (a) To get the dividing point between the two generation-$(n+1)$ circles, one must reflect the point in the generation-$(n-1)$ circle, not on the generation-n circle in that circle; in the figure, that is the rightmost point in the second largest circle. The distance of that point in ∂C_{n-1} from the center of the dimension-n circle is $d_{n-1} - \frac{1}{2}d_n$, so the reflected point is a distance $(\frac{1}{2}d_n)^2/(d_{n-1} - \frac{1}{2}d_n)$ from the center, which immediately implies (8.3.12).

(b) (8.3.13) follows from (8.3.12) by straightforward algebra. $\square$

Theorem 8.3.5. *R_n has 2^n circles with maximal diameter exactly* $1/(n+1)$.

Remarks. 1. This is well-illustrated in Figure 8.3.2.

2. In fact, it follows from the more detailed analysis in Problem 6 except for the circles at the extreme, the maximal diameter is bounded by $1/2(n+1)$.

Proof. We first claim that $x_n \le n/(n+1)$ and for all choices plus, we have equality. For this holds when $n = 1$ ($d_0 = 1$, $d_1 = \frac{1}{2}$) and if it holds for n, by (8.3.13),

$$x_{n+1}^- \le x_{n+1}^+ \le \frac{1}{2-\frac{n}{n+1}} = \frac{n+1}{n+2} \tag{8.3.14}$$

This shows that if $x_n = n/(n+1)$, then $x_{n+1} = (n+1)/(n+2)$, so for all plus, that is, left-most disk, we have all $x_j = j/(j+1)$. Then, with $d_0 = 1$,

$$d_n = x_n x_{n+1} \dots x_1 1 \le \frac{n}{n+1}\,\frac{n-1}{n} \cdots \frac{2}{3}\,\frac{1}{2}\,1 = \frac{1}{n+1}$$

with equality on the left-most disk. $\square$

Proof of Theorem 8.3.1. By induction, we can extend λ to each $C_n \subset S_0$ since it is real inductively on ∂R_{n-1} and the 2^{n-1} hyperbolic triangles in C_n are each the reflection of a hyperbolic triangle in C_{n-1} in a circle.

By Theorem 8.3.5, $\bigcup_{j=0}^n C_n \supset \{z \in S_0 \mid \operatorname{Im} z > 1/(n+1)\}$, so $\bigcup_{j=1}^\infty C_n = S_0$, which says we can analytically continue λ to S_0. Then $\mathcal{R}_0$ defines λ on S_1 and $(\mathcal{R}_1 \circ \mathcal{R}_0)^n$ to all of $\mathbb{C}_+$. Since λ is obtained from $\lambda \restriction \widetilde{\mathcal{F}}$ by reflection and $\mathbb{C} \setminus \{0,1\}$ is conjugation invariant, we see (i) of the theorem, and (ii) follows from the definition via reflection and the fact that $\lambda' \neq 0$ on $\widetilde{\mathcal{F}}$.

Let $\Gamma(2)$ be the group of FLTs which are products of an even number of $\mathcal{R}_0, \mathcal{R}_{1/2}, \mathcal{R}_1$. Since $\lambda(\mathcal{R}_j z) = \overline{\lambda(z)}$ and $\mathcal{R}_j \equiv \overline{\mathcal{R}_j z}$, we see that

$$\lambda(\gamma(z)) = \lambda(z) \qquad \text{for all } \gamma \in \Gamma(2) \tag{8.3.15}$$

To begin the converse, we need to note a property of $\Gamma(2)$: for any $w \in \mathbb{C}_+$, we claim there is $\gamma \in \Gamma(2)$ and $z \in \widetilde{\mathcal{F}}$, so $w = \gamma(z)$. Since $\Gamma(2)$ includes $z \to z+2$ (8.3.10), there is $z_1 \in S_0 \cup S_{-1}$ and γ_0, so $w = \gamma_0(z_1)$. Suppose $z_1 \in S_0$. Then by the construction of the region C_j and the fact we proved $\bigcup_{j=1}^\infty C_j = S_0$, there are reflections $r_1, \dots, r_\ell$ in circles in ∂R_k for some k so that $r_1 \dots r_\ell(z_1) \in C_0$.

Since each r_j is a product of $\mathcal{R}_0$, $\mathcal{R}_{1/2}$, or $\mathcal{R}_1$ (Problem 3), we can find $\mathcal{R}^{(1)}, \dots, \mathcal{R}^{(m)}$, each one of the three generators, so $\mathcal{R}^{(1)} \dots \mathcal{R}^{(m)}(z_1) \in C_0 \subset \widetilde{\mathcal{F}}$. If $z_1 \in S_{-1}$ by looking at $\mathcal{R}_0(z_1)$, we can do the same, so this is true for any $z_1 \in S_0 \cup S_{-1}$. If m is odd, look at (see Remark 1 below) $\mathcal{R}_0\mathcal{R}^{(1)} \dots \mathcal{R}^{(m)}(z_1) \in \mathcal{R}_0(C_1) \subset \widetilde{\mathcal{F}}$. Thus, there is $\gamma_2 \in \Gamma(2)$ and $z \in \widetilde{\mathcal{F}}$, so $\gamma_2(z_1) = z$. It follows that if $\gamma = \gamma_0\gamma_2^{-1}$, then $\gamma(z) = w$ as claimed.

Now suppose $\lambda(w) = \lambda(z)$. Find $z_1, z_2 \in \widetilde{\mathcal{F}}$, $\gamma_1, \gamma_2 \in \Gamma(2)$, so $w = \gamma_1(z_1)$, $z = \gamma_2(z_2)$. Then invariance of λ implies $\lambda(z_1) = \lambda(z_2)$. Since λ is bijective on $\widetilde{\mathcal{F}}$, $z_1 = z_2$ and $w = \gamma_1\gamma_2^{-1}(z)$. $\square$

Remarks. 1. There was a small cheat in the above proof. If $z \in \{\tau \mid \operatorname{Im}\tau > 0, \operatorname{Re}\tau = 1 \text{ or } |\tau - \frac{1}{2}| = \frac{1}{2}\}$, $\mathcal{R}_0(z)$ will not be in $\widetilde{\mathcal{F}}$. But $\mathcal{R}_1(z)$ or $\mathcal{R}_{1/2}(z)$ will be and we can use that.

2. The proof shows that for each w, there is a unique $z \in \mathcal{F}$ with $w = \gamma(z)$ for some $\gamma \in \Gamma(2)$ and it is not hard to see (Problem 1) that γ is unique also.

Finally, for our needs later, we want to say something about the class of groups to which $\Gamma(2)$ belongs.

Definition. A *Fuchsian group*, Γ, is a group of FLTs that all leave fixed the same disk or half-plane and which is discrete in that for each $R > 0$, $\{T \in \mathbb{SL}(2, \mathbb{C}) \mid f_T \in \Gamma, \, \|T\| \leq R\}$ is finite.

Lemma 8.3.6. *Let $T \in \mathbb{SU}(1,1)$. Then with $\|(\begin{smallmatrix} \alpha & \beta \\ \gamma & \delta \end{smallmatrix})\|_{\mathrm{HS}}^2 = |\alpha|^2 + |\beta|^2 + |\gamma|^2 + |\delta|^2$, we have*

$$(1 - |f_T(0)|)^{-1} \leq 2\|T\|_{\mathrm{HS}}^2 \leq 16(1 - |f_T(0)|)^{-1} \tag{8.3.16}$$

Remark. $\|\cdot\|_{\mathrm{HS}}$ is the Hilbert–Schmidt norm. Of course, all norms on 2×2 matrices are comparable.

Proof. $T = (\begin{smallmatrix} \alpha & \gamma \\ \bar{\gamma} & \bar{\alpha} \end{smallmatrix})$ with $|\alpha|^2 - |\gamma|^2 = 1$. Thus, $f_T(0) = \gamma/\bar{\alpha}$ and

$$(1 - |f(0)|^2) = \frac{|\alpha|^2 - |\gamma|^2}{|\alpha|^2} = \frac{1}{|\alpha|^2} \tag{8.3.17}$$

This plus

$$1 - |f(0)| \leq 1 - |f(0)|^2 = (1 - |f(0)|)(1 + |f(0)|) \leq 2(1 - |f(0)|)$$

plus (since $|\gamma|^2 = |\alpha|^2 - 1 \leq |\alpha|^2$)

$$|\alpha|^2 \leq \|T\|_{\mathrm{HS}}^2 \leq 4|\alpha|^2$$

implies (8.3.16). □

Theorem 8.3.7. *Let Γ be a group of FLTs leaving some disk, Ω, fixed. Then the following are equivalent:*

(1) *Γ is Fuchsian.*
(2) *For one $z_0 \in \Omega$, $\{\gamma(z_0) \mid \gamma \in \Gamma\}$ is discrete in Ω, that is, it has no limit point in Ω.*
(3) *For all $z_0 \in \Omega$, $\{\gamma(z_0) \mid \gamma \in \Gamma\}$ is discrete in Ω.*

Proof. Since all disks are equivalent under an FLT, without loss we can suppose $\Omega = \mathbb{D}$ and $\Gamma \subset \mathbb{PSU}(1,1)$. By conjugation, we can also suppose $z_0 = 0$ (conjugation doesn't preserve norms, but since $\|\gamma_0 \gamma \gamma_0^{-1}\| \leq \|\gamma_0\| \, \|\gamma_0^{-1}\| \, \|\gamma\|$, discreteness of Γ and one of its conjugates is equivalent).

The lemma immediately implies (1) $\Leftrightarrow$ (2). Clearly, (3) $\Rightarrow$ (2), and by the conjugacy argument, (1) $\Rightarrow$ (2) for $z_0 = 0$ means (1) $\Rightarrow$ (2) for any z_0, and so (3). □

Corollary 8.3.8. *$\Gamma(2)$ is a Fuchsian group.*

Proof. Since λ is not identically constant, for any z_0, $\{z \mid \lambda(z) = \lambda(z_0)\}$ is discrete. But that set is $\{\gamma(z_0) \mid \gamma \in \Gamma(2)\}$, so orbits are discrete. $\square$

Notes and Historical Remarks. $\Gamma(2)$ will appear again in Section 10.2 where it will be defined as the set of elements of $\mathbb{SL}(2,\mathbb{Z})$ (invertible matrices with integral coefficient) where $T = \left(\begin{smallmatrix} \alpha & \beta \\ \gamma & \delta \end{smallmatrix}\right)$ with α, δ odd and β, γ even. In Problem 2 of Section 10.2, the reader will prove that $\Gamma(2)$, defined in that manner, is generated by the FLTs associated to $\pm\left(\begin{smallmatrix} 1 & 2 \\ 0 & 1 \end{smallmatrix}\right)$ and $\left(\begin{smallmatrix} 1 & 0 \\ 2 & 1 \end{smallmatrix}\right)$, and in Problem 3 of the current section that the group of even numbers of reflections is generated by $f_1(z) = z + 2$ and $f_2(z) = z/(2z+1)$.

It is easy to see (Problem 7) that λ has a natural boundary on $\mathbb{R}$.

There are remarkable number-theoretic properties of the points of intersection of the semicircles in ∂C_n with $\mathbb{R}$. In particular, they are all rational (as is easy to see inductively from (8.3.12)) and, moreover, which is not as easy to see, every rational in $(0,1)$ occurs; see Problem 6. If $p/q < p'/q'$ are two successive points at a level n, written in lowest rational form (i.e., ends of a semicircle in ∂C_n), then

$$p'q - pq' = 1 \tag{8.3.18}$$

and if p''/q'' is the point in between them at level $n+1$, then (see Problem 5)

$$p'' = p + p', \qquad q'' = q + q' \tag{8.3.19}$$

(which proves (8.3.18) inductively if we start with $p = 0$, $q = p' = q' = 1$). This also shows inductively that the first circle has edges $0/1$ and $1/n$. p''/q'' is called the *mediant* of p/q and p'/q'.

John Farey (1766–1826) was a British geologist who noted in 1816 [**185**] that if one wrote down all fractions in $[0,1]$ with denominator at most n in order, then any three successive ones obeyed (8.3.19). As he put it:

> If all the possible vulgar fractions of different values, whose greatest denominator (when in their lowest terms), does not exceed any given number, be arranged in the order of their values or quotients; then if both the numerator and denominator of any fraction therein be added to the numerator and denominator respectively, of the fraction next but one to it (on either side) the sums will give the fraction next to it; although perhaps not in its lowest terms.

This had been noted fourteen years earlier by Haros [**241**] (who actually, unlike Farey, proved it!), but Cauchy learned the result from Farey's note and named the series, Farey series—and the name stuck! Our sequence at level n is not even the full Farey series, but a subset that preserves (8.3.19). Nevertheless, the cover of $\mathbb{C}_+$ by hyperbolic triangles that we study here has extensively been called the Farey tesselation. The Farey tesselation will

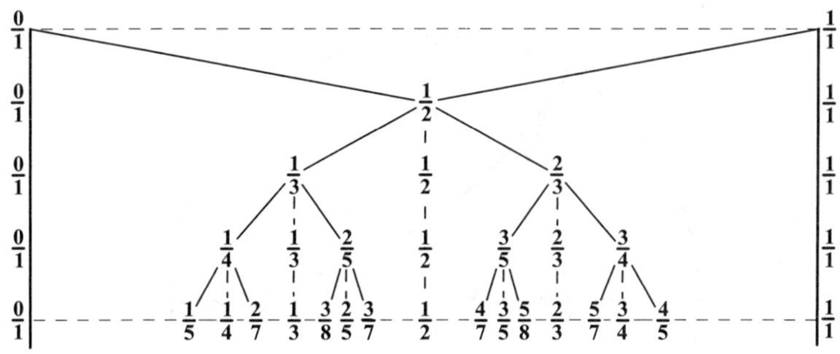

Figure 8.3.4. The Stern–Brocot tree.

appear again in the Notes to Section 2.8 of Part 3 in the context of the relation of geodesic flow and ergodic properties of continued fractions.

Related to this is a binary tree formed by singling out the new points at level n and joining them to their level $n-1$ neighbor; see Figure 8.3.4. This is the *Stern–Brocot tree.*

Moritz Stern (1807–94) was Gauss' replacement and noted for being the first Jewish German professor not required to convert for his position. Achille Brocot (1817–78) was a French clockmaker. Stern [**538**] was motivated by continued fraction considerations and Brocot [**76**] by sequences of gears! In terms of continued fractions, the children of $[a_1, \dots, a_n]$ with $a_n \geq 2$ are $[a_1, \dots, a_n + 1]$ and $[a_1, \dots, a_n - 1, 2]$. (Since $[a_1, \dots, a_j, 1] = [a_1, \dots, a_j{+}1]$, every rational in $(0, 1)$ has a continued fraction representation not ending in 1; see (7.5.6).) For the history and connection to clock-making, see Austin [**26**] or Hayes [**247**]. For more on number-theoretic aspects of Farey tesselations, see [**63, 214, 240, 264, 288, 306, 517**]. Rademacher [**464**] and Bonahon [**63**] discuss the connection to the Ford circles discussed in the Notes to Section 7.5 and below. For other textbook constructions of the elliptic modular function as a Riemann map extended by reflections, see Ullrich [**559**] and Veech [**562**].

Ford [**192**] looked at disks in $\mathbb{C}_+$ tangent to $\mathbb{R}$ at a rational p/q (in lowest order) of diameter $1/q^2$. Remarkably, the interiors of these disks are all disjoint and two, associated with $p_1/q_1 < p_2/q_2$, have touching boundaries if and only if $p_2q_1 - q_2p_1 = 1$. In that case, the only Ford circle tangent to these two is the one associated with the mediant, $(p_1 + p_2)/(q_1 + q_2)$; see Rademacher [**464**]. Rademacher considered this in connection with his work on partitions, the subject of Example 15.4.7 of Part 2B.

As we'll see in Section 12.2 of Part 2B, geodesics in the hyperbolic metric on $\mathbb{C}_+$ are precisely images under the FLT from $\mathbb{D}$ onto $\mathbb{C}_+$ of diameters of $\mathbb{D}$. It is not hard to see these are precisely those circles that intersect

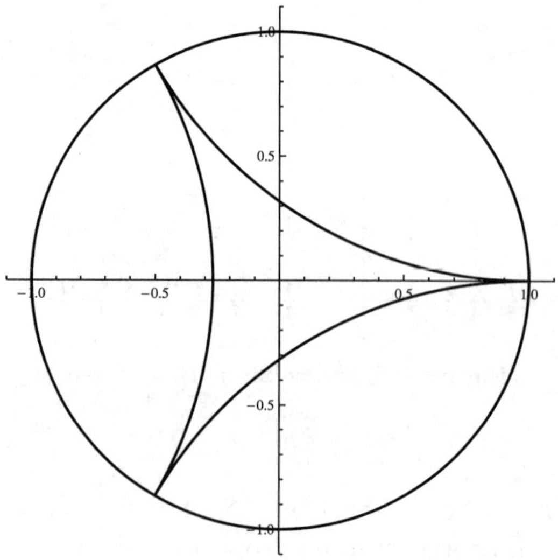

Figure 8.3.5. An image of $\mathcal{F}^\sharp$ in $\mathbb{D}$.

$\mathbb{R}$ orthogonally. Thus, our hyperbolic triangles have three sides which are hyperbolic geodesics.

$\mathcal{F}^\sharp$ doesn't look very symmetric, but Figure 8.3.5 shows an image in $\mathbb{D}$ under an FLT that takes $0, 1, \infty$ to the cube roots of unity.

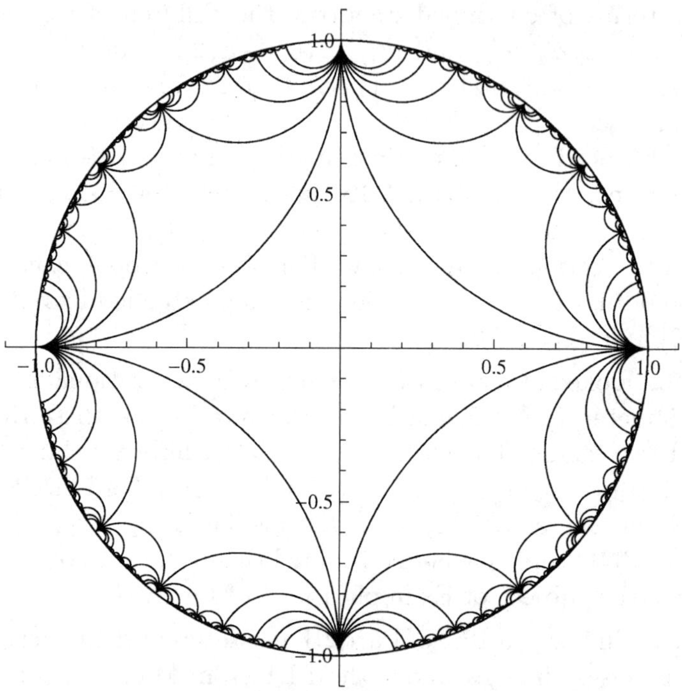

Figure 8.3.6. An image of $\widetilde{\mathcal{F}}$ and its reflections in $\mathbb{D}$.

Figure 8.3.6 shows the image of $\widetilde{\mathcal{F}}$ under a map that takes $0, 1, \infty$ to $-i, 1, i$ (and -1 to -1) and the action of the image of $\Gamma(2)$ under that map. These figures are very symmetric tesselations of $\mathbb{D}$. For a lovely online gallery of hyperbolic tesselations, see [**243**].

Fuchsian groups are also called *discontinuous groups* because of the distinction from continuous orbits under continuous groups. A discrete group of $\mathbb{SL}(2, \mathbb{C})$ matrices that doesn't leave a disk fixed is called a *Kleinian group.* See [**397**] for the spectacular geometry and history of Fuchsian and Kleinian groups. Fuchsian groups are extensively discussed in Beardon [**34**], Ford [**193**], Katok [**300**], in a section of Simon [**525**], and, in a historical context, in Gray [**217**].

Problems

1. (a) Prove that no element of $\Gamma(2)$ (defined via reflection) other than $\mathbb{1}$ leaves any points of $\mathbb{C}_+$ fixed. (*Hint*: Prove if $\gamma(z_0) = z_0$ and $\gamma(z) - z$ has a zero at z_0, then $\lambda'(z_0) = 0$.)

   (b) Prove if $z, w \in \mathbb{C}_+$, with $\lambda(z) = \lambda(w)$, then the γ in $\Gamma(2)$ with $\gamma(z) = w$ is unique.

2. Prove that if $\widetilde{\lambda}$ maps $\overline{\widetilde{\mathcal{F}}^\sharp} \cup \{\infty\}$ conformally and bijectively to either $\mathbb{C}_+ \cup \{\infty\}$ or $\overline{\mathbb{C}}_+ \cup \{\infty\}$ and $\widetilde{\lambda}(\infty) = 0$, $\widetilde{\lambda}(0) = 1$, $\widetilde{\lambda}(1) = \infty$, then $\widetilde{\lambda}$ goes to $\mathbb{C}_+ \cup \{\infty\}$ because $\widetilde{\lambda}$ is orientation preserving.

3. (a) Prove that $\Gamma(2)$ is generated by $\mathcal{R}_1 \circ \mathcal{R}_0$ and $\mathcal{R}_{1/2} \circ \mathcal{R}_0$ (and their inverses). (*Hint*: Write $\mathcal{R}^{(1)} \circ \mathcal{R}^{(2)} = \mathcal{R}^{(1)} \circ \mathcal{R}_0 \circ \mathcal{R}_0 \circ \mathcal{R}^{(2)}$.)

   (b) Prove $\mathcal{R}_1 \circ \mathcal{R}_0$ is the FLT generated by $\left(\begin{smallmatrix} 1 & 2 \\ 0 & 1 \end{smallmatrix}\right)$ and $\mathcal{R}_{1/2} \circ \mathcal{R}_0$ by $\left(\begin{smallmatrix} 1 & 0 \\ 2 & 1 \end{smallmatrix}\right)$.

   (c) Conclude that if $\left(\begin{smallmatrix} \alpha & \beta \\ \gamma & \delta \end{smallmatrix}\right) \in \Gamma(2)$, then α, δ are odd integers and β, γ are even integers.

   (d) Prove that reflection in each circle in ∂R_n is a conjugate of $\mathcal{R}_{1/2}$ by some element of $\Gamma(2)$.

4. (a) Prove the 32 diameters at level 6 (i.e., of the semidisks in R_5) are the following sixteen numbers followed by the same string in reverse order: $\frac{1}{6}, \frac{1}{30}, \frac{1}{45}, \frac{1}{36}, \frac{1}{44}, \frac{1}{77}, \frac{1}{70}, \frac{1}{30}, \frac{1}{33}, \frac{1}{88}, \frac{1}{104}, \frac{1}{65}, \frac{1}{60}, \frac{1}{84}, \frac{1}{63}, \frac{1}{18}$.

   (b) Using (8.3.12), show that the dividing points along $[0, 1]$ of these circles at generation six are: $0, \frac{1}{6}, \frac{1}{5}, \frac{2}{9}, \frac{1}{4}, \frac{3}{11}, \frac{2}{7}, \frac{3}{10}, \frac{1}{3}, \frac{4}{11}, \frac{3}{8}, \frac{5}{13}, \frac{2}{5}, \frac{5}{12}, \frac{3}{7}, \frac{4}{9}, \frac{1}{2}, \frac{5}{9}, \frac{4}{7}, \frac{7}{12}, \frac{3}{5}, \frac{8}{13}, \frac{5}{8}, \frac{7}{11}, \frac{2}{3}, \frac{7}{10}, \frac{5}{7}, \frac{8}{11}, \frac{3}{4}, \frac{7}{9}, \frac{4}{5}, \frac{5}{6}, 1$.

   (c) Obtain the sequence in (b) by starting instead with $\frac{0}{1}$ and $\frac{1}{1}$ and repeatedly using (8.3.19).

   (d) Verify (8.3.18) for the sequence in (b).

5. (a) Let $p_1q_0 - p_0q_1 = 1$. Let $x_0 = p_0/q_0$, $x_1 = p_1/q_1$, and $x_{1/2} = (p_0+p_1)/(q_0+q_1)$. Define $d_{n-1} = x_1 - x_0$, $d_n = x_{1/2} - x_0$, and let d_{n+1}^+ be given by (8.3.12). Prove that $d_{n+1}^+ = 1/q_0(2q_0+q_1)$.

   (b) Prove that $x_0 + d_{n+1}^+ = (2p_0+p_1)/(2q_0+q_1)$.

   (c) Prove inductively, starting with the fact that the endpoints in the circles of R_1 are $(0, \frac{1}{2}, 1)$, that the endpoints of the circles of R_n are given by the left side of the Stern–Brocot tree in Figure 8.3.4, that is, are given by repeated use of the Farey relation (8.3.19).

6. The purpose of this problem is to prove that if $0 < a < b$ are relatively prime integers, then a/b appears as one of the numbers in level $b-1$ of the Stern–Brocot tree so that every rational appears in the tree. (*Note*: It may appear also in earlier levels of the tree.)

   (a) Prove inductively that the new elements at level n of the Stern–Brocot tree have the form p/q with $q \geq n+1$.

   (b) Suppose a/b is not one of the rationals at level k so $p_0/q_0 < a/b < p_1/q_1$ with $p_1q_0 - p_0q_1 = 1$ with either p_0/q_0 or p_1/q_1 new at level k. Prove that

$$\frac{a}{b} - \frac{p_0}{q_0} \geq \frac{1}{bq_0}, \qquad \frac{p_1}{q_1} - \frac{a}{b} \geq \frac{1}{bq_1}, \qquad \frac{p_1}{q_1} - \frac{p_0}{q_0} = \frac{1}{q_0q_1}$$

   (c) Prove that $b \geq q_0 + q_1$ so $b \geq k+2$. Thus, for $k = b-1$, a/b must be one of the rationals at level k.

   (d) Prove again that all circles at level n of R_n have diameter at least $n+1$.

7. (a) Show that inside each semicircle in each R_n that λ takes all values in $\mathbb{C} \setminus \{0,1\}$.

   (b) If $x \in \mathbb{R}$, prove that arbitrarily close to x lies a complete semicircle in some R_n.

   (c) Prove that for each $x \in \mathbb{R}$, $\lim_{z \to x,\, z \in \mathbb{C}_+} \lambda(z)$ does not exist and conclude that $\mathbb{R}$ is a natural boundary of λ.

8.4. Some Explicit Conformal Maps

Like contour integrals, conformal maps are not only theoretically important but important in engineering and other applications—in which case, explicit formulae are important. So this section will present many examples, some where the Riemann map is a composition of FLTs, $\exp(\cdot)$, $\log(\cdot)$ and z^α. In several cases, the map will be given by an explicit integral and one of the integrals will be an elliptic integral. To keep the formulae simple-looking, we'll often write compositions and have $\varphi = \varphi_k \circ \varphi_{k-1} \dots \varphi_1$ without doing the composition in closed form. We'll generally use φ for maps of Ω to $\mathbb{C}_+$ or $\mathbb{D}$ and ψ for the inverses.

Example 8.4.1 (Lens Regions). A region between two intersecting circles is sometimes called a *lens region* since something like the region on the left of Figure 8.4.1 looks like a lens but not the one on the right (recall a straight line and circle look "the same" to FLTs).

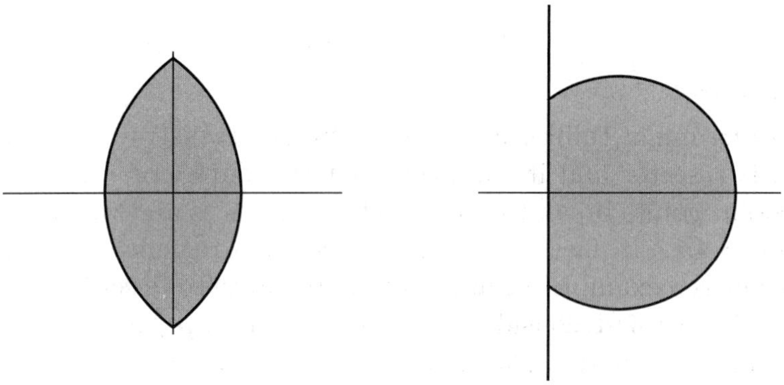

Figure 8.4.1. Two lens regions.

Let Ω be a lens region. Let z_0, z_1 be the two intersection points; let α be the angle formed by the tangents at z_0, and β the angle between the line from z_0 to z_1 and the tangent to the circle on the right (see Figure 8.4.2). Let

$$\varphi_1(z) = e^{i\beta}\,\frac{z_0 - z}{z - z_1}, \qquad \varphi_2(z) = z^{\pi/\alpha}, \qquad \varphi_3(z) = \frac{z-i}{z+i} \tag{8.4.1}$$

φ_1 maps z_0 to 0, z_1 to ∞, and $\frac{1}{2}(z_0+z_1)$ to $e^{i\beta}$. It follows that Ω is mapped to $\{z \mid 0 < \arg(z) < \alpha\}$. φ_2 maps this region to $\mathbb{C}_+$ and φ_3 is the canonical map of $\mathbb{C}_+$ to $\mathbb{D}$. Thus, $\varphi_3 \circ \varphi_2 \circ \varphi_1$ maps Ω to $\mathbb{D}$. An example of a lens region is the semicircular region $\mathbb{D} \cap \mathbb{C}_+$. Problem 1 will find the map of that to $\mathbb{D}$.

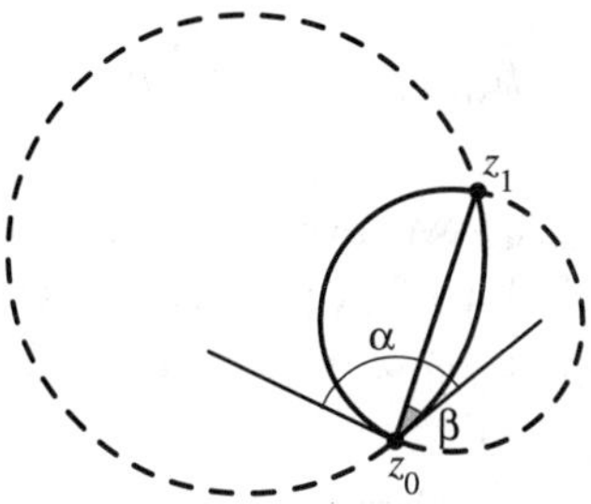

Figure 8.4.2. A lens with labels.

$\square$

Example 8.4.2 (Slit Plane; Koebe Function). This is an example of a degeneracy of the lens region. If we take one vertex of a lens to infinity, we get a sector, which we'll take as

$$\Omega_\alpha = \left\{ z \mid |\arg z| < \frac{\alpha}{2}\,;\, |z| > 0 \right\} \tag{8.4.2}$$

To be a lens, we need $0 < \alpha < 2\pi$, but we'll be especially interested in $\alpha = 2\pi$, that is,

$$\Omega_{2\pi} = \mathbb{C} \setminus (-\infty, 0] \tag{8.4.3}$$

the cut plane.

The reader might think the region "between" two straight lines through the origin is a sector and its negative, but two circles or lines with a pair of intersection points break the plane into four pieces and we only took one in Example 8.4.1. In fact, if $\alpha > \pi$, Ω_α is three regions. The analogy is imperfect but the example natural and interesting. To describe the map ψ_α from $\mathbb{D}$ to Ω_α, we first consider $z \mapsto \frac{1+z}{1-z}$ which maps $\mathbb{D}$ to $\mathbb{H}_+$, the right half-plane $\{z \mid \operatorname{Re} z > 0\} = \Omega_\pi$. The α/π power maps Ω_π to Ω_α. Thus,

$$\psi_\alpha(z) = \left(\frac{1+z}{1-z} \right)^{\alpha/\pi} \tag{8.4.4}$$

maps $\mathbb{D}$ biholomorphically and bijectively to Ω_α. In particular, $\psi_{2\pi}$ maps $\mathbb{D}$ to $\mathbb{C} \setminus (-\infty, 0]$.

Since $\frac{1+z}{1-z} = 1 + 2z + 2z^2 + \dots$, if we want maps on $\mathbb{D}$ normalized by $f(0) = 0$, $f'(0) = 1$, it will be natural to take $\frac{1}{4}(\psi_\alpha - 1)$, that is,

$$f_{\text{Koebe}}(z) = \frac{1}{4}\left(\frac{1+z}{1-z} \right)^2 - \frac{1}{4} \tag{8.4.5}$$

$$= z(1-z)^{-2} \tag{8.4.6}$$

the *Koebe function* a bijection of $\mathbb{D}$ to $\mathbb{C} \setminus (-\infty, -\frac{1}{4}]$.

Notice since $(1-z)^{-2} = \frac{d}{dz}(1-z)^{-1} = \frac{d}{dz}(1 + z + z^2 + \dots)$, we see that

$$f_{\text{Koebe}}(z) = \sum_{n=1}^{\infty} n z^n \tag{8.4.7}$$

As a final formula for f_{Koebe}, we note $1/f_{\text{Koebe}}$ maps $\mathbb{D}$ to $\mathbb{C} \setminus [-4, 0]$, so $2 + f_{\text{Koebe}}^{-1}$ is a bijection of $\mathbb{D}$ to $\mathbb{C} \setminus [-2, 2]$. In the next example, we'll see this is the Joukowski map, $z \mapsto z + z^{-1}$ (and the normalization is right). Thus,

$$f_{\text{Koebe}}(z) = (z + z^{-1} - 2)^{-1} \tag{8.4.8}$$

as can easily be checked directly. The Koebe function will play a starring role in Section 16.1 of Part 2B. □

Example 8.4.3 (Joukowski Map). The Joukowski (also Joukowsky or Zhukovsky) map is the map from $\widehat{\mathbb{C}}$ to itself by

$$\psi(z) = z + z^{-1} \tag{8.4.9}$$

It is of degree 2, that is, $\psi(z) = w$ has two solutions if $w \neq \pm 2$. It is one-one on $\mathbb{D}$ and maps $\mathbb{D}$ bijectively to $\widehat{\mathbb{C}} \setminus [-2, 2]$. Its inverse (to $\mathbb{D}$) is given by

$$\varphi(z) = \frac{z - \sqrt{z^2 - 4}}{2} \tag{8.4.10}$$

It is of considerable theoretical use. For example, the classical orthogonal polynomials on $[-1, 1]$, like Legendre and Chebyshev, have asymptotics for $z \in \mathbb{C} \setminus [-1, 1]$,

$$P_n(z) \sim c_n d(z) \big[z + \sqrt{z^2 - 1}\,\big]^n \tag{8.4.11}$$

The factor in $[\dots]$ is the conformal map of $\widehat{\mathbb{C}} \setminus [-1, 1]$ to $\mathbb{C} \setminus \overline{\mathbb{D}}$, closely related to (8.4.10).

Figure 8.4.3 shows the image under ψ of a circle $|z - z_0| = |1 - z_0|$ for z_0 near but shifted from 0 (chosen to be $-0.08 + i0.08$). It is no coincidence that this looks like an idealization of the cross-section of an airplane wing. This is the Joukowski airfoil, useful especially because one can compute airflow around a circle and conformally map to airflow around this airfoil.

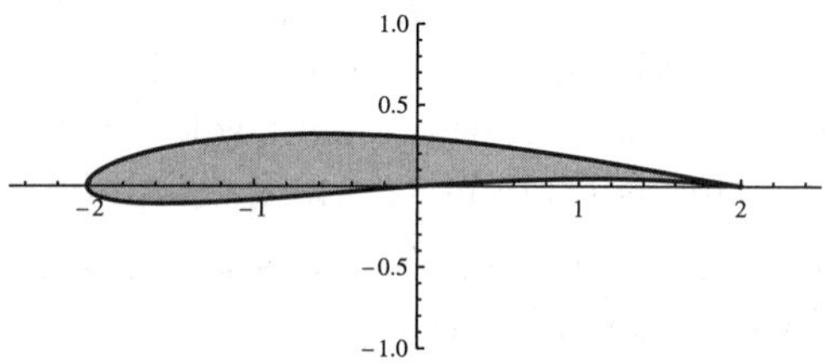

Figure 8.4.3. The Joukowski airfoil.

Note that this has a cusp at the end which produces instabilities. For that reason, airplanes used the Kármán–Trefftz airfoil shown in Figure 8.4.4,

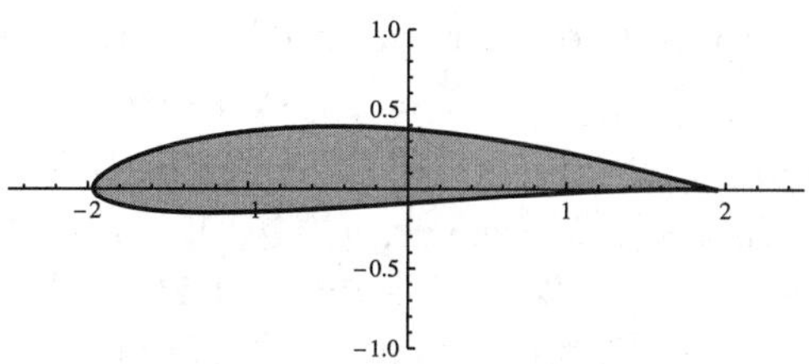

Figure 8.4.4. The Kármán–Trefftz airfoil.

where ψ is replaced by

$$\psi_\beta(z) = \beta\,\frac{(1+z^{-1})^\beta + (1-z^{-1})^\beta}{(1+z^{-1})^\beta - (1-z^{-1})^\beta} \tag{8.4.12}$$

(the figure shows the same z_0 and $\beta = 1.94$). □

Example 8.4.4 (Half-Strip). Let

$$\Omega = \left\{ z \;\middle|\; \operatorname{Im} z > 0,\ -\frac{\pi}{2} < \operatorname{Re} z < \frac{\pi}{2} \right\} \tag{8.4.13}$$

a half-strip. We begin by looking at maps from Ω to $\mathbb{C}_+$ (or vice-versa) since it is easy to get from $\mathbb{C}_+$ to $\mathbb{D}$. The reader might pause—and using some of the functions they are familiar with—seek a function that maps Ω to $\mathbb{C}_+$. (*Hint*: Try mapping $\partial\Omega$ to $\mathbb{R}$.)

We begin by focusing on a map of $\mathbb{C}_+$ to Ω or, rather, a monotone map of $\mathbb{R}$ to $\partial\Omega$. Let's normalize it by mapping ± 1 to $\pm\pi/2$. If Ω were the quarter-plane with $0 < \operatorname{Re} z < \infty$, the proper map from $\mathbb{C}_+$ would be $z \mapsto \sqrt{z-1}$, so one's first guess for the actual Ω might be $z \mapsto \sqrt{1-z^2}$. That doesn't work because $\arg(1-z)$ is not constant on $\operatorname{Re} z = -1$. Put differently, the arg of $w - \pi$ is not constant on $\operatorname{Re} w = \pi$—it's not $\arg(w)$ but $\arg(\frac{dw}{dz})$ that is constant. Some thought about this suggests that the map we want to take is

$$w = \psi(z) = \int_0^z \frac{dx}{\sqrt{1-x^2}} \tag{8.4.14}$$

or a constant multiple. We want to define this in $\mathbb{C}_+$ where it is clearly analytic since $\frac{dw}{dz} = (1-z^2)^{-1/2}$. And since $(\sqrt{1-x^2})^{-1}$ is even integrable at $x = \pm 1$, it has continuous boundary values on all of $\mathbb{R}$.

Of course, $\psi(z) = \arcsin(z)$ and the inverse function is $\varphi(w) = \sin(w)$. One can use this directly to solve the problem of mapping $\mathbb{C}_+$ to Ω (see Problem 3), but since we'll later have analogs which are not expressible in terms of elementary functions, we focus on analyzing the integral directly.

We need to know

$$\int_0^1 \frac{dx}{\sqrt{1-x^2}} = \frac{\pi}{2} \tag{8.4.15}$$

This is usually done with trigonometric substitution but it can also be done via the Joukowski map and contour integration (Problem 4).

The ψ defined by (8.4.14) maps $(0,1)$ to $(0,\pi/2)$, but at $x = 1$, $\arg(\sqrt{1-x^2})$ becomes $\pi/2$ (since $\arg(1-x)$ goes from 0 to $-\pi$ as x goes from $1-\varepsilon$ to $1+\varepsilon$ in $\mathbb{C}_+$). Moreover, $\int_1^\infty dx|1-x^2|^{-1/2} = \infty$, so ψ maps $(1,\infty)$ "monotonically" to $(\pi/2, \pi/2 + i\infty)$. If $\Omega \subset \mathbb{C}_+$ is mapped to $\mathbb{D}$ in the usual way for $\mathbb{C}_+$ to $\mathbb{D}$, the image of $\mathbb{R}$ under ψ gets mapped to the curve on the exterior of Figure 8.4.5. Since ψ maps $\mathbb{R}$ monotonically around the curve, winding number arguments (Problem 5) show that ψ is a one-one

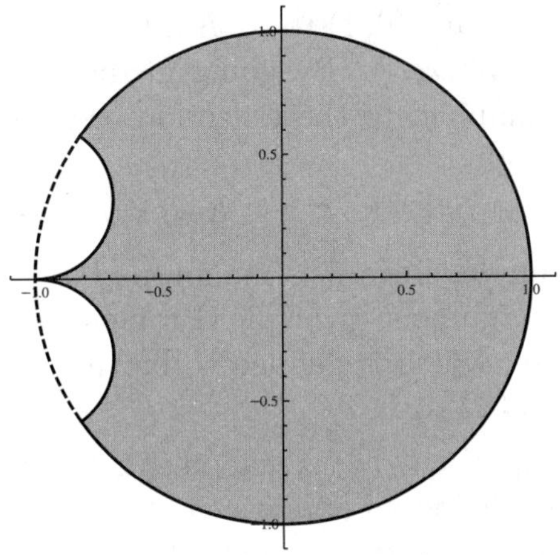

Figure 8.4.5. Image of the boundary of the half-strip in the disk.

map of $\mathbb{C}_+$ to Ω, solving the problem of finding a conformal map of the half-strip to $\mathbb{C}_+$, and from there to $\mathbb{D}$.

Notice also that the inverse function, φ, defined on $\overline{\Omega}$ with Ω given by (8.4.16) is real on $\partial\Omega$. So, by the repeated use of reflection, we can define it as a map onto $\mathbb{C}$, first on $-\pi/2 < \operatorname{Re} z < \pi/2$ and then on all of $\mathbb{C}$. Thus, φ is an entire function, and since two complex conjugations are the identity, $\varphi(z+2\pi) = \varphi(z)$. Thus, we've used conformal mapping to prove the inverse function to the integral is periodic! $\square$

Example 8.4.5 (Conformal Map of a Rectangle via Elliptic Integrals). Let $0 < k < 1$. The *elliptic integral of the first kind* is defined by

$$w = \psi(z) = \int_0^z \frac{dt}{\sqrt{(1-t^2)(1-k^2t^2)}} \tag{8.4.16}$$

Several special values are relevant. Define $K(k)$, the *complete elliptic integral of the first kind* by

$$K(k) = \int_0^1 \frac{dt}{\sqrt{(1-t^2)(1-k^2t^2)}} \tag{8.4.17}$$

and also

$$k' = \sqrt{1-k^2}, \qquad K'(k) = K(k') \tag{8.4.18}$$

which enter because one can show (Problem 6) that

$$\int_1^{k^{-1}} \frac{dt}{\sqrt{(t^2-1)(1-k^2t^2)}} = K(k') \tag{8.4.19}$$

$$\int_{k^{-1}}^{\infty} \frac{dt}{\sqrt{(t^2-1)(k^2t^2-1)}} = K(k) \tag{8.4.20}$$

This implies that ψ maps $(0,1)$ to $(0,K)$, then $(1,k^{-1})$ to $(K,K+iK')$, and (k^{-1},∞) to $(K+iK',iK')$. By going in the opposite direction along $\mathbb{R}$, we see $(-\infty,0)$ is mapped to the reflection, that is, ψ maps $\mathbb{R}$ into the boundary of the rectangle

$$\Omega = \{z = x+iy \mid -K < x < K,\, 0 < y < K'\} \tag{8.4.21}$$

As in the last example, since ψ is monotone (as) a real-valued function of $\mathbb{R}\cup\{\infty\}$ to $\partial\Omega$, the argument principle (Problem 5) shows ψ is a bijection of $\mathbb{C}_+$ to Ω. Its inverse, which is defined to be the *Jacobi elliptic function*, $\operatorname{sn}(z,k)$, is a bijection of Ω to $\mathbb{C}_+$.

As in the last example, $\operatorname{sn}(\,\cdot\,,k)$ is real on $\partial\Omega$, so we can use the reflection principle repeatedly to define an analytic continuation of $\operatorname{sn}(z,k)$ to $z\in\mathbb{C}$, and this continuation will be doubly periodic with periods $4K$ and $2iK'$. Doubly periodic functions are called elliptic functions and are the subject of Chapter 10.

Here the point is that $\operatorname{sn}(\,\cdot\,,k)$ gives a conformal map of the rectangle (8.4.21) to $\mathbb{C}_+$ and then $z\mapsto(\operatorname{sn}(z,k)-i)/(\operatorname{sn}(z,k)+i)$ to $\mathbb{D}$. To be sure that one can handle all rectangles, one needs that, as k runs from 0 to 1, $4K/2K'$ runs monotonically from 0 to ∞ (Problem 7).

In Problem 15 of Section 10.5, we'll prove that $|\operatorname{sn}(x+\frac{iK'}{2},k)| = k^{-1/2}$ for x real. Thus, $\operatorname{sn}(z,k)$ maps the bottom and side edges of the rectangle $\widetilde{\Omega} = \{z = x+iy \mid |x| < K,\, 0 < y < K'/2\}$ to $(-k^{-1/2},k^{-1/2})$ in $\mathbb{R}$ and the top edge to $\mathbb{C}_+\cup\partial\mathbb{D}_{k^{-1/2}}(0)$, so $\operatorname{sn}(z,k)$ maps Ω to all of $\mathbb{C}_+$ and $\widetilde{\Omega}$, the bottom half of Ω, to the semidisk $\mathbb{C}_+\cap\{z \mid |z| < k^{-1/2}\}$.

You might hope to use this to construct a map of a rectangle to a disk (directly, rather than via a map to $\mathbb{C}_+$)—map a half-disk to a rectangle and reflect in an edge of the rectangle, but that doesn't work because the diameter of the disk goes to three sides of the rectangle rather than one. This "defect" will be useful in Example 8.4.14.

Besides the Jacobi sn function, there is a Weierstrass $\wp$-function among elliptic functions. Problem 13 of Section 10.4 shows that

$$z \to \wp\left(\frac{z}{2a};\tau = \frac{ib}{a}\right) \tag{8.4.22}$$

maps $\{z = x+iy \mid 0 < x < a,\, 0 < y < b\}$ conformally onto $\mathbb{C}_+$.

We discussed sn rather than $\wp$ because it fits with Examples 8.4.4 and (8.4.6). $\square$

Example 8.4.6 (Schwarz–Christoffel Transformations). Motivated by the last two examples, we'll consider maps from $\mathbb{C}_+$ given by

$$w = \psi = \int_0^z \prod_{j=1}^n (x_j - x)^{\alpha_j - 1}\,dx \tag{8.4.23}$$

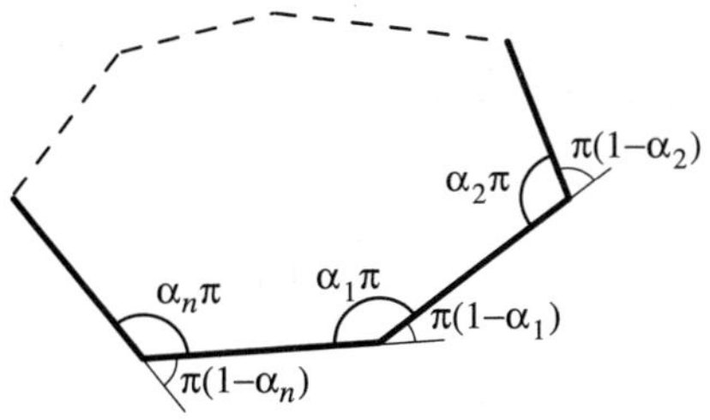

Figure 8.4.6. Angles on polygon.

where $x_1 < \cdots < x_n$ and

$$0 < \alpha_j < 1 \tag{8.4.24}$$

and

$$\sum_{j=1}^{n} (1 - \alpha_j) = 2 \tag{8.4.25}$$

We'll see in a moment the geometric meaning of these conditions (and explore in Problem 10 what happens if (8.4.25) is replaced by $\sum_{j=1}^{n}(1-\alpha_j) < 2$ and discuss replacing (8.4.24) by $0 < \alpha_j < 2$ below).

Notice that (8.4.13) is $n = 4$, $x_j = (-k^{-1}, -1, 1, k^{-1})$ and $\alpha_j = \frac{1}{2}$.

For $x < x_1$, each $x_j - x$ is positive and we can pick the integrand so that it is positive there. With this choice,

$$\arg(\psi'(z)) = \beta_j \text{ on } [x_j, x_{j+1}] \qquad \beta_{j+1} - \beta_j = \pi(1 - \alpha_j)$$

so $\psi[\mathbb{R}]$ describes a polygon with interior angles $\alpha_j \pi$ (see Figure 8.4.6).

The condition (8.4.24) implies that the polygon is convex and (8.4.25) that with these images of $(-\infty, x_1)$ and (x_n, ∞) are parallel (Problem 8). (8.4.25) also implies the integrand is $O(x^{-2})$ at ∞, so $\lim_{x\to\pm\infty} \psi(x)$ exist, and by closing a contour (Problem 9), $\psi(\infty) = \psi(-\infty)$, so the image is a closed polygon.

Since the enclosed polygon is convex, the boundary is simple, and so the argument principle implies ψ defines a conformal map of $\mathbb{C}_+$ to the interior of the polygon. $\square$

This is nice as a set of conformal maps, but it avoids a main issue: given a polygonal region, is there a ψ of the form (8.4.23) mapping $\mathbb{C}_+$ to that region. That there is, is the content of the following:

Theorem 8.4.7 (Schwarz–Christoffel Theorem). *Let Ω be a polygonal region, that is, a Jordan region whose boundary is a finite number of linear pieces. Let $\varphi\colon \overline{\Omega} \to \overline{\mathbb{C}}_+ \cup \{\infty\}$ be a bijection guaranteed by the Riemann mapping theorem (as extended in Theorem 8.2.3), chosen so no vertex of $\overline{\Omega}$ goes into $\{\infty\}$. Let $x_1, \dots, x_n$ be the images in $\mathbb{R}$ of the vertices $\zeta_1, \dots, \zeta_n$*

and $\alpha_1, \dots, \alpha_n$ *the interior angles at the vertices (which obey* (8.4.25)*). Let* $\psi\colon \overline{\mathbb{C}}_+ \cup \{\infty\} \to \overline{\Omega}$ *be the inverse to* φ*. Then for suitable constants* A *and* C*,*

$$w = \psi(z) = A + C \int_0^z \prod_{j=1}^n (x_j - x)^{\alpha_j - 1}\, dx \tag{8.4.26}$$

Proof. By the Riemann mapping theorem as extended (Theorem 8.2.3), ψ is continuous on $\overline{\mathbb{C}}_+ \cup \{\infty\}$ and analytic in $\mathbb{C}_+$, and by the reflection principle (Theorem 5.5.1), it has a continuation across each interval (x_j, x_{j+1}). Each reflection defines on $\mathbb{C}_-$ a bijection to the reflection, $\widetilde{\Omega}$, of Ω in the edge $[\zeta_j, \zeta_{j+1}]$, which is also a polygonal region.

This continuation can be continued by reflection back to $\mathbb{C}_+$ across each interval. In this way, by repeated reflections, we see that ψ defines a multi-valued analytic function.

Let $\tilde{\psi}$ be a result of two reflections yielding a function on $\mathbb{C}_+$. In the w plane, this is a result of two reflections (in an edge of Ω and in one of $\widetilde{\Omega}$), which is a rotation and translation (see Figure 8.4.7). Thus, for suitable constant, μ, λ,

$$\tilde{\psi}(z) = \mu\psi(z) + \lambda \tag{8.4.27}$$

so if

$$h(z) = \frac{\psi''(z)}{\psi'(z)} \tag{8.4.28}$$

then

$$\tilde{h}(z) \equiv \frac{\mu\psi''(z)}{\mu\psi'(z)} = h(z) \tag{8.4.29}$$

so h has a single-valued continuation to $\widehat{\mathbb{C}} \setminus \{x_1, \dots, x_n\}$, that is, h is analytic and single-valued in punctured neighborhoods of each x_j.

Note next, since ψ is analytic at ∞ (since $\psi(\infty)$ is an interior point of the edge $[\zeta_n, \zeta_1]$ of Ω), so near ∞, $\psi(z) = a_0 + a_1 z^{-1} + a_2 z^{-2} + \dots$. Since ψ is a bijection near ∞, $a_1 \neq 0$, and thus, $\psi'(z) = a_1 z^{-2} + O(z^{-3})$, $\psi''(z) = 2z^{-3} + O(z^{-4})$, so

$$h(z) = -2z^{-1} + O(z^{-2}) \quad \text{near } \infty \tag{8.4.30}$$

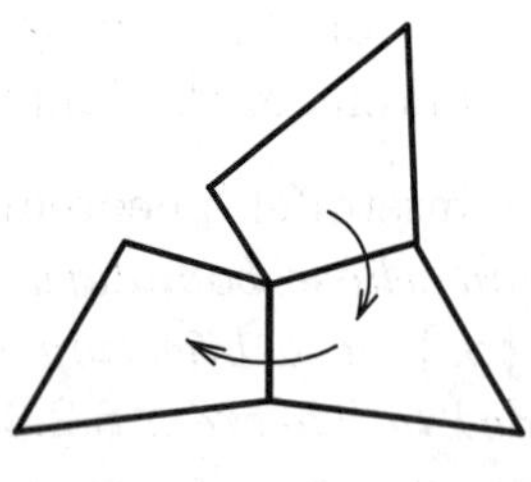

Figure 8.4.7. Two reflections are a translation plus rotation.

If φ is the inverse of ψ, Theorem 5.5.8 implies, near ζ_j, φ has a convergent Puiseux expansion

$$\varphi(w) = x_j + a_j^{(1)}(w - \zeta_j)^{1/\alpha_j} + a_j^{(2)}(w - \zeta_j)^{2/\alpha_j} + \dots \tag{8.4.31}$$

with $a_j^{(1)} \neq 0$. Thus, near x_j, ψ has a convergent expansion

$$\psi(z) = \zeta_j + b_j^{(1)}(z - x_j)^{\alpha_j} + b_j^{(2)}(z - x_j)^{2\alpha_j} + \dots \tag{8.4.32}$$

with $b_j^{(1)} \neq 0$, which means that

$$\psi'(z) = b_j^{(1)}\alpha_j(z - x_j)^{\alpha_j - 1} + O(|z - x_j|^{2\alpha_j - 1})$$
$$\psi''(z) = b_j^{(1)}\alpha_j(\alpha_j - 1)(z - x_j)^{\alpha_j - 2} + O(|z - x_j|^{2\alpha_j - 2})$$

which implies, near x_j,

$$h(z) = \frac{\alpha_j - 1}{z - x_j} + O(1) \tag{8.4.33}$$

We have thus proven that $h(z)$ is an analytic function from $\widehat{\mathbb{C}}$ to $\widehat{\mathbb{C}}$, so

$$h(z) = \sum_{j=1}^{n} \frac{\alpha_j - 1}{z - x_j} \tag{8.4.34}$$

since the difference is entire and vanishing at infinity by (8.4.30).

Integrating twice (h involves ψ'') yields (8.4.26) (Problem 10). $\square$

Example 8.4.8 ($\mathbb{C}_+$ with a missing slit). Let (Figure 8.4.8)

$$\Omega = \{z \mid \operatorname{Im} z > 0\} \setminus \{z \mid \operatorname{Re} z = 0,\ 0 < \operatorname{Im} y \leq h\} \tag{8.4.35}$$

This is especially interesting as a degenerate Schwarz–Christoffel example and also as one where the integrals can be done explicitly. Ω doesn't look like a polygon, but it can be thought of an one with two degeneracies—its boundary includes the point at infinity and it has an interior angle of all of 2π! Thus, we seek a map of $\mathbb{R}$ so that $(-\infty, -1)$ goes to $(-\infty, 0)$, $(-1, 0)$

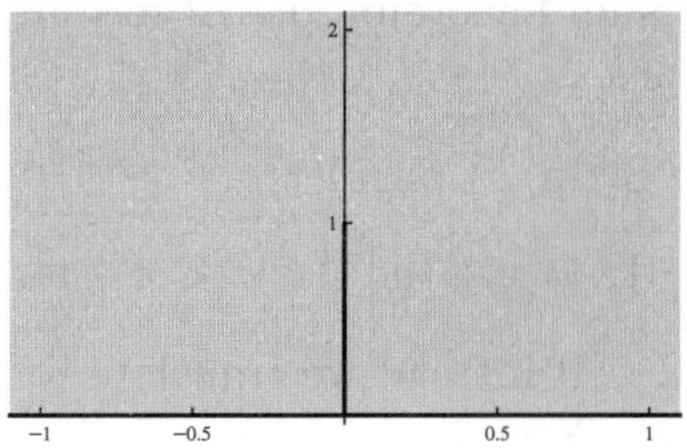

Figure 8.4.8. $\mathbb{C}_+$ with missing slit.

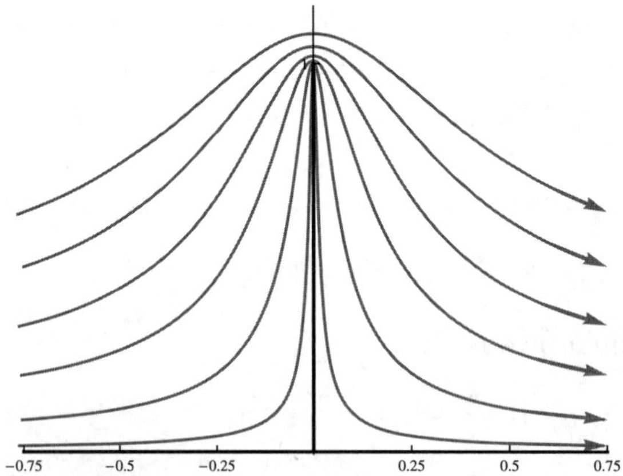

Figure 8.4.9. Airflow around a slit.

and $(0,1)$ go to $i(0,h)$, and $(1,\infty)$ goes to $(0,\infty)$. The angles divided by π at $(x_1,x_2,x_3)=(-1,0,1)$ are $(\alpha_1,\alpha_2,\alpha_3)=(\frac{1}{2},2,\frac{1}{2})$. Thus, in (8.4.26),

$$\begin{aligned} w=\psi(z) &= A+C\int_0^z \frac{x}{(x^2-1)^{1/2}}\,dx \\ &= A'+C\sqrt{z^2-1} \end{aligned} \tag{8.4.36}$$

since the integrand is an elementary perfect derivative. To take $z=\pm 1$ to 0, we want $A'=0$, and for $z=0$ to go to ih, we want $C=h$. Thus,

$$w=\psi(z)=h\sqrt{z^2-1} \tag{8.4.37}$$

maps $\mathbb{R}$ to $\partial\Omega$, and with the right branch of the square root, $\mathbb{C}_+$ to Ω. Figure 8.4.9 shows the streamlines for fluid flow over such a simple barrier. □

Example 8.4.9 (Elliptic Modular Function). In Section 8.3, we found a map λ of the region $\mathcal{F}^\sharp$ in Figure 8.3.1 to $\mathbb{C}_+$. We mention that Section 10.6 has an explicit formula for λ in terms of an elliptic function, $\wp$, called the Weierstrass $\wp$-function and Theorem 10.4.12 has an explicit formula for $\wp$ as an infinite sum of sines. □

Example 8.4.10 (The Strip). Consider

$$\Omega=\{z=x+iy \mid 0<x<\pi,\ y\in\mathbb{R}\} \tag{8.4.38}$$

$\varphi(z)=e^{iz}$ maps Ω to $\mathbb{C}_+$. If instead we translate to

$$\widetilde{\Omega}=\left\{z=x+iy \,\middle|\, -\frac{\pi}{2}<x<\frac{\pi}{2},\ y\in\mathbb{R}\right\} \tag{8.4.39}$$

then e^{iz} maps $\Omega = \mathbb{H}_+$. Since $w \to \frac{1}{i}\frac{w-1}{w+1}$ maps $\mathbb{H}_+$ to $\mathbb{D}$, we see that

$$\varphi(z) = \tan\left(\frac{z}{2}\right) \tag{8.4.40}$$

maps $\widetilde{\Omega}$ to $\mathbb{D}$. □

Example 8.4.11 (Slice of Annulus). Let (Figure 8.4.10)

$$\Omega = \{z = re^{i\theta} \mid r_1 < r < r_2,\ |\theta| \le \beta < \pi\} \tag{8.4.41}$$

Then $\varphi_1(z) = \log(z)$ maps Ω to the rectangle $\{w = u + iv \mid \log(r_1) < u < \log(r_2),\ -\beta < v < \beta\}$, so composing φ with the map, sn, of Example 8.4.11 maps this Ω to $\mathbb{C}_+$. □

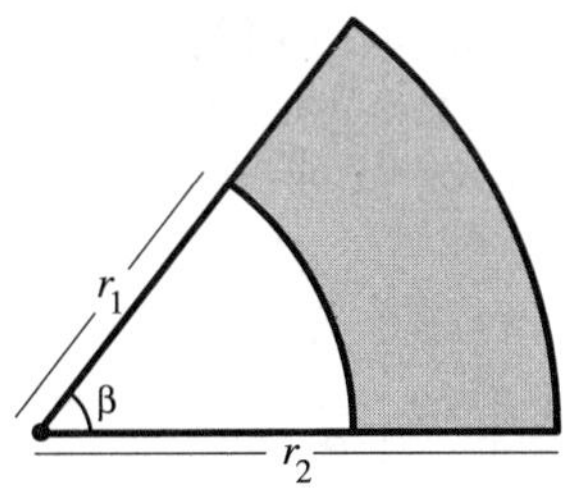

Figure 8.4.10. Slice on an annulus.

Example 8.4.12 (Hyperbolic Region). Consider the region

$$\Omega = \{x + iy \mid x > 0,\ y > 0,\ |xy| < 1,\ |x^2 - y^2| < 1\} \tag{8.4.42}$$

the set in the upper quadrant inside three hyperboles (Figure 8.4.11). As you might expect, since conic sections are quadratic, this is connected to the function z^2. Indeed, since

$$z \in \Omega \Leftrightarrow 0 < \operatorname{Im} z^2 < 1,\ |\operatorname{Re} z^2| < 1$$

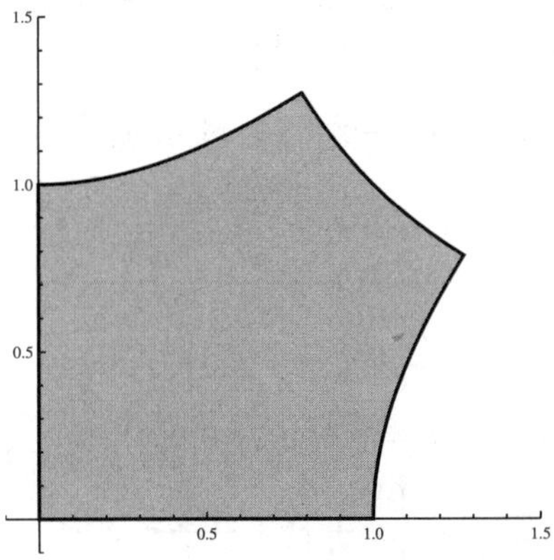

Figure 8.4.11. A hyperbolic region.

the map $\varphi_1(z) = z^2$ maps Ω to a rectangle bijectively, and then the elliptic map, sn, of Example 8.4.5 will go to $\mathbb{C}_+$. $\square$

Example 8.4.13 (Exterior of an Ellipse and Semi-Ellipse). Let $a > b$ and

$$\Omega_0 = \left\{ z = x + iy \,\middle|\, \frac{x^2}{a^2} + \frac{y^2}{b^2} < 1 \right\} \tag{8.4.43}$$

which is an ellipse with foci at $\pm c$ where $c^2 = a^2 - b^2$. In this example, we'll take $c = 2$. Let

$$\Omega_1 = \Omega_0 \cap \mathbb{C}_+, \qquad \Omega_2 = \widehat{\mathbb{C}} \setminus \Omega_0 \tag{8.4.44}$$

Here we'll see that the inverse Joukowski map provides Riemann maps or components of such a map for Ω_1 and Ω_2, but for Ω_0 we'll need to do more (see the next example). Recall the Joukowski map is (8.4.9). Let $r < 1$. Let $a = r + r^{-1}$, $b = r^{-1} - r$, so $a^2 - b^2 = 4$. In Problem 12, the reader will prove that

$$\psi(\partial\mathbb{D}_r(0)) = \left\{ z = x + iy \,\middle|\, \frac{x^2}{a^2} + \frac{y^2}{b^2} = 1 \right\} \qquad \psi(\mathbb{D}_r(0)) = \Omega_2$$

Thus, $\varphi = r^{-1}\psi^{-1}$ maps $\Omega_2 \to \mathbb{D}$ is the Riemann map for Ω_2.

Moreover, ψ maps $\mathbb{A}_{r,1}$ to $\Omega_0 \backslash [-2, 2]$ and $\mathbb{C}_- \cap \mathbb{A}_{r,1}$ to Ω_1. Since $\mathbb{C}_- \cap \mathbb{A}_{r,1}$ is a slice of an annulus, the map of Example 8.4.20 combined with ψ can be used to map Ω_1 to $\mathbb{D}$. $\square$

Example 8.4.14 (Ellipse). $2\cos$ is the composition of $z \mapsto e^{iz}$ and the Joukowski map, so the inverse is a composition of the inverse Joukowski map and log. Since log maps an annulus into a rectangle, the above map of Ω_1 is essentially arcsin. Put differently, we expect sin to map a suitable rectangle into a half-ellipse. This will be extended to mapping a full ellipse. To avoid the factors of 2, our ellipses in this example will have foci at ± 1. Thus, $a^2 - b^2 = 1$. y_0 will be defined by

$$\cosh y_0 = a, \qquad \sinh y_0 = b \tag{8.4.45}$$

$$\Omega = \left\{ u + iv \,\middle|\, \frac{u^2}{a^2} + \frac{v^2}{b^2} \le 1 \right\} \tag{8.4.46}$$

Since

$$\sin(x + iy) = u + iv \Leftrightarrow u = (\cosh y)(\sin x), \qquad v = (\sinh y)(\cos x)$$

$$\arcsin(u + iv) \colon \Omega \cap \mathbb{C}_+ \to \left\{ x + iy \,\middle|\, -\frac{\pi}{2} < x < \frac{\pi}{2},\ 0 < y < y_0 \right\} \equiv \mathcal{R} \tag{8.4.47}$$

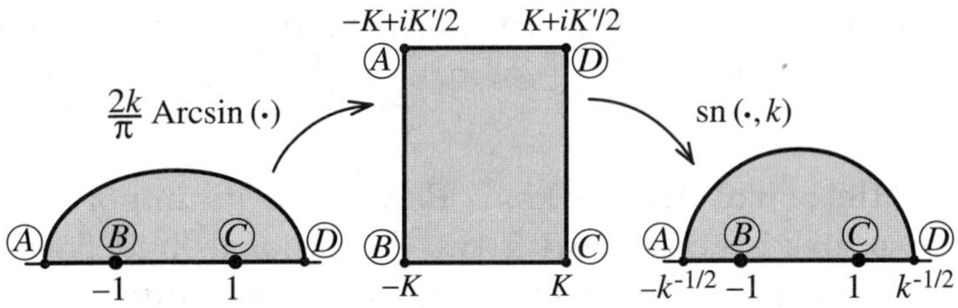

Figure 8.4.12. Two maps where an interval is mapped to three sides of a rectangle.

As we saw in Example 8.4.5, if k is chosen so that (in terms of the τ of Sections 10.4–10.5, this is $\tau = 2iy_0/\pi$)

$$\frac{K'}{K} = \frac{y_0}{\frac{\pi}{2}} \tag{8.4.48}$$

then $\text{sn}(\frac{2K}{\pi}\,\cdot)$ maps $\mathcal{R}$ to a half-disk of radius $1/\sqrt{k}$. But now the "defect" mentioned at the end of that example is a virtue. Figure 8.4.12 shows how points are mapped. The diameter of both the circle and ellipse map to the same three sides of the rectangle,

$$\varphi(z) = \sqrt{k}\,\text{sn}\biggl(\frac{2K}{\pi}\,\arcsin(z)\biggr) \tag{8.4.49}$$

with k given to solve (8.4.48) with y_0 picked by (8.4.45) maps the diameter to the diameter. Thus, using Schwarz reflection, φ maps the Ω of (8.4.46) to $\mathbb{D}$. $\square$

Example 8.4.15. We end with an example of a map among doubly connected sets rather than the simply connected examples discussed previously in this section. Returning to Example 8.4.13, we see the Joukowski map takes $\mathbb{A}_{r,1}$ to the ellipse Ω_0 of (8.4.37) (with $a = r + r^{-1}$) but with $(-2,2)$ removed. If we follow this with the map (8.4.49), we'll get a map of $\mathbb{A}_{r,1}$ to a disk with a suitable interval $[-\alpha, \alpha]$ removed.

Turning around now to get to arcsin, the use of arcsin and the Joukowski map can be replaced by a use of log. Namely, $w \in \mathbb{A}_{r,1} \cap \mathbb{H}_+$ mapped by $i\log(w)$ to $\{z = x + iy \mid -\frac{\pi}{2} < x < \frac{\pi}{2},\ 0 < y < \log(r)\}$. Picking the $\text{sn}(\cdot)$ that maps this to a circle of radius $k^{-1/2}$, we see the image of $\mathbb{A}_{r,1} \cap \mathbb{R}$ goes to $(-k^{-1}, 1) \cup (1, k^{-1})$

$$\varphi(w) = \sqrt{k}\,\text{sn}\biggl(\frac{2Ki}{\pi}\,\log(w)\biggr) \tag{8.4.50}$$

maps $\mathbb{A}_{r,1}$ to $\mathbb{D} \setminus (-k, k)$. If instead we take k so the rectangle goes to all of $\mathbb{C}_+$ and we drop the i, we get a map to $\widehat{\mathbb{C}} \setminus (-k^{-1}, -1) \cup (1, k^{-1})$. In

Problem 13 of Section 10.4, we'll see the Weierstrass $\wp$-function, composed with log, maps an annulus to $\mathbb{C}\setminus(-\infty,\alpha]\cup[\beta,\gamma]$ for suitable $\alpha<\beta<\gamma$. $\square$

Notes and Historical Remarks. Because of its importance in engineering applications connected with two-dimensional fluid flow (including aerodynamics) and potential theory, there is a huge literature on explicit and theoretical conformal maps.

Kober [**321**] is nothing but a listing of conformal maps between many different regions and a "standard" region like $\mathbb{D}$ or $\mathbb{C}_+$ (or, once maps from a strip to $\mathbb{C}_+$ are found, to a strip). Interestingly enough, it was originally published by the British admiralty! Books with more of a theoretical leaning include Bieberbach [**50**], Nehari [**405**], and Schinzinger–Laura [**504**]. Books that focus on computational aspects include Driscoll–Trefethen [**150**], Ivanov–Trubetskov [**278**], and Kythe [**334**].

Nikolai Egorovich Zhukovsky (1847–1921), usually transliterated as Joukowski, is often referred to as the father of Russian aviation. He worked on both practical and theoretical aspects of the subject. He found the map $z\mapsto z+z^{-1}$ for use in airfoil design in 1910 [**294**]. The Kármán–Trefftz map was found by them in 1918 [**567**].

For the use of the Joukowski map in the theory of orthogonal polynomials, see [**550, 525**].

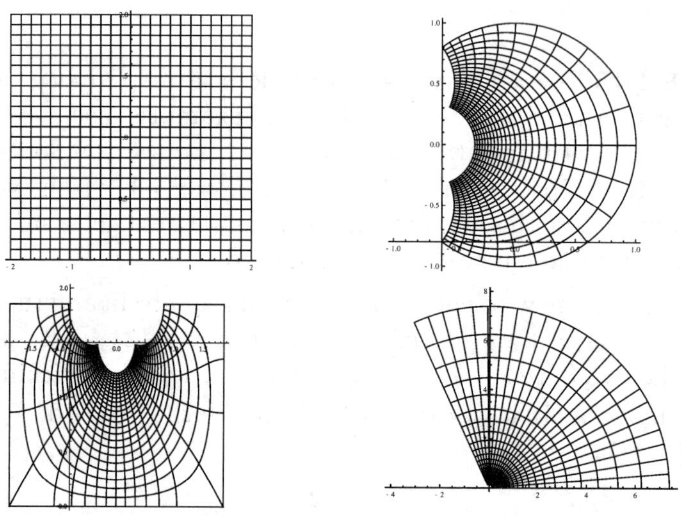

Figure 8.4.13. Three examples of conformal grid maps. (a) original grid. (b) $z\to(i-z)/(i+z)$. (c) $z\to\operatorname{sn}^{-1}(z,k)$ with $k=1/\sqrt{2}$. (d) $z\to e^z$.

The Schwarz–Christoffel maps appeared first in an 1867 paper of Christoffel [**116**] and shortly thereafter (and apparently independently) in Schwarz [**509, 510**]. Schwarz focused on triangles and quadrilaterals[2]. His work was connected to what he was doing on monodromy of hypergeometric functions (see the Notes to Section 14.5 of Part 2B). The book of Driscoll–Trefethen [**150**] is entirely on Schwarz–Christoffel transforms, especially computer code for them. Ablowitz–Fokas [**4**] has lots of examples of Schwarz–Christoffel mappings. Crowdy [**127**] has recent work on Schwarz–Christoffel maps for multiply connected domains.

Fluid flow in $\mathbb{C}_+$ is parallel to the boundary, namely $\mathbb{R}$, so one cares about the image under a conformal map for $\mathbb{C}_+$ with a coordinate grid. Figure 8.4.13 shows a rectangle mesh in $\mathbb{C}_+$ and its image under three conformal maps as indicated in the caption. Figure 8.4.13 (c) is a slice of the map of Example 8.4.5 to a rectangle. At the corner points, the map is not conformal but halves angles, which is why a 90° angle becomes 45°.

Problems

1. Using the ideas of Example 8.4.1, find an explicit formula for $\varphi\colon \{z \mid |z| < 1, \operatorname{Im} z > 0\} \to \mathbb{D}$.

2. Prove that the Kármán–Trefftz map, ψ_β, given by (8.4.12), becomes the Joukowski map, $\psi(z) = z + z^{-1}$ when $\beta = 2$.

3. Prove directly that $\varphi(z) = \sin z$ is a bijection from (8.4.13) to $\mathbb{C}_+$. (*Hint*: If $\sin(x+iy) = u+iv$, prove that $v > 0$ and $u^2/\cosh^2 y + v^2/\sinh^2 y = 1$, and that for y fixed, $u+iv$ is the entire set of (u,v) with this property.)

4. By changing variables from x to $z \in \partial\mathbb{D}$ by $\frac{1}{2}(z+z^{-1}) = x$, show that
$$2\int_{-1}^{1} \frac{dx}{\sqrt{1-x^2}} = \frac{1}{i}\oint_{|z|=1} \frac{dz}{z} = 2\pi$$

5. Let Ω be a Jordan region with boundary curve $\{\gamma(z)\}_{0\le t\le 1}$. Suppose that you know there are curves $\{\gamma_s(t)\}_{0\le s\le 1,\, 0\le t\le 1}$ so that, uniformly in t, $\gamma_s(t) \to \gamma(t)$ as $s \uparrow 1$ with $\gamma_s(t) \in \Omega$ for $s < 1$, and that f is a function analytic on Ω, continuous on $\bar\Omega$ with $f\circ\gamma$ a simple closed curve in $\mathbb{C}$. Prove that f is a bijection of Ω and the interior of $f\circ\gamma$. (*Hint*: Use the argument principle.)

6. (a) Prove (8.4.20). (*Hint*: $t = k^{-1}s^{-1}$.)
   (b) Prove (8.4.19). (*Hint*: $t = (1-(k')^2s^2)^{-1/2}$.)

[2]One way suggests this is how Schwarz squared the circle.

7. Let $0 < k < 1$, $K(k) = \int_0^1 [(1-t^2)(1-k^2t^2)]^{-1/2}\,dt$, and $K'(k) = K(\sqrt{1-k^2}\,)$.

   (a) Prove that $\lim_{k\to 0} K(k) = \pi/2$ and $\lim_{k\to 1} K(k) = \infty$.

   (b) Prove that $K(k)$ is strictly monotone increasing in k.

   (c) Prove that $4K/2K'$ is strictly monotone in k and runs from 0 to ∞ as k runs from 0 to 1.

8. (a) Let $\gamma_j\pi$ be the exterior angle at vertex j of a polygon and $\alpha_j\pi$ the interior angle (see Figure 8.4.6) so $\alpha_j + \gamma_j = 1$. Prove that a closed polygon has $\sum_{j=1}^n \gamma_j = 2$ (so that (8.4.25) implies the polygon is closed).

   (b) Prove that if $0 < \alpha_j < 1$, then the polygon is convex.

   (c) Prove that if one starts at a point and turns at vertices $\alpha_1, \dots, \alpha_n$ obeying (8.4.25), then the initial and final segments are at least parallel.

9. (a) Prove that if (8.4.25) and (8.4.24) hold, then $\int_{-\infty}^{\infty} \prod_{j=1}^n (x_j - x)^{\alpha_j - 1}\,dx = 0$. (*Hint*: Close the contour.)

   (b) Show that when (8.4.24) and (8.4.25) hold, $\psi(z)$, given by (8.4.24) for $-\infty < z < \infty$, describes a closed polygon (minus one point which is $\psi(\infty)$).

10. Suppose (8.4.24) holds for $0 < \alpha_j < 1$ but that $1 < \sum_{j=1}^n (1-\alpha_j) < 2$. Prove that ψ, given by (8.4.23), describes a closed $(n+1)$-edge polygon with vertices at $\{\psi(x_j)\}_{j=1}^n$ plus at $\psi(\infty)$.

11. If $\psi''/\psi' = h$ is given by (8.4.34), prove that $\psi'(x) = C\prod_{j=1}^n (x_j - x)^{\alpha_j - 1}$ for some constant C.

12. Let $\psi(z) = z + z^{-1} = u + iv$. Prove that $u(re^{i\theta}) = a\cos\theta$, $v(re^{i\theta}) = -b\sin\theta$, where $a = r + r^{-1}$, $b = r^{-1} - r$, and that $u^2/a^2 + v^2/b^2 = 1$ for r fixed.

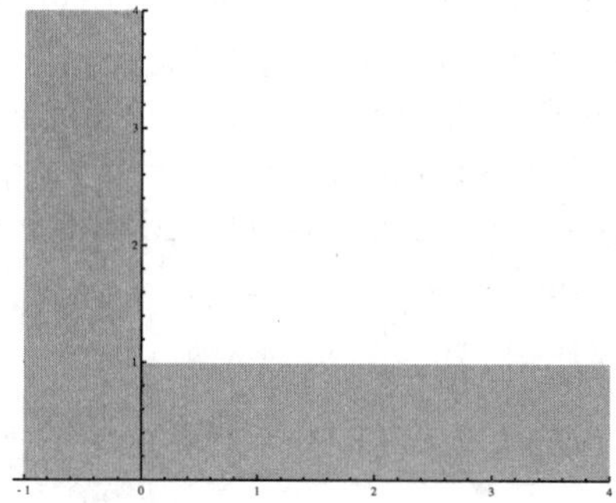

Figure 8.4.14. An L-shaped region.

13. Show that $w = \frac{2}{\pi}\,[\tanh^{-1}\sqrt{z} - \tan^{-1}\sqrt{z}]$ maps $\mathbb{C}_+$ to $\{w = u + iv \mid -1 < u < \infty,\ 0 < v < 1 \text{ or } v > 0,\ -1 < u < 0\}$, an L-shaped region (see Figure 8.4.14).

14. Show that $w = \tan^2 \frac{\pi}{4}\sqrt{z}$ maps the parabolic region $\{z = x + iy \mid y^2 = 4(1 - x)\}$ to the unit disk.

8.5. Bonus Section: Covering Map for General Regions

Note: The reader will need to look carefully at Section 1.7 before this bonus section and before bonus Section 8.7.

The Riemann mapping theorem says for any simply connected proper region, Ω, there exists $f\colon \mathbb{D} \to \Omega$ which is analytic, onto Ω, and globally one-one. Since the existence of such a map implies Ω is simply connected, we can't hope for such a map for a general region but suppose, instead of requiring f to be globally one-one, we only demand that it be locally one-one, that is, $f'(z) \neq 0$ for all z. There might be such a map, and indeed, we'll see, with one exceptional case, there is.

But even more is so. If f is locally one-one, for any $z, w \in \mathbb{D}$, distinct points, with $f(z) = f(w)$, then there is a neighborhood U of $f(z)$ and disjoint neighborhoods V_1, V_2 of z and w, so f is a bijection of V_j and U. But if we fix z and vary w, we may need to shrink U, and so V_1. Clearly, it would be nice to have a single U so $f^{-1}[U]$ is a disjoint union of open sets, each mapped bijectively to U. That's exactly saying that f is a covering map (see Section 1.7), and since $\mathbb{D}$ is simply connected, it will be the universal cover. As a covering map, there will be a group of deck transformations, so we'll actually prove the following:

Theorem 8.5.1. *For any region $\Omega \subset \mathbb{C}$ where $\mathbb{C}\setminus\Omega$ has at least two points, there exist an analytic map $f\colon \mathbb{D} \to \Omega$ and a Fuchsian group, Γ, of hyperbolic and/or parabolic elements of $\mathbb{A}\mathrm{ut}(\mathbb{D})$ so that*

(i) *f is onto Ω.*
(ii) *f is locally one-one, that is, $f'(z) \neq 0$ for all $z \in \mathbb{D}$.*
(iii) *$f(z) = f(w)$ if and only if there exists $\gamma \in \Gamma$ with $\gamma(z) = w$.*

Moreover, for any $w \in \Omega$, we can find f so $f(0) = w$ and $f'(0) > 0$ and the f that does this is unique.

The "moreover" follows as in the proof of the Riemann mapping theorem from the use of the structure of $\mathbb{A}\mathrm{ut}(\mathbb{D})$, so we'll say nothing more about it. Remarkably, the proof of this result will be almost identical to the Rado proof of the Riemann mapping theorem—indeed, it is from the same paper! As we'll explain in the Notes (and Section 8.7), the result is a corollary of the uniformization theorem of Koebe–Poincaré, which we'll prove in Section 8.7

(see Theorem 8.7.1). To put this theorem in context, we begin with the general issue of covering maps of Riemann surfaces.

Theorem 8.5.2. *Let $\mathcal{S}$ be a Riemann surface and $\pi\colon \mathcal{U} \to \mathcal{S}$ a covering map. Then there is a unique (up to compatibility) analytic structure on $\mathcal{U}$ so that π is analytic, and in that structure, all the deck transformations are analytic bijections of $\mathcal{U}$ (i.e., elements of $\mathbb{A}\mathrm{ut}(\mathcal{U})$). Moreover, if $\pi_1\colon \mathcal{U}_1 \to \mathcal{S}$ and $\pi_2\colon \mathcal{U}_2 \to \mathcal{S}$ are covering maps and $g\colon \mathcal{U}_1 \to \mathcal{U}_2$ obeys $\pi_2 \circ g = \pi_1$, then g is analytic in the associated analytic structures. In particular, two universal covers have analytic structures in which the unique homeomorphism between them (given choice of base point) is analytic.*

Proof. This is immediate from the definitions. Given $z_0 \in \mathcal{U}$, pick an open neighborhood U so $\pi \restriction U$ is a bijection of U and an open neighborhood, U_1, of $\pi(z_0)$. Pick V an open neighborhood of $\pi(z_0)$ contained in U_1 and $f\colon V \to \mathbb{C}$ a coordinate map. If π is to be analytic, $f \circ \pi\colon \pi^{-1}[V] \to \mathbb{C}$ must be a coordinate map, so there is a unique (up to compatibility) possible analytic structure. It is easy to confirm all these coordinate maps are consistent.

One can see (Problem 1) that the fundamental group, $\pi_1(\mathcal{S})$, is countable. Pick a countable family, $\{U_n, g_n\}_{n=0}^{\infty}$, of coordinate maps covering $\mathcal{S}$ so that for each U_n, $\pi^{-1}(U_n)$ is a family of disjoint sets. The number of disjoint sets is countable since $\pi_1(\mathcal{S})$ and the fibers are in one-one correspondence with a subgroup of $\pi_1(\mathcal{S})$. This gives a countable family of coordinate maps, so $\mathcal{U}$ is a Riemann surface.

Uniqueness of the analytic structure proves the deck transformations are analytic and the other statements in the theorem. $\square$

The uniqueness of the analytic structure of universal covers proves if $\pi_j\colon \mathcal{U}_j \to \mathcal{S}_j$ for $j = 1, 2$ and $f\colon \mathcal{S}_1 \to \mathcal{S}_2$ is an analytic bijection, then the map $g\colon \mathcal{U}_1 \to \mathcal{U}_2$ with $\pi_2 \circ g = f \circ \pi_1$ is an analytic bijection. From this, it is easy to see that:

Theorem 8.5.3. *Let $\mathcal{S}_1$ and $\mathcal{S}_2$ be two Riemann surfaces. Let $\mathcal{U}_1, \mathcal{U}_2$ be their universal covering spaces and Γ_1, Γ_2 the groups of deck transformations. Then $\mathcal{S}_1$ is analytically equivalent to $\mathcal{S}_2$ if and only if there is an analytic equivalence $g\colon \mathcal{U}_1 \to \mathcal{U}_2$ so that $\Gamma_2 = g\Gamma_1 g^{-1}$.*

Proof. If such a g exists, given $z \in \mathcal{S}_1$, pick $\tilde{z} \in \mathcal{U}_1$ with $\pi_1(\tilde{z}) = z$ and set $f(z) = \pi_2(g(\tilde{z}))$. If $\tilde{z}' \in \mathcal{U}_1$ is another point in $\mathcal{U}_1$ with $\pi_1(\tilde{z}_1') = z$, then there is $\gamma_1 \in \Gamma_1$, with $\tilde{z}' = \gamma_1(\tilde{z})$, so if $\gamma_2 = g\gamma_1 g^{-1}$, then $\gamma_2(g(\tilde{z})) = g\gamma_1(\tilde{z}) = g(\tilde{z}')$, so $\pi_2(g(\tilde{z})) = \pi_2(g(\tilde{z}')$ since $\gamma_2 \in g\Gamma_1 g^{-1}$. Thus, f is well-defined and so analytic. Applying the same argument to g^{-1} shows f is a bijection. Thus, the existence of g yields an analytic equivalence.

Conversely, if $f\colon \mathcal{S}_1 \to \mathcal{S}_2$ is an analytic equivalence, $f\circ\pi_1\colon \mathcal{U}_1 \to \mathcal{S}_2$ is a covering map, so by uniqueness and analyticity of covering maps, there is an analytic bijection $g\colon \mathcal{U}_1 \to \mathcal{U}_2$ so that $\pi_2 g = f\pi_1$. Tracking fibers and deck transformations shows that $\Gamma_2 = g\Gamma_1 g^{-1}$. □

Thus, the issue of analytic equivalence of Riemann surfaces is reduced to the study of simply connected Riemann surfaces (which we'll settle in bonus Section 8.7), the discrete subgroups of their fixed point free automorphisms, and the conjugacy problem for them. This will be the major theme of Sections 8.6, 8.7, and 10.7.

One can interpret the existence of the elliptic modular function, λ, of Section 8.3 as saying that $\mathbb{C}_+$ is the universal covering space of $\mathbb{C}\setminus\{0,1\}$ (or since it is analytically equivalent to $\mathbb{C}_+$, we could say $\mathbb{D}$ is the covering space). For one shows that λ is a covering map as follows: Given $z_0 \in \mathbb{C}\setminus\{0,1\}$, pick $w \in \widetilde{\mathcal{F}}$ with $\lambda(w) = z_0$. If $w \in \widetilde{\mathcal{F}}^{\text{int}}$, let $\varepsilon = \text{dist}(w_0, \mathbb{C}_+ \setminus \widetilde{\mathcal{F}})$ and $U = \{w_1 \mid \text{dist}(w, w_0) < \varepsilon\}$. For any $\gamma\in\Gamma$ and $w_1 \in U$, $\gamma(w_1) \notin \mathcal{F}$, so $\gamma[U]\cap U = \emptyset$, which implies for all γ_1,γ_2 that $\gamma_1[U]\cap\gamma_2[U] = \emptyset$. Thus, if $V = \lambda[U]$, $\lambda^{-1}[V] = \bigcup_{\gamma\in\Gamma}\gamma[U]$ is a disjoint union of open sets on which λ is a bijection. A separate argument (see Problem 2) works on $\partial\mathcal{F}$.

We now turn to the proof of Theorem 8.5.1. We'll require a theorem (the monodromy theorem, Theorem 11.2.1) only proven in Section 11.2. If f is defined and analytic in a neighborhood of a point z_0 of a simply connected Riemann surface, $\mathcal{U}$, and f has an analytic continuation in a neighborhood of any simple path starting at z_0. then f has an analytic continuation to all of $\mathcal{U}$.

We need to find an analog of the set, $\mathcal{R}$, of functions in (8.1.1) used to prove the Riemann mapping theorem. We'll assume $\Omega\subset\mathbb{C}$ is missing at least two points of $\mathbb{C}$ and let $\pi\colon\mathcal{U}\to\Omega$ be its universal cover. Pick $z_0\in\mathcal{U}$. Define $\mathcal{R}$ by

$$\begin{aligned}\mathcal{R} = \{f\colon \mathcal{U}\to\mathbb{D} \mid\ & f \text{ is analytic, } f(z_0) = 0,\\ & f'(z_0) > 0,\ f(z) = f(w) \Rightarrow \pi(z) = \pi(w)\}\end{aligned} \tag{8.5.1}$$

Here f' is defined in the local coordinates inherited from Ω using π.

Lemma 8.5.4. *$\mathcal{R}$ is nonempty.*

Proof. Suppose that $a\neq b$ with $a,b\notin\Omega$. Let h be an FLT that maps ∞ to ∞, 0 to a, and 1 to b. Let λ be the elliptic modular function. Then $\Lambda = h\circ\lambda$ is a covering map of $\mathbb{C}_+$ to $\mathbb{C}\setminus\{a,b\}$, so any curve in $\mathbb{C}\setminus\{a,b\}$ can be lifted. In particular, if γ is a curve in $\mathcal{U}$ starting at z_0, we can find $\tilde\gamma$ a curve in $\mathbb{C}_+$, so $\Lambda\circ\tilde\gamma(z) = \pi\circ\gamma(z)$. We can define g in a neighborhood of $\gamma(z)$ as $\Lambda^{-1}\circ\pi$ since Λ is one-one at each point in $\tilde\gamma(z)$.

By the monodromy theorem, g defined as $\Lambda^{-1} \circ \pi$ in a small neighborhood of z_0 extends to all of $\mathcal{U}$. If $g(z) = g(w)$, then $\Lambda(g(z_0)) = \Lambda(g(w))$, so $\pi(z) = \pi(w)$.

Using a conformal map, $\tilde{h}$, of $\mathbb{C}_+$ to $\mathbb{D}$, we get $f = \tilde{h} \circ g$ so $f(z_0) = 0$, $f'(z_0) > 0$, and since $\tilde{h}$ is invertible, $f(z) = f(w) \Rightarrow g(z) = g(w) \Rightarrow \pi(z) = \pi(w)$. Thus, $f \in \mathcal{R}$. $\square$

Proof of Theorem 8.5.1. Define $\mathcal{R}$ by (8.5.1). If we find $g \in \mathcal{R}$ which is onto $\mathbb{D}$, define $f\colon \mathbb{D} \to \Omega$ as follows: Given $\zeta \in \mathbb{D}$, find $z \in \mathcal{U}$ so $g(z) = \zeta$ and define $f(\zeta) = \pi(z)$. If $g(w) = \zeta$ also, by (8.5.1), $\pi(z) = \pi(w)$, so f is well-defined. Since π is a covering map, it is easy to see that f, and thus, by uniqueness of the universal covering space, $\mathbb{D}$ is the universal covering space, and the deck transformations are the required Γ. Indeed, one can see (Problem 3) that this implies g is a bijection and Γ is the image of the deck transformation on $\mathcal{U}$ under g.

By a Hurwitz theorem argument (Problem 4), $\mathcal{R} \cup \{0\}$ is compact in the topology of uniform convergence, so we can find $g \in \mathcal{R}$ maximizing $g'(z_0)$.

Since $\mathcal{U}$ is simply connected, it obeys the square root property, so the argument of Lemma 8.1.6 applies, and the g maximizing $g'(z_0)$ must be onto $\mathbb{D}$. $\square$

Notes and Historical Remarks. As noted, this proof is from Radó's 1923 paper [**466**], although the result follows from the uniformization theorem which was proven earlier in 1907; see Section 8.7.

Problems

1. Prove that the fundamental group for any Riemann surface is countable. (*Hint*: Pick a countable dense set and prove that for any n, any γ is homotopic to a curve with values in this dense set for $s = j/n$, $j = 0, 1, \dots, n$.)

2. Let $w \in \partial\widetilde{\mathcal{F}}$. Prove there is a neighborhood U of w with $\gamma[U] \cap U = \emptyset$ for all $\gamma \in \Gamma_2$ with $\gamma \neq \mathbb{1}$.

3. Prove that the map g in the first paragraph of the proof of Theorem 8.5.1 is a bijection.

4. Prove that any uniform limit on compact subsets of $\mathcal{U}$ of functions in $\mathcal{R}$ obeys either $g \equiv 0$ or $g(w) = g(z) \Rightarrow \pi(w) = \pi(z)$.

5. Using Theorem 8.5.3, prove that if $\lambda, \widetilde{\lambda}$ obey the properties of Theorem 8.3.1, then for some $\gamma \in \mathbb{A}\mathrm{ut}(\mathbb{C}_+)$, we have $\widetilde{\lambda} = \lambda \circ \gamma$.

8.6. Doubly Connected Regions

The keystone result in this chapter says that (except for $\mathbb{C}$) all simply connected regions are analytically isomorphic to $\mathbb{D}$. In this section, we'll determine the isomorphism classes for doubly connected regions, that is, Ω so that $\mathbb{C}\setminus\Omega$ has exactly one bounded component; equivalently (by Theorem 4.5.1), $\widehat{\mathbb{C}}\setminus\Omega$ has exactly two components.

Recall the annulus, $\mathbb{A}_{r,R}$, is defined for $0 \le r < R \le \infty$ by

$$\mathbb{A}_{r,R} = \{z \mid r < |z| < R\} \tag{8.6.1}$$

We have some degenerate cases: $r = 0$, $R = \infty$, is the punctured plane, $\mathbb{C}^\times$, and $r = 0$, $R < \infty$ a punctured disk, $\mathbb{D}^\times$. Notice under $z \to z^{-1}$, $\mathbb{A}_{r,\infty}$ maps to $\mathbb{A}_{0,r^{-1}}$, so if we care only up to analytic isomorphism, we need not consider $R = \infty$, $r > 0$. Here are the two results we'll prove:

Theorem 8.6.1. *$\mathbb{A}_{r,R}$ and $\mathbb{A}_{r',R'}$ are analytically isomorphic if and only if*

$$\frac{r}{R} = \frac{r'}{R'} \tag{8.6.2}$$

with the special rule for $r/R = 0$ that $r = 0$, $R = \infty$ is in its own class and all $(r,R) \in \{(r',R') \mid r' = 0,\ R' < \infty\} \cup \{(r',R') \mid r' > 0, R' = \infty\}$ are isomorphic.

Theorem 8.6.2. *Every doubly connected region is analytically isomorphic to some annulus $\mathbb{A}_{r,R}$.*

Our (first) proof of Theorem 8.6.1 will use nothing more than the reflection principle—essentially the arguments used in Problem 7 of Section 7.4 to find $\mathrm{Aut}(\mathbb{A}_{r,R})$. Theorem 8.6.2 relies though on some heavy machinery—the covering space results of the last section—and so, in some sense, all the current section can be thought of as a bonus section. One can also use that machinery to prove Theorem 8.6.1 (see the remark after Theorem 8.6.3).

Proof of Theorem 8.6.1. If $0 < r < R < \infty$ and $0 < r' < R' < \infty$ and (8.6.2) holds, then there is $\lambda \in \mathbb{C}^\times$ so that $r' = \lambda r$, $R' = \lambda R$, and then $z \to \lambda z$ maps $\mathbb{A}_{r,R}$ bijectively to $\mathbb{A}_{r',R'}$ so they are analytically isomorphic.

If $0 < r < R = \infty$ and $0 < r' < R' = \infty$, pick $\lambda = r'/r$ and $z \to \lambda z$ maps $\mathbb{A}_{r,R}$ to $\mathbb{A}_{r',R'}$ and similarly, if $r = r' = 0$ and $R, R' < \infty$, let $\lambda = R'/R$. Finally, if $0 = r < R < \infty$ and $0 < r' < R' = \infty$, $z \to r'R/z$ maps $\mathbb{A}_{r,R}$ to $\mathbb{A}_{r',R'}$. Thus, all the claimed isomorphisms hold, and we need the converse.

Suppose $\varphi\colon \mathbb{A}_{r,R} \to \mathbb{A}_{r',R'}$ is an analytic bijection. If $R < \infty$, let $\rho = \frac{1}{2}(r+R)$, and if $R = \infty$, let $\rho = r+1$. $\gamma(s) = \varphi(\rho e^{2\pi i s})$ for $0 \le s \le 1$ is a closed analytic Jordan curve, so by the Jordan curve theorem, γ breaks $\mathbb{C}$, and so $\mathbb{A}_{r',R'}$, into an inside and outside. By continuity, φ must take

$\{z \mid r < |z| < \rho\}$ entirely into the inside of γ or entirely into the outside. If into the outside, replace $\varphi(z)$ by $1/\varphi(z)$ and $\mathbb{A}_{r',R'}$ by $\mathbb{A}_{1/R',1/r'}$. Thus, we can suppose φ takes the inside to the inside.

As in the proof of Theorem 8.2.1, as $|z| \to r$ or R, $|\varphi(z)|$ must approach r' or R'. Since φ takes the inside of $|z| = \rho$ to the inside of γ, we must have

$$\lim_{|z|\downarrow r} |\varphi(z)| = r', \qquad \lim_{|z|\uparrow R} |\varphi(z)| = R' \tag{8.6.3}$$

We first claim that if $r > 0$, then $r' > 0$, for if $r' = 0$, then the Laurent coefficients of φ are all zero by taking $\kappa \downarrow r$ in $a_n = (2\pi i) \oint_{|z|=\kappa} z^{-n-1}\varphi(z)\, dz$, which is impossible. Similarly, looking at $1/\varphi$, we see that if $R < \infty$, then $R' < \infty$.

By the reflection principle, if $r > 0$, we can extend φ to $\mathbb{A}_{r^2/R,r}$ by $\varphi(z) = (r')^2\, \overline{\varphi(r^2/z)}^{-1}$ and this extended φ is a bijection of $\mathbb{A}_{r^2/R,R}$ to $\mathbb{A}_{(r')^2/R',R'}$.

By iterating and doing the same near infinity, if $R < \infty$, we extend φ to $\mathbb{C}^\times$ and the extended φ is a bijection to $\mathbb{C}^\times$ and bounded near 0, so 0 is a removable singularity and φ is a bijection of $\mathbb{C}$ to $\mathbb{C}$ with $\varphi(0) = 0$. So by Theorem 7.3.4, $\varphi(z) = \lambda z$ for some λ.

Restoring the prescribed replacement of φ by $1/\varphi$, we see the original φ is either λz or λ/z. In either case, (8.6.3) holds. □

* * * * *

As a preliminary for the proof of Theorem 8.6.2, we need to identify the covering space and, when $\mathbb{D}$, the Fuchsian group for the universal cover of $\mathbb{A}_{r,R}$. We'll use the $\mathbb{C}_+$ model instead of $\mathbb{D}$.

Theorem 8.6.3. (a) *The map*

$$E(z) = e^{2\pi i z} \tag{8.6.4}$$

from $\mathbb{C}$ onto $\mathbb{C}^\times \equiv \mathbb{A}_{0,\infty}$ is a universal covering map with deck transformation, for $n \in \mathbb{Z}$,

$$f_n(z) = z + n \tag{8.6.5}$$

(b) *The maps*

$$E_R(z) = R\, e^{2\pi i z} \tag{8.6.6}$$

from $\mathbb{C}_+$ to $\mathbb{A}_{0,R}$ are universal covering maps with (parabolic) deck transformations given by (8.6.5).

(c) *The maps*

$$\widetilde{E}_r(z) = r\, e^{-2\pi i z} \tag{8.6.7}$$

from $\mathbb{C}_+$ to $\mathbb{A}_{r,\infty}$ are universal covering maps with (parabolic) deck transformations given by (8.6.5).

(d) *Let* $L\colon \mathbb{C}_+ \to \{x+iy \mid x \in \mathbb{R},\, 0<y<\pi\}$ *by*

$$L(z) = \log(z) = \log(|z|) + i\arg(z), \qquad 0 < \arg(z) < \pi \tag{8.6.8}$$

For $0 < r < \infty$, *let*

$$\alpha = \frac{1}{\pi}\log\left(\frac{R}{r}\right) \tag{8.6.9}$$

The map

$$\varphi_{r,R}(z) = r\, e^{-i\alpha L(z)} \tag{8.6.10}$$

is a universal covering map of $\mathbb{C}_+$ *onto* $\mathbb{A}_{r,R}$ *with deck transformations given by*

$$f_n(z) = \lambda_{r,R}^n z \tag{8.6.11}$$

where

$$\lambda_{r,R} = \exp\left(\frac{2\pi}{\alpha}\right) \tag{8.6.12}$$

Proof. (a)–(c) are all straightforward calculations. For (d), note first that L maps $\mathbb{C}_+$ to the indicated region and

$$L(f_n(z)) = L(z) + \frac{2\pi n}{\alpha} \tag{8.6.13}$$

Since $e^{-i\alpha(i\pi)} = R/r$, $\varphi_{r,R}$ maps the strip to $\mathbb{A}_{r,\mathbb{R}}$, and by (8.6.13), $\varphi_{r,R}(z) = \varphi_{r,R}(w) \Leftrightarrow \exists n$ so that $w = f_n(z)$. $\square$

Notice first that this provides another proof of Theorem 8.6.1 since if $r'/R' \neq r/R$, the Fuchsian groups are not conjugate. So by Theorem 8.5.3, there cannot be an analytic bijection. Also note:

Corollary 8.6.4. *As* R *runs through* $(1,\infty]$, *the Fuchsian groups for* $\mathbb{A}_{1,R}$ *include one from each conjugacy class of fixed point free one-parameter Fuchsian subgroups of* $\mathbb{A}\mathrm{ut}(\mathbb{C}_+)$.

Proof. As R/r runs from 1 to ∞, α in (8.6.9) runs from 0 to ∞, so $\lambda_{r,R}$ runs from 1 to ∞, so $\lambda_{r,R} + \lambda_{r,R}^{-1}$ runs from 2 to ∞, and thus, every possible hyperbolic class occurs, so by Theorem 7.4.5, $R=\infty$, $r=1$ is parabolic. While there are two parabolic classes, they are inverses, so there is only one group with parabolic generators. $\square$

The key to the proof of Theorem 8.6.2 is thus to show that for any doubly connected region Ω, the fundamental group, $\pi_1(\Omega)$, is $\mathbb{Z}$. One can deduce this from purely topological considerations, but we'll exploit some complex structure. By the Riemann mapping theorem, except for a special degenerate case by a preliminary conformal map, we'll be able to take $\mathbb{C}\setminus\mathbb{D}$ as the unbounded component.

Lemma 8.6.5. *Let $K \subset \mathbb{D}$ be connected and compact with $0 \in K$ so that $\Omega = \mathbb{D} \setminus K$ is connected. Let E be the map* (8.6.4). *Then $E^{-1}(\mathbb{D} \setminus K)$ is a connected and simply connected set.*

Proof. Note first that since E is a covering map of $\mathbb{C}^\times$, E is a covering map of $\mathbb{D} \setminus K$, except perhaps for connectedness of $E^{-1}(\mathbb{D} \setminus K)$.

By compactness, $\sup_{z \in K} |z| = \rho < 1$, so

$$Q \equiv E^{-1}(\mathbb{D} \setminus K) \tag{8.6.14}$$

obeys

$$\{z \mid \log(\rho) < \operatorname{Re} z < 0\} \subset Q \subset \{z \mid \operatorname{Re} z < 0\} \tag{8.6.15}$$

Since Ω is connected and open, it is arcwise connected, so there is a curve γ for any $z_0 \in \Omega$ to $\frac{1}{2}(\rho + 1)$. Since E is a covering map, given any $z_1 \in Q$, let $z_0 = E(z_1)$ and $\tilde{\gamma}$ the left of γ to Q. Thus, any z_1 can be connected by an arc in Q to the strip on the extreme left of (8.6.15). This strip is connected, so Q is.

If Q is not simply connected, by the proof of Theorem 4.5.1, we can find $A \subset \mathbb{C} \setminus Q$ compact, Q' open with

$$A \subset Q' \subset Q \cup A \tag{8.6.16}$$

We'll prove that $E[A]$ is closed and relatively open in K. Then by connectedness of K, $E[A] = K$. Since $0 \notin E[A]$, this is impossible, and the contradiction implies that Q is simply connected.

A must be disjoint from $E^{-1}(\mathbb{C} \setminus \mathbb{D})$ since the latter is connected and A is both closed, bounded, and relatively open. Thus, $E[A]$ is closed and contained in K. Since E is a covering map from $\mathbb{C}$ to $\mathbb{C}^\times$, it takes open sets to open sets, and so $E[Q']$ is open. Since $Q = E^{-1}(\mathbb{D} \setminus K)$, $E[Q'] \subset (\mathbb{D} \setminus K) \cup A$, that is, A is relatively open in K. $\square$

Proof of Theorem 8.6.2. Let $\Omega \subset \mathbb{C}$ be a doubly connected region and let K_1, K_2 be the two components of $\widehat{\mathbb{C}} \setminus \Omega$. If each has one point, $\Omega = \mathbb{C} \setminus \{z_0\}$ for some z_0, so clearly, Ω is analytically equivalent to $\mathbb{A}_{0,\infty}$, an annulus. If not, by an FLT, we can map Ω to $\widetilde{\Omega}$ in $\mathbb{C}$, where $\widetilde{K}_1, \widetilde{K}_2$ are such that $\#(\widetilde{K}_2) \geq 2$.

It follows that $\widetilde{\Omega} \cup K_1$ is simply connected and not all of $\mathbb{C}$, so by the Riemann mapping theorem, $\widehat{\mathbb{C}}$, and so Ω, is conformally equivalent to $\mathbb{D} \setminus K$ for some compact K. By an element of $\mathbb{A}\mathrm{ut}(\mathbb{D})$, we can arrange $0 \in K$. Thus, by a preliminary conformal map, Ω is taken to $\widetilde{\Omega}$ of the form where Lemma 8.6.5 is applicable.

Thus, we find $\mathcal{U}$ connected and simply connected, a universal covering map of the original Ω so that the deck transformations are given by

$z \to z + 2\pi i n$. By the Riemann mapping theorem, we can map $\mathcal{U}$ to $\mathbb{C}_+$ and so suppose that $\mathbb{C}_+$ is the universal cover of Ω and the sets of deck transformations are $\{f^{[n]}\}_{n\in\mathbb{Z}}$ for some fixed point free element of $\mathbb{A}\mathrm{ut}(\mathbb{C}_+)$.

By Corollary 8.6.4, every such group is the Fuchsian group for some $\mathbb{A}_{1,R}$. So by Theorem 8.5.3, Ω is analytically equivalent to some $\mathbb{A}_{1,R}$, $R \in [1,\infty]$, or else to $\mathbb{A}_{0,\infty}$. $\square$

We end with a single result about n-connected regions:

Theorem 8.6.6. *Let Ω be an open region of $\widehat{\mathbb{C}}$ which is n-connected (i.e., $\widehat{\mathbb{C}} \setminus \Omega$ has exactly n components) and nondegenerate in that no component of $\widehat{\mathbb{C}} \setminus \Omega$ is a single point. Then, there exist $C_1, \dots, C_{n-1}$, disjoint analytic Jordan curves in $\mathbb{D}$, so that Ω is conformally equivalent to $\mathbb{D}\setminus\bigcup_{j=1}^{n-1} \widetilde{C}_j$ where $\widetilde{C}_j$ is C_j and the points inside it.*

Remarks. 1. If there are $n-k$ components of $\widehat{\mathbb{C}} \setminus \Omega$ which are points and $k \geq 1$ which are not, the same proof shows that there are $C_1, \dots, C_{k-1}$ disjoint analytic curves and $w_1, \dots, w_{n-k}$ distinct points in $\mathbb{D} \setminus \bigcup_{j=1}^{k-1} \widetilde{C}_k$ so that Ω is conformally equivalent to $\mathbb{D} \setminus \big[\{z_j\}_{j=1}^{n-k} \cup \bigcup_{j=1}^{k-1} \widetilde{C}_j\big]$.

2. This implies that for purely internal issues (i.e., not involving boundary behavior) there is no loss in considering Ω with an analytic and so C^∞ boundary!

3. This result is sometimes stated with all occurrences of $\mathbb{D}$ replaced by $\widehat{\mathbb{C}}\setminus\overline{\mathbb{D}}$.

Proof. If $n = 1$, Ω is simply connected and $\neq \mathbb{C}$, so the conclusion is just the Riemann mapping theorem.

Suppose $n > 1$ and we have the result for $(n-1)$-connected regions and $\Omega = \widehat{\mathbb{C}} \setminus \bigcup_{j=1}^{n} \mathfrak{e}_j$ with each $\mathfrak{e}_j$ compact and these sets pairwise disjoint. Let $\Omega_0 = \widehat{\mathbb{C}} \setminus \bigcup_{j=1}^{n-1} \mathfrak{e}_j$. By the induction hypothesis, there is $f_0\colon \Omega_0 \to \mathbb{D} \setminus \bigcup_{j=1}^{n-2} \widetilde{C}_j^{(0)}$ a biholomorphic bijection. Let $\tilde{e}_n = f_0\,[e_n]$. Then $\tilde{e}_n$ is connected so $\widehat{\mathbb{C}} \setminus \tilde{e}_n$ is simply connected and there exists $g\colon \widehat{\mathbb{C}} \setminus \tilde{e}_n \to \mathbb{D}$, a biholomorphic bijection. Then $g \circ \big[f_0 \restriction (\Omega_0 \setminus e_n)\big]$ is a biholomorphic bijection of Ω and $\mathbb{D} \setminus \bigcup_{j=1}^{n-1} \widetilde{C}_j$ where $C_j = g\big[C_j^{(0)}\big]$ for $j = 1, \dots, n-2$ and $C_{n-1} = g\big[\partial\mathbb{D}\big]$. Since g is analytic, each C_j is an analytic Jordan curve. $\square$

Notes and Historical Remarks. One way of thinking about the results of this section is to note that all doubly connected sets are homeomorphic, so we've determined all possible nonequivalent analytic structures on this topological space $\mathbb{C}^\times$. They are parametrized by a single real number $\alpha = r/R$ that runs from 0 to 1 with two extra points ($(r,R) = (1,\infty)$ and $(0,\infty)$), which one can adjoin to $(0,1)$ in some way if one wishes. (To adjoin $(1,\infty)$ as $\alpha = 0$ is natural—since $(0,\infty)$ has a nonequivalent universal cover, it is natural to leave it out.)

A set of natural parameters for classifying analytic structures is called a *modular space* or *space of moduli*, and a number like α is called a modulus. In Sections 10.6 and 10.7, we'll study the moduli for complex tori.

For n-connected sets, one method of classification would be to look at possible conjugacy classes of finitely generated Fuchsian groups with no elliptic elements. Another method tries to find canonical sets like $\mathbb{A}_{\alpha,1}$ with every n-connected set equivalent to 1. For n-connected sets, the canonical sets are $\mathbb{A}_{\alpha,1}$ with $n-2$ arcs removed, that is, sets of the form $\{re^{i\theta} \mid r = r_0,\ \theta_0 < \theta < \theta_1\}$. There are $n-2$ r's, θ_0 and θ_1 yielding $3(n-2)$ parameters, but one overall rotational symmetry (so adding the same β to all θ_0's and θ_1's gives an equivalent set). Of course, α is an additional parameter to make up for the one lost by rotations. This shows the modular space for n-connected regions with $n \geq 3$ is $(3n-6)$-dimensional. Ahlfors [**9**], Krantz [**331**], and Nehari [**405**] discuss this further as well as a representation in terms of parallel slits. This work, as well as the more common analysis of the doubly connected case, depends on the use of potential theory, aka the theory of harmonic functions, rather than the approach we have used here.

8.7. Bonus Section: The Uniformization Theorem

Here we want to discuss and partially prove the following remarkable theorem:

Theorem 8.7.1 (Uniformization Theorem). *The only simply connected Riemann surfaces (up to equivalence) are $\widehat{\mathbb{C}}$, $\mathbb{C}$, and $\mathbb{D}$.*

This result of Koebe and Poincaré can be viewed as a strengthening of the Riemann mapping theorem—it implies the Riemann mapping theorem once one proves that any simply connected $\Omega \subsetneq \mathbb{C}$ has a bounded analytic function, so it cannot be analytically equivalent to all of $\mathbb{C}$ (and since Ω is not compact, it can't be equivalent to $\widehat{\mathbb{C}}$).

The proof of this theorem relies on methods of potential theory—essentially the existence of certain harmonic functions with prescribed singularities. We'll quote the results needed below but only prove them in Part 3 (see Section 3.8 of that part).

Before turning to the proof, we want to note an important consequence:

Theorem 8.7.2. *All Riemann surfaces have $\mathbb{D}$ as their universal covering space with the following exceptions:*

(a) $\widehat{\mathbb{C}}$, *which is the only Riemann surface with universal cover* $\widehat{\mathbb{C}}$

(b) $\mathbb{C}$
(c) $\mathbb{C}\setminus\{0\}$
(d) *Tori* $\mathcal{J}_{\tau_1,\tau_2}$

Remarks. 1. Tori are constructed in Example 7.1.3.

2. We mean here, of course, up to analytic equivalence. For example, we can put a natural complex structure on the infinite cylinder $\{z \mid 0 \le \operatorname{Re} z < 1\}$ with iy and $1+iy$ made equivalent for all y. This is analytically isomorphic to $\mathbb{C}^\times$ (Problem 1).

3. Section 10.7 will determine which $\mathcal{J}_{\tau_1,\tau_2}$ are nonequivalent.

Proof. The deck transformations are groups of fixed point free elements of $\mathbb{A}\mathrm{ut}(\mathcal{U})$. $\widehat{\mathbb{C}}$ has no fixed point free automorphisms, so only $\widehat{\mathbb{C}}$ has $\widehat{\mathbb{C}}$ as universal cover.

The only fixed point free automorphisms of $\mathbb{C}$ are translations $z \to z+a$. Thus, we are interested in additive subgroups, Γ, of $\mathbb{C}$ which are discrete in that $\{a \in \Gamma\}$ has no limit points. In Theorem 10.2.1, we'll prove every such group is of the form $\{n\tau\}_{n\in\mathbb{Z}}$ or $\{n_1\tau_1 + n_2\tau_2\}_{n_1,n_2\in\mathbb{Z}}$ for τ,τ_1,τ_2 in $\mathbb{C}$ nonzero with $\tau_2/\tau_1 \notin \mathbb{R}$. All $\{n\tau\}_{n\in\mathbb{Z}}$ are conjugate (Problem 2) and $\widehat{\mathbb{C}}\setminus\{0\}$ has covering map, E, given by (8.6.4) with $\Gamma = \{2\pi i n\}_{n\in\mathbb{C}}$. $\mathcal{J}_{\tau_1,\tau_2}$, by construction, has covering map $\mathbb{C}$ with deck transformations $\{n_1\tau_1 + n_2\tau_2\}$. $\square$

Once one has uniformization, many questions can be tackled on a case-by-case basis. For example, in Problem 6, the reader will prove:

Theorem 8.7.3. *Every Riemann surface has a non-abelian fundamental group with the following exceptions:*

(i) $\widehat{\mathbb{C}}$, $\mathbb{C}$, $\mathbb{D}$.
(ii) $\mathbb{C}\setminus\{0\}$ *and the tori* $\mathcal{L}_{\tau_1,\tau_2}$ *of Example* 7.1.3.
(iii) $\mathbb{D}\setminus\{0\}$ *and the annulli* $\mathbb{A}_{r,R}$ *with* $0<r<R<\infty$.

Remarks. 1. Of course, we mean these exceptions up to analytic equivalence.

2. $\mathcal{L}_{\tau_1,\tau_2}$ is a family of Riemann surfaces since only some are analytically equivalent as τ_1,τ_2 vary. The equivalence problem is solved in Section 10.7.

3. $\mathbb{A}_{r,R}$ has an equivalence problem solved in Theorem 8.6.1. What matters is r/R. We use $0<r<R<\infty$ rather than $0\le r<R\le\infty$, since $\mathbb{A}_{0,\infty}\simeq\mathbb{C}\setminus\{0\}$ and for any $R<\infty$ or $0<r$, $\mathbb{A}_{r,\infty}\simeq\mathbb{A}_{0,R}\simeq\mathbb{D}\setminus\{0\}$.

Notice that Theorem 8.7.1 provides another immediate proof of Theorem 8.5.1 if we note the only regions of $\mathbb{C}$ in the list of exceptions are $\mathbb{C}$ and $\mathbb{C}^\times$. The covering map and group of deck transformations provide the necessary map. Notice also that Theorem 8.7.1 provides a second construction of the elliptic modular function: $\mathbb{D}$ is the universal cover of $\mathbb{C}\setminus\{0,1\}$ so, since there is analytic bijection of $\mathbb{C}_+$ and $\mathbb{D}$, we get $\mathbb{C}_+$ as universal cover. The covering map and Fuchsian group of deck transformations provide the λ and Γ of Theorem 8.3.1.

We now turn to the proof of Theorem 8.7.1. As we noted, it relies on the use of harmonic functions and associated analytic functions. A function, f, on a Riemann surface, $\mathcal{S}$, is called *harmonic* if locally it is given as the real part of an analytic function. We begin with a warmup:

Proposition 8.7.4. *Let u be a harmonic function on a simply connected Riemann surface, $\mathcal{U}$. Then there is an analytic function, f, on $\mathcal{U}$ so $u = \operatorname{Re} f$.*

Proof. In local coordinates, $z = x + iy$, consider the one-form

$$\begin{aligned}\tilde{du} &= \left(\frac{\partial u}{\partial x} - i\frac{\partial u}{\partial y}\right)dx + \left(\frac{\partial u}{\partial y} + i\frac{\partial u}{\partial x}\right)dy \qquad (8.7.1)\\ &= \left(\frac{\partial u}{\partial x} - i\frac{\partial u}{\partial y}\right)(dx + idy) = 2(\partial u)\,dz\end{aligned}$$

whose expression we pick, since if $f = u + iv$ is analytic, $\frac{\partial f}{\partial x}dx + \frac{\partial f}{\partial y}dy$ is given by (8.7.1) because of the Cauchy–Riemann equation. It is easy to see (Problem 3(a)) that as a one-form, $g\,dz$, this is coordinate-system independent, and because u is harmonic, $d(\tilde{du}) = 0$ (Problem 3(b)). Define

$$f(z) = u(z_0) + \int_{z_0}^{z} \tilde{du} \qquad (8.7.2)$$

Because $\tilde{du}$ obeys $d(\tilde{du}) = 0$ and $\mathcal{U}$ is simply connected, the integral is path-independent. One can check that f is analytic and $\operatorname{Re} f = u$ (Problem 3(c)). $\square$

We want to extend this to allow local singularities.

Definition. We say u is harmonic near $p \in \mathcal{S}$, a Riemann surface, up to a *polar singularity* at p with *charge*, $\eta \in \mathbb{C}_+$, if there is an open neighborhood, N, of p with u harmonic in $N \setminus \{p\}$ and in some local coordinate z with $z(p) = 0$, $u(q) + \eta\log(|z(q)|)$ is harmonic near p.

It is easy to see this notion is independent of local coordinate system.

Theorem 8.7.5. *Let $\mathcal{U}$ be a simply connected Riemann surface and $p_1, \dots, p_\ell$ a finite set of distinct points in $\mathcal{U}$. Let u be a real-valued function on $\mathcal{U}\setminus\{p_1,\dots,p_\ell\}$ so that u is harmonic on that set with polar singularities at the p_j with charges $\eta_j \in \mathbb{Z}\setminus\{0\}$. Then there is a meromorphic function f*

on $\mathcal{U}$ with poles only at $\{p_j \mid \eta_j < 0\}$ and zeros only at $\{p_j \mid \eta_j > 0\}$, where $|\eta_j|$ is the order of pole or zero and

$$|f(p)| = e^{-u(p)} \tag{8.7.3}$$

Proof. As in the last theorem, form $\tilde{du}$. This is a differential form on $\mathcal{U} \setminus \{p_1, \dots, p_\ell\}$. Near p_ℓ, by using local coordinates, it looks like $-\eta_j\, dz/z(p_j)$, which means that its integral is not path-independent, but the values along different paths differ by $2\pi i$ times an integer, so $e^{-u(z_0)} \exp(-\int_{z_0}^{z} \tilde{du})$ is path-independent and defines an analytic single-valued function on $\mathcal{U}\setminus\{p_1, \dots, p_\ell\}$ and it obeys (8.7.3).

Near p_j, the explicit form of the singularity shows $f(q) \sim C(z - z(p))^{\eta_j}$, so the singularity is a zero (if $\eta_j > 0$) or pole (if $\eta_j < 0$). □

We need two special kinds of functions:

Definition. Let $\mathcal{U}$ be a simply connected Riemann surface. A *Green's function*, $g(q;p)$, for a point $p \in \mathcal{S}$ is a function harmonic on $\mathcal{S} \setminus \{p\}$ with the following properties:

(i) p has a polar singularity with charge 1

(ii) $g(q;p) > 0 \quad$ for all q (8.7.4)

(iii) Let $\varphi(q,p)$ be an analytic function with

$$|\varphi(q,p)| = e^{-g(q;p)} \tag{8.7.5}$$

Then if $\psi(q)$ is any other analytic function

$$|\psi(q)| < 1 \qquad \psi(p) = 0 \quad \Rightarrow \quad |\psi(q)| \le |\varphi(q,p)| \tag{8.7.6}$$

for all q.

For the more usual definition, which makes sense for general Riemann surfaces, (iii) is replaced by a different maximum condition, but it is equivalent (see Section 3.8 of Part 3).

Definition. Let $\mathcal{S}$ be a Riemann surface. A *bipolar Green's function* $\beta(q;p_0,p_1)$ defined for $p_0, p_1 \in \mathcal{S}$ is a function harmonic on $\mathcal{S} \setminus \{p_0, p_1\}$ with polar singularities at p_0 and p_1, with charges 1 at p_0 and -1 at p_1, and with β bounded on each $\mathcal{S} \setminus N_0 \cup N_1$, where N_j is an arbitrary neighborhood of p_j.

Here are four basic facts we'll prove in Part 3 (see Section 3.8 of Part 3). In that discussion, we'll define Green's functions for any Riemann surface, so we state the results in the general context. $\mathcal{S}$ will denote a general Riemann surface and $\mathcal{U}$ one that is simply connected.

Fact 1. If a Green's function $g(q;p)$ exists for one $p \in \mathcal{S}$, it exists for all $p \in \mathcal{S}$.

Fact 2. If $\mathcal{S}$ has a nonconstant bounded analytic function, then a Green's function exists. Conversely, if $\mathcal{U}$ is simply connected and a Green's function exists, then $\mathcal{U}$ has bounded analytic functions.

Note: Indeed, $\varphi(q,p)$ is such a function.

Fact 3. If $\mathcal{S}$ has a Green's function, then

$$g(p;q) = g(q;p) \tag{8.7.7}$$

Fact 4. Bipolar Green's functions exists for any distinct q_0, q_1 in any $\mathcal{S}$.

We now have the tools to prove Theorem 8.7.1. If $\mathcal{U}$ has a Green's function, we'll prove $\mathcal{U}$ is analytically equivalent to $\mathbb{D}$, and if it does not, to either $\widehat{\mathbb{C}}$ or $\mathbb{C}$.

Theorem 8.7.6. *If $\mathcal{U}$ is a simply connected Riemann surface with a Green's function, then $\mathcal{U}$ is conformal to $\mathbb{D}$.*

Proof. We need to find $f\colon \mathcal{U} \to \mathbb{C}$, an analytic bijection. If we show that $\text{Ran}(f)$ is not all of $\mathbb{C}$, then by the Riemann mapping theorem, $\text{Ran}(f)$ and so $\mathcal{U}$ are conformal to $\mathbb{D}$.

Pick $p_0 \in \mathcal{U}$. We'll show that $\varphi(q,p_0)$ is a bijection. Since φ is analytic and $|\varphi(q,p_0)| < 1$, $\text{Ran}(f)$ is not all of $\mathbb{C}$. Given $p_1 \in \mathcal{U}$, let

$$\psi(q) = \frac{\varphi(q,p_0) - \varphi(p_1,p_0)}{1 - \overline{\varphi(p_1,p_0)}\,\varphi(q,p_0)} \tag{8.7.8}$$

Then $|\psi(q)| \le 1$ and $\psi(p_1) = 0$, so by the definition of Green's function, $|\psi(q)| \le e^{-g(q;p_1)}$.

Thus,

$$\left|\frac{\psi(q)}{\varphi(q,p_1)}\right| \le 1 \tag{8.7.9}$$

On the other hand,

$$\begin{aligned}\left|\frac{\psi(p_0)}{\varphi(p_0,p_1)}\right| &= \frac{|\varphi(p_1,p_0)|}{|\varphi(p_0,p_1)|} \\ &= e^{g(p_1;p_0)-g(p_0;p_1)} = 1\end{aligned}$$

by the symmetry of the Green's functions.

By the maximum principle, $|\psi(q)| = |\varphi(q;p_1)| = e^{-g(q;p_1)}$, so $\psi(q)$ only vanishes at $q = p$. By (8.7.8),

$$\varphi(q,p_0) = \varphi(p_1,p_0) \Leftrightarrow q = p_1 \tag{8.7.10}$$

We've thus proven $\varphi(\,\cdot\,,p_0)$ is one-one. $\square$

Proof of Theorem 8.7.1. We've shown that if there is a Green's function, $\mathcal{U}$ is conformal to $\mathbb{D}$. So suppose there is no Green's function, and so, by Fact 2, no nonconstant bounded analytic function.

Pick $p_0 \neq p_1$, so a bipolar Green's function exists and so, by Theorem 8.7.5, a meromorphic function $\varphi(q; p_0, p_1)$ with a simple zero at p_0, simple pole at p_1, and bounded away from any neighborhood of p_1. Suppose that we show $\varphi(\,\cdot\,; p_0, p_1)$ is one-one. Then $\text{Ran}(\varphi)$ is a simply connected subset of $\widehat{\mathbb{C}}$, but not one conformal to $\mathbb{D}$ (because if it were, $\mathcal{U}$ would have a Green's function). Since any simply connected subset, U, of $\widehat{\mathbb{C}}$ with $\#(\widehat{\mathbb{C}} \setminus U) \geq 2$ is conformal to $\mathbb{D}$, by the Riemann mapping theorem, we conclude $\#(\widehat{\mathbb{C}} \setminus U) = 0$ or 1, that is, $\text{Ran}(\varphi)$ is $\widehat{\mathbb{C}}$ or $\mathbb{C}$. Thus, it suffices to prove that φ is one-one.

Pick $p_2 \in \mathcal{U}$. Since $\varphi(q; p_0, p_1)$ and $\varphi(q; p_2, p_1)$ have simple poles at p_1, for some constant c_1,

$$\varphi(q; p_0, p_1) - c_1 \varphi(q; p_2, p_1)$$

has a removable singularity at p_1. Since both are bounded away from p_1, this function is bounded, hence a constant c_2, that is,

$$\varphi(q; p_0, p_1) = c_1 \varphi(q; p_2, p_1) + c_2 \tag{8.7.11}$$

In particular,

$$\varphi(p_2; p_0, p_1) = c_2 \tag{8.7.12}$$

If $\varphi(q; p_0, p_1) = \varphi(p_2; p_0, p_1)$, then by (8.7.11) and (8.7.12),

$$\varphi(q; p_0, p_1) = c_2 \Rightarrow \varphi(q; p_2, p_1) = 0 \Rightarrow q = p_2$$

that is, φ is one-one. $\square$

Notes and Historical Remarks.

> A significant mathematical problem, like the uniformization problem which appears as No. 22 on Hilbert's list, is never solved only once. Each generation of mathematicians, as if obeying Goethe's dictum, rethinks and reworks solutions discovered by their predecessors, and fits these solutions into the current conceptual and notational framework. Because of this, proofs of important theorems become, as if by themselves, simpler and easier as time goes by—as Ahlfors observed in his 1938 lecture on uniformization. Also, and this is more important, one discovers that solved problems present further questions.
>
> —*Lipman Bers* [**45**]

Our approach to uniformization is based on that of Gamelin [**199**].

Uniformization has a rich and complicated history, so much so that when the theorem is given names, they can be Koebe, Koebe–Poincaré, or Klein–Poincaré. [**45, 3**] have more about these issues. The original idea is that the set of solutions of an equation of two variables could be written as pairs

$(f_1(z), f_2(z))$ of functions of a single variable. $(\wp'(z), \wp(z))$ for elliptic curves is the motivating example. A further refinement of Poincaré is that these functions should be automorphic, that is, invariant under a discrete group, for example, lattice translations in the elliptic case.

What we have called the uniformization theorem may not look like that, but it is. Each of $\widehat{\mathbb{C}}, \mathbb{C}, \mathbb{D}$ is described by a single variable, and π, the universal covering map is automorphic under the group of deck transformations. If the original surface is embedded in $\mathbb{C}^2$, π followed by the coordinate maps provide the pair of automorphic functions.

Klein [**311**] first had the idea of uniformization holding in great generality and presented a proof for algebraic functions. He informed Poincaré of his result who announced he knew that and had a stronger result [**440**]. Klein's proof relied on some topological facts that he did not prove and, indeed, the techniques he would have needed to prove them were only invented forty years later. It is likely that Hilbert didn't accept his proof (see below). In the aftermath, Schwarz, whom Klein consulted, invented the notion of covering space, in part to explain uniformization.

Almost twenty years after the Klein–Poincaré work, a new burst of activity was initiated by Hilbert, listing uniformization for analytic surfaces among his famous 1900 list of problems [**259**]—it is the 22nd problem. In describing the background for the problem, Hilbert mentioned Poincaré's work on the algebraic case, but not that of Klein, his colleague at Göttingen, leading to the assumption that he had doubts about the validity of Klein's proof. Yandell [**598**], in his work on Hilbert's problems, has lots on the lives of Poincaré and Koebe.

Uniformization for analytic surfaces was accomplished in 1907, independently by Poincaré [**444**] and by Koebe [**324**]. The modern proof we use is highly influenced by work of Hilbert [**261**]. All three authors used potential theory ideas (which go back to Riemann) and Hilbert used the dipoles (aka bipolar) we do. It was Weyl [**590**] who emphasized that one should view uniformization as an issue of general Riemann surfaces, not merely as something involving solutions of equations of two variables. Weyl used the following poetic description (as translated in [**323**]):

> Here we enter the temple in which the Divinity (if I may be allowed this metaphor) is released from the earthly prison of its individual realizations: in the symbol of the two-dimensional non-Euclidean crystal, the archetype of the Riemann surface itself appear (as far as this is possible) pure and free of all obscurities and inessentials.

Paul Koebe [1882–1945] was a German mathematician and student of Schwarz. He spent his entire career working on conformal mapping, with important contributions to the proofs of the Riemann mapping theorem and to uniformization. He also found the class of one-one maps that saturate the bounds in the Bieberbach conjecture, which was motivated by Koebe's work. Both Yandell [**598**] and Reid's biography of Courant [**474**] paint unflattering portraits of Koebe's personality; in particular, Courant always felt that Koebe had stolen the ideas in Courant's thesis.

There is another aspect, quite remarkable, of the uniformization theorem we should note. Each of the three universal models, $\widehat{\mathbb{C}}, \mathbb{C}, \mathbb{D}$, supports a homogeneous Riemann metric, that is, one for which there are isometries which act transitively, in that for any z, w, there is an isometry, ρ, with $\rho(z_0) = w$. Indeed, these isometries are also orientation preserving and every one is an analytic automorphism. On $\widehat{\mathbb{C}}$, it is the spherical metric $(dz)^2/(1+|z|^2)^2$ of Section 6.5, on $\mathbb{C}$ the usual Euclidean metric, and on $\mathbb{D}$, the hyperbolic metric $(dz)^2/(1-|z|^2)^2$ of Section 12.2 of Part 2B. These metrics each have constant curvatures (1, 0, and -1, respectively). There are approaches to uniformization that depend on this constant curvature property.

One can complain that our comment that the uniformization theorem implies the Riemann mapping theorem is silly because our proof of the uniformization theorem uses the Riemann mapping theorem! But there are other proofs of uniformization that do not use the Riemann mapping theorem, so the comment can be reasonable.

Uniformization is also a key to understanding $\mathbb{A}\mathrm{ut}(\Omega)$ for a region, Ω, or more generally for a Riemann surface. If $\widetilde{\Omega}$ is the universal cover of Ω, $\pi\colon \widetilde{\Omega} \to \Omega$ the covering map, and Γ the subgroup of $\mathbb{A}\mathrm{ut}(\widetilde{\Omega})$ which are the deck transformations, the universal lifting property shows that if $f \in \mathbb{A}\mathrm{ut}(\Omega)$, there is $g \in \mathbb{A}\mathrm{ut}(\widetilde{\Omega})$ so that $f\pi = \pi g$. Moreover, if $\tilde{g}$ is another such automorphism of $\widetilde{\Omega}$, we have $\tilde{g} = hg$ for $h \in \Gamma$.

A little more work shows g induces such an f if and only if f is in the normalizer of Γ, that is, $f\Gamma f^{-1} = \Gamma$ setwise. So $\mathbb{A}\mathrm{ut}(\Omega)$ is the quotient of the normalizer by Γ. Because Γ is discrete, the connected component of $\mathbb{1}$ in $\mathbb{A}\mathrm{ut}(\Omega)$ is generated by these f in the centralizer of Γ, that is, f so that $fg = gf$ for all $g \in \Gamma$.

For $\mathbb{A}\mathrm{ut}(\Omega)$ to act transitively, the centralizer must be two-dimensional. It is not hard to see that if $\Gamma \subset \mathbb{A}\mathrm{ut}(\mathbb{D})$ is a nontrivial discrete group of fixed point free maps, its centralizer is never more than one-dimensional. But if $\Gamma \subset \mathbb{A}\mathrm{ut}(\mathbb{C})$ is a discrete subgroup of translations, its centralizer is two-parameter. Thus, other than $\mathbb{D}$, $\mathbb{C}$, $\widehat{\mathbb{C}}$, the only Riemann surfaces whose automorphism groups can possibly be transitive are those with $\mathbb{C}$ as universal cover, that is, $\mathbb{C} \setminus \{0\}$ or a torus—and they do, indeed have transitive automorphism groups. In particular, the only regions in

$\mathbb{C}$ with transitive automorphism groups are equivalent to one of $\mathbb{D}$, $\mathbb{C}$, or $\mathbb{C} \setminus \{0\}$.

Problems

1. (a) Prove that $\mathbb{C}^\times$ and the cylinder $\{z \mid 0 \le \operatorname{Re} z \le 1\}$ with $iy \cong 1 + iy$ are isomorphic Riemann surfaces. (*Hint*: $e^{2\pi i z}$.)

   (b) Prove this by looking at the universal covers and groups of deck transformations.

2. Given $\tau \in \mathbb{C}^\times$, find an element, f, of $\mathbb{A}\mathrm{ut}(\mathbb{C})$ so that if $g_w(z) = z + w$, then $f g_\tau f^{-1} = g_1$.

3. (a) Prove the form $\tilde{du}$ of (8.7.1) is independent of coordinate choice.

   (b) Prove that $d(\tilde{du}) = 0$.

   (c) Check that the function, f, of (8.7.2) is analytic.

4. (a) Prove that $\log|z|^{-1}$ is a Green's function for $\mathbb{D}$ and $p = 0$.

   (b) Prove that $\log|z|^{-1}$ is a bipolar Green's function for $\widehat{\mathbb{C}}$ with charge 1 at 0 and -1 at ∞.

   (c) Find the Green's functions for $\mathbb{D}$ for arbitrary $p \in \mathbb{D}$.

   (d) Find the bipolar Green's function for $\mathbb{C}$ and for $\widehat{\mathbb{C}}$ for arbitrary pairs of points.

   (e) In terms of a Riemann map, find the Green's function for an arbitrary simply connected proper region of $\mathbb{C}$.

   (f) Find the bipolar Green's function in terms of the Green's function if a Green's function exists.

   (g) Find the Green's function for $\mathbb{A}_{r,R}$, $0 < r < R < \infty$.

5. Recall (see the Notes to Section 8.1) that we say $\Omega \subset \mathbb{C}$ has a classical Green's function at $z_0 \in \Omega$ if $G(z, z_0)$ is harmonic on $\Omega \setminus \{z_0\}$ with a pole of 1 at z_0 and $\lim_{z \to \partial\Omega} G(z, z_0) = 0$. Prove that G is then a Green's function as defined in this section. (*Hint*: Apply the maximum principle for analytic functions to ψ/φ.)

6. This problem will prove Theorem 8.7.3

   (a) Prove that the classification of Riemann surfaces with abelian fundamental group is equivalent to finding all abelian subgroups of $\mathrm{Aut}(U)$ with U simply connected and all elements of the subgroup other than e being free of fixed points in U. The subgroups which are conjugate are not considered distinct.

(b) For $U = \mathbb{C}$, show that the subgroups must be either $\{n\tau \mid n \in \mathbb{Z}\}$ or $\{n_1\tau_1 + n_2\tau_2 \mid \operatorname{Im}(\overline{\tau_1}\tau_2) \neq 0, n_1, n_2 \in \mathbb{Z}\}$ and that the corresponding Riemann surface are $\widehat{\mathbb{C}} \setminus \{0\}$ and $\mathcal{L}_{\tau_1,\tau_2}$.

(c) Let $f, g \in \mathbb{A}\mathrm{ut}(\mathbb{D})$ both be parabolic and/or hyperbolic. If $fg = gf$, prove that either both are parabolic or both hyperbolic and they must have the same fixed points.

(d) Prove that any subgroup $G \subset \mathbb{A}\mathrm{ut}(\mathbb{C}_+)$ that is abelian and discrete and has no elliptic elements is of the form g^n for a single g.

(e) Up to conjugates, prove G is either $\{\tau_n \mid n \in \mathbb{Z}; \tau_n(z) = z + n\}$ or $\{\tau_n \mid n \in \mathbb{Z} \mid \tau_n(z) = \lambda^n z, \lambda \in (1, \infty)\}$ and conclude that these correspond to the Riemann surfaces $\mathbb{D} \setminus \{0\}$ and $\mathbb{A}_{r,R}$ with $\lambda = R/r$.

8.8. Ahlfors' Function, Analytic Capacity and the Painlevé Problem

We proved the Riemann mapping theorem for $\Omega \subset \mathbb{C}$, $\Omega \neq \mathbb{C}$ and simply connected by maximizing $\operatorname{Re} f'(z_0)$ among $f\colon \Omega \to \mathbb{D}$ with $f(z_0) = 0$. We also imposed that f be one-one and then found a unique maximizer. In fact, using the Schwarz lemma, we showed (see Problems 7 and 8 in Section 8.1) that we could drop $f(z_0) = 0$ and the one-one condition and still had a unique maximizer.

Remarkably, there is a unique maximizer for any $\Omega \subset \widehat{\mathbb{C}}$. While not too closely connected to the main theme of this chapter, we will prove this result and show its connection to the issue of which sets are removable in the sense of the Riemann removable singularities theorem. If $z_0 = \infty$, we write $f(z) = a + bz^{-1} + O(z^{-2})$ near infinity and interpret b as $f'(\infty)$.

Theorem 8.8.1. *Let Ω be a region in $\widehat{\mathbb{C}}$ and $z_0 \in \Omega$. Suppose $\mathfrak{A}(\Omega)$ has nonconstant bounded analytic functions. Then, there is a unique $f\colon \Omega \to \mathbb{D}$ with*

$$f'(z_0) = \sup\{\operatorname{Re} g'(z_0) \mid g\colon \Omega \to \mathbb{D}\} \tag{8.8.1}$$

and it has $f(z_0) = 0$.

Remarks. 1. If the only bounded functions are constants, $g'(z_0) = 0$ for all $g\colon \Omega \to \mathbb{D}$ and if we demand $g(z_0) = 0$, the maximizer is unique.

2. The proof shows that the maximizer is an extreme point in the analytic functions from Ω to $\mathbb{D}$.

Proof. The set of $f \in \mathfrak{A}(\Omega)$ with $\|f\|_\infty \leq 1$ is compact by Montel's theorem and $f \to \operatorname{Re} f'(z_0)$ is continuous so there is a maximizer. Since $\mathfrak{A}(\Omega)$ has nonconstant functions, it is easy to see (Problem 1) that there are f's with $f'(z_0) > 0$, with f nonconstant, and with $\operatorname{Ran} f \subset \mathbb{D}$. Since we can pick

$e^{i\theta}$ so that $e^{i\theta}f'(z_0) = |f'(z_0)|$, the maximizer has $f'(z_0) > 0$ so (8.8.1) holds. If $g(z) = (f(z) - f(z_0))(1 - \overline{f_0(z_0)}f(z))^{-1}$, then $\|g\|_\infty \leq 1$ while $g'(z_0) = f'(z_0)(1-|f(z_0)|^2)^{-1}$ so f must have $f(z_0) = 0$. Thus we need only prove uniqueness.

Suppose f_1 and f_2 are two maximizers and define

$$f = \tfrac{1}{2}(f_1 + f_2), \quad k = \tfrac{1}{2}(f_1 - f_2) \tag{8.8.2}$$

so f is a maximizer and $\|f \pm k\|_\infty \leq 1$. Thus

$$|f|^2 + |k|^2 = \tfrac{1}{2}(|f+k|^2 + |f-k|^2) \leq 1 \tag{8.8.3}$$

Let $g = k^2/2$. Then

$$|g| \leq \frac{1-|f|^2}{2} = (1-|f|)\left(\frac{1+|f|}{2}\right) \leq 1 - |f| \tag{8.8.4}$$

i.e.,

$$|g| + |f| \leq 1 \tag{8.8.5}$$

We are heading towards a proof that $g \equiv 0$. For simplicity of notation, suppose $z_0 = 0$. Suppose first that $g(0) \neq 0$. Let

$$h = f(1 + \overline{g(0)}\,|g(0)|^{-1}g) \tag{8.8.6}$$

Then $|h| \leq |f| + |g| \leq 1$ and $h(0) = 0$ so h maps Ω to $\mathbb{D}$. Since $f(0) = 0$,

$$h'(0) = f'(0)(1 + |g(0)|) > f'(0) \tag{8.8.7}$$

violating maximality. Thus $g(0) = 0$.

If $g \not\equiv 0$, we can write for $|z|$ small

$$g(z) = \sum_{k=\ell}^{\infty} a_k z^k, \quad a_\ell \neq 0 \tag{8.8.8}$$

for some $\ell \geq 1$ (since $g(0) = 0$). We'll take

$$h(z) = f(z) + \varepsilon \overline{a_\ell}\, z^{-(\ell-1)}\, g(z) \tag{8.8.9}$$

where $\varepsilon > 0$ will be chosen shortly.

Since $|f(0)| + |z^{-(\ell-1)}\, g(z)|_{z=0} = 0$, we can pick $R > 0$ so that

$$0 < |z| < R \Rightarrow |f(z)| + |\bar{a}_\ell\, z^{-(\ell-1)}\, g(z)| < 1 \tag{8.8.10}$$

Pick ε so that

$$0 < \varepsilon < \min(1, |a_\ell|^{-1}\, R^{(\ell-1)}) \tag{8.8.11}$$

Then, (8.8.10) says that $|h(z)| < 1$ if $|z| < R$ and (8.8.11) says if $|z| \geq R$, then

$$\begin{aligned} |h(z)| &\leq |f(z)| + \varepsilon\, |a_\ell|\, R^{-(\ell-1)}\, |g(z)| \\ &\leq |f(z)| + |g(z)| \leq 1 \end{aligned} \tag{8.8.12}$$

by (8.8.5). Thus, $\|h\|_\infty \leq 1$, so since $h(0) = 0$, $h\colon \Omega \to \mathbb{D}$.

By (8.8.9),

$$h'(0) = f'(0) + \varepsilon\,|a_\ell|^2 > |f'(0)| \tag{8.8.13}$$

contradicting maximality. Thus $g \equiv 0 \Rightarrow k = 0 \Rightarrow f_1 = f_2$ proving uniqueness. □

The maximizer, f, is called the *Ahlfors function* for (Ω, z_0). If Ω has no nonconstant bounded analytic functions, we set the Ahlfors function to $f \equiv 0$. If $\mathfrak{e} \subset \mathbb{C}$ is compact, we say that the Ahlfors function for $\mathfrak{e}$ is the Ahlfors function of $(\Omega, z = \infty)$ where Ω is the unbounded component of $\widehat{\mathbb{C}} \setminus \mathfrak{e}$. $f'(\infty)$ is called the *analytic capacity* of $\mathfrak{e}$, $A(\mathfrak{e})$, i.e.,

$$A(\mathfrak{e}) = \sup\{f'(\infty) \mid f \in \mathfrak{A}(\Omega), \|f\|_\infty \le 1, f(\infty) = 0\} \tag{8.8.14}$$

The name comes from the connection of $A(\mathfrak{e})$ to the potential theoretic capacity, $C(\mathfrak{e})$, discussed in Chapter 3, especially Section 3.6, of Part 3. There are two conventions for the normalization—in most of Chapter 3 of Part 3, we define Coulomb energy as $\mathcal{E}(\mu) = (2\pi)^{-1} \int \log|x-y|^{-1} d\mu(x)\, d\mu(y)$ since $(2\pi)^{-1} \log|x|^{-1}$ is the fundamental solution for $-\Delta$, but it is common when dealing with $\mathbb{C}$ (as opposed to general $\mathbb{R}^\nu$) to drop the $(2\pi)^{-1}$ which we do. With this latter convention which we'll use, $C(\mathfrak{e}) = e^{-R(\mathfrak{e})}$ where $R(\mathfrak{e})$ is the minimum of $\mathcal{E}(\mu)$ (without the $(2\pi)^{-1}$) over all $\mu \in \mathcal{M}_{+,1}(\mathfrak{e})$.

One can prove (and we will prove something equivalent in Section 3.6 of Part 3), that

$$\begin{aligned} C(\mathfrak{e}) = \sup\{\lim_{|z|\to\infty} |z|\, g(z) \mid g(z) = e^{u(z)}, u \text{ subharmonic on } \Omega,& \\ u \le 0, u(\infty) = -\infty, \text{ or } u \equiv -\infty\}& \end{aligned} \tag{8.8.15}$$

Since one can take $u(z) = \log|f(z)|$ for $f \in \mathfrak{A}(\Omega)$ with $\|f\|_\infty \le 1$, we'll see that $A(\mathfrak{e}) \le C(\mathfrak{e})$. In Section 3.6 of Part 3, we'll also show that if $\mathfrak{e}$ and $\widehat{\mathbb{C}} \setminus \mathfrak{e}$ are both connected, then $A(\mathfrak{e}) = C(\mathfrak{e})$. In the rest of this section, we want to discuss the relation of $A(\mathfrak{e})$ to the Riemann removable singularities theorem and compute the Ahlfors function and $A(\mathfrak{e})$ when $\mathfrak{e} \subset \mathbb{R}$.

Definition. Let $\mathfrak{e} \subset \Omega \subset \mathbb{C}$ where $\mathfrak{e}$ is compact and Ω a domain. We say that $\mathfrak{e}$ is *removable* for Ω if $f \in \mathfrak{A}(\Omega \setminus \mathfrak{e})$ and $\|f\|_\infty < \infty$ implies f is the restriction to $\Omega \setminus \mathfrak{e}$ of a function analytic in all of Ω.

The canonical example is $\mathfrak{e} = \{z_0\}$ for any Ω with $z_0 \in \Omega$ which expresses the Riemann removable singularities theorem.

Theorem 8.8.2. *Let $\mathfrak{e} \subset \mathbb{C}$ be compact. Then the following are equivalent:*

(1) $A(\mathfrak{e}) = 0$.
(2) *Every bounded analytic function on $\mathbb{C} \setminus \mathfrak{e}$ is constant.*
(3) *$\mathfrak{e}$ is removable for any domain Ω with $\mathfrak{e} \subset \Omega$.*
(4) *$\mathfrak{e}$ is removable for $\mathbb{C}$.*

Proof. We saw (1) $\Leftrightarrow$ (2) above and (3) $\Rightarrow$ (4) is trivial. We'll show (2) $\Rightarrow$ (3) and (4) $\Rightarrow$ (2).

(2) $\Rightarrow$ (3). Suppose (2) holds and f is bounded and analytic on $\Omega \setminus \mathfrak{e}$. By Theorem 4.1.3, $f = f_+ + f_-$ where f_+ is analytic and bounded on $\mathbb{C} \setminus \mathfrak{e}$ and

$$\lim_{|z|\to\infty} f_+(z) = 0 \tag{8.8.16}$$

and f_- is analytic and bounded on Ω. By (2), f_+ is constant and then by (8.8.16), $f_+ = 0$ so $f = f_-$ is analytic on all of Ω.

(4) $\Rightarrow$ (2). Suppose (4) holds and f is analytic and bounded on $\mathbb{C} \setminus \mathfrak{e}$. By (4), f extends to an entire function and by the maximum principle, this extension is bounded. Thus, f is constant. $\square$

As we've seen, $A(\mathfrak{e}) = 0 \Rightarrow C(\mathfrak{e}) = 0$ and in Section 3.6 of Part 3, we'll show that any compact set in $\mathbb{C}$ with $C(\mathfrak{e}) = 0$ is totally disconnected. Thus, removable sets are totally disconnected.

The *Painlevé problem* is to find a geometric characterization of removable sets. We'll say more about it in the Notes.

Finally, we turn to $\mathfrak{e} \subset \mathbb{R}$ ($\widehat{\mathbb{C}} \setminus \mathfrak{e}$ is then called a *Denjoy domain*).

Theorem 8.8.3 (Pommerenke's Theorem). *Let $\mathfrak{e} \subset \mathbb{R} \subset \mathbb{C}$ be compact. Then the Ahlfors function for $\mathfrak{e}$ is*

$$F(z) = \tanh\biggl(\frac{1}{4}\int_{\mathfrak{e}} \frac{dx}{z-x}\biggr) \tag{8.8.17}$$

and

$$A(\mathfrak{e}) = \tfrac{1}{4}|\mathfrak{e}| \tag{8.8.18}$$

where $|\cdot|$ is the Lebesgue measure on $\mathbb{R}$.

Proof. If $F\colon \widehat{\mathbb{C}} \setminus \mathfrak{e} \to \mathbb{D}$, then so does $Q(z) = \overline{F(\bar{z})}$ (since $\bar{\mathfrak{e}} = \mathfrak{e}$) and $Q'(\infty) = \overline{F'(\infty)}$, so by uniqueness of the Ahlfors function, if F is the Ahlfors function, $F = Q$, i.e., F is real on $\mathbb{R} \setminus \mathfrak{e}$.

Define

$$G(z) = \frac{1-F(z)}{1+F(z)} \tag{8.8.19}$$

since $w \mapsto \frac{1-w}{1+w}$ maps $\mathbb{D}$ to $\mathbb{H}_+$, $\operatorname{Re} G(z) > 0$ for all $z \in \widehat{\mathbb{C}} \setminus \mathfrak{e}$. Thus, we can define a single-valued log,

$$H(z) = \log G(z) \tag{8.8.20}$$

which is analytic on $\mathbb{C}_+ \subset \widehat{\mathbb{C}} \setminus \mathfrak{e}$ with

$$-\frac{\pi}{2} < \operatorname{Im} H(z) < \frac{\pi}{2} \tag{8.8.21}$$

there. Since $F(\infty)=0$, $G(\infty)=1$ and thus

$$\lim_{|z|\to\infty} H(z)=0 \tag{8.8.22}$$

As a function on $\mathbb{C}_+$, $H(z)$ is analytic with $\|\operatorname{Im} H\|_\infty<\infty$. Such functions are analyzed in Theorem 5.9.2 of Part 3 which implies for Lebesgue a.e. $x\in\mathbb{R}$, $\lim_{\varepsilon\downarrow 0} H(x+i\varepsilon)\equiv H^*(x)$ exists and for all $z\in\mathbb{C}_+$, we have that

$$H(z)=\alpha+\frac{1}{\pi}\int\frac{\operatorname{Im} H^*(x)dx}{x-z} \tag{8.8.23}$$

for some $\alpha\in\mathbb{R}$. Since F is real on $\mathbb{R}\setminus\mathfrak{e}$, so is G and thus so is H, i.e., $\operatorname{Im} H^*$ is supported on $\mathfrak{e}$. Thus, $\operatorname{Im} H^*\in L^1$ which implies $\alpha=0$ given (8.8.22).

We can go backwards; any $h(x)$ on $\mathfrak{e}$ with $|h(x)|\le\frac{\pi}{2}$ yields $H(z)$ by (8.8.23) with $\operatorname{Im} H^*(x)=h(x)$ and $\alpha=0$ obeying $|\operatorname{Im} H|\le\frac{\pi}{2}$ and (8.8.22). Then $G=e^{H(z)}$ has $\operatorname{Re} G>0$ and $G(\infty)=1$. (8.8.19) is equivalent to

$$F(z)=\frac{1-G(z)}{1+G(z)} \tag{8.8.24}$$

and $\operatorname{Ran} F\in\mathbb{D}$, $F(\infty)=0$.

Thus (8.8.23) with $\alpha=0$ sets up a one-one correspondence between h supported on $\mathfrak{e}$ with $|h(x)|\le\frac{\pi}{2}$ and F's mapping $\widehat{\mathbb{C}}\setminus\mathfrak{e}$ to $\mathbb{D}$ and $F(\infty)=0$. Let $\beta=\frac{1}{\pi}\int h(x)dx$. Then near ∞,

$$H(z)=-\tfrac{\beta}{z}+O(|z|^{-2}),\ G(z)=1-\tfrac{\beta}{z}+O(|z|)^{-2},\ F(z)=\tfrac{\beta}{2z}+O(|z|^{-2}) \tag{8.8.25}$$

To maximize $F'(\infty)$, we want to maximize β which given $\operatorname{supp} h\subset\mathfrak{e}$ and $\|h\|_\infty\le\frac{\pi}{2}$, means $h=\frac{\pi}{2}\chi_{\mathfrak{e}}$, i.e., the Ahlfors function comes from

$$F'(\infty)=\tfrac{1}{4}|\mathfrak{e}|,\quad H(z)=\frac{1}{2}\int_{\mathfrak{e}}\frac{dx}{x-z} \tag{8.8.26}$$

Thus

$$F(z)=\frac{1-e^H}{1+e^H}=\frac{e^{-H/2}-e^{H/2}}{e^{-H/2}+e^{H/2}}=\tanh\left(-\frac{H}{2}\right) \tag{8.8.27}$$

so (8.8.26) yields (8.8.17) □

Corollary 8.8.4. *$\mathfrak{e}\subset\mathbb{R}$ is a removable set if and only if $|\mathfrak{e}|=0$.*

Corollary 8.8.5. *Let $\mathfrak{e}\subset\mathbb{R}$ be compact. Then*

$$A(\mathfrak{e})=C(\mathfrak{e}) \tag{8.8.28}$$

if $\mathfrak{e}$ is an interval and

$$A(\mathfrak{e})<C(\mathfrak{e}) \tag{8.8.29}$$

for any $\mathfrak{e}$ which is not a single interval.

Remark. In Section 3.6 of Part 3,we'll prove that $A(\mathfrak{e})\le C(\mathfrak{e})$ for any compact $\mathfrak{e}$ in $\mathbb{C}$ with equality if $\mathfrak{e}$ is connected.

Proof. As noted, (8.8.28) holds for any connected $\mathfrak{e}$. In Problem 2, the reader will show that if $a \notin \mathfrak{e}$ and $\mathfrak{e}_- = \mathfrak{e} \cap (-\infty, a)$, $\mathfrak{e}_+ = \mathfrak{e} \cap (a, \infty)$ are both nonpolar, then

$$C(\mathfrak{e}) > \tfrac{1}{4}|\mathfrak{e}| = A(\mathfrak{e}) \tag{8.8.30}$$

□

Theorem 8.8.6. *Let $\alpha_1 < \beta_1 < \alpha_2 < \cdots < \alpha_\ell < \beta_\ell$ in $\mathbb{R}$. Let*

$$\mathfrak{e} = \bigcup_{j=1}^{\ell} [\alpha_j, \beta_j] \tag{8.8.31}$$

Then

(a) *The Ahlfors function, F, is given by (8.8.24) where*

$$G(z) = \prod_{j=1}^{\ell} \sqrt{\frac{z - \beta_j}{z - \alpha_j}} \tag{8.8.32}$$

for $z \in \widehat{\mathbb{C}} \setminus \mathfrak{e}$ where the branch of square root which is 1 at $z = \infty$ is taken.

(b) *The function*

$$\Delta(z) = F(z) + F(z)^{-1} \tag{8.8.33}$$

is a rational function of the form

$$\Delta(z) = P(z)/Q(z) \tag{8.8.34}$$

with $\deg(P) = \ell$, $\deg(Q) = \ell - 1$ and Q has a single simple zero, c_j, in each interval (β_j, α_{j+1}), $j = 1, \dots, \ell - 1$ so that Δ has poles exactly at ∞ and at $\{c_j\}_{j=1}^{\ell-1}$.

(c) $$\mathfrak{e} = \Delta^{-1}([-2, 2]) \tag{8.8.35}$$

Moreover

$$\Delta(\alpha_j) = -2, \quad \Delta(\beta_j) = 2 \tag{8.8.36}$$

and for some $\{r_j\}_{j=1}^{\ell-1}$ in $(0, \infty)$ and $B \in \mathbb{R}$:

$$\Delta(z) = \frac{z}{A(\mathfrak{e})} + B + \sum_{j=1}^{\ell-1} \frac{r_j}{c_j - z} \tag{8.8.37}$$

(d) *$F\colon \widehat{\mathbb{C}} \setminus \mathfrak{e} \to \mathbb{D}$ is an ℓ to 1 map onto $\mathbb{D}$ in the sense that each $w \in \mathbb{D}$ is taken ℓ times counting multiplicity. We have*

$$F(z) = \frac{\Delta(z)}{2} - \sqrt{\left(\frac{\Delta(z)}{2}\right)^2 - 1} \tag{8.8.38}$$

Remarks. 1. F can't be a local bijection since $\mathbb{D}$ doesn't have any nontrivial covering spaces. That is, there have got to be points with $F'(z) = 0$, i.e., where multiplicity matters,

2. Any Δ of the form (8.8.37) with $A(\mathfrak{e}) > 0$, $r_j > 0$ is the Δ associated to $\mathfrak{e}$ given by (8.8.35).

3. $\operatorname{Im}\Delta(z) > 0$ for $z \in \mathbb{C}_+$ and it is easy to see that Δ is the unique rational function with this property and (8.8.35).

4. There is an analog of (8.8.32) for any compact set $\mathfrak{e} \subset \mathbb{R}$. For one can write (with a different labeling of the α's and β's from the finite gap case) $\mathbb{R} \setminus \mathfrak{e} = (-\infty, \alpha_0) \cup (\beta_0, \infty) \cup \bigcup_{\ell=1}^{L} (\beta_\ell, \alpha_l)$ (where L is infinite if $\mathfrak{e}$ is not of the form (8.8.31)). Here the α's aren't ordered but the $(\beta_\ell, \alpha_\ell)$ are all the bounded connected components of $\mathbb{R}\setminus\mathfrak{e}$. Since $\sum_{\ell=1}^{L} |\alpha_\ell - \beta_\ell| < |\beta_0 - \alpha_0| < \infty$, the product in (8.8.32) with j running from 0 to L converges and the Ahlfors function is still given by (8.8.24). Unlike the finite gap case, G need not have square-root behavior at the edges of the gaps. This kind of analysis goes back at least to Craig [**126**].

Proof. (a) We have for $z \in \mathbb{R} \setminus \mathfrak{e}$,

$$H(z) = \frac{1}{2} \int_{\mathfrak{e}} \frac{dx}{x - z} = \frac{1}{2} \sum_{j=1}^{\ell} \log\left(\frac{z - \beta_j}{z - \alpha_j}\right) \tag{8.8.39}$$

so $G = e^H$ is given by (8.8.32) initially for $z \in \mathbb{R} \setminus \mathfrak{e}$, but then by analyticity for $z \in \widehat{\mathbb{C}} \setminus \mathfrak{e}$. $G(\infty) = 1$ is true for any G given by (8.8.19) with $F(\infty) = 0$.

(b) By (8.8.24), Δ given by (8.8.33) has

$$\Delta(z) = \frac{2G^2 + 2}{1 - G^2} \tag{8.8.40}$$

which clearly is a ratio P/Q with $\deg(P) = \ell$ and $\deg(Q) \le \ell - 1$. By (8.8.39), H, and so G, is strictly monotone in each interval (β_j, α_{j+1}) and runs from $G(\beta_j) = 0$ to $G(\alpha_{j+1}) = \infty$, so there is exactly one point where $G = 1$ and thus $\Delta = \infty$. This accounts for $\ell - 1$ zeros of Q so $\deg Q = \ell - 1$. Since $G(\infty) = 1$, we also have a pole at ∞.

(c) By Example 8.4.3, since F maps from $\widehat{\mathbb{C}} \setminus \mathfrak{e}$ to $\mathbb{D}$, Δ maps from $\widehat{\mathbb{C}} \setminus \mathfrak{e}$ to $\widehat{\mathbb{C}} \setminus [-2, 2]$, so $\Delta^{-1}([-2, 2]) \subset \mathfrak{e}$. On the other hand, by (8.8.32), G has pure imaginary boundary values on $\mathfrak{e}$ so F has boundary values on $\partial\mathbb{D}$ so Δ has values in $[-2, 2]$. Since $G(\alpha_j) = \infty$, $G(\beta_j) = 0$, (8.8.40) implies (8.8.36). Thus, $\Delta \in (2, \infty)$ on (β_j, c_j) and $\Delta \in (-\infty, -2)$ on (c_j, α_{j+1}) which means the r_j in the partial fraction expansion are positive. Since $F(z) = \frac{A(\mathfrak{e})}{z} + O(z^{-2})$, we see near ∞, $\Delta(z) = \frac{z}{A(\mathfrak{e})} + O(1)$.

(d) Every rational function on $\widehat{\mathbb{C}}$ has a fixed degree and since there are ℓ simple poles, Δ has degree ℓ. Thus every value in $\widehat{\mathbb{C}}$ is taken ℓ times counting multiplicity. Since (8.8.35) says $\Delta^{-1}(\widehat{\mathbb{C}} \setminus [-2,2]) = \widehat{\mathbb{C}} \setminus \mathfrak{e}$, we get the multiplicity ℓ result. (8.8.38) follows from (8.4.10). $\square$

Notes and Historical Remarks. The foundational paper for the themes of this section is a 1947 paper of Ahlfors [**6**] in which he considered n-connected regions (i.e., $\widehat{\mathbb{C}} \setminus \Omega$ has n-connected components) so that each component of the complement is bounded by a smooth curve. By Theorem 8.6.6, any n-component region so that each component of $\widehat{\mathbb{C}} \setminus \Omega$ is more than a single point is biholomorphically equivalent to such a region with smooth boundaries. For such regions, Ahlfors proved uniqueness of the maximizer and that this maximizer, f, was onto all of $\mathbb{D}$ and each point in $\mathbb{D}$ was taken n times counting multiplicity.

The uniqueness proof that works for any region and proves that the Ahlfors function is an extreme point of the unit ball of $H^\infty(\Omega)$ is from Fisher [**190**].

Pommerenke's theorem is due to him in [**450**]. The proof I use was shown to me by P. Yuditski who also showed me the calculations of Theorem 8.8.6.

While our calculation in Theorem 8.8.6 and Ahlfors' for general n-connected sets shows that the range of the Ahlfors map is all of $\mathbb{D}$ in the n-connected case, the range may not be all of $\mathbb{D}$ in general. It can be shown (see Havinson [**246**]) that $\mathbb{D} \setminus \operatorname{Ran} f$ has zero analytic capacity. Röding [**489**], Minda [**381**], and Yamada [**596, 597**] have examples with nontrivial missing sets—the second Yamada paper even has an omitted set with $C(\mathbb{D} \setminus \operatorname{Ran} f) > 0$.

For domains with smooth boundary, there are explicit formulae for the Ahlfors function in terms of some reproducing kernels (Szegő kernel and Garabedian kernel)—this is discussed in Bell's book [**38**] and references therein.

Garnett [**204**] and Pajot [**424**] are two sets of lecture notes on analytic capacity. In particular, Pajot [**424**] discusses results on removable sets including that a general $\mathfrak{e} \subset \mathbb{C}$ with zero Hausdorff measure, $h^1(\mathfrak{e}) = 0$, is removable (but there are removable sets $\mathfrak{e}$ with $h^1(\mathfrak{e}) > 0$).

Problems

1. (a) Suppose $z_0 \in \Omega \subset \widehat{\mathbb{C}}$ where Ω is open and connected. If there are nonconstant bounded analytic functions on Ω, and $z_0 \neq \infty$, show that there is one with $f(z_0) = 0$, $f'(z_0) > 0$. (*Hint*: If g is bounded, initially take $\big(f(z) - f(z_0)\big)(z - z_0)^{-\ell}$ for a suitable ℓ.)

   (b) Do the same when $z_0 = \infty$.

2. (a) Let $\mathfrak{e} \subset \mathbb{R}$ be compact with $a \in \mathbb{R} \setminus \mathfrak{e}$. Let $\mathfrak{e}_- = (-\infty, a) \cap \mathfrak{e}$ and $\mathfrak{e}_+ = (a, \infty) \cap \mathfrak{e}$. Suppose neither $\mathfrak{e}_+$ nor $\mathfrak{e}_-$ is polar. Let $\mathfrak{e}(\lambda) = \mathfrak{e}_- \cup (\mathfrak{e}_+ + \lambda)$ for $\lambda > 0$. Prove that $C\bigl(\mathfrak{e}(\lambda)\bigr)$ is strictly monotone increasing in λ. (*Hint*: Generalize the fact that if $g(x, y)$ is a smooth function of two variables so that there is a unique $x(y)$ with $g\bigl(x(y), y\bigr) = \min g(x, y)$, then $\frac{d}{dy} g\bigl(x(y), y\bigr) = \frac{\partial}{\partial y} g\bigl(x(y), y\bigr)$.)

(b) Prove that $C(\mathfrak{e}) > \frac{1}{4}|\mathfrak{e}|$.

Chapter 9

Zeros of Analytic Functions and Product Formulae

I am so impressed by the importance of your theorem, that on a closer examination I can't conceal that the establishment of the theorem must be made significantly shorter and simpler if it is to find its place in the elements of analysis, where it belongs. The large apparatus of formulas that you apply would, I fear, scare off many readers; in any case it makes it harder to penetrate into the essence of the matter. I therefore believed that it did not lie in your interest for the memoir to be published in the present form, and wanted to suggest to you that you give me permission to present the theorems in question in a free treatment to our Academy, the more so since they are already published and your property rights to them are in any case secured.

—*Weierstrass*, as translated in [**557**][1]

Big Notions and Theorems: Absolutely Convergent Products, Euler Product Formula, Partial Fraction Expansion of cot and $\csc^2$, Bernoulli Numbers, Euler Numbers, Mittag-Leffler Theorem, $\bar{\partial}$-Problem, Weierstrass Factor, Weierstrass Product Theorem, Meromorphic $\equiv$ Field of Quotients, Natural Boundary, Domain of Holomorphy, Gamma Function, Wielandt's Theorem, Bohr–Mollerup Theorem, Beta Function, Euler Reflection Formula, Legendre Duplication Formula, Gauss Multiplication Formula, Euler–Maclaurin Series, Stirling's Formula, Bernoulli Polynomials, Jensen Formula, Poisson–Jensen Formula, Blaschke Products, Müntz–Szász Theorem, Cauchy Formula, Gram Determinant, Order of an Entire Function, Type of a Function of Finite Order, Hadamard Product Formula, Genus of a Function

[1]Letter to Mittag-Leffler, June 7, 1880.

The two themes of this chapter are the determination of an analytic function by its zeros and the connected issue of product representations.

A key early development was Euler's 1735 formula for $\sin(\pi x)$. Euler started with the representation of a polynomial, P, in terms of its zeros $\{z_j\}_{j=1}^n$, normally written as $c\prod_{j=1}^n(z - z_j)$, but which Euler preferred to write as (so long as no $z_j = 0$)

$$P(z) = P(0)\prod_{j=1}^{n}\left(1 - \frac{z}{z_j}\right) \tag{9.0.1}$$

Since $\sin(\pi z)$ has zeros as $0, \pm 1, \pm 2, \dots$, and $\sin(\pi z)/\pi z$ is 1 at $z = 0$, he reasoned by analogy that

$$\frac{\sin(\pi z)}{\pi z} = \prod_{n=1}^{\infty}\left(1 - \frac{z^2}{n^2}\right) \tag{9.0.2}$$

He treated this as obviously true and, in the spirit of his time, didn't worry about the issue of convergence. We don't have that luxury (nor, in the end, did Euler—see the historical notes to Section 9.2). Section 9.1 will deal with convergence of infinite products, and Section 9.2 will have two proofs of Euler's formula: one looks at the ratio and uses the fact that a nonvanishing entire function is of the form $e^{h(z)}$. This theme will recur throughout the chapter. The second proof will involve a second recurring theme—the use of partial fractions. We'll prove first that

$$\frac{\pi^2}{\sin^2(\pi z)} = \sum_{n=-\infty}^{\infty}\frac{1}{(z-n)^2} \tag{9.0.3}$$

and relate this to (9.0.2) by noting if $f(z) = \sin(\pi z)$, then $\frac{d^2}{dz^2}\log(f) = \pi^2/\sin^2(\pi z)$.

In carrying over the Euler idea to general functions, one faces the fact that, in general, $\prod_{j=1}^{\infty}(1 - z/z_j)$ will not converge. In 1876, Weierstrass had one of the great ideas in science if we realize that it was the precursor to renormalization in quantum field theory—he subtracted out the first few terms in $\log(1 - z/z_j)$ by using

$$E_n(z) = (1-z)\exp\left(\sum_{j=1}^{n}\frac{z^j}{j}\right) \tag{9.0.4}$$

With this extra factor, he could prove, for any $z_1, z_2, \dots,$ converging to infinity, that $\prod_{n=1}^{\infty} E_n(z/z_n)$ converged and vanished exactly at $z = z_n$. This paper had enormous impact on his contemporaries since it opened up a new way to actually construct analytic functions.

Renormalization for partial fractions, a result known as the Mittag-Leffler theorem, is somewhat more straightforward, so we present it first in Section 9.3, followed by the Weierstrass product formula in Section 9.4. These sections present results on all of $\mathbb{C}$; Section 9.5 does the analogs for an arbitrary region that, in particular, show for any region, Ω, there is $f \in \mathfrak{A}(\Omega)$ with $\partial\Omega$ as a natural boundary.

Sections 9.6 and 9.7 discuss the gamma function—they are in this chapter because we use the Weierstrass definition as an infinite product rather than either of the Euler definitions, which we show are equivalent to the product definition. That said, Section 9.7, on the proof of Stirling's formula, is essentially a real-variables discussion.

The final three sections are closely related and concern the relation between zeros and growth. Section 9.8 has a general formula, Jensen's formula, that expresses this quantitatively and leads us to consider specialized products for situations where one has growth restrictions: Blaschke products for $\mathbb{D}$ in Section 9.9 and, in Section 9.10, the Hadamard product formula for entire finite f's obeying $|f(z)| \leq D\exp(C|z|^m)$ (functions of finite order).

9.1. Infinite Products

In this section, we present criteria for convergence of infinite products $\prod_{n=1}^\infty f_n(z)$ of analytic functions. The criterion will be similar to the one presented for sums in Problem 1 of Section 6.1 as the Weierstrass M-test. There, for $\sum_{n=1}^\infty g_n(z)$ to converge, we required

$$\sum_{n=1}^\infty \sup_{z\in K} |g_n(z)| < \infty \tag{9.1.1}$$

for each compact K. Here we'll need

$$\sum_{n=1}^\infty \sup_{z\in K} |1 - f_n(z)| < \infty \tag{9.1.2}$$

Since this may seem like a strong condition, we begin by showing it is implied by a seemingly weaker condition explaining why it is virtually the only condition ever used.

Proposition 9.1.1. *Let Ω be a region. Let $\{g_n\}_{n=1}^\infty \subset \mathfrak{A}(\Omega)$ so that for every compact $K \subset \Omega$, we have*

$$\sup_{z\in K}\left(\sum_{n=1}^\infty |g_n(z)|\right) < \infty \tag{9.1.3}$$

Then (9.1.1) *holds.*

Proof. By Theorems 4.1.2 and 4.4.1 we can find Γ a chain in Ω so that $\mathrm{Ran}(\Gamma) \cap K = \emptyset$, and for $z \in K$,

$$f(z) = \frac{1}{2\pi i} \oint_\Gamma \frac{f(w)}{w-z}\, dw \tag{9.1.4}$$

and, in particular,

$$\sup_{z \in K} |f(z)| \le (2\pi)^{-1} \mathrm{dist}(K, \mathrm{Ran}(\Gamma))^{-1} \int_\Gamma |f(w)|\, d|w| \tag{9.1.5}$$

Let

$$a_n = \int_\Gamma |g_n(w)|\, d|w| \tag{9.1.6}$$

By (9.1.3) for Γ,

$$\sum_{n=1}^\infty |a_n| < \infty \tag{9.1.7}$$

This and (9.1.5) imply (9.1.1). □

Lemma 9.1.2. *Let $\{w_j\}_{j=1}^N$ be a finite set in $\mathbb{C}$. Then*

(a)
$$\biggl|\prod_{j=1}^N (1+w_j)\biggr| \le \exp\biggl(\sum_{j=1}^N |w_j|\biggr) \tag{9.1.8}$$

(b)
$$\biggl|\biggl[\prod_{j=1}^N (1+w_j)\biggr] - 1\biggr| \le \biggl(\sum_{j=1}^N |w_j|\biggr) \exp\biggl(1 + \sum_{j=1}^N |w_j|\biggr) \tag{9.1.9}$$

Proof. (a) For $x \ge 0$, $e^x = 1 + x + \frac{x^2}{2!} + \cdots \ge 1 + x$, so $1 + |w_j| \le e^{|w_j|}$, which implies (9.1.8).

(b) If $f(\lambda)$ is an entire function, by the maximum principle and a Cauchy estimate for $\lambda \in [0,1]$,

$$|f'(\lambda)| \le R^{-1} \sup_{|z|=R+1} |f(z)| \tag{9.1.10}$$

so

$$|f(1) - f(0)| \le R^{-1} \sup_{|z|=R+1} |f(z)| \tag{9.1.11}$$

Let

$$f(z) = \prod_{n=1}^N (1 + zw_j) \tag{9.1.12}$$

Then, by (9.1.8),

$$|f(1) - f(0)| \le R^{-1} \exp\biggl((1+R) \sum_{j=1}^N |w_j|\biggr) \tag{9.1.13}$$

Pick $R = (\sum_{j=1}^N |w_j|)^{-1}$ to get (9.1.9). □

Theorem 9.1.3. *Let Ω be a region and $\{f_n\}_{n=1}^\infty \subset \mathfrak{A}(\Omega)$. Suppose for each $K \subset \Omega$, (9.1.2) holds. Then as $N \to \infty$, $F_N \equiv \prod_{n=1}^N f_n$ converges in the $\mathfrak{A}(\Omega)$ topology to an analytic function, F (which we'll write $\prod_{n=1}^\infty f_n$). Moreover,*

$$F(z_0) = 0 \Leftrightarrow \exists n\, f_n(z_0) = 0$$

and the order of a zero of F is the sum of the orders of the zeros of those f_n which vanish at that zero.

Remarks. 1. By (9.1.2) at any z_0, only finitely many f_n can vanish.

2. When (9.1.2) holds, we say $\prod_{n=1}^\infty f_n$ is an *absolutely convergent product.*

Proof. By (9.1.8) and (9.1.9), if $M \geq N$,

$$\|F_N - F_M\|_K \leq \left(\sum_{j=N+1}^{M} \|f_j - 1\|_K \right) \exp\left(1 + \sum_{j=1}^{M} \|f_j - 1\|_K \right) \tag{9.1.14}$$

since

$$F_M - F_N = \left[\left(\prod_{j=N+1}^{M} f_j \right) - 1 \right] \prod_{j=1}^{N} f_j \tag{9.1.15}$$

Since, by (9.1.2),

$$\lim_{N\to\infty} \sum_{j=N+1}^{\infty} \|f_j - 1\|_K = 0 \tag{9.1.16}$$

we see F_N is Cauchy, so the existence of a limit in $\mathfrak{A}(\Omega)$ follows from the Weierstrass convergence theorem.

Because $e^{6/5} \approx 3.32 < 5$, if $0 < y < \frac{1}{5}$, then $ye^{1+y} < 1$. It follows from (9.1.9) if $\sum_{j=1}^N |w_j| \leq \frac{1}{5}$, then $|\prod_{j=1}^N w_j - 1| < 1$, so $\prod_{j=1}^N w_j \neq 0$. Thus, given K, find N so $\sum_{j=N+1}^\infty \|f_j\|_K \leq \frac{1}{5}$ and conclude that $\prod_{N+1}^\infty f_j$ is non-vanishing on K. Thus, $F = F_N \prod_{N+1}^\infty f_j$ vanishes if and only if F_N vanishes if and only if one of $\{f_j\}_{j=1}^N$ vanishes. This proves the final statement in the theorem. □

Notes and Historical Remarks. The usual method of controlling infinite products when (9.1.2) holds is to use log. It shows the power of Cauchy estimates to see that they lead directly to (9.1.9) without the need to estimate logs.

One can also control zeros in Theorem 9.1.3 using Hurwitz's theorem.

There are, of course, also conditionally convergent products, for example, $\prod_{n=1}^\infty (1 - a_n z)$ where $a_{2n} = n^{-1}$ and $a_{2n-1} = -n^{-1}$ which conditionally converges to $\prod_{n=1}^\infty (1 - z^2/n^2)$ even though $\sum |a_n z| = \infty$. We'll see general results on conditionally convergent products in Problem 6 of Section 9.4.

An important aspect of absolutely convergent products is that they are rearrangement invariant (Problem 2).

Problems

1. (a) Using the power series expansion of $\log(1+z)$, prove that

$$|z| \le \tfrac{1}{2} \Rightarrow |\log(1+z) - z| \le |z|^2 \tag{9.1.17}$$

so that

$$|z| \le \tfrac{1}{2} \Rightarrow \tfrac{1}{2}\,|z| \le |\log(1+z)| \le \tfrac{3}{2}\,|z| \tag{9.1.18}$$

(b) Suppose $\{w_j\}_{j=1}^\infty$ obeys $|w_j| \le \frac{1}{2}$. Prove that $\prod_{j=1}^\infty (1+w_j)$ is an absolutely convergent product if and only if $\sum_{j=1}^\infty w_j$ is an absolutely convergent sum.

(c) Prove Theorem 9.1.3 by applying the M-test to the $\sum_{j=N}^\infty \log(f_j)$ for N sufficiently large (and K-dependent).

2. A rearrangement of a sequence $\{a_j\}_{j=1}^\infty$ is a sequence $b_j = a_{\pi(j)}$ where π is a bijection of $\{1, 2, \dots\}$.

(a) If $\sum_{j=1}^\infty |a_j| < \infty$, prove for any rearrangement, π, that $\lim_{N\to\infty} \sum_{j=1}^N a_j = \lim_{N\to\infty} \sum_{j=1}^N b_j$.

(b) Prove that if $\sum_{j=1}^\infty |1-w_j| < \infty$ and $z_j = w_{\pi(j)}$, for a rearrangement, π, then $\lim_{N\to\infty} \prod_{j=1}^N w_j = \lim_{N\to\infty} \prod_{j=1}^N z_j$.

(c) Prove that if $a_j \to 0$, $\sum_{j=1}^\infty |a_j| = \infty$ but $\lim_{N\to\infty} \sum_1^N a_j$ exists and is finite, then for any $x \in \mathbb{R} \cup \{\infty\} \cup \{-\infty\}$, there is a rearrangement with $\lim_{N\to\infty} \sum_1^N b_j = x$.

(d) Prove that if $w_j \in (0,\infty)$, $w_j \to 1$, $\sum_{j=1}^\infty |1-w_j| = \infty$ but $\lim_{N\to\infty} \prod_{j=1}^N w_j$ exists and lies in $(0,\infty)$, then for any $x \in [0,\infty) \cup \{\infty\}$, there is a rearrangement with $\lim_{N\to\infty} \prod_{j=1}^N z_j = x$.

3. (a) Let $f_N(z) = \prod_{n=1}^N (1+z^{2^n})$. Where does $f_N(z)$ have zeros?

(b) Do you expect $\lim_{N\to\infty} f_N(z)$ to have a natural boundary on $\partial\mathbb{D}$?

(c) Compute $\prod_{n=1}^\infty (1+z^{2^n})$ for $z \in \mathbb{D}$. (*Hint*: Compute $(1-z)f_N(z)$ inductively.)

4. (a) Let $z_n = \frac{i}{n}, n = 1, 2, \dots$. Prove that $\prod_{n=1}^N (1+z_n)$ does not converge but that $\prod_{n=1}^N |1+z_n|$ does.

(b) Let $z_n = (-1)^n n^{-1/2}$. Prove that $\sum_{n=1}^N z_n$ converges but that $\prod_{n=1}^N (1+z_n)$ does not.

9.2. A Warmup: The Euler Product Formula

Euler's decisive step was daring. In strict logic, it was an outright fallacy: he applied a rule to a case for which the rule was not made, a rule about algebraic equations to an equation which is not algebraic. In strict logic, Euler's step was not justified. Yet it was justified by analogy, by the analogy of the most successful achievements of a rising science that he called himself a few years later the "Analysis of the Infinite." Other mathematicians, before Euler, passed from finite differences to infinitely small differences, from sums with a finite number of terms to sums with an infinity of terms, from finite products to infinite products. And so Euler passed from equations of finite degree (algebraic equations) to equations of infinite degree, applying the rules made for the finite to the infinite.

—*G. Pólya* (1887–1985), in [**449**, pg. 21]

In this section, we'll prove a remarkable product formula of Euler for $\sin(\pi x)$. This will, first of all, allow us to introduce several themes of import in the rest of this chapter, including the use of improved Cauchy estimates (Theorem 3.2.2) and the relation between zeros and partial fractions. But the formula and the related sum formulae for $\cot(\pi z)$ and $\csc^2(\pi z)$ will also be of interest for their own sake and as input to the study of two other special functions that appear later: the Euler gamma function in Section 9.6 and the Weierstrass $\wp$-function in Section 10.4.

Theorem 9.2.1 (Euler Product Formula). *For all $z \in \mathbb{C}$,*

$$\frac{\sin(\pi z)}{\pi z} = \prod_{n=1}^{\infty}\left(1 - \frac{z^2}{n^2}\right) \tag{9.2.1}$$

Corollary 9.2.2. *We have*

$$\sum_{n=1}^{\infty}\frac{1}{n^2} = \frac{\pi^2}{6} \tag{9.2.2}$$

Proof. We give the formal argument here, leaving some technical details and the extension to the formula for $\sum_{n=1}^{\infty} 1/n^{2\ell}$ to the problems (see Problems 1–6). The Taylor series for $\sin z$ shows

$$\frac{\sin(\pi z)}{\pi z} = 1 - \frac{(\pi z)^2}{6} + O(z^4) \tag{9.2.3}$$

On the other hand, expanding the product (and here's where we leave out an argument),

$$\prod_{n=1}^{\infty}\left(1 - \frac{z^2}{n^2}\right) = 1 - z^2\left(\sum_{n=1}^{\infty}\frac{1}{n^2}\right) + O(z^4) \tag{9.2.4}$$

(9.2.2) comes from identifying the $O(z^2)$ coefficients. $\square$

We'll give two proofs of Theorem 9.2.1 (and sketch a third in Problems 10 and 11). The first concerns the notion that if two entire functions have the same zeros, their ratio is a nonvanishing entire function, so of the form $e^{h(z)}$. Bounds will limit the possibilities for h. To get upper bounds on the ratio, we need an upper bound on the numerator and lower bound on the denominator. Upper bounds will be relatively easy, but lower bounds are subtle and require more detailed information. In the general case, treated in Section 9.10, we'll put the product in the denominator since we have more information about it than an arbitrary function. But here, $\sin(\pi x)$ is so explicit, we'll put it in the denominator. We begin with the upper bound on the product.

Lemma 9.2.3. *For any $\varepsilon > 0$, there is a constant C_ε so that for all $z \in \mathbb{C}$,*

$$\left|\pi z \prod_{n=1}^{\infty}\left(1 - \frac{z^2}{n^2}\right)\right| \le C_\varepsilon \exp(\varepsilon|z|^2) \tag{9.2.5}$$

Remark. It isn't hard to show that in (9.2.5) a bound still holds when $|z|^2$ is replaced by $|z|^{1+\delta}$ for any $\delta > 0$ (see Problem 8). Of course, once we have (9.2.1), there is a bound by $e^{\pi|z|}$, but this is a question of a priori bounds.

Proof. Fix $\varepsilon > 0$. Since $\sum_{n=1}^\infty n^{-2} < \infty$, find N so that

$$\sum_{n=N+1}^{\infty} n^{-2} \le \frac{\varepsilon}{2} \tag{9.2.6}$$

which, by (9.1.6), implies that

$$\left|\prod_{n=N+1}^{\infty}\left(1 - \frac{z^2}{n^2}\right)\right| \le \exp\left(\frac{\varepsilon}{2}|z|^2\right) \tag{9.2.7}$$

On the other hand, to give a crude estimate,

$$\begin{aligned}|z|\prod_{n=1}^{N}\left|1 + \frac{|z|^2}{n^2}\right| &\le (1+|z|^2)^{N+1} \le 2^{N+1}(1+|z|^{2N+2}) \\ &\le \left(\frac{\varepsilon}{2}\right)^{-N-1} 2^{N+1}(N+1)!\left(1 + \frac{(\frac{\varepsilon}{2})^{N+1}|z|^{2N+2}}{(N+1)!}\right) \\ &\le \left(\frac{\varepsilon}{2}\right)^{-N-1} 2^{N+1}(N+1)!\,\exp\left(\frac{\varepsilon}{2}|z|^2\right) \end{aligned} \tag{9.2.8}$$

This plus (9.2.7) implies (9.2.5). □

Lemma 9.2.4. *For $n = 0, 1, 2, \dots$, let Γ_n be the square (four edges) with corners $\pm(n + \frac{1}{2}) \pm i(n + \frac{1}{2})$. Then for all n and all $z \in \Gamma_n$,*

$$|\sin(\pi z)| \ge 1 \tag{9.2.9}$$

Proof. If $z = (n + \frac{1}{2} + iy)$, since

$$|\sin(\pi z)| = |\cosh(\pi y)| \geq 1 \tag{9.2.10}$$

(9.2.9) holds on the vertical edges. On the horizontal edges,

$$|\sin(x \pm i(n + \tfrac{1}{2}))| \geq \tfrac{1}{2}(e^{\pi(n+1/2)} - 1) \geq 1 \tag{9.2.11}$$

since $e^{\pi/2} > 3$. Thus, (9.2.9) holds. □

First Proof of Theorem 9.2.1. Since

$$\sup_{|z|\leq R} \left(\sum_{n=1}^{\infty} \frac{|z|^2}{n^2} \right) < \infty \tag{9.2.12}$$

the product in (9.2.1) defines an entire analytic function by Theorem 9.1.3. Define

$$g(z) = \frac{\pi z \prod_{n=1}^{\infty}(1 - \frac{z^2}{n^2})}{\sin(\pi z)} \tag{9.2.13}$$

Since the numerator and denominator have the same zeros, g is a nonvanishing entire function with $g(0) = 1$.

Thus, there is an entire function h with

$$e^{h(z)} = g(z), \qquad h(0) = 0 \tag{9.2.14}$$

By (9.2.9) and (9.2.5),

$$\sup_{z\in\Gamma_n} |g(z)| \leq C_\varepsilon \exp(2\varepsilon(n + \tfrac{1}{2})^2) \tag{9.2.15}$$

By the maximum principle, this is true for z inside Γ_n and, in particular, in $\mathbb{D}_{(n+1/2)}(0)$. It follows that

$$\sup_{|z|\leq n+1/2} |g(z)| \leq C_\varepsilon \exp(2\varepsilon(n + \tfrac{1}{2})^2) \tag{9.2.16}$$

so that

$$|g(z)| \leq C_\varepsilon \exp(2\varepsilon(|z| + 1)^2) \tag{9.2.17}$$

By (9.2.14),

$$\operatorname{Re} h(z) \leq \log(C_\varepsilon) + 2\varepsilon(|z| + 1)^2 \tag{9.2.18}$$

so, by (3.2.2), if

$$h(z) = \sum_{n=0}^{\infty} a_n z^n \tag{9.2.19}$$

then for all R and k,

$$|a_k| \leq R^{-k}[\log(C_\varepsilon) + 2\varepsilon(R + 1)^2] \tag{9.2.20}$$

Taking $R \to \infty$, and then, if $k = 2$, $\varepsilon \downarrow 0$, we see that

$$a_k = 0 \qquad \text{for } k \geq 2 \tag{9.2.21}$$

that is,

$$g(z) = e^{a+bz} \tag{9.2.22}$$

for some complex a, b. By (9.2.14), $a = 0$.

Since $\sin(\pi z)$ and $z\prod_{n=1}^{\infty}(1 - z^2/n^2)$ are both odd under $z \to -z$, $g(-z) = g(z)$, which means $b = 0$, so $g(z) \equiv 1$, that is, (9.2.1) holds. $\square$

For the second proof, we note that (9.2.1), if it holds, implies

$$\log(\sin(\pi z)) = \log(\pi z) + \sum_{n=1} h_n(z) \tag{9.2.23}$$

with

$$h_n(z) = \log(n^2 - z^2) - \log(n)^2 \tag{9.2.24}$$

$$= \log(n - z) + \log(n + z) - \log(n)^2 \tag{9.2.25}$$

Thus,

$$-h_n''(z) = \frac{1}{(n-z)^2} + \frac{1}{(n+z)^2} \tag{9.2.26}$$

while the negative second derivative of $\log(\sin(\pi z))$ is $\pi^2/\sin^2(\pi z)$.

So, formally,

$$\frac{\pi^2}{\sin^2(\pi z)} = \sum_{n=-\infty}^{\infty} \frac{1}{(z-n)^2} \tag{9.2.27}$$

We say formally because we haven't worried about branches of log, sums, and we don't want to assume (9.2.1) but prove it! The idea of the second proof will be to first prove (9.2.27) and then integrate to get (9.2.1). We begin with

Theorem 9.2.5 (Partial Fraction Expansion of $\pi^2\csc^2(\pi z)$). *For any $z \in \mathbb{C}\setminus\mathbb{Z}$, the sum on the right side of* (9.2.27) *converges and defines an analytic function, and* (9.2.27) *holds.*

Proof. Let

$$g_n(z) = \frac{1}{(z-n)^2} \tag{9.2.28}$$

Then for $N = 1, 2, \dots,$

$$|z| < N < n \Rightarrow |g_n(z)| \le |n - N|^{-2} \tag{9.2.29}$$

so

$$\sum_{N<|n|<\infty} \sup_{|z|\le N} |g_n(z)| < \infty \tag{9.2.30}$$

so, by the M-test,

$$g(z) \equiv \sum_{n=-\infty}^{\infty} g_n(z) \tag{9.2.31}$$

defines a meromorphic function. Moreover, since $g_n(z+1) = g_{n-1}(z)$,

$$\sum_{n=-N}^{N} [g_n(z+1) - g_n(z)] = g_{-N-1}(z) - g_N(z)$$

which goes to zero as $N \to \infty$ for z fixed, so

$$g(z+1) = g(z) \tag{9.2.32}$$

Moreover, for x real in $[0,1]$ and $y > 0$,

$$|g_n(x+iy)| \le [y^2 + (\max(|n|-1,0))^2]^{-1} \tag{9.2.33}$$

so

$$\lim_{|y|\to\infty} \sup_{x\in[0,1]} |g(x+iy)| = 0 \tag{9.2.34}$$

Since $\pi^2 \csc^2(z)$ is also periodic, has the same poles and principal parts as g, and obeys

$$\lim_{|y|\to\infty} \sup_{x\in[0,1]} \pi^{-2} \csc^2(x+iy) = 0 \tag{9.2.35}$$

we conclude that

$$H(z) = \pi^2 \csc^2(z) - g(z) \tag{9.2.36}$$

is entire, periodic, and has

$$\lim_{|y|\to\infty} \sup_{x\in[0,1]} H(x+iy) = 0 \tag{9.2.37}$$

Since H is entire and periodic, by Theorem 3.10.3, there is G analytic in $\mathbb{C}^\times$ so that

$$H(z) = G(e^{2\pi iz}) \tag{9.2.38}$$

By (9.2.37), $\lim_{|w|\to 0} G(w) = \lim_{|w|\to\infty} G(w) = 0$, so the singularity at zero is removable, G is entire, and then, by Liouville's theorem, $G \equiv 0$. Thus, H is zero and (9.2.27) holds. $\square$

One integral of (9.2.27) is particularly simple:

Theorem 9.2.6 (Partial Fraction Expansion of $\pi\cot(\pi z)$). *We have*

$$\pi\cot(\pi z) = \frac{1}{z} + \sum_{n=1}^{\infty} \frac{2z}{z^2 - n^2} \tag{9.2.39}$$

in that the sum on the right is uniformly convergent on compact subsets of $\mathbb{C}\setminus\mathbb{Z}$ to the left side.

Proof. An estimate like (9.2.29) shows that the sum is convergent as claimed. Continuity of derivatives in the topology of $\mathfrak{A}(\Omega)$ and

$$\frac{d}{dz}\left(\frac{2z}{z^2-n^2}\right)=\frac{d}{dz}\left(\frac{1}{z-n}+\frac{1}{z+n}\right)=-\frac{1}{(z-n)^2}-\frac{1}{(z+n)^2} \tag{9.2.40}$$

show that

$$\begin{aligned}\frac{d}{dz}\,(\text{RHS of (9.2.39)}) &= -\text{RHS of (9.2.27)}\\ &= -\text{LHS of (9.2.27)}\\ &= \frac{d}{dz}\,(\text{LHS of (9.2.39)})\end{aligned}$$

so LHS of (9.2.39) – RHS of (9.2.39) is constant. But since each side is odd under $z \to -z$, the constant must be odd, that is, it must be zero. □

The next integration involves logarithmic derivatives, and so is slightly more subtle. We need:

Proposition 9.2.7. *Let $\{f_j\} \subset \mathfrak{A}(\Omega)$ obey (9.1.2) so that*

$$F(z)=\prod_{j=1}^{\infty} f_j(z) \tag{9.2.41}$$

is an absolutely convergent product. Suppose that no f_j is identically zero. Then for z not in the zeros of F,

$$\frac{F'(z)}{F(z)}=\sum_{j=1}^{\infty}\frac{f_j'(z)}{f_j(z)} \tag{9.2.42}$$

where the sum is absolutely and uniformly convergent on compact subsets of $\Omega \setminus \{z \mid F(z)=0\}$.

Proof. See Problem 9. □

Second Proof of Theorem 9.2.1. Let $F(z)$ be πz times the infinite product on the right side of (9.2.1) and $G(z)=\sin(\pi z)$. If $f_n(z)=1-z^2/n^2$, then

$$\frac{f_n'(z)}{f_n(z)}=\frac{2z}{z^2-n^2} \tag{9.2.43}$$

so, by (9.2.42), $F'(z)/F(z) = $ RHS of (9.2.39).

Clearly, by (9.2.39), $G'(z)/G(z) = $ LHS of (9.2.39), so we have

$$F'(z)G(z)-G'(z)F(z)=0 \tag{9.2.44}$$

But that implies $(F/G)'=0$ so that

$$\frac{F(z)}{G(z)}=\lim_{z\to 0}\frac{F(z)}{G(z)}=1 \tag{9.2.45}$$

which is what we wanted to prove. □

While we used Proposition 9.2.7 to go from (9.2.39) to (9.2.1) and so complete our second proof of (9.2.1), we can use it to go in the other direction: first prove (9.2.1) as we did at the start of the section, and then derive (9.2.39) and so also (9.2.27) from (9.2.1).

Notes and Historical Remarks.

> And so is satisfied the burning desire of my brother, who, realizing that the investigation of the sum was more difficult than anyone would have thought, openly confessed that all his zeal had been mocked. If only my brother were alive now.
>
> —*Johann Bernoulli*, quoted in Young [**599**] [2]

The Euler product formula appeared in his 1734/35 note [**172**] on the Basel problem and included his proof of (9.2.2) by these methods (although he appears to have found other proofs of (9.2.2) earlier; he also reported on these earlier proofs). Johann Bernoulli, who had been Euler's mentor, complained that the proof required that one know that $\sin(\pi z)$ have no zeros in $\mathbb{C}\setminus\mathbb{R}$. Spurred by this, Euler studied complex exponentials leading to $e^{i\theta}=\cos\theta+i\sin\theta$ and, in 1741, to a second proof of the product formula relying on $e^z=\lim_{n\to\infty}(1+z/n)^n$ (this proof is sketched in Problem 10). In Section 9.6, we'll get the product formula as a byproduct of our analysis of the gamma function (although, traditionally, the formula for the gamma we prove is obtained by first proving the Euler product formula). This is a variant of our second proof. In Section 9.10, we'll provide a fifth proof that is a variant of our first—deriving Euler's formula from the Hadamard product formula. Apostol [**16**] has a summary of various elementary proofs of this formula and Apostol [**17**], another proof.

An amazing application of the Euler formula for $\sum_{n=1}^{\infty}n^{-2}$ is to compute

$$\int_0^1 \frac{\log y}{1+y}\,dy=-\frac{\pi^2}{12} \tag{9.2.46}$$

(obtained by writing $y=e^{-x}$ and expanding $1+e^{-x}$ in a geometric series). Since we'll need it in Section 2.8 of Part 3, we'll leave the details to Problem 8 of that section.

The term "Euler product formula" is also used for his unrelated formula

$$\sum_{n=1}^{\infty}n^{-s}=\prod_{p\text{ prime}}\left(1-\frac{1}{p^s}\right)^{-1} \tag{9.2.47}$$

that will reappear in Chapter 13 of Part 2B, where we'll call it the Euler factorization theorem.

[2] writing to his former pupil, Euler, when Euler found the Basel sum which had so fascinated Jakob Bernoilli.

By 1740, Euler also had the partial fraction expansion of $\pi\cot(\pi z)$ which appeared in his 1748 book on calculus [**178**]. An interesting proof relying on work of Schröter, Schottky, and Herglotz (from the 1890s) is sketched in Problem 14.

Leonhard Euler (1707–83) was a Swiss mathematician who spent most of his career in St. Petersburg (1727–41 and 1766–83) and Berlin (1741–66) as a member of the Russian and Prussian academies. His father and maternal grandfather were ministers in the Protestant church and Euler was expected to go into the family business. Fortunately, his mathematical talents were discovered by Johann Bernoulli who knew Euler's father from their years in school, and Bernoulli persuaded the father to let Euler follow his talents. Interestingly enough, Bernoulli's father was also a pastor who had expected his older son, Jakob (the most famous of the Bernoulli mathematicians), to become a pastor.

Bernoulli also helped Euler in finding his initial position in Russia; indeed, Euler replaced one of Bernoulli's sons who died of appendicitis, and lived with another son, Daniel, until Euler's marriage. Euler had thirteen children. Both the shift to Berlin and back had political roots. Indeed, in Berlin, he had conflicts with Voltaire (perhaps caused partly by Euler's deep religious beliefs and Voltaire's free thinking) and Frederick the Great. Another factor was that Catherine the Great had ascended to the throne and wanted to get Euler back. Euler lost vision in one eye in 1735 and the other in 1766 but remained productive with the help of scribes; for example, in 1775, at age 68, he produced over fifty papers! Euler's prodigious output can be seen by two facts: it has been estimated that one-third of all published work in mathematics and mechanics in 1726–1800 is Euler's and the Russian Academy was posthumously publishing his papers for fifty years!

Euler, widely regarded as the greatest mathematician of the eighteenth century, was also a great physicist, making important contributions to acoustics, hydrodynamics, and mechanics. An indication of the significance of his work is that, in a poll *The Mathematical Intelligencer* did in 1999, Euler had three of the top five formulae of all time ($e^{\pi i} = -1$, his relation of edges, faces, and vertices of polyhedra, and $\sum_{n=1}^{\infty} n^{-2} = \pi^2/6$). He is also responsible for much of modern notation, including the use of sigma for sum, $f(\,\cdot\,)$ notation, "e" for the base of the natural logs (it is generally thought that he picked "e" for exponential), π for pi, i for $\sqrt{-1}$, and the symbols for the trigonometric functions. There is an online archive at the MAA[3] of hundreds of articles on Euler, including translations of many of his key papers. Euler has long fascinated mathematicians. Two charming books

[3]`http://eulerarchive.maa.org/`

that combine biography and a look at his mathematics are Dunham [**154**] and Varadarajan [**560**].

Besides the results of this section, we'll see Euler again in the sections on the gamma and beta functions (Sections 9.6 and 9.7) and in the study of elliptic functions (see Section 10.4) and of the zeta function (see Section 13.3 of Part 2B). His book [**178**] had enormous influence, establishing and codifying much of what we call calculus. His book [**177**] was one of the first successful popular science books.

Problems

1. Complete the proof of (9.2.1) by justifying (9.2.4).

2. (a) Prove that

$$\pi\cot(\pi z) = \sum_{n=0}^{\infty}\left[\frac{1}{z+n} - \frac{1}{z-(n+1)}\right] \tag{9.2.48}$$

(b) By taking $z = \frac{1}{3}$, prove that

$$\frac{1}{1\cdot 2} + \frac{1}{4\cdot 5} + \frac{1}{7\cdot 8} + \frac{1}{10\cdot 11} + \cdots = \frac{\pi}{3\sqrt{3}} \tag{9.2.49}$$

3. Prove the following

(a) $\pi\tan(\pi z) = \sum\limits_{n=0}^{\infty}\dfrac{2z}{(n+\frac{1}{2})^2 - z^2}$

(b) $\pi\csc(\pi z) = \dfrac{1}{z} + \sum\limits_{n=1}^{\infty}\dfrac{(-1)^n(2z)}{z^2-n^2}$

(c) $\pi\sec(\pi z) = \sum\limits_{n=0}^{\infty}\dfrac{(-1)^n(2n+1)}{(n+\frac{1}{2})^2 - z^2}$

4. The Bernoulli numbers, B_n, are defined by (3.1.48). You proved them rational in Problem 11 of Section 3.1. Prove that for $\ell = 1, 2, \dots,$

$$\sum_{n=1}^{\infty}\frac{1}{n^{2\ell}} = (-1)^{\ell-1}\frac{(2\pi)^{2\ell}}{2(2\ell)!}B_{2\ell} \tag{9.2.50}$$

Hint: Use the partial fraction expansion of $\pi\cot(\pi z)$ plus

$$\pi\cot(\pi z) = z^{-1} + \sum_{\ell=1}^{\infty}(-1)^{\ell}\frac{(2\pi)^{2\ell}}{(2\ell)!}B_{2\ell}z^{2\ell-1}$$

which you should prove, given (3.1.51). The first four of the sums $S_\ell = \sum_{n=1}^\infty n^{-2\ell}$ are

$$S_1 = \frac{\pi^2}{6}, \qquad S_2 = \frac{\pi^4}{90}, \qquad S_3 = \frac{\pi^6}{945}, \qquad S_4 = \frac{\pi^8}{9450}$$

Note. Bernoulli numbers were found by Jakob Bernoulli, the older brother of Euler's teacher, Johann. Bernoulli numbers are discussed further in Section 9.7 and Section 13.3 of Part 2B.

5. The Euler numbers, E_n, are defined by

$$(\cosh z)^{-1} = \sum_{n=0}^\infty E_n z^n/n! \tag{9.2.51}$$

Prove that for $\ell = 1, 2, \dots$, one has

$$\sum_{n=0}^\infty \frac{(-1)^n}{(2n+1)^{2\ell+1}} = \frac{(-1)^\ell \pi^{2\ell+1}}{2^{2\ell+2}(2\ell)!} E_{2\ell} \tag{9.2.52}$$

(*Hint*: Develop a partial fraction expansion for $[\cosh(\pi z)]^{-1}$.)

Remark. Some sources define E_n with $(\cos z)^{-1}$ not $(\cosh z)^{-1}$. These $\widetilde{E}_n$ are related to our E_n by $\widetilde{E}_{2n} = (-1)^n E_{2n}$. All $\widetilde{E}_n > 0$.

6. In Problem 24 of Section 5.7, you extended Theorem 5.7.13. Use this extension to prove (9.2.50). Similarly, prove an extended version of the result in Problem 19 of Section 5.7 and use it to prove (9.2.52).

7. Recover Wallis' formula, (5.7.80), from the Euler product formula. (*Hint*: Take $z = \frac{1}{2}$.)

8. (a) Prove for $0 < \alpha < 1$, we have that, for $x > 0$, $(1+x)^\alpha \le (1 + x^\alpha)$. (*Hint*: Prove $\frac{d}{dx}[(1+x^\alpha) - (1+x)^\alpha] \ge 0$.)

(b) Prove that for $0 < \alpha < 1$,

$$\left| \prod_{n=N}^\infty \left(1 - \frac{z^2}{n^2}\right) \right| \le \exp\left(\alpha^{-1} |z|^{2\alpha} \sum_{n=N}^\infty n^{-2\alpha} \right)$$

and conclude that (9.2.5) is valid if $|z|^2$ is replaced by $|z|^{1+\delta}$.

9. (a) Suppose g is analytic in a neighborhood of $\overline{\mathbb{D}_\delta(z_0)}$ and $\sup_{|w_0 - z_0| = \delta} |g(w) - 1| < \frac{1}{2}$. Prove that

$$\left| \frac{g'(z_0)}{g(z_0)} \right| \le 2\delta' \sup_{|w - z_0| = \delta} |g(w) - 1|$$

(b) Under the hypothesis of Proposition 9.2.7, prove the sum in (9.2.42) is uniformly and absolutely convergent.

(c) Prove for finite products the analog of (9.2.42) and then deduce (9.2.42).

10. (a) Prove that $\sin z = \lim_{n\to\infty}(2i)^{-1}q_n(iz/n)$ where $q_n(z) = (1+z)^n - (1-z)^n$.

(b) Show $q_{2m}(w) = 0$ if and only if $w = 0$ or $w = \pm i\tan j\pi/m$ for $j = 1, \dots, m-1$, and conclude that

$$q_{2m}(z) = 4mz\prod_{j=1}^{m-1}\left(1 + z^2\cot^2\left(\frac{j\pi}{m}\right)\right)$$

(c) Using the fact that $\lim_{z\to 0,\, z\neq 0} z\cot(z) = 1$, prove the Euler product formula.

Remark. The interchange of product and limit is tricky because of the fact that the number of factors is increasing. It may help to prove the result only for iz real and small (and then use Vitali) and to use the inequality $\tan x \geq x$ for $0 < x < \pi/2$. See Eberlein [**158**].

11. This is a variant of the approach of Problem 10, following Ebbinghaus et al. [**157**].

(a) Show that the Chebyshev polynomial of the second kind, $U_n(x)$, of Problem 8 of Section 3.1 is even for n even, and conclude that for $n = 0, 1, 2, \dots$, $\sin((2n+1)x) = \sin x\, S_{2n}(\sin x)$ for a polynomial S_{2n} of degree $2n$.

(b) Prove that $S_{2n}(y) = 0$ if and only if $y = \sin(j\pi/(2n+1))$, $j = \pm 1, \pm 2, \dots, \pm n$, and that $S_{2n}(0) = 2n+1$. Use this to conclude that

$$\sin z = (2n+1)\sin\left(\frac{z}{2n+1}\right)\prod_{j=\pm 1,\dots,\pm n}\left[1 - \frac{\sin(\frac{z}{2n+1})}{\sin(\frac{j\pi}{2n+1})}\right]$$

(c) By taking $n\to\infty$, prove the Euler product formula.

12. This will provide another proof of (9.2.2) that Euler found.

(a) Prove that

$$\int_0^1 \frac{\arcsin(t)}{\sqrt{1-t^2}}\,dt = \frac{\pi^2}{8}$$

(*Hint*: Look at $\frac{d}{ds}\frac{1}{2}\arcsin^2(s)$.)

(b) Prove $\arcsin(t) = \sum_{n=0}^{\infty} t^{2n+1}\frac{1\cdot 3\cdots(2n-1)}{2\cdot 4\cdots(2n)}\frac{1}{2n+1}$, where the coefficient for $n = 0$ is 1.

(c) Let $J_n = \int_0^1 \frac{t^n}{\sqrt{1-t^2}}\,dt$ for $n = 1, 2, \dots$. Prove that $J_{n+1} = \frac{n}{n+1} J_{n-1}$ and $J_1 = 1$.

(d) Prove that $\frac{\pi^2}{8} = \sum_{n=0}^{\infty} \frac{1}{(2n+1)^2}$ and then (9.2.2).

13. This has yet another elementary proof of (9.2.2) due to Beukers et al. [**48**].

(a) Write $\zeta(2) = \sum_{n=1}^{\infty} n^{-2}$. Prove that

$$\sum_{n=0}^{\infty} (2n+1)^{-2} = \tfrac{3}{4}\zeta(2) \tag{9.2.53}$$

(b) Prove that

$$\int_0^1 \int_0^1 \frac{dx\,dy}{1 - x^2y^2} = \sum_{n=0}^{\infty} (2n+1)^{-2} \tag{9.2.54}$$

(*Hint*: Geometric series.)

(c) Map $\Omega = \{(u,v) \mid u > 0,\ v > 0,\ u + v < \frac{\pi}{2}\}$ to $[0,1] \times [0,1]$ by $(x,y) = (\frac{\sin u}{\cos v}, \frac{\sin v}{\cos u})$. Prove that this map is a bijection with Jacobian $\frac{\partial(x,y)}{\partial(u,v)} = 1 - x^2y^2$.

(d) Prove that the left-hand side of (9.2.54) is the area of Ω which is $\pi^2/8$ and conclude that $\zeta(2) = \pi^2/6$.

14. (a) Let h be an entire function obeying $4h(2z) = h(z) + h(z + \frac{1}{2})$. Prove $h(z) = 0$. (*Hint*: Look at $\sup_{|z| \le 2} |h(z)|$.)

(b) Let g be an entire function obeying

$$2g(2z) = g(z) + g(z + \tfrac{1}{2}) \tag{9.2.55}$$

Prove that g is constant. (*Hint*: Let $h(z) = g'(z)$.)

(c) Let $S_N(z) = z^{-1} + \sum_{n=1}^{N}((z+n)^{-1} + (z-n)^{-1})$. Prove that $S_N(z) + S_N(z + \frac{1}{2}) = 2S_{2N}(2z) + (2z + 2N + 1)^{-1}$.

(d) Prove that both sides of (9.2.39) obey (9.2.55), and so conclude that (9.2.39) is valid.

15. Let Γ_n be the contour in Lemma 9.2.4 and $f(z) = \pi\cot(\pi z) - z^{-1}$.

(a) Show $\sup_{n, z \in \Gamma_n} |f(z)| < \infty$.

(b) For $z \in \mathbb{C} \setminus \mathbb{Z}$, compute $1/2\pi i \oint_{\Gamma_n} f(\zeta) d\zeta / \zeta(\zeta - z)$ via residues, and thereby find another proof of (9.2.39).

16. (a) Prove the following formula of Euler:

$$\frac{\sin(\pi z)}{\pi z} = \prod_{j=1}^{\infty} \cos\left(\frac{\pi z}{2^j}\right)$$

(*Hint*: $\sin(2w) = 2\sin(w)\cos(w)$.)

(b) Prove the following formula of F. Viète (1540–1603):

$$\frac{2}{\pi} = \frac{\sqrt{2}}{2} \cdot \frac{\sqrt{2+\sqrt{2}}}{2} \cdot \frac{\sqrt{2+\sqrt{2+\sqrt{2}}}}{2} \cdots$$

(*Hint*: $z = \frac{1}{2}$.)

Remark. This formula provides a rapidly convergent way to compute π. Define x_n inductively by $x_{n+1} = \sqrt{2+x_n}$ and then $\pi = \lim_{n\to\infty} 2^{n+1}(\prod_{j=1}^{n} x_j)^{-1}$. The error is $O(2^{-n})$.

17. Prove that

$$e^{\pi z} - 1 = \pi z e^{\pi z/2} \prod_{n=1}^{\infty} \left(1 + \frac{z^2}{4n^2}\right)$$

18. Prove Euler's formula for $\cos(\pi z)$,

$$\cos(\pi z) = \prod_{j=1}^{\infty} \left(1 - \frac{4z^2}{(2j-1)^2}\right)$$

(*Hint*: See the hint for Problem 16.)

9.3. The Mittag-Leffler Theorem

Our goal in this section is to prove

Theorem 9.3.1 (Mittag-Leffler Theorem). *Let $\{z_n\}_{n=1}^{\infty}$ be a sequence of distinct points with $\lim_{n\to\infty}|z_n| = \infty$. Let $P_1, P_2, \dots$ be nonzero polynomials with $P_j(0) = 0$. Then there exists an entire meromorphic function, f, whose only poles are at $\{z_n\}_{n=1}^{\infty}$ with principal parts at z_n equal to $P_n(1/(z-z_n))$.*

The obvious way to try to prove this result is to set

$$f(z) = \sum_{n=1}^{\infty} P_n((z-z_n)^{-1}) \tag{9.3.1}$$

The problem is there is no reason for the sum to converge! Since analytic "corrections" don't impact principal parts, why not subtract an analytic correction to force convergence? This is exactly what we'll do.

Proof of Theorem 9.3.1. We can add on $P(z^{-1})$ by hand, so we'll suppose each $z_n \neq 0$. Since $P_n((z-z_n)^{-1})$ is analytic in a neighborhood of $\overline{\mathbb{D}_{|z_n|/2}(0)}$, its Taylor series converges uniformly there, so we can find polynomials $Q_n(z)$ (the start of the Taylor series about zero) so that, with $f_n(z) = P_n((z-z_n)^{-1}) - Q_n(z)$, we have

$$\sup_{|z|\le\frac{1}{2}|z_n|} |f_n(z)| \le 2^{-n} \tag{9.3.2}$$

By (9.3.2),

$$f(z) = \sum_{n=1}^{\infty} f_n(z) \tag{9.3.3}$$

converges uniformly on compact subsets of $\mathbb{C}\setminus\{z_n\}_{n=1}^{\infty}$ and, by construction, $f(z) - P_n((z-z_n)^{-1})$ has a removable singularity at z_n. Thus, f is entire meromorphic with the prescribed principal parts. □

Notes and Historical Remarks. Gösta Leffler (1846–1927) was a Swedish mathematician who added his mother's maiden name to his own while a student at Uppsala. His career was greatly impacted by a fellowship to study abroad in 1873–76. He went first to Paris to study with Hermite, who recommended that he go to Berlin to learn with Weierstrass, which he did. Motivated by Weierstrass' work on infinite products (see the next section), Mittag-Leffler found a partial version of his theorem with an involved proof in 1876–77 [**384**]. In an 1880 note [**586**], Weierstrass gave the complete result with the now standard proof.

Mittag-Leffler was notable for founding and leading the journal *Acta Mathematica* and for his support of Cantor's work on set theory. There is a biographical note by Yngve Domarat on Mathematical research during the first decades of the University of Stockholm. There is a full biography of Mittag-Leffler by Stubhaug [**547**]. Persistent stories claim there is no Nobel prize in mathematics because of bad blood between Nobel and Mittag-Leffler. One version says this was because Mittag-Leffler had an affair with Nobel's wife. In fact, Nobel was never married and [**203**] debunks the rumor (see also Alex Lopez-Ortiz[4]).

Problems

1. Recall that in Problem 11 in Section 5.4, you constructed solutions of $\bar\partial f = g$ for $g \in C_0^\infty(\mathbb{C})$. Here you'll use the strategy of the proof of this section to solve $\bar\partial f = g$ for any $g \in C^\infty(\mathbb{C})$.

   (a) Prove that any g can be written $\sum_{n=0}^\infty g_n$ where g_n is C^∞ and $\text{supp}(g_n) \subset \{z \mid n-1 < |z| < n+1\}$.

[4]`http://www.cs.uwaterloo.ca/~alopez-o/math-faq/node50.html`

(b) Show that for $n \geq 3$, there exists f_n solving $\bar{\partial} f_n = g_n$ with $\sup_{|z|\leq n/2} |f_n^{(k)}(z)| \leq 2^{-n}$ for $k = 1, 2, \dots, n$. (*Hint*: f_n is analytic in $|z| \leq n/2$. Use the strategy of this section.)

(c) Show that $f = \sum_{n=0}^\infty f_n$ defines a C^∞ function solving $\bar{\partial} f = g$.

2. This problem will lead to a proof of the Mittag-Leffler theorem, assuming you know you can solve $\bar{\partial} f = g$ for any C^∞ function, g, in $\mathbb{C}$.

(a) Given $\{z_n\}_{n=1}^\infty$ distinct points, let $r_n = \min_{j\neq n} |z_j - z_n|$. Show there exist functions $h_n \in C_0^\infty(\mathbb{C})$ so that $\text{supp}(h_n) \subset \mathbb{D}_{\frac{1}{2} r_n}(z_n)$ and $h_n \equiv 1$ in $\mathbb{D}_{\frac{1}{4} r_n}(z_n)$.

(b) Show that there is a C^∞ function g so that for $z \in \mathbb{C} \setminus \{z_n\}$,

$$\bar{\partial}\bigg(\sum_{n=1}^\infty h_n(z) P_n((\cdot - z_n)^{-1})\bigg) = -g(\cdot)$$

(c) Let f solve $\bar{\partial} f = g$. Show that $f + \sum_{n=1}^\infty h_n(\cdot) P_n((\cdot - z_n)^{-1})$ is an entire meromorphic function that solves the Mittag-Leffler problem.

3. Two functions, $f, g \in \mathfrak{A}(\mathbb{C})$, are called *relatively prime* if and only if they have no common zero. Prove Wedderburn's lemma [**581**]: If f, g are relatively prime, there exist $a, b \in \mathfrak{A}(\Omega)$ so that

$$1 = af + bg \tag{9.3.4}$$

(*Hint*: Use the Mittag-Leffler theorem to show $(fg)^{-1} = a_1 + b_1$ where a_1 has poles only at the zeros of g and b_1 only at the poles of f.)

9.4. The Weierstrass Product Theorem

Our goal in this section is to prove:

Theorem 9.4.1. *Let $\{z_j\}_{j=1}^\infty$ be a distinct set of points in $\mathbb{C}$ obeying $|z_j| \to \infty$ as $j \to \infty$. Let $n_1, n_2, \dots$ be strictly positive integers. Then there exists an entire function f with zeros precisely at $\{z_j\}_{j=1}^\infty$ and with the order of the zero at z_j equal to n_j.*

As with the last section, the obvious first guess is $\prod_{j=1}^\infty (1 - z/z_j)^{n_j}$, but unless $\sum (R/|z_j|)^{n_j} < \infty$ for all R, that won't converge. So we need to renormalize with "subtractions," but to preserve zeros, these subtractions have to be multiplicative. Since $\log(1 - x) = -x - x^2/2 - x^3/3 - \dots$, the natural form is to use:

Definition. The *Weierstrass factors*, $E_n(z)$, are defined by

$$E_0(z) = 1 - z \tag{9.4.1}$$

$$E_n(z) = (1 - z) \exp\bigg(\sum_{j=1}^n \frac{z^j}{j}\bigg) \qquad \text{for } n \geq 1 \tag{9.4.2}$$

Here is the key bound:

Proposition 9.4.2. *For $|z| < 1$, one has*

$$|1 - E_n(z)| \leq |z|^{n+1} \tag{9.4.3}$$

Proof. We compute

$$-E_n'(z) = \left[1 - (1-z)\sum_{j=1}^{n} z^{j-1}\right] \exp\left(\sum_{j=1}^{n} \frac{z^j}{j}\right) \tag{9.4.4}$$

$$= z^n \exp\left(\sum_{j=1}^{n} \frac{z^j}{j}\right) \tag{9.4.5}$$

$$\equiv \sum_{m=0}^{\infty} z^{n+m} b_{n,m} \tag{9.4.6}$$

for suitable $b_{n,m} \geq 0$ (by expanding and using the positivity of the Taylor coefficients of e^z). The sum is convergent for all z. Thus, integrating,

$$(1 - E_n(z)) = \sum_{m=0}^{\infty} z^{n+m+1}(n+m+1)^{-1} b_{n,m} \tag{9.4.7}$$

Thus, for $|z| \leq 1$,

$$\begin{aligned} |1 - E_n(z)| &\leq |z|^{n+1} \sum_{m=0}^{\infty} (n+m+1)^{-1} b_{n,m} \\ &= |z|^{n+1}(1 - E_n(1)) \\ &= |z|^{n+1} \end{aligned} \tag{9.4.8}$$

□

The following immediately implies Theorem 9.4.1 (given that one can accommodate $z_j = 0$ with a prefactor of z^ℓ and $n_j > 1$ by repeating z_n's below):

Theorem 9.4.3 (Weierstrass Product Theorem). *Let $\{z_n\}_{n=0}^{\infty}$ be a sequence of numbers in $\mathbb{C}^\times$ with $|z_n| \to \infty$ (but not necessarily distinct). Then for each $R < \infty$,*

$$\sum_{n=0}^{\infty} \sup_{|z| \leq R} \left| 1 - E_n\left(\frac{z}{z_n}\right)\right| < \infty \tag{9.4.9}$$

so

$$f(z) = \prod_{n=0}^{\infty} E_n\left(\frac{z}{z_n}\right) \tag{9.4.10}$$

defines an entire function whose zeros are precisely at the z_n with multiplicity of $z_n = \#\{j \mid z_j = z_n\}$.

Remark. This is sometimes called the *Weierstrass factorization theorem.*

Proof. By (9.4.2), since $\#\{z_n \mid |z_n| < 2R\}$ is finite and $|z/z_n| < 1$ if $|z| \leq R \leq 2R \leq |z_n|$, (9.4.9) is implied by

$$\sum_{\{z_n||z_n|\geq 2R\}} \left(\frac{R}{|z_n|}\right)^{n+1} < \infty \tag{9.4.11}$$

But if $|z_n| > 2R$, $R/|z_n| < \frac{1}{2}$ and (9.4.11) is implied by $\sum_{n=0}^{\infty} 2^{-n-1} < \infty$.

Once one has (9.4.9), we get the convergence of the product and zero structure from Theorem 9.1.3. □

Corollary 9.4.4. *If f is an entire meromorphic function, then $f = h/g$ with h, g entire analytic functions with no common zero.*

Proof. If f has poles at $\{z_j\}_{j=1}^{\infty}$ with orders n_j, let g be a function with zeros there of exactly those orders, and let $h = fg$. □

Corollary 9.4.5. *Let $\{z_n\}_{n=1}^{\infty}$ be a set of distinct points with $|z_n| \to \infty$ and let $\{w_n\}_{n=1}^{\infty} \in \mathbb{C}$. Then there is an entire function with*

$$f(z_n) = w_n \tag{9.4.12}$$

Proof. Let g be an entire function with simple zeros at $\{z_n\}_{n=1}^{\infty}$. Let h be a meromorphic function, as guaranteed by the Mittag-Leffler theorem, with simple poles at $\{z_n\}_{n=1}^{\infty}$ and principal parts

$$(z - z_n)^{-1}\left[\frac{w_n}{g'(z_n)}\right] \tag{9.4.13}$$

Let $f = hg$. Then f is entire and obeys (9.4.12). □

Notes and Historical Remarks. Weierstrass published his theorem in 1876 [**585**]. (He originally announced the result at the Berlin Akademie on December 10, 1874.) He explained that his motivation was to understand the gamma function in terms of its poles. His resulting definition of Γ will be a major piece of our presentation of Γ in Section 9.6.

An interesting precursor of Weierstrass is work of Enrico Betti (1823–92)[5] in 1860 [**47**] on defining elliptic functions in terms of products. Betti showed if $\inf_{i\neq j}|w_i - w_j| > 0$ and $\inf_j |w_j| > 0$, then $\sum_j |w_j|^{-\alpha} < \infty$ for $\alpha > 2$ (see Problem 1 of Section 10.4) and then defined functions with zeros at those points by

$$\prod_{j=1}^{\infty}\left(1+\frac{1}{w_j}\right)^z\left(1+\frac{1}{2w_j^2}\right)^z\left(1+\frac{1}{2w_j^2}\right)^{z^2}\left(1-\frac{z}{w_j}\right)$$

[5]Betti was influenced enormously by a visit to Riemann in 1858. In turn, he was the father of a lively school of analysis in Pisa including Dini, Arzelà, Ricci, and Volterra.

Karl Weierstrass (1815–97) was the son of a Prussian government functionary. Weierstrass dropped out of college in 1838 because of a conflict with his father who wanted him to pursue studies in finance with an eye to becoming a government functionary. The resolution was his entering training to be a gymnasium teacher, which he became in 1842. During this period, he did copious but unpublished work on power series. During the summer break of 1853, he wrote a manuscript on abelian functions [**583**]). In the words of E. T. Bell [**37**, pp. 406–432]:

> The memoir on Abelian functions published in *Crelle's Journal* in 1854 created a sensation. Here was a masterpiece from the pen of an unknown schoolmaster in an obscure village nobody in Berlin had ever heard of. This in itself was sufficiently astonishing. But what surprised those who could appreciate the magnitude of the work even more was the almost unprecedented fact that the solitary worker had published no preliminary bulletins announcing his progress from time to time, but with admirable restraint had held back everything till the work was completed.

As a result, Weierstrass got numerous offers, and even honorary degrees, leading to a professorship at the University of Berlin in late 1856. He suffered from ill health and, in later years, he lectured in a wheelchair with an assistant writing on the blackboard for him. His students and others strongly influenced by him included Cantor, Frobenius, Hölder, Hurwitz, Killing, Klein, Lie, Minkowski, Mittag-Leffler, Schönflies, Schottky, Schwarz, von Mangoldt, and Sofia Kovalevskaya, for whom he helped arrange a professorship in Stockholm. There is an online collection of Weierstrass' papers[6].

Besides his work on products and elliptic functions discussed in this chapter and the next, Weierstrass is noted for championing rigor in analysis and is responsible for systematizing the ε-δ definition of limit, the notion of uniform convergence, the construction of the first continuous but nowhere differentiable functions, his polynomial approximation theorem discussed in Section 2.4 of Part 1, and the Weierstrass preparation theorem, a fundamental result in power series in several variables.

The bound (9.4.3) goes back to Fejér, according to a footnote on page 227 in Hille [**262**]: "This proof was communicated to me some forty years ago by my teacher Marcel Riesz. If I remember correctly, he ascribed it to Fejér. The proof does not seem to have been published." It is remarkable that it doesn't appear in all books on the subject—the majority settle for the weaker $|1 - E_n(z)| \leq 2|z|^{n+1}$ if $|z| < \frac{1}{2}$ (see Problem 1). While this suffices for applications, it is a shame that the more elegant (9.4.3) isn't more common, although it is in a substantial minority of texts.

[6]`http://bibliothek.bbaw.de/bibliothek-digital/digitalequellen/schriften/autoren/weierstr/`

Another, less explicit, proof of Theorem 9.4.1 uses the Mittag-Leffler theorems (see Problems 2 and 3). It was popular in texts of the early twentieth century, causing the following outburst from Pringsheim [**461**], as quoted in Remmert [**477**]: "This topsy-turvy way of doing things should not be sanctioned by anyone who sees mathematics as something other than a disordered heap of mathematical results."

For other approaches to the interpolation result (Corollary 9.4.5), see Davis [**133**], Eidelheit [**160**], and Luecking–Rubel [**369**, Ch. 18].

Problems

1. Prove from

$$\log(1-z)+\sum_{j=1}^{n}\frac{z^j}{j}=-\sum_{j=n+1}^{\infty}\frac{z^j}{j}$$

that for $|z|\le\frac{1}{2}$, $|E_n(z)-1|\le 2|z|^{n+1}$. (*Hint*: You'll need to do $n=0$ separately, and prove that if $0\le x\le\frac{1}{2}$, then $e^x\le 1+2x$.)

2. Let h be an entire meromorphic function. Suppose all poles of h are simple and have residues equal to a positive integer. Prove there is an entire analytic function, f, with $h=f'/f$ where f has zeros exactly at the poles of h with the order of a zero equal to the residue of h at that point. (*Hint*: Show that $\exp(\oint_{z_0}^{z_1}h(z)\,dz)$ is independent of contour γ with $\gamma(0)=z_0$, $\gamma(1)=z_1$, and $\operatorname{Ran}(\gamma)\subset\Omega\setminus\{\text{poles of } h\}$.)

3. Use the theorem from Problem 2 and the Mittag-Leffler theorem to prove Theorem 9.4.1.

4. Let $\{z_j\}_{j=1}^{\infty}$ be a sequence of distinct points in $\mathbb{C}$. Let $\{n_j\}_{j=1}^{\infty}$ be a sequence of positive integers and $\{a_k^{(j)}\}_{k=1,\dots,n_j;\,j=1,2,\dots}$ a multisequence of complex numbers. By following the proof of Corollary 9.4.5, prove that there is an entire function f with $f^{(k)}(z_j)=a_{k+1}^{(j)}$ for $k=0,2,\dots,n_j-1$ and $j=1,\dots$ (where $f^{(0)}(z_j)\equiv f(z_j)$).

5. This problem will lead you through a different solution of the interpolation result of Problem 4. Pick $\varepsilon_j=\frac{1}{4}\min_{k\ne j}|z_k-z_j|$. Let ψ_j be C^∞ functions with $0\le\psi_j\le 1$, $\operatorname{supp}(\psi_j)\subset\mathbb{D}_{2\varepsilon_j}(z_j)$, and $\psi_j\equiv 1$ in $\mathbb{D}_{\varepsilon_j}(z_j)$. Let g be an entire function with a zero of order n_j at each z_j.

   (a) Prove that

$$\varphi(z)=g(z)^{-1}\sum_{j=1}^{\infty}\biggl(\sum_{k=0}^{n_j-1}\frac{a_{k+1}^{(j)}(z-z_j)^k}{k!}\biggr)\psi_j(z)$$

   is a C^∞ function on $\mathbb{C}$.

(b) Let η be the solution of $\bar\partial\eta = -\varphi$ guaranteed by Problem 1 of Section 9.3. Prove that

$$f(z) = \eta(z)g(z) + \sum_{j=1}^{\infty}\biggl(\sum_{k=0}^{n_j-1} \frac{a_{k+1}^{(j)}(z-z_j)^k}{k!}\biggr)\psi_j(z)$$

is an entire function that solves the interpolation question of Problem 4.

6. Let $\{a_j\}$ be a sequence of complex numbers going to zero so that, for some fixed n, $\sum_{j=1}^{\infty}|a_j|^{n+1} < \infty$, and for $\ell = 1, 2, \dots, n$, $\lim_{N\to\infty}\sum_{j=1}^{N} a_j^{\ell}$ exists (and is finite). Prove that $\lim_{N\to\infty}\prod_{j=1}^{N}(1+a_j)$ exists and, so long as no a_j is -1, is nonzero.

7. Suppose a_j are real, $\lim_{N\to\infty}\sum_{j=1}^{N} a_j$ exists (and is finite), and that $\sum_{j=1}^{\infty} a_j^2 = \infty$. Prove that $\lim_{N\to\infty}\prod_{j=1}^{N}(1+a_j) = 0$.

8. (a) Let $f_1, \dots, f_n \in \mathfrak{A}(\mathbb{C})$, none of them identically zero. Show that they have a *greatest common divisor*, that is, a function $f \in \mathfrak{A}(\mathbb{C})$, so that each f_j/f is analytic, and if $h \in \mathfrak{A}(\mathbb{C})$ is such that f_j/h is analytic, so is f/h.

   (b) Prove that if f is a greatest common divisor of $f_1, \dots, f_n \in \mathfrak{A}(\mathbb{C})$, then there exist $a_1, \dots, a_n \in \mathfrak{A}(\mathbb{C})$, so $f = \sum_{j=1}^{n} a_j f_j$. (*Hint*: Use induction and Wedderburn's lemma; see Problem 3 of Section 9.3.)

   (c) Prove Helmer's theorem [**251**]: Every finitely generated ideal in $\mathfrak{A}(\mathbb{C})$ is a principal idea.

   (d) Let $h_n(z) = \sin(\pi z)/z\prod_{j=1}^{n}(z^2 - j^2)$. Let $\mathfrak{I}$ be the set of all functions of the form fh_n for some n and some entire function f. Prove that $\mathfrak{I}$ is an ideal in $\mathfrak{A}(\mathbb{C})$ that is not a principal ideal.

9.5. General Regions

Here we'll prove analogs of the Mittag-Leffler and Weierstrass theorems for a general region, $\Omega \subset \mathbb{C}$.

Theorem 9.5.1. *Let $\{z_j\}_{j=1}^{\infty}$ be a sequence of distinct points in Ω with all limit points in $\partial\Omega \cup \{\infty\}$. Let $\{P_j\}_{j=1}^{\infty}$ be a sequence of nonzero polynomials with $P_j(0) = 0$. Then there exists a meromorphic function, f, on Ω whose only poles are at $\{z_j\}_{j=1}^{\infty}$ with principal part at z_j given by $P_j((z-z_j)^{-1})$.*

Proof. For $n = 1, 2, \dots$, let K_n be given by

$$K_n = \biggl\{z \in \Omega \mid \mathrm{dist}(z, \partial\Omega) \geq \frac{1}{n};\ |z| \leq n\biggr\} \tag{9.5.1}$$

Let

$$g_n(z) = \sum_{\{j \mid z_j \in K_{n+1} \setminus K_n\}} P_j((z - z_j)^{-1}) \tag{9.5.2}$$

Clearly, g_n is analytic in a neighborhood of K_n, so by Theorem 4.7.4, there exist functions h_n analytic in Ω so that

$$\|g_n - h_n\|_{K_n} \le 2^{-n} \tag{9.5.3}$$

Let

$$f(z) = g_0(z) + \sum_{n=1}^{\infty} (g_n(z) - h_n(z)) \tag{9.5.4}$$

where

$$g_0(z) = \sum_{\{j \mid z_j \in K_1\}} P_j((z - z_j)^{-1}) \tag{9.5.5}$$

By (9.5.3), the sum converges uniformly on compact subsets of $\Omega \setminus \{z_j\}_{j=1}^{\infty}$ and $f(z) - P_j((z - z_j)^{-1})$ has a removable singularity at z_j. □

There is also a proof of the Weierstrass theorem using a suitable Runge-type theorem, namely, the improved version of Problem 3 of Section 4.7; see Problem 1. We'll instead construct suitable renormalized products.

We need one piece of preparation for the proof of the Weierstrass theorem using products. Since the z_j's can have limit points in $\partial\Omega$, $1 - E_n(z/z_n)$ may not be small as n gets large with z fixed in Ω. If $z_{n_j} \to z_\infty \in \partial\Omega$, we need to use $E_{n_j}((z_{n_j} - z_\infty)/(z - z_\infty))$. The problem is that different z_j's may approach infinity or points in $\partial\Omega$, so we have to find a way to choose between the two. That is the purpose of this lemma:

Lemma 9.5.2. *Let Ω be a region in $\mathbb{C}$ and $\{z_j\}_{j=1}^{\infty}$ a sequence in Ω. Then we can write*

$$\{1, 2, \dots\} = N_1 \cup N_2 \quad \textit{with} \quad N_1 \cap N_2 = \emptyset \tag{9.5.6}$$

so that there is for any $j \in N_1$, $w_j \in \partial\Omega$, and so that

$$\lim_{\substack{n\to\infty \\ n \in N_1}} |z_n - w_n| = 0, \qquad \lim_{\substack{n\to\infty \\ n \in N_2}} |z_n|^{-1} = 0 \tag{9.5.7}$$

Proof. Let K_n be given by (9.5.1). For each m, eventually $z_m \in K_n$, so let

$$n(m) = \begin{cases} 1, & z_m \in K_1 \\ \max(n \mid z_m \notin K_n), & z_m \notin K_1 \end{cases} \tag{9.5.8}$$

We claim that

$$\lim_{m\to\infty} n(m) = \infty \tag{9.5.9}$$

for $n(m) < N$ implies $z_m \in K_N$, and for m large, $z_m \notin K_N$.

If $z_m \in K_1$, put $m \in N_2$ (this is a finite set and so irrelevant for (9.5.7)). If $z_m \notin K_1$, since $z_m \notin K_{n(m)}$, either $\operatorname{dist}(z_m, \partial\Omega) < 1/n(m)$ or $|z_m| > n(m)$ or both. If the former, put $m \in N_1$; if not, put $m \in N_2$. If $m \in N_1$, pick $w_m \in \partial\Omega$ with $\operatorname{dist}(z_m, \partial\Omega) = |z_m - w_m|$

$$m \in N_1 \Rightarrow |z_m - w_m| \leq \frac{1}{n(m)} \to 0 \qquad \text{by (9.5.9)}$$

$$m \in N_2 \Rightarrow |z_m| \geq n(m) \to \infty \qquad \text{by (9.5.1)}$$

so we have proven (9.5.7). □

Theorem 9.5.3. *Let $\{z_j\}_{j=1}^\infty$ be a set of distinct points in a region, Ω, with no limit points in Ω. Let $n_1, n_2, \dots$ be a sequence of strictly positive integers. Then there exists a function $f \in \mathfrak{A}(\Omega)$ so that f vanishes precisely at $\{z_j\}_{j=1}^\infty$ and the order of the zero at z_j is n_j.*

Proof. As in the proof of (9.4.1), we can suppose no z_j is 0 and ignore n_j, but allow nondistinct z_j. Let N_1, N_2, and $\{w_j\}_{j\in N_1}$ be as given by Lemma 9.5.2.

We claim that for each m, with K_m given by (9.5.1),

$$\sup_{z\in K_m} \left[\sum_{j\in N_1} \left| \frac{z_j - w_j}{z - w_j} \right|^j + \sum_{j\in N_2} \left| \frac{z}{z_j} \right|^j \right] < \infty \tag{9.5.10}$$

For eventually for $j \in N_1$, $|z_j - w_j| \leq 1/2m \leq 1/m \leq |z - w_j|$, and for $j \in N_2$, $|z_j| \geq 2m \geq m \geq |z|$, so eventually the summand is uniformly in K bounded by $(\frac{1}{2})^j$.

By the same argument, for each m, uniformly in z in K_m, eventually the factors in $|\ |$ in (9.5.10) are less than $\frac{1}{2}$, so we can apply (9.4.3) and conclude

$$\sup_{z\in K_m} \sum_{j\in N_1} \left| 1 - E_j\left(\frac{z_j - w_j}{z - w_j} \right) \right| + \sum_{j\in N_2} \left| 1 - E_j\left(\frac{z}{z_j} \right) \right| < \infty \tag{9.5.11}$$

so

$$f(z) = \prod_{j\in N_1} E_j\left(\frac{z_j - w_j}{z - w_j} \right) \prod_{j\in N_z} E_j\left(\frac{z}{z_j} \right) \tag{9.5.12}$$

converges in $\mathfrak{A}(\Omega)$ to a function in Ω with the required zeros. □

Corollary 9.5.4. *For any region Ω, there is a function $f \in \mathfrak{A}(\Omega)$ for which $\partial\Omega$ is a natural boundary, that is, f cannot be continued into any disk about any point $z_0 \in \partial\Omega$.*

Proof. Pick w_m a dense sequence in $\partial\Omega$ (for isolated points of $\partial\Omega$, repeat the point infinitely often) and then $z_m \in \Omega$ with $|z_m - w_m| < 1/m$ so all of $\partial\Omega$ are limit points of $\{z_m\}_{m=1}^\infty$. Let f be analytic in Ω with zeros precisely at the $\{z_m\}_{m=1}^\infty$. Then f cannot be analytic in a disk $\mathbb{D}_\delta(z_\infty)$ with $z_\infty \in \Omega$ since z_∞ is a limit point of zeros of f, but f is not identically zero. $\square$

The following corollaries have identical proofs to their $\mathbb{C}$ analog in Section 9.4:

Corollary 9.5.5. *Let Ω be a region. Any meromorphic function on Ω is a ratio of analytic functions on Ω.*

Corollary 9.5.6. *Given any sequence $\{z_n\}_{n=1}^\infty$ of distinct points in Ω with no limit point in Ω and any $\{a_n\}_{n=1}^\infty \in \mathbb{C}$, there is a function $f \in \mathfrak{A}(\Omega)$ with $f(z_n) = a_n$.*

Notes and Historical Remarks. Picard seems to be first to have used $E_n((z_n - w_n)/(z - w_n))$-type factors in dealing with zeros for general regions. The full result appeared in Mittag-Leffler [**385**]. He also had the Mittag-Leffler theorem for general regions in this paper.

Corollary 9.5.4, which is due to Mittag-Leffler [**385**] and Runge [**498**] using their respective theorems, is especially interesting because its analog fails in higher dimensions; see Section 11.5. A region with a function that cannot be continued any further is called a *domain of holomorphy.*

The method of Problems 2 and 3 of Section 9.4 to get the Weierstrass theorem from the Mittag-Leffler theorem works for any simply connected region. Once one has the result for simply connected regions, one can handle finitely connected regions by hand. For example, for any annulus, $\mathbb{A}_{a,b}$, one can break the potential zeros into those with $|z_n| \geq (a+b)/2$ and $|z_n| \leq (a+b)/2$. For the former, find a function, g, on $\mathbb{D}_b(0)$ with zeros at those z_n, and then a function, h, on $\mathbb{D}_{a^{-1}}(0)$ with zeros at the z_n^{-1} from the second set of zeros. Then $f(z) = g(z)h(z^{-1})$ solves the $\mathbb{A}_{a,b}$ problem.

Wedderburn's lemma (Problem 3 of Section 9.3) and Helmer's theorem (Problem 8 of Section 9.4) extend to $\mathfrak{A}(\Omega)$ for any region $\Omega \subset \mathbb{C}$.

Problems

1. This will provide an alternate proof of Theorem 9.5.3 using a Runge-type theorem. Let K_n be the set of the form (4.7.4) with $R = n$ and $\varepsilon = 1/n$. Let $F_n(z) = \prod_{z_j \in K_n}(z - z_j)$.

   (a) Prove there exist functions $h_n \in \mathfrak{A}(\Omega)$ nonvanishing on Ω so
$$\sup_{z \in K_n} \left| \frac{F_{n+1}}{F_n} h_n - 1 \right| \leq 2^{-n}$$
   Hint: Use Problem 3 of Section 4.7 to pick h_n so
$$\sup_{z \in K_n} \left| h_n - \frac{F_n}{F_{n+1}} \right| \leq 2^{-n} \left[\sup_{x \in K} \left| \frac{F_n(z)}{F_{n+1}(z)} \right| \right]^{-1}$$

(b) Let G_n be defined on K_n by

$$G_n(z) = F_n(z) \prod_{m \geq n} \left(\frac{F_{m+1}(z)}{F_m(z)} h_m(z) \right) h_1(z) \dots h_{n-1}(z)$$

Prove $G_{n+1} \restriction K_n = G_n$ so the G's define a function in $\mathfrak{A}(\Omega)$ with the right zeros.

9.6. The Gamma Function: Basics

Trigonometric functions are at least medieval and the log and exponential go back at least to the Renaissance. In many ways, the gamma function is the first modern special function, the result of the flowering of analysis following Newton and Leibniz and of the genius of Euler. In this section, we'll present the basic properties and, and in the next, the asymptotics as $z \to \infty$ (with $|\arg(z)| \leq \pi - \varepsilon$ for some $\varepsilon > 0$).

There are three natural possible definitions:

Euler:

$$\Gamma(z) = \int_0^\infty e^{-t} t^{z-1}\, dt \qquad (\operatorname{Re} z > 0) \tag{9.6.1}$$

Euler–Gauss:

$$\Gamma(z) = \lim_{n \to \infty} \frac{n! n^{z-1}}{z(z+1) \dots (z+n-1)}, \qquad z \in \mathbb{C} \setminus \{0, -1, -2, \dots\} \tag{9.6.2}$$

Weierstrass:

$$\frac{1}{\Gamma(z)} = z e^{\gamma z} \prod_{n=1}^\infty \left(1 + \frac{z}{n}\right) e^{-z/n}, \qquad \text{all } z \in \mathbb{C} \tag{9.6.3}$$

In (9.6.3), γ is the Euler–Mascheroni constant

$$\gamma = \lim_{n \to \infty} \left(\sum_{k=1}^n \frac{1}{k} \right) - \log(n) \tag{9.6.4}$$

The reader is asked in Problem 1 to prove the limit exists. Its value is $0.5772156649\dots$. Any approach starts with one of these definitions and shows their equivalence. We'll use the Weierstrass definition which fits into our discussion of products.

Proposition 9.6.1. *The infinite product in* (9.6.3) *converges uniformly on* $\mathbb{C}$ *to an entire function, so the* gamma function, $\Gamma(z)$, *defined by* (9.6.3) *is an analytic function in* $\mathbb{C} \setminus \{0, -1, -2, \dots\}$ *with simple poles at* $z = 0, -1, -2, \dots$ *and which is nowhere vanishing.*

Proof. The product is $\prod_{n=1}^\infty E_1(z/n)$, so by (9.4.2),

$$|z| \leq n \Rightarrow \left| E_1\left(\frac{z}{n}\right) - 1 \right| \leq \frac{|z|^2}{|n|^2} \tag{9.6.5}$$

which means, for every R,

$$\sup_{|z|\le R} \sum_{j=1}^{\infty} \left| E_1\left(\frac{z}{j}\right) - 1 \right| < \infty \tag{9.6.6}$$

proving convergence of the product. □

Proposition 9.6.2. *For $z \notin \{0,-1,-2,\dots\}$, the limit on the right of (9.6.2) exists and is $\Gamma(z)$. Moreover, for all $x, y \in \mathbb{R}$,*

$$|\Gamma(x+iy)| \le |\Gamma(x)| \tag{9.6.7}$$

Proof. Define

$$\Gamma_n(z) = \frac{n! n^{z-1}}{z(z+1)\dots(z+n-1)} \tag{9.6.8}$$

Then

$$\begin{aligned} \Gamma_n(z)^{-1} &= z\left(1+\frac{z}{1}\right)\left(\dots\right)\left(1+\frac{z}{n-1}\right)e^{-z\log(n)} \\ &= z\left[\prod_{j=1}^{n-1}\left(1+\frac{z}{j}\right)e^{-z/j}\right]\exp\left(z\left(\sum_{j=1}^{n}\frac{1}{j}-\log(n)\right)\right)e^{-z/n} \end{aligned} \tag{9.6.9}$$

Given the convergence of the product in (9.6.3) and the existence of the limit in (9.6.4), we see this converges to $\Gamma(z)^{-1}$.

To get (9.6.7), we note that for x, y real, $|x+j|^{-1} \ge |x+j+iy|^{-1}$ for $j = 0, \dots, n-1$, and that $|n^{iy}| = 1$, so

$$|\Gamma_n(x+iy)| \le |\Gamma_n(x)| \tag{9.6.10}$$

from which (9.6.7) follows. □

We prove below the crucial *functional equation for* Γ:

$$\Gamma(z+1) = z\Gamma(z) \tag{9.6.11}$$

We begin by analyzing solutions of this equation:

Proposition 9.6.3. *Let f be defined and analytic on the right half-plane, $\mathbb{H}_+$, and obey*

$$f(z+1) = zf(z) \tag{9.6.12}$$

Then f has a meromorphic continuation to $\mathbb{C}$ with poles only possible at $0, -1, \dots$. We define

$$u(z) = f(z)f(1-z) \tag{9.6.13}$$

which is analytic on $\mathbb{C}\setminus\mathbb{Z}$ and meromorphic on $\mathbb{C}$. Then

(a) *u obeys*

$$u(z+1) = -u(z), \qquad u(1-z) = u(z) \tag{9.6.14}$$

$$u(z+2) = u(z), \qquad u(-z) = -u(z) \tag{9.6.15}$$

(b) (Wielandt's Theorem) *If* f *is bounded on* $\{z \mid 1 \le \operatorname{Re} z \le 2\}$, *then*

$$f(z) = f(1)\Gamma(z) \tag{9.6.16}$$

Remarks. 1. We use (9.6.11) proven in the next theorem.

2. A similar result to (a) holds if f is only defined on $(0, \infty)$.

3. f need only be defined and analytic on $\{z \mid a < \operatorname{Re} z < b\}$ for $a, b \in (0, \infty)$ with $b - a > 1$ and only (9.6.12) for $\{z \mid a < \operatorname{Re} z < b - 1\}$.

4. One doesn't need boundedness of f in $\{z \mid 1 \le \operatorname{Re} z \le 2\}$. $|f(x+iy)| = o(|y|^{-\frac{1}{2}+x}\exp(\frac{3}{2}\pi|y|))$ suffices there; see Problem 11.

Proof. The functional equation (9.6.12) can be used for the continuation. Define

$$f_1(z) = z^{-1} f(z+1) \tag{9.6.17}$$

This is meromorphic on $\{z \mid \operatorname{Re} z > -1\}$ and agrees with $f(z)$ on $\{z \mid \operatorname{Re} z > 0\}$ by (9.6.12). Then (9.6.17) on the extended f extends to $\{z \mid \operatorname{Re} z > -2\}$. By iteration, we get a meromorphic continuation to $\mathbb{C}$.

(a) Let

$$g(z) = f(1-z) \tag{9.6.18}$$

Then

$$g(z+1) = f(-z) = (-z)^{-1} f(-z+1) = (-z)^{-1} g(z) \tag{9.6.19}$$

From this and (9.6.12), the first equation in (9.6.14) follows.

Since $g(1-z) = f(z)$, the second equation in (9.6.14) holds. Iterating $u(z+1) = -u(z)$ yields $u(z+2) = u(z)$ and, finally, $u(-z) = -u(1-z) = -u(z)$, proving (9.6.15).

(b) Let

$$\widetilde{f}(z) = f(z) - f(1)\Gamma(z) \tag{9.6.20}$$

Clearly, $\widetilde{f}$ obeys the functional equation (9.6.12), so it has a meromorphic continuation. But since $\widetilde{f}(1) = 0$, $\widetilde{f}(0)$ is regular at $z = 0$, and then by the functional equation, at $-1, -2, \dots$. So $\widetilde{f}$ is an entire analytic function.

By (a), $\tilde{u}$ is periodic with $\tilde{u}(z+1) = -\tilde{u}(z)$, and by assumption and (9.6.7), $\tilde{u}$ is bounded on $\{z \mid 1 \le \operatorname{Re} z \le 2\}$. By the antiperiodicity, $\tilde{u}$ is globally bounded, so constant. Since $\tilde{u}(0)$, $\tilde{u} \equiv 0$. But then, either $\widetilde{f}$ or $\widetilde{f}(1 - \cdot)$ is identically zero, so $\widetilde{f} \equiv 0$ and (9.6.16) holds. $\square$

Here are the main properties of Γ, other than Stirling's formula and Binet's formula to be discussed later.

Theorem 9.6.4. (a) $\Gamma(z)$ *is meromorphic on all of* $\mathbb{C}$ *with singularities at* $\{0, -1, -2, \dots\}$ *with simple poles there. It is nonvanishing at all points.*

(b) *For all* z *not in the poles,*

$$\Gamma(z+1) = z\Gamma(z) \tag{9.6.21}$$

(c) $\Gamma(1) = 1$*; for* $n = 1, 2, \dots$ *(here and below,* $0! = 1$*),*

$$\Gamma(n) = (n-1)! \tag{9.6.22}$$

and

$$\Gamma(\tfrac{1}{2}) = \sqrt{\pi} \tag{9.6.23}$$

(d) *For* $n = 0, 1, 2, \dots,$

$$\operatorname{Res}(\Gamma; -n) = \frac{(-1)^n}{n!} \tag{9.6.24}$$

(e) (Euler Reflection Formula) *For* $z \notin \mathbb{Z}$,

$$\Gamma(z)\Gamma(1-z) = \frac{\pi}{\sin(\pi z)} \tag{9.6.25}$$

(f) (Legendre Duplication Formula) *For* $z \notin \{0, -\frac{1}{2}, -1, -\frac{3}{2}, \dots\}$,

$$\Gamma(2z)\Gamma(\tfrac{1}{2}) = 2^{2z-1}\Gamma(z)\Gamma(z+\tfrac{1}{2}) \tag{9.6.26}$$

(g) (Gauss Multiplication Formula) *For* $m = 2, 3, \dots$ *and* $z \notin \{0, -\frac{1}{m}, -\frac{2}{m}, \dots\}$,

$$\Gamma(mz)(2\pi)^{(m-1)/2} = m^{mz-1/2}\Gamma(z)\Gamma\left(z+\frac{1}{m}\right)\dots\Gamma\left(z+\frac{m-1}{m}\right) \tag{9.6.27}$$

(h) *For* $z \notin \{0, -1, \dots\}$,

$$\frac{\Gamma'(z)}{\Gamma(z)} = -\gamma - \frac{1}{z} - \sum_{n=1}^{\infty}\left(\frac{1}{z+n} - \frac{1}{n}\right) \tag{9.6.28}$$

and

$$\frac{d}{dz}\left(\frac{\Gamma'(z)}{\Gamma(z)}\right) = \sum_{n=0}^{\infty}\frac{1}{(z+n)^2} \tag{9.6.29}$$

(i) *For* $x, y \in (0, \infty)$ *and* $0 \le \theta \le 1$,

$$\Gamma(\theta x + (1-\theta)y) \le \Gamma(x)^{\theta}\Gamma(y)^{1-\theta} \tag{9.6.30}$$

(j) *For* $\operatorname{Re} z > 0$,

$$\Gamma(z) = \int_0^{\infty} e^{-t}t^{z-1}\,dt \tag{9.6.31}$$

Remarks. 1. The reader should check (9.6.26) by hand when $z = 1, 2, \dots$.

2. (9.6.30) holds for $|\Gamma(\theta x + (1-\theta)y)|$ for $x, y \in (-n, -n+1)$ for $n = 1, 2, \dots$ (Problem 10).

3. (9.6.21) will be easy, given (9.6.2). In Problem 3, the reader is asked to prove it directly from (9.6.3).

4. The condition (9.6.30) is called *log convexity*. We'll say more about it later. Figures 9.6.1–9.6.3, plot Γ and $\log|\Gamma|$ on $\mathbb{R}$ and $|\Gamma|$ on $\mathbb{C}$ and illustrate the convexity of $\log(\Gamma)$ and of Γ (see Problem 5).

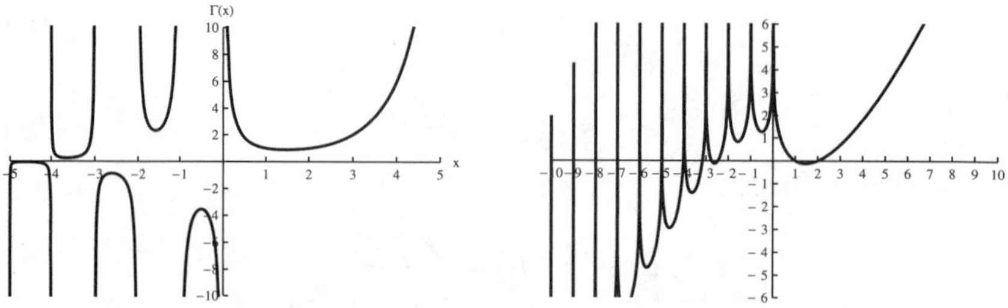

Figure 9.6.1. $\Gamma(x)$ for $-5 < x < 5$ and $\log|\Gamma(x)|$ for $-10 < x < 10$.

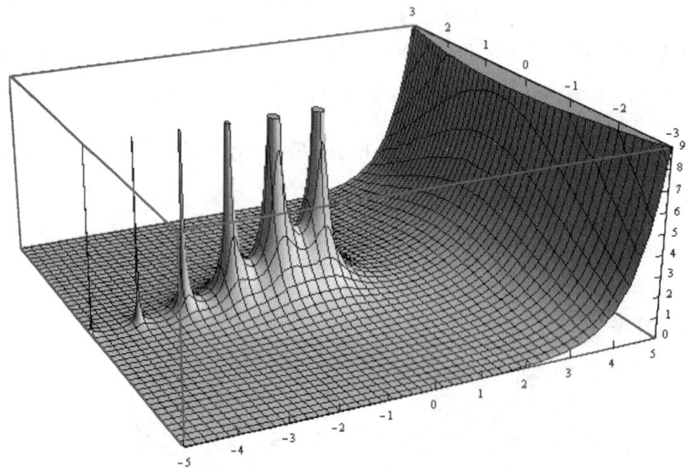

Figure 9.6.2. Surface plot of $|\Gamma(x+iy)|$; $-5 < x < 5$, $-3 < y < 3$.

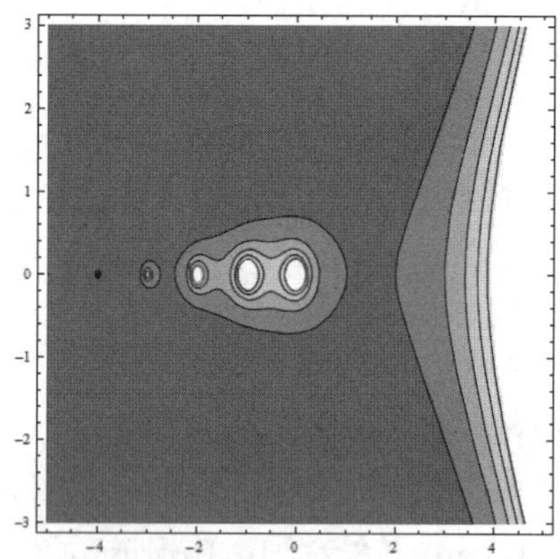

Figure 9.6.3. Contour plot of $|\Gamma(x+iy)|$; $-5 < x < 5$, $-3 < y < 3$.

5. See Problem 32 for another way to see the residues at $n = 0, -1, -2, \ldots$.

6. (9.6.28) is due to Gauss.

Proof. (a) is immediate from (9.6.3) and Proposition 9.6.1.

(b) By (9.6.8),

$$\Gamma_n(z+1) = z\Gamma_{n+1}(z)\left(\frac{n}{n+1}\right)^z \tag{9.6.32}$$

Since z is fixed and, as $n \to \infty$, $n/(n+1) \to 1$, $(n/(n+1))^z \to 1$, so (9.6.21) follows from taking $n \to \infty$.

(c) By (9.6.3),

$$\lim_{\substack{z\to 0\\ z\neq 0}} \frac{1}{z\Gamma(z)} = 1 \tag{9.6.33}$$

so, by (9.6.21), $\lim_{z\to 0}\Gamma(z+1) = 1$ which, given the analyticity, implies $\Gamma(1) = 1$. (9.6.22) then follows inductively using (9.6.21).

By (9.6.8),

$$\begin{aligned}\Gamma(\tfrac{1}{2})^2 &= \lim_{n\to\infty} \frac{n(n-1)(n-1)(n-2)(n-2)\ldots(1)(1)}{(\frac{1}{2})(\frac{1}{2})(\frac{3}{2})(\frac{3}{2})\ldots(n-\frac{1}{2})(n-\frac{1}{2})} \\ &= \lim_{n\to\infty} 2\left[\prod_{j=1}^{n} \frac{(2j)^2}{(2j+1)(2j-1)}\right]\frac{2n+1}{2n}\end{aligned} \tag{9.6.34}$$

converges to π by Wallis' formula, (5.7.80).

(d) By (9.6.21),

$$[(z+1)+(n-1)]\Gamma(z+1) = z[z+n]\Gamma(z) \tag{9.6.35}$$

which, taking $z \to -n$ implies

$$\text{Res}(\Gamma; -(n-1)) = (-n)\text{Res}(\Gamma; -n) \tag{9.6.36}$$

Moreover, since $\Gamma(1) = 1$, $\lim_{z\to 0} z\Gamma(z) = 1$, so (9.6.24) follows by induction.

(e) Let $u(z) = \Gamma(z)\Gamma(1-z)$. By Proposition 9.6.3, $u(z)$ is entire meromorphic of period 2, so by an obvious extension of Theorem 3.10.2 to meromorphic functions, we have that

$$u(z) = h(e^{\pi i z}) \tag{9.6.37}$$

where h is meromorphic on $\mathbb{C}^\times$.

Since u has only simple poles at $z = 0, 1$, $h(w)$ has simple poles at $w = \pm 1$. Since $u(1+z) = -u(z)$,

$$h(-w) = -h(w) \tag{9.6.38}$$

By Proposition 9.6.2, Γ is bounded on $\{z \mid 1 \le \text{Re}\, z \le 2\}$, so since $\Gamma(z-1) = (z-1)^{-1}\Gamma(z)$, Γ is also bounded on $\{z \mid -1 \le \text{Re}\, z \le 1,\ |\text{Im}\, z| >$

1}. Thus, u is also bounded there and thus, h is bounded on $\{w \mid |w| > e^{\pi} \text{ or } |w| < e^{-\pi}\}$. It follows that h has removable singularities at 0 and ∞. Thus,

$$h(w) = \frac{a}{1-w} + \frac{b}{1+w} + c \tag{9.6.39}$$

But (9.6.38) implies $c = 0$ and $b = -a$, that is,

$$h(w) = a\left(\frac{1}{1-w} - \frac{1}{1+w}\right) = \frac{2aw}{1-w^2} = -\frac{2a}{w - w^{-1}} \tag{9.6.40}$$

or

$$u(z) = \frac{d}{\sin(\pi z)} \tag{9.6.41}$$

Since $z\Gamma(z)\Gamma(1-z) = \Gamma(1+z)\Gamma(1-z) \to 1$ as $z \to 0$, we must have $d = \pi$, that is, we have proven (9.6.25).

(f),(g) See Problem 6 or 15 or 21.

(h) follows from (9.6.3) and (9.2.42).

(i) (9.6.29) implies $\frac{d^2}{dx^2}\log(\Gamma(x)) > 0$, so $\log(\Gamma(x))$ is convex (see Section 5.3 of Part 1), which is what (9.6.30) says.

(j) Call the integral $\widetilde{\Gamma}(z)$. By Theorem 3.1.6, $\widetilde{\Gamma}(z)$ is analytic in $\mathbb{H}_+$. By an elementary estimate, the integral converges and obeys $|\widetilde{\Gamma}(x+iy)| \le \widetilde{\Gamma}(x)$ for $x > 0$, so $\sup_{1 \le \operatorname{Re} z \le 2}|\widetilde{\Gamma}(z)| < \infty$. Clearly, $\widetilde{\Gamma}(1) = 1$ and, by $\frac{d}{dt}e^{-t} = -\frac{d}{dt}e^{-t}$ and an integration by parts, $\widetilde{\Gamma}(z+1) = z\widetilde{\Gamma}(z)$. Thus, by Wielandt's theorem, $\widetilde{\Gamma}(z) = \widetilde{\Gamma}(1)\Gamma(z) = \Gamma(z)$. □

Corollary 9.6.5.

$$\prod_{n=1}^{\infty}\left(1 - \frac{z^2}{n^2}\right) = \frac{\sin(\pi z)}{\pi z} \tag{9.6.42}$$

Proof.

$$\frac{1}{\Gamma(1-z)} = -\frac{1}{z\Gamma(-z)} = e^{-\gamma(z)}\prod_{n=1}^{\infty}\left(1 - \frac{z}{n}\right)e^{z/n}$$

so

$$\text{LHS of (9.6.42)} = \frac{1}{z\Gamma(z)\Gamma(1-z)} = \text{RHS of (9.6.42)}$$

by (9.6.25). □

In addition to this new proof of the Euler product formula, (9.6.25) implies $\Gamma(\frac{1}{2})^2 = \pi/\sin(\pi/2) = \pi$, so running (9.6.35) backwards, a new proof of Wallis' formula.

Finally, we turn to using Γ to evaluate a large number of different integrals. One key is the *beta function* defined for $\operatorname{Re} x > 0$, $\operatorname{Re} y > 0$ by

$$B(x,y) = \int_0^1 t^{x-1}(1-t)^{y-1}\,dt \tag{9.6.43}$$

Theorem 9.6.6. *For* $\operatorname{Re} x > 0$, $\operatorname{Re} y > 0$, *we have*

$$B(x,y) = \frac{\Gamma(x)\Gamma(y)}{\Gamma(x+y)} \tag{9.6.44}$$

Proof. We sketch the proof, leaving the details to Problem 16. By the change of variables $t = \sin^2\theta$,

$$\int_0^{\pi/2} \sin^{2x-1}\theta\cos^{2y-1}\theta\,d\theta = \tfrac{1}{2}\,B(x,y) \tag{9.6.45}$$

Next, letting $t = u^2$, $s = v^2$,

$$\begin{aligned}\Gamma(x)\Gamma(y) &= \int_0^\infty\int_0^\infty t^{x-1}s^{y-1}e^{-(t+s)}\,dtds \\ &= 4\int_0^\infty\int_0^\infty u^{2x-1}v^{2y-1}e^{-(u^2+v^2)}\,dudv \end{aligned}\tag{9.6.46}$$

From this and a shift to polar coordinates, we get (9.6.44) from (9.6.45). □

Problem 20 provides an alternate proof of (9.6.44) using Wielandt's theorem.

Remark. (9.6.45) implies $B(\frac{1}{2},\frac{1}{2}) = \pi$ which, by (9.6.44), proves once again that $\Gamma(\frac{1}{2}) = \sqrt{\pi}$. Since the above is a variant of the polar coordinate evaluation of Gaussian integrals, this is not surprising.

A huge number of integrals can be expressed in terms of Γ and B. First, using $t^\alpha = s$ (Problem 23),

$$\int_0^\infty e^{-t^\alpha}\,dt = \frac{1}{\alpha}\Gamma\left(\frac{1}{\alpha}\right) \tag{9.6.47}$$

which shows once again that $\int_{-\infty}^\infty e^{-t^2}\,dt = \Gamma(\frac{1}{2}) = \sqrt{\pi}$.

Next, using $s = t/(t+1)$ (Problem 23)

$$\int_0^\infty \frac{t^{x-1}}{(1+t)^{x+y}}\,dt = B(x,y) \tag{9.6.48}$$

which implies (Problem 23; also Example 5.7.8 and Problem 10 of Section 5.7)

$$\int_0^\infty \frac{t^{x-1}}{1+t}\,dt = \frac{\pi}{\sin(\pi x)} \tag{9.6.49}$$

In the beta integral, set $y = \frac{1}{2}$, $x = \frac{m}{n}$, and let $s = t^{1/n}$ to get (Problem 23)

$$\int_0^1 \frac{s^{m-1}}{\sqrt{1-s^n}}\, ds = \frac{\sqrt{\pi}\,\Gamma(\frac{m}{n})}{n\Gamma(\frac{m}{n}+\frac{1}{2})} \tag{9.6.50}$$

so, for example, an elliptic integral, called the *lemniscate integral*,

$$\int_0^1 \frac{dt}{\sqrt{1-t^4}} = \frac{(\Gamma(\frac{1}{4}))^2}{\sqrt{32\pi}} \tag{9.6.51}$$

Here are some additional integrals for the reader to confirm (see Problem 24):

$$\int_0^\infty \sin x^n\, dx = \frac{1}{n}\Gamma\left(\frac{1}{n}\right)\sin\left(\frac{\pi}{2n}\right), \qquad n = 2,3,\dots \tag{9.6.52}$$

$$\int_0^\infty \frac{\sin x}{x^\alpha}\, dx = \frac{\pi}{2\Gamma(\alpha)\sin(\frac{\pi\alpha}{2})}, \qquad 0 < \operatorname{Re}\alpha < 2 \tag{9.6.53}$$

$$\int_0^\infty \frac{\cos x}{x^\alpha}\, dx = \frac{\pi}{2\Gamma(\alpha)\cos(\frac{\pi\alpha}{2})}, \qquad 0 < \operatorname{Re}\alpha < 1 \tag{9.6.54}$$

$$\int_0^\pi \sin^n x\, dx = \frac{\sqrt{\pi}\,\Gamma(\frac{n+1}{2})}{\Gamma(\frac{n+2}{2})}, \qquad n = 0,1,2,\dots \tag{9.6.55}$$

$$\int_0^{\pi/2} \sin^n x\, dx = \frac{1}{2}B\left(\frac{n+1}{2},\frac{1}{2}\right), \qquad n = 0,1,2,\dots \tag{9.6.56}$$

$$\int_0^{\pi/2} \sqrt{\tan x}\, dx = \frac{\pi}{\sqrt{2}} \tag{9.6.57}$$

$$\int_0^\infty e^{-ax}x^{s-1}e^{ibx}\, dx = \frac{\Gamma(s)}{(a^2+b^2)^{s/2}}\, e^{is\varphi}, \qquad \tan\varphi = \frac{b}{a} \tag{9.6.58}$$

$$\int_0^\infty t^{z-1}e^{-t}\log(t)\, dt = \Gamma'(z), \qquad \operatorname{Re} z > 0 \tag{9.6.59}$$

$$\int_0^1 (1-x^a)^{b-1}\, dx = \frac{1}{a}B\left(\frac{1}{a},b\right), \qquad \operatorname{Re} a > 0,\ \operatorname{Re} b > 0 \tag{9.6.60}$$

$$\int_0^\infty \frac{dt}{(1+t^2)^b} = \frac{1}{2}B\left(\frac{1}{2},b-\frac{1}{2}\right), \qquad \operatorname{Re} b > \frac{1}{2} \tag{9.6.61}$$

$$\int_0^1 \frac{x^{a-1}(1-x)^{b-a}}{(x+\beta)^{a+b}}\, dx = \frac{\Gamma(a)\Gamma(b)}{\Gamma(a+b)}\,\frac{1}{(1+\beta)^a\beta^b}, \qquad a > 0,\ b > 0,\ \beta > 0 \tag{9.6.62}$$

$$\int_0^1 \left[\log\left(\frac{1}{t}\right)\right]^{z-1} dt = \Gamma(z), \qquad \operatorname{Re} z > 0 \tag{9.6.63}$$

Notes and Historical Remarks. Euler developed the gamma function in his early twenties in response to a question of Daniel Bernoulli and Goldbach to find a natural interpolation formula for $n!$. In a letter to Goldbach dated October 13, 1729, Euler solved the interpolation problem using (9.6.2). The reader should confirm by hand that (9.6.2) gives $(m-1)!$ when $z=m$. In a letter dated January 8, 1730, Euler showed this definition was equivalent to $\Gamma(z+1)=\int_0^1(-\log(t))^z\,dt$ for real z with $z>-1$. These results were published in [**171**]. Euler's original definition, for $x>0$, was (9.6.2). We have associated Gauss' name with this definition because he emphasized the definition for complex z, but also because Gauss is often given credit for this definition, for example, by Weierstrass [**585**] and in Artin's book [**23**] discussed below. The name "gamma" was chosen by Legendre in 1809, who also reformulated Euler's $(-\log(t))^k$ integral as (9.6.1). The name "beta" is due to Binet [**51**].

While Euler was motivated by interpolating $n!$, Hadamard found an entire analytic interpolation (i.e., no poles in $\mathbb{C}$)! (see Problem 34). Hölder [**265**] has proven that Γ does not obey any differential equation with rational coefficients.

In 1739, Euler [**176**] studied beta integrals, and by iterating the functional relation that he proved (see Problem 20(a))

$$B(x,y)=\frac{x+y}{y}\,B(x,y+1) \tag{9.6.64}$$

he was able to find (9.6.44) by using his definition (9.6.2) of Γ. He then deduced (9.6.1) from this relation to B. The modern approach, via a change of variables in a double integral, is from Poisson [**445**] and Jacobi [**282**]. Some of the older literature, following Legendre, calls B the Eulerian integral of the first kind and Γ the Eulerian integral of the second kind.

In 1944, Alte Selberg (1917–2007) [**516**] found a spectacular generalization of the beta integral in terms of gamma functions, namely, if $\operatorname{Re} a>0$, $\operatorname{Re} b>0$, $\operatorname{Re} c>\max(-\frac{1}{n},-\frac{\operatorname{Re} a}{n-1},-\frac{\operatorname{Re} b}{n-1})$, then

$$\begin{aligned}\int_0^1\cdots\int_0^1\prod_{i<1}^{n}x_i^{a-1}(1-x_i)^{b-1}\prod_{1\le j<k\le n}|x_j-x_k|^{2c}\,d^nx&\\ =\prod_{j=0}^{n-1}\frac{\Gamma(a+jc)\Gamma(b+jc)\Gamma(1+(j+1)c)}{\Gamma(a+b+(n-1+j)c)\Gamma(1+c)}&\end{aligned} \tag{9.6.65}$$

For a proof of this Selberg integral formula, see Beals–Wong [**32**] or Andrews et al. [**15**, Ch. 8].

As mentioned earlier, Weierstrass [**585**] developed his product formula in part to understand (9.6.2). His 1876 formula had appeared earlier in an 1843 paper of Schlömilch [**505**] and an 1848 paper of Newman [**410**].

The Euler–Mascheroni constant has developed a following, as seen, for example, in a long and informative Wikipedia article[7] and by an entire book [**245**] on the subject. It is not known if it is rational or not (although its continued fraction expansion has at least 470,000 elements). It occurs not only in the study of the gamma function, but also the zeta function. Key to many occurrences is (9.6.28) which implies (see Problem 27)

$$\lim_{\substack{z\to 0\\ z\neq 0}}\left(\Gamma(z)-\frac{1}{z}\right)=-\gamma,\qquad \Gamma'(1)=-\gamma \tag{9.6.66}$$

and leads to various integral relations (see Problem 28). Lorenzo Mascheroni (1750–1800) was born fifteen years after Euler discovered γ; his name is associated to the constant because he investigated ways of computing γ. He found the first 19 digits (he claimed 32, but only 19 are correct).

Helmut Wielandt, whose published work is almost entirely in finite group theory, found his theorem and simple proof in 1939, but he did not publish it. His friend, Konrad Knopp, included the result with a footnote about Wielandt in the second volume of the 1941 fifth edition of his *Theory of Functions* book. Alas, the widely available English translation [**319**] is based on the fifth edition of Volume 1 [**316**] but only the fourth edition of Volume 2 [**317**]. Remmert in his complex analysis book [**477**] and in an article in the *Monthly* [**476**] has championed the result, and I learned of it from him. It is unfortunate that this result which, after all, has been available in German since 1941 and which I regard as deeper and more elegant than Bohr–Mollerup is not more widely known.

An interesting way of understanding the naturalness of Euler's extension of $n!$ is the following theorem of Bohr and Mollerup [**60**]:

Theorem 9.6.7 (Bohr–Mollerup Theorem). *If f is a continuous function from $(0,\infty)$ to $(0,\infty)$ and f obeys*

(a) $f(x+1)=xf(x)$ (9.6.67)

(b) *For $x,y>0$ and $0\le\theta\le 1$,*

$$f(\theta x+(1-\theta)y)\le f(x)^\theta f(y)^{1-\theta} \tag{9.6.68}$$

Then

$$f(x)=f(1)\Gamma(x) \tag{9.6.69}$$

where $\Gamma(x)$ is given by (9.6.2).

The reader is led through Artin's proof of this in Problem 7. The Bohr in Bohr–Mollerup is Harald Bohr (1887–1951), who was the brother of the

[7] `http://en.wikipedia.org/wiki/Euler-Mascheroni_constant`

famous physicist. In his youth, he was more famous than his older brother since he was on the medal-winning Danish Olympic soccer team.

As his title suggests, Srinivasan [**535**] has an eclectic collection of facts (that he proves) about Γ, including two I'll mention: a formula of Stern that for $z \in \mathbb{H}_+$,

$$\frac{\Gamma'(z)}{\Gamma(z)} = -\gamma + \sum_{j=1}^{\infty} \frac{(z-j)(z-j+1)\dots(z-1)}{j(j!)} \tag{9.6.70}$$

and *Schobloch's reciprocity formula* (that generalizes the Gauss multiplication formula (which is $p = m$, $q = 1$)):

If, for $p, q \in (1, 2, \dots)$,

$$\Gamma_{p,q}(z) = (2\pi)^{-q/2} q^{z+(pq-q-p)/2} \prod_{j=0}^{q-1} \Gamma\left(\frac{z+pj}{q}\right) \tag{9.6.71}$$

then

$$\Gamma_{p,q}(z) = \Gamma_{q,p}(z) \tag{9.6.72}$$

For extensive discussions of the gamma function, see the charming books of Artin [**23**] and Havil [**245**], the first chapter of Andrews–Askey–Roy [**15**], and Davis' historical note [**132**]. Havil's book is on γ but has a lot on Γ. Artin uses the Bohr–Mollerup theorem as the organizing principle for most of his presentation.

Problems

1. (a) Let

$$\gamma_n = \sum_{k=1}^{n} \frac{1}{k} - \log(n) \tag{9.6.73}$$

Prove that $|\gamma_{n+1} - \gamma_n| \le C(n+1)^{-2}$ and deduce the limit (9.6.4) exists.

(b) Prove $\gamma_{n+1} - \gamma_n \ge 0$ and conclude $\gamma > 0$.

(c) Prove that

$$\gamma = \int_1^{\infty} \left(\frac{1}{[x]} - \frac{1}{x}\right) dx \tag{9.6.74}$$

(d) Prove that

$$\gamma_n = 1 - \log(2) + \frac{1}{n} + \sum_{k=2}^{\infty} \frac{(-1)^k}{k} \left(\sum_{j=2}^{n-1} \frac{1}{j^k}\right) \tag{9.6.75}$$

(e) Define the Riemann zeta function, $\zeta(s)$, for $\operatorname{Re} s > 1$ by

$$\zeta(s) = \sum_{k=1}^{\infty} k^{-s} \tag{9.6.76}$$

Prove that

$$\zeta(s) \leq 1 + 2^{-(s-1)}(1 - 2^{-(s-1)})^{-1} \tag{9.6.77}$$

(*Hint*: Group $\sum_{j=2}^{\infty} j^{-k}$ into $2+4+8+\dots$ terms.)

(f) Use this for another proof that $\lim \gamma_n = \gamma$ exists and

$$\gamma = 1 - \log(2) + \sum_{k=2}^{\infty} \frac{(-1)^k}{k} [\zeta(k) - 1] \tag{9.6.78}$$

(g) Similarly, prove that

$$\gamma = 1 - \sum_{k=2}^{\infty} \frac{1}{k} [\zeta(k) - 1] \tag{9.6.79}$$

Remark. $\gamma - \gamma_n \geq (2(n+1))^{-1}$, so the convergence of γ_n to γ is very slow, but (9.6.78) and (9.6.79) are exponentially rapidly converging approximations. However, $\zeta(k)$ isn't known in closed form for k an odd integer, so accurate decimal approximations of γ rely on the method of the next section (see Problem 13 of Section 9.7).

2. For $n = 1, 2, \dots$, prove for $x, y \in (-n, -n+1)$ and $0 \leq \theta \leq 1$,

$$|\Gamma(\theta x + (1-\theta)y)| \leq |\Gamma(x)|^{\theta} |\Gamma(y)|^{1-\theta} \tag{9.6.80}$$

3. Let

$$G(z) = \prod_{n=1}^{\infty} \left(1 + \frac{z}{n}\right) e^{-z/n} \tag{9.6.81}$$

(a) Prove that

$$\frac{G'(z-1)}{G(z-1)} = \frac{1}{z} + \frac{G'(z)}{G(z)} \tag{9.6.82}$$

(b) Conclude that for some constant γ,

$$G(z-1) = e^{\gamma}(zG(z)) \tag{9.6.83}$$

and then, by setting $z = 1$, that γ is given by (9.6.4).

(c) If Γ is given by (9.6.3), conclude that $\Gamma(z+1) = z\Gamma(z)$.

4. Show that

$$(1-z)\left(1 + \frac{z}{2}\right)\left(1 - \frac{z}{3}\right) \cdots = \frac{\sqrt{\pi}}{\Gamma(1 + \frac{1}{2}z)\Gamma(\frac{1}{2} - \frac{1}{2}z)}$$

5. If $f > 0$ and C^2 on an interval (a, b), compute $(\log(f))''$ and conclude that f log-convex $\Rightarrow$ f convex.

6. (a) Starting from (9.6.2), prove that for $m = 2, 3, \ldots,$

$$\Gamma(mz) = Cm^{mz-1}\Gamma(z)\Gamma\left(z + \frac{1}{m}\right) \ldots \Gamma\left(z + \frac{m-1}{m}\right) \tag{9.6.84}$$

where

$$C = \left[\Gamma\left(\frac{1}{m}\right) \ldots \Gamma\left(\frac{m-1}{m}\right)\right]^{-1} \tag{9.6.85}$$

(b) Using the Euler reflection formula, prove that

$$C^2 = \frac{\pi^{m-1}}{\sin(\frac{\pi}{m})\sin(\frac{2\pi}{m}) \ldots \sin(\frac{(m-1)\pi}{m})} \tag{9.6.86}$$

(c) Show the product of sines in (9.6.86) is $m/2^{m-1}$. (*Hint*: $m = \lim_{x\to 1,\, x\neq 1}(x^m - 1)/(x-1)$ and $x^m - 1$ has roots $e^{2\pi j/m}$.)

(d) Conclude the Gauss multiplication formula.

7. Here you'll prove the Bohr–Mollerup theorem (Theorem 9.6.7), following Artin [**23**].

(a) If f obeys (9.6.68), then for $n = 1, 2, \ldots$ and $0 < x < 1$, we have $x + n = (1-x)n + x(n+1)$, so

$$f(x+n) \le n^x f(n) = f(1)n^x(n-1)! \tag{9.6.87}$$

(b) If f obeys (9.6.68), then for $n = 1, 2, \ldots,$ and $0 < x < 1$, we have $n + 1 = x(n+x) + (1-x)(n+1+x)$, so

$$f(n) = f(1)(n-1)! \le f(x+n)(x+n)^{1-x}n^{-1} \tag{9.6.88}$$

(c) With Γ_n given by (9.6.8), prove for $n = 1, 2, \ldots,$

$$\Gamma_n(x)\left(\frac{n}{x+n}\right)^{1-x} \le \frac{f(x)}{f(1)} \le \Gamma_n(x) \tag{9.6.89}$$

and conclude

$$\frac{f(x)}{f(1)} = \lim_{n\to\infty} \Gamma_n(x) \tag{9.6.90}$$

which is (9.6.69) for $x \in (0, 1)$.

(d) Conclude (9.6.69) for all x in $(0, \infty)$. (*Hint*: Functional equation.)

(e) From (9.6.87), (9.6.88), prove that for $x \in (0, \infty)$,

$$\frac{\Gamma(x+n)}{\Gamma(n)n^x} \to 1 \qquad \text{as } n \to \infty \tag{9.6.91}$$

8. This will provide another proof of the Euler reflection formula, (9.6.25).

   (a) Prove that for any $y > 0$ and $0 < s < 1$,

$$\Gamma(1-s) = y \int_0^\infty e^{-yx}(xy)^{-s}\,dx$$

   (b) Using (a), prove that for $0 < s < 1$,

$$\begin{aligned}\Gamma(1-s)\Gamma(s) &= \int_0^\infty \int_0^\infty e^{-t(1+x)}x^{-s}\,dxdt \\ &= \int_0^\infty \frac{x^{-s}}{1+x}\,dx \qquad (9.6.92)\end{aligned}$$

   (c) Prove (9.6.25) for $0 < s < 1$ (*Hint*: See (5.7.30)) and then for all of $s \notin \mathbb{Z}$.

9. Show that for $y \neq 0$ real, one has

$$|\Gamma(iy)| = \left(\frac{\pi}{y \sinh \pi y}\right)^{1/2} \qquad (9.6.93)$$

10. Prove (9.6.31) using the Bohr–Mollerup theorem plus analyticity. (*Hint*: You'll need Hölder's inequality to verify log convexity.)

11. This will improve Wielandt's theorem by a priori allowing considerably more growth than the boundedness assumption. It will require the leading terms in Stirling's formula (Theorem 9.7.2) of the next section. S will denote the strip $\{z \mid z = x + iy;\ 1 \leq x \leq 2\}$.

    (a) Let $g(z)$ be an entire function obeying

$$g(z+1) = g(z); \qquad \lim_{\substack{|y|\to\infty \\ z\in S}} |g(z)|e^{-2\pi|y|} = 0 \qquad (9.6.94)$$

    Prove that g is a constant. (*Hint*: See Theorem 3.10.1.)

    (b) For some $c \in (0,\infty)$ and all $z = x + iy \in S$, prove that

$$c \leq \frac{|\Gamma(z)|}{[|z|^{x-\frac{1}{2}} \exp(-\frac{1}{2}\pi|y|)]} \leq c^{-1} \qquad (9.6.95)$$

    (c) Let f defined on $\mathbb{H}_+$ obey $f(1) = 1$, $f(z+1) = zf(z)$. Prove that $g(z) = f(z)/\Gamma(z)$ can be extended to an entire function obeying $g(1) = 1$, $g(z+1) = g(z)$.

    (d) Prove that if f is defined on $\mathbb{H}_+$ with $f(1) = 1$, $f(z+1) = zf(z)$, and for $z \in S$, we have as $|y| \to \infty$,

$$|f(z)||z|^{-\frac{1}{2}-x} \exp(-\tfrac{3}{2}\pi|y|) = o(1) \qquad (9.6.96)$$

    then $f(z) = \Gamma(z)$.

    Remark. This extension of Wielandt's theorem is due to Fuglede [**196**].

12. (a) Prove that if f is defined on $(0,\infty)$, $f(x+1)=xf(x)$, and

$$\lim_{n\to\infty}\frac{f(x+n)}{n!\,n^{x-1}}=1 \tag{9.6.97}$$

then $f(x)=\Gamma(x)$.

(b) Let $g(x)$ be a function on $(0,\infty)$ with $\lim_{x\to\infty}|g(x)|=\infty$ and $|f'(x)|\le Cx^{-1}|f(x)|$ for x large. Prove that

$$\lim_{n\to\infty}\frac{\int_0^\infty g(x)x^n e^{-x}\,dx}{g(n)\int_0^\infty x^n e^{-x}\,dx}=1 \tag{9.6.98}$$

(*Hint*: This will require some detailed analysis to show that if $a_n/n^{1/2}\to\infty$, then the probability measure $x^n e^{-x}dx/\int_0^\infty x^n e^{-x}dx$ is concentrated on $(n-a_n,n+a_n)$.)

(c) If $\widetilde{\Gamma}(z)$ is the integral in (9.6.31), prove $\widetilde{\Gamma}(x+n)/n^x\to 1$ and conclude another proof that (9.6.31) holds.

13. This will lead to yet another proof of (9.6.31). Define

$$\Gamma_{n,k}(z)=\int_0^n t^{z-1}\left(1-\frac{t}{n}\right)^k dt \tag{9.6.99}$$

(a) Prove that

$$\lim\Gamma_{n,n}(z)=\int_0^\infty t^{z-1}e^{-t}\,dt \tag{9.6.100}$$

(*Hint*: Look at Problem 14 of Section 2.3.)

(b) Prove that $\Gamma_{n,0}(z)=z^{-1}n^z$, and that for $k\ge 1$,

$$\Gamma_{n,k}(z)=\frac{k}{nz}\Gamma_{n,k-1}(z+1) \tag{9.6.101}$$

(*Hint*: $t^{z-1}=\frac{1}{z}\frac{d}{dt}t^z$), and then that

$$\Gamma_{n,n}(z)=\left(\frac{n}{n+1}\right)^z\Gamma_{n+1}(z) \tag{9.6.102}$$

(c) Conclude that (9.6.31) holds.

14. This goes through still another proof of (9.6.31)—this one with a more complex variable flavor.

(a) Prove that $|\Gamma(1+iy)|^{-2}=\sinh(\pi y)/\pi y$ based on the Weierstrass definition of $(z\Gamma(z))^{-1}$.

(b) Prove $\max(|\Gamma(1+iy)|^{-1},|\Gamma(2+iy)|^{-1})\le e^{\pi|y|/2}$.

(c) If g is C^2 on $[0,1]$, prove that

$$\sup_{0\le x\le 1}|g(x)|\le\max(|g(0)|,|g(1)|)+\tfrac{1}{8}\sup_{0\le x\le 1}|g''(x)|$$

(*Hint*: If g is real-valued and $C = -\inf_{x\in[0,1]} g''(x)$, then $h(x) = g(x) - \frac{1}{2}Cx(1-x)$ is convex.)

(d) Prove $\sup_{1\le x\le 2}|\Gamma(x+iy)|^{-1} \le Ce^{\pi|y|/2}$ for a suitable constant C. (*Hint*: Apply (c) to $\log|\Gamma(x+iy)|^{-1}$.)

(e) If F is entire and periodic with period 1 and

$$|F(x+iy)| \le Ce^{\alpha|y|} \tag{9.6.103}$$

with $\alpha < 2\pi$, prove F is constant. (*Hint*: See Section 3.10.)

(f) Let $F(z) = \widetilde{\Gamma}(z)/\Gamma(z)$ where $\widetilde{\Gamma}$ is the integral in (9.6.31). Prove F is entire and periodic and obeys (9.6.103) with $\alpha = \pi/2$ so $F = 1$. Thus, (9.6.31) holds.

15. (a) Using the Bohr–Mollerup theorem for

$$f(x) = m^{x-1}\Gamma\left(\frac{x}{m}\right)\Gamma\left(\frac{x+1}{m}\right)\dots\Gamma\left(\frac{x+m-1}{m}\right) \tag{9.6.104}$$

prove (9.6.84). (C can be evaluated as in Problem 6.)

(b) Use Wielandt's theorem in place of Bohr–Mollerup.

16. Provide the details of the proof of Theorem 9.6.6.

The next three problems will study Γ using (9.6.28)/(9.6.29) as the basic objects, following in part Srinivasan [**535**].

17. Let

$$g(z) = \sum_{n=0}^{\infty} \frac{1}{(z+n)^2} \tag{9.6.105}$$

for $z \in \mathbb{C} \setminus \{0, -1, -2, \dots\}$.

(a) Prove $L(z) = \log(\Gamma(z))$ solves

$$L''(z) = g(z), \qquad L(1) = 0, \quad L'(1) = -\gamma \tag{9.6.106}$$

and is the unique such solution.

(b) Let $E(z) = L(z) + L(1-z)$. Prove that

$$E''(z) = \frac{\pi^2}{\sin^2(\pi z)} \tag{9.6.107}$$

(*Hint*: Use (9.2.27).)

(c) Use (b) to prove that for some A, B,

$$\Gamma(z)\Gamma(1-z) = \frac{e^{A+Bz}\pi}{\sin(\pi z)} \tag{9.6.108}$$

(d) Using $\lim_{z\to 0} z\Gamma(z) = 1$, prove that $A = 0$.

(e) Using $z \to 1 - z$ symmetry, prove that $B = 0$ and thereby obtain another proof of the Euler reflection formula.

18. This will provide another proof of the Legendre duplication formula using (9.6.29).

(a) Let $L(z) = \log(\Gamma(z))$. using (9.6.106), prove that

$$L'(z) - \tfrac{1}{2}\,L'(\tfrac{1}{2}\,z) - \tfrac{1}{2}\,L'(\tfrac{1}{2}\,(z+1)) = \lim_{N\to\infty} \sum_{n=N+1}^{2N+1} \frac{1}{n+z} \tag{9.6.109}$$

(b) Prove $\lim_{N\to\infty} \sum_{n=N+1}^{2N+1} [\frac{1}{n+z} - \frac{1}{n}] = 0$ and that

$$\lim_{N\to\infty} \sum_{n=N+1}^{2N+1} \frac{1}{n} = \log 2 \tag{9.6.110}$$

(*Hint*: (9.6.4).)

(c) Conclude that, for some C,

$$\Gamma(z) = C 2^z \Gamma(\tfrac{1}{2}\,z)\Gamma(\tfrac{1}{2}\,(z+1)) \tag{9.6.111}$$

Taking $z = 1$, conclude that $C = 1/2\sqrt{\pi}$.

(d) By the same argument, prove that for $k = 3, 4, \dots,$

$$\Gamma(z) = C_k k^z \prod_{j=0}^{k-1} \Gamma\left(\frac{z+j}{k}\right) \tag{9.6.112}$$

for a constant C_k.

19. This problem will use (9.6.29) to prove there is a Stirling-type expansion for $\log(\Gamma(z))$ with $z \in (0, \infty)$, but will only compute the leading term explicitly. For $\ell > 1$ and $z \in (0, \infty)$, define

$$g_\ell(z) = \sum_{n=1}^{\infty} (n+z)^{-\ell} \tag{9.6.113}$$

(a) For $0 < t < 1$, show there are constants, $d_{\ell,j}$ and $C_{\ell,J}$, so for $J = 0, 1, 2, \dots,$

$$\left| (z+n+t)^{-\ell} - \sum_{j=0}^{J} d_{\ell,j} (z+n)^{-\ell-j} t^j \right| \le C_{\ell,J} (z+n)^{-\ell-J} \tag{9.6.114}$$

Here $d_{\ell,0} = 1$.

(b) Prove that for $z \in (1, \infty)$,

$$|g_\ell(z)| \le C_\ell |z|^{-\ell+1} \tag{9.6.115}$$

(c) Prove that

$$\frac{1}{\ell-1}z^{-\ell+1}=\sum_{j=0}^{J}(j+1)^{-1}d_{\ell,j}g_{\ell+j}(z)+O(|z|^{-\ell-J})$$

(d) For constants $q_{\ell,j}$ with $q_{\ell,0}=(\ell-1)^{-1}$, prove that

$$g_\ell(z)=\sum_{j=0}^{J}q_{\ell,j}z^{-\ell+1-j}+O(|z|^{-J-\ell})$$

(e) Using (9.6.29), prove that for $z\in(0,\infty)$,

$$\log(\Gamma(z+1))=z\log z+c_{-1}\log z+\sum_{j=0}^{J}c_jz^{-j}+O(|z|^{-J-1})$$

for suitable constants $\{c_j\}_{j=-1}^{\infty}$.

(f) Evaluate c_{-1}.

20. This problem will lead you through an alternate proof of (9.6.44).
(a) Prove that for $\operatorname{Re}x>0$, $\operatorname{Re}y>0$,

$$B(x+1,y)=\frac{x}{x+y}B(x,y)$$

(*Hint:* $\int_0^1\frac{d}{dt}(t^x(1-t)^y),dt=0$ and $x(1-t)-yt=x-(x+y)t$.)
(b) Prove $|B(z,w)|\le B(\operatorname{Re}z,\operatorname{Re}w)$ and $B(1,1)=1$.
(c) By applying Wielandt's theorem to $\Gamma(x+y)B(x,y)$, prove (9.6.44).

21. By looking at $\int_0^{\pi/2}\sin^{2x-1}(2\theta)\,d\theta$, prove that

$$2^{2x-1}B(x,x)=B(x,\tfrac{1}{2}) \tag{9.6.116}$$

and deduce the Legendre duplication formula, (9.6.26), from this.

22. (a) Show that

$$B(z,z)B\left(z+\frac{1}{2},z+\frac{1}{2}\right)=\frac{\pi}{2^{4z-1}z}$$

(b) Show that

$$B(z,w)B(z+w,u)=B(w,u)B(w+u,z)$$

23. Using the changes of variable indicated in the text, verify (9.6.47)–(9.6.50).

24. Verify (9.6.52)–(9.6.63).

25. Using (9.6.48) and (5.7.30), prove that for $0 < \operatorname{Re} z < 1$, $B(z, 1-z) = \pi/\sin(\pi z)$, and then using (9.6.44) obtain a new proof of the Euler reflection formula.

26. (a) Compute $\int_0^1 x^m (\log(x))^m \, dx$. (*Hint:* $x = e^{-t}$.)

    (b) Prove the following result of Johann Bernoulli:

$$\int_0^1 x^x \, dx = 1 - \tfrac{1}{2^2} + \tfrac{1}{3^3} - \tfrac{1}{4^4} + \cdots$$

27. (a) Using (9.6.28), verify that the Laurent expansion of $\Gamma(z)$ at $z = 0$ is

$$\Gamma(z) = \frac{1}{z} - \gamma + O(z) \tag{9.6.117}$$

    (b) Prove $\Gamma'(1) = -\gamma$.

28. Confirm that the following expressions all equal γ:

    (a) $\displaystyle\int_0^\infty \left(\frac{1}{t+1} - e^{-t}\right) \frac{dt}{t}$

    (b) $\displaystyle -\int_0^\infty e^{-t} \log(t) \, dt$

    (c) $\displaystyle\int_0^1 \log\left(\log\left(\frac{1}{t}\right)\right) dt$

    (d) $\displaystyle\int_0^\infty \left(\frac{1}{1-e^{-x}} - \frac{1}{x}\right) e^{-x} \, dx$

    (e) $\displaystyle\int_0^1 \left[[\log(x)]^{-1} + (1-x)^{-1}\right] dx$

    (f) $\displaystyle\int_0^1 \int_0^1 \frac{x-1}{(1-xy)\log(xy)} \, dx dy$

    (g) $\displaystyle\lim_{x \downarrow 1} \sum_{n=1}^\infty \left(\frac{1}{n^x} - \frac{1}{x^n}\right)$

    (h) $\displaystyle\frac{1}{2(2!)} - \frac{1}{4(4!)} + \frac{1}{6(6!)} - \int_1^\infty \frac{\cos x}{x} \, dx$

29. (a) Using (9.6.29), prove that

$$\Gamma''(1) = \gamma^2 + \frac{\pi^2}{6} \tag{9.6.118}$$

    (b) Prove that

$$\int_0^\infty e^{-t} [\log(t)]^2 \, dt = \gamma^2 + \frac{\pi^2}{6} \tag{9.6.119}$$

30. Prove that

$$\int_0^\infty e^{-t^2} \log(t)\, dt = \tfrac{1}{4}\,(\gamma + 2\log(2))\sqrt{\pi} \tag{9.6.120}$$

31. Prove that (with ζ given by (9.6.76)) for $\operatorname{Re} x > 1$

$$\zeta(x)\Gamma(x) = \int_0^\infty \frac{t^{x-1}}{e^t - 1}\, dt \tag{9.6.121}$$

(*Hint*: Expand $(1 - e^{-t})^{-1}$ in a geometric series. Note that this formula can be used as a basis for the analytic continuation of ζ; see Problem 3 of Section 13.3 of Part 2B and Problem 5 of Section 14.7 of Part 2B.)

32. (a) By using $e^{-t} = \sum_{n=0}^\infty (-t)^n/n!$ and justifying change of integral and sum, prove Prym's decomposition [**462**] that in $\operatorname{Re} z > 0$,

$$\Gamma(z) = \sum_{n=0}^\infty \frac{(-1)^n}{n!(z+n)} + \int_1^\infty e^{-t} t^{z-1}\, dt \tag{9.6.122}$$

(b) Show that the sum in (9.6.122) converges in all of $\mathbb{C} \setminus \{0, -1, -2, \dots\}$ and even at $-m$ if the $n = m$ term is dropped and that the integral defines an entire function, so (9.6.122) represents Γ on all of $\mathbb{C}$ as a partial fraction expansion with remainder.

33. Prove the formula of Saalschütz [**499**] that for $k = 0, 1, 2, \dots$ and $-k-1 < \operatorname{Re} z < k$, one has

$$\Gamma(z) = \int_0^\infty t^{z-1}\left(e^{-t} - \sum_{j=0}^k \frac{(-1)^j t^j}{j!}\right) dt \tag{9.6.123}$$

(*Hint*: Induction in k and integration by parts in $t^{z-1} = z^{-1}\frac{d}{dt}\, t^z$.)

34. Following Hadamard [**231**], define

$$F(z) = \frac{1}{\Gamma(1-z)}\, \frac{d}{dz} \log\left[\frac{\Gamma(\frac{(1-z)}{2})}{\Gamma(1-\frac{z}{2})}\right]$$

(a) Prove that $F(1) = 1$.

(b) Prove that $F(z+1) = zF(z) + 1/\Gamma(1-x)$ and conclude that for $n = 1, 2, \dots$, $F(n) = \Gamma(n) = (n-1)!$.

(c) Prove that F is an entire function.

9.7. The Euler–Maclaurin Series and Stirling's Approximation

In this section, which is essentially on real, not complex, analysis, we study asymptotics of $\Gamma(z)$. The classical Stirling approximation for $n!$ says that

Theorem 9.7.1 (Stirling's Approximation). *As $n \to \infty$,*

$$n! \sim \sqrt{2\pi}\, n^{n+1/2} e^{-n} \tag{9.7.1}$$

in the sense that the ratio goes to 1.

We'll prove this below as well as a vast generalization for gamma functions. We used this in our discussion of the law of the iterated logarithm in Section 7.2 of Part 1. Recall that the Bernoulli numbers, B_n, were discussed in Problem 11 of Section 3.1 and will be discussed further below.

Theorem 9.7.2. *As $|z| \to \infty$ away from $(-\infty, 0)$,*

$$\log(\Gamma(z)) \sim (z - \tfrac{1}{2}) \log(z) - z + \tfrac{1}{2} \log(2\pi) + \sum_{j=1}^{\infty} \frac{B_{2j}}{2j(2j-1)} \frac{1}{z^{2j-1}} \tag{9.7.2}$$

as an asymptotic series. To be precise, for each N, if the sum is truncated at $j = N$, the difference is $O(|z|^{-(2N+1)})$ uniformly in each sector $\{z \mid |\arg(z)| < \pi - \delta\}$ with $\delta > 0$.

Remarks. 1. In both $\log(\Gamma(z))$ and $\log(z)$, the branch is taken with $\log(\cdot) > 0$ for $z > 0$.

2. The leading error term on the right in (9.7.1) is thus $e^{-1/12n}$.

3. In fact, this result holds in each region $\mathbb{C} \setminus \{z = x + iy \mid x < 0, |y| \le |x|^\alpha\}$ for any $\alpha > 0$. And the leading order (i.e., up to $\frac{1}{2}\log(2\pi)$) holds if $\operatorname{Re} z \to -\infty$, so long as $|\operatorname{Im} z| \to \infty$ no matter how slowly (see Problem 3). Moreover, we'll have an "explicit" formula for the error after truncating at N.

Our approach to these theorems in this section will be via the Euler–Maclaurin series. There is another approach later, using asymptotics of integrals. A Gaussian approximation (see Example 15.2.6 of Part 2B) will get (9.7.1). Knowing that, in Problem 19 of this section, you'll get an integral representation, known as Binet's formula, for $\log(\Gamma(z)) - (z - \frac{1}{2})\log(z) + z - \frac{1}{2}\log(2\pi)$ whose asymptotics can be determined by the methods of Section 15.2 of Part 2B; see Problem 4 of that section.

We begin with a proof of (9.7.1) which we use in proving Theorem 9.7.2. As (9.7.2) suggests, Stirling's approximation should be thought of as a relation of logs, so (9.7.1) says

$$\sum_{j=1}^{n} \log(j) = n \log(n) - n + \tfrac{1}{2} \log(n) + \log\bigl(\sqrt{2\pi}\bigr) \tag{9.7.3}$$

Since

$$\int_1^n \log(x)\, dx = n \log(n) - n + 1 \tag{9.7.4}$$

this is a relation of sums to integrals. One can even understand the $\frac{1}{2}\log(n)$ since

$$\int_n^{n+1} g(x)\,dx \sim \tfrac{1}{2}\,(g(n)+g(n+1)) \tag{9.7.5}$$

the integral should be near $\sum_{j=2}^{n-1}\log(j)+\frac{1}{2}[\log(1)+\log(n)]$. Thus, we focus on finding the errors in (9.7.5).

If g is C^1 on $[0,1]$, we start with

$$\int_0^1 [(x-\tfrac{1}{2})g(x)]'\,dx = \tfrac{1}{2}\,(g(1)+g(0)) \tag{9.7.6}$$

or

$$\tfrac{1}{2}\,(g(0)+g(1)) = \int_0^1 g(x)\,dx + \int_0^1 (x-\tfrac{1}{2})g'(x)\,dx \tag{9.7.7}$$

giving us an exact error in (9.7.5). By summing such formula (recall $\{x\}$ is given in Section 1.1 as $x-[x]$):

Theorem 9.7.3 (First-Order Euler–Maclaurin Series). *If g is C^1 on $[1,n]$, then*

$$\tfrac{1}{2}\,(g(1)+g(n)) + \sum_{j=2}^{n-1} g(j) = \int_1^n g(x)\,dx + \int_1^n (\{x\}-\tfrac{1}{2})g'(x)\,dx \tag{9.7.8}$$

This leads to

Proposition 9.7.4. *We have, as $n\to\infty$,*

$$\lim_{n\to\infty}(\log(n!) - [(n+\tfrac{1}{2})\log(n)-n]) = 1 + \int_1^\infty \frac{(\{x\}-\frac{1}{2})}{x}\,dx \tag{9.7.9}$$

$$= 1 - \int_1^\infty \frac{(\{x\}-\frac{1}{2})^2}{x([x]+\frac{1}{2})}\,dx \tag{9.7.10}$$

Remark. The integral in (9.7.9) is only conditionally convergent.

Proof. By (9.7.8) with $g(x)=\log(x)$ and (9.7.4),

$$\log(n!) - [(n+\tfrac{1}{2})\log(n)-n] = 1 + \int_1^n \frac{(\{x\}-\frac{1}{2})}{x}\,dx \tag{9.7.11}$$

Since

$$\int_j^{j+1} \frac{(\{x\}-\frac{1}{2})}{x}\,dx = \int_0^1 \frac{(x-\frac{1}{2})}{x+j}\,dx \quad\text{and}\quad \int_0^1 (x-\tfrac{1}{2})\,dx = 0$$

we have

$$\int_j^{j+1} \frac{(\{x\}-\frac{1}{2})}{x}\,dx = \int_0^1 (x-\tfrac{1}{2})\left[\frac{1}{x+j}-\frac{1}{j+\frac{1}{2}}\right]dx$$

$$= -\int_0^1 \frac{(x-\frac{1}{2})^2}{(x+j)(j+\frac{1}{2})}\, dx \tag{9.7.12}$$

so

$$\text{RHS of (9.7.11)} = 1 - \int_1^n \frac{(\{x\}-\frac{1}{2})^2}{x([x]+\frac{1}{2})}\, dx \tag{9.7.13}$$

This integrand is bounded by $\frac{1}{4}[x]^2$, which is integrable at infinity, so the limit exists. □

Proof of Theorem 9.7.1. The last proposition says

$$C_n \equiv \frac{n!}{n^{n+\frac{1}{2}}e^{-n}} \tag{9.7.14}$$

obeys $C_n \to C_\infty$, a finite number, and the proof shows

$$C_1 \geq C_n \geq C_\infty \tag{9.7.15}$$

Wallis' formula, (5.7.80), can be rewritten

$$\lim_{n\to\infty} \frac{2^{2n}(n!)^4}{(2n)!(2n+1)!} = \frac{\pi}{2} \tag{9.7.16}$$

which says $C_\infty^2 = 2\pi$. □

$C_1 = e$ and $e/\sqrt{2\pi} \cong 1.08$, so by (9.7.15) over the full range of n, the Stirling approximation is off by no more than about 8%. That said, their ratio only goes to one with leading error like $1/12n$.

For later purposes, we note that we have proven that

$$\lim_{n\to\infty} \left[1 + \int_1^n \frac{(\{x\}-\frac{1}{2})}{x}\, dx\right] = \log(\sqrt{2\pi}) \tag{9.7.17}$$

To go beyond first order, we'll want to push (9.7.7) further. Let $b_1(x) \equiv (x-\frac{1}{2})$. If we find b_2 so $b_2'(x) = b_1(x)$, then $b_2(1) - b_2(0) = \int_0^1 b_1(x)\, dx = 0$. So if we integrate by parts,

$$\int_0^1 b_1(x)g'(x)\, dx = -\int_0^1 b_2(x)g''(x)\, dx + b_2(1)(g'(1) - g'(0)) \tag{9.7.18}$$

then the boundary terms telescope when we sum over multiple intervals. Therefore, to continue this process, we want to arrange b_2 so $\int_0^1 b_2(x)\, dx = 0$ also. This determines b_2 and then, by induction, all b_n. It turns out to be easiest to pick the generating functions for the $b_n(x)$ out of a hat, but one can "deduce" the choice if one wants. We note though that, clearly, $b_n(x)$ is a polynomial with leading term $x^n/n!$. It is traditional to use B_n for monic polynomials, so $b_n(x) = B_n(x)/n!$.

Definition. The *Bernoulli polynomials*, $B_n(x)$, are defined by

$$\frac{ze^{zx}}{e^z - 1} = \sum_{n=0}^{\infty} B_n(x)\, \frac{z^n}{n!} \tag{9.7.19}$$

Here are the basic properties:

Proposition 9.7.5. (a) *B_n is a polynomial of degree n.*

(b) $B_n'(x) = nB_{n-1}(x)$ (9.7.20)

(c) $B_n(x) = x^n +$ *lower order* (9.7.21)

(d) $\displaystyle\int_0^1 B_n(x)\, dx = \delta_{n0}$ (9.7.22)

(e) *$B_n(0)$ is the nth Bernoulli number, B_n. In particular, $B_1 = -\frac{1}{2}$, $B_{2k+1} = 0$ for $k \geq 1$ (see (3.1.50)).*

Proof. (a) $z(e^z - 1)^{-1} = 1 + \sum_{m=1}^{\infty} B_m z^m$ and $e^{zx} = \sum_{n=1}^{\infty} (zx)^n/n!$, so the product is a sum of terms of the form $(zx)^n z^m$ and the coefficient of z^{n-m} is a sum of x^n with $n \leq n_m$.

(b) Let $F(x,z)$ denote the left side of (9.7.19). Then by a direct calculation,

$$\frac{\partial F}{\partial x} = zF \tag{9.7.23}$$

which leads directly to (9.7.20) plus $B_0' = 0$.

(c) This follows inductively from $B_0(x) = 1$ and (9.7.20).

(d) By a direct calculation, $\int_0^1 F(x,z)\, dx = 1$, which implies (9.7.22).

(e) $F(0,z)$ is the generating function for the Bernoulli numbers. □

Theorem 9.7.6 (General Euler–Maclaurin Series). *For $k = 1, 2, \dots$, let g be a C^k function on $[1,n]$. Then*

$$\tfrac{1}{2}\,(g(1) + g(n)) + \sum_{j=1}^{n-1} g(j) = \int_1^n g(x)\, dx + \sum_{\ell=1}^{[k/2]} \frac{B_{2\ell}}{(2\ell)!}\, [g^{(2\ell-1)}(n) - g^{(2\ell-2)}(1)] + \frac{(-1)^{k+1}}{k!} \int_1^n B_k(\{x\}) g^{(k)}(x)\, dx \tag{9.7.24}$$

Proof. The proof is by induction in k. $k = 1$ is (9.7.8). Given (9.7.20) for k, we write $\int_1^n$ and dx as $\sum_{j=1}^{n-1} \int_j^{j+1} \dots dx$ and write $\int_j^{j+1} B_k(\{x\}) g^{(k)}(x)\, dx$

as $\int_j^{j+1} B_k(x)g^{(k)}(j+x)\,dx$, and then

$$\begin{aligned}\frac{(-1)^{k+1}}{k!}&\int_j^{j+1} B_k(x)g^{(k)}(j+x)\,dx\\ &=\frac{(-1)^{k+1}}{(k+1)!}\int \frac{d}{dx}B_{k+1}(x)g^{(k)}(j+x)\,dx\\ &=\frac{(-1)^{k+2}}{(k+1)!}\int B_{k+1}(x)g^{(k+1)}(j+x)+(-1)^{k+1}\frac{B_{k+1}(0)}{(k+1)!}(g^{(k)}(j+1)-g^{(k)}(j))\end{aligned} \tag{9.7.25}$$

If $k+1$ is odd, then $B_{k+1}(0)=0$, and so we get (9.7.24) for $k+1$. $\square$

The following implies Theorem 9.7.2:

Theorem 9.7.7. *For $z\in\mathbb{C}\setminus(-\infty,0]$ and any L,*

$$\begin{aligned}\log(\Gamma(z))&=(z-\tfrac{1}{2})\log(z)-z+\tfrac{1}{2}\log(2\pi)\\ &\qquad+\sum_{\ell=1}^{L}\frac{B_{2\ell}}{2\ell(2\ell-1)}\frac{1}{z^{2\ell-1}}-\frac{1}{2L}\int_0^\infty\frac{B_{2L}(\{x\})}{(x+z)^{2L}}\,dx\end{aligned} \tag{9.7.26}$$

Proof. We begin by claiming that it suffices to prove

$$\log(\Gamma(z))=(z-\tfrac{1}{2})\log(z)-z+\tfrac{1}{2}\log(2\pi)-\int_0^\infty\frac{B_1(\{x\})}{x+z}\,dx \tag{9.7.27}$$

where, as in going from (9.7.9) to (9.7.10), the integral is conditionally convergent and bounded by

$$\int_0^\infty\frac{|B_1(\{x\})|^2}{|x+z||[x]+\frac{1}{2}+z|}\,dx \tag{9.7.28}$$

For once we have (9.7.27), we write the integral as a sum of $\int_j^{j+1}$ and integrate by parts repeatedly, as in the proof of Theorem 9.7.6. Since $\frac{d^2}{dx^2}(x+z)^{-2L}=(2L)(2L+1)(x+z)=2L-2$, this leads from (9.7.26) for L to (9.7.26) for $(L+1)$ (and a single integration by parts goes from (9.7.27) to (9.7.26) for $L=1$).

Problem 17 has an interesting proof of (9.7.27) by Wielandt using his theorem. We start our proof of (9.7.27) by writing (Γ_n is given by (9.6.8))

$$\Gamma_n(z)=n^{z-1}\prod_{j=1}^{n}\left(\frac{j}{z+j-1}\right) \tag{9.7.29}$$

so with

$$f(x) = \log(z - 1 + x) - \log(x) \tag{9.7.30}$$

$$\log(\Gamma_n(z)) = (z - 1)\log(n) - \sum_{j=1}^{n} f(j) \tag{9.7.31}$$

We note the following properties of f:

$$\int_1^n f(x)\,dx = (n + z - 1)\log(n + z - 1) - n\log(n) - z\log(z) \tag{9.7.32}$$

$$f'(x) = \frac{1}{x + z - 1} - \frac{1}{x} \tag{9.7.33}$$

$$\tfrac{1}{2} f(1) = \tfrac{1}{2}\log(z), \qquad \tfrac{1}{2} f(n) = \tfrac{1}{2}\log\biggl(\frac{n + z - 1}{n}\biggr) \to 0 \tag{9.7.34}$$

Thus, by (9.7.8) and (9.7.28),

$$\log(\Gamma_n(z)) = -\tfrac{1}{2} f(n) + (z - \tfrac{1}{2})\log(z) + R_n(z) - \int_1^n B_1(\{x\})\biggl[\frac{1}{x + z - 1} - \frac{1}{x}\biggr] dx \tag{9.7.35}$$

where

$$\begin{aligned} R_n(z) &= (z - 1)\log(n) - (n + z - 1)\log(n + z - 1) - n\log(n) \\ &= -(n + z - 1)[\log(n + z - 1) - \log(n)] \\ &= -(n + z - 1)\biggl[\log\biggl(1 + \frac{z - 1}{n}\biggr)\biggr] \\ &= -(z - 1) + O\biggl(\frac{1}{n}\biggr) \end{aligned} \tag{9.7.36}$$

Thus,

$$\log(\Gamma(z)) = (z - \tfrac{1}{2})\log(z) - z + \biggl(1 + \int_1^\infty \frac{B_1(\{x\})}{x}\biggr) - \int_1^\infty \frac{B_1(\{x\})}{x + z - 1}\,dx \tag{9.7.37}$$

By (9.7.17), this is (9.7.27). $\square$

One defines the function $\mu(z)$ on $\mathbb{C} \setminus (-\infty, 0]$ by

$$\Gamma(z) = \sqrt{2\pi}\, z^{z-1/2} e^{-z} e^{\mu(z)} \tag{9.7.38}$$

so that (9.7.27) becomes

$$\mu(z) = -\int_0^\infty \frac{B_1(\{x\})}{x + z}\,dx \tag{9.7.39}$$

In Problem 19, the reader will prove *Binet's formula:*

$$\mu(z) = \int_0^\infty \biggl(\frac{1}{e^t - 1} - \frac{1}{t} + \frac{1}{2}\biggr) e^{-zt}\,\frac{dt}{t} \tag{9.7.40}$$

Notes and Historical Remarks. In 1730, in rapid order, Abraham de Moivre (1667–1754) [**134**] found the leading $n^{n+1/2}e^{-n}$ behavior of $n!$ for large n; James Stirling (1692–1770) [**544**] found the $\sqrt{2\pi}$, and then de Moivre provided a proof and the full asymptotic formula [**135**]. As explained in Section 7.2 of Part 1, de Moivre worked on these asymptotics in connection with his discovery of the central limit theorem for the binomial distribution. For a review article on proofs of Stirling's formula, see Dominici [**148**].

The calculation of the constant in Stirling's formula comes one way or another from either Wallis' formula, (5.7.80), or the closely related value of $\Gamma(\frac{1}{2})$ (which, since it is essentially the Gaussian integral, can come from any method of computing that). The quickest way of getting the $\sqrt{2\pi}$ in Stirling's formula is to first compute $\Gamma(\frac{1}{2})$ from the Euler duplication formula $(\Gamma(\frac{1}{2})^2 = \pi/\sin(\pi/2))$, and then via the Legendre duplication formula, get the constant assuming Stirling's formula holds up to the constant (see Problem 1). Another way to get the constant is to use Stirling's formula in binomial coefficients and take the limit as $n \to \infty$ in $\sum_{j=0}^{n} \binom{j}{n} = 2^n$. However, this requires a priori knowledge of the Gaussian integral!

As we noted, Stirling's formula is accurate even when $n = 1$. Indeed, one can show that for $n = 1, 2, \dots,$

$$\frac{1}{12n+1} \le \log(n!) - (n + \tfrac{1}{2}) \log(n) - n + \tfrac{1}{2} \log(2\pi) \le \frac{1}{12n} \tag{9.7.41}$$

Bernoulli numbers were introduced by Jakob Bernoulli (1654–1705) in his posthumously published *Ars Conjectandi* [**43**], who noticed that there was a regularity in the coefficients of the polynomials $Q_k(n)$ for $\sum_{j=1}^{n} j^k$ (see Problem 4). The famous Japanese mathematician, Seki Takakazu (1642–1708) also discovered Bernoulli numbers independently at the same time. His work was published posthumously in 1712 [**551**], a year before Bernoulli's publication. Since (9.2.50) holds and $\lim_{\ell\to\infty} \sum_{n=1}^{\infty} n^{-2\ell} = 1$, we get

$$(-1)^{\ell-1} B_{2\ell} \sim \frac{2(2\ell)!}{(2\pi)^{2\ell}} \sim 4\sqrt{\pi\ell} \left(\frac{\ell}{\pi e}\right)^{2\ell} \tag{9.7.42}$$

by Stirling's formula.

Figure 9.7.1 shows $B_n(x)$ for $0 < x < 1$ for $n = 3, 5, 7, 9$ and for $n = 4, 6, 8, 10$. They have similar shapes for those two sets of values.

The Bernoullis had many mathematicians in their family, most notably Jakob (1654–1705), Johann (1667–1748), and Daniel (1700–82). The family was Swiss, although several generations before had fled Belgium (during a time of persecution of Protestants) and Daniel was born in the Netherlands where Johann was teaching. Johann was Jakob's younger brother and the father of Daniel, notable for his strong competitions with his brother and son.

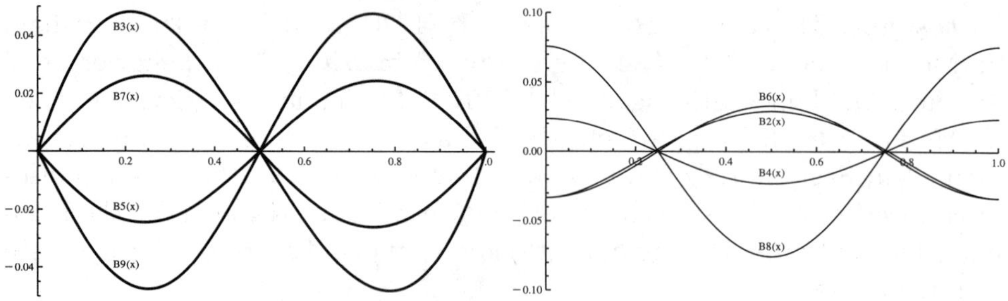

Figure 9.7.1. $B_n(x)$; $0 < x < 1$, $n = 3, \ldots, 10$.

The brothers were involved in a controversy over the isoperimetric problem, and shortly before his death in 1705, Jakob was worried that Johann was after his chair at Basel which Johann got after his death. Most remarkably, Johann wrote his book on hydrodynamics after his son's book on the same subject, but he predated it and claimed his son's ideas in it were his own! l'Hôpital published a book on calculus that brought him some fame, but the book was almost verbatim copies of lectures Johann had sent him as part of lessons, for which l'Hôpital, a French nobleman, paid handsomely. In particular, what is called l'Hôpital's rule is due to Johann Bernoulli.

Jakob's contributions are the most notable, including Bernoulli numbers and polynomials, fundamental work in probability theory (Bernoulli distribution), and early work on elliptic integrals. Johann gained fame for solving the catenary problem. Daniel is known for Bernoulli's principle in hydrodynamics.

The Euler–Maclaurin series was found around 1734 independently by Euler and Colin Maclaurin (1698–1746), a Scottish mathematician. Euler found it [**173, 174**] as part of his concerted work that also led to the product formula and the solution of the Basel problem. Indeed, he used the series to compute $\sum_{n=1}^{\infty} 1/n^2$ to twenty decimal places, presumably as part of his finding/checking that the sum was $\pi^2/6$. Euler wrote to Stirling in 1736 about his work and heard back in 1738 that Maclaurin had the same results. Maclaurin only published his results in 1742, as part of his landmark text, *A Treatise on Fluxions* [**371**], one of the first English language texts on calculus.

Euler further developed his ideas in a 1755 book [**180**]; there are online excerpts in English[8] that amply demonstrate Euler's fascination with values.

Darboux [**129**] found a generalization of the Euler–Maclaurin series that has a further arbitrary function; see, for example, the discussion in [**592**, Ch. VII].

[8] `http://www.math.nmsu.edu/~davidp/institutionescalcdiff.pdf`

An interesting insight to the form of Euler–Maclaurin series and nature of the B_n's can be seen by the following formal considerations. Let

$$I_j = \int_1^n f^{(j)}(x)\,dx \tag{9.7.43}$$

$$S_j = \sum_{k=1}^{n-1} f^{(j)}(k) \tag{9.7.44}$$

Then, since $\int_0^1 y^j\,dy = (j+1)!$, formally using $f(k+y) = \sum_{j=0}^\infty f^{(j)}(k)\frac{y^j}{j!}$ on $[0,1]$, we see

$$I^{(j)} = \sum_{k=0}^\infty \frac{S^{(j+k)}}{(k+1)!} = \sum_{\ell=-\infty}^\infty a_{j\ell} S^{(\ell)} \tag{9.7.45}$$

where a is the upper-triangular Toeplitz matrix

$$a_{j\ell} = \begin{cases} 0, & \ell \le j-1 \\ \frac{1}{(\ell-j+1)!}, & \ell \ge j \end{cases} \tag{9.7.46}$$

Upper-triangular Toeplitz matrices appeared already in Problem 22 of Section 2.3. Their inverse is given in terms of numerical inverse power series. That is, (a_{jk}) has the inverse matrix, $b_{j\ell}$, where

$$b_{j\ell} = \begin{cases} 0, & \ell \le j-1 \\ c_{\ell-j}, & \ell \ge j \end{cases} \tag{9.7.47}$$

where $\sum_{j=0}^\infty c_j z^j = 1/\big[\sum_{j=0}^\infty z^j/(j+1)!\big]$. But

$$\sum \frac{z^j}{(j+1)!} = \frac{e^z - 1}{z}$$

which means that the c's are the Bernoulli numbers divided by $n!$ That is, we've shown, again formally, that

$$\begin{aligned} S^{(0)} &= \sum_{k=0}^\infty \frac{B_k}{k!} I^{(k)} \\ &= \int_1^n f(x)\,dx + \sum_{k=1}^\infty \frac{B_k}{k!}\Big[f^{(k-1)}(n) - f^{(k-1)}(1)\Big] \end{aligned} \tag{9.7.48}$$

since $I_k = f^{(k-1)}(n) - f^{(k-1)}(1)$ if $k \ge 1$. Taking into account that $B_1 = -\frac{1}{2}$ and $B_{2\ell+1} = 0$ if $\ell \ge 1$, we see that formally,

$$\tfrac{1}{2}(f(1)+f(n)) + \sum_{j=2}^{n-1} f(j) = \int_1^n f(x)\,dx + \sum_{\ell=1}^\infty \frac{B_{2\ell}}{(2\ell)!}[f^{(2\ell-1)}(n) - f^{(2\ell-1)}(1)] \tag{9.7.49}$$

which is precisely the formal limit of (9.7.24)!

The Stirling expansion (9.7.2) is a good example of an asymptotic series. We'll describe such a series at 0, although (9.7.2) is at infinity. If f is a function defined in a sector

$$\Omega = \{z \mid \alpha \le \arg(z) \le \beta,\, 0 \le |z| < \varepsilon\}$$

we say $\sum_{n=0}^\infty a_n z^n$ is an asymptotic series for f if and only if, for all N,

$$\lim_{\substack{|z|\to 0 \\ z\in\Omega}} \frac{|f(z) - \sum_{n=0}^N a_n z^n|}{|z|^N} = 0 \tag{9.7.50}$$

It can happen that the series diverges for $z \ne 0$ but (9.7.50) holds. Indeed, while not every power series is a convergent series, every one is asymptotic for some f so long as $|\beta - \alpha| < 2\pi$. We discuss these issues further in Section 15.1 of Part 2B.

The definition (9.7.38) of $\mu(z)$ as a tool in getting refined bounds in Stirling's formula is due to Stieltjes [**540**] in 1889. He proved that

$$\begin{aligned} |\mu(z)| &\le \frac{1}{12}\,\frac{1}{\operatorname{Re} z} && \text{if } \operatorname{Re} z > 0 \\ |\mu(iy)| &\le \frac{1}{6}\,\frac{1}{|y|} && \text{if } y \in \mathbb{R} \\ |\mu(z)| &\le \frac{1}{8}\,\frac{1}{\cos^2(\frac{1}{2}\varphi)}\,\frac{1}{|z|}, && z = |z|e^{i\varphi} \in \mathbb{C}\setminus(-\infty,0] \end{aligned}$$

The expansion of the error given in Problem 20 is from Gudermann [**224**] in 1845.

Problems

1. From $\Gamma(z+1) = z\Gamma(z)$, deduce that for $n = 1, 2, \dots$ that

$$n!\,\Gamma(n+\tfrac{1}{2}) = 2^{-2n}(2n)!\,\Gamma(\tfrac{1}{2}) \tag{9.7.51}$$

and from this, show that if $\Gamma(z) \sim Cz^{z-1/2}e^{-z}$ as $z \in (0,\infty)$ goes to infinity, then $C = \sqrt{2\pi}$.

Remarks. 1. (9.7.51) is, of course, a special case of the Legendre duplication formula, (9.6.26), which also implies $C = \sqrt{2\pi}$.

2. The reader might need to first solve the puzzle that, while $\Gamma(z) \sim \sqrt{2\pi}\, z^{z-1/2}e^{-z}$, we have $\Gamma(z+1) \sim \sqrt{2\pi}\, z^{z+1/2}e^{-z}$, not $e^{-(z+1)}$.

2. (a) From (9.6.91), prove that (9.7.1) implies for $x \in (0,\infty)$ that $\log(\Gamma(x)) \sim (x - \frac{1}{2})\log(x) - x + \frac{1}{2}\log(2\pi)$.

(b) Prove directly that

$$\int_{n-1/2}^{n+1/2} \log(t)\,dt = \log(n) + O\Big(\frac{1}{n^2}\Big)$$

and deduce (9.7.1) with $\sqrt{2\pi}$ replaced by an unknown constant C. (*Hint*: $\log(n+z)+\log(n-z)=2\log(n)+O(t^2/n^2)$.)

3. (a) Prove that the remainder term in (9.7.26) for $\operatorname{Re} z<0$, $|\operatorname{Im} z|>1$, is bounded by $|\operatorname{Im} z|^{-2L+2}$.

   (b) Prove that for any C,α,L in $\mathbb{C}\setminus\{z=x+iy,\ x<0,\ |y|\le C|x|^\alpha\}$, we have

$$\log(\Gamma(z)) = (z-\tfrac{1}{2})\log(z)-z+\tfrac{1}{2}\log(2\pi) + \sum_{\ell=1}^{L}\frac{B_{2\ell}}{2\ell(2\ell-2)}\,\frac{1}{z^{2\ell-1}}+O(z^{2\ell-3})$$

   (*Hint*: Go to much higher order than L in (9.7.26).)

   (c) If $|\operatorname{Im} z_n|\to\infty$, prove that uniformly in $\operatorname{Re} z_n$,

$$\log(\Gamma(z))-(z_n-\tfrac{1}{2})\log(z_n)-z_n\to\tfrac{1}{2}\log(2\pi)$$

   Note. This includes, for example, $z_n=-n+i\log(n)$ not included in (b).)

4. (a) Prove that in terms of Bernoulli numbers,

$$\sum_{j=1}^{n} j^k = \frac{1}{k+1}n^{k+1}+\frac{1}{2}n^k+\sum_{\ell=1}^{[k/2]}\frac{1}{2\ell}\binom{k}{2\ell-1}B_{2\ell}n^{k+1-2} \tag{9.7.52}$$

   (b) Prove that the Bernoulli polynomials obey

$$B_n(1)-B_n(0)=\delta_{n1} \tag{9.7.53}$$

   (c) Assuming that

$$B_k(x+1)-B_k(x)=kx^{k-1} \tag{9.7.54}$$

   for some $k\ge 1$, prove it for $k+1$. (*Hint*: Use (9.7.20) and (9.7.53).)

   (d) Conclude that (9.7.54) holds for all $k\ge 1$.

   (e) Prove that

$$\sum_{j=0}^{n} j^k = \frac{B_{k+1}(n+1)-B_{k+1}(0)}{k+1} \tag{9.7.55}$$

Remarks. 1. Note that (9.7.52) and (9.7.55) show the coefficients of $B_n(x)$ are expressible in terms of $\binom{n+1}{2\ell}$ and $B_{2\ell}(0)$.
2. Part (a) can be obtained from the generating function for B_n but also from a terminating Euler–Maclaurin expansion.
3. (9.7.55) can also be proven by looking at generating functions (i.e., if $\delta_k(n)$ is the difference, look at $\sum_{k=0}^\infty z^k\delta_k(n)/k!$).

5. (a) Using generating functions, prove that for any n, x, y,

$$B_n(x+y) = \sum_{k=0}^{n} \binom{n}{k} y^k B_{n-k}(x) \tag{9.7.56}$$

(b) Prove (9.7.56) using Taylor series.

6. (a) Define

$$B_n^\sharp(m) = \int_0^1 B_n(x) e^{-2\pi i m x}\, dx \tag{9.7.57}$$

For any $n \geq 1$ and $m \in \mathbb{Z}$, prove that (you'll need (9.7.53))

$$n B_{n-1}^\sharp(m) = (2\pi i m) B_n^\sharp(m) + \delta_{n1} \tag{9.7.58}$$

(b) Prove for $n \geq 1$ that $B_n^\sharp(0) = 0$ and conclude for $n \geq 1$ that

$$B_n^\sharp(m) = -\frac{n!}{(2\pi i)^n} \begin{cases} 0 & \text{if } m = 0 \\ \frac{1}{m^n} & \text{if } m \neq 0 \end{cases} \tag{9.7.59}$$

(c) For $n \geq 1$, prove that (for $n = 1$, $x \notin \mathbb{Z}$)

$$B_n(\{x\}) = -\frac{n!}{(2\pi i)^n} \sum_{m \neq 0} \frac{e^{2\pi i m x}}{m^n} \tag{9.7.60}$$

$$B_{2n}(\{x\}) = (-1)^{n+1} \frac{2(2n)!}{(2\pi)^{2n}} \sum_{m=1}^{\infty} \frac{\cos(2\pi m x)}{m^{2n}} \tag{9.7.61}$$

$$B_{2n-1}(\{x\}) = (-1)^n \frac{2(2n-1)!}{(2\pi)^{2n-1}} \sum_{m=1}^{\infty} \frac{\sin(2\pi m x)}{m^{2n-1}} \tag{9.7.62}$$

Remarks. 1. For (c), you'll need results on convergence of Fourier series; see Section 3.5 of Part 1.

2. In particular, (9.7.60) for $n = 2\ell$ and $x = 0$ proves (9.2.50) yet again.

7. This problem will use the Euler–Maclaurin series for $k = 2$ to prove the Poisson summation formula, Theorem 6.6.10 of Part 1. We'll prove it for $f \in \mathcal{S}(\mathbb{R})$. One can go from that to less regular f's by suitable approximation.

(a) For $f \in \mathcal{S}(\mathbb{R})$, prove that

$$\sum_{n=-\infty}^{\infty} f(n) = \int_{-\infty}^{\infty} f(x)\, dx - \tfrac{1}{2} \int_{-\infty}^{\infty} B_2(\{x\}) f''(x)\, dx \tag{9.7.63}$$

(b) Using (9.7.60), prove that

$$\int_{-\infty}^{\infty} B_2(\{x\}) f''(x) = -\frac{2}{(2\pi i)^2} \sum_{m \neq 0} \int_{-\infty}^{\infty} \frac{e^{2\pi i m x}}{m^2} f''(x)\, dx \tag{9.7.64}$$

$$= -2 \sum_{m \neq 0} \int_{-\infty}^{\infty} e^{2\pi i m x} f(x)\, dx \tag{9.7.65}$$

(c) Conclude that

$$\sum_{n=-\infty}^{\infty} f(n) = (2\pi)^{1/2} \sum_{n=-\infty}^{\infty} \widehat{f}(2\pi n) \tag{9.7.66}$$

which is equivalent to the Poisson summation formula.

Remark. The idea of the connection is from Butzer–Stens [**83**].

8. Prove that

$$\frac{1}{n} \sum_{k=1}^{n} \binom{n}{k} B_k B_{n-k} = -B_n - B_{n-1} \tag{9.7.67}$$

9. For $n \geq 1$, prove that one has

$$\sum_{k=0}^{n} \binom{n}{k} B_k(x) = B_n(x) + n x^{n-1} \tag{9.7.68}$$

Remarks. 1. This generalizes (3.1.49).

2. This is often written

$$\sum_{k=0}^{n} \binom{n+1}{k} B_k(x) = (n+1) x^n, \qquad n \geq 2 \tag{9.7.69}$$

10. The Euler polynomials, $E_n(x)$, are defined by the generating function

$$\frac{2e^{zx}}{e^z + 1} = \sum_{n=0}^{\infty} E_n(x)\, \frac{z^n}{n!} \tag{9.7.70}$$

(a) If E_n is given by (9.2.51), prove that

$$E_n = 2^n E_n(\tfrac{1}{2}) \tag{9.7.71}$$

(b) Prove that

$$E_n(1) = -E_n(0) + 2\delta_{n0} \tag{9.7.72}$$

(c) Prove that $\{e^{(2n+1)i\pi x}\}_{n=-\infty}^{\infty}$ is an orthonormal basis for $L^2([0,1], dx)$. (*Hint*: Multiplication by $e^{i\pi x}$ is unitary.)

(d) Define $E_n^\flat(m)$ by

$$E_n^\flat(m) = \int_0^1 e^{-(2m+1)i\pi x} E_n(x)\, dx \tag{9.7.73}$$

Prove that for $n \geq 1$,

$$n E_{n-1}^\flat(m) = (2m+1)(i\pi) E_n^\flat(m) \tag{9.7.74}$$

(e) Prove that

$$E_0^\flat(m) = \frac{2}{(i\pi)(2m+1)} \tag{9.7.75}$$

(f) Prove that

$$E_n^\flat(m) = \frac{2(n!)}{[(2m+1)i\pi]^{n+1}} \tag{9.7.76}$$

(g) For $0 \le x \le 1$, prove that

$$E_n(x) = \frac{2n!}{(i\pi)^{n+1}} \sum_{m=-\infty}^{\infty} \frac{e^{(2m+1)i\pi x}}{(2m+1)^{n+1}} \tag{9.7.77}$$

(h) Prove (9.2.52).

Remark. Just as $B_n(x)$ extended from $[0,1]$ to be periodic is C^{n-1}, $E_n(x)$ has to be extended to be antiperiodic (i.e., $x \to x+1$ multiplies by -1) to be C^n.

11. Prove that

$$\lim_{m\to\infty} \sum_{j=1}^{m-1} \left(\frac{j}{m}\right)^m = \frac{1}{e-1} \tag{9.7.78}$$

(see Spivey [**532**])

12. Let $F(x,z)$ be the left side of (9.7.19).

(a) Prove that

$$F\left(x=0, \frac{z}{2}\right) = \tfrac{1}{2}\left[F(\tfrac{1}{2}, z) + F(0,z)\right] \tag{9.7.79}$$

(b) Prove that

$$B_n(\tfrac{1}{2}) = -B_n(0)\left[1 - \frac{1}{2^{n-1}}\right] \tag{9.7.80}$$

13. (a) Use an Euler–Maclaurin series with $g(n) = n^{-1}$ to prove, for each L, that there is a constant C_L so

$$1 + \frac{1}{2} + \cdots + \frac{1}{n} - \log(n) = C_L + \frac{1}{2n} - \sum_{\ell=1}^{L} \frac{B_{2\ell}}{(2\ell)n^{2\ell}} + O(n^{-2L-2})$$

(b) Prove $C_L = \gamma$, the Euler–Mascheroni constant, and so deduce that for fixed L,

$$\gamma = \sum_{j=1}^{n} \frac{1}{j} - \log(n) - \frac{1}{2n} + \sum_{\ell=1}^{L} \frac{B_{2\ell}}{(2\ell)n^{2\ell}} + O(n^{-2L-2})$$

This formula was used by Euler to compute γ to fifteen decimal places. He took $n = 10$, but had to choose L carefully, since for n fixed the series diverges as $L \to \infty$!

14. Use the Euler–Maclaurin series to make several digit approximations for

(a) γ (b) $\displaystyle\sum_{n=1}^{\infty} \frac{1}{n^2}$ (c) $\displaystyle\sum_{n=1}^{\infty} \frac{1}{n^3}$

15. Prove that for $n \geq 2$, $B_n(1-x) = (-1)^n B_n(x)$.

16. This problem will prove

$$T_{2n-1} = \frac{(-1)^{n-1} 4^n (4^n - 1) B_{2n}}{2n} \tag{9.7.81}$$

is a positive integer.

(a) Let

$$G(x,z) = \frac{\sin z + x \cos z}{\cos z - x \sin z} = \sum_{n=0}^{\infty} \frac{T_n(x)}{n!} z^n$$

Prove that $(1+x^2)\frac{\partial}{\partial x} G(x,z) = \frac{\partial}{\partial z} G(x,z)$.

(b) Prove that $T_{n+1}(x) = (1+x^2) T_n'(x)$ and $T_0(x) = x$.

(c) Prove that $T_n(x) = \sum_{j=0}^{n+1} t_j^{(n)} x^j$ where each $t_j^{(n)}$ is an integer and $t_{n+1}^{(n)}, t_{n-1}^{(n)}, t_{n-3}^{(n)}, \dots > 0$ with $t_n^{(n)} = t_{n-2}^{(n)} = \dots = 0$.

(d) Prove that T_{2n-1} given by (9.7.81) is $T_{2n-1}(0)$ and that this quantity is a nonnegative integer. (*Hint*: See Problem 11 of Section 3.1.)

17. This will provide Wielandt's proof of (9.7.27). There is a partly alternate proof in the next problem.

(a) With η given by the right side of (9.7.39), define

$$\widetilde{\Gamma}(z) = e^{\eta(z)} \Gamma_0(z), \qquad \Gamma_0(z) = z^{z-1/2} e^{-z}$$

Prove that $|\Gamma_0(z)|$ and $|e^{\eta(z)}|$ are bounded in the vertical strip $\{z \mid 1 \leq \operatorname{Re} z \leq 2\}$.

(b) Prove that η obeys the functional equation $\eta(z) - \eta(z+1) = (z + \frac{1}{2}) \log(1 + \frac{1}{z}) - 1$.

(c) Prove that $\widetilde{\Gamma}(z+1) = z\widetilde{\Gamma}(z)$ and conclude, by Wielandt's theorem (Proposition 9.6.3), that $\widetilde{\Gamma}(z) = c\Gamma(z)$.

(d) By taking $z = 1, 2, \dots$ to infinity and using Stirling's formula for $n!$, conclude $\widetilde{\Gamma}(z) = (\sqrt{2\pi})^{-1} \Gamma(z)$, that is, that (9.7.27) holds.

18. This problem will prove that if η obeys

$$\eta(z) - \eta(z+1) = \left(z + \frac{1}{2}\right)\log\left(1 + \frac{1}{z}\right) - 1, \qquad \eta(z) \to 0 \quad \text{as } z \to \infty \text{ on } \mathbb{R} \tag{9.7.82}$$

for $z \in \mathbb{R}$ sufficiently large, then for z real and large, we have $\eta(z) = \mu(z)$, the function given by (9.7.38). Thus, if η is also analytic, we get this for all z in the region of analyticity.

(a) Prove the functional equation for $\Gamma(z)$ implies that μ, given by (9.7.38), obeys

$$\mu(z) - \mu(z+1) = \left(z + \frac{1}{2}\right)\log\left(1 + \frac{1}{z}\right) - 1 \tag{9.7.83}$$

(b) Using (9.7.1) for n in $(0, \infty)$ and $n!$ is $\Gamma(n+1)$, show that $\mu(z) \to 0$ as $z \to \infty$ along $\mathbb{R}$ on account of (9.7.1).

(c) For $z \in (0, \infty)$, prove that $\mu(z+n) - \eta(z+n) \to 0$ as $n \to \infty$.

(d) If $\eta(z) - \eta(z+1) = \mu(z) - \mu(z+1)$, conclude $\mu(z) - \eta(z) = \mu(z+n) - \eta(z+n) \to 0$, so $\mu(z) = \eta(z)$ for z real and positive.

(e) Use this observation to prove (9.7.39), and so (9.7.27).

19. This will prove Binet's formula, (9.7.40).

(a) For $x \in (1, \infty)$, prove that one has the convergent series

$$\mu(x) - \mu(x+1) = \sum_{n=1}^{\infty} (-1)^n \left(\frac{n-1}{(n+1)(2n)}\right) x^{-n} \tag{9.7.84}$$

(*Hint*: See (9.7.83).)

(b) The goal of the rest of this problem is to prove that η given for $\operatorname{Re} z > 0$ by

$$\eta(z) = \int_0^\infty F(t) e^{-zt}\, dt \tag{9.7.85}$$

$$F(t) = [(e^t - 1)^{-1} - t^{-1} + \tfrac{1}{2}]\, t^{-1} \tag{9.7.86}$$

has $\eta(z) = \mu(z)$. By analyticity, it suffices to prove that $\eta(x) = \mu(x)$ for $x \in (1, \infty)$. Prove that for all $t \in [0, \infty)$,

$$G(t) \equiv F(t)(1 - e^{-t})$$

$$G(t) = \sum_{n=1}^{\infty} (-1)^n \frac{n-1}{(n+1)(2n)} \frac{1}{(n-1)!} t^{n-1} \tag{9.7.87}$$

(c) For all $t \in [0,\infty)$ and $n = 0,1,2,\ldots$, prove that

$$\sum_{k=n}^{\infty} \frac{t^k}{k!} \le \frac{t^n}{n!} e^t \tag{9.7.88}$$

(*Hint*: Taylor's series with remainder.)

(d) Let $b_n = (-1)^n \frac{1}{2}(n-1)[(n+1)!]^{-1}$. For t real in $(0,\infty)$, prove that

$$\left| G(t) - \sum_{n=1}^{N} b_n t^{n-1} \right| \le \frac{1}{2} \frac{t^N}{N!} e^t \tag{9.7.89}$$

and conclude that for $x \in (1,\infty)$, we have

$$\int_0^\infty \left| G(t) - \sum_{n=1}^{N} b_n t^{n-1} \right| e^{-xt}\, dt \le \tfrac{1}{2}\,(x-1)^{-N-1} \tag{9.7.90}$$

(e) For $x \in (1,\infty)$, prove that

$$\eta(x) - \eta(x+1) = \sum_{n=1}^{\infty} (-1)^n \left(\frac{n-1}{(n+1)(2n)} \right) x^{-n} \tag{9.7.91}$$

with a convergent power series.

(f) Conclude that for $x \in (1,\infty)$, $\eta(x) = \mu(x)$, and so Binet's formula. (*Hint*: See Problem 18.)

Remarks. 1. Binet's formula is due to Binet [**51**] in 1838. For alternate proofs, see Sasvári [**501**] and Zheng et al. [**602**]. Jacques Philippe Marie Binet (1786–1856) is best known for his formula for the nth Fibonacci number

$$F_n = \big(\sqrt{5}\big)^{-1}(\omega_+^n - \omega_-^n), \qquad \omega_\pm = \frac{1 \pm \sqrt{5}}{2}$$

also known as Binet's formula. There is also another formula of Binet for $\mu(z)$, namely,

$$\mu(z) = 2 \int_0^\infty \frac{\arctan(\frac{t}{z})}{e^{2\pi t} - 1}\, dt \tag{9.7.92}$$

2. We return to Binet's formula to prove Theorem 9.7.2 from an analysis of the asymptotics of the integral in Problem 4 of Section 15.2 of Part 2B.

20. (a) With μ given by (9.7.39), prove that

$$\mu(z) - \mu(z+n+1) = \sum_{j=0}^{n} \left[\left(z + j + \frac{1}{2} \right) \log\left(1 + \frac{1}{z+j} \right) - 1 \right]$$

(b) For fixed z, prove that in $\mathbb{C} \setminus (-\infty, 0)$

$$\left| \left(z + j + \frac{1}{2} \right) \log\left(1 + \frac{1}{z+j} \right) - 1 \right| \le O\left(\frac{1}{j^2} \right)$$

(c) Prove Gudermann's series

$$\mu(z) = \sum_{j=0}^{\infty} \left[\left(z + j + \frac{1}{2} \right) \log\left(1 + \frac{1}{z+j} \right) - 1 \right]$$

converging uniformly on $\mathbb{C} \setminus (-\infty, 0]$.

9.8. Jensen's Formula

In the last three sections of this chapter, we study the issue of lots of zeros forcing the growth of a function. This section has a main tool; the next section, the consequences for functions on the disk; and in the final section, consequences for entire functions. In particular, we'll answer the following two questions:

(1) If $f \in \mathfrak{A}(\mathbb{D})$ is bounded, what can one say about the zeros of f?
(2) If f is entire and

$$|f(z)| \le C \exp(D(|z|^k)) \qquad (9.8.1)$$

what does that say about zeros?

In each case, we'll also have canonical forms for the products of the zeros. In the case of (9.8.1), we'll also analyze the nonproduct pieces in Section 9.10. In the disk, we'll only turn to the behavior of the nonproduct pieces in Chapter 5 of Part 3.

One can ask why one should expect the zeros to force growth—after all, f is small at zeros, so the intuition might be that they make f small, not large! One big hint is seen by polynomials. Their number of zeros determines their degree which determines the rate of growth at infinity. Another hint is that, for entire functions, more zeros means the need for higher-order Weierstrass factors and, of course, an individual $E_n(z/z_j)$ growth like $\exp(C|z|^n)$ near infinity. Of course, this growth might be shielded by other $E_n(z/z_k)$ or by an $e^{h(z)}$ term. One of our goals is to prove this can't happen. The argument principle gives us a further hint. The more zeros in a disk, the more the argument varies for $f(re^{i\theta})$ and the greater the cancellations in $\int f(re^{i\theta})\frac{d\theta}{2\pi}$. Since this integral is $f(0)$, $\sup_\theta|f(re^{i\theta})|$ will need to grow to overcome the cancellation.

We'll need to remove the zeros without changing the size of $|f(re^{i\theta})|$ for some fixed r. Focus first on $r = 1$. We'd clearly like a function g with prescribed zeros and $|g(e^{i\theta})| = 1$. We studied such functions in Problem 5 of Section 3.6. We'll modify the building blocks used there by changing the phase.

Definition. Let $w \in \mathbb{D}$. The *Blaschke factor*, $b(z,w)$, is defined as

$$b(z,w) = \begin{cases} z & \text{if } w = 0 \\ \frac{|w|}{w}\,\frac{w-z}{1-\bar{w}z}, & \text{otherwise} \end{cases} \tag{9.8.2}$$

Remarks. 1. $b(\,\cdot\,,w)$ can be viewed either as an analytic function on $\mathbb{D}$, a meromorphic one on $\mathbb{C}$, or as a regular function on $\overline{\mathbb{D}}$.

2. The extra phase $|w|/w$ makes $b(\,\cdot\,,w)$ discontinuous in w at $w=0$ (e.g., $\lim_{r\downarrow 0} b(\frac{1}{2}, re^{i\theta}) = -\frac{1}{2}e^{-i\theta}$). This may seem like a strange choice, but to minimize $|1-b(z,w)|$ for product considerations, it will be essential to make $b(0,w)>0$ for $w\neq 0$, which is what the $|w|/w$ factor does. In most of this section, we could dispense with $|w|/w$, but since we need it late in this and in the next section, we put it in now.

Notice that since

$$\left|\frac{w-e^{i\theta}}{1-\bar{w}e^{i\theta}}\right| = \left|\frac{w-e^{i\theta}}{\bar{w}-e^{-i\theta}}\right| = \left|\frac{w-e^{i\theta}}{\overline{w-e^{i\theta}}}\right| = 1$$

we have

$$|b(e^{i\theta},w)| = 1 \tag{9.8.3}$$

Here's one main tool we'll need in the remainder of this chapter:

Theorem 9.8.1 (Jensen's Formula). *Fix $R>0$. Let f be analytic in a neighborhood of $\overline{\mathbb{D}_R(0)}$ and not identically zero. Then f has finitely many zeros, $\{z_j\}_{j=1}^N$, counting multiplicity, in $\mathbb{D}_R(0)$. If $f(0)\neq 0$, then*

$$\int_0^{2\pi} \log|f(Re^{i\theta})|\,\frac{d\theta}{2\pi} = \log|f(0)| + \sum_{j=1}^N \log\left(\frac{R}{|z_j|}\right) \tag{9.8.4}$$

Remarks. 1. There is a variant in case $f(0)=0$. If $f(z) = \eta(f)z^{n_0} + O(z^{n_0+1})$ for $\eta(f)\neq 0$, by applying (9.8.4) to $R^{n_0}f(z)/z^{n_0}$, we get an analog of (9.8.4) with $\log|f(0)|$ replaced by

$$\log|\eta(f)| + n_0\log(R) \tag{9.8.5}$$

2. In particular, if f is a polynomial of degree N, this implies the integral is $O(\log|R|^N)$ as $R\to\infty$. Indeed, even the constant $\log(|f(0)|/\prod_{j=1}^N |z_j|)$ is right.

3. Problem 3 provides an alternate proof.

Proof. Analyticity implies finiteness of the number of zeros. By applying the argument below to $g(z) = f(z/R)$. we lose nothing by supposing $R=1$, which we do. We'll also suppose that f is nonvanishing on $\partial\mathbb{D}$. For if we prove the result for $f_r(z) = f(rz)$ with $r\in(1-\varepsilon,1)$, we can get the result for f by taking a limit as $r\uparrow 1$. Here we use the fact that even if f vanishes at some finite number of points on $\partial\mathbb{D}$, $\int_0^{2\pi} \log|f(re^{i\theta})| \frac{d\theta}{2\pi}$ is continuous as $r \uparrow 1$ (Problem 1).

Let

$$g(z) = \frac{f(z)}{\prod_{j=1}^N b(z, z_j)} \tag{9.8.6}$$

so

$$g(0) = \frac{f(0)}{\prod_{j=1}^N |z_j|} \tag{9.8.7}$$

g is analytic and nonvanishing in a neighborhood of $\overline{\mathbb{D}}$. Thus, by Theorem 2.6.1, $\log(g(z))$ is analytic in a neighborhood of $\overline{\mathbb{D}}$ and the Cauchy formula and $\log(|a|) = \operatorname{Re}\log(a)$ imply

$$\log|g(0)| = \int \log|g(e^{i\theta})| \frac{d\theta}{2\pi} \tag{9.8.8}$$

By (9.8.3), $|f(e^{i\theta})| = |g(e^{i\theta})|$, and then by (9.8.7), we have that (9.8.8) is (9.8.4). □

Of course, we not only have (9.8.8), we have the full-blown, complex Poisson formula of (5.3.2). If $f(z) \neq 0$, we define g by (9.8.6). Then (5.3.2) says

$$\log(g(z)) = i\psi + \int \frac{e^{i\theta} + z}{e^{i\theta} - z} \log(|g(e^{i\theta})|) \frac{d\theta}{2\pi} \tag{9.8.9}$$

where $\psi = \operatorname{Im}\log(g(0)) = \operatorname{Im}\log(f(0)) = \arg(f(0))$, since $\operatorname{Im}\log(\prod_{j=1}^N |z_j|) = 0$. Exponentiating and scaling to $\mathbb{D}_R(0)$, we get

Theorem 9.8.2 (Poisson–Jensen Formula). *Fix $R > 0$. Let f be analytic in a neighborhood of $\overline{\mathbb{D}_R(0)}$ and not identically zero. Then f has finitely many zeros, $\{z_j\}_{j=1}^N$, in $\mathbb{D}_R(0)$. If $f(0) \neq 0$, then for $z \in \mathbb{D}_R(0)$,*

$$f(z) = \frac{f(0)}{|f(0)|} \prod_{j=1}^N b\Big(\frac{z}{R}, \frac{z_j}{R}\Big) \exp\bigg(\int \frac{Re^{i\theta} + z}{Re^{i\theta} - z} \log|(f(Re^{i\theta}))| \frac{d\theta}{2\pi}\bigg) \tag{9.8.10}$$

If $f(0) = 0$, let $z_1 = z_2 = \cdots = z_k = 0$ and $z_j \neq 0$ for $j = k+1, \dots, N$. Then (9.8.10) holds if $\prod_{j=1}^N b(z/R, z_j/R)$ is replaced by $z^k \prod_{j=k+1}^N b(z/R, z_j/R)$ and if $f(0)/|f(0)|$ is replaced by $f^{(k)}(0)/|f^{(k)}(0)|$.

Proof. Given that $e^{i\psi} = f(0)/|f(0)|$, this is immediate from (9.8.9). If $f(0) = 0$, the "corrected" formula is just (9.8.10) for the function $h(z) = f(z)/z^k$. □

Notes and Historical Remarks. Jensen's formula is normally attributed to Jensen [**284**] in 1899, although it was stated and proven for polynomials already in 1827 by Jacobi [**282**]. As noted by Landau [**342**], Jacobi's proof works for general analytic functions. The Poisson–Jensen formula, including its name, is due to Nevanlinna [**408**] who exploited it in his value distribution theory, which we discuss in Chapter 17 of Part 2B.

Problems

1. (a) Prove that for all $e^{i\theta} \in \partial\mathbb{D}$ and $r \in [0,1]$,

$$\tfrac{1}{2}\,|1 - re^{i\theta}| \geq \tfrac{1}{4}\,|1 - e^{i\theta}| \tag{9.8.11}$$

(b) Let $g_r(\theta) = \log(\frac{1}{2}|1-re^{i\theta}|)$. Use the dominated convergence theorem to prove that

$$\lim_{r\uparrow 1} \int g_r(\theta)\,\frac{d\theta}{2\pi} = \int g(\theta)\,\frac{d\theta}{2\pi} \tag{9.8.12}$$

(c) For any f analytic in a neighborhood of $\overline{\mathbb{D}}$ and not identically zero, prove that

$$\lim_{r\uparrow 1} \int \log|f(re^{i\theta})|\,\frac{d\theta}{2\pi} = \int \log|f(e^{i\theta})|\,\frac{d\theta}{2\pi} \tag{9.8.13}$$

2. Find a version of Jensen's formula, including the variant indicated in (9.8.5), that allows poles in $\overline{\mathbb{D}_R(0)}$ also.

3. This will provide an alternate proof of Jensen's formula. Suppose $f(0) \neq 0$.
(a) Let $n(r)$ be the number of zeros in $\mathbb{D}_r(0)$ of a function $f(z)$ analytic in a neighborhood of $\overline{\mathbb{D}_r(0)}$. If f has no zeros on $\partial\mathbb{D}_{r_0}(0)$, prove that

$$n(r_0) = r\,\frac{d}{dr}\left[\int_0^{2\pi} \log|f(re^{i\theta})|\,\frac{d\theta}{2\pi}\right]\Bigg|_{r=r_0} \tag{9.8.14}$$

(*Hint*: Argument principle and (2.1.24).)
(b) If $\{z_j\}_{j=1}^N$ are the zeros of f in $\mathbb{D}_{r_0}(z)$, prove that

$$\sum_{j=1}^N \log\left(\frac{r_0}{|z_j|}\right) = \int_0^{r_0} \frac{n(r)}{r}\,dr \tag{9.8.15}$$

(c) Prove Jensen's formula, (9.8.4).

9.9. Blaschke Products

In this section, we study analytic functions on $\mathbb{D}$, especially uniformly bounded ones. We'll prove the following:

Theorem 9.9.1. *Let $f \in H^\infty(\mathbb{D})$, the analytic functions on $\mathbb{D}$ with $\sup_{z\in\mathbb{D}}|f(z)| < \infty$. Let $\{z_n\}_{n=1}^\infty$ be the zeros of f counting multiplicity. Then*

$$\sum_{n=1}^\infty (1-|z_n|) < \infty \tag{9.9.1}$$

Conversely, let $\{z_n\}_{n=1}^\infty$ obey (9.9.1). *Then there is a function $f \in H^\infty(\mathbb{D})$ where zeros are precisely $\{z_n\}_{n=1}^\infty$.*

(9.9.1) is called a *Blaschke condition.* We'll actually prove that (9.9.1) holds under a lot weaker conditions than $f \in H^\infty(\mathbb{D})$. Note that Jensen's formula shows if $k \ge 0$ is chosen so that $(f(z)/z^k)$ is nonvanishing at 0, then

$$\int_0^{2\pi} \log|f(re^{i\theta})|\,\frac{d\theta}{2\pi} \ge \log(r^k) + \log\biggl(\biggl[\frac{|f(z)|}{|z|^k}\biggr]\bigg|_{z=0}\biggr) \tag{9.9.2}$$

Define

$$\log_\pm(x) = \max(\pm\log(x), 0) \tag{9.9.3}$$

Definition. A function f on $\mathbb{D}$ is said to lie in the *Nevanlinna class, N,* if and only if f is analytic and

$$\sup_{0\le r\le 1} \int \log_+|f(re^{i\theta})|\,\frac{d\theta}{2\pi} < \infty \tag{9.9.4}$$

Remark. It can be shown (see Theorem 5.1.2 of Part 3) that the integral in (9.9.4) is monotone in r.

By (9.9.2), if (9.9.4) holds, we have a finiteness in r also for $\log_-$ or $|\log|f(\cdot)||$. Here is half of Theorem 9.9.1:

Theorem 9.9.2. *Let $f \in N$. Then its zeros obey* (9.9.1).

Proof. Since we can replace $f(z)$ by $f(z)/z^k$, we can suppose $f(0) \ne 0$. Then, by Jensen's inequality, for any $r < 1$ and any k,

$$\begin{aligned}\sum_{\substack{|z_j|\le r\\ j\le k}} \log\biggl(\frac{r}{|z_j|}\biggr) + \log|f(0)| &\le \int_0^{2\pi} \log|f(re^{i\theta})|\,\frac{d\theta}{2\pi} \\ &\le \int_0^{2\pi} \log_+|f(re^{i\theta})|\,\frac{d\theta}{2\pi}\end{aligned} \tag{9.9.5}$$

So taking $r \uparrow 1$,

$$\prod_{j=1}^k |z_j| \ge |f(0)|\exp\biggl(-\sup_{0\le r\le 1}\int \log_+|f(re^{i\theta})|\,\frac{d\theta}{2\pi}\biggr)$$

Thus, $\prod_{j=1}^\infty |z_j| > 0$, which implies (9.9.1) (Problem 1). ☐

For the opposite direction, we need a preliminary:

Lemma 9.9.3. *Let $b(z,w)$ be given by (9.8.2). For any $z,w \in \mathbb{D}$, we have*

$$|1 - b(z,w)| \le \frac{(1-|w|)(1+|z|)}{1-|z|} \tag{9.9.6}$$

Proof. We note that

$$b(e^{i\theta}z, e^{i\theta}w) = b(z,w) \tag{9.9.7}$$

so, without loss, we can suppose $w \in (0,1)$, in which case,

$$1 - b(z,w) = 1 - \frac{w-z}{1-wz} = \frac{(1-w)(1+z)}{1-wz} \tag{9.9.8}$$

from which (9.9.6) is immediate. □

Theorem 9.9.4. *Let $\{z_j\}_{j=1}^\infty$ be a sequence in $\mathbb{D}$. Then $\lim_{N\to\infty} \prod_{j=1}^N b(z,z_j)$ converges uniformly on compacts. If (9.9.1) holds, the product is an absolutely convergent product and the limit*

$$B_\infty(z, \{z_j\}_{j=1}^\infty) = \prod_{j=1}^\infty b(z,z_j) \tag{9.9.9}$$

called a Blaschke product, *vanishes exactly at $\{z_j\}_{j=1}^\infty$ (counting multiplicity). If (9.9.1) fails, the limit is identically zero.*

Proof. Suppose first that (9.9.1) holds. Then, by (9.9.6),

$$\sum_{j=1}^\infty |1 - b(z,z_j)| \le \frac{1+|z|}{1-|z|} \sum_{j=1}^\infty (1-|z_j|) \tag{9.9.10}$$

converges uniformly on each $\{z \mid |z| \le r < 1\}$ so that the product is absolutely and uniformly convergent and has the requisite zeros by Theorem 9.1.3.

If (9.9.1) fails, then $\{\prod_{j=1}^N b(\cdot\,, z_j)\}$ is uniformly bounded on $\mathbb{D}$. So, by Montel's theorem, it suffices to show that any limit point, call if F, is 0.

Suppose first that infinitely many z_j are zero. Then for any m, eventually $|\prod_{j=1}^N b(z,z_j)| \le |z|^m$, so for all m, $|F(z)| \le |z|^m$, and thus, $F \equiv 0$.

If there are infinitely many nonzero z_j's converging to 0, since $F(z_j) = 0$, zero is not an isolated zero, so again $F \equiv 0$.

If zero is not a limit point of the z_j's and no $z_j = 0$, then $\prod_{j=1}^N b(z,z_j)$ has no zero near $z = 0$. So, by Hurwitz's theorem, either $F(0) \neq 0$ or $F \equiv 0$. Since (9.9.1) fails, $\prod_{j=1}^N b(0,z_j) = \prod_{j=1}^N |z_j| \to 0$ by Problem 1. Thus, $F \equiv 0$.

If finitely many z_j's are zero, say $\{z_{j_k}\}_{k=1}^\ell$, and zero is not a limit point of $\{z_j\}$, then for $N \ge \max(j_k)$, $\prod_{j=1}^N b(z,z_j) = z^\ell \prod_{1\le j\le N, j\neq j_k} b(z,z_j)$ goes to zero by the prior case. Thus, in all cases, $F \equiv 0$, so the limit is zero. □

We'll study B_∞ extensively in Chapter 5 of Part 3. In particular, B_∞ has boundary values $\lim_{r\uparrow 1} B_\infty(re^{i\theta})$ for Lebesgue a.e. θ and these boundary values have absolute value 1.

Proof of Theorem 9.9.1. By Theorem 9.9.2, if $f \in H^\infty(\mathbb{D})$, then (9.9.1) holds.

Conversely, if (9.9.1) holds, then B_∞ given by (9.9.9) is in $H^\infty(\mathbb{D})$ (since $|\prod_{j=1}^N b(z,z_j)| \le 1$) and has the requisite zeros. □

Here is an interesting application of Theorem 9.9.1. Recall that the Weierstrass approximation theorem (see Theorem 2.4.1 of Part 1) says that the span of $\{x^n\}_{n=1}^\infty$ is dense in $C([0,1])$. But since any continuous function of x is also a continuous function of x^2, so is $\{x^{2n}\}_{n=0}^\infty$, so lots of sets of powers span $C([0,1])$. Here is a result that follows from Theorem 9.9.1 (the direction that follows from Jensen's formula):

Theorem 9.9.5 (Müntz–Szász Theorem). *Let $t_0 \equiv 0 < t_1 < \cdots < t_n < \cdots$ be a sequence of reals so that*

$$\sum_{j=1}^\infty t_j^{-1} = \infty \tag{9.9.11}$$

Then $\{1, x^{t_1}, x^{t_2}, \dots\}$ are a total subset of $C([0,1])$.

Remarks. 1. By total, we mean finite linear combinations are dense.

2. The converse is also true (and in the original papers; see, e.g., Rudin [**496**, Sect. 15.25]). See also Problem 8.

3. We emphasize the t_j need not be integers.

4. We'll need to know some facts from real analysis; namely, that if $\{f_j\}_{j=1}^\infty$ do not span $C([0,1])$, then there is a linear functional on $L \in C([0,1])$ with $L(f_j) = 0$ but with L not identically zero (Hahn–Banach theorem, see Theorem 5.5.5 of Part 1). We also need to know that the linear functions are of the form $L(f) = \int_0^1 f(x)\,d\mu(x)$ for a signed measure on $[0,1]$ (Riesz–Markov theorem; see Theorem 4.8.8 of Part 1).

Proof. Let μ be a signed measure on $[0,1]$. Define for $\operatorname{Re} w > 0$,

$$F(w) = \int_0^1 e^{w\log(x)}\,d\mu(x) \tag{9.9.12}$$

which, by Theorem 3.1.6, is analytic in $\{w \mid \operatorname{Re} w > 0\}$.

Let $T\colon \mathbb{D} \to \{w \mid \operatorname{Re} w > 0\}$,

$$T(z) = \frac{1-z}{1+z} \tag{9.9.13}$$

with inverse map

$$T^{-1}(w) = \frac{1-w}{1+w} \tag{9.9.14}$$

Then since F is bounded by $\|\mu\|$,

$$g(z) = F(T(z)) \tag{9.9.15}$$

lies in $H^\infty(\mathbb{D})$.

Let

$$z_j = T^{-1}(t_j) \tag{9.9.16}$$

then

$$1 - |z_j| = \frac{2\min(1,t_j)}{1+t_j} \geq \frac{2\min(1,t_1)}{t_1^{-1}t_j + t_j} = \frac{2t_1\min(1,t_1)}{1+t_1}\, t_j^{-1} \tag{9.9.17}$$

so (9.9.11) implies $\sum_{j=1}^\infty 1 - |z_j| = \infty$. By Theorem 9.9.1, $g \equiv 0$, so $F \equiv 0$, so $F(n) = 0$ for $n = \{1, 2, \dots\}$.

Thus, given (9.9.11),

$$\int x^{t_j}\, d\mu(x) = 0, \text{ all } j \text{ including } t_j = 0 \Rightarrow \int x^n\, d\mu(x) = 0, \quad n = 0, 1, \dots \tag{9.9.18}$$

$$\Rightarrow \int f\, d\mu = 0$$

for all f since the Weierstrass theorem says $\{x^n\}_{n=0}^\infty$ are total. Thus, by the Hahn–Banach theorem, $\{x^{t_j}\}_{j=0}^\infty$ span $C([0,1])$. □

Notes and Historical Remarks. The argument to go from Jensen's inequality to bounds like (9.9.1) is due to Carathéodory and Fejér [**91**]. Blaschke products were constructed by Wilhelm Blaschke (1885–1966) in 1915 [**53**]. With this exception, almost all of Blaschke's work was in geometry. He was Austrian having been raised in Graz where his father was a mathematics professor but Wilhelm spent the bulk of his career in Hamburg. He is best known for his 1916 book, *Kreis und Kugel* (circle and ball) on isoperimetric problems. During the earlier years of the Hitler regime, he defended Jewish mathematicians, most notably Reidemeister, but he had a change of heart in 1936.

While Blaschke was not as bad as Bieberbach or Teichmüller, he was on the side of oppressors in the Nazi era, becoming a party member in the late 1930s. The story is told of Szegő, who had a reputation of a gentle soul, teaching complex analysis at Stanford after the war using the term "products," and when asked by a student why he didn't call them "Blaschke products" like others, replying: "I will not speak that man's name."

The Müntz–Szász theorem is named after their discovery of it in 1914–16 [**398, 548**].

The H in H^∞ is for Hardy. Hardy spaces, that is, those functions analytic on $\mathbb{D}$ with

$$\sup_{0<r<1} \int |f(re^{i\theta})|^p \frac{d\theta}{2\pi} < \infty$$

and the Nevanlinna class will be heavily studied in Chapter 5 of Part 3.

Blaschke products for matrix-valued functions are subtle because of commutativity issues. They are due to Potapov [**457**]; see Kozhan [**329**] for a modern exposition.

Problems

1. Let $0 < x_j < 1$. Prove that

$$\prod_{j=1}^\infty x_j > 0 \Leftrightarrow \sum_{j=1}^\infty (-\log|x_j|) < \infty \Leftrightarrow \sum_{j=1}^\infty (1 - |x_j|) < \infty$$

2. Let f be bounded and analytic in the right half-plane with $f(n) = 0$ for $n = 1, 2, \dots$. Prove that $f = 0$.

3. Let $\{x_j\}_{j=1}^\infty \subset (0,1)$, $\{\omega_j\}_{j=1}^\infty \subset \partial\mathbb{D}$.

   (a) Prove that

$$b(z, (1-x)\omega) = \frac{E_0(\frac{x}{1-\omega^{-1}z})}{E_0(\frac{-x\omega^{-1}z}{1-\omega^{-1}z})}$$

   (b) Prove that if $\sum_{n=1}^\infty |x_n|^{m+1} < \infty$, then if

$$b_n(z, (1-x)\omega) = \frac{E_n(\frac{x}{1-\omega^{-1}z})}{E_n(\frac{-x\omega^{-1}z}{1-\omega^{-1}z})}$$

   we have that $\prod_{j=1}^\infty b_m(z, (1-x_j)\omega_j)$ converges and has zeros precisely at $\{(1-x_n)\omega_n\}$.

   (c) How do b_n and b differ for $n > 1$? (*Hint*: Look at $\lim_{r\uparrow 1}|b_n(r\omega, (1-x)\omega)|$. Is $\prod b_m$ bounded on $\mathbb{D}$?)

   Remark. In (c), you are not being asked to prove something for arbitrary choices of ω and x, but for suitable examples.

4. Let $\{f_n\}_{n=1}^\infty$ be a sequence of analytic functions on $\mathbb{D}$ with $\sup_{n,z\in\mathbb{D}}|f_n(z)| < \infty$. Let $\{z_k\}_{k=1}^\infty$ be a sequence of distinct points in $\mathbb{D}$ with $\sum_{k=1}^\infty (1 - |z_k|) < \infty$. Suppose for each k, $f_n(z_k)$ exists. Prove that for some function, $f \in H_\infty$, $f_n \to f$ uniformly on each $\mathbb{D}_r(0)$ with $r < 1$.

5. (a) Prove for $w \in \mathbb{D}$, $z \in \mathbb{C}^\times$,

$$|1 - b(z,w)| \le \frac{(1-|w|)(1+|z|)}{|z|\operatorname{dist}(z^{-1},\bar{w})}$$

(b) Suppose $\{z_j\}_{j=1}^\infty \subset \mathbb{D}$ obeys a Blaschke condition (9.9.1), and for some interval $I \subset \partial\mathbb{D}$, $\inf_{z\in I,\, j=1,2,\dots}|z_j - z| > 0$. Prove $B_\infty(z)$ has an analytic continuation across I and defines a function meromorphic on $\mathbb{C}\setminus(\partial\mathbb{D}\setminus I)$ with poles at $\{\bar{z}_j^{-1}\}_{j=1}^\infty$.

6. In this problem and the next, we discuss results not directly related to Blaschke products but needed for an alternate approach to the Müntz–Szász theorem. This problem will prove the *Cauchy determinant formula* [**106**] that for $\{a_1,\dots,a_n\}$ and $\{b_1,\dots,b_n\}$, $2n$ complex numbers so that for all i,j : $a_i \ne a_j$, $b_i \ne b_j$, $a_i + b_j \ne 0$ (where the first two require $i \ne j$), we have

$$\det\left(\frac{1}{a_i+b_j}\right)_{1\le i,j\le n} = \frac{\prod_{1\le i<j\le n}(a_i-a_j)(b_i-b_j)}{\prod_{i,j=1,\dots,n}(a_i+b_j)} \tag{9.9.19}$$

(a) Fix $\{a_j\}_{j=1}^n$, $\{b_j\}_{j=1}^{n-1}$ and replace b_n by z and let $R_n(z;a_1,\dots,a_n,b_1,\dots,b_{n-1})$ be the right side of (9.9.19), Prove that $R_n(z) = O(\frac{1}{z})$ at ∞ and that if $\rho_1,\dots,\rho_n$ are the residues of the poles at $z = -a_1,\dots,-a_n$, then

$$R_n(z) = \sum_{j=1}^n \frac{\rho_j}{a_j+z} \tag{9.9.20}$$

(b) Prove that

$$\rho_n = R_{n-1}(b_{n-1};a_1,\dots,a_{n-1};b_1,\dots,b_{n-2}) \tag{9.9.21}$$

(c) Prove (9.9.19) inductively.

7. Let $x_1,\dots,x_n \in \mathcal{H}$, a Hilbert space, be linearly independent. Their *gramian* (determinant) is

$$G(x_1,\dots,x_n) = \det\big(\langle x_i,x_j\rangle_{1\le i,j\le n}\big)$$

Given y_1,y_2, let $\widetilde{G}(x_1,\dots,x_n;y_1,y_2)$ be the $(n+1)\times(n+1)$ determinant with $\langle y_1,x_j\rangle$ plus $\langle y_1,y_2\rangle$ added as an extra row on the bottom and $\langle x_j,y_2\rangle$ plus $\langle y_1,y_2\rangle$ as an extra column on the right.

(a) Prove that for $x_1,\dots,x_n$ fixed, $\widetilde{G}$ is a sesquilinear form in y_1,y_2. Define

$$Q(y_1,y_2) = \widetilde{G}(x_1,\dots,x_n;y_1,y_2)/G(x_1,\dots,x_n) \tag{9.9.22}$$

(b) Prove $Q(y_1, y_2) = 0$ if y_1 and/or y_2 are linear combinations of $x_1, \dots, x_n$.

(c) Let P be the projection onto the span of $x_1, \dots, x_n$. Prove that

$$\|(1-P)y\|^2 = Q(y,y) \tag{9.9.23}$$

8. This problem will prove an L^2-version of the Müntz-Szász theorem (Note that the requirement that $t_0 = 0$ is missing in this L^2 version!). Fix real numbers $-1 < s_1 < s_2 < \dots$ with $s_n \to \infty$. Let Q_n be the $L^2([0,1], dx)$ projection onto the orthogonal of the complement span of $\{x^{s_j}\}_{j=1}^n$.

(a) Let $m \in \mathbb{N}$, $m \neq s_j$ for any s_j. Prove that

$$\langle x^m, x^{s_j} \rangle = (m+1+s_j)^{-1} \tag{9.9.24}$$

(b) Prove that

$$Q_n(x^m) = (2m+1)^{-1} \prod_{j=1}^n \left(1 - \frac{2m+1}{m+s_j+1}\right)^2 \tag{9.9.25}$$

(c) Prove that for all $m \notin \{s_j\}_{j=1}^\infty$, x^m is in the L^2-closure of the span of the x^{s_j} if and only if

$$\sum_{s_j > 0} \frac{1}{s_j} = \infty \tag{9.9.26}$$

(d) Prove that $\{x^{s_j}\}_{j=1}^\infty$ is complete in $L^2([-1,1], dx)$ if and only if (9.9.26) holds.

(e) Prove a fortiori that if (9.9.26) fails, then $\{x^{s_j}\}_{j=1}^\infty$ is not $L^\infty([0,1], dx)$ complete.

9. This will provide an alternate proof of the rest of the Müntz-Szász theorem.

(a) Let $\{f_j\}_{j=1}^\infty$ be a family of C^1 functions on $(0,1)$, so that $g_j \equiv f_j'$ is in $L^2([0,1], dx)$ and $\{g_j\}_{j=1}^\infty$ are total in $L^2([0,1], dx)$. Prove that $\{\mathbb{1}\} \cup \{f_j\}_{j=1}^\infty$ are total in $C([0,1])$. (*Hint*: $h(x) = h(0)\,\mathbb{1} + \int_0^x h'(y)\,dy$.)

(b) Prove if $0 = t_0 < t_1 < \dots$ and $t_n \to \infty$, then $\{x^{t_j}\}_{j=0}^\infty$ are total in $C([0,1])$.

Remark. Problem 8 is essentially Müntz's original proof [**398**]. He had a more involved argument to go from L^2 to $C([0,1])$ spanning sets. Problem 9 is Szász' method to get to $C([0,1])$ [**548**].

9.10. Entire Functions of Finite Order and the Hadamard Product Formula

An entire function, f, is said to have *finite order* if for constants C, D and some positive real α,

$$|f(z)| \le C\exp(D|z|^\alpha) \tag{9.10.1}$$

Our main goal in this section is to prove a representation for such functions of the form

$$f(z) = e^{P(z)} z^\ell \prod_{j=1}^{\infty} E_m\left(\frac{z}{z_j}\right) \tag{9.10.2}$$

with a fixed m and P a polynomial. We want to relate m and $\deg(P)$ to the optimal α in (9.10.1). We begin by defining this optimal α.

Definition. Let f be a nonconstant entire function. The *order* of f, $\rho(f)$, is defined by

$$\rho(f) = \limsup_{R\to\infty} \sup_{\theta\in[0,2\pi)} \frac{\log(\log|f(Re^{i\theta})|)}{\log(R)} \tag{9.10.3}$$

If $\rho(f)$ is finite, we say f has *finite order.*

It is not hard to see that (9.10.1) holds for any $\alpha > \rho(f)$ and fails if $\alpha < \rho(f)$. It might or might not hold if $\alpha = \rho(f)$. It is also easy to see that one can replace $\sup|f(Re^{i\theta})|$ by averages (see Problem 1). Since f is not constant, $\lim_{R\to\infty}\sup_\theta|f(Re^{i\theta})| = \infty$, so $0 \le \rho(f) \le \infty$. If $0 < \rho(f) < \infty$, we define

Definition. If $0 < \rho(f) < \infty$, the *type* of f is defined by

$$\tau(f) = \limsup_{r\to\infty} \sup_{\theta\in[0,2\pi)} R^{-\rho(f)} \log|f(Re^{i\theta})| \tag{9.10.4}$$

Notice $0 \le \tau(f) \le \infty$. If $\tau(f) = 0$, we say f has *minimal type*, if $\tau(f) \in (0,\infty)$, of *normal type*, and if $\tau = \infty$, of *maximal type.*

Except for some results in the Problems, we'll say nothing more about type, except to note that $\sin(az)$ has order 1 and type a, that by Stirling's formula, $\Gamma(z)^{-1}$ has order 1 and maximal type (see Problem 13), and that Problem 14 has an "explicit" function of order 1 and minimal type. The problems (see Problems 2 and 3) have criteria for the order and type in terms of the Taylor coefficients; for example, if $f(w)$ is given by (2.3.1), then

$$\rho(f) = \limsup_{n\to\infty} \frac{n\log(n)}{\log(|a_n|^{-1})} \tag{9.10.5}$$

and, in particular,

$$F(z) = \sum_{n=0}^{\infty} \frac{z^n}{(n!)^{1/\rho}} \tag{9.10.6}$$

has order ρ.

For a representation like (9.10.2) to converge, we need the zeros to obey $\sum_{j=1}^\infty |z_j|^{-m-1} < \infty$. We thus define

Definition. The *exponent of convergence* of the zeros of f is defined to be

$$\sigma = \inf\left\{\alpha > 0 \,\middle|\, \sum_{j=1}^\infty |z_j|^{-\alpha} < \infty\right\} \tag{9.10.7}$$

where $\{z_j\}_{j=1}^\infty$ are the zeros of f other than 0, counting multiplicity.

We are heading towards a proof that

$$\sigma(f) \le \rho(f) \tag{9.10.8}$$

a compact way of expressing quantitatively that zeros force growth. It will help to have two enumerative functions for the zeros:

$$n(r;f) = \#\text{ of zeros of } f \text{ in } \overline{\mathbb{D}_r(0)} \tag{9.10.9}$$

$$N(r;f) = \int_0^r \frac{n(s)-n(0)}{s}\,ds \tag{9.10.10}$$

We note that if $r_0 > 0$ is picked so f has no zeros in $\overline{\mathbb{D}_{r_0}(0)} \setminus \{0\}$, then

$$\sum_{\{j \mid |z_j| \le R\}} |z_j|^{-\alpha} = \int_{r_0}^R r^{-\alpha}\,dn(r) \tag{9.10.11}$$

since dn is a point measure with pure points at $|z_j|$ and mass equal to the number of zeros with $|z_j| = r$. If $f(0) \ne 0$, we can take $r_0 = 0$. The following is elementary and left to Problem 4:

Proposition 9.10.1. *The following are equivalent for each $\alpha > 0$:*

(1) $$\sum_{j=1}^\infty |z_j|^{-\alpha} < \infty \tag{9.10.12}$$

(2) $$\int_{r_0}^\infty r^{-1-\alpha} n(r)\,dr < \infty \tag{9.10.13}$$

(3) $$\int_{r_0}^\infty r^{-1-\alpha} N(r)\,dr < \infty \tag{9.10.14}$$

We next rewrite Jensen's formula:

Proposition 9.10.2. *Let f be an entire function with $f(0) \ne 0$. Then for any $r > 0$,*

$$\frac{1}{2\pi}\int_0^{2\pi} \log|f(re^{i\theta})|\,d\theta = \log|f(0)| + N(r;f) \tag{9.10.15}$$

Proof. Clearly,

$$\log\left(\frac{r}{|z_j|}\right) = \int_{|z_j|}^{r} \frac{ds}{s} \tag{9.10.16}$$

So, given the definition of n and N,

$$N(r;f) = \sum_{|z_j|<r} \log\left(\frac{r}{|z_j|}\right) \tag{9.10.17}$$

and (9.10.15) is (9.8.4). □

Theorem 9.10.3. *For any entire function,*

$$\sigma(f) \le \rho(f) \tag{9.10.18}$$

In particular, if $m = [\rho(f)]$, the integral part of $\rho(f)$, and $\{z_j\}_{j=1}^{\infty}$ are the zeros of f away from $z=0$,

$$\prod_{j=1}^{\infty} E_m\left(\frac{z}{z_j}\right) \tag{9.10.19}$$

is an absolutely convergent product.

Proof. Suppose first that $f(0) \neq 0$. For $\varepsilon > 0$, $|f(z)| \le C\exp(D|z|^{\rho+\frac{1}{2}\varepsilon})$, so

$$\text{LHS of (9.10.15)} \le \log(C) + Dr^{\rho+\frac{1}{2}\varepsilon}$$

Thus, by (9.10.15),

$$\int_0^\infty N(r;f)r^{-1-\rho-\varepsilon}\,dr < \infty \tag{9.10.20}$$

By Proposition 9.10.1,

$$\sum_{j=1}^{\infty} |z_j|^{-\rho-\varepsilon} < \infty \tag{9.10.21}$$

which implies (9.10.18).

If $f(0)=0$, pick ℓ so $g(z) = f(z)/z^\ell$ is nonvanishing and finite at $z=0$ and note that $\rho(g) = \rho(f)$ and $\sigma(g) = \sigma(f)$.

Since $m+1 > \rho(f) \ge \sigma(f)$ if $m = [\rho(f)]$, we have that

$$\sum_{j=1}^{\infty} |z_j|^{-m-1} < \infty \tag{9.10.22}$$

which, by (9.4.3), implies

$$\sup_{|z|\le r} \sum_{j=1}^{\infty} \left|1 - E_m\left(\frac{z}{z_j}\right)\right| < \infty \tag{9.10.23}$$

and yields the convergence of the product. □

We need the following elementary fact left to Problem 5:

Proposition 9.10.4. *Let f, g be two entire functions of finite order. Then fg has finite order and*

$$\rho(fg) \le \max(\rho(f), \rho(g)) \tag{9.10.24}$$

Remark. Using the Hadamard product formula, one can prove equality holds in (9.10.24) if $\rho(f) \ne \rho(g)$; see Problem 5.

Given a sequence $\{z_k\}_{k=1}^\infty$ in $\mathbb{C}^\times$ with $|z_k| \to \infty$ and exponent of convergence σ, if m is an integer with $m+1 > \sigma$, or both $m+1 = \sigma$ and

$$\sum_{j=1}^\infty |z_j|^{-\sigma} < \infty \tag{9.10.25}$$

the product

$$\prod_{j=1}^\infty E_m\left(\frac{z}{z_j}\right) \tag{9.10.26}$$

converges. If we pick m to be the smallest integer for which $\sum_{j=1}^\infty |z_j|^{-m-1} < \infty$, then (9.10.26) is called a *canonical product* and m is called its *genus*.

Lemma 9.10.5. *Let $m \ge 1$ and define*

$$C_m = 1 + \sum_{j=1}^m \frac{1}{j} \tag{9.10.27}$$

Then, for $|z| \ge 1$,

$$|E_m(z)| \le \exp(C_m |z|^m) \tag{9.10.28}$$

Proof. Since $|1-z| \le 1 + |z| < \exp(|z|)$ and for $|z| \ge 1$ and $j \le m$, $|z|^j \le |z|^m$, we have for $|z| \ge 1$,

$$\begin{aligned} |E_m(z)| &\le \exp\biggl(|z| + \biggl(\sum_{j=1}^m \frac{1}{j}|z|^j\biggr)\biggr) \\ &\le \exp(C_m |z|^m) \end{aligned}$$

□

Proposition 9.10.6. *Let g be a canonical product of genus m with exponent of convergence of zeros σ. Then*

$$\rho(g) = \sigma(g) \tag{9.10.29}$$

Proof. Since $\sigma \le \rho$ for any function, we only need $\rho(g) \le \sigma(g)$. By definition of canonical product, we have that $m \le \sigma \le m+1$. If $|z| \le 1$, by (9.4.3),

$$|E_m(z)| \le 1 + |z|^{m+1} \le \exp(|z|^{m+1}) \tag{9.10.30}$$

so, by the lemma,

$$|g(z)| \le \exp\biggl(\sum_{|z_j|\ge|z|} \biggl(\frac{|z|}{|z_j|}\biggr)^{m+1} + C_m \sum_{|z_j|\le|z|} \biggl(\frac{|z|}{|z_j|}\biggr)^{m} \biggr) \tag{9.10.31}$$

If $\sigma = m+1$, we have for $|z| \ge |z_j|$, that $|z/z_j|^m \le |z/z_j|^{m+1}$, so

$$|g(z)| \le \exp\biggl(C_m |z|^{m+1} \sum_{j=1}^\infty |z_j|^{-m-1} \biggr) = \exp(C|z|^\sigma) \tag{9.10.32}$$

by (9.10.25). Thus, $\rho \le \sigma$.

If $\sigma < m+1$, pick σ' with $m \le \sigma < \sigma' < m+1$. For $|z| > |z_j|$, $|z/z_j|^m \le |z/z_j|^{\sigma'}$, and for $|z| \le |z_j|$, $|z/z_j|^{m+1} \le |z/z_j|^{\sigma'}$, so

$$|g(z)| \le \exp\biggl(C_m |z|^{\sigma'} \sum_{j=1}^\infty |z_j|^{-\sigma'} \biggr) = \exp(C|z|^{\sigma'}) \tag{9.10.33}$$

Thus, $\rho(g) \le \sigma'$. Since σ' is arbitrary in $(\sigma, m+1)$, we get $\rho(g) \le \sigma(g)$. □

Remark. If $m = 0$, (9.10.28) and so (9.10.31) do not hold. In (9.10.31), replace $(|z|/|z_j|)^m$ by $\log\bigl[1+|z|/|z_j|\bigr]$. Since $0 < 1/k < \sigma'$ and $w > 1$ implies $1 + w \le \exp((k)^{1/k} w^{\sigma'})$, the proof goes through.

We need one final preliminary before turning to the main result of this section:

Lemma 9.10.7. *For all $m = 0, 1, 2, \dots,$*

$$|z| \le \tfrac{1}{2} \Rightarrow |E_m(z)| \ge \exp(-2|z|^{m+1}) \tag{9.10.34}$$

$$|z| \ge 2 \Rightarrow |E_m(z)| \ge \exp(-2|z|^m) \tag{9.10.35}$$

Proof. If $x \in [0, \frac{1}{2}]$,

$$\frac{\log(1-x)}{x} = -1 - \frac{x}{2} - \frac{x^2}{3} - \cdots \ge -1 - \frac{1}{2} - \frac{1}{4} - \frac{1}{8} = -2 \tag{9.10.36}$$

so

$$0 \le x \le \tfrac{1}{2} \Rightarrow 1 - x \ge e^{-2x} \tag{9.10.37}$$

Thus,

$$\begin{aligned} |z| \le \tfrac{1}{2} \Rightarrow |E_m(z)| &\ge 1 - |z|^{m+1} \qquad \text{(by (9.4.3))} \\ &\ge \exp(-2|z|^{m+1}) \end{aligned}$$

On the other hand, $|z| \ge 2 \Rightarrow |1-z| \ge |z| - 1 \ge 1$, while $-|z| - \frac{|z|^2}{2} - \cdots - \frac{|z|^m}{m} \ge -|z|^m(1 + \frac{1}{2} + \cdots + (\frac{1}{2})^{m-1}) \ge -2|z|^m$, so

$$\begin{aligned} |z| \ge 2 \Rightarrow |E_m(z)| &\ge |1-z| \exp\biggl(-|z| - \frac{|z|^2}{2} - \cdots - \frac{|z|^m}{m} \biggr) \\ &\ge \exp(-2|z|^m) \end{aligned}$$

□

Definition. Let f be an entire function of finite order and $\sigma(f)$ the exponent of convergence of the zeros. If $\sigma \in \mathbb{Z}$ and $\sum_{j=1}^{\infty} |z_j|^{-\sigma} < \infty$, set $m = \sigma - 1$ and otherwise $m = [\sigma]$. The associated canonical product, (9.10.26), is called the *canonical product for* f.

We now turn to the main result of this section:

Theorem 9.10.8 (Hadamard Product Formula). *Let f be an entire function of finite order. Let $\rho(f)$ be its order, $\sigma(f)$ the exponent of convergence of its zeros, ℓ the order of the zero at $z = 0$ if any (otherwise $\ell = 0$), and m the genus in its canonical product. Then*

$$f(z) = e^{P(z)} z^{\ell} \prod_{j=1}^{\infty} E_m\left(\frac{z}{z_j}\right) \tag{9.10.38}$$

where P is a polynomial of degree q with

$$q \leq [\rho(f)] \tag{9.10.39}$$

Moreover,

$$\rho(f) = \max(q, \sigma(f)) \tag{9.10.40}$$

Remarks. 1. $\max(q, m)$ is called the *genus* of f but this term has fallen into disuse. Normally, the genus is $[\rho(f)]$, but it can happen that $\rho(f) \in \mathbb{Z}$ and $m = \rho(f) - 1$ is the genus.

2. Since q is an integer, if $\rho(f) \notin \mathbb{Z}$, it must be that $\sigma(f) = \rho(f)$. We'll discuss the consequences of this below.

3. This is sometimes called the *Hadamard factorization theorem*.

Proof. If $\ell \neq 0$, apply the theorem to $f(z)/z^{\ell}$ to get the general case, so without loss, we may suppose $\ell = 0$. We set

$$g(z) = \prod_{j=1}^{\infty} E_m\left(\frac{z}{z_j}\right) \tag{9.10.41}$$

Clearly, $f(z)/g(z)$ is an entire, nonvanishing function, so of the form $e^{P(z)}$. The whole issue is proving that P is a polynomial of degree bounded by $\rho(f)$. We'll then see that (9.10.40) is easy.

The key is getting a bound on $|e^{P(z)}|$ and the potential problem is related to zeros of $g(z)$. Since zeros of g are cancelled by those of f, at first sight this wouldn't seem to be a problem. But on second sight, borrowing part of f makes bounding f difficult. There is, however, a beautiful trick of Landau, using the maximum principle that finesses the problem!

Define for each $R > 0$,

$$g_R(z) = \prod_{|z_n| \le 2R} E_m\left(\frac{z}{z_n}\right) \tag{9.10.42}$$

$$h_R(z) = \prod_{|z_n| > 2R} E_m\left(\frac{z}{z_n}\right) \tag{9.10.43}$$

On the circle $|z| = R$, $h_R(z)$ has no nearby zeros, so bounding $1/h_R(z)$ is easy. If $|z_n| > 2R$ and $|z| = R$, $|z/z_n| < \frac{1}{2}$. So by (9.10.34),

$$\sup_{|z|=R} |h_R(z)|^{-1} \le \exp\left(2 \sum_{|z_n|>2R} \left(\frac{R}{|z_n|}\right)^{m+1}\right) \tag{9.10.44}$$

Alas, $g_R(z)$ can have lots of zeros on or near $|z| = R$, but $f(z)/g_R(z)$ is an entire function, so

$$\sup_{|z|=R} \left|\frac{f(z)}{g_R(z)}\right| \le \sup_{|z|=4R} \left|\frac{f(z)}{g_R(z)}\right| \tag{9.10.45}$$

$$= \frac{\sup_{|z|=4R} |f(z)|}{\inf_{|z|=4R} |g_R(z)|} \tag{9.10.46}$$

Since f has order ρ, for any $\varepsilon > 0$, there are $C_\varepsilon, D_\varepsilon$ with

$$|f(z)| \le D_\varepsilon \exp(C_\varepsilon |z|^{\rho+\varepsilon}) \tag{9.10.47}$$

so,

$$\sup_{|z|=4R} |f(z)| \le D_\varepsilon \exp(4^{\rho+\varepsilon} C_\varepsilon R^{\rho+\varepsilon}) \tag{9.10.48}$$

If $|z_n| \le 2R$ and $|z| = 4R$, then $|z|/|z_n| \ge 2$, so (9.10.35) implies that

$$\inf_{|z|=4R} |g_R(z)| \ge \exp\left(-2 \sum_{|z_n|\le 2R} \left(\frac{4R}{|z_n|}\right)^{m}\right) \tag{9.10.49}$$

Putting together $f/g = (f/g_R)(1/h_R)$, (9.10.44), (9.10.46), (9.10.48), and (9.10.49) yields

$$\sup_{|z|=R} |e^{P(z)}| \le D_\varepsilon \exp\left[4e^{\rho+\varepsilon} C_\varepsilon R^{\rho+\varepsilon} + 2 \sum_{|z_n|>2R} \left(\frac{R}{|z_n|}\right)^{m+1} + 2 \sum_{|z_n|\le 2R} \left(\frac{4R}{|z_n|}\right)^{m}\right] \tag{9.10.50}$$

Now pick σ' with $\sigma' = m+1$ if $m = \sigma - 1$ (in which case, $\sum_{n=1}^\infty |z_n|^{-m-1} < \infty$), and otherwise $m \le \sigma < \sigma' < m+1$. Since $m \le \sigma' \le m+1$, we can replace m by σ' in the second sum (since $4R/|z_n| \ge 2$) and we can replace $m+1$ by σ' in the first sum (since $R/|z_n| \le \frac{1}{2}$) and, use $\sum_{n=1}^\infty |z_n|^{-\sigma'} < \infty$, to get that

$$\text{RHS of (9.10.50)} \le D_\varepsilon \exp(C(R^{\rho+\varepsilon} + R^{\sigma'}))$$

By Theorem 3.2.1, P is a polynomial of degree q, at most $\max(\rho+\varepsilon, \sigma')$. Taking ε to zero, σ' to σ, and using $\sigma \le \rho$, we get $q \le \rho$. This yields (9.10.39).

Finally, we turn to (9.10.40). Since $\sigma \le \rho$ and $q \le [\rho]$, we have $\max(q, \sigma(f)) \le \rho$. On the other hand, $\rho(e^P) = q$ and $\rho(g) = \sigma$ (by Proposition 9.10.6). Thus, by Proposition 9.10.4, $\rho(f) \le \max(q, \sigma(f))$. □

There is one specialized case (but which arises in application; see Section 3.12 in Part 4), where one can improve (9.10.39):

Theorem 9.10.9 (Refined Hadamard Product Formula). *Let f be a function of finite order with:*

(i) $k = \rho(f) \in \mathbb{Z}$; $k > 0$
(ii) $\tau(f)$, *the type, is* 0
(iii) $\sigma(f) = k$ *and*

$$\sum_{n=1}^{\infty} |z_n|^{-k} < \infty \tag{9.10.51}$$

Then (9.10.39) *holds with* $m = k - 1$ *and, moreover,*

$$q \le k - 1 \tag{9.10.52}$$

In particular, if $k = 1$ *and* $f(0) \ne 0$,

$$f(z) = f(0) \prod_{j=1}^{\infty} \left(1 - \frac{z}{z_n}\right) \tag{9.10.53}$$

Remark. We already know that (9.10.51) and $k \in \mathbb{Z}$ implies that $m = k-1$. The new element is (9.10.52).

Proof. As remarked, we need only prove (9.10.52). Since the type is 0, (9.10.47) can be replaced with

$$|f(z)| \le D_\varepsilon \exp(\varepsilon |z|^k) \tag{9.10.54}$$

so (9.10.50) becomes

$$\sup_{|z|=R} |e^{P(z)}| \le D_\varepsilon \exp\left(\varepsilon 4^k R^k + 2 \sum_{|z_n| \ge 2R} \left(\frac{R}{z_n}\right)^k + 2 \sum_{|z_n| \le 2R} \left(\frac{4R}{|z_n|}\right)^{k-1}\right) \tag{9.10.55}$$

Fix N and in the final sum replace $k-1$ by k only if $n \ge N$. The result is

$$\sup_{|z_n| \le R} |e^{P(z)}| \le D_\varepsilon \exp(c_N R^k + d_N R^{k-1}) \tag{9.10.56}$$

where

$$d_N = 2 \sum_{n=1}^{N-1} 4^{k-1} |z_n|^{-(k-1)}, \qquad c_N = \varepsilon 4^k + 2(4)^k \sum_{n=N}^{\infty} |z_n|^{-k} \tag{9.10.57}$$

While d_N may go to ∞ as $N \to \infty$, it is finite for N finite, and in using (9.3.2) irrelevant in estimating the $O(z^k)$ term in P. The result is

$$P(z) = a_k z^k + O(z^{k-1}) \tag{9.10.58}$$

where (taking $R \to \infty$)

$$|a_k| \le 4c_N \tag{9.10.59}$$

Since $c_N \to 0$ as first $\varepsilon \downarrow 0$ and then $N \to \infty$, we see $a_k = 0$, that is, $\deg(P) \le k - 1$. □

Example 9.10.10. Here is yet another proof of the Euler product formula. Let

$$f(z) = \frac{\sin(\pi\sqrt{z})}{\pi\sqrt{z}} \tag{9.10.60}$$

Since $z \to \sin(\pi z)/\pi z$ is invariant under $z \to -z$, it has only even terms in its Taylor series, so f is an entire function. Since

$$\begin{aligned} \sin(\pi w) &= \int_0^1 \frac{d}{dt} \sin(\pi w t)\, dt \\ &= \pi w \int_0^1 \cos(\pi w t)\, dt \end{aligned} \tag{9.10.61}$$

and, for all $w \in \mathbb{C}$, $|\cos(w)| \le e^{|w|}$, we see that

$$|f(z)| \le e^{\pi |z|^{1/2}} \tag{9.10.62}$$

Thus, f has order $\frac{1}{2}$ and zeros exactly at $\{n^2\}_{n=1}^\infty$. Thus, by the Hadamard product formula ($\deg(P) \le \frac{1}{2} \Rightarrow P$ is constant),

$$f(z) = f(0) \prod_{j=1}^\infty \left(1 - \frac{z}{j^2}\right) \tag{9.10.63}$$

Since $f(0) = 1$, this is the Euler product formula. □

Finally, we want to show that Hadamard's theorem implies the little Picard theorem for entire functions of finite order and (more than) the great Picard theorem for entire functions of nonintegral order. Of course, we'll eventually prove this (in several ways) for all entire functions (and even functions only analytic near infinity), but those proofs are more involved.

Theorem 9.10.11. *If f is an entire, nonconstant function of finite order, f takes every value, with perhaps one exception. If $\rho(f) \notin \mathbb{Z}$ or if $\rho(f) = 0$ and f is not a polynomial, f takes every value infinitely often.*

Remark. In Problem 10, we'll boost this argument to show the great Picard theorem for all entire functions of finite order.

Proof. If $f(z)$ never takes the value a, by the Hadamard theorem,

$$f(z) - a = e^{P(z)} \tag{9.10.64}$$

with $P(z)$ a nonconstant polynomial. By the fundamental theorem of algebra, $P(z)$ takes every value. In particular, for any $b \neq a$, if $b-a = |b-a|e^{i\varphi}$, then there are points, z_b, with $P(z_b) = \log|b-a| + i\varphi$. In that case, we have that $f(z_b) = b$. Since there are infinitely many φ with $b - a = |b - a|e^{i\theta}$ (differing by $2\pi\mathbb{Z}$), $f(z) = b$ infinitely often.

If $f(z) - a$ has only finitely many zeros, say $z_1, \dots, z_k$, by the Hadamard theorem applied to $(f(z) - a)/\prod_{j=1}^k (z - z_k)$,

$$f(z) - a = \prod_{j=1}^{k} (z - z_k) e^{P(z)} \tag{9.10.65}$$

for a polynomial, P. If $\deg(P) = 0$, $\rho(f - a) = 0$, and if $\deg(P) \neq 0$, $\rho(f - a) = \deg(P) \in \mathbb{Z}$. Since $\rho(f) = \rho(f - a)$ (see Problem 5), $\rho(f) \in \mathbb{Z}$. It follows that if $\rho(f) \notin \mathbb{Z}$, then, for each a, $f(z) - a$ has infinitely many zeros.

If $\rho(f) = 0$ and (9.10.65) holds, then f is a polynomial, so if $\rho(f) = 0$ and f is not a polynomial, each value must get taken infinitely often. □

Notes and Historical Remarks. Hadamard's theorem is named after his partial result in [**230**] (see below). Landau's trick of breaking g into two products, estimating on circles of different size, and then using the maximum principle is from [**340**].

It should be emphasized that virtually any entire function that enters in applications is of finite order—indeed, almost always of order at most 1. Thus, the results of this section and the special situations in the problems (e.g., of real zeros) are extremely important. That said, it should be pointed out that Pólya [**447**] has proven that compositions are "usually" not of finite order. He has proven that if f and g are entire and $f(g(z))$ is of finite order, then either f is a polynomial and g of finite order or f is of finite order and g is of order 0.

The key to the proof of Hadamard's theorem is to show that $h(z) = \log(f(z)/g(z))$ is a polynomial. I am aware of two other widely used proofs besides the Landau trick. One (Problem 8) uses a partial fraction decomposition of $h'(z)$, and the other, a detailed analysis of zeros (Problem 9). Each has a slightly messy calculation. To my taste, the Landau proof is by far the most elegant and simple, with the partial fraction proof in second place. Remarkably, in the many texts that prove the Hadamard theorem, only one (Titchmarsh [**556**], from which I learned it) has the Landau proof, and five (Ash [**24**], Conway [**120**], Greene–Krantz [**220**] Hahn–Epstein [**232**], and Segal [**514**]) the partial fraction proof. All the many others use the detailed analysis. (There are a few other "one-off" proofs, e.g., Berenstein–Gay [**42**] use potential and subharmonic function theory.) I'm hoping this is because the slicker proofs are not known, but there could be another reason. The detailed analysis is where brute force considerations lead one and some will prefer "straightforward" but grubby arguments to some "trick that happens to work." It should also be mentioned that the ideas behind the detailed analysis, while not necessary to get Hadamard's theorem, can be useful in other contexts. The moral is that multiplicities of style are an advantage!

Jacques Hadamard (1865–1963) proved his theorem in 1892–93 (while he was a high school teacher) in part to understand Picard's theorem and, in particular, found the application to prove that result for functions of finite order. It was also key in his celebrated proof four years later of the prime number theorem (done independently by de la Vallée Poussin, both based on work of Riemann). We prove the prime number theorem in Section 13.5 of Part 2B.

While the Weierstrass product formula was German, its refinement, as discussed in this section, is due to a group of French mathematicians. In 1882, Laguerre [**336**] singled out what we'd now call canonical products of genus 1 as entire functions that look like polynomials (see his theorem proven in Problem 11). Poincaré [**437, 441**] took up this subject almost immediately, defined genus n (actually, the early papers used the name *genre* rather than genus and related genus to the decrease of the coefficients of Taylor coefficients).

Hadamard [**230**], in 1893, then studied these notions further. In particular, he established a product formula for the Riemann zeta function (in modern terminology, he proved that the function $\xi(s)$ of (13.3.20) of Part 2B is of genus zero), a fact which earned him a grand prix of the French academy and established his reputation. He had the key relation (9.10.39) but didn't have the notion of exact exponent of convergence nor (9.10.40).

Borel [**66**] introduced the exponent of convergence and his book on the subject [**67**] had the full results including (9.10.40). Borel is the first to use the name "order" although he used that for the exponent of convergence and called what we call order, the "apparent order." In the book, he also proved that the exponent of convergence for the zeros, $\sigma(f-a)$, is equal to $\rho(f)$ for all but at most one value of a. For $\rho(f) \notin \mathbb{Z}$, our argument shows that, but the case $\rho(f) \in \mathbb{Z}_+$ is more subtle. We'll prove Borel's theorem as Theorem 17.3.6 of Part 2B. Chapter 17 of Part 2B discusses much more subtle extensions. We note that Jensen's formula was proven after Hadamard's work. In later life, Hadamard regarded his not finding this result as a missed opportunity!

Hadamard lived a long and eventful, sometimes tragic, life. His family moved to Paris when he was three and there suffered the effects of the siege of Paris in the 1870 Franco-Prussian war. Two of his sons were killed in the First World War and a third in the Second World War. He and his wife were of Jewish descent and he was very active in supporting Dreyfus in the struggle to get him freed and vindicated. He fled France for the U.S. at the beginning of the Second World War. Ironically, because of his left-wing politics and campaigning for world peace, he was almost denied re-entry to the U.S. for the 1950 International Congress of Mathematics.

Hadamard was prolific mathematically over a long career, not only in complex analysis but also in the study of PDEs, matrix theory, and geometry. Hadamard was a student of Picard and, in turn, Hadamard's students include Fréchet, Lévy, and Weil (the last shared with Picard). There is an interesting mathematical biography of Hadamard [**377**].

Edmund Landau (1877–1938) was a German Jewish mathematician and student of Frobenius. On his father's side, he was descended from the famous eighteenth-century rabbi, the Noda b'Yehuda (Yechezkel Landau). His father was from a minor banking family and his mother from a major banking family, so Landau had enormous inherited wealth.

Landau spent most of his career in Göttingen where he was Minkowski's successor. He had a strong and arrogant personality; his published criticisms of the work of Blaschke and Bieberbach produced lifelong enemies. His arrogance was probably a factor in a development that turned out to be tragic[9]. He visited the newly founded Hebrew University in 1927 and accepted a professorship while there. Magnes, the founder and first chancellor, offered him the position of Rector, but Magnes hadn't discussed this appointment with the other members of the triumvirate that ran things (the others from afar): Albert Einstein and Chaim Weizmann. The offer was withdrawn in the resulting power struggle. The cause was partly a fight over control of the university, but it is likely that Landau's arrogance was a factor. In any event, Landau then resigned and returned to Göttingen—in retrospect, an unwise decision for a prominent Jew. Landau was an early victim of the anti-Jewish feeling with a famous incident of a group of students, organized by Teichmüller, barring attendance to a Landau lecture. Landau retired to Berlin and died in 1938.

Landau was noted for his breadth of knowledge in analytic number theory. In 1903, he greatly simplified the Hadamard–de la Vallée Poussin proof of the prime number theorem and he wrote the first systematic text on analytic number theory. His writing and proofs were marked by elegance and clarity.

[9]`http://www.ma.huji.ac.il/~landau/landuniv.html`

Problems

1. This will show that $\sup_\theta$ in (9.10.3) can be replaced by an average. Given an entire function, f, define

$$M_f^{(\infty)}(r) = \sup_{\theta\in[0,2\pi)} |f(re^{i\theta})|, \quad M_f^{(1)}(r) = \int_0^{2\pi} |f(re^{i\theta})| \frac{d\theta}{2\pi} \tag{9.10.66}$$

(a) Prove that $M_f^{(1)}(r) \le M_f^{(\infty)}(r)$ for all r.

(b) Let $C = \sup_\theta P_{1/2}(\theta)$ where $P_r(\theta)$ is the Poisson kernel $\frac{1-r^2}{1+r^2-2r\cos\theta}$, i.e., $C = 3$. Prove that for all r

$$M_f^{(\infty)}(r) \le CM_f^{(1)}(2r) \tag{9.10.67}$$

Prove that $\rho(f)$ given by (9.10.12) obeys

$$\rho(f) = \lim\sup_{R\to\infty} \frac{\log\bigl[\log M_f^{(1)}(R)\bigr]}{\log(R)} \tag{9.10.68}$$

2. (a) Suppose $|f(z)| \le De^{C|z|^\alpha}$ and $f(z) = \sum_{n=0}^\infty a_n z^n$. Prove that

$$\limsup_{n\to\infty} |a_n|^{1/n} n^{1/\alpha} < \infty \tag{9.10.69}$$

(*Hint*: Optimize a Cauchy estimate.)

(b) Fix $c > 0$ and $\rho > 0$. Consider the function

$$f(z) = \sum_{n=0}^\infty \frac{c^n}{n^{n/\rho}} z^n \tag{9.10.70}$$

Prove f has order ρ. (*Hint*: Compute $\max_y[c^y r^y / y^{y/\rho}]$ as a function of r.)

(c) Prove that if (9.10.69) holds, then $|f(z)| \le D_\varepsilon \exp(C_\varepsilon |z|^{\alpha+\varepsilon})$ for any $\varepsilon > 0$.

(d) Verify that if $f(z) = \sum_{n=0}^\infty a_n z^n$, then $\rho(f)$ is given by (9.10.5).

Note. The function in (9.10.70) is a close relative to the function $\sum_{n=0}^\infty z^n/\Gamma(\alpha n + 1)$, which Mittag-Leffler noted has order α^{-1}.

3. (a) Compute the type of the function in (9.10.70).

(b) Prove that if $f(z) = \sum_{n=0}^\infty a_n z^n$ has order $\rho > 0$, then its type is given by

$$\tau(f) = (e\rho)^{-1} \limsup n|a_n|^{\rho/n} \tag{9.10.71}$$

4. (a) Prove (1) $\Leftrightarrow$ (2) in Proposition 9.10.1. (*Hint*: Integration by parts in (9.10.11).)

(b) Prove (2) $\Leftrightarrow$ (3) in Proposition 9.10.1. (*Hint*: Write $n(r) - n(0) = r\frac{dN}{dr}$ and integrate by parts.)

5. Suppose that f and g are entire functions of finite order.

 (a) Prove that $\rho(fg) \leq \max(\rho(f), \rho(g))$ and $\rho(f+g) \leq \max(\rho(f), \rho(g))$.

 (b) If $\rho(f) > \rho(g)$, prove that $\rho(f+g) = \rho(f)$.

 (c) Find examples where $\rho(fg) < \max(\rho(f), \rho(g))$ and $\rho(f+g) < \max(\rho(f), \rho(g))$.

 (d) Prove that $\sigma(f) \leq \rho(fg)$.

 (e) If $\rho(fg) < \rho(f)$, prove first that $\sigma(f) < \rho(f)$ and then that $\rho(f) \in \mathbb{Z}_+$ and that the e^P in the Hadamard theorem has $\deg(P) = \rho(f)$.

 (f) If $\rho(fg) < \max(\rho(f), \rho(g))$, prove that $\rho(f) = \rho(g) \in \mathbb{Z}_+$ and $\deg(P_f + P_g) < \deg(P_f)$. In particular, if $\rho(f) > \rho(g)$, then $\rho(fg) = \rho(f)$.

6. Let $\{z_j\}_{j=1}^\infty \subset \mathbb{C}^\times$ so that σ given by (9.10.7) is finite. Suppose $m \in \mathbb{Z}$ has $m > \sigma$. Prove that the order of the product in (9.10.19) (which is no longer a canonical product) is m. (*Hint*: Pick $m' = [\sigma]$ and write $\prod_{j=1}^\infty E_m(\frac{z}{z_j}) = e^{P(z)} \prod_{j=1}^\infty E_{m'}(\frac{z}{z_j})$ where $\deg P = m$ and then use Theorem 9.10.8 and Problem 5(d).)

7. Let $M_r(f) = \sup_{|z| \leq r} |f(z)|$.

 (a) Prove that $M_r(f') \leq M_{r+1}(f)$.

 (b) Prove that $M_r(f) \leq |f(0)| + rM_r(f')$.

 (c) Show that $\rho(f) = \rho(f')$.

 Remark. We mentioned in the Notes that genus was used earlier than order. One of the difficulties with genus was the relation of the genus of f and of f'.

8. This problem provides an alternate to the essence of the proof of the Hadamard factorization theorem.

 (a) By using the Poisson–Jensen formula (Theorem 9.8.2), prove that if f is an entire function and $\rho(f) < m+1$ with $m \in \mathbb{Z}$, then

$$\frac{d^m}{dz^m}\left(\frac{f'(z)}{f(z)}\right) = -m! \sum_{j=1}^\infty \frac{1}{(z_j - z)^{m+1}} \tag{9.10.72}$$

 where $\{z_j\}_{j=1}^\infty$ are the zeros of f including any possible zeros at $z = 0$.

 (b) Let $g(z) = z^\ell \prod_{z/z_j \neq 0} E_m(z/z_j)$. Prove that

$$\frac{d^m}{dz^m}\left(\frac{g'(z)}{g(z)}\right) = -m! \sum_{j=1}^\infty \frac{1}{(z_j - z)^{m+1}} \tag{9.10.73}$$

 (c) Prove $f/g = e^P$ where P is a polynomial of degree at most m.

9. This problem provides an alternate to the essence of the proof of the Hadamard factorization theorem, the one most commonly used in textbook presentations. It has much in common with the proof we gave, except that in place of Landau's trick, it estimates $[\prod_{j=1}^{n}(1-z/z_j)]^{-1}$ away from small circles about the z_j. This proof supposes that $m = [\sigma]$ (and so doesn't handle the special case where $\sigma = m-1$ and $\sum_{j=1}^{\infty}|z_j|^{-\sigma} < \infty$).

(a) Prove that for any $\sigma < \sigma'$,

$$n(r) \leq \left(\sum_{j=1}^{\infty} |z_j|^{-\sigma'}\right)|r|^{\sigma'} \tag{9.10.74}$$

(b) Prove that

$$\sum_{|z_j|\leq 2|z|} \log|z_j| \leq C_{\sigma'}|r|^{\sigma'} \tag{9.10.75}$$

for any $\sigma' > \sigma$.

(c) Consider small circles of radius $|z_j|^{-m-1}$ about z_j removed from $\mathbb{C}$, that is, let

$$\Omega = \mathbb{C}\Big\backslash \bigcup_{j=1}^{\infty} \mathbb{D}_{|z_j|^{-m-1}}(z_j) \tag{9.10.76}$$

On Ω, prove that

$$\prod_{|z_j|\leq 2|z|}\left|1-\frac{z}{z_j}\right| \geq \exp(-(m+2)C_{\sigma'}|z|^{\sigma'}) \tag{9.10.77}$$

(d) With g given by (9.10.41), prove on Ω that

$$|g(z)| \geq \exp(-D_{\sigma'}|z|^{\sigma'}) \tag{9.10.78}$$

for any $\sigma' \in (\sigma, m+1)$.

Remark. Follow the strategy of our proof but don't move the circle $|z| = R$ to $|z| = 4R$.

(e) Prove there exists $R_k \to \infty$ so $\partial\mathbb{D}_{R_k}(0) \subset \Omega$ for all k. (*Hint*: $\bigcup_{j=1}^{\infty}[z_j - |z_j|^{-m-1}, z_j + |z_j|^{-m-1}]$ has finite total measure in $\mathbb{R}$.)

(f) Complete the proof of the Hadamard theorem.

10. (a) Let g be analytic in $\{z \mid |z| > R\}$ with $|g(z)|/|z|^N \to 0$ as $|z| \to \infty$. Prove that if P is a polynomial of degree N, there exists b so that for every w with $|w| > b$, there is a z with $|z| > R$ and $P(z) + g(z) = w$. (*Hint*: Rouché.)

(b) Prove that if f is an entire function of integral order at least 1, then f takes every value with one possible exception infinitely often. (*Hint*: Use (9.10.65).)

11. The purpose of this problem is for the reader to prove the following theorem of Laguerre [**336**]:

    Theorem 9.10.12. *Let f be an entire function of genus* 0 *or* 1 *(i.e., $q \le 1$, $m \le 1$) with $f(\bar{z}) = \overline{f(z)}$ and with all its zeros on $\mathbb{R}$. Then all the zeros of f' are real and interlace those of f.*

    (a) Prove $\operatorname{Im} f'/f > 0$ on $\mathbb{C}_+$ and conclude that the zeros of f' are also real.

    (b) Prove $(f'/f)' > 0$ on $\mathbb{R}$ and f'/f has simple zeros and conclude that the zeros interlace.

12. Assume that $f(z)$ has genus zero so that

$$f(z) = z^m \prod_n \left(1 - \frac{z}{a_n}\right)$$

    Compare $f(z)$ with

$$g(z) = z^m \prod_n \left(1 - \frac{z}{|a_n|}\right)$$

    and show that the maximum modulus $\max_{|z|=r}|f(z)|$ is $\le$ the maximum modulus of g, and that the minimum modulus of f is $\ge$ the minimum modulus of g.

13. (a) Prove that $f(z) = 1/\Gamma(z)$ has order 1. (*Hint*: Proposition 9.10.6.)

    (b) Prove that $f(-(n+\frac{1}{2})) = \frac{(-1)^{n+1}}{2^{2n+1}} \frac{(2n+1)!}{n!}$.

    (c) Prove that f has maximal type.

14. Let $f(z) = \prod_{n=2}^{\infty}(1 + \frac{z}{n(\log(n))^2})$.

    (a) Prove that f has order 1. (*Hint*: Proposition 9.10.6.)

    (b) Prove for any ε, there exist N and C so that $|f(z)| \le C(1+|z|)^N e^{\varepsilon|x|}$, and conclude that f has minimal type.

Chapter 10

Elliptic Functions

> Scarcely had my work seen the light of day, scarcely could its title have become known to scientists abroad, when I learned with as much astonishment as satisfaction that two young geometers, MM. Jacobi of Königsberg and Abel of Christiania, had succeeded in their own individual work in considerably improving the theory of elliptic functions at its highest points.
>
> —*Adrien-Marie Legendre (1752-1833)* as quoted in [**69**]

Big Notions and Theorems: Elliptic Functions, Lattice, Jacobi's Theorem on Dimension of Lattices, Basis, Minimal Basis, $\mathbb{SL}(2,\mathbb{Z})$, $\Gamma(2)$, Fundamental Region, $\mathcal{F}$, $\widetilde{\mathcal{F}}$, Liouville's Theorems, Abel's Theorem, Principal Part Theorem, Legendre Relation, Weierstrass σ-function, Weierstrass ζ-function, Weierstrass $\wp$-function, Addition Theorem for $\wp$, Elliptic Integrals, Chord-Tangent Construction, Lemniscate Integral, Jacobi θ-functions, Triple Product Formula, Jacobi Elliptic Functions: sn, cn, dn, q-ology, Elliptic Modular Function, J-invariant, Equivalence of Complex Tori

This chapter discusses elliptic functions which are entire meromorphic functions that are doubly periodic, that is,

$$f(z+\tau_j) = f(z) \tag{10.0.1}$$

for $j = 1,2$, where τ_1, τ_2 are "independent" (which will mean $\tau_2/\tau_1 \notin \mathbb{R}$). One can ask why we don't look at doubly periodic analytic functions but, as we'll see momentarily, the only such functions are constant.

While this problem seems to be specialized, it turns out to have a broad impact because of its relevance to computation of integrals (elliptic integrals), to certain nonlinear ordinary differential equations including the one for the classical pendulum, to certain completely integrable systems, to certain conformal maps (such as a rectangle to $\mathbb{C}_+$ in Example 8.4.5 and an ellipse to $\mathbb{D}$ in Example 8.4.14; indeed, we saw directly in Example 8.4.5

that the inverse function to the elliptic integral given by (8.4.16) is doubly periodic), to number theory (see Section 13.1 of Part 2B), and to the study of certain algebraic curves (elliptic curves)—we will explore some of these below and, in depth, later in the chapter. That said, we emphasize that the material in this chapter is only used in the rest of the nonbonus parts of this book in one place—where we need the elliptic modular function, and we've constructed that by alternate means in Section 8.3 so that this chapter can be skipped.

Mind you, I do not recommend that this chapter be skipped. A hundred years ago, elliptic functions were regarded as a central—maybe even *the* central—part of the analysis curriculum. Nowadays, it seems as if the algebraic geometers and number theorists have claimed ownership of the subject (given, e.g., the connection of Wiles' proof of the Fermat conjecture to elliptic curves, their interest is certainly understandable) and, on the other side, analysts tend to ignore the subject. This is part of the woeful trend of many analysts to dismiss special functions. But fashions are cyclical—in 1928, Felix Klein [**314**] complained, "When I was a student, abelian functions were, as an effect of the Jacobian tradition, considered the uncontested summit of mathematics, and each of us was ambitious to make progress in this field. And now? The younger generation hardly knows abelian functions"—and we can hope these particular special functions regain their rightful place in the analysis curriculum.

In Section 3.10, we studied periodic functions by showing (Theorem 3.10.2) that if $f(z+\tau)=f(z)$, then there is $g\colon \mathbb{C}^\times \to \mathbb{C}$, analytic, so that $f(z)=g(e^{2\pi i z/\tau})$. This relied on the fact that $\pi(z)=e^{2\pi i z/\tau}$ is a covering map with $z\to z+\tau$ the deck transformations.

Similarly, in the current setup, if f obeys (10.0.1) for $j=1,2$ and $\mathcal{T}_{\tau_1,\tau_2}$ is the complex torus described in Example 7.1.3, then there is a meromorphic function g on $\mathcal{T}_{\tau_1,\tau_2}$ so

$$f(z)=g(\pi(z)) \tag{10.0.2}$$

where π mapping $\mathbb{C}$ to $\mathcal{T}_{\tau_1,\tau_2}=\mathbb{C}/\mathcal{L}_{\tau_1,\tau_2}$ is the canonical projection. For (10.0.1) implies that g defined by (10.0.2) is well-defined and g is meromorphic since π is locally an analytic bijection.

This implies that the study of elliptic functions is the same as the study of meromorphic functions on tori. The torus is the second simplest compact Riemann surface, so we start in Section 10.1 with a quick summary of the situation for the simplest such surface, $\widehat{\mathbb{C}}$. Basically, for $\widehat{\mathbb{C}}$, the only restriction on zeros and poles is that there are equal numbers. The surprise is that for the torus, there are additional restrictions.

Section 10.2 is a preliminary for the study of tori—the structure of lattices $\mathcal{L}_{\tau_1,\tau_2}$; see Figure 7.1.1. Sections 10.3 and 10.4 are the central ones in the chapter. Section 10.3 will prove the basic theorems of Liouville and state Abel's theorem, which will be reduced to the construction of certain analytic functions which are doubly periodic up to a linear exponential.

Section 10.4, on the Weierstrass elliptic functions, will construct the needed analytic functions as Weierstrass products and use them to both prove Abel's theorem and to construct the Weierstrass $\wp$-function, an elliptic function with the remarkable property that

$$(\wp'(z))^2 = 4(\wp(z) - e_1)(\wp(z) - e_2)(\wp(z) - e_3) \tag{10.0.3}$$

Here e_1, e_2, e_3 are three special values of $\wp(z)$ that depend on τ_1, τ_2. This will link elliptic functions both to elliptic integrals and to cubic curves.

Bonus Section 10.5 will present Jacobi's alternate approach based on Θ functions and the related sn and cn functions. Section 10.6 will define the elliptic modular function in terms of the values of e_1, e_2, e_3 for $\tau_1 = 1$, $\tau_2 = \tau$ and prove its basic properties. This is the same function we constructed by other means in Section 8.3. Finally, Section 10.7 will classify complex tori—that is, settle when τ_1, τ_2 and $\tilde{\tau}_1, \tilde{\tau}_2$ generate tori which are biholomorphically equivalent.

This chapter will only scratch the surface of the subject of elliptic functions, on which there are many complete books [**10, 21, 71, 102, 110, 156, 236, 237, 273, 315, 322, 344, 378, 409, 473, 520, 577, 589**]. In particular, we'll say almost nothing about their applications to number theory (however, see Section 13.1 of Part 2B).

Notes and Historical Remarks.

> December 23, 1751 is "the birth day of the theory of elliptic functions."
>
> —*Jacobi* as quoted in [**478**][1]

For purposes of presentation, the central theme in the chapter is the study of doubly periodic functions, but in terms of both history and interest in the results, this gives a slanted picture. Consider how one might view the trigonometric functions:

(a) As periodic functions. If $g(z) = \cos(\pi z)$, then by Problem 1 of Section 3.10, every periodic function, f, has a unique expansion

$$f(z) = \sum_{n=0}^{\infty} a_n g(z)^n + g'(z) \sum_{n=0}^{\infty} b_n g(z)^n \tag{10.0.4}$$

[1]discussing Euler's beginning of work that led to his discovery of addition formulae for the lemniscate integral.

In that way, there will be an analog for the Weierstrass $\wp$-functions. We'll see (Theorem 10.4.9) that every doubly periodic function, f, can be written uniquely as

$$f(z) = \prod_{j=1}^{n} \frac{\wp(z) - w_j}{\wp(z) - y_j} + \wp'(z) \prod_{k=1}^{m} \frac{\wp(z) - r_k}{\wp(z) - s_k} \tag{10.0.5}$$

and from the point of view of the study of doubly periodic functions per se, this is a way to understand the importance of $\wp$, but just as (10.0.4) is not the real reason one should care about $\cos(\pi z)$, (10.0.5) is only part of why $\wp$ is important.

(b) As the inverse of an integral. Consider the formula

$$\int_0^z \frac{dx}{\sqrt{1-x^2}} = \arcsin(z) \tag{10.0.6}$$

Suppose one discovered sin by noting the integral on the left of (10.0.6) was the inverse of a periodic function and concluded this inverse function might be an object worth studying. Given how sin and cos entered in real history, this would seem fairly bizarre. But historically, the analog is exactly how elliptic functions were discovered!

Just as the reason (10.0.6) works is that $u(z) = \sin(z)$ obeys $u'(z)^2 = 1 - u(z)^2$, so the reason the Weierstrass $\wp$-function is connected to integrals of $(\sqrt{Q(z)})^{-1}$ for Q, a cubic polynomial, is that $\wp$ obeys

$$\wp'(z)^2 = 4(\wp(z) - e_1)(\wp(z) - e_2)(\wp(z) - e_3) \tag{10.0.7}$$

for suitable $\{e_j\}_{j=1}^3$.

(c) As uniformization functions. Suppose we are interested in a quadratic curve $\{(w,z) \in \mathbb{C}^2 \mid w^2 + z^2 = 1\}$ and what the topology of its compactification (as described in Section 7.1) is. If we note $w = \cos(\zeta)$, $z = \sin(\zeta)$ not only solves $w^2 + z^2 = 1$, but also describes all solutions. In fact, if we add two points at infinity (since $w + iz$ is missing the values 0 and ∞), we see the compactification is topologically a sphere. In the same way, because of (10.0.7), $\zeta \mapsto (\wp'(\zeta), \wp(\zeta))$ uniformizes the cubic $w^2 + 4(e_1 - z)(e_2 - z)(e_3 - z) = 0$ and shows the compactification is a torus.

(d) As arithmetic functions given addition formula. Suppose you are looking for Pythagorean triples, that is, integers p, q, r so $p^2 + q^2 = r^2$. If we let $x = p/r$, $y = q/r$, this is equivalent to looking for rational solutions of $x^2 + y^2 = 1$ in the first quadrant. Draw a straight line between $(-1, 0)$ and (x, y) and you'll have a rational slope $t = n/m$ where the geometry (see Figure 10.0.1) has $t < 1$, that is, $n < m$.

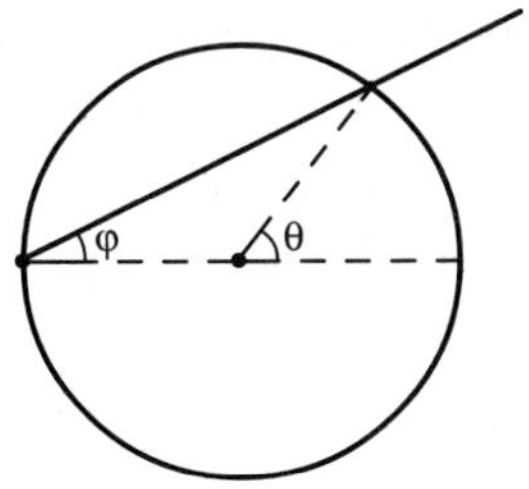

Figure 10.0.1. Chord-tangent construction for a quadratic curve.

Thus, $y = t(x+1)$, so $y^2 + x^2 = 1$ is the quadratic equation

$$x^2(1+t^2) + 2t^2x + t^2 - 1 = 0 \tag{10.0.8}$$

This has the obvious solution $x = -1$, so looking at the product of the roots, we see the other solution is $x = (1-t^2)/(1+t^2)$, and then $y = 2t/(1+t^2)$. Thus, solutions are

$$p = (m^2 - n^2)k, \qquad q = (2mn)k, \qquad r = (m^2+n^2)k \tag{10.0.9}$$

a well-known parametrization of the Pythagorean triples. This geometric construction is believed to go back to Diophantus. (The formulae appeared even earlier in Book 10 of Euclid's *Elements*; Diophantus did his work about 250 AD, Euclid about 300 BC.) The geometry of this situation shows that the key to this algebra is actually the double-angle formula for trigonometric functions $\theta = 2\varphi$ (in Figure 10.0.1), and (10.0.9) is essentially $\sin(\theta) = 2\cos(\varphi)\sin(\varphi)$; $\cos(\theta) = \cos^2(\varphi) - \sin^2(\varphi)$. It explains why Jacobi called December 23, 1751, the day Euler started working on what led to addition formulae for elliptic integrals, the birthday of elliptic function theory. We'll only touch on these addition formulae briefly in Section 10.4 and not at all on other number theoretic aspects, but these calculations should illustrate one reason why there are number theoretic aspects of elliptic function theory.

One of the early successes of elliptic function theory was Jacobi's two- and four-squares theorems, discussed in Section 13.1 of Part 2B. Another was the proof of cubic and quartic reciprocity laws by Jacobi and Eisenstein (see Lemmermeyer [**353**]). Other significant nineteenth-century applications include Hermite's solution in 1858 [**256**] of quintic equations in terms of elliptic functions (see King [**307**]) and the finding of several completely integrable systems, including geodesic flow on an ellipsoid by Jacobi, and nutation of tops, including the Kovaleskaya top (see Arnol'd [**22**] and Bolsinov–Fomenko [**61**]). The last four chapters of [**21**] have lots of applications of elliptic functions and integrals, including solutions of quintics.

To get a feel for what elliptic functions have to do with solving quintics, we note, following Hermite, that $x^3 - 3x + 2a = 0$ can be solved by finding θ so that $\sin\theta = a$ and noting that, by triple-angle formulae, the solutions of the cubic are $2\sin((2\pi_j+\theta)/3)$ for $j = 0, 1, 2$. Quintics use Jacobi θ functions and elliptic modular functions in place of sin and arcsin. Klein [**312**] developed a link of this approach to Galois theory by using the symmetry group of the icosahedron.

In the end, as a first pass, we'll focus on double periodicity per se, but these Notes should give the reader some flavor of why the full theory is so much more and why it has led to multiple books.

10.1. A Warmup: Meromorphic Functions on $\widehat{\mathbb{C}}$

Theorem 7.1.9 says that two meromorphic functions on a compact Riemann surface, $\mathcal{S}$, are determined up to a multiplicative constant by their zeros and poles (including their orders) and are also determined up to an additive constant by their principal parts. So it is a natural question to ask what sets of zeros and poles and what principal parts can occur.

One obvious restriction comes from Theorem 7.1.10—the total numbers of zeros and of poles (counting orders) must agree. The main result in this warmup section is that, for $\mathcal{S} = \widehat{\mathbb{C}}$, this is the only restriction.

Recall that a *rational function* is a ratio of polynomials $Q(z)/P(z)$, that we suppose has no common zero. The following has been implicit in some places earlier in this book:

Theorem 10.1.1. *Every meromorphic function on $\widehat{\mathbb{C}}$ is a rational function. The number of zeros and poles (counting orders) agree.*

Remark. Of course, the converse of the first sentence is also true and the last sentence follows from Theorem 7.1.10, but we provide a simple direct proof.

Proof. Let f be the meromorphic function on $\widehat{\mathbb{C}}$ and $\{p_j\}_{j=1}^{N_0}$ the set of poles in $\mathbb{C}$, counting orders. Let

$$P(z) = \prod_{j=1}^{N_0}(z - p_j) \tag{10.1.1}$$

Then $Q(z) \equiv f(z)P(z)$ is an entire function on $\mathbb{C}$.

Since f is meromorphic at ∞, near infinity, for some k,

$$|f(z)| \le C|z|^k \tag{10.1.2}$$

so, $|Q(z)| \le C|z|^{k+N_0}$ and Q is a polynomial by Theorem 3.1.9. Thus, f is rational.

If $f = Q/P$ is rational, the order of the zero or pole at infinity is $\deg(P)-\deg(Q)$ (with pole orders counted as negative numbers and f nonvanishing and regular at ∞ if this number is 0). Since $\deg(P)$ counts the sum of the orders of poles in $\mathbb{C}$ and $\deg(Q)$ of the zeros, we get the claimed equality on the total number of zeros and poles. $\square$

Theorem 10.1.2. *Let $\{z_j\}_{j=1}^{N_z}$ and $\{p_j\}_{j=1}^{N_p}$ be finite sequences of points in $\widehat{\mathbb{C}}$ with $p_j \neq z_k$ for all j,k, and suppose $N_z = N_p$. Then there exists a rational function whose zeros (counting orders) is precisely $\{z_j\}_{j=1}^{N_z}$ and poles $\{p_j\}_{j=1}^{N_p}$.*

Proof. By renumbering, we can suppose for some $N_z^{(0)}, N_p^{(0)}$ that $z_j = \infty$ for $j \geq N_z^{(0)} + 1$ and $z_j \neq \infty$ for $j \leq N_z^{(0)}$, and similarly for the poles (by hypothesis, either $N_z^{(0)} = N_z$ or $N_p^{(0)} = N_p$ or both).

$$f(z) = \frac{\prod_{j=1}^{N_z^{(0)}} (z - z_j)}{\prod_{j=1}^{N_z^{(0)}} (z - p_j)} \tag{10.1.3}$$

has the required poles and zeros in $\mathbb{C}$ and the order of its pole or zero at infinity is, by $N_p = N_z$, precisely the number of times ∞ occurs in the sets. $\square$

Theorem 10.1.3. *Let $\{p_j\}_{j=1}^{\widetilde{N}_p}$ be a set of distinct points in $\widehat{\mathbb{C}}$ and $\{P_j\}_{j=1}^{\widetilde{N}_p}$ potential principal parts of a pole at p_j. Then there exists a meromorphic function, f, on $\widehat{\mathbb{C}}$ so that the poles and principal parts are precisely $\{p_j\}_{j=1}^{\widetilde{N}_p}$ and $\{P_j\}_{j=1}^{\widetilde{N}_p}$.*

Remark. The principal part at ∞ of a meromorphic function on $\widehat{\mathbb{C}}$ is the negative terms in the Taylor series in z^{-1}, so a polynomial in z with zero constant term.

Proof. Let $\widetilde{P}_j(z) = P_j(1/z - p_j)$ if $p_j \neq \infty$ and $P_j(z)$ if $p_j = \infty$. Take

$$f(z) = \sum_{j=1}^{\widetilde{N}_p} \widetilde{P}_j(z) \tag{10.1.4}$$

Since $f(z) - \widetilde{P}_j$ is analytic at p_j, f has the requisite principal parts. $\square$

10.2. Lattices and $\mathbb{SL}(2,\mathbb{Z})$

Elliptic functions are left invariant by adding to their arguments elements of a subset $\mathcal{L} \equiv \{t\} \subset \mathbb{C}$ of those t with $f(z+t) = f(z)$ for all z. Clearly, if $t_1, t_2 \in \mathcal{L} \Rightarrow t_1 + t_2 \in \mathcal{L}$ and $-t \in \mathcal{L}$. For f to be nonconstant, $\mathcal{L}$ cannot have finite limit points. If $t_n \to t_\infty$, $t_{n+1} - t_n \to 0$, so it suffices that 0 is not a limit point. This leads to the natural definition:

Definition. A *lattice*, $\mathcal{L} \subset \mathbb{C}$, is a subset that obeys

(i) $t_1, t_2 \in \mathcal{L} \Rightarrow t_1 - t_2 \in \mathcal{L}$

(ii) $\inf_{t\in\mathcal{L},\, t\neq 0}|t| > 0$

In other words, a lattice is a discrete additive subgroup. We'll always suppose $\mathcal{L}$ is nontrivial, that is, $\mathcal{L}$ has at least two (and so, an infinity of) points.

This section, as a preliminary, will study lattices in $\mathbb{C}$ and their geometry. We'll begin by defining bases showing the existence of what we'll call minimal bases. We'll see the set of all bases with $\operatorname{Im} \tau_2/\tau_1 > 0$ for a lattice, $\mathcal{L}$, is naturally a transitive homogeneous space for the group, $\mathbb{SL}(2,\mathbb{Z})$, and use the existence of minimal bases to find a fundamental domain for the action of $\mathbb{SL}(2,\mathbb{Z})$ on $\mathbb{C}_+$. There will be a subtle interplay between the group and the set of lattices. Finally, we'll study $\frac{1}{2}\mathcal{L} \supset \mathcal{L}$ and an associated natural subgroup, $\mathbf{\Gamma}(2)$, of $\mathbb{SL}(2,\mathbb{Z})$ and find a fundamental domain for it.

We begin by showing that every lattice as a group is isomorphic to $\mathbb{Z}$ or $\mathbb{Z}^2$.

Theorem 10.2.1. *Let $\mathcal{L}$ be a lattice in $\mathbb{C}$ and pick*

$$\tau_1 \in \{t \in \mathcal{L} \mid |t| = \inf_{s\in\mathcal{L}\setminus\{0\}} |s|\} \tag{10.2.1}$$

which is a nonempty set. Then, either

(i) $\mathcal{L} \subset \mathbb{R}\tau_1 = \{xt_1 \mid x \in \mathbb{R}\}$ (10.2.2)

in which case,

$$\mathcal{L} = \mathbb{Z}\tau_1 = \{n\tau_1 \mid n \in \mathbb{Z}\} \tag{10.2.3}$$

or

(ii) $\mathcal{L} \not\subset \mathbb{R}\tau_1$

in which case, if

$$\tau_2 \in \{t \in \mathcal{L}\setminus\mathbb{R}\tau_1 \mid |t| = \inf_{s\in\mathcal{L}\setminus\mathbb{R}\tau_1} |s|\} \tag{10.2.4}$$

which is nonempty, then

$$\mathcal{L} = \mathbb{Z}\tau_1 + \mathbb{Z}\tau_2 = \{n_1\tau_1 + n_2\tau_2 \mid n_j \in \mathbb{Z}\} \tag{10.2.5}$$

and any $t \in \mathcal{L}$ has a unique representation $n_1\tau_1 + n_2\tau_2$.

Proof. Let $R = \inf_{t\in\mathcal{L},\, t\neq 0}|t| > 0$, by hypothesis. If the set (10.2.1) were empty, there must be $t_n \in \mathcal{L}$ with $|t_n| > R$ but $|t_n| \to R$ and then, by compactness, distinct $t_n \to t_\infty$, which is forbidden since $|t - s| \geq R$ for all $t, s \in \mathcal{L}$ with $t \neq s$.

Suppose $\mathcal{L} \subset \mathbb{R}\tau_1$. Then every $t \in \mathcal{L}$ is of the form $x\tau_1$ with $x \in \mathbb{R}$. If $x \notin \mathbb{Z}$, $\{x\}\tau_1 = t - [x]\tau_1 \in \mathcal{L}$ and $0 < |\{x\}\tau_1| < |\tau_1| = R$, which is impossible. Thus, $\mathcal{L} \subset \mathbb{Z}\tau_1$. Clearly, $\mathbb{Z}\tau_1 \subset \mathcal{L}$, so (10.2.3) holds.

If $\mathcal{L} \not\subset \mathbb{R}\tau_1$, let R_1 be the inf in the set in (10.2.4). Since points in $\mathcal{L}$ are a distance at least R apart if there are k points in $\mathcal{L} \cap \mathbb{D}_{2R_1}(0)$, there are k disjoint disks of radius $R/2$ in $\mathbb{D}_{2R_1+R/2}$, so $k \le ((2R_1 + \frac{1}{2}R)/(\frac{1}{2}R))^2 < \infty$. Since there are finitely many points in $\mathcal{L} \cap \mathbb{D}_{2R_1} \setminus \mathbb{R}\tau_1$ and there are some points (by definition of R_1), there must be one of smallest norm, so the set in (10.2.4) is nonempty.

Since $\tau_2 \notin \mathbb{R}\tau_1$, τ_1 and τ_2 are a vector space basis for $\mathbb{C}$ as a vector space over $\mathbb{R}$, so any $t \in \mathcal{L}$ can be written

$$t = x_1(t)\tau_1 + x_2(t)\tau_2 \tag{10.2.6}$$

Picking integers with $|x_j(t) - n_j| \le \frac{1}{2}$, we see

$$s \equiv t - n_1\tau_1 - n_2\tau_2 = y_1\tau_1 + y_2\tau_2 \tag{10.2.7}$$

with $|y_j| \le \frac{1}{2}$. If $y_2 \neq 0$, $s \in \mathcal{L} \setminus \mathbb{R}\tau_1$ with

$$|s| < \tfrac{1}{2}\,|\tau_1| + \tfrac{1}{2}\,|\tau_2| \le |\tau_2| \tag{10.2.8}$$

(10.2.8) is obvious if $y_1 = 0$ and is true if $y_1 \neq 0$, since nonparallel vectors obey a strict triangle inequality. This is a contradiction, so $y_2 \equiv 0$. But then, since as above, $\mathcal{L} \cap \mathbb{R}\tau_1 = \mathbb{Z}\tau_1$, $y_1 = 0$. We conclude $s = 0$ so $t \in \mathbb{Z}\tau_1 + \mathbb{Z}\tau_2$.

Uniqueness of representation follows from the fact that τ_1 and τ_2 are linearly independent over $\mathbb{R}$. $\square$

We call $\mathcal{L}$ *one-dimensional* if $\mathcal{L} = \mathbb{Z}\tau_1$ and *two-dimensional* otherwise. Henceforth, we only consider the two-dimensional case. Two vectors, t_1, t_2, in $\mathcal{L}$, a two-dimensional lattice, are called a *basis* of $\mathcal{L}$ if and only if

$$\mathcal{L} = \{n_1t_1 + n_2t_2 \mid n_1, n_2 \in \mathbb{Z}\} \tag{10.2.9}$$

Since $\mathcal{L} \not\subset \mathbb{R}t_1$, we have that $\operatorname{Im}(t_2/t_1) \neq 0$. Clearly, if (t_1, t_2) are a basis, so are $(\pm t_1, \pm t_2)$ for the four independent choices of $\pm$. We can partly remove this multiplicity by demanding $\operatorname{Im}(t_2/t_1) > 0$, which can be arranged if necessary by replacing t_2 by $-t_2$. We'll call such a basis an *oriented basis* and $\tau = t_2/t_1$, the *ratio of the basis.* If (τ_1, τ_2) is an oriented basis which obeys (10.2.1) and (10.2.4), we call it a *minimal basis.*

$\mathbb{SL}(2,\mathbb{Z})$ is the set of elements in $\mathbb{SL}(2,\mathbb{C})$ with integral coefficients. The inverse of such a matrix is again in $\mathbb{SL}(2,\mathbb{Z})$ by Cramer's rule for the inverse. In addition, we note that if $A \in \mathbb{GL}(2,\mathbb{Z})$ with $A^{-1} \in \mathbb{GL}(2,\mathbb{Z})$, $\det(A) = \pm 1$, since $\det(A)\det(A^{-1}) = 1$ but both have to be integers. Any $A \in \operatorname{hom}(\mathbb{C}^2)$, the 2×2 complex matrices, induces a map $\mathcal{B}_A\colon (t_1, t_2) \to (s_1, s_2)$

on pairs of vectors by

$$\mathcal{B}_A(t_1,t_2) = (a_{11}t_1 + a_{12}t_2, a_{21}t_1 + a_{22}t_2) \tag{10.2.10}$$

Proposition 10.2.2. (a) *$\mathcal{B}_A$ maps $\mathcal{L}\times\mathcal{L}$ into $\mathcal{L}\times\mathcal{L}$ for any $A\in\hom(\mathbb{Z}^2)$, the 2×2 matrices with integral coefficients.*

(b) *If t_1,t_2 is an oriented basis for $\mathcal{L}$, then $\mathcal{B}_A(t_1,t_2)=(s_1,s_2)$ is an oriented basis if and only if $A\in\mathbb{SL}(2,\mathbb{Z})$.*

(c) *Given any two oriented bases of $\mathcal{L}$, (t_1,t_2) and (s_1,s_2), there is a unique $A\in\mathbb{SL}(2,\mathbb{Z})$ with*

$$\mathcal{B}_A(t_1,t_2) = (s_1,s_2) \tag{10.2.11}$$

Proof. (a) is obvious.

(b) By definition of basis, if (t_1,t_2) is a basis, then for any $(s_1,s_2)\in\mathcal{L}\times\mathcal{L}$ can be written (10.2.11) for $A\in\hom(\mathbb{Z}^2)$. If s_1,s_2 is also a basis, then $(t_1,t_2)=\mathcal{B}_B(s_1,s_2)$ where B also has integrable elements. Thus, $BA=\mathbb{1}$, so since $\det(A)\det(B)\in\mathbb{Z}$, $\det(A)=\pm 1$. But a simple calculation shows that

$$\operatorname{Im}\left(\frac{s_2}{s_1}\right) = \frac{\det(A)\operatorname{Im}(\frac{t_2}{t_1})}{|a_{11}+a_{12}\frac{t_2}{t_1}|^2} \tag{10.2.12}$$

so if both bases are oriented, $\det(A)>0$, that is, $A\in\mathbb{SL}(2,\mathbb{Z})$. Running these calculations backwards shows, conversely, that $\mathcal{B}_A$ for $A\in\mathbb{SL}(2,\mathbb{Z})$ maps oriented bases to oriented bases.

(c) Since (s_1,s_2) and (t_1,t_2) are bases, there exist A,B with integral elements so that $\mathcal{B}_A(t_1,t_2)=(s_1,s_2)$ and $\mathcal{B}_B(s_1,s_2)=(t_1,t_2)$. Thus, $AB=\mathbb{1}$, so $\det(A)=\pm 1$ and (10.2.12) implies $\det(A)>0$, so $A\in\mathbb{SL}(2,\mathbb{Z})$. $\square$

If $A=\left(\begin{smallmatrix} a_{11} & a_{12}\\ a_{21} & a_{22}\end{smallmatrix}\right)$ and

$$g_A(z) = \frac{a_{22}z + a_{21}}{a_{12}z + a_{11}} \tag{10.2.13}$$

(which, in terms of f_A of Section 7.3, is f_C with $C=\left(\begin{smallmatrix} 0 & i\\ i & 0\end{smallmatrix}\right)A\left(\begin{smallmatrix} 0 & i\\ i & 0\end{smallmatrix}\right)^{-1}$), if (10.2.11) holds, and if $\tau=t_2/t_1$, $\sigma=s_2/s_1$, then

$$\sigma = g_A(\tau) \tag{10.2.14}$$

Thus, $\mathbb{SL}(2,\mathbb{Z})$ acts on the set of all bases in such a way that the induced action on basis ratios is by FLTs, albeit a conjugate of the "usual" action.

We want to ask what ratios arise for minimal bases. Define (see Figure 10.2.1)

$$\mathcal{F}^{\text{int}} = \{z\in\mathbb{C}_+ \mid |z|>1,\ |\operatorname{Re} z|<\tfrac12\} \tag{10.2.15}$$

$$\mathcal{C}_+ = \{z\in\mathbb{C}_+ \mid |z|=1,\ 0\le\operatorname{Re} z\le\tfrac12\} \tag{10.2.16}$$

$$\mathcal{C}_- = \{z\in\mathbb{C}_+ \mid |z|=1,\ -\tfrac12\le\operatorname{Re} z\le 0\} \tag{10.2.17}$$

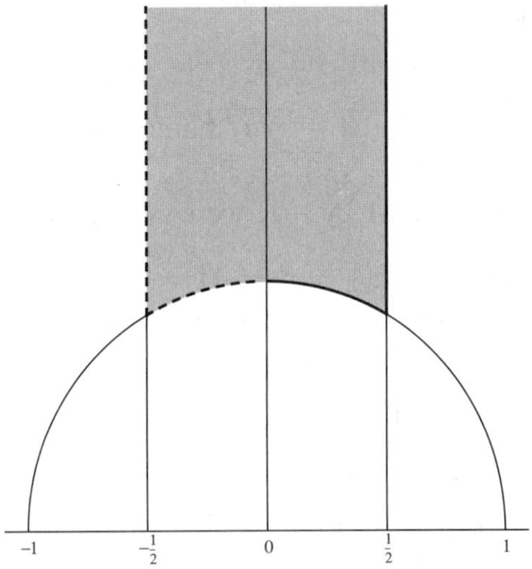

Figure 10.2.1. The region $\mathcal{F}$.

$$\mathcal{E}_\pm = \{z \in \mathbb{C}_+ \mid |z| \geq 1, \operatorname{Re} z = \pm\tfrac{1}{2}\} \tag{10.2.18}$$

$$\mathcal{F} = \mathcal{F}^{\text{int}} \cup \mathcal{C}_+ \cup \mathcal{E}_+, \qquad \overline{\mathcal{F}} = \mathcal{F} \cup \mathcal{C}_- \cup \mathcal{E}_- \tag{10.2.19}$$

A special role is played by the single point in $\mathcal{C}_+ \cap \mathcal{E}_+$, that is

$$\omega_0 = \tfrac{1}{2} + i\sqrt{\tfrac{3}{4}} \tag{10.2.20}$$

The lattice with $\tau_1 = 1$, $\tau_2 = \omega_0$ has a symmetry under rotations by 60°. Notice that ω_0 is a sixth root of unity. A less special role involves the point in $\mathcal{C}_+ \cap \mathcal{C}_-$, that is, $\tau = i$. The lattice with $\tau_1 = 1$, $\tau_2 = i$ has a symmetry under 90° rotations.

Theorem 10.2.3. *Let $A \in \mathbb{SL}(2,\mathbb{Z}), \tau, \sigma \in \mathcal{F}$ with $g_A(\tau) = \sigma$. Then one of the following holds:*

(i) $A = \pm\mathbb{1}$

(ii) $\tau = \sigma = i$; $A = \pm\left(\begin{smallmatrix} 0 & 1 \\ -1 & 0 \end{smallmatrix}\right)$

(iii) $\tau = \sigma = \omega_0$; $A = \pm\left(\begin{smallmatrix} 0 & 1 \\ -1 & 1 \end{smallmatrix}\right)$ *or* $A = \pm\left(\begin{smallmatrix} -1 & 1 \\ -1 & 0 \end{smallmatrix}\right)$

In all cases, $\tau = \sigma$.

Proof. Since $\tau = g_{A^{-1}}(\sigma)$, by interchanging τ and σ and replacing A by A^{-1}, we can suppose $\operatorname{Im}\sigma \geq \operatorname{Im}\tau$. In this equivalence, it is important to note that up to $\pm\mathbb{1}$, the A in (ii) is its own inverse and the A's in (iii) are each other's inverse. Now, since $\det(A) = 1$,

$$\sigma = \frac{a_{22}\tau + a_{21}}{a_{12}\tau + a_{11}}, \qquad \operatorname{Im}\sigma = \frac{\operatorname{Im}\tau}{|a_{12}\tau + a_{11}|^2} \tag{10.2.21}$$

so $\operatorname{Im}\sigma \geq \operatorname{Im}\tau$ implies

$$|a_{12}\tau + a_{11}|^2 \leq 1 \tag{10.2.22}$$

This implies $|a_{12}| \leq 1/|\operatorname{Im}\tau| \leq \sqrt{4/3}$ since the minimum value of $\operatorname{Im}\tau$ in $\mathcal{F}$ is at $\tau = \omega_0$ with $\operatorname{Im}\tau = \sqrt{3/4}$. Thus, $a_{12} = 0$ or ± 1. So long as we multiply the possibilities for A by -1, we can suppose $a_{12} = 0$ or $a_{12} = 1$.

Case 1: $a_{12} = 0$. Then $a_{11}a_{22} = \det(A) = 1$, so $a_{11} = a_{22} = \pm 1$ and, by (10.2.21), $\sigma = \tau \pm a_{21}$. Since $\mathcal{F} \cap (\mathcal{F} + n) = \emptyset$ for $n \in \mathbb{Z} \setminus \{0\}$, we must have $a_{21} = 0$, that is, $A = \pm\mathbb{1}$. So we have (i).

Case 2: $a_{12} = 1$, $a_{11} \neq 0$. Then

$$1 \geq |a_{11} + \operatorname{Re}\tau|^2 \geq (|a_{11} - \tfrac{1}{2}|)^2 \tag{10.2.23}$$

so $a_{11} = \pm 1$ and so, $|\operatorname{Im}\tau|^2 \leq 1 - \frac{1}{4}$ which implies $\tau = \omega_0$ since the only $\tau \in \mathcal{F}$ with $\operatorname{Im}\tau \leq \sqrt{3/4}$ is ω_0. Thus, $\operatorname{Re}\tau = \frac{1}{2}$ and (10.2.23) implies $a_{11} = -1$. Thus, $|a_{12}\tau + a_{11}| = |\omega_0 - 1| = 1$ and $\operatorname{Im}\sigma = \operatorname{Im}\tau$, so $\sigma = \omega_0$ also. Thus,

$$a_{22}\omega_0 + a_{21} = \omega_0(\omega_0 - 1) = \omega_0^3 = -1 \tag{10.2.24}$$

so $a_{22} = 0$, $a_{21} = -1$, that is, $A = \left(\begin{smallmatrix} -1 & 1 \\ -1 & 0 \end{smallmatrix}\right)$ and $\sigma = \tau = \omega_0$, which is one of the possibilities of (iii).

Case 3: $a_{12} = 1$, $a_{11} = 0$. Then $\det(A) = 1$ implies $a_{21} = -1$ and $\sigma = a_{22} - \tau^{-1}$. (10.2.22) implies $|\tau| \leq 1$, so since $\tau \in \mathcal{F}$, $|\tau| = 1$, so $\tau \in \mathcal{C}_+$ and $-\tau^{-1} \in \mathcal{C}_-$. Only two pairs $\mu_+ \in \mathcal{C}_+$, $\mu_- \in \mathcal{C}_-$ have $\mu_+ - \mu_- \in \mathbb{Z}$. $\mu_+ = \mu_- = i$ and $a_{22} = 0$ (which is when (ii) holds) or $\mu_+ = \omega_0$, $a_{22} = 1$ and $1 - \omega_0^{-1} = \omega_0$, which is the other possibility where (iii) holds. □

Theorem 10.2.4. (a) *For any minimal basis of a lattice, $\mathcal{L}$, the ratio τ lies in $\overline{\mathcal{F}}$.*

(b) *Every lattice has a minimal basis whose ratio lies in $\mathcal{F}$.*

(c) *If a lattice has two oriented bases whose ratios both lie in $\mathcal{F}$, the ratios are the same.*

(d) *If a basis has a ratio in $\overline{\mathcal{F}}$, then it is a minimal basis.*

(d) *Every $\tau \in \mathcal{F}$ is the ratio of a minimal basis of some lattice.*

Remark. Problem 1 will explore the number of distinct minimal bases a lattice has.

Proof. (a) Let τ_1, τ_2 be a minimal basis with $\tau_2/\tau_1 = \tau$. $|\tau_2| \geq |\tau_1|$ implies $|\tau| \geq 1$. Since it is oriented, $\operatorname{Im}\tau > 0$. Since $(\tau_1, \tau_2 + \tau_1)$ is also a basis, $|\tau_2 \pm \tau_1| \geq |\tau_2| \Rightarrow |1 \pm \tau|^2 \geq |\tau|^2 \Rightarrow 1 \pm 2\operatorname{Re}\tau \geq 0 \Rightarrow |\operatorname{Re}\tau| \leq \frac{1}{2}$. Thus, $\tau \in \{\tau \mid |\tau| \geq 1, \operatorname{Im}\tau > 0, |\operatorname{Re}\tau| \leq \frac{1}{2}\} = \overline{\mathcal{F}}$.

(b) Let (τ_1, τ_2) be a minimal basis with ratio τ. If $\tau \in \mathcal{E}_-$, $(\tau_1, \tau_2 + \tau_1)$ is an oriented basis and $|\tau_2 + \tau_1|/\tau_2 = |\tau + 1| = |\tau|$ since $\tau \in \mathcal{E}_-$, so $(\tau_1, \tau_2 + \tau_1)$ is a basis with ratio $1+\tau \in \mathcal{E}_+$. If $\tau \in \mathcal{C}_-$, $(\tau_2, -\tau_1)$ is an oriented basis, which is also minimal (since $|\tau_2| = |\tau_1|$), and the ratio for $(\tau_2, -\tau_1)$ is $-\tau^{-1} \in \mathcal{C}_+$. Thus, if $\tau \in \overline{\mathcal{F}} \setminus \mathcal{F}$, there is also a minimal basis with ratio in $\mathcal{F}$.

(c) Let (t_1, t_2) and (s_1, s_2) both be bases whose ratios τ, σ lie in $\mathcal{F}$. Then there is $A \in \mathbb{SL}(2,\mathbb{C})$ so $(s_1, s_2) = \mathcal{B}_A(t_1, t_2)$. Thus, $\sigma = g_A(\tau)$, so by Theorem 10.2.3, $\sigma = \tau$.

(d) Let (t_1, t_2) be the basis with ratio $\tau \in \overline{\mathcal{F}}$. By the construction in the proof of (b), there is another oriented basis with $|t_1'| = |t_1|$, $|t_2'| = |t_2|$, and ratio $\tau' \in \mathcal{F}$. By (b), there is a minimal basis (s_1, s_2) with ratio, σ, in $\mathcal{F}$. Thus, by (c), $\tau = \sigma$. Moreover, by Theorem 10.2.3, either the change of basis, A, is $\pm\mathbb{1}$, or A is one of the two special cases where all four lattice vectors (t_1', t_2', s_1, s_2) have the same length. Thus, t' and so t is minimal.

(e) Let $\tau \in \mathcal{F}$. Let $\mathcal{L} = \{n_1 + n_2\tau \mid n_1, n_2 \in \mathbb{Z}\}$. $1, \tau \in \mathcal{L}$ are clearly a basis. Their ratio is in $\mathcal{F}$, so it is a minimal basis. □

Definition. Let $\mathbb{G}$ be a group of analytic automorphisms of a region $\Omega \subset \mathbb{C}$. A *fundamental domain*, $\mathcal{G}$, is a subset of Ω that contains exactly one point of each orbit of $\mathbb{G}$, that is, for all $z \in \Omega$, there is a unique $w \in \mathcal{G}$ and some $F \in \mathbb{G}$ so $F(w) = z$.

Usually, one wants $\mathcal{G}$ to be connected and also that $\overline{\mathcal{G}^{\text{int}}} = \overline{\mathcal{G}}$.

Theorem 10.2.5. *$\mathcal{F}$ is a fundamental domain for the action $A \mapsto g_A$ of $\mathbb{SL}(2,\mathbb{Z})$ on $\mathbb{C}_+$.*

Remark. Since there is $T \in \mathbb{SL}(2,\mathbb{Z})$ with $f_A = g_{TAT^{-1}}$ so the orbits $\{f_A(Tw) \mid A \in \mathbb{SL}(2,\mathbb{C})\}$ are the orbits $\{g_A(w) \mid A \in \mathbb{SL}(2,\mathbb{C}\}$, $\mathcal{F}$ is also a fundamental domain for the action $A \mapsto f_A$.

Proof. Let $\tau \in \mathbb{C}_+$. Let $\mathcal{L} = \{n_1 + n_2\tau \mid n_1, n_2 \in \mathbb{Z}\}$. $\mathcal{L}$ is a lattice with basis $(1, \tau)$. By Theorem 10.2.4, it has a minimal basis with ratio, σ, in $\mathcal{F}$ so $\tau = g_A(\sigma)$ for some $A \in \mathbb{SL}(2,\mathbb{C})$. So each orbit contains a point in $\mathcal{F}$. Uniqueness follows from Theorem 10.2.3. □

The relation of the lattice $\frac{1}{2}\mathcal{L} = \{\frac{1}{2}z \mid z \in \mathcal{L}\}$ to $\mathcal{L}$ will play an important role, especially in Section 10.6, and related to this is the subgroup of $\mathbb{SL}(2,\mathbb{C})$

$$\Gamma(2) = \left\{ \begin{pmatrix} \alpha & \beta \\ \gamma & \delta \end{pmatrix} \;\middle|\; \alpha, \delta \text{ odd}; \beta, \gamma \text{ even} \right\} \tag{10.2.25}$$

called the *principal conjugacy subgroup of level* 2. As a test, the reader might guess the answer to the question of what its index is as a subgroup of $\mathbb{SL}(2,\mathbb{Z})$, that is, how many left cosets it has. We'll answer it shortly.

$\mathcal{L}$ is a subgroup of index 4 in $\frac{1}{2}\mathcal{L}$—its four cosets are naturally labeled *ee*, *eo*, *oe*, and *oo*. For example, $oe = \{\frac{1}{2}(n_1\tau_1 + n_2\tau_2) \mid n_1 \text{ odd and } n_2 \text{ even}\}$.

If we pick a basis τ_1, τ_2 of $\mathcal{L}$, $\mathbb{SL}(2,\mathbb{Z})$ acts on $\frac{1}{2}\mathcal{L}$ by $\mathcal{A}_A\colon \frac{1}{2}\mathcal{L} \to \frac{1}{2}\mathcal{L}$ by

$$\mathcal{A}_A(\tfrac{1}{2}n_1\tau_1 + \tfrac{1}{2}n_2\tau_2) = \tfrac{1}{2}n_1(a_{11}\tau_1 + a_{12}\tau_2) + \tfrac{1}{2}n_2(a_{21}\tau_1 + a_{22}\tau_2) \quad (10.2.26)$$

then each $\mathcal{A}_A$ takes ee to ee, that is, $\mathcal{A}_A$ maps $\mathcal{L}$ to $\mathcal{L}$, and so it maps cosets of $\mathcal{L}$ in $\frac{1}{2}\mathcal{L}$ to (possibly distinct) cosets, that is, there is a map $\widetilde{\mathcal{A}}_A\colon \frac{1}{2}\mathcal{L}/\mathcal{L} \to \frac{1}{2}\mathcal{L}/\mathcal{L}$ by

$$\widetilde{\mathcal{A}}_A([t]) = [\mathcal{A}_A(t)] \quad (10.2.27)$$

Since

$$\mathcal{A}_A(\tfrac{1}{2}n_1\tau_1 + \tfrac{1}{2}n_2\tau_2) = \tfrac{1}{2}\tau_1(a_{11}n_1 + a_{21}n_2) + \tfrac{1}{2}\tau_2(a_{12}n_1 + a_{22}n_2) \quad (10.2.28)$$

we see $a_{11}n_1+a_{21}n_2 \equiv n_1 \pmod 2$, $a_{21}n_1+a_{22}n_2 \equiv n_2 \pmod 2$ for all n_1, n_2 if and only if all of $a_{11}-1, a_{21}, a_{21}, a_{22}-1$ are even, that is, $A \in \ker(\widetilde{\mathcal{A}}_A) \equiv \{A \mid \mathcal{A}_A([t]) = [t] \text{ for all } t\}$ if and only if $A \in \Gamma(2)$. Thus, $\mathbb{SL}(2,\mathbb{Z})/\Gamma(2)$ acts on $\frac{1}{2}\mathcal{L}/\mathcal{L}$ and this action is one-one. The image maps all map the coset $[\mathcal{L}] \equiv ee$ to itself and permute the other three cosets oo, oe, and eo in some way. $A = \left(\begin{smallmatrix} 0 & 1 \\ 1 & 0 \end{smallmatrix}\right)$ leaves oo fixed but permutes oe and eo, while by (10.2.28), $A = \left(\begin{smallmatrix} 1 & 1 \\ -1 & 0 \end{smallmatrix}\right)$ maps oo to eo, eo to oe, and oe to oo. These two elements of S_3, the six-element permutation group on three objects, generate the whole group, so $\text{Ran}(\widetilde{\mathcal{A}}_A)$ has six elements, that is, the index of $\Gamma(2)$ in $\mathbb{SL}(2,\mathbb{Z})$ is six. Thus, we've proven that:

Theorem 10.2.6. $\Gamma(2)$ *has index* 6 *in* $\mathbb{SL}(2,\mathbb{Z})$.

Now, let $\widetilde{\mathcal{F}}$ be the set

$$\widetilde{\mathcal{F}} = \{z \mid -1 < \text{Re}\, z \le 1;\ \text{Im}\, z > 0;\ |z - \tfrac{1}{2}| \ge \tfrac{1}{4},\ |z + \tfrac{1}{2}| > \tfrac{1}{4}\} \quad (10.2.29)$$

which we've seen in Section 8.3; see Figure 8.3.1.

Theorem 10.2.7. $\widetilde{\mathcal{F}}$ *is a fundamental region for the group* $\Gamma(2)$ *under the action* g_A *of* $\Gamma(2)$.

Remark. In Section 8.3, we proved that $\widetilde{\mathcal{F}}$ is a fundamental region for the group of FLTs which were products of an even number of reflections in $|z - \frac{1}{2}| = \frac{1}{2}$, in $\text{Re}\, z = 0$, and in $\text{Re}\, z = 1$. In Problem 2, the reader will show this group is $\Gamma(2)$ and so provide another proof of this theorem.

Proof. It's proof by picture!

Figure 10.2.2 divides $\widetilde{\mathcal{F}}$ in twelve regions, six shaded and six not. The two regions labeled $\boxed{1}$ and $\boxed{7}$ form $\mathcal{F}$. There are five elements, $C_2, \dots, C_6$, in $\mathbb{SL}(2,\mathbb{Z})$, one from each nontrivial coset of $\mathbb{SL}(2,\mathbb{Z})/\Gamma(2)$, that map region $\boxed{1}$ to $\boxed{2}$–$\boxed{6}$ and five different ones, $C_8, \dots, C_{12}$, that map $\boxed{7}$ to $\boxed{8}$–$\boxed{11}$. Set $C_1 = C_7 = \mathbb{1}$. Given any $z \in \mathbb{C}_+$, find $B \in \mathbb{SL}(2,\mathbb{C})$ with $g_B(z) \in \mathcal{F}$. If in region $\boxed{1}$, find the one of $C_1, \dots, C_6$ so $C_jB \in \Gamma(2)$. Then $g_{C_jB}(z) \in \widetilde{\mathcal{F}}$,

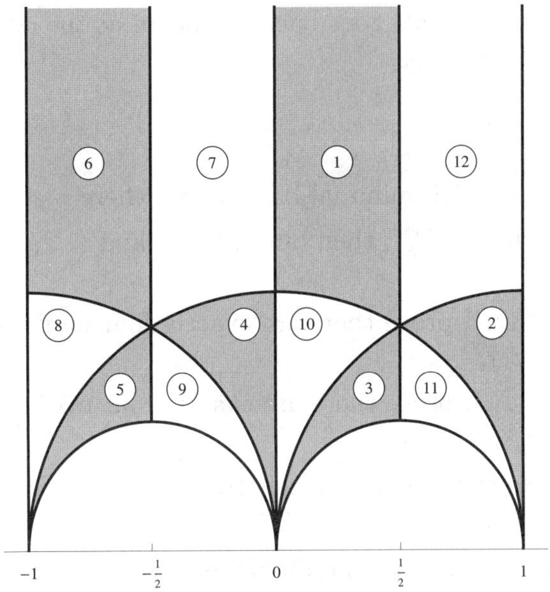

Figure 10.2.2. Using $\mathcal{F}$ to cover $\widetilde{\mathcal{F}}$.

showing each orbit intersects $\widetilde{\mathcal{F}}$ and the uniqueness of the element of $\mathbb{SL}(2,\mathbb{Z})$ and that only one of $C_jB \in \Gamma(2)$ for $j = 1, \dots, 6$ imply uniqueness. □

We haven't worried about proving proper counting of boundary points. We leave that to Problem 4.

Notes and Historical Remarks. The lattices associated with the symmetry points $\tau = \omega_0$ and $\tau = i$ have six-fold and four-fold rotational symmetries. It is no coincidence that four and six occur and not, say, 10 or eight (because of $\tau \to -\tau$ symmetry, only even orders occur). For in the τ_1, τ_2 basis, the implementation of the rotations lies in $\mathbb{SL}(2,\mathbb{Z})$, so their trace is in $\mathbb{Z}$. But a rotation of angle at θ has trace $2\cos(\theta)$. Since $\cos(\theta) \in [-1,1]$, this is in $\mathbb{Z}$ only for $\theta = 0, \pi j/3$ ($j = 1, \dots, 5$), $\pi j/2$ ($j = 1, 2, 3$), so only two-, three-, four-, six-fold symmetries occur.

These same arguments imply that the only elements of $\mathbb{SL}(2,\mathbb{Z})$ that have a fixed point in $\mathbb{C}_+$ are of order two or three in $\mathbb{PSL}(2,\mathbb{Z})$ (as we've seen, elements of $\mathbb{A}\text{ut}(\mathbb{D})$ with a fixed point in $\mathbb{D}$ are elliptic, so that is true of $\mathbb{C}_+$). Note the special A's in Theorem 10.2.3 are order four and six and induce elements of $\mathbb{PSL}(2,\mathbb{Z})$ of order two and three.

One consequence of the theorem that a lattice is spanned by a two-element basis is that if a function has three periods that are independent in that no nontrivial integral combination of them is zero, then f is constant (see Problem 5). This result goes back to Jacobi at the dawn of the theory of doubly periodic functions.

Problems

1. Let $\mathcal{L}$ be a lattice with minimal basis τ_1, τ_2 where $\tau = \tau_2/\tau_1 \in \mathcal{F}$.

   (a) Prove that if $\tau \in \mathcal{F}^{\text{int}}$, then only (τ_1, τ_2) and $(-\tau_1, -\tau_2)$ are minimal bases.

   (b) If $\tau \in \mathcal{E}_+ \setminus \{\omega_0\}$, prove there are exactly four minimal bases $\pm(\tau_1, \tau_2)$ and $\pm(\tau_1, \tau_2 - \tau_1)$.

   (c) If $\tau \in \mathcal{C}_+ \setminus \{\omega_0\}$, prove there are exactly four minimal bases $\pm(\tau_1, \tau_2)$ and $\pm(\tau_2, -\tau_1)$.

   (d) If $\tau = \omega_0$, prove there are exactly twelve minimal bases $(\tau_1\omega_0^\ell, \tau_1\omega_0^{\ell+1})$ and $(\tau_1\omega_0^\ell, \tau_2\omega_0^{\ell+2})$ for $\ell = 0, 1, \dots, 5$.

   (e) Prove that the number of minimal bases with ratio τ is two if $\tau \neq i, \omega_0$, four if $\tau = i$, and six if $\tau = \omega_0$.

2. This will lead you through a proof that $\Gamma(2)$ is generated by $\pm\left(\begin{smallmatrix} 1 & 2 \\ 0 & 1 \end{smallmatrix}\right)$ and $\left(\begin{smallmatrix} 1 & 0 \\ 2 & 1 \end{smallmatrix}\right)$. Let $\left(\begin{smallmatrix} a & b \\ c & d \end{smallmatrix}\right) \in \Gamma(2)$ and let G be the subgroup generated by $\pm\tau = \pm\left(\begin{smallmatrix} 1 & 2 \\ 0 & 1 \end{smallmatrix}\right)$ and $\sigma = \left(\begin{smallmatrix} 1 & 0 \\ 2 & 1 \end{smallmatrix}\right)$.

   (a) Prove that if $c = 0$, then $\left(\begin{smallmatrix} a & b \\ c & d \end{smallmatrix}\right) \in G$.

   (b) Let $X = \Gamma(2) \setminus G$. Prove that $x \in X$ and $y \in G \Rightarrow yx \in X$.

   (c) Suppose X is nonempty and $x_0 = \left(\begin{smallmatrix} a & b \\ c & d \end{smallmatrix}\right) \in X$ is chosen so that $|c|$ is minimal among $x \in X$, and among those x with this value of $|c|$, $|a|$ is minimal. By considering $\tau^{\pm 1}x$, prove that $|a \pm 2c| \geq |a|$ or $|a| \leq |c|$. Note that, by (a), $|c| \geq 2$, and by a odd, $|a| \geq 1$. By considering $\sigma^{\pm 1}x$, prove that $|c \pm 2a| \geq |c|$ or $|c| \leq |a|$. Get a contradiction and conclude that $G = \Gamma(2)$.

   (d) Prove that the group generated by even numbers of reflections in $\operatorname{Re} z = 0$, $\operatorname{Re} z = 1$, and $|z - \frac{1}{2}| = \frac{1}{2}$ is $\Gamma(2)$.

3. Find explicit formulae for the $\mathbb{SL}(2, \mathbb{Z})$ matrices $C_1, \dots, C_{12}$ of the proof of Theorem 10.2.7.

4. Prove that boundary points are properly counted so that each orbit in $\Gamma(2)$ contains exactly one point of $\widetilde{\mathcal{F}}$. (*Hint*: You'll need to treat ω_0 and i differently from other points! Why?)

5. (a) If τ_1, τ_2 is a basis of a lattice, $\mathcal{L}$, prove that if σ_1, σ_2 are any independent points in $\mathcal{L}$ (in that for no $(m_1, m_2) \in \mathbb{Z}^2$ other than $(0, 0)$ is $m_1\sigma_1 + m_2\sigma_2 = 0$), then τ_1 and τ_2 are rational linear combinations of σ_1 and σ_2.

(b) Prove for any $\sigma_1, \sigma_2, \sigma_3 \in \mathcal{L}$, there are integers m_1, m_2, m_3 not all zero so that $m_1\sigma_1 + m_2\sigma_2 + m_3\sigma_3 = 0$.

(c) If $\sigma_1, \sigma_2, \sigma_3$ are three points in $\mathbb{C}$ independent over $\mathbb{Z}$, prove that the integral combinations of $\sigma_1, \sigma_2, \sigma_3$ are dense in $\mathbb{C}$. (This is a result of Jacobi.)

10.3. Liouville's Theorems, Abel's Theorem, and Jacobi's Construction

In this section, we'll prove a series of four theorems of Liouville that delineate what possibilities there are for elliptic functions and state, but only begin the proof of the basic existence theorems that say the necessary conditions found by Liouville are also sufficient. We'll reduce, following Jacobi, the existence theorems to the existence of a single nonelliptic function with suitable properties. Later, once in Section 10.4 and once in Section 10.5, we'll construct the required function in two ways. We begin by stating Liouville's theorems together. f will be nonconstant and have periods τ_1, τ_2 with $\operatorname{Im}(\tau_2/\tau_1) > 0$, $\{z_j\}_{j=1}^{N_z}$ and $\{p_j\}_{j=1}^{N_p}$ will be its zeros and poles in

$$\{x\tau_1 + y\tau_2 \mid 0 \le x < 1,\ 0 \le y < 1\} \tag{10.3.1}$$

counting multiplicity. (10.3.1) is called the *fundamental cell* or fundamental period parallelogram; see Figure 10.3.1.

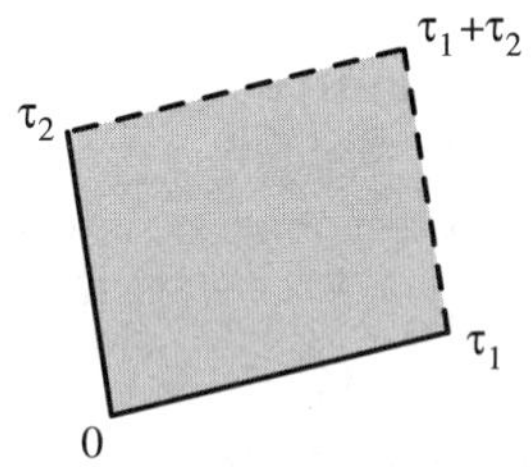

Figure 10.3.1. A fundamental cell.

Theorem 10.3.1 (Liouville's First Theorem). *f has poles, that is, $N_p \ge 1$.*

Theorem 10.3.2 (Liouville's Second Theorem).

$$N_p = N_z \tag{10.3.2}$$

Theorem 10.3.3 (Liouville's Third Theorem). *The sum of the residues at the poles is zero, that is,*

$$\sum_{j=1}^{N_p} \operatorname{Res}(f; p_j) = 0 \tag{10.3.3}$$

Corollary 10.3.4. $N_p \ge 2$

Proof. By the first theorem, $N_p \geq 1$. If $N_p = 1$, there is a single pole and it is first order, so its residue is nonzero. This contradicts (10.3.3). □

Theorem 10.3.5 (Liouville's Fourth Theorem). *We have*

$$\sum_{j=1}^{N_z} (z_j - p_j) \in \mathcal{L}_{\tau_1,\tau_2} \tag{10.3.4}$$

Remarks. 1. The numbering first–fourth is not canonical. We presented them in logical order, although our proofs will be in a rather different order.

2. We have, of course, seen the first and second theorems before—they are special cases of Theorems 7.1.7 and 7.1.10 but, following Liouville, we'll give "direct" proofs that don't use Liouville's theorem on bounded analytic functions or the notion of degree. Corollary 10.3.4 which, given what we know about degree, says $\deg(f) \geq 2$ may be surprising at first sight, but it is obvious from general consideration. If $\deg(f) = 1$, f is a bijection, so $\mathcal{J}_{\tau_1,\tau_2}$ and $\widehat{\mathbb{C}}$ would be analytically equivalent, but they aren't even topologically homeomorphic!

3. The most interesting results here are those that do not have analogs for $\widehat{\mathbb{C}}$, namely, (10.3.3) and (10.3.4).

Proofs of Liouville's Four Theorems. We'll use contour integrals around the contour, Γ, shown in Figure 10.3.2 for $\tau_1 = 1$, $\tau_2 = \tau$, that is, it goes counterclockwise around the boundary of the fundamental cell, 0 to τ_1 to $\tau_1 + \tau_2$ to τ_2 to 0. Because $\operatorname{Im}(\tau_2/\tau_1) > 0$, this contour is counterclockwise. We'll suppose f has no zeros or poles on Γ. If it does then, for small positive ε, $f(z + \varepsilon(\tau_1 + \tau_2))$ has no zeros or poles on Γ and apply the argument to it.

If g is any doubly periodic function with no poles on Γ, $\oint_\Gamma g(z)\,dz = 0$ because the integrals over Γ_1 and Γ_3 and over Γ_2 and Γ_4 cancel. Thus,

$$\oint_\Gamma f(z)\,dz = 0 = \oint_\Gamma \frac{f'(z)}{f(z)}\,dz \tag{10.3.5}$$

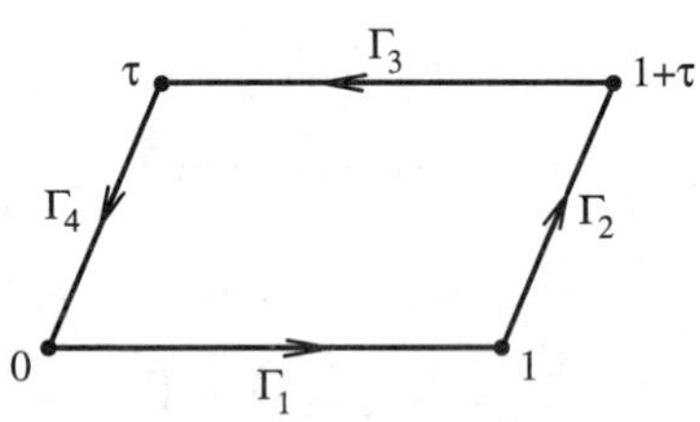

Figure 10.3.2. Contour for Liouville's theorems.

By the residue calculus, the integral on the left is the left side of (10.3.3), proving the third theorem; and by the argument principle, the integral on the right is $2\pi i(N_z - N_p)$, proving the second theorem.

The second theorem implies the first, for since f is nonconstant, $f(z)-a$ has isolated zeros for any a, so $N_z(f-a) = N_p(f-a)$ implies $N_p \neq 0$. Indeed, this argument shows that $N_z(f-a)$ is constant for all a, that is, reproves the constancy of degree.

(10.3.4) is the most subtle. We consider

$$Q = \frac{1}{2\pi i}\oint_\Gamma z\,\frac{f'(z)}{f(z)}\,dz \tag{10.3.6}$$

On the one hand, by the fact that f'/f has poles exactly at the z_j and p_j with residues $\pm$ (the order of zeros or poles)

$$Q = \sum_j (z_j - p_j)$$

There is no cancellation between Γ_1 and Γ_3 and Γ_2 and Γ_4; $f'(z+\tau_j)/f(z+\tau_j) = f'(z)/f(z)$ but the $z+\tau_j$ doesn't cancel z but leaves τ_j, so

$$Q = -\frac{\tau_2}{2\pi i}\int_{\Gamma_1}\frac{f'(z)}{f(z)}\,dz + \frac{\tau_1}{2\pi i}\int_{\Gamma_2}\frac{f'(z)}{f(z)}\,dz \tag{10.3.7}$$

Now, $\frac{1}{2\pi i}\int_{\Gamma_1} f'/f\,dz$ is the change of the argument of f from 0 to 1. Since $f(0) = f(1)$, $f[\Gamma_1]$ is a closed curve and the integral is its winding number about 0, that is, the integral is an integer, $-n_2$. Similarly, the second integral in (10.3.7) is an integer, n_1, so

$$Q = n_1\tau_1 + n_2\tau_2 \tag{10.3.8}$$

□

These theorems place two restrictions, namely, (10.3.2) and (10.3.4) on the zeros and poles, and one, namely, (10.3.3) on the principal parts. The next results say these are the only restrictions:

Theorem 10.3.6 (Abel's Theorem). *Let $\{z_j\}_{j=1}^{N_z}$ and $\{p_j\}_{j=1}^{N_z}$ be two collections of points in the fundamental cell of a lattice $\mathcal{L}_{\tau_1,\tau_2}$ with $z_j \neq p_k$ for all j,k. Then a necessary and sufficient condition for there to exist a doubly periodic function with period lattice $\mathcal{L}_{\tau_1,\tau_2}$ with precisely those zeros and poles in the fundamental cell is that* (10.3.2) *and* (10.3.4) *hold.*

Theorem 10.3.7 (Principal Part Theorem). *Let $\{p_j\}_{j=1}^{\widetilde{N}_p}$ be a finite collection of distinct points in the fundamental cell and $\{P_j(z)\}_{j=1}^{\widetilde{N}_p}$ a collection of* *nonzero polynomials with* $P_j(0) = 0$. *Then a necessary and sufficient condition for there to exist a doubly periodic function with period lattice* $\mathcal{L}_{\tau_1,\tau_2}$ *with precisely these poles and principal part* $P_j(1/(z-p_j))$ *at* p_j *is that*

$$\sum_{j=1}^{\widetilde{N}_p} P_j'(0) = 0 \tag{10.3.9}$$

Remarks. 1. It is not claimed that there couldn't be a larger period lattice; indeed, if the zeros and poles are invariant under adding $\tau_j/2$, the period lattice will include $\mathcal{L}_{\frac{1}{2}\tau_1,\frac{1}{2}\tau_2}$.

2. By Theorem 7.1.9, these functions are unique up to one overall constant.

3. Of course, all the zeros and poles are $\{z_j+t\}_{j=1,\,t\in\mathcal{L}}^{N_z}$ and $\{p_j+t\}_{j=1,\,t\in\mathcal{L}}^{N_p}$.

4. (10.3.9) is just (10.3.3).

5. We use $\widetilde{N}_p$ in the second theorem because there, $\{p_j\}_{j=1}^{\widetilde{N}_p}$ are distinct, that is, multiplicity is encoded in the $\deg(P_j)$, not in the set of $\{p_j\}_{j=1}^{\widetilde{N}_p}$.

6. The fact that $N_p \geq 2$ is forced by Abel's theorem, because if z, p are in the fundamental cell and $z \neq p$, then $z - p \notin \mathcal{L}$.

How might one go about constructing such a set of functions? Suppose we could find a doubly periodic analytic function, g, with a single zero in the fundamental cell at $z = 0$. Of course, such functions do not exist! But ignore that for a moment. Then

$$f(z) = \frac{\prod_{j=1}^{N_z} g(z - z_k)}{\prod_{k=1}^{N_p} g(z - p_k)} \tag{10.3.10}$$

is doubly periodic and has precisely the right zeros and poles. Of course, that would be too good since it violates both (10.3.2) and (10.3.4)!

For f to be doubly periodic, it isn't necessary that g be doubly periodic. Suppose instead that

$$g(z + \tau_j) = e^{\alpha_j} g(z) \tag{10.3.11}$$

for constants $e^{\alpha_1}, e^{\alpha_2}$. $g(0) = 0$ still forces $g(t) = 0$ for all $t \in \mathcal{L}$. Then $f(z)$ will obey $f(z + \tau_j) = f(z)$ so long as $N_p = N_z$. For the constants cancel in the ratio.

It isn't quite obvious we can't arrange (10.3.11). After all,

$$g(z) = C_1 e^{C_2 z} \tag{10.3.12}$$

does obey (10.3.11), but it doesn't have the required zero. Again, it would be too good if this g existed since it could produce f's, violating (10.3.4). In fact (see Problem 1), (10.3.12) are the only functions that obey (10.3.11).

But we've made some progress. Suppose, instead, that for some $\alpha_1, \alpha_2, \omega_1, \omega_2 \in \mathbb{C}$, g obeys

$$g(z+\tau_j) = e^{\alpha_j + \omega_j z} g(z) \tag{10.3.13}$$

If $\sum_{j=1}^{N_z}(z_j - p_j) = 0$, then, in fact, g given by (10.3.10) will be doubly periodic since, if (10.3.13) holds,

$$\prod_{k=1}^{N_z} g(z+\tau_j - z_k) = e^{\alpha_j N_z} e^{\omega_j (\sum_{k=1}^{N_z} z_k)} \prod_{k=1}^{N_z} g(z - z_k) \tag{10.3.14}$$

But what can we do if $N_z = N_p$ and $\sum_{j=1}^{N_z}(z_j - p_j) \equiv t_0 \in \mathcal{L}$ but is not zero?

Define

$$\tilde{z}_1 = z_1 - t_0, \quad \tilde{z}_j = z_j, \quad j = 2, \dots, N_z; \qquad \tilde{p}_j = p_j, \quad j = 1, \dots, N_z \tag{10.3.15}$$

and $g(z - \tilde{z}_j)$ has zeros precisely at $\{\tilde{z}_j + t\}_{t \in \mathcal{L}}^{N_z} = \{z_j + t\}_{t \in \mathcal{L}}^{N_z}$ since $t_0 \in \mathcal{L}$. Thus,

$$f(z) = \frac{\prod_{j=1}^{N_z} g(z - \tilde{z}_j)}{\prod_{j=1}^{N_z} g(z - \tilde{p}_j)} \tag{10.3.16}$$

is doubly periodic with the right zeros and poles. We have thus proven:

Theorem 10.3.8 (Jacobi's Construction). *If there exists an entire analytic function, g, obeying* (10.3.13) *for some $\alpha_1, \alpha_2, \omega_1, \omega_2$, and if the only zero of g in the unit cell is at $z = 0$, then* (10.3.16) *solves the existence part of Abel's theorem and proves that part of Theorem* 10.3.6.

Of course, Liouville's theorems include the other half.

We'll construct the required g twice: in the next section as a Weierstrass product (calling g, σ) and in the section after that, using Jacobi's theta functions (calling g, θ). g is not unique (see Problem 5) and, indeed, the ω's for σ and θ are different.

As an extra bonus, the existence of g also proves Theorem 10.3.7.

Proof of Theorem 10.3.7 if g exists. Let

$$h_1(z) = \frac{g'(z)}{g(z)}; \qquad h_n(z) = -\frac{1}{n-1}\frac{d}{dz} h_{n-1}(z), \qquad n = 2, 3, \dots \tag{10.3.17}$$

Then, since $g(z)$ vanishes to first order at $z = 0$, $h_1(z) = \frac{1}{z} + O(1)$ at $z = 0$, and then, by induction,

$$h_n(z) = \frac{1}{z^n} + O(1) \tag{10.3.18}$$

since $(h_1(z) - \frac{1}{z})$ is analytic at $z = 0$.

Moreover, by (10.3.13),

$$h_1(z+\tau_j) = h_1(z) + \omega_j, \qquad h_n(z+\tau_j) = h_n(z) \quad n \geq 2 \tag{10.3.19}$$

where the second equation holds by induction. Write

$$P_k(z) = \sum_{m=1}^{\ell_k} a_{km} z^m \tag{10.3.20}$$

By (10.3.18),

$$f(z) = \sum_{km} a_{km} h_m(z - z_k) \tag{10.3.21}$$

has the requisite principal parts.

Since (10.3.9) says that $\sum_k a_{k1} = 0$, we have that

$$\begin{aligned}\sum_{k,\ell} a_{kl} h_\ell(z+\tau_j - z_k) &= \sum_{k,\ell} a_{kl} h_\ell(z - z_k) + \tau_j \sum_k a_{k1} \\ &= \sum_{k,\ell} a_{kl} h_\ell(z - z_k)\end{aligned}$$

so f is doubly periodic. □

The doubly periodic functions, f, for $\mathcal{L}_{\tau_1,\tau_2}$ that one constructs from g either have second derivatives (e.g., $h_1 = (g'/g)'$) or since $\deg(f) \geq 2$, a total of four g's. But one can construct doubly periodic functions from only two g's but with periods, τ_1 and $2\tau_2$. Since the fundamental cell is now twice as big, in that cell, one has two zeros and two poles. The functions will be antiperiodic when only τ_2 is added, so in the fundamental cell for $\mathcal{L}_{\tau_1,2\tau_2}$, there will be zeros at $0, \tau_2$ and poles at $z_0, z_0 + \tau_2$. For this to obey (10.3.4) for $\mathcal{L}_{\tau_1,2\tau_2}$, we will need $2z_0 \in \mathcal{L}_{\tau_1,2\tau_2}$, so we'll take $z_0 = \tau_1/2$. Thus, we want a ratio $g(z)/g(z+\tau_1/2)$ with a phase adjustment. We'll suppose

$$g(-z) = -g(z) \tag{10.3.22}$$

By Problems 3 and 5, this can always be arranged if any g obeying (10.3.13) exists. When (10.3.22) holds, we have that (Problem 3)

$$g(z+\tau_j) = -e^{\omega_j(z+\tau_j/2)} g(z) \tag{10.3.23}$$

Here is the result:

Theorem 10.3.9. *Let g obey (10.3.13) and (10.3.22). Define for $j = 1, 2$,*

$$g_j(z) = e^{-\omega_j z/2} g\left(z + \frac{\tau_j}{2}\right) \tag{10.3.24}$$

Then with $\tilde{\tau}_j = \tau_{3-j}$ for $j = 1, 2$, the function

$$f_j(z) = \frac{g(z)}{g_j(z)} \tag{10.3.25}$$

has poles in $\tau_j/2 + \mathcal{L}_\tau$ *and zeros in* $\mathcal{L}_\tau$ *and obeys*

$$f_j(z+\tau_j) = f_j(z), \qquad f_j(z+\tilde{\tau}_j) = -f_j(z) \tag{10.3.26}$$

Proof. Problem 6 proves (10.3.26). □

One way of understanding the phase in (10.3.24) is to plug $z - \tau_j/2$ for z in(10.3.23) to see

$$g\left(z+\frac{\tau_j}{2}\right)e^{-\omega_j z} = g\left(z-\frac{\tau_j}{2}\right) \tag{10.3.27}$$

so

$$g_j^2(z) = g\left(z+\frac{\tau_j}{2}\right)g\left(z-\frac{\tau_j}{2}\right) \tag{10.3.28}$$

and

$$f_j(z)^2 = \frac{g(z)^2}{g(z+\frac{\tau_j}{2})g(z-\frac{\tau_j}{2})}$$

is the proper choice (using a shift like in (10.3.16) to arrange $\sum_j \tilde{z}_j - \tilde{p}_j = 0$) to get an elliptic function with poles at $\frac{1}{2}\tau_j + \mathcal{L}_\tau$ and zeros at $\mathcal{L}_\tau$.

Theorem 10.3.9 is not only interesting from an abstract point of view, but will be used for our construction of the Jacobi elliptic functions $\mathrm{sn}(z), \mathrm{cn}(z), \mathrm{dn}(z)$ in Section 10.5. In that regard, we note that τ_1, τ_2 can be replaced by any basis s_1, s_2 with $\mathrm{Im}(s_2/s_1) > 0$ (since that is all we use for τ_2/τ_1 and g obeys (10.3.13) for τ_1, τ_2 if and only if it obeys it for s_1, s_2 with different α_j, ω_j). In particular, if

$$g_3(z) = e^{-(\omega_1+\omega_2)z/2} g\left(z+\frac{\tau_1+\tau_2}{2}\right) \tag{10.3.29}$$

then

$$f_3(z) = \frac{g(z)}{g_3(z)} \tag{10.3.30}$$

has

$$f_3\left(z+\frac{\tau_1+\tau_2}{2}\right) = f_3(z), \qquad f_3(z+\tau_1) = -f_3(z) \tag{10.3.31}$$

since $(\tau_1, \tau_1+\tau_2)$ is also a basis.

Notes and Historical Remarks. While Liouville's results are natural as part of a pedagogic presentation, they not only have much less mathematical depth than the work of Abel, Jacobi, and Weierstrass, but had much less historical significance. Indeed, as explained in the Notes to Section 3.1, while Liouville discussed this work already in 1844 (almost twenty years after the work of Abel and Jacobi), it was only published in 1880 [**64**].

Abel and his contemporaries would certainly not recognize what we (and others) call Abel's theorem. Abel was studying elliptic integrals, notably the lemniscate integral $\int_0^x (1-y^4)^{-1/2}\,dy = u$ and made the critical observation that the inverse function $F(u) = x$ was, as a function of a complex variable, doubly periodic [**2**]. More generally, he looked at $\int_{x_0}^{x} R(z,w)\,dz$ where R is a rational function and (z,w) obeys a polynomial equation $P(z,w) = 0$. These are now called *abelian integrals* (some authors require $\deg(P) > 4$ while others include elliptic integrals among the Abelian integrals). Unlike the elliptic case, this integral may not define a one-one multivalued map of x to u so there is not an inverse function, but the integral is still multiple valued with periods. Abelian integrals provide existence results on what are essentially meromorphic functions on certain Riemann surfaces, so the name Abel's theorem to the more general result (see Section 11.6) is appropriate, albeit the form and language we use here is much more modern.

In his major work on elliptic functions [**280**], Jacobi constructed elliptic functions as ratios of products of certain nonelliptic functions which obey (10.3.13), so the name we give Theorem 10.3.8 is appropriate. Jacobi's construction is described in bonus Section 10.5. The Notes to that section and to Section 10.4 say more about the historical context of the work of Abel and Jacobi, and the Notes to Section 10.5 include a capsule biography of Jacobi.

Functions obeying (10.3.13) are sometimes called theta functions. We'll reserve that name for what is often called Jacobi's theta function after the functions Jacobi used in his later work (a slight variant on his first theta functions). We will sometimes refer to "general theta functions" for functions obeying (10.3.13). The term *quasi-elliptic* (or even quasi-periodic, a terrible name, given its use for something very different; see Section 6.6 of Part 4) is sometimes used for functions obeying (10.3.13)) or ones like h_1 of (10.3.17) that are doubly periodic up to some simple "error term."

The term Legendre relations (see Problem 2) is based on the special case he proved in his work on elliptic functions [**350, 351**]; see Problem 13 of Section 10.5. They were generalized to the general doubly periodic case by Weierstrass in his first work on elliptic integrals [**583**] and are sometimes called the Legendre–Weierstrass relations.

Niels Henrik Abel (1802–29) was a Norwegian mathematician whose life was so melodramatic that were it a Hollywood movie, it would be dismissed as too tragic to be true! Among other aspects, his father, like Euler's, a minister, and family were adversely impacted by fallout from the Napoleonic wars (specifically, a side conflict between England and Denmark), his final illness was precipitated by a trip through the snow to his fiancée, and the letter informing him that he finally had a position as a professor arrived two days after his death (like that of Riemann) from a lung ailment, in his case, consumption. A year after he was unable to find employment, he not only had that offer, but a posthumous grand prize (jointly with Jacobi) from the French Academy. Earlier, Cauchy lost a manuscript that Abel had submitted to the French Academy. (Galois also had a manuscript lost through a combination of actions by Cauchy and Fourier!) See [**11, 528, 529**] for more on Abel's life.

Abel's first great work showed the impossibility of solving quintics in terms of radicals, and eventually led to the notion of group—"abelian groups" comes from this work. His celebrated mémoire on elliptic functions, as we've noted, found the double periodicity and, in more general contexts, abelian integrals, abelian functions, and abelian periods. Livio [**364**] has an interesting biography in his book. Pesic [**427**] describes Abel's proof for the insolvability of the quintic by radicals.

Joseph Liouville (1809–82) was a French mathematician who was active in the political turmoil around 1848, including an elective office. His contributions include not only his results on elliptic functions but also the construction of the first transcendental numbers (e.g., $\sum_{n=1}^{\infty} 10^{-n!}$, sometimes called Liouville's constant; see Problem 6 of Section 5.4 of Part 1), important contributions to mechanics (his result on constancy of phase space volumes and on completely integrable systems including the introduction of action-angle variables), and his work with Sturm on ordinary differential equations and their eigenfunction expansions (see Section 3.2 of Part 4) and the Liouville–Green approximation (see Section 15.5 of Part 2B). He is also noteworthy for appreciating and publicizing Galois' work ten years after Galois' death. Liouville served as a member of the moderate republican majority of the National Assembly in 1848 but was defeated for reelection because he was too moderate for the electorate after the revolution of that year. Lützen [**370**] has a biography of Liouville.

Liouville was a mentor to Charles Hermite (1822–1901) who (see Belhoste [**40**]) seems to have been the first to make systematic use of the Cauchy integral theorem in the study of elliptic functions.

Problems

1. Suppose g is an entire function that obeys (10.3.11).

   (a) Prove that g'/g is doubly periodic.

   (b) Prove that g'/g is constant. (*Hint*: Since g has no poles, what do you know about the residues of g'/g?)

   (c) Prove g has the form (10.3.12).

2. Suppose that g obeys (10.3.13) and has a single zero in the unit cell. By considering $\oint (g'/g)\, dz$ around a Liouville contour, prove the *Legendre*

relation (see also Problem 13 of Section 10.5)

$$\omega_1\tau_2 - \omega_2\tau_1 = 2\pi i \tag{10.3.32}$$

3. Let g be entire and obey (10.3.13).

(a) Prove that

$$g(z-\tau_j) = e^{(\omega_j\tau_j-\alpha_j)-\omega_j z}g(z) \tag{10.3.33}$$

(b) Let $h(z) = g(z)/g(-z)$. Prove that

$$h(z+\tau_j) = e^{2\alpha_j-\omega_j\tau_j}h(z) \tag{10.3.34}$$

(c) If $g(-z) = -g(z)$, prove that

$$g(z+\tau_j) = -e^{\omega_j(z+\tau_j/2)}g(z) \tag{10.3.35}$$

(*Hint*: Prove first (10.3.35) holds where $-$ is a priori ± 1 and then look at $z = -\tau_j/2$.)

(d) Returning to general g, suppose the only zeros of g occur at $z \in \mathcal{L}$. Prove that $h(z) = g(z)/g(-z)$ obeys $h(z) = e^{bz}$ for some b, and then that $\tilde{g}(z) = e^{-bz/2}g(z)$ has $\tilde{g}(-z) = \pm\tilde{g}(z)$, and that if the zeros are also simple, that $\tilde{g}(-z) = -\tilde{g}(z)$. (*Hint*: Prove that h is entire and doubly periodic.)

4. Let $g, \tilde{g}$ obey (10.3.13), be entire, and both have only simple zeros and exactly at $z \in \mathcal{L}$. Prove that $\tilde{g}(z) = e^{az^2+bz+c}g(z)$ for some a, b, c. (*Hint*: Prove that if $h = g/\tilde{g}$, then $(h'/h)''$ is doubly periodic.)

5. Let g obey (10.3.13). Let $\tilde{g}(z) = e^{az^2+bz+c}g(z)$.

(a) Prove $\tilde{g}$ also obeys a formula like (10.3.13) with

$$\tilde{\alpha}_j = \alpha_j + a\tau_j^2 + b\tau_j \tag{10.3.36}$$
$$\tilde{\omega}_j = \omega_j + 2a\tau_j \tag{10.3.37}$$

(b) Prove if there is an entire g that obeys (10.3.15), there is always a second $\tilde{g}$ with $\tilde{g}(-z) = -\tilde{g}(z)$, $\omega_1 = 0$, and $\omega_2 = -2\pi i/\tau_1$.

(c) Prove that α_1, α_2 can be picked (they are each only determined mod $2\pi i$) so that

$$\alpha_1\tau_2 - \alpha_2\tau_1 = \tfrac{1}{2}(\omega_2-\omega_1)\tau_1\tau_2 + i\pi(\tau_2-\tau_1) \tag{10.3.38}$$

6. Let g be an entire function obeying (10.3.13) and $g(z) = -g(-z)$ so (10.3.35) holds. Fix $a, b \in \mathbb{C}$ and define

$$g_{a,b}(z) = e^{bz}g(z+a) \tag{10.3.39}$$

(a) Prove that

$$g_{a,b}(z+\tau_j) = -e^{a\omega_j+b\tau_j}e^{\omega_j(z+\tau_j/2)}g_{a,b}(z) \tag{10.3.40}$$

(b) Let

$$h_{a,b}(z) = \frac{g(z)}{g_{a,b}(z)} \tag{10.3.41}$$

Prove that

$$h_{a,b}(z+\tau_j) = e^{-(a\omega_j+b\tau_j)}h_{a,b}(z) \tag{10.3.42}$$

(c) Pick $a = \tau_1/2$, $b = -\omega_1/2$. Prove that

$$h_{a,b}(z+\tau_1) = h_{a,b}(z), \qquad h_{a,b}(z+\tau_2) = -h_{a,b}(z) \tag{10.3.43}$$

(*Hint*: Use Problem 2.)

7. Fix $p_1, \dots, p_\ell$ in the fundamental cell. Let $V(p_1, \dots, p_\ell; n_1, \dots, n_\ell)$ for $n_j \geq 1$ be the vector space of all elliptic functions whose only poles are at $\{p_j\}_{j=1}^{\ell}$ with order at most n_j at p_j. Prove that

$$\dim(V(p_1, \dots, p_\ell; n_1, \dots, n_\ell)) = \sum_{j=1}^{\ell} n_j \tag{10.3.44}$$

(*Hint*: Don't forget to include the constant functions. (10.3.44) is the *Riemann–Roch theorem* for genus 1.)

10.4. Weierstrass Elliptic Functions

We'll begin this section by finding a g that completes the proof of Abel's theorem begun in the last section. The function g has to obey (10.3.13) and have a zero at $z = 0$ and so at all $z \in \mathcal{L}$. Since this is the only zero in the fundamental cell, the points in $\mathcal{L}$ are the only zeros in $\mathbb{C}$. We know from Chapter 9 how to construct functions with prescribed zeros as products of Weierstrass factors (see the definition of $E_n(z)$ in (9.4.2)). Such factors don't automatically obey (10.3.13), but if E_2 works (and it does), the quadratic exponential and the fact that the difference of a quadratic and its translate is linear is suggestive that (10.3.13) will hold. So we'll begin with implementing this idea—to define the Weierstrass σ-function, and so implement a general formula for elliptic functions in terms of σ and its translates.

Then we'll study the $\wp$-function, the simplest doubly periodic function formed from σ via

$$\wp = -\left(\frac{\sigma'}{\sigma}\right)' \tag{10.4.1}$$

This $\wp$-function will be the star of this section and, in many ways, the chapter—we'll reformulate the form of general elliptic functions in terms of

$\wp$ and $\wp'$, prove a cubic differential equation for $\wp$, and so relate $\wp$ to elliptic integrals and to elliptic curves, and state general addition formulae for $\wp$.

Two preliminary remarks on the choice of periods: The first involves scaling. If $\lambda \in \mathbb{C}$ and

$$\tilde{\tau}_1 = \lambda\tau_1, \qquad \tilde{\tau}_2 = \lambda\tau_2 \tag{10.4.2}$$

and f and g are functions on $\mathbb{C}$ related by

$$f(z) = g(\lambda z) \tag{10.4.3}$$

then

$$f(z + \tau_j) = g(\lambda z + \tilde{\tau}_j) \tag{10.4.4}$$

so (10.4.3) sets up a one-one correspondence between elliptic functions f with periods τ_j and g with periods $\tilde{\tau}_j$. We will often pick $\lambda = \tau_1^{-1}$, that is, after dropping the tilde, we'll consider the case where the periods are 1 and $\tau \in \mathbb{C}_+$. We note that in specific situations, one wants to multiply by constants in (10.4.4). For example, the $\wp$-function is normalized by $\wp(z) = 1/z^2 + O(1)$ near $z = 0$, so making τ_j-dependence explicit, what we have is

$$\wp(z \mid \tau_1\, \tau_2) = \lambda^2 \wp(\lambda z \mid \lambda\tau_1\, \lambda\tau_2) \tag{10.4.5}$$

with an extra λ^2.

The other remark concerns the convention about how one labels the basis of the period lattices. We will take τ_1 and τ_2 as the basis for the period lattice. There is another convention that uses $2\tau_1$ and $2\tau_2$ (or $2\omega_j$) for the basis. A variant uses $2\tau_1$ and $2\tau_3$ (and defines $\tau_2 = \tau_1 + \tau_3$). This basis is chosen because $\frac{1}{2}\mathcal{L}_{\tau_1,\tau_2}$ is important, so where we use $\frac{1}{2}\tau_j$ below, those using the other convention can use τ_j. Historically, the $2\tau_j$ convention was almost universally used a hundred years ago. The overwhelming majority of recent general complex analysis texts use our convention (exceptions include Markushevich [**372**] and Segal [**514**]), but books specializing on elliptic functions are split, with the $2\tau_j$ convention in a slight majority. I've no doubt many of their authors, thinking of the traditions of their specialty, rail about the barbarians who use our τ_j convention. In any event, the reader should be aware of the differing conventions.

Since our basic function is going to be zero on $\mathcal{L}$, we need to know about the index of convergence of these zeros:

Lemma 10.4.1. *We have that*

$$\alpha(|\tau_1^2| + |\tau_2|^2)(n_1^2 + n_2^2) \le |n_1\tau_1 + n_2\tau_2|^2 \le (|\tau_1|^2 + |\tau_2|^2)(n_1^2 + n_2^2) \tag{10.4.6}$$

where

$$\alpha = \frac{\operatorname{Im}(\tau^2)}{1 + |\tau|^2}, \qquad \tau = \frac{\tau_2}{\tau_1} \tag{10.4.7}$$

Remark. The upper bound also comes from the Cauchy–Schwarz inequality.

Proof. Let $\tau_1 = 1$, $\tau_2 = \tau$. Let $A = \left(\begin{smallmatrix} 1 & \operatorname{Re}\tau \\ \operatorname{Re}\tau & |\tau|^2 \end{smallmatrix}\right)$. Then

$$|n_1\tau_1 + n_2\tau_2|^2 = (n_1 n_2) A \begin{pmatrix} n_1 \\ n_2 \end{pmatrix} \tag{10.4.8}$$

$\operatorname{Tr}(A) = 1 + |\tau|^2$, $\det(A) = |\operatorname{Im}\tau|^2$. Since both are positive, A is positive, so the two eigenvalues, $\lambda_- \le \lambda_+$, obey $\lambda_+ \le \operatorname{Tr}(A)$ and $\lambda_- = \det(A)/\lambda_+ \ge |\operatorname{Im}\tau|^2/\operatorname{Tr}(A)$.

We thus have (10.4.6) if $\tau_1 = 1$, $\tau_2 = \tau$. Since everything scales quadratically under $\tau_1 \to \lambda\tau_1$, $\tau_2 \to \lambda\tau_2$, we get the result in general. □

Proposition 10.4.2. *For any two-dimensional lattice, $\mathcal{L}$, we have that*

$$\sum_{t\in\mathcal{L}\setminus\{0\}} |t|^{-\alpha} < \infty \tag{10.4.9}$$

if and only if $\alpha > 2$.

Remark. See Problems 1 and 2 for alternate proofs.

Proof. For $\alpha = 2$, we have by (10.4.6) that

$$\begin{aligned}
\sum_{t\in\mathcal{L}\setminus\{0\}} |t|^{-2} &\ge C \sum_{(n_1,n_2)\neq 0} (n_1^2 + n_2^2)^{-1} \\
&\ge 8C \sum_{n_1=1}^{\infty} \sum_{n_2=1}^{n_1} (2n_1^2)^{-1} \\
&= 8C \sum_{n_1=1}^{\infty} \tfrac{1}{2} n_1^{-1} \\
&= \infty
\end{aligned}$$

proving divergence if $\alpha \le 2$.

For $\alpha > 2$, again using (10.4.6),

$$\begin{aligned}
\sum_{t\in\mathcal{L}\setminus\{0\}} |t|^{-2} &\le \widetilde{C} \sum_{(n_1,n_2)\neq 0} (n_1^2 + n_2^2)^{-\alpha/2} \\
&\le 4\widetilde{C} \sum_{n=1}^{\infty} n^{-\alpha} + 4\widetilde{C} \sum_{n_1,n_2\ge 1} (2n_1n_2)^{-\alpha/2} \\
&\le 4\widetilde{C} \sum_{n=1}^{\infty} n^{-\alpha} + 4\widetilde{C} \left(\sum_{n=1}^{\infty} n^{-\alpha/2} \right)^2 \\
&< \infty
\end{aligned}$$

by the integral test. We used here that $n_1^2 + n_2^2 \ge 2n_1n_2$. □

We define the *Eisenstein series*, $S_k(\mathcal{L})$, for $k \geq 3$ by

$$S_k(\mathcal{L}) = \sum_{t \in \mathcal{L}\setminus\{0\}} t^{-k} \tag{10.4.10}$$

For k odd, $S_k = 0$, since then $(-t)^{-k} = -(t)^{-k}$. So we'll mainly look at k even. When $\mathcal{L} = \{n + m\tau\}_{n,m\in\mathbb{Z}}$, we write $S_k(\tau)$.

By Proposition 10.4.2, we cannot use E_1, but we can use E_2. We thus define the *Weierstrass sigma-function*,

$$\sigma(z \mid \tau_1\, \tau_2) = z \prod_{t \in \mathcal{L}_{\tau_1,\tau_2}\setminus\{0\}} E_2\left(\frac{z}{t}\right) \tag{10.4.11}$$

where E_2 is the Weierstrass factor $E_2(z) = (1-z)e^{(z+z^2/2)}$. By the estimate (9.4.3), we get the first sentence in

Theorem 10.4.3. *The product in* (10.4.11) *converges uniformly and absolutely on compacts of* $\mathbb{C}$ *to an entire function of order* 2. *We have that*

$$\sigma(-z \mid \tau_1\, \tau_2) = -\sigma(z \mid \tau_1\, \tau_2) \tag{10.4.12}$$

Moreover, for suitable ω_1, ω_2,

$$\sigma(z+\tau_j \mid \tau_1\, \tau_2) = -e^{\omega_j(z+\tau_j/2)}\sigma(z \mid \tau_1\, \tau_2) \tag{10.4.13}$$

Also,

$$\sigma(0) = 0, \qquad \sigma'(0) = 1 \tag{10.4.14}$$

Proof. As noted, (9.4.3) implies convergence and analyticity. (10.4.12) follows from $E_2(-z/-t) = E_2(z/t)$ and $-\mathcal{L}\setminus\{0\} = \mathcal{L}\setminus\{0\}$. Since $E_2(0) = 1$, we have (10.4.14).

We turn to (10.4.13). Fix $j = 1$ or 2. For $t \in \mathcal{L}$, we let

$$g_t(z) = \begin{cases} E_2(\frac{z+\tau_j}{t+\tau_j})/E_2(z/t), & t \neq 0, -\tau_j \\ E_2(\frac{z+\tau_j}{\tau_j})/z, & t = 0 \\ (z+\tau_j)/E_2(z/-\tau_j), & t = -\tau_j \end{cases}$$

Since $1 - (z+\tau_j)/(t+\tau_j) = [t/(t+\tau_j)](1 - z/t)$, we see each $g_t(z) =$ exp(second degree polynomial). By the absolute convergence of the product, which lets us rearrange terms,

$$\frac{\sigma(z+\tau_j)}{\sigma(z)} = \prod_{t\in\tilde{\mathcal{L}}} g_t(z) = \exp(\text{second degree polynomial}) \tag{10.4.15}$$

The point, of course, is to prove that the quadratic terms cancel, and one can do this by controlling the boundary terms in the sum (see Problem 3), but instead we'll use a trick.

Let ζ, the *Weierstrass zeta-function*, be given by

$$\zeta(z) = \frac{\sigma'(z)}{\sigma(z)} \tag{10.4.16}$$

(*Warning*: This has nothing to do with the Riemann zeta-function. We'll only use this Weierstrass ζ in this section and the next.) Since the derivative of a quadratic polynomial is linear, (10.4.15) implies for some β_j, ω_j,

$$\zeta(z+\tau_j) - \zeta(z) = \beta_j z + \omega_j \tag{10.4.17}$$

Since σ is odd, σ' is even, so

$$\zeta(-z) = -\zeta(z) \tag{10.4.18}$$

Taking $z = -3\tau_j/2$ in (10.4.17) and using (10.4.18) implies

$$-\zeta(\tfrac{1}{2}\,\tau_j) + \zeta(\tfrac{3}{2}\,\tau_j) = \beta_j(\tfrac{3}{2}\,\tau_j) + \omega_j \tag{10.4.19}$$

Comparing this to (10.4.17) for $z = \frac{1}{2}\tau_j$ implies

$$(-\tfrac{3}{2}\,\tau_j)\beta_j + \omega_j = (\tfrac{1}{2}\,\tau_j)\beta_j + \omega_j \tag{10.4.20}$$

so $\beta_j = 0$, that is

$$\zeta(z+\tau_j) = \zeta(z) + \omega_j \tag{10.4.21}$$

This implies

$$\sigma(z+\tau_j) = \sigma(z)e^{(\alpha_j+\omega_j z)} \tag{10.4.22}$$

By (10.4.12) and (10.3.22) $\Rightarrow$ (10.3.23), we get (10.4.13). □

The proofs of Theorems 10.3.6 and 10.3.7 are thus complete. Let us be explicit about the form of the first:

Theorem 10.4.4. *Let $\{z_j\}_{j=1}^{N_z}$ and $\{p_j\}_{j=1}^{N_z}$ be collections of points in the fundamental cell so that $z_j \neq p_k$ for all j,k and*

$$\sum_{j=1}^{N_z}(z_j - p_j) \in \mathcal{L} \tag{10.4.23}$$

Let $\tilde{z}_j, \tilde{p}_j$ be chosen so that $z_j - \tilde{z}_j$, $p_j - \tilde{p}_j \in \mathcal{L}$ and

$$\sum_{j=1}^{N_z}(\tilde{z}_j - \tilde{p}_j) = 0 \tag{10.4.24}$$

Then

$$F(z) = \frac{\prod_{j=1}^{N_z}\sigma(z-\tilde{z}_j)}{\prod_{j=1}^{N_p}\sigma(z-\tilde{p}_j)} \tag{10.4.25}$$

is an elliptic function whose zeros are $\{z_j+\mathcal{L}\}_{j=1}^{N_z}$ and poles are $\{p_j+\mathcal{L}\}_{j=1}^{N_z}$.

Since, using $f'/f = (\log f)'$, $E_2'(z/w)/E_2(w/z) = (z-w)^{-1}+1/w+z/w$, we see, by Proposition 9.2.7, that

$$\zeta(z) = \frac{1}{z} + \sum_{w\in\mathcal{L}\setminus\{0\}} \left(\frac{1}{z-w} + \frac{1}{w} + \frac{z}{w^2}\right) \tag{10.4.26}$$

Expanding $1/(z-w) = -\sum_{n=0}^{\infty} z^n/w^{n+1}$, we see that $\zeta(z)$ has a Laurent expansion about $z=0$ given by

$$\zeta(z) = \frac{1}{z} - \sum_{\ell=2}^{\infty} S_{2\ell}(\mathcal{L}_{\tau_1,\tau_2})z^{2\ell-1} \tag{10.4.27}$$

We now define the *Weierstrass* $\wp$*-function* by

$$\wp(z\mid\tau_1\,\tau_2) = -\zeta'(z\mid\tau_1\,\tau_2) \tag{10.4.28}$$

$$= \frac{1}{z^2} + \sum_{w\in\mathcal{L}\setminus\{0\}} \left[\frac{1}{(z-w)^2} - \frac{1}{w^2}\right] \tag{10.4.29}$$

One often wants to consider $\tau_1 = 1$, so we use

$$\wp(z\mid\tau) \equiv \wp(z\mid 1\,\tau) \tag{10.4.30}$$

By the lattice scaling (10.4.5), which is easy to verify,

$$\wp(z\mid\tau_1\,\tau_2) = \tau_1^{-2}\wp\left(\frac{z}{\tau_1}\,\middle|\,\frac{\tau_2}{\tau_1}\right) \tag{10.4.31}$$

We give the first set of its properties:

Theorem 10.4.5. (a) *$\wp$ is an elliptic function with periods τ_1,τ_2, that is,*

$$\wp(z+\tau_j\mid\tau_1\,\tau_2) = \wp(z\mid\tau_1\,\tau_2) \tag{10.4.32}$$

$\wp$ is jointly analytic in z,τ_1,τ_2 in $\{(z,\tau_1,\tau_2)\mid z\notin\mathcal{L}_{\tau_1,\tau_2},\ \tau_1\neq 0\neq \tau_2,\ \mathrm{Im}(\tau_2/\tau_1)>0\}$.

(b) *$\wp$ is even, that is,*

$$\wp(-z\mid\tau_1\,\tau_2) = \wp(z\mid\tau_1\,\tau_2) \tag{10.4.33}$$

More generally, for any $t\in\mathcal{L}$,

$$\wp_{(t)}(z) = \wp\left(z-\frac{t}{2}\right) \tag{10.4.34}$$

obeys

$$\wp_{(t)}(-z) = \wp_{(t)}(z) \tag{10.4.35}$$

(c) *$\wp(z)$ has the Laurent expansion about zero*

$$\wp(z\mid\tau_1\,\tau_2) = \frac{1}{z^2} + \sum_{\ell=1}^{\infty}(2\ell+1)S_{2\ell+2}(\mathcal{L}_{\tau_1,\tau_2})z^{2\ell} \tag{10.4.36}$$

(d) *$\wp(z)$ has degree two. For any z_0 in the fundamental cell other than $\frac{1}{2}\tau_1, \frac{1}{2}\tau_2, \frac{1}{2}\tau_3$ where*

$$\tau_3 = \tau_1 + \tau_2 \tag{10.4.37}$$

$\wp(z) - \wp(z_0)$ has a simple zero, and the only other solution of $\wp(z) = \wp(z_0)$ in the fundamental cell is $z = \tau_3 - z_0$, that is,

$$z \in \text{unit cell}, \qquad \wp(z) = \wp(z_0) \Leftrightarrow z = z_0 \text{ or } \tau_3 - z_0 \tag{10.4.38}$$

The only zeros of $\wp'(z)$ in the unit cell are at $\frac{1}{2}\tau_j$, $j = 1, 2, 3$. Define

$$e_1 = \wp(\tfrac{1}{2}\tau_1), \qquad e_2 = \wp(\tfrac{1}{2}\tau_3), \qquad e_3 = \wp(\tfrac{1}{2}\tau_2)$$

(e) *$\wp$ obeys the following nonlinear differential equations:*

$$\wp'(z)^2 = 4(\wp(z) - e_1)(\wp(z) - e_2)(\wp(z) - e_3) \tag{10.4.39}$$

$$= 4\wp^3(z) - g_2\wp(z) - g_3 \tag{10.4.40}$$

$$g_2 = 60S_4(\mathcal{L}_{\tau_1,\tau_2}), \qquad g_3 = 140S_6(\mathcal{L}_{\tau_1,\tau_2}) \tag{10.4.41}$$

Remarks. 1. It is unfortunate that the close to standard convention takes e_3 to be the value at $\frac{1}{2}\tau_2$ and e_2 at the centroid. This is the reason that some authors use τ_1, τ_3 for the basis and save τ_2 for their $\frac{1}{2}(\tau_1 + \tau_3)$.

2. (10.4.39) and (10.4.40) imply, of course,

$$e_1 + e_2 + e_2 = 0 \tag{10.4.42}$$

$$-4(e_1e_2 + e_1e_3 + e_2e_3) = g_2 \tag{10.4.43}$$

$$4e_1e_2e_3 = g_3 \tag{10.4.44}$$

3. By taking higher derivatives, we can get relations linear in the highest derivative and polynomial in the lower ones. Canceling $\wp'$ from both sides of (10.4.40) implies

$$\wp''(z) = 6\wp^2(z) - \tfrac{1}{2}g_2 \tag{10.4.45}$$

$$\wp'''(z) = 12\wp(z)\wp'(z) \tag{10.4.46}$$

4. $z \to \tau_3 - z_0$ maps the interior, but not the edge, of the fundamental cell to itself (e.g., the analog for the bottom edge is $\tau_1 - z_0$). Thus, (10.4.38) is, strictly speaking, only true on the interior.

5. Figure 10.4.1 shows a contour plot of $|\wp(z)|$ for $g_2 = 4$, $g_3 = 0$ ($\tau_1 = -i\tau_2$ approximately $\frac{8}{3}$).

Proof. (a) In (10.4.17), we proved $\beta_j = 0$, so ζ' is doubly periodic. Joint analyticity is immediate from the joint analyticity of $E_2(z/(n_1\tau_1 + n_2\tau_2))$ for each n_1, n_2 and the uniform convergence.

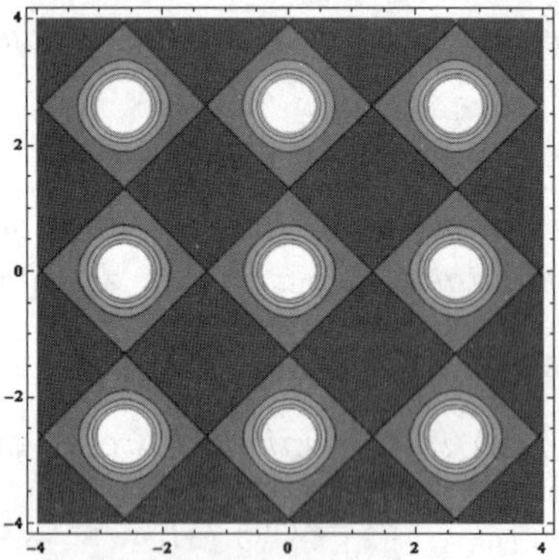

Figure 10.4.1. Contour plot of the Weierstrass $\wp$-function for $\tau = 1$.

(b) (10.4.33) is immediate from (10.4.18). That then implies

$$\begin{aligned}\wp_{(t)}(-z) &= \wp\left(\frac{t}{2} - z\right) \\ &= \wp\left(z - \frac{t}{2}\right) \\ &= \wp\left(z - \frac{t}{2} + t\right) \qquad (10.4.47) \\ &= \wp_{(t)}(z)\end{aligned}$$

where (10.4.47) uses periodicity of $\wp$.

(c) is immediate from (10.4.27).

(d) For each $a \in \mathbb{C}$, since $\wp - a$ has a single double pole and no other poles in the unit cell, $\wp - a$ has either two simple zeros or one double zero (by Theorem 10.3.2). Since $\wp^{(\tau_j)}$ is even, $\wp(\tau_j - z) = \wp(z)$, so if z_0 is not such that $2z_0 \in \mathcal{L}$, that is, if $z_0 \notin \frac{1}{2}\mathcal{L}$, $\wp(\tau_j - z_0) = \wp(z_0)$, and we must have two simple zeros if $a = \wp(z_0)$. By the evenness of $\wp^{(\tau_j)}$, the derivative of $\wp$ vanishes at τ_j $(j = 1, 2, 3)$, so there are double values at those points.

(e) Both sides of (10.4.42) have a sixth-order pole at $z = 0$, have double zeros at $\frac{1}{2}\tau_j$ $(j = 1, 2, 3)$, so by the uniqueness result, they are multiples of each other.

By (10.4.36),

$$\wp'(z) = -\frac{2}{z^3} + 6S_4 z + 20S_6 z^3 + O(z^5) \qquad (10.4.48)$$

while, by (10.4.36),

$$\wp^3(z) = \frac{1}{z^6} + \frac{9S_4}{z^2} + 15S_6 + O(z^2) \qquad (10.4.49)$$

so the constant in (10.4.39) must be 4.

By (10.4.48) and (10.4.49),

$$\begin{aligned}\wp'(z)^2 - 4\wp^3(z) &= \frac{4}{z^6} - \frac{24S_4}{z^2} - 80S_6 - \left(\frac{4}{z^6} + \frac{36S_4}{z^2} + 60S_6\right) + O(z^2) \\ &= -60S_4\wp(z) - 140S_6 = -g_2\wp(z) - g_3\end{aligned}$$

thus, the difference of the two sides of (10.4.47) is entire, doubly periodic, and 0 at $z = 0$, so 0. □

Here's a corollary that shows that $\wp$ uniformizes certain elliptic curves:

Corollary 10.4.6. *Let τ_1, τ_2 be given and let g_2, g_3 be given by (10.4.41). Let $\mathcal{C}$ be the Riemann surface of pairs $(\zeta, w) \in \mathbb{C}^2$ so*

$$\omega^2 = 4\zeta^3 - g_2\zeta - g_3 \tag{10.4.50}$$

with a single point added at ∞. Map the torus $\mathcal{J}_{\tau_1,\tau_2}$ to $\mathbb{C}^2 \cup \{\infty\}$ by

$$\mathcal{U}(z) = (\wp(z), \wp'(z)) \tag{10.4.51}$$

Then $\mathcal{U}$ is an analytic bijection of $\mathcal{J}_{\tau_1,\tau_2}$ onto $\mathcal{C}$.

Remark. The way we defined uniformization in Section 8.7, to get uniformization we define $\mathcal{U}$ on all of $\mathbb{C}$, which is the universal cover of $\mathcal{J}$.

Proof. The map is clearly analytic and it is onto $\mathcal{C}$ by (10.4.40). It is one-one since, by (d) of the theorem, $\wp(z_0) = \wp(z_1)$ implies either $z_1 = z_0$ or $z_1 = \tau_3 - z_0$. Since $\wp'$ is odd about inversion in $\frac{1}{2}\tau_3$, $\wp'(\tau_3 - z_0) = -\wp'(z_0)$. So $\wp'(z_0) \neq \wp'(z_1)$ if $z_0 \neq z_1$ since the only points in the unit cell where $\wp'$ is 0 (or ∞) are in $\frac{1}{2}\mathcal{L}$, and for such point, $z_1 \in z_0 + \mathcal{L}$.

To see the map is onto, let $(\zeta, \omega) \in \mathcal{C}$. First, find a z_0 so $\wp(z_0) = \zeta$, which we can do since $\wp$ takes every value. By the fact that (ζ, ω) solve (10.4.50), we see that $\wp'(z_0) = \pm\omega$, so if $z_1 = \tau_3 - z_0$, either $\mathcal{U}(z_0) = (\zeta, \omega)$ or $\mathcal{U}(z_1) = (\zeta, \omega)$. □

We say a function, $f(z)$, obeys an addition formula if $f(z + w)$ is an algebraic function of $f(z)$ and $f(w)$. For example, if $f(z) = \sin(z)$,

$$\begin{aligned}\sin(z + w) &= \sin(z)\cos(w) + \sin(w)\cos(z) \\ &= \sin(z)\sqrt{1 - \sin^2(w)} + \sin(w)\sqrt{1 - \sin^2(z)}\end{aligned}$$

provides the necessary formula. The $\wp$-function also obeys such an addition formula. The key will be that, in the language of the map $\mathcal{U}$, $\mathcal{U}(z)$, $\mathcal{U}(w)$, and $\mathcal{U}(-z - w)$ lie on a straight line.

Theorem 10.4.7 (Addition Formula for $\wp$). *Let z, w be in the unit cell of the periodic lattice, $\mathcal{L}$, for $\wp(z) \equiv \wp(z \mid \tau_1\tau_2)$. Suppose that $z, w, z+w \notin \mathcal{L}$ and $z \neq w$. Then*

$$\wp(z+w) = -\wp(z) - \wp(w) + \frac{1}{4}\left(\frac{\wp'(z) - \wp'(w)}{\wp(z) - \wp(w)}\right)^2 \tag{10.4.52}$$

Remarks. 1. Since $\wp'(u) = \sqrt{4\wp(u)^3 - g_2\wp(u) - g_3}$, this gives $\wp(z+w)$ as an algebraic function of $\wp(z)$ and $\wp(w)$, so (10.4.52) is called the *addition formula* for $\wp$.

2. The intermediate formula (10.4.59) below is sometimes called the addition formula.

3. See Problem 4 for another proof.

4. The conditions, $z, w, z+w \notin \mathcal{L}$, say that none of $\wp(z)$, $\wp(w)$, $\wp(-z-w)$ are infinite. Suitably interpreted, one can often regard (10.4.52) as true even when one of the values of $\wp$ is infinite. For example, if $z, w \notin \mathcal{L}$, $z+w \in \mathcal{L}$ means $\wp(z) = \wp(w)$, so both sides of (10.4.52) are infinite.

5. By taking a limit as $w \to z$, we obtain the *duplication formula* for $z \notin \frac{1}{2}\mathcal{L}$,

$$\wp(2z) = -2\wp(z) + \frac{1}{4}\left(\frac{\wp''(z)}{\wp'(z)}\right)^2 \tag{10.4.53}$$

In terms of the geometry of our proof, where (10.4.52) involves chords to the curve associated to $\wp$, (10.4.53) involves tangents; see Figure 10.4.2.

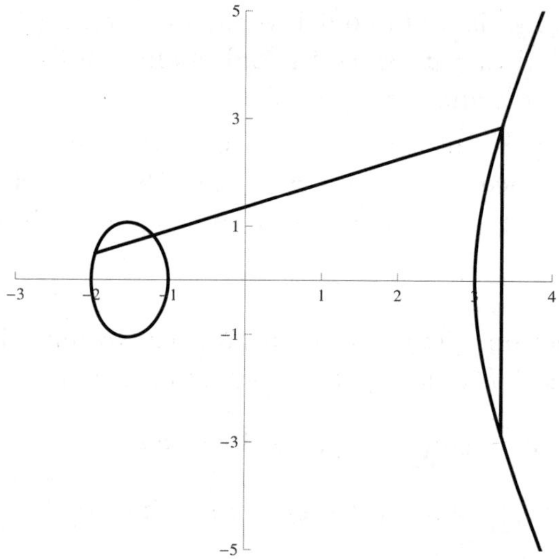

Figure 10.4.2. Chord-tangent construction for a cubic curve.

Proof. Let $(\zeta_1, \omega_1) \equiv (\wp(z), \wp'(z))$, $(\zeta_2, \omega_2) \equiv (\wp(w), \wp'(w))$. There is a straight line through these points in $\mathbb{C}^2$ of the form

$$\omega = m\zeta + b \tag{10.4.54}$$

As usual, the slope m is given by

$$m = \frac{\omega_1 - \omega_2}{\zeta_1 - \zeta_2} = \frac{\wp'(z) - \wp'(w)}{\wp(z) - \wp(w)} \tag{10.4.55}$$

This won't work if $\wp(z) = \wp(w)$, but that only happens if $z = w$ or $z+w \in \mathcal{L}$, which we are supposing doesn't occur.

Consider the function

$$f(u) = \wp'(u) - m\wp(u) - b \tag{10.4.56}$$

This has a third-order pole at $u = 0$ and zeros at $u = z$ and $u = w$. By Theorem 10.3.6, its third zero must be at a point η with $\eta + z + w \in \mathcal{L}$, that is, $\eta = -z - w$.

Thus,

$$\zeta_3 = \wp(-z - w) = \wp(z + w) \tag{10.4.57}$$

$$\omega_3 = \wp'(-z - w) = -\wp'(z + w) \tag{10.4.58}$$

also lies on the line (10.4.54), which we can summarize by the vanishing of the determinant

$$\begin{vmatrix} \wp(z) & \wp'(z) & 1 \\ \wp(w) & \wp'(w) & 1 \\ \wp(z+w) & -\wp'(z+w) & 1 \end{vmatrix} = 0 \tag{10.4.59}$$

From the relation of $\wp'$ and $\wp$ and the definition of f,

$$(f(u) + m\wp(u) + b)^2 = 4\wp^3(u) - g_2\wp(u) - g_3 \tag{10.4.60}$$

By the fact that f vanishes at $u = z, w, -z - w$, we see that

$$(m\zeta + b)^2 = 4\zeta^3 - g_2\zeta - g_3 \tag{10.4.61}$$

has $\zeta_1, \zeta_2, \zeta_3$ as its three roots. That means that

$$4(\zeta - \zeta_1)(\zeta - \zeta_2)(\zeta - \zeta_3) = 4\zeta^3 - m^2\zeta^2 - (g_2 + 2mb)\zeta - (g_3 + b^2) \tag{10.4.62}$$

Identifying ζ^2 terms,

$$\zeta_1 + \zeta_2 + \zeta_3 = \frac{m^2}{4} \tag{10.4.63}$$

which is (10.4.52). □

(10.4.25) shows any elliptic function can be written as a ratio of products of translates of σ. But σ itself is not elliptic. However, if F is both elliptic and even, one can use ratios of translates of $\wp$.

Theorem 10.4.8. *If F is an entire meromorphic function and F obeys*

$$F(z+\tau_j) = F(z), \quad j=1,2; \qquad F(z) = F(-z) \tag{10.4.64}$$

Then there exist c, $a_1, \dots, a_n$, and $b_1, \dots, b_m$ in $\mathbb{C}$ so that

$$F(z) = c\,\frac{\prod_{j=1}^n (\wp(z \mid \tau_1\tau_2) - a_j)}{\prod_{k=1}^m (\wp(z \mid \tau_1\tau_2) - b_k)} \tag{10.4.65}$$

Remarks. 1. This is an analog of the result in Problem 1 of Section 3.10.

2. Unlike (10.4.25), we do not require $n=m$. The equality in (10.4.25) was forced by total degree. Here, that is enforced by the fact that $\wp$ is elliptic.

Proof. Since F is even, the order of any pole or zero at 0 is even. But also,

$$F\left(-z+\frac{\tau_j}{2}\right) = F\left(z-\frac{\tau_j}{2}\right) = F\left(z+\frac{\tau_j}{2}\right) \tag{10.4.66}$$

so F is even about $\tau_j/2$ and about $(\tau_1+\tau_2)/2$, so these are also even order. Otherwise, zeros and poles come in pairs, since if $z_j+\mathcal{L}$ are zeros, so is $-z_j+\mathcal{L}$ and $-z_j+\mathcal{L} \neq z_j+\mathcal{L}$ if $z_j \notin \frac{1}{2}\mathcal{L}$.

This means that the nonzero zeros and poles can each be broken in two sets, that is, there exist $\{\tilde{z}_j\}_{j=1}^n$, $\{\tilde{p}_j\}_{j=1}^m$ in the fundamental cell so $\{\tilde{z}_j\}_{j=1}^n \cup \{-\tilde{z}_j\}_{j=1}^n$ are all the zeros away from $z=0$, counting multiplicity (for half periods, we take only half the order in the $\tilde{z}_j$).

It follows that with $a_j = \wp(\tilde{z}_j)$, $b_j = \wp(\tilde{p}_j)$, then F and the right side of (10.4.65) have the same zeros and poles, except perhaps for $z=0$, so the ratio is an elliptic function with no zeros and poles, except perhaps at $\mathcal{L}$. But this ratio can't have both zeros and poles at zero, so the ratio is constant. □

Theorem 10.4.9. *Any elliptic function, F, can be written*

$$F(z) = F_1(z) + \wp'(z)F_2(z) \tag{10.4.67}$$

where F_j are elliptic and even, and so F is a rational function of $\wp$ and $\wp'$.

Remark. Only first orders of $\wp'$ are needed because $(\wp')^2$ is a polynomial in $\wp$, and so $(\wp')^{-1} = \wp'/(\wp')^2$ is $\wp'$ times a rational function of $\wp$. Thus, $\wp, \wp'$ generate the field of elliptic functions subject to the single relation (10.4.40).

Proof. Let $F_e(z) = \frac{1}{2}(F(z)+F(-z))$, $F_o(z) = \frac{1}{2}(F(z)-F(-z))$. Obviously, F_e (respectively, F_o) is even (respectively, odd) and $F = F_e + F_o$. Clearly, $F_2 = F_o/\wp'$ is even as a ratio of odd functions, so (10.4.67) holds with $F_1 = F_e$. □

The differential equation also lets us express certain integrals, which are certain elliptic integrals, in terms of the inverse function to $\wp$ just as arcsin can be used to integrate via

$$\int_0^x \frac{dy}{\sqrt{1-y^2}} = \arcsin(x) \tag{10.4.68}$$

Theorem 10.4.10. *Let $\wp(u \mid \tau_1\tau_2)$ be a Weierstrass $\wp$-function with associated parameters g_1, g_2, etc. Suppose*

$$u = \int_{\alpha_0}^{\alpha} \frac{dy}{\sqrt{4y^3 - g_2 y - g_3}} \tag{10.4.69}$$

Let u_0 solve $\wp(u_0) = \alpha_0$. Then

$$\wp(u + u_0) = \alpha \tag{10.4.70}$$

Remarks. 1. What seems at first a defect in this is actually a virtue. Namely, the integral in (10.4.69) has a branch cut, so the integral is a multivalued function of α. But (10.4.70) has multiple solutions. If u_0 solves (10.4.70), so does $u_0 + t$ for any $t \in \mathcal{L}$. Of course, $\arcsin(\cdot)$ is also multivalued. Abel's great discovery is that the inverse function to the elliptic integral (10.4.69) is doubly periodic.

2. Formally, if $y = \wp(x)$, then

$$\frac{dy}{\sqrt{4y^3 - g_2 y - g_3}} = \frac{dy}{\wp'(x)} = \frac{\wp'(x)\,dx}{\wp'(x)} = dx$$

so the integral is $\int_{\wp^{-1}(\alpha_0)}^{\wp^{-1}(\alpha)} dx$, so $u = \wp^{-1}(\alpha) - \wp^{-1}(\alpha_0)$ and $\alpha = \wp(u + \wp^{-1}(\alpha_0))$, which is (10.4.70).

Proof. We'll suppose $\alpha_0 \neq e_j$ for $j = 1, 2, 3$, or $\alpha_0 = \infty$, so the integrand is nonsingular for α near α_0. Once one has the formula in that case, one even gets it for $\alpha_0 = e_j$ or $\alpha_0 = \infty$, either by taking limits or by using the additive structure of integrals. We'll also prove it only for α near α_0 and appeal to analytic continuation to get the general case.

Let $I(\alpha)$ denote the integral. Clearly, $I(\alpha)$ is analytic for α near α_0 and $I'(\alpha_0) \neq 0$, so I is a local bijection and there is local inverse. For u small, define

$$g(u) = I(\wp(u + u_0)) \tag{10.4.71}$$

Clearly, $g(0) = 0$. By the chain rule,

$$g'(u) = I'(\wp(u + u_0))\wp'(u + u_0) \tag{10.4.72}$$

Since $I'(\alpha) = (4\alpha^3 - g_2\alpha - g_3)^{-1/2}$, (10.4.72) and (10.4.40) imply that $g'(u) = 1$. It follows that $g(u) = u$. Thus, I is a local inverse of $\wp(\,\cdot\, + u_0)$, so since I is a local bijection, $\wp(\,\cdot\, + u_0)$ is a local inverse of I, that is, (10.4.70) holds. $\square$

Of course, when studying integrals, we are not normally given τ_1, τ_2, but rather g_2, g_3, so the natural question that arises concerns existence. Not all (g_2, g_3) can occur since the roots of $(\wp - e_1)(\wp - e_2)(\wp - e_3)$ are distinct. The condition for the roots of the cubic $4y^3 - g_2 y - g_3$ to be distinct is that the *discriminant*

$$\Delta(\tau, \tau') = g_2^3 - 27g_3^2 \tag{10.4.73}$$

be nonzero. The reader will check (see Problem 7) that

$$\Delta = 16(e_1 - e_2)^2(e_2 - e_3)^2(e_1 - e_3)^2 \tag{10.4.74}$$

There is a simple scaling relation that reduces the problem of finding two variables τ_1, τ_2 in terms of two variables g_2, g_3 to a single function. We note that if $Q(y, g_2, g_3) = y^3 - g_2 y - g_3$, then

$$Q(\mu y, \mu^2 g_2, \mu^3 g_3) = \mu^3 Q(y, g_2, g_3) \tag{10.4.75}$$

This is connected to the scaling relation (10.4.5) with $\mu = \lambda^2$. We are thus interested in the scale invariant quantity (often multiplied by 1728; see the Notes to Section 10.6)

$$\widetilde{J}(\tau_1, \tau_2) = \frac{g_2^3}{g_2^3 - 27g_3^2} \tag{10.4.76}$$

Tracking through the scaling, (10.4.75) and (10.4.5), shows $\widetilde{J}$ is only a function of $\tau = \tau_2/\tau_1$, so there is a single function, the *J-invariant*

$$J(\tau) = \widetilde{J}(1, \tau) \tag{10.4.77}$$

The problem of studying the range of J is the same as confirming all nondegenerate cubics (i.e., distinct roots) arise for some (τ_1, τ_2); this is called the *modular problem* since $\tau = \tau_2/\tau_1$ is called the *modulus* for the elliptic integral. Given the similarity of names, the reader must suspect a connection to the elliptic modular function of Section 8.3, and we will see there is. This set of ideas is the subject of Section 10.6.

This completes the main discussion of $\wp$. We want to end with a brief indication of two further topics, one involving $\sqrt{\wp(z) - e_j}$, something of relevance in Section 10.5, and one involving writing $\wp$ as a sum of $\csc^2$, relevant to Section 10.6.

For the next discussion, for notational convenience, it will be helpful to write

$$\begin{aligned} \tilde{\tau}_1 &= \tau_1, & \tilde{\tau}_2 &= \tau_1 + \tau_2, & \tilde{\tau}_3 &= \tau_2 \\ \tilde{\omega}_1 &= \omega_1, & \tilde{\omega}_2 &= \omega_1 + \omega_2, & \tilde{\omega}_3 &= \omega_2 \end{aligned} \tag{10.4.78}$$

so σ obeys

$$\sigma(z + \tilde{\tau}_j) = -e^{\tilde{\omega}_j(z + \tilde{\tau}_j/2)}\sigma(z) \tag{10.4.79}$$

Define the *associated σ-functions*, $\sigma_j(z)$, $j = 1, 2, 3$, by

$$\sigma_j(z) = \frac{\sigma(z + \frac{1}{2}\tilde{\tau}_j)}{\sigma(\frac{1}{2}\tilde{\tau}_j)}\, e^{-\tilde{\omega}_j z/2} \tag{10.4.80}$$

These are essentially the functions g_j of (10.3.24) for $g = \sigma$ multiplied by the constant $\sigma(\frac{1}{2}\tilde{\tau}_j)^{-1}$ so that

$$\sigma_j(0) = 1 \tag{10.4.81}$$

By (10.3.28),

$$\frac{\sigma_j(z)^2}{\sigma(z)^2} = \frac{\sigma(z + \frac{1}{2}\tilde{\tau}_j)\sigma(z - \frac{1}{2}\tilde{\tau}_j)}{\sigma(z)^2\sigma(\frac{1}{2}\tilde{\tau}_j)^2} \tag{10.4.82}$$

and by (10.3.26), this function is elliptic with periods τ_1, τ_2. The function has double poles at $\mathcal{L}$ and double zeros at $\frac{1}{2}\tau_j + \mathcal{L}$, and so does $\wp(z) - e_j$, so they must agree up to a multiplicative constant. Since $\sigma'(0) = 1$ and (10.4.81) holds, near 0,

$$\frac{\sigma_j(z)^2}{\sigma(z)^2} = \frac{1}{z^2} + O\left(\frac{1}{z}\right) \tag{10.4.83}$$

so the constant is 1, that is, we have proven the first part of

Theorem 10.4.11. *In terms of associated σ-functions, we have*

(a) $$\left(\frac{\sigma_j(z)}{\sigma(z)}\right)^2 = \wp(z) - e_j \tag{10.4.84}$$

Thus, $\sqrt{\wp(z) - e_j}$ is an entire function periodic under adding $\tilde{\tau}_j$ and antiperiodic upon adding $\tilde{\tau}_k$ for $k \neq j$.

(b) $$\wp'(z) = -2\,\frac{\sigma_1(z)\sigma_2(z)\sigma_3(z)}{\sigma(z)^3} \tag{10.4.85}$$

Remark. We already know $\wp(z) - e_j$ has only double poles and double zeros, so it has a square root which is entire, which it is easy to see, and must be periodic or antiperiodic under adding $\tilde{\tau}_k$. What the explicit formula

$$\sqrt{\wp(z) - e_j} = \frac{\sigma_j(z)}{\sigma(z)} \tag{10.4.86}$$

shows is that it is antiperiodic in $\tilde{\tau}_k$, $k \neq j$, and periodic in $\tilde{\tau}_j$.

Proof. (a) As noted, we proved (10.4.84). The (anti)-periodicity result follows from Theorem 10.3.9. That theorem only proves (anti)-periodicity for two of the $\tilde{\tau}_j$, but since the other is a sum or difference of this $\tilde{\tau}_k$ and $\tilde{\tau}_j$, we get antiperiodicity for it.

(b) Both sides of (10.4.85) are elliptic (since two $\sigma_j(z)/\sigma(z)$ are antiperiodic and the other one periodic under adding any $\tilde{\tau}_k$) and they have the same zeros and poles, so they agree up to a multiplicative constant. Since both sides are $-2/z^3 + O(1/z^2)$ at 0, the constant is 1. □

Finally, we note that since $\sum_{n=-\infty}^{\infty}(z+w-n)^{-2}$ is $\pi^2/\sin^2(z+w)$, one can rewrite the double sums in (10.4.29) as single sums of $\sin^{-2}(\cdot)$. It is convenient to write the formula for $\tau_1 = 1$, $\tau_2 = \tau$.

Theorem 10.4.12. *For $z \notin \mathcal{L}_\tau$,*

$$\wp(z \mid \tau) = \frac{\pi^2}{\sin^2(\pi z)} - \frac{\pi^2}{3} + \pi^2 \sum_{m\neq 0} \left[\frac{1}{\sin^2(\pi z - m\pi\tau)} - \frac{1}{\sin^2(m\pi\tau)}\right] \tag{10.4.87}$$

Proof. $\wp$ is an absolutely convergent sum over m and n of $(z-n-m\tau)^{-2} - (n+m\tau)^{-2}$ without the second term for $n = m = 0$. We fix m and sum over n. Since, for m fixed, the sums of the two terms are separately convergent in n, we can use (9.2.22) to do those sums to obtain (10.4.87) if we use $\sum_{n=1}^{\infty} 1/n^2 = \pi^2/6$. □

(10.4.87) allows one to compute asymptotics of $e_j(1,\tau) - e_k(1,\tau)$ as $\operatorname{Im}\tau \to \infty$, with $\operatorname{Re}\tau$ fixed. See Problem 8. We'll also use (10.4.87) in Section 13.1 of Part 2B.

Notes and Historical Remarks.

> It is well known that the reluctance of Gauss to publish his discoveries was due to the rejection of his *Disquisitiones arithmeticae* by the French Academy, the rejection being accompanied by a sneer which, as Rouse Ball has said, would have been unjustifiable even if the work had been as worthless as the referees believed. It is the irony of fate that, but for this sneer, the *Traite des fonctions elliptiques*, the work of a Frenchman, might have assumed a different and vastly more valuable form, and Legendre might have been spared the pain of realizing that many years of his life had been practically wasted, had the method of inversion come to be published when Legendre's age was fifty instead of seventy-six.
>
> —*G. N. Watson* [**580**]

Let's begin with a summary of the history of elliptic functions, except for theta functions (which are discussed in the next section). Major developments in the prehistory include Wallis' first encounter with an elliptic integral in 1655, when he tried to compute the arclength of an ellipse and got an answer that involved (in modern notation) $\int_0^1 \sqrt{(1-k^2x^2)/(1-x^2)}\,dx$ for $0 \le k < 1$ (see Problem 9). This was the first example of an elliptic integral—the name came from the ellipse whose arclength it was measuring. In 1694, Jakob Bernoulli computed the arclength of the *lemniscate* (given by $(x^2+y^2)^2 = x^2 - y^2$) as $4\int_0^1 dx/\sqrt{1-x^4}$, a famous example called the lemniscate integral (see Problem 10). He conjectured that this could not be expressed in terms of direct or inverse trigonometric functions. A final piece of the prehistory is Fagnano's 1718 paper [**184**] on the partial lemniscate integral $\int_0^x dy/\sqrt{1-y^4}$, which included what we now regard as special cases of the addition formula.

The major developments prior to the birth of the theory of elliptic functions in the hands of Abel and Jacobi in the period 1825–35 are due to three giants: Euler, Legendre, and Gauss. We have already alluded to Euler's work on addition formulas. Beginning in 1786, Adrien-Marie Legendre (1752–1833) spent a significant part of his efforts on what he named elliptic functions—what we now call elliptic integrals, rather than their inverses, which is what we call elliptic functions. In the 1820s, he produced a major multivolume compendium [**351**]: Volume 1 in 1825 and Volume 2 in 1826. Shortly before his death, he published some addenda reacting to the new developments of Abel and Jacobi which he championed. His encyclopedic work was made largely obsolete by these new developments.

The two key ideas that were necessary for the new developments were to look at the inverse function and to realize this function is doubly periodic. These ideas were first found by Carl Friedrich Gauss (1777–1855) in work that appeared in his notebooks beginning in 1796 (see the Notes to Section 14.4 of Part 2B for a capsule biography). But as with the cases of Cauchy's theorem, hyperbolic geometry, and the prime number theorem, Gauss did not publish this work. It has been conjectured [**28**] that Gauss' reticence was due to the poor reception his *Disquisitiones Arithmeticae* received from the French Academy and Legendre in particular. As noted by Watson and others, it is ironic that if Legendre had not caused Gauss to be reluctant to publish, he might have been spared realizing after twenty-five years that his masterwork, written after Gauss's discovery, was largely obsolete.

It was Abel and Jacobi who independently published the key discoveries (it is clear that Jacobi was already working on these things when Abel sent him his work, but Jacobi may have gotten some hints on what to do—in any event, Jacobi always admitted that Abel was first). We'll discuss Jacobi's work on theta functions in the next section. For more on the history of the work of Abel and Jacobi, see Bottazzini–Gray [**69**] and Markushevich [**373**]. As Bottazzini–Gray emphasize, it is a surprising fact that contour integrals and Cauchy's theorem play no role in the initial work of Abel and Jacobi.

Given that, in many ways, the main elements of the theory were settled in the initial 1825–35 period, it is remarkable that Weierstrass' work many years later revolutionized the subject. It should be mentioned that earlier, in 1847 [**163**], Gotthold Eisenstein (1823–52) defined what is essentially the $\wp$-function. He did not make the Weierstrass subtraction but rather looked at $\sum_{m=-\infty}^{\infty}(\sum_{n=-\infty}^{\infty}(z-n-m\tau)^{-2})$ which is convergent in that the sum over n for each fixed m is absolutely convergent and the resulting sums then converge absolutely! (The reader will check this, for $z = 0$ with the $(n,m) = (0,0)$ term dropped, in Problem 11). Weierstrass, with his concern about convergence issues, always complained about this approach. Weil [**589**] has championed Eisenstein's cause. Like Abel, Eisenstein died in his twenties from consumption. The Eisenstein series is named after this work.

Weierstrass began lecturing on his σ and $\mathcal{P}$ functions as early as 1862. Only in 1881 did his student Schwarz convince him that a collection of formulas and theorems on this approach would be useful and Schwarz edited such a booklet appearing in 1883 [**588**] in German and in French translation.

The idea of using elliptic functions to uniformize cubic curves is due to Clebsch [**118**]. Geometrically, the addition theorem associated to $P_1 = (\zeta_1, \omega_1)$, $P_2 = (\zeta_2, \omega_2)$ on a cubic, a new point, $P_3 = (\zeta_3, \omega_3)$, by taking the chord through P_1 and P_2 and seeing its third point, P_4, of intersection and then flipping the sign of ω_3. The flip means that iteration is possible, yielding additional points. This procedure gives a group operation to the points on the cubic which is, of course, just using the map $\mathcal{U}$ of (10.4.51) to move the natural addition on the torus to $\mathcal{C}$ (as seen by the $u, w \mapsto u + w$ rule); see Figure 10.4.2.

If $\mathcal{C}$ is a rational curve, that is, g_2, g_3 are in $\mathbb{Q}$, and if (ζ_1, ω_1), (ζ_2, ω_2) are rational, then the line will have rational coefficients and the third point will also be rational, so this produces new rational solutions from existing ones and is related to the Pythagorean triple construction mentioned in the Notes at the start of the chapter. This procedure is called the *chord-tangent construction* (the tangent case finds a second point from a single point) and, as a way to generate integral solutions, goes back to Diophantus almost eighteen-hundred years ago (later developments are due to Fermat and Newton). See Stillwell [**543**] for more of the history and Husemoller [**273**] for more on the subject. One major result in this subject is a theorem of Mordell [**394**] that the subgroup of all rational solutions of a rational cubic is finitely generated. For a textbook presentation of this theorem and other elements of the number theory associated to elliptic curves, see [**522**].

Problems

1. (a) Let $\{z_\beta\}_{\beta \in I}$ be a collection of points in the plane all nonzero, with $N(R) = \#\{\beta \mid |z_\beta| < R\}$ finite for all R. Prove that

$$\sum_{\{\beta \mid |z_\beta| < R\}} |z_\beta|^{-\alpha} = R^{-\alpha} N(R) + \int_0^R \alpha s^{-\alpha-1} N(s)\, ds \tag{10.4.88}$$

and that

$$\sum_{\beta} |z_\beta|^{-\alpha} < \infty \Leftrightarrow \int_0^\infty s^{-\alpha-1} N(s)\, ds < \infty \tag{10.4.89}$$

(b) Prove Betti's result [**47**] that if $\min_{\beta\in I}|z_\beta| > 0$ and $\min_{\gamma\neq\beta}|z_\gamma - z_\beta| = r > 0$, then for all $\alpha > 2$, $\sum|\zeta_\beta|^{-\alpha} < \infty$. (*Hint*: Prove that $N(R) \leq (R+r)^2/r^2$.)

(c) For $\operatorname{Im}(\tau_2/\tau_1) > 0$, $\{z_\beta\} = \{n_1\tau_1 + n_2\tau_2\}$, prove $N(R) \geq cR^2$ and conclude that $\sum_\beta |z_\beta|^{-2} = \infty$.

2. In $\mathcal{L}_{\tau_1,\tau_2}$, let $S_k = \{n_1\tau_1 + n_2\tau_2 \mid \max(|n_1|,|n_2|) = k\}$.

   (a) Prove that $\#(S_k) = 8k$.

   (b) Let $\widetilde{S}_1 = \{x_1\tau_1 + x_2\tau_2 \mid \max(|x_1|,|x_2|) = 1,\ x_j \in \mathbb{R}\}$ and let $m_\pm = {\max \atop \min}\{|s| \mid s \in \widetilde{S}_1\}$. Prove that $\min_{S_k}|s| \geq km_-$ and $\max_{S_k} \leq km_+$.

   (c) Conclude that $\sum_{t\neq 0,\, t\in\mathcal{L}} |t|^{-\alpha} < \infty$ if and only if $\alpha > 2$.

3. In terms of the notation of Problem 2, let $T_k = \bigcup_{j=1}^k S_j$.

   (a) Prove that for $j = 1, 2$, $T_k \Delta (T_k + \tau_j) \subset S_k \cup S_{k+1}$.

   (b) For $j = 1, 2$, prove that if

$$\Delta_k(j) = \sum_{\substack{t\in T_k+\tau_j \\ t\neq 0}} t^{-2} - \sum_{\substack{t\in T_k \\ t\neq 0}} t^{-2}$$

   then $|\Delta_k(j)| \to 0$ as $k \to \infty$.

   (c) Conclude that the constants β_j in (10.4.17) are zero.

4. This will provide another proof of the addition formula for $\wp$.

   (a) Prove that

$$\wp(z) - \wp(w) = -\frac{\sigma(z-w)\sigma(z+w)}{\sigma^2(z)\sigma^2(w)} \tag{10.4.90}$$

   (b) Prove that

$$\frac{\wp'(z)}{\wp(z) - \wp(w)} = \zeta(z-w) + \zeta(z+w) - 2\zeta(z) \tag{10.4.91}$$

   (c) Prove that

$$\frac{1}{2}\frac{\wp'(z) - \wp'(w)}{\wp(z) - \wp(w)} = \zeta(z+w) - \zeta(z) - \zeta(w) \tag{10.4.92}$$

(d) Prove that

$$\wp''(z) = 6\wp(z)^2 - \tfrac{1}{2}\, g_2 \tag{10.4.93}$$

(e) By differentiating (10.4.92) and using (10.4.93), prove (10.4.52).

5. If $z_1 + z_2 + z_3 = 0$, prove that

$$[\zeta(z_1) + \zeta(z_2) + \zeta(z_3)]^2 + \zeta'(z_1) + \zeta'(z_2) + \zeta'(z_3) = 0$$

6. Prove that (j and k are the other two of $1, 2, 3$)

$$(\wp(z) - e_i)(\wp(z + \tfrac{1}{2}\,\tilde{\tau}_i) - e_i) = (e_j - e_i)(e_k - e_i)$$

(*Hint*: Look at the zeros and poles of the left side and then evaluate it at $\frac{1}{2}\tilde{\tau}_j$.)

Note. See Problem 15 of Section 10.5 for a related result.

7. This problem will prove that Δ defined by (10.4.73) obeys (10.4.74). Let $f(z) = 4z^3 - g_2 z - g_3 = 4(z - e_1)(z - e_2)(z - e_3)$.

(a) Prove that for j, k, ℓ distinct among $1, 2, 3$ that $f'(e_j) = 4(e_j - e_k)(e_j - e_k)$.

(b) Prove that

$$\begin{aligned} 16(e_1 - e_2)^2(e_2 - e_3)^2(e_1 - e_3)^2 &= -\tfrac{1}{4}\, f'(e_1) f'(e_2) f'(e_3) \\ &= -\tfrac{1}{4}\,(12e_1^2 - g_2)(12e_2^2 - g_2)(12e_3^2 - g_3) \end{aligned} \tag{10.4.94}$$

(c) Using (10.4.42) and (10.4.43), prove that

$$e_1^2 + e_2^2 + e_3^2 = \frac{g_2}{2}$$

$$e_1^2 e_2^2 + e_2^2 e_3^2 + e_1^2 e_3^2 = \tfrac{1}{2}\,(e_1^4 + e_2^4 + e_3^4) = \frac{g_2^2}{16}$$

(d) Using (10.4.94), part (c), and (10.4.44), prove (10.4.74).

8. For $0 \le \operatorname{Re}\tau \le 2$ as $\operatorname{Im}\tau \to \infty$, prove that ($\tau_1 = 1$, $\tau_2 = \tau$)

$$e_1 = \frac{2\pi^2}{3} + O(e^{-2\pi \operatorname{Im}\tau}) \tag{10.4.95}$$

$$e_2 = -\frac{\pi^2}{3} + 8\pi^2 e^{i\pi\tau} + O(e^{-3\pi \operatorname{Im}\tau}) \tag{10.4.96}$$

$$e_3 = -\frac{\pi^2}{3} - 8\pi^2 e^{i\pi\tau} + O(e^{-3\pi \operatorname{Im}\tau}) \tag{10.4.97}$$

$$\lambda(\tau) \equiv \frac{e_2 - e_3}{e_1 - e_3} = 16 e^{i\pi\tau} + O(e^{-3\pi \operatorname{Im}\tau}) \tag{10.4.98}$$

(*Hint*: See (10.4.87).)

9. If $y = f(x)$ is a C^1 curve, the arclength of the curve from α to β is $\int_\alpha^\beta \sqrt{1+(df/dx)^2}\,dx$. Let $0 < b < a$ and $0 < \beta < a$. Prove that the arclength of the ellipse $x^2/a^2 + y^2/b^2 = 1$ from $x = 0$ to $x = \beta$ (and $y > 0$) is (with $k^2 = 1 - b^2/a^2$)
$$L = a\int_0^{\beta/a} \sqrt{\frac{1-k^2u^2}{1-u^2}}\,du$$
This is an elliptic integral of the second kind; see the discussion in the Notes to Section 10.5.

10. A lemniscate has the form $(x^2+y^2)^2 = (x^2-y^2)$.

(a) Show, in polar coordinates, the lemniscate is $r^2 = \cos(2\theta)$.

(b) In polar coordinates, arclength is given by $\int_\alpha^\beta [r^2 + (dr/d\theta)^2]^{1/2}\,d\theta$. Show that the total arclength of the lemniscate is $4\int_0^{\pi/2} d\theta/r$.

(c) Prove $d\theta/dr = r/\sqrt{1-r^4}$, so the total arclength is $4\int_0^1 dr/\sqrt{1-r^4}$.

(d) Prove that this total length is $\beta(\frac{1}{2}, \frac{1}{4})$ where β is the Euler beta function.

(e) Prove that this total length is $\Gamma(\frac{1}{4})^2/\sqrt{2\pi}$.

11. (a) Let $\operatorname{Im}\tau > 0$ and let $Q_m = \sum_{n=-\infty}^{\infty}(n - m\tau)^{-2}$. Prove that $|Q_m| \le 4e^{-m\operatorname{Im}\tau}/(1-e^{-2m\operatorname{Im}\tau})$. (*Hint*: See (10.4.87).)

(b) Conclude that $\sum_{m\neq 0} Q_m$ is absolutely convergent.

(c) Let $Q_m^{(k)} = \sum_{n=-\infty}^{\infty}(n-m\tau)^{-k}$ for any $k \ge 2$. Prove that $|Q_m^{(k)}| \le C_k e^{-m(\operatorname{Im}\tau)}$ for any k and $\operatorname{Im}\tau > 1$.

(d) Let $\zeta(2k) = \sum_{n=1}^{\infty} n^{-2k}$ for $k = 1, 2, \dots$. Prove that $\lim_{y\to\infty} S_{2k}(1, iy) = 2\zeta(2k)$ for any $k = 2, 3, \dots$.

12. Because of (d) of Problem 11, one defines the normalized Eisenstein series $E_{2k}(\tau_1, \tau_2) = S_{2k}(\tau_1,\tau_2)/2\zeta(2k)$. Prove that $J(\tau) = E_4(1,\tau)^3/(E_4(1,\tau)^3 - E_6(1,\tau)^2)$. (*Hint*: You need the values of $\zeta(4)$ and $\zeta(6)$ in Problem 4 of Section 9.2.)

13. (a) Let $\tau = iy_0$ with $y_0 > 0$. Prove that $\wp(z \mid \tau)$ is real when $\operatorname{Im} z = \frac{n}{2}\tau$ or $\operatorname{Re} z = \frac{m}{2}$ for $n, m \in \mathbb{Z}$. (*Hint*: First prove it for $m = 0$, then using covariance under $z \to iz$, for $n = 0$, and then using conjugate covariance and translation invariance, for $m = 1$ and $n = 1$.)

(b) Let $\widetilde{\Omega} = \{x + iy \mid 0 < x < \frac{1}{2}, 0 < y < \frac{1}{2}y_0\}$ so $\wp$ is real on $\partial\widetilde{\Omega} \setminus \{0\}$. Prove that $\wp$ is one-one to $\mathbb{R}$ on $\partial\widetilde{\Omega}\setminus\{0\}$.

(c) Prove for $\lim_{x\downarrow 0}\wp(x,\tau) \to \infty$, $\lim_{y\downarrow 0}\wp(iy,\tau) = -\infty$ and show $\wp$ is strictly monotone decreasing as one goes counterclockwise around $\partial\widetilde{\Omega} \setminus \{0\}$ and that $e_1 > e_2 > e_3$.

(d) Prove that $\wp$ is a bijection of $\widetilde{\Omega}$ to $\mathbb{C}_+$. This provides another solution to Example 8.4.5.

(e) Prove that $\wp$ is a bijection of $\{x + iy \mid 0 < x < 1,\, 0 < y < \frac{1}{2}y_0\}$ to $\mathbb{C} \setminus (-\infty, e_2] \cup [e_1, \infty)$.

(f) Let $0 < r < 1$ and let $\tau - 2\pi i/\log(r^{-1})$. Prove that $\wp(\log(w)/\log(r))$ maps $\mathbb{A}_{r,1}$ conformally to $\mathbb{C} \setminus (-\infty, e_3] \cup [e_2, e_1]$.

10.5. Bonus Section: Jacobi Elliptic Functions

Having seen Weierstrass elliptic functions, we turn to another class of both historical and practical significance and, in particular, emphasize a remarkable formula (the triple product identity) that is useful in analysis and in number theory. In fact, we'll discuss two rather distinct families of functions: the Jacobi theta functions and the Jacobi sn, cn, dn functions. Jacobi used the theta functions to understand the others, but we'll be able to use the Weierstrass σ-function in its stead—we'll leave the connections between the two to the Problems, so the two halves of this section are almost disconnected!

The theta functions will be a replacement for σ, providing an alternate tool to complete the proof of the existence part of Abel's theorem. The Jacobi elliptic functions, sn, etc., arise as inverses of particularly simple elliptic integrals involving square roots of quartic rather than cubic polynomials; for example, $\text{sn}(x,k)$ obeys

$$x = \int_0^{\text{sn}(x,k)} \frac{dt}{\sqrt{(1-t^2)(1-k^2t^2)}} \tag{10.5.1}$$

so for $k = 0$, it is just $\sin(x)$. As you can imagine, in many applications it is integrals like this that enter. It should also be recalled that in Example 8.4.5, we saw that $\text{sn}(x,k)$ provides a conformal map of a rectangle to $\mathbb{C}_+$.

* * * * *

We begin with the θ-function, an analog of the σ-function in that it will obey (10.3.13). We'll suppose that $\tau_1 = 1$, $\tau_2 = \tau$ with $\operatorname{Im}\tau > 0$. For reasons that will eventually be clear, we want to place zeros at $\frac{1}{2} + \frac{1}{2}\tau + \mathcal{L}$ rather than $\mathcal{L}$. We'll also shift to calling the variable w, using z for something else. We'll call the resulting function θ (it will eventually be θ_3). The starting idea will be to make θ periodic under $w \to w + 1$, that is, we expect

$$\theta(w+1 \mid \tau) = \theta(w \mid \tau), \qquad \theta(w+\tau \mid \tau) = e^{-2\pi i(w+\frac{1}{2}\tau)}\theta(w \mid \tau) \tag{10.5.2}$$

We've built two things into this formula. We expect θ to have the same relation between ω_j and α_j as σ does since that comes from a natural symmetry, but since σ is odd and we guess that θ will be even, there is no minus sign. Secondly, we used the Legendre relation (10.3.32) when $\omega_1 = 0$, $\tau_1 = 1$ to conclude $\omega_2 = -2\pi i$. Of course, this is only an expectation that we will need to prove.

Since θ is periodic, Theorem 3.10.3 applies, and we can write θ as a function of $z = e^{2\pi i w}$ where this new function is analytic in $\mathbb{C}^\times$. The condition about $w \to w + \tau$ is now a multiplicative one, $z \to z e^{2\pi i \tau}$. The fact that $\tau/2$ often appears suggests the right parameter is

$$q = e^{\pi i \tau} \tag{10.5.3}$$

The condition $\operatorname{Im} \tau > 0$ is equivalent to

$$0 < |q| < 1 \tag{10.5.4}$$

Thus, we want a function $\Theta(z \mid q)$ and

$$\theta(w \mid \tau) = \Theta(e^{2\pi i w} \mid q = e^{\pi i \tau}) \tag{10.5.5}$$

For $\theta(w)$ to be zero if $w = \frac{1}{2} + \frac{1}{2}\tau(2n+1)$, we need Θ to be zero if $z = e^{2\pi i w} = (-1) q^{2n+1}$. Thus, a first guess for Θ might be $\prod_{n=-\infty}^{\infty}(1 + q^{2n-1} z)$ which has zeros at the right places. Since $|q| < 1$, the product $\prod_{n=1}^{\infty}(1 + q^{2n-1} z)$ is fine since $\sum_{n=1}^{\infty} |q^{2n-1} z| < \infty$. But as $n \to -\infty$, q^{2n-1} blows up and no kind of Weierstrass trick will help. However, $q^{2n-1} z = -1$ if and only if $q^{1-2n} z^{-1} = -1$ and as $|n|$ increases with $n < 0$, $|q^{1-2n}|$ goes to zero. Of course, $1 + q^{1-2n} z^{-1}$ is singular at $z = 0$, but Θ is only required to be analytic on $\mathbb{C}^\times$. Indeed, with zeros at $q^{-(2n-1)}$ with $n \to \infty$, it has to be singular at 0. This leads to the following definition:

$$\Theta(z \mid q) = \prod_{n=1}^{\infty} (1 - q^{2n})(1 + q^{2n-1} z)(1 + q^{2n-1} z^{-1}) \tag{10.5.6}$$

The preliminary factor $\prod_{n=1}^{\infty}(1 - q^{2n})$ is an overall constant—irrelevant to the next few theorems, but important later when we identify the Laurent expansion for Θ. Because of the three factors, this is called the *triple product.* Here are the main properties of Θ, short of the Laurent expansion to come below in Theorem 10.5.4 (see (10.5.20)).

Theorem 10.5.1. (a) *For $q \in \mathbb{D}$, $\Theta(z \mid q)$ defined by (10.5.6) is an absolutely and uniformly convergent product for z in compact subsets of $\mathbb{C}^\times$, and the limit is jointly analytic in (z, q) in $\mathbb{C}^\times \times \mathbb{D}$.*

(b) *For $q \neq 0$,*

$$\Theta(zq^2 \mid q) = (zq)^{-1} \Theta(z \mid q) \tag{10.5.7}$$

(c) *For* $q \neq 0$, $\Theta(z \mid q) = 0$ *if and only if* $z = -q^{\pm 1}, -q^{\pm 3}, -q^{\pm 5}, \dots$ *and these zeros are all simple.*

Proof. (a), (c) For $|q| \le Q < 1$ and $r < |z| < r^{-1}$ with $r < 1$,

$$\sum_{n=1}^{\infty} (|q^{2n}| + |q^{2n-1}|\,|z| + |q^{2n-1}|\,|z|^{-1}) \le (Q + 2r^{-1})\,\frac{Q}{1-Q^2}$$

proving uniform and absolute convergence of the products. Analyticity follows from the Weierstrass convergence theorem and the zero condition comes from $q^{2n-1}z = -1$ or $q^{2n-1}z^{-1} = -1$.

(b)
$$\frac{\Theta(z \mid q)}{\prod_{n=1}^{\infty}(1-q^{2n})} = [(1+qz)(1+q^3 z)\dots][(1+qz^{-1})(1+q^3 z^{-1})\dots] \tag{10.5.8}$$

If z is replaced by $q^2 z$, the first product in $[\dots]$ is $[(1+q^3 z)\dots]$ while the second is $[(1+q^{-1}z^{-1})(1+qz^{-1})\dots]$, so

$$\frac{\Theta(zq^2 \mid q)}{\Theta(z \mid q)} = \frac{1+q^{-1}z^{-1}}{1+qz} = \frac{1}{qz} \tag{10.5.9}$$

proving (10.5.7). □

We now define θ on $\mathbb{C} \times \mathbb{C}_+$ by

$$\theta(w \mid \tau) = \Theta(e^{2\pi i w} \mid e^{\pi i \tau}) \tag{10.5.10}$$

The following is a direct translation of the last theorem if we note that when $z = e^{2\pi i w}$, $q = e^{\pi i \tau}$, then

$$(zq)^{-1} = e^{-2\pi i (w + \frac{1}{2}\tau)}, \qquad zq^2 = e^{2\pi i (w+\tau)} \tag{10.5.11}$$

the expected prefactor in (10.5.2) under the result of $w \to w + \tau$.

Theorem 10.5.2. θ *defined by* (10.5.10) *defines an entire function of* $(w, \tau) \in \mathbb{C} \times \mathbb{C}_+$ *and*

(a) θ *obeys* (10.5.2).
(b) $\theta(w \mid \tau) = 0 \Leftrightarrow w = \frac{1}{2} + \frac{1}{2}\tau + m + n\tau$ *for* $m, n \in \mathbb{Z}$.
(c) $\theta(-w \mid \tau) = \theta(w \mid \tau)$

Proof. As noted, (a) is just $e^{2\pi i (w+1)} = e^{2\pi i w}$ and (10.5.7). To get (b), note that $w = \frac{1}{2} + \frac{1}{2}\tau + m + n\tau \Leftrightarrow z = e^{2\pi i \omega} = (-1)q^{2m+1}$. (c) is immediate from the formula (10.5.6), which implies $\Theta(z^{-1} \mid q) = \Theta(z \mid q)$. □

Thus,

$$g(w) \equiv \theta(w - \tfrac{1}{2} - \tfrac{1}{2}\tau \mid \tau) \tag{10.5.12}$$

has the properties required in Theorem 10.3.8 and allows a new proof of the existence parts of Abel's theorem and of the principal part theorem. Of course, by Problem 4 of Section 10.3, θ and σ_2 have a simple relation (see Problem 1).

Before leaving θ and Θ, we prove a striking formula that, because it relates a Laurent series to the triple product (10.5.6), is called the *triple product formula.* As with the gamma function, a key role is played by uniqueness implied by a functional equation.

Theorem 10.5.3. *Fix $q \in \mathbb{D}^\times$ and let g be analytic in a neighborhood of $\bar{\mathbb{A}}_{q,q^{-1}}$. Suppose that*

(i) *If $|z| = q^{-1}$, then*

$$g(q^2 z) = (qz)^{-1} g(z) \tag{10.5.13}$$

(ii) $g(-q) = 0$ (10.5.14)

Then g has an analytic continuation to $\mathbb{C}^\times$ and

$$g(z) = C\Theta(z \mid q) \tag{10.5.15}$$

for some constant C.

Remark. We emphasize that, a priori, g is allowed to have zeros at points in $\bar{\mathbb{A}}_{q,q^{-1}}$ other than $z = -q, q^{-1}$. That it does not is explained in Problem 2.

Proof. The functional equation holds by analyticity for $||z| - q^{-1}| < \varepsilon$ for some small ε. Thus, if we use (10.5.13) to define $g(z)$ for $z \in \bar{\mathbb{A}}_{q^3,q}$, we get an analytic continuation to a neighborhood of $\bar{\mathbb{A}}_{q^3,q}$ and repeating this to all of $\mathbb{C}^\times$. Moreover, (10.5.13) then holds by analyticity on all of $\mathbb{C}^\times$. The functional equation also implies $g(z) = 0$ for $z \in \{-q^{2n+1}\}_{n=-\infty}^{\infty}$. Let

$$h(z) = \frac{g(z)}{\Theta(z \mid q)} \tag{10.5.16}$$

By the functional equation for g and Θ, h obeys

$$h(q^2 z) = h(z) \tag{10.5.17}$$

Moreover, since g vanishes where Θ does and Θ has only first-order zeros, h is analytic in all of $\mathbb{C}^\times$. In particular,

$$\sup_{z \in \bar{\mathbb{A}}_{q,q^{-1}}} |h(z)| = C < \infty \tag{10.5.18}$$

But then, by (10.5.17),

$$\sup_{z \in \mathbb{C}^\times} |h(z)| = C \tag{10.5.19}$$

so the singularity at 0 is removable and, by Liouville's theorem, h is a constant. $\square$

Theorem 10.5.4. *For any $z \in \mathbb{C}^\times$, $q \in \mathbb{D}$,*

$$\Theta(z \mid q) = \sum_{n=-\infty}^{\infty} q^{n^2} z^n \tag{10.5.20}$$

Remark. It is for this to be true that we took $\prod_{n=1}^\infty (1-q^{2n})$ in defining Θ.

Proof. Let $\widetilde{\Theta}(z \mid q)$ denote the right side of (10.5.20). We'll show first that $\widetilde{\Theta}$ obeys the hypotheses of Theorem 10.5.4 and then compute the constant.

Fix $q \in \mathbb{D}$. Note first that since $|q^{n^2}|^{1/n} = |q|^n \to 0$, by the Cauchy radius formula, $\sum_{n=0}^\infty q^{n^2} z^n$ converges and defines a function analytic in z in all of $\mathbb{C}$, and similarly, $\sum_{n=-\infty}^{-1} q^{n^2} z^n$ in $\mathbb{C}^\times$, so $\widetilde{\Theta}$ is analytic in $\mathbb{C}^\times$ (we note though that for $z \neq 0$ fixed, by the Fabry gap theorem, $\widetilde{\Theta}$ has a natural boundary at $|q| = 1$). Next,

$$\begin{aligned}\widetilde{\Theta}(q^2 z \mid q) &= \sum_{n=-\infty}^{\infty} q^{n^2+2n} z^n \\ &= (qz)^{-1} \sum_{n=-\infty}^{\infty} q^{(n+1)^2} z^{n+1} &\text{(10.5.21)} \\ &= (qz)^{-1} \widetilde{\Theta}(z \mid q) &\text{(10.5.22)}\end{aligned}$$

by changing $n+1 \to n$ in the summation.

Moreover,

$$\begin{aligned}\widetilde{\Theta}(-q \mid q) &= \sum_{n=-\infty}^{\infty} (-1)^n q^{n^2+n} \\ &= \sum_{m=-\infty}^{\infty} (-1)^{-1-m} q^{m^2+m} &\text{(10.5.23)} \\ &= -\widetilde{\Theta}(-q \mid q) &\text{(10.5.24)}\end{aligned}$$

so $\widetilde{\Theta}(-q \mid q) = 0$. In (10.5.23), we noted that if $n = -1 - m$, then

$$n^2 + n = n(n+1) = (-1-m)(-m) = m(m+1) \tag{10.5.25}$$

Thus, by Theorem 10.5.3,

$$\widetilde{\Theta}(z \mid q) = C_q \Theta(z \mid q) \tag{10.5.26}$$

and we need to prove that $C_q = 1$. This is a subtle fact with no really simple proof, so there are many alternatives (see Problem 7 and the Notes for other proofs). We evaluate Θ and $\widetilde{\Theta}$ at $z = -1$ and at $z = i$. Clearly,

$$\widetilde{\Theta}(-1 \mid q) = \sum_{n=-\infty}^{\infty} (-1)^n q^{n^2}, \qquad \Theta(-1 \mid q) = \prod_{n=1}^{\infty} (1 - q^n)(1 - q^{2n-1}) \tag{10.5.27}$$

since

$$\prod_{n=1}^{\infty} (1 - q^n) = \prod_{n=1}^{\infty} (1 - q^{2n}) \prod_{n=1}^{\infty} (1 - q^{2n-1}) \tag{10.5.28}$$

Since $(-i)^n = -(i)^{-n}$ for n odd, the odd n terms in $\widetilde{\Theta}$ cancel. Thus,

$$\widetilde{\Theta}(i \mid q) = \sum_{n=-\infty}^{\infty} (-1)^n q^{4n^2}, \qquad \Theta(i \mid q) = \prod_{n=1}^{\infty} (1 - q^{4n})(1 - q^{8n-4}) \tag{10.5.29}$$

since

$$(1 - q^{2n})(1 + iq^{2n-1})(1 - iq^{2n-1}) = (1 - q^{2n})(1 + q^{4n-2}) \tag{10.5.30}$$

so that rearranging absolutely convergent products,

$$\begin{aligned} \prod_{n=1}^{\infty} (1 - q^{2n})(1 + q^{4n-2}) &= \prod_{n=1}^{\infty} (1 - q^{4n}) \prod_{n=1}^{\infty} (1 - q^{4n-2})(1 + q^{4n-2}) \\ &= \prod_{n=1}^{\infty} (1 - q^{4n})(1 - q^{8n-4}) \end{aligned} \tag{10.5.31}$$

Comparing (10.5.27) and (10.5.29) shows

$$\widetilde{\Theta}(i \mid q) = \widetilde{\Theta}(-1 \mid q^4), \qquad \Theta(i \mid q) = \Theta(-1 \mid q^4) \tag{10.5.32}$$

so that

$$C_{q^4} = C_q \tag{10.5.33}$$

On the other hand,

$$\lim_{q \to 0} \Theta(z \mid q) = 1 = \lim_{q \to 0} \widetilde{\Theta}(z \mid q) \Rightarrow \lim_{q \to 0} C_q = 1 \tag{10.5.34}$$

so, iterating (10.5.31),

$$C_q = C_{q^4} = C_{q^{16}} = C_{q^{64}} = \cdots = 1 \tag{10.5.35}$$

□

We will apply $\Theta(1, q)$ to number theory in Section 13.2 of Part 2B and say something about general q-ology in the Notes.

Just as there are four σ's, there are four θ's defined by (θ_3 is our θ):

$$\theta_3(w \mid \tau) \equiv \Theta(e^{2\pi i w} \mid q = e^{\pi i \tau}) \tag{10.5.36}$$

$$\theta_4(w \mid \tau) \equiv \theta_3(w + \tfrac{1}{2} \mid \tau) = \Theta(-e^{2\pi i w} \mid q) \tag{10.5.37}$$

$$\theta_1(w \mid \tau) \equiv -iq^{1/4}e^{\pi iw}\theta_3(w+\tfrac{1}{2}+\tfrac{1}{2}\tau \mid \tau) = -iq^{1/4}e^{\pi iw}\Theta(-qe^{2\pi iw} \mid q) \tag{10.5.38}$$

$$\theta_2(w \mid \tau) \equiv \theta_1(w+\tfrac{1}{2} \mid \tau) = q^{1/4}e^{\pi iw}\Theta(qe^{2\pi iw} \mid q) \tag{10.5.39}$$

These have both product forms and sums; the sums are:

$$\theta_1(w \mid \tau) = 2\sum_{n=0}^{\infty} q^{(n+\frac{1}{2})^2}(-1)^n \sin((2n+1)\pi w) \tag{10.5.40}$$

$$\theta_2(w \mid \tau) = 2\sum_{n=0}^{\infty} q^{(n+\frac{1}{2})^2} \cos((2n+1)\pi w) \tag{10.5.41}$$

$$\theta_3(w \mid \tau) = 1+2\sum_{n=1}^{\infty} q^{n^2} \cos(2\pi nw) \tag{10.5.42}$$

$$\theta_4(w \mid \tau) = 1+2\sum_{n=1}^{\infty} (-1)^n q^{n^2} \cos(2\pi nw) \tag{10.5.43}$$

In Problems 1(b) and 3, the reader will confirm these last four formulae and show that if $\sigma \equiv \sigma_0$, then up to a quadratic exponential, σ_j and θ_{j+1} agree. The choice of indices on σ and θ varies among authors, and some authors replace πw by w in their definition of θ (so their τ_1 is π, not 1). We will use σ to express the relation of sn, cn, dn below to theta-type functions, leaving the (original) connection to Jacobi theta functions to the Problems.

Before leaving the subject of theta functions, we want to note a remarkable symmetry that has applications to number theory (Riemann's original proof of the functional equation for the zeta function and, as discussed in Section 13.1 of Part 2B, to counting representations as sums of squares):

Theorem 10.5.5. *Define for* $\operatorname{Im}\tau > 0$ *and* $w \in \mathbb{C}$,

$$\theta(w \mid \tau) = \sum_{n=-\infty}^{\infty} e^{\pi i n^2 \tau} e^{2\pi i w n} \tag{10.5.44}$$

$$= \Theta(z = e^{2\pi iw} \mid q = e^{\pi i\tau}) \tag{10.5.45}$$

Then

$$\theta\left(w \,\middle|\, -\frac{1}{\tau}\right) = \sqrt{\frac{\tau}{i}}\, e^{\pi i\tau w^2}\theta(w\tau \mid \tau) \tag{10.5.46}$$

In particular, taking $w=0$, $\tau = ix$, *if*

$$\theta_0(x) = \sum_{n=-\infty}^{\infty} e^{-\pi n^2 x} \tag{10.5.47}$$

then

$$\theta_0\left(\frac{1}{x}\right) = \sqrt{x}\,\theta_0(x) \tag{10.5.48}$$

Proof. (10.5.46) for $\tau = ix$ with $x \in (0, \infty)$ was proven in Part 1, both as Example 6.6.11 and in Problem 11 of Section 6.9. Since $\theta(w \mid \tau)$ is analytic in τ for $\tau \in \mathbb{C}_+$, we get (10.5.46) for all $\tau \in \mathbb{C}_+$. $\square$

* * * * *

We turn now to the Jacobi elliptic functions and associated elliptic integrals. It has the advantage of depending on a single parameter k in the integral whereas (10.4.69) depends on two g_2, g_3 (which, in turn, do depend on only a single τ plus scaling). In this theory, other basic parameters are k', K, and K' given by

$$(k')^2 = 1 - k^2 \tag{10.5.49}$$

$$K = \int_0^1 \frac{dt}{\sqrt{(1-t^2)(1-k^2t^2)}} \tag{10.5.50}$$

$$K' = \int_0^1 \frac{dt}{\sqrt{(1-t^2)(1-(k')^2t)}} \tag{10.5.51}$$

The reader will show (Problem 4) that $\mathrm{sn}(\,\cdot\,, k)$ has periods $4K$ and $2iK'$. This was also done earlier in Example 8.4.5. The second period is Jacobi's (and Abel's) great discovery. These functions are mainly used when $0 < k^2 < 1$, so also $0 < (k')^2 < 1$ and, of course, $K > 0$, $K' > 0$. Below we assume $0 < k^2 < 1$. We'll show in the next section (see Theorem 10.6.6) that for any such k, there is a unique τ on the positive imaginary axis solving (10.5.55) below, and in that case,

$$e_1 > e_2 > e_3 \tag{10.5.52}$$

(see also Problem 13 of Section 10.4 for (10.5.52)). Theorem 10.5.6 is stated for this case, although once one has it, one can analytically continue sn, etc. by using the formulae of this theorem as definitions for all k. The integral formulae are valid with suitable branch choices in the square root.

The other functions that arise are $\mathrm{cn}(x)$, $\mathrm{dn}(x)$ given by

$$\mathrm{sn}^2(x) + \mathrm{cn}^2(x) = 1 \tag{10.5.53}$$

$$\mathrm{dn}^2(x) + k^2\mathrm{sn}^2(x) = 1 \tag{10.5.54}$$

cn and dn are determined by requiring them to be analytic near 0 with value at $x = 0$ with $\mathrm{dn}(0) = \mathrm{cn}(0) = 1$. That they extend to entire functions follows from Theorem 10.5.6. Essentially, the required double zeros of $1 - \mathrm{sn}^2(x)$ or $1 - k^2\mathrm{sn}^2(x)$ come from the double zeros of $\wp(z) - e_j$.

Figures 10.5.1, 10.5.2, and 10.5.3 plot the Jacobi elliptic functions.

Here is the key relation of the Jacobi elliptic functions to σ and σ_j. We'll prove this and leave the relation to θ and other connections to the Problems.

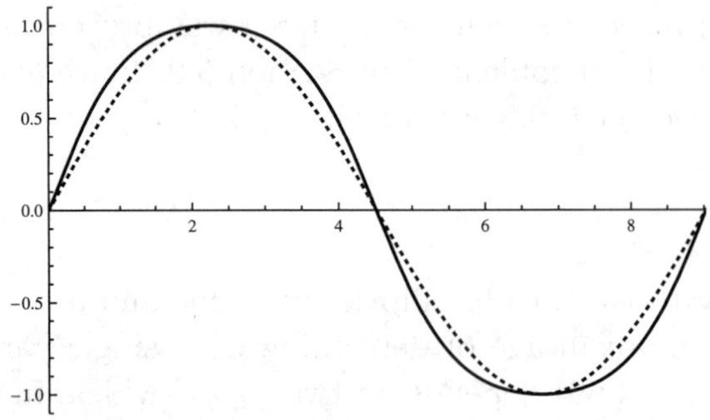

Figure 10.5.1. $\mathrm{sn}(x,k)$ and $\sin(x)$ (scaled to have the same x-axis period) for $k^2 = \frac{4}{5}$. sn is solid; sin dashed.

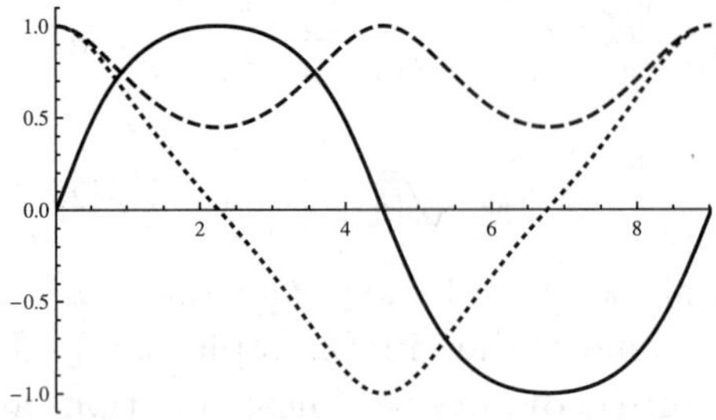

Figure 10.5.2. $\mathrm{sn}(x,k)$ (solid), $\mathrm{cn}(x,k)$ (dotted), $\mathrm{dn}(x,k)$ (dashed) for $k^2 = \frac{4}{5}$.

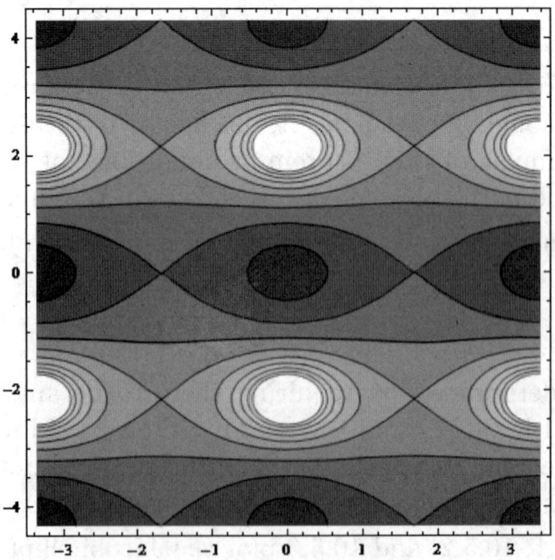

Figure 10.5.3. Contour plot of $|\mathrm{sn}(x+iy,k)|$ for $k = \frac{1}{2}$. This illustrates the double periodicity.

Theorem 10.5.6. *Given k in $(0,1)$, let $\tau \in i(0,\infty)$ be determined by*

$$k^2 = \frac{e_2(\tau) - e_3(\tau)}{e_1(\tau) - e_3(\tau)} \tag{10.5.55}$$

where $e_j(\tau)$ are the values of $\wp(\cdot \mid \tau)$ on the half-lattice which obey $e_1 > e_2 > e_3$. Let

$$\mu = \sqrt{e_1 - e_3} \tag{10.5.56}$$

Then

$$\operatorname{sn}(x,k) = \frac{\mu\sigma(x/\mu)}{\sigma_3(x/\mu)} \tag{10.5.57}$$

$$\operatorname{cn}(x,k) = \frac{\sigma_1(x/\mu)}{\sigma_3(x/\mu)} \tag{10.5.58}$$

$$\operatorname{dn}(x,k) = \frac{\sigma_2(x/\mu)}{\sigma_3(x/\mu)} \tag{10.5.59}$$

$$K = \frac{\mu}{2}, \qquad K' = -\frac{i\mu\tau}{2} \tag{10.5.60}$$

Proof. We start with (10.4.69) by picking $\alpha_0 = \infty$, and since $\wp(u) = \wp(-u)$, flipping signs to get

$$u = \int_\alpha^\infty \frac{dy}{\sqrt{4y^3 - g_2 y - g_3}} \Leftrightarrow \wp(u) = \alpha \tag{10.5.61}$$

In (10.5.1), we change variables to $s^2 = \beta$, $t^2 = z$ to get (since $dz = 2t\,dt = 2\sqrt{z}\,dt$)

$$x = \int_0^\beta \frac{dz}{2\sqrt{z(1-z)(1-k^2 z)}} \Leftrightarrow \beta = \operatorname{sn}(x)^2 \tag{10.5.62}$$

The square root in (10.5.61) has zeros at $y = e_1, e_2, e_3$, while the one in (10.5.62), at $z = 1,\ k^{-2},\ \infty$. This suggests we pick a fractional linear transformation that takes e_1 to 1, e_3 to ∞, and e_2 to k^{-2}, that is,

$$z = \frac{e_1 - e_3}{y - e_3} \tag{10.5.63}$$

does the right thing to e_1 and e_3. If we pick τ so that (10.5.55) holds, then (10.5.63) moves the zeros to the right place.

By (10.5.63) with μ given by (10.5.56),

$$y - e_3 = \frac{\mu^2}{z}, \qquad y - e_1 = \mu^2\,\frac{(1-z)}{z} \tag{10.5.64}$$

$$y - e_2 = \mu^2\,\frac{1 - k^2 z}{z}, \qquad dz = -\frac{z^2}{\mu^2}\,dy \tag{10.5.65}$$

Thus,

$$\begin{aligned}\sqrt{4y^3 - g_2 y - g_3} &= 2\sqrt{(y-e_1)(y-e_2)(y-e_3)} \\ &= \frac{2\mu^3}{z^2}\sqrt{z(1-z)(1-k^2 z)}\end{aligned} \tag{10.5.66}$$

so

$$\frac{dy}{\sqrt{4y^3 - g_2 y - g_3}} = -\frac{1}{\mu}\frac{dz}{2\sqrt{z(1-z)(1-k^2 z)}} \tag{10.5.67}$$

Taking into account the change of variables in limits and the inversion of limits, (10.5.61), (10.5.62), and (10.5.67) translate to

$$\alpha = \wp(u), \qquad \beta = \text{sn}(x)^2 \quad \Leftrightarrow \quad u = \frac{x}{\mu}, \qquad \beta = \frac{\mu^2}{\alpha - e_3} \tag{10.5.68}$$

or

$$\text{sn}(x) = \frac{\mu}{\sqrt{\wp(x/\mu) - e_3}} \tag{10.5.69}$$

Given (10.4.86) for $j = 3$, this translates to (10.5.57). This plus (10.5.53) plus (10.4.86) for $j = 1$ and 3 implies (10.5.58), and similarly, (10.5.69) plus (10.5.54) plus (10.5.55) plus (10.4.86) for $j = 2$ and 3 implies (10.5.59).

K is determined by $\text{sn}(K) = 1$ which, given $\wp(\frac{1}{2}) = e_1$ and (10.5.56), yields the first equation in (10.5.60).

Making the k- and τ-dependence explicit, we see if $\tau(k)$ solves (10.5.55), then

$$K(k) = \tfrac{1}{2}\sqrt{e_1(\tau(k)) - e_3(\tau(k))} \tag{10.5.70}$$

If k^2 obeys (10.5.55), then $k'^2 = (e_2 - e_1)/(e_3 - e_1)$. Thus $\tau(k')$ obeys

$$\frac{e_2(\tau(k')) - e_3(\tau(k'))}{e_1(\tau(k')) - e_3(\tau(k'))} = \frac{e_2(\tau(k)) - e_1(\tau(k))}{e_3(\tau(k)) - e_1(\tau(k))}$$

Using $e_1 + e_2 + e_3 = 0$, it is easy to see this is equivalent to

$$\frac{e_1(\tau(k'))}{e_3(\tau(k'))} = \frac{e_3(\tau(k))}{e_1(\tau(k))}$$

Thus, we need to interchange the two lattice generators and replace one by the negative. Using the invariance of e_j/e_k under scaling τ_1 and τ_2, this implies

$$\tau(k') = -\tau(k)^{-1} \tag{10.5.71}$$

Restoring the second element of the basis,

$$\begin{aligned}e_1(-\tau^{-1}) &= e_1(1, -\tau^{-1}) = e_3(\tau^{-1}, 1) \\ &= \tau^2 e_3(1,\tau) = \tau^2 e_3(\tau)\end{aligned} \tag{10.5.72}$$

where we used the scale covariance (10.4.5) with $\lambda = \tau$. There is a similar formula for e_3.

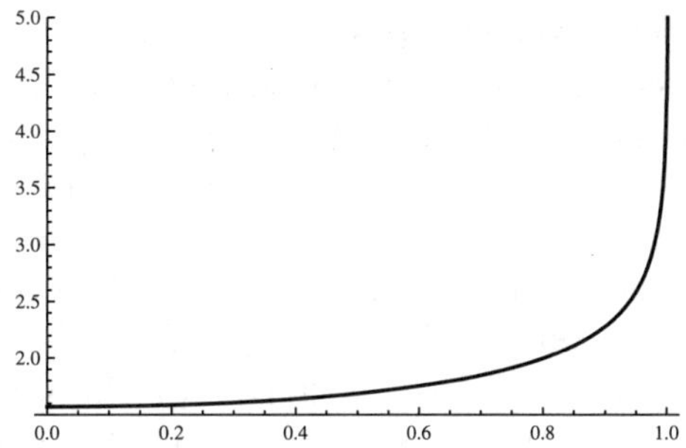

Figure 10.5.4. $K(k)\colon 0 < k < 1$.

Thus, (10.5.11) and (10.5.71) imply

$$\begin{aligned} K' = K(k') &= \tfrac{1}{2}\sqrt{e_1(-\tau^{-1}(k)) - e_3(-\tau^{-1}(k))} \\ &= \tfrac{1}{2}\sqrt{\tau^2(e_3(\tau) - e_1(\tau))} \\ &= -i\tau K(k) \qquad (10.5.73) \end{aligned}$$

verifying the second half of (10.5.60). Since $\tau \in \mathbb{C}_+$ with $\operatorname{Re}\tau = 0$ and $e_3 < e_1$, the argument of the final square root is positive. We take $-i$ in (10.5.73) to arrange that $K' > 0$. $\square$

Figure 10.5.4 shows $K(k)$ for $0 < k < 1$. Note that $\lim_{k\downarrow 0} K(k) = \pi/2$ and there is a logarithmic divergence at $k = 1$.

Notes and Historical Remarks. Bernoulli [**43**] considered the sums $\sum_{n=0}^\infty x^{n^2}$ and $\sum_{n=0}^\infty x^{n(n+1)}$ and Euler [**178**] has several results on such sums (and even a variant of triple product identity) and considered the θ-like product $\prod_{n=1}^\infty (1 - x^n z)^{-1}$. Gauss' unpublished works on the lemniscate integral contain objects we would now recognize as θ functions, both as products and sums. He found this in part by looking at the arithmetic-geometric mean (start with a_0, b_0, form a_{n+1}, b_{n+1} inductively by $a_{n+1} = \frac{1}{2}(a_n + b_n)$, $b_{n+1} = \sqrt{a_n b_n}$; they have a common limit called the *arithmetic geometric mean*), which he related to the lemniscate integral and wrote in terms of theta functions. Cox [**125**] discusses both this mean and the history.

It was Jacobi, though, who first fully presented θ functions in his fundamental work in 1829 on elliptic functions [**281**]. The four-fold θ_j is from his 1835 paper [**283**]. For more on the history, see Hellegouarch [**250**, Ch. 5]. For a recent book on theta functions, see Farkas–Kra [**187**].

As we have seen, Θ and θ obey a variety of remarkable relations. As Hellegouarch [**250**] remarks:

> The originator of the famous wisecrack about the five operations in arithmetic—addition, subtraction, multiplication, division and

> modular functions—would appear to have been the mathematician M. Eichler. As we will see, the fifth and last "operations" are actually functions, which have so incredibly many symmetries that one may really wonder how they can possibly exist.

Theta functions, as generalized by Riemann [**484**] to higher dimensions ((10.5.20) is replaced by $\Theta(\vec{z}) = \sum_{n \in \mathbb{Z}^\nu} \exp(-(\vec{n}, A\vec{n})) z_1^{n_1} \dots z_\nu^{n_\nu}$ for a suitable positive define matrix A) are the basis for the study of meromorphic functions on higher-genus Riemann surfaces and have many other applications. See [**274, 446, 472**] and Section 11.6.

The reader should be warned that while our definitions of θ_j are very common, there are many, many other normalizations of functions called "theta"—as remarked in the Wikipedia article[2] on "Jacobi theta-functions (notational variations)" (yes, there is an entire article on that subject!): "The warning in Abramowitz and Stegun, 'There is a bewildering variety of notations ... in consulting books caution should be exercised,' may be viewed as an understatement."

The triple product formula appeared in Jacobi's 1829 book [**281**] and it is often called the Jacobi triple product theorem, but there are the usual suspects on the prehistory. Gauss also knew it and a special case appeared in Euler. [**15**] point out that it follows from what they call the q-binomial theorem found by Rothe in 1811 [**492**] (see Problem 6). They also interpret the triple product formula as an Euler duplication formula for q-Gamma functions.

Carl Gustav Jacob Jacobi (1804–51) was, like Landau later, the son of a Jewish banker based in Berlin, although, unlike Landau, he had to convert to Christianity to be eligible for academic appointment, and due to economic turmoil in Germany, he lost his family inheritance. In his later years, unable to teach because of ill health (diabetes), he depended on a royal stipend. In 1851, at age only 47, he suffered first from influenza and then smallpox and passed away.

He is best known for his work on elliptic functions, but also for continued fractions (and the related Jacobi matrices), determinant theory (Jacobians), mechanics (Hamilton–Jacobi equations), number theory (much of it based on his work on elliptic functions), special functions (Jacobi polynomials), and differential geometry (variational equations for geodesics).

The mention of q-binomial and q-Gamma function points out that θ functions are a part of what is sometimes called *q-ology*. While the formulae work

[2] `http://en.wikipedia.org/wiki/Jacobi_theta_functions_(notational_variations)`

for $q \in \mathbb{D}$, the intuition comes from $q \in (0,1)$. The *q-integral* is defined by

$$\int_0^a f(x)\, d_q x = \sum_{n=0}^{\infty} f(aq^n)(aq^n - aq^{n+1}) \tag{10.5.74}$$

the *q-difference*

$$(\Delta_q f)(x) = \frac{f(x) - f(qx)}{x - qx} \tag{10.5.75}$$

and the q-factorial and q-binomial coefficient

$$n!_q = \frac{(1-q)(1-q^2)\dots(1-q^n)}{(1-q)^n} \tag{10.5.76}$$

$$\begin{bmatrix} n \\ k \end{bmatrix}_q = \frac{n!_q}{k!_q(n-k)!_q} \tag{10.5.77}$$

The point is that one uses geometrically spaced values and defines objects that, as $q \to 1$, approach their standard form (the reader should check this for the four q-objects just defined). Analogy to $q = 1$ gives one remarkable identities and relations. There are q-Gamma and q-beta functions, q-orthogonal polynomials, and a q-ology interpretation of the triple product formula. [**15**, Ch. 10] is a good introduction to q-ology.

As noted, there are many other evaluations of what we called C_q. See Andrews [**13, 14**] for two analytic calculations and Cheema [**113**], Wright [**595**], and [**15**, Ch. 11] for combinatorial proofs. Simon [**524**, Sec. 1.6] has an interesting proof based on setting up a family of orthogonal polynomials on the unit circle (Rodgers–Szegő polynomials) and analyzing them in terms of q-binomial coefficients. Szegő's theorem for this set of polynomials evaluates C_q.

For suitable values of parameters, θ obeys a heat equation. To get rid of annoying π's, define

$$H(x,t) = \theta_3\left(\frac{x}{2\pi}\,\middle|\,\frac{it}{\pi}\right) = \sum_{n=-\infty}^{\infty} e^{-n^2 t} e^{inx} \tag{10.5.78}$$

and note that H obeys the heat equation for $x = \mathbb{R}$, $t > 0$,

$$\frac{\partial}{\partial t} H(x,t) = \frac{\partial^2}{\partial x^2} H(x,t) \tag{10.5.79}$$

Indeed, it is the fundamental solution as periodic functions in x with $H(\cdot \mid t = 0) = c\delta(x)$.

At first, some are puzzled about why there are four θ's, four σ's, and three Jacobi elliptic. Why not just use a single function and ratios and translates? But, after all, we use not only sin but also cos and tan and .... In fact, there are standardly not three Jacobi functions but twelve! Going back to Glaisher [**209**], they have easy-to-remember symbols, for example, $\mathrm{cd}(x) = \mathrm{cn}(x)/\mathrm{dn}(x)$. And they actually enter in integrals, for example,

$$x = \int_{\mathrm{cd}(x)}^{1} (1-t^2)^{-1/2}(1-k^2t^2)^{-1/2}\,dt$$

See Armitage–Eberlein [**21**, Ch. 2] for a summary. There are, of course, dozens of formulae for sums of arguments, algebraic relations, and integrals. The Problems have a selection of some of them.

Integrals of the form $\int \frac{R(z)}{\sqrt{P(z)}}\,dz$, where P is a polynomial of degree 3 or 4 and R is a rational function, are called *elliptic integral.* Legendre reduced such integrals to normal forms. As we did in the proof of Theorem 10.5.6 for the special case (10.5.61), one can use fractional linear transformations and quadratic change of variable to reduce to the special case $P(z) = [(1-z^2)(1-k^2z^2)]$. By further use of integration by parts and that for this case of P, if $R(z) = zf(z^2)$, we have a perfect derivative, Legendre reduced to three normal forms:

$$\int \frac{dz}{\sqrt{(1-z^2)(1-k^2z^2)}}, \qquad \int \frac{\sqrt{1-k^2z^2}}{\sqrt{1-z^2}}\,dz, \qquad \int \frac{dz}{(z^2-b)\sqrt{(1-z^2)(1-k^2z^2)}} \tag{10.5.80}$$

called, respectively, *elliptic integrals of the first kind*, *second kind*, and *third kind*. Hancock [**236, 237**] from 1910 and 1917 are the standard "modern" references on elliptic integrals. Section 11.1 of Beals–Wong [**32**] has a textbook presentation of Legendre's theorem that all elliptic integrals can be reduced to the three in (10.5.80).

Problems

1. (a) Prove that

$$\sigma(z) = e^{\omega_1 z^2/2\tau_1} \frac{\theta_1(z)}{\theta_1'(0)} \tag{10.5.81}$$

(b) Prove that for $j = 1, 2, 3$,

$$\sigma_j(z) = e^{\omega_1 z^2/2\tau_1} \frac{\theta_{j+1}(z)}{\theta_{j+1}(0)} \tag{10.5.82}$$

2. One of the surprises in Theorem 10.5.3 is that one doesn't assume a priori that $z = -q$ is the only zero in $\bar{\mathbb{A}}_{q,q^{-1}}$. This problem will take the mystery out of that.

(a) If f obeys (10.5.14) and $f(z) = g'(z)/g(z)$, prove that $f(q^2z) = f(z) - z^{-1}$.

(b) Use the argument principle to show that if g obeys (10.5.14), then g has exactly one zero in any $\mathbb{A}_{q^2a,a}$ with g nonvanishing on $\partial\mathbb{D}_{q^2a}$.

3. Given the definitions of $\theta_j(z)$ in (10.5.36)–(10.5.39), prove that (10.5.41)–(10.5.43) are valid.

4. Given (10.5.57), prove that

$$\operatorname{sn}(x+2K,k) = -\operatorname{sn}(x,k), \qquad \operatorname{sn}(x+2iK',k) = \operatorname{sn}(x,k) \tag{10.5.83}$$

What are the analogous periodicities of cn and dn?

5. Prove the q-binomial coefficients given by (10.5.77)/(10.5.76) obey

$$\begin{bmatrix} n+1 \\ k \end{bmatrix}_q = q^k \begin{bmatrix} n \\ k \end{bmatrix}_q + \begin{bmatrix} n \\ k-1 \end{bmatrix}_q = \begin{bmatrix} n \\ k \end{bmatrix}_q + q^{n+1-k} \begin{bmatrix} n \\ k-1 \end{bmatrix}_q$$

6. This problem will lead the reader through a proof of the q-binomial theorem, that for $|q| < 1$, $|z| < 1$, and $a \in \mathbb{C}$,

$$\sum_{n=0}^{\infty} \frac{(a;q)_n}{(q;q)_n} z^k = \frac{(az;q)_\infty}{(z;q)_\infty} \tag{10.5.84}$$

where (including for $n = \infty$)

$$(a;q)_n = \prod_{j=1}^{n} (1 - aq^{j-1}) \tag{10.5.85}$$

(a) Let $F(z)$ be the right-hand side of (10.5.84). Prove that F is analytic for q, a fixed and $z \in \mathbb{D}$. Write $F(z) = \sum_{n=0}^{\infty} b_n z^n$.

(b) Prove that $(1-z)F(z) = (1-az)F(qz)$.

(c) Prove that $b_1 = 1$ and $b_n = [(1-aq^{n-1})/(1-q^n)]b_{n-1}$ and conclude that $b_n = (a;q)_n/(q;q_n)$.

(d) Prove *Rothe's formula* that

$$\sum_{k=0}^{N} \begin{bmatrix} N \\ k \end{bmatrix} (-1)^k q^{k(k-1)/2} z^k = (z;q)_N \tag{10.5.86}$$

(*Hint*: Take $a = q^{-N}$.)

7. This will deduce the Jacobi triple product formula from Rothe's formula. Take $z \in \mathbb{C}^\times$.

(a) Prove that

$$(zq^{-n};q)_{2n} = (-1)^n z^n q^{-n^2+n(n-1)/2} (q/z;q)_n (z;q)_n \tag{10.5.87}$$

(b) Take $N = 2n$ in (10.5.86), replace z by zq^{-n}, and use (10.5.87) to prove that

$$(q/z;q)_n(z;q) = \sum_{k=-n}^{n} \frac{(q;q)_{2n}(-1)^k q^{k(k-1)/2} z^k}{(q;q)_{n+k}(q;q)_{n-k}}$$

(c) Take $n \to \infty$ to obtain

$$(z;q)_\infty (q/z;q)_\infty (q;q)_\infty = \sum_{k=-\infty}^{\infty} (-1)^k q^{k(k-1)/2} z^k \tag{10.5.88}$$

(d) Replace q by q^2 and z by $-qz$ to get

$$(-qz;q^2)_\infty (-q/z;q^2)_\infty (q^2;q^2) = \sum_{n=-\infty}^{\infty} q^{n^2} z^n \tag{10.5.89}$$

and verify that this is just the triple product formula (10.5.20).

8. For $q \in \mathbb{D}$, prove that

(a) $\displaystyle\sum_{n=-\infty}^{\infty} q^{n^2} = \prod_{n=1}^{\infty} (1-q^{2n})(1+q^{2n-1})^2$

(b) $\displaystyle\sum_{n=-\infty}^{\infty} (-1)^n q^{n^2} = \prod_{n=1}^{\infty} (1-q^{2n})(1-q^{2n-1})^2$

(c) $\displaystyle\sum_{n=-\infty}^{\infty} (-1)^n q^{n(3n+1)/2} = \prod_{n=1}^{\infty} (1-q^n)$

(d) $\displaystyle\sum_{n=0}^{\infty} q^{n(n+1)/2} = \prod_{n=1}^{\infty} \left[\frac{1-q^{2n}}{1-q^{2n-1}}\right]$

(*Hint*: Use the triple product formula. For (c), replace q by $q^{3/2}$ and $z = -q^{1/2}$, and for (d), replace q by $q^{1/2}$ and take $z = q^{1/2}$.)

9. Let $g(x) = \int_0^x f(y)\,d_q y$ and let Δ_q be given by (10.5.75). Prove that $(\Delta_q g)(x) = f(x)$.

10. (a) Prove that for $0 < k < 1$, $\varphi \in (-\pi/2, \pi/2)$, we have, by a change of variables, that

$$\int_0^\varphi \frac{d\psi}{\sqrt{1-k^2\sin^2(\psi)}} = \int_0^{\sin(\varphi)} \frac{dy}{\sqrt{(1-k^2y^2)(1-y^2)}}$$

This is another way that elliptic integrals and functions arise.

(b) Prove that

$$K(k) = \int_0^{\pi/2} \frac{d\psi}{\sqrt{1-k^2\sin^2(\psi)}} \tag{10.5.90}$$

(c) Fix $k \in (0,1)$. For $x \in (-K, K)$, define $\varphi(x) \in (-\pi/2, \pi/2)$ to be the unique solution of

$$x = \int_0^{\varphi(x)} \frac{d\psi}{\sqrt{1-k^2\sin^2(\psi)}}$$

Prove that

$$\operatorname{sn}(x,k) = \sin(\varphi(x)), \qquad \operatorname{cn}(x,k) = \cos(\varphi(x)) \tag{10.5.91}$$

(d) Prove that

$$\frac{d\varphi}{dx} = \sqrt{1-k^2\sin^2(\varphi(x))} = \operatorname{dn}(x,k) \tag{10.5.92}$$

and from that, prove that

$$\frac{d}{dx}\operatorname{sn}(x) = \operatorname{cn}(x)\operatorname{dn}(x) \tag{10.5.93}$$

(e) Prove that $\frac{d^2\varphi}{dx^2} = -\frac{1}{2}k^2\sin(2\varphi)$ so that $\eta = 2\varphi$ for k suitable solves the *pendulum differential equation* $\frac{d^2\eta}{dx^2} = -\omega^2\sin(\eta)$. (*Hint*: Apply $\frac{d}{dx}$ to $(\frac{d\varphi}{dx})^2 = 1 - k^2\sin^2(\varphi)$.)

Remark. The function φ defined by $\operatorname{sn}(x,k) = \sin(\varphi(x))$ is sometimes called the *Jacobi amplitude function*, written $am(x,k)$ as in $\operatorname{sn}(x,k) = \sin(am(x,k))$.

11. (a) Using the product formula for Θ, prove that

$$\Theta(1 \mid q)\,\Theta(-1 \mid q) = \prod_{n=1}^{\infty}(1-q^{2n})^2(1-q^{4n-2})^2 \tag{10.5.94}$$

(b) Prove that

$$\Theta(1 \mid q)\,\Theta(-1 \mid q)\,\Theta(q \mid q) = 2\prod_{n=1}^{\infty}(1-q^{2n})^3 \tag{10.5.95}$$

(c) Prove that

$$\Theta'(-q \mid q) = q^{-1}\prod_{n=1}^{\infty}(1-q^{2n})^3 \tag{10.5.96}$$

(d) Define four constants that are in (10.5.81) and (10.5.82),

$$\theta_j = \theta_j(0), \quad j = 2,3,4; \qquad \theta_1' = \theta_1'(0) \tag{10.5.97}$$

where θ_j is related to Θ by (10.5.36)–(10.5.39). Prove that

$$\theta_1' = \pi\theta_2\theta_3\theta_4 \tag{10.5.98}$$

12. This problem will establish the main formulae that write the Jacobi elliptic functions in terms of $\{\theta_j(w)\}_{j=1}^4$ and the constants of (10.5.97). You need Theorem 10.5.6, (10.5.81)/(10.5.82), and (10.5.98).

(a) Prove that

$$\sqrt{\wp(z) - e_j} = \frac{\theta_1'}{\theta_{j+1}} \frac{\theta_{j+1}(z)}{\theta_1(z)}$$

and conclude that

$$\sqrt{e_1 - e_j} = \frac{\theta_1'}{\theta_{j+1}} \frac{\theta_{j+1}(\frac{1}{2})}{\theta_1(\frac{1}{2})}$$

so that

$$\sqrt{e_1 - e_2} = \frac{\theta_1'\theta_4}{\theta_3\theta_2} = \pi\theta_4^2 \quad \text{and} \quad \mu \equiv \sqrt{e_1 - e_3} = \pi\theta_3^2 \tag{10.5.99}$$

(b) Similarly, prove that

$$\sqrt{e_2 - e_3} = \pi\theta_2^2 \tag{10.5.100}$$

(c) Using (a) + (b), prove that

$$\theta_2^2 + \theta_4^2 = \theta_3^2 \tag{10.5.101}$$

(d) Prove (10.5.101) using the product formulae.

(e) Prove that

$$k = \frac{\theta_2^2}{\theta_3^2}, \qquad k' = \frac{\theta_4^2}{\theta_3^2}, \qquad K = \frac{\pi}{2}\,\theta_3^2, \qquad K' = -\frac{i\pi\tau}{2}\,\theta_3^2 \tag{10.5.102}$$

(f) Prove that

$$\operatorname{sn}(2Kw) = \frac{1}{\sqrt{k}} \frac{\theta_1(w)}{\theta_4(w)}, \qquad \operatorname{cn}(2Kw) = \sqrt{\frac{k'}{k}} \frac{\theta_2(w)}{\theta_4(w)}, \qquad \operatorname{dn}(2Kw) = \sqrt{k'}\,\frac{\theta_3(w)}{\theta_4(w)} \tag{10.5.103}$$

(g) Prove that

$$\operatorname{sn}(K) = 1 \tag{10.5.104}$$

13. This problem will verify the original relation that Legendre proved that is equivalent to what we've called the Legendre relation (see (10.3.32)),

although it looks quite different. The complete elliptic integrals of the second kind are

$$E(k) = \int_0^{\pi/2} \sqrt{1 - k^2 \sin^2(\psi)}\, d\psi = \int_0^1 \sqrt{\frac{1-k^2t^2}{1-t^2}}\, dt \tag{10.5.105}$$

If $E = E(k)$ and $E' = E(k')$, then the Legendre relations that you'll prove say that

$$EK' + E'K - KK' = \frac{\pi}{2} \tag{10.5.106}$$

(a) Prove that

$$E = \int_0^K (\mathrm{dn})^2(x)\, dx \tag{10.5.107}$$

(*Hint*: Use (10.5.91), (10.5.92), and (10.5.104).)

(b) Prove that

$$E = \mu \int_0^{1/2} 1 + \left(\frac{e_3 - e_2}{\wp(y) - e_3}\right) dy \tag{10.5.108}$$

(*Hint*: Use (10.5.59) and (10.4.84).)

(c) Prove that

$$E = \frac{\mu}{2} - \frac{1}{\mu} \int_0^{1/2} (\wp(y+\tau) - e_3)\, dy \tag{10.5.109}$$

(*Hint*: Use Problem 6 of Section 10.4.)

(d) Prove that (ω_1 is given in (10.4.13))

$$E = \frac{e_1 + \omega_1}{\mu} \tag{10.5.110}$$

(*Hint*: Use (10.4.25) and (10.4.28).)

(e) Similarly, prove that

$$E' = \frac{i(\tau e_3 + \omega_3)}{\mu} \tag{10.5.111}$$

(f) Prove the Legendre relation (10.5.106). (*Hint*: Use (10.5.60) and (10.3.32).)

14. With K given by (10.5.50), E by (10.5.105), and k' by (10.5.49), prove that

$$\frac{dE}{dk}(k) = \frac{E(k) - K(k)}{k}, \quad \frac{dK}{dk}(k) = \frac{E(k) - (k')^2 K(k)}{k(k')^2} \tag{10.5.112}$$

15. Let $0 < k < 1$. This problem will prove for x real,

$$\left|\operatorname{sn}\left(x + \frac{iK'}{2}\right)\right| = k^{-1/2} \tag{10.5.113}$$

a fact used in Examples 8.4.5 and 8.4.14.

(a) Using (10.5.60) and (10.5.69), prove that

$$\operatorname{sn}(K) = 1, \qquad \operatorname{sn}(K + iK') = k^{-1} \tag{10.5.114}$$

(b) Using (10.5.60) and (10.5.69), prove that $\operatorname{sn}(z + iK')$ and $\operatorname{sn}(z)$ have zeros and poles reversed and conclude that

$$\operatorname{sn}(z)\operatorname{sn}(z + iK') = k^{-1} \tag{10.5.115}$$

(*Hint*: First prove the product is constant and then evaluate it at $z = K$.)

(c) Prove (10.5.113). (*Hint*: sn is real on $\mathbb{R}$ so $\operatorname{sn}(\bar{z}) = \overline{\operatorname{sn}(z)}$.)

Note. See Problem 7 of Section 10.4 for a related result.

10.6. The Elliptic Modular Function

This section will study the function

$$\lambda(\tau) = \frac{e_2(\tau) - e_3(\tau)}{e_1(\tau) - e_3(\tau)} \tag{10.6.1}$$

called the *elliptic modular function*. From the point of view of this chapter, this will allow the solution of elliptic integral inversion problems, specifically the questions of what values of k^2 in (10.5.1) can occur and what g_2, g_3 in (10.4.81) can occur as we vary τ or τ_1, τ_2. From this standpoint, the main results of this section are:

(1) λ is a bijection of $\{iy \mid 0 < y < \infty\} \equiv \mathcal{K}$ and $(0,1)$, and for $\tau \in \mathcal{K}$, $e_1(\tau) > e_2(\tau) > e_3(\tau)$. This was used in the last section and, in particular, says that all classical elliptic integrals of the form (10.5.1) fall within the purview of Theorem 10.5.6.

(2) λ maps $\mathbb{C}_+$ to $\mathbb{C}\setminus\{0,1\}$, is locally one-one, and its restriction to $\widetilde{\mathcal{F}}$, the fundamental domain for $\Gamma(2)$, is a bijection to $\mathbb{C}\setminus\{0,1\}$. (Note that this bijection is continuous, but its inverse is not in the topology as a subset of $\mathbb{C}$; if we think of $\mathcal{F} \cong \mathbb{C}\setminus\Gamma(2)$ with the quotient topology, it is.) Thus, the extension of Theorem 10.5.6 to complex k^2 handles all cases $k^2 \neq 0, 1$. Of course, the elliptic integrals for $k = 0, 1$ are trivially doable without recourse to elliptic functions ($k = 0$ is arcsin and $k = 1$ is arctanh).

(3) For every g_2, g_3, with $\Delta \neq 0$, there is a τ_1, τ_2 in $\mathbb{C}^\times \times \mathbb{C}^\times$, with $\tau_2/\tau_1 \notin \mathbb{R}$, so that $g_2(\tau_1, \tau_2) = g_2$, $g_3(\tau_1, \tau_2) = g_3$ (again, if $\Delta = 0$, the integral is doable in terms of inverse trigonometric functions). The map $\tau \to J(\tau)$

is a bijection of $\mathcal{F}$ and $\mathbb{C}$, although not a local bijection at some points in $\partial\mathcal{F}$, that is, $J\colon \mathbb{C}_+ \to \mathbb{C}$ is not everywhere a local bijection.

From a perspective that goes beyond elliptic functions:

$$(1)\ \tau \to \frac{\lambda(\tau) - i}{\lambda(\tau) + i} \tag{10.6.2}$$

is a Riemann map of $\{\tau \mid 0 < \operatorname{Re}\tau < 1;\ |\tau - \frac{1}{2}| > \frac{1}{2}\}$ to $\mathbb{D}$, so an interesting example in an atlas of conformal maps. See Example 8.4.9.

(2) $\lambda\colon \mathbb{C}_+ \to \mathbb{C} \setminus \{0,1\}$ is onto, a local bijection, and $\lambda(z) = w$ if and only if there exists $\gamma \in \Gamma(2)$ so that $z = \gamma(w)$. This will implement the function needed for the proofs of Picard's theorems (see Section 10.3). It is the third construction of this map (see Sections 8.3 and 8.5).

Recall what we learned in Problem 8 of Section 10.4:

$$\lambda(\tau) = 16e^{i\pi\tau} + O(e^{-3|\operatorname{Im}\tau|}) \tag{10.6.3}$$

as $\operatorname{Im}\tau \to \infty$ with $\operatorname{Re}\tau \in [0,2]$ (we'll see shortly, uniformly in all real τ) and, in particular,

$$\lambda(\tau) \to 0 \qquad \text{as } \operatorname{Im}\tau \to \infty \tag{10.6.4}$$

Moreover, again by Problem 8 of Section 10.4, for $\operatorname{Im}\tau$ large, once we know $e_j(\tau)$ are real on $i(0,\infty)$, we have

$$e_1 > e_2 > e_3 \tag{10.6.5}$$

and by continuity and (10.6.7) below, this holds for all $\tau \in i(0,\infty)$; see also Problem 13 of Section 10.4.

Finally, since $\wp(z) - e_j$ has only one (double) zero in the fundamental cell,

$$e_1 \neq e_2 \neq e_3 \neq e_1 \tag{10.6.6}$$

We begin this section's analysis with the covariance properties of λ under the action of $\mathbb{SL}(2,\mathbb{Z})$.

Theorem 10.6.1. (a) *If* $\left(\begin{smallmatrix} a & b \\ c & d \end{smallmatrix}\right) \in \Gamma(2)$, *then*

$$\lambda\left(\frac{a\tau + b}{c\tau + d}\right) = \lambda(\tau) \tag{10.6.7}$$

$$\text{(b)}\ \lambda\left(-\frac{1}{\tau}\right) = 1 - \lambda(\tau) \tag{10.6.8}$$

$$\text{(c)}\ \lambda(\tau + 1) = -\lambda(\tau)(1 - \lambda(\tau))^{-1} \tag{10.6.9}$$

Proof. (a) If $\tilde\tau_1 = d + c\tau$, $\tilde\tau_2 = b + a\tau$, then $\tilde\tau = (a\tau + b)/(c\tau + d) = \tilde\tau_2/\tilde\tau_1$ is the τ for the new lattice. Not only does this preserve $\mathcal{L}_{1,\tau}$, and so $\wp(z)$, but if a, d are odd and b, c even, $\wp(\tilde\tau_1/2) = \wp(\frac{1}{2})$, $\wp(\tilde\tau_1/2) = \wp(\tau/2)$ since $\tilde\tau_1/2 - \frac{1}{2}$, $\tilde\tau_2/2 - \tau/2 \in \mathcal{L}_{1,\tau}$. Thus, $e_j(\tilde\tau_1, \tilde\tau_2) = e_j(1,\tau)$ and λ is unchanged.

(b) If $\tilde{\tau}_1 = \tau$, $\tilde{\tau}_2 = -1$, then $\tilde{\tau} = -1/\tau$ and e_2 is unchanged (since $\frac{1}{2}(\tilde{\tau}_1 + \tilde{\tau}_2) - \frac{1}{2}(1+\tau) = -1 \in \mathcal{L}_{1,\tau}$), but e_1 and e_3 are interchanged. Thus,

$$\lambda\biggl(-\frac{1}{\tau}\biggr) = \frac{e_2(\tau) - e_1(\tau)}{e_3(\tau) - e_1(\tau)} = 1 - \lambda(\tau)$$

(c) Similarly, $\tilde{\tau}_1 = 1$, $\tilde{\tau}_2 = 1 + \tau$ has $\tilde{\tau} = 1 + \tau$ and e_1 is unchanged, but e_2 and e_3 are interchanged. Thus,

$$\lambda(1+\tau) = \frac{e_3 - e_2}{e_1 - e_2} = -\frac{\lambda}{1-\lambda} \qquad \square$$

We said the action of all of $\mathbb{SL}(2,\mathbb{Z})$, but $\mathbb{SL}(2,\mathbb{Z})/\Gamma(2)$, has six elements not two. However, using cycle notation (see the Notes), the two permutations we look at, namely, (13) and (23), generate the whole group, so the full action is easy to obtain, summarized in the following table (see Problem 1):

	$\mathbb{1}$	(123)	(132)	(12)	(13)	(23)
$\tilde{\tau}$	τ	$-1/(\tau+1)$	$-1-1/\tau$	$\tau/(1+\tau)$	$-1/\tau$	$\tau+1$
$\lambda(\tilde{\tau})$	λ	$(1-\lambda)^{-1}$	$1-\lambda^{-1}$	λ^{-1}	$1-\lambda$	$-\lambda(1-\lambda)^{-1}$

Theorem 10.6.2. (a) *λ is real if τ is contained in* $\partial\mathcal{F}^\sharp \equiv \{\tau \mid \operatorname{Re}\tau = 0$ *or* $\operatorname{Re}\tau = 1$ *or* $|\tau - \frac{1}{2}| = \frac{1}{2}\} \cap \mathbb{C}_+$.

(b) $\lambda \to 0$ *uniformly in* $\operatorname{Re}\tau$ *as* $\operatorname{Im}\tau \to \infty$.

(c) $\lambda \to 1$ (*respectively,* ∞) *as* $\tau \to 0$ (*respectively,* 1) *in* $\mathbb{C}_+$ *with* $\arg(\tau)$ (*respectively,* $\arg(\tau - 1)$) *in* $(\varepsilon, \pi - \varepsilon)$ *for any* $\varepsilon > 0$.

(d) *For all* $\tau \in \mathbb{C}_+$, $\lambda(\tau)$ *is never* 0, 1, *or* ∞.

(e) $\lambda[\{iy \mid y \in (0,\infty)\}] = (0,1)$, $\lambda[\{\frac{1}{2} + \frac{1}{2}e^{i\theta} \mid \theta \in (0,\pi)\}] = (1,\infty)$, $\lambda[\{1 + iy \mid y \in (0,\infty)\}] = (-\infty, 0)$.

Remarks. 1. We'll see shortly that λ takes every value in $\mathbb{C} \setminus \{0,1\}$ and that λ is one-one on each of the sets in (v).

2. As our proof shows, $\lambda(\tau) \to 1$ as $\operatorname{Im}(-1/\tau) \to \infty$. This is not the same as $|\tau| \to 0$, $\operatorname{Im}\tau > 0$. For example, if $\tau = \varepsilon + i\varepsilon^2$, then $\operatorname{Im}(1/\tau) = \varepsilon^2/(\varepsilon^2 + \varepsilon^4) \to 1$ as $\varepsilon \downarrow 0$.

Proof. (a) Suppose first $\operatorname{Re}\tau = 0$, so $\tau = iy$ and $\mathcal{L}_\tau = \{n_1 + in_2 y\}$. Since $\bar{\mathcal{L}}_\tau = \mathcal{L}_\tau$, the formula (10.4.29) for $\wp$ proves that

$$\wp(\bar{z} \mid 1\, iy) = \overline{\wp(z \mid 1\, iy)} \tag{10.6.10}$$

Since always $\wp(-z \mid 1\,\tau) = \wp(z \mid 1\,\tau)$, we have

$$\wp(-\bar{z} \mid 1\,iy) = \overline{\wp(z \mid 1\,iy)}$$

Thus, e_1 and e_3 are real, and since $e_1 + e_2 + e_3 = 0$, e_2 is also real, so λ is real. By (10.6.9), λ is real on $\operatorname{Re}\tau = 1$, and similarly, λ is real if $\operatorname{Re}\tau = -1$. $\tau \to -1/\tau$ maps -1 to 1 and ∞ to 0 and maps $\mathbb{R}$ to $\mathbb{R}$, $\{\tau \mid \operatorname{Re}\tau = -1\}$ is mapped to a circle orthogonal to $\mathbb{R}$, so the circle $|\tau - \frac{1}{2}| = \frac{1}{2}$ (as algebra also shows) and $\mathbb{C}_+ \cap \{\tau \mid \operatorname{Re}\tau = -1\}$ is mapped to $\mathbb{C}_+ \cap \{\tau \mid |\tau - \frac{1}{2}| = \frac{1}{2}\}$. Thus, by (10.6.8), λ is real on $\mathbb{C}_+ \cap \{\tau \mid |\tau - \frac{1}{2}| = \frac{1}{2}\}$.

(b) Since $\left(\begin{smallmatrix} 1 & 2 \\ 0 & 1 \end{smallmatrix}\right) \in \Gamma(2)$, $\lambda(\tau+2) = \lambda(\tau)$, so (10.6.3) uniformly for $\operatorname{Re}\tau \in [0,2]$ implies it for all $\operatorname{Re}\tau$.

(c) By (10.6.8), $\lambda \to 1$ as $\operatorname{Im}(-1/\tau) \to \infty$, and with the restriction on $\arg(\tau)$, $|\tau| \to 0 \Rightarrow \operatorname{Im}(-1/\tau) \to \infty$. Then, by (10.6.9), we get from $\tau \to 0$ to $\tau \to 1$.

(d) This is immediate from $e_1 \neq e_2 \neq e_3 \neq e_1$.

(e) Since λ is real, never 1 or 0, and $\lambda(iy) \to 1$ as $y \to \infty$ and $\lambda(iy) \to 1$ as $y \to 0$, it must take all values and only those values on $\{iy \mid y \in (0,\infty)\}$. (10.6.9) and $\lambda(\tau - 2) = \lambda(\tau)$ imply λ takes all values in $(-\infty, 0)$ on $\pm 1 + iy$ and thus, (10.6.8) yields the result on $\{z \mid |z - \frac{1}{2}| = \frac{1}{2}\} \cap \mathbb{C}_+$. $\square$

Here is (part (a) below) the most subtle argument in the analysis of $\lambda(\tau)$. Let

$$\mathcal{F}^\sharp = \{\tau \mid \operatorname{Im}\tau > 0,\ 0 < \operatorname{Re}\tau < 1,\ |\tau - \tfrac{1}{2}| > \tfrac{1}{2}\} \tag{10.6.11}$$

Theorem 10.6.3. (a) *λ is a bijection of $\mathcal{F}^\sharp$ to $\mathbb{C}_+$.*
(b) *λ is a bijection of $\widetilde{\mathcal{F}}$ to $\mathbb{C} \setminus \{0,1\}$.*
(c) *$\lambda'(z) \neq 0$ for all z in $\mathbb{C}_+$.*
(d) *$\lambda(z) = \lambda(w) \Leftrightarrow z = (aw+b)/(cw+d)$ for some $\left(\begin{smallmatrix} a & b \\ c & d \end{smallmatrix}\right) \in \Gamma(2)$.*

Proof. (a) Fix ε and let Γ_ε be the following contour in $\overline{\mathcal{F}^\sharp}$ (Figure 10.6.1).

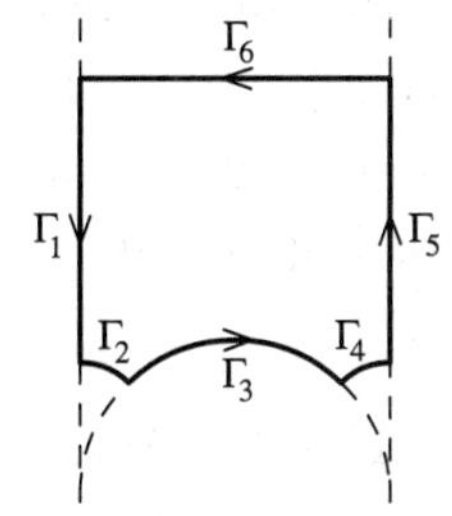

Figure 10.6.1. Contour around $\widetilde{\mathcal{F}}$ cutoff.

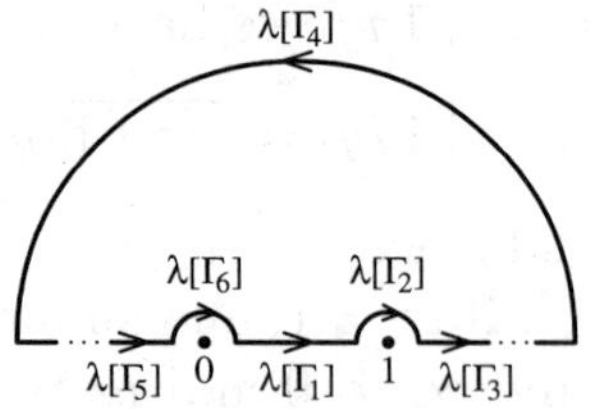

Figure 10.6.2. Image of Γ under the elliptic modular function.

Start at $i\varepsilon^{-1}$ and go down to $i\varepsilon$ (call this Γ_1), then go along the curve $\Gamma_2(x) = (x - i\varepsilon^{-1})^{-1} = (x + i\varepsilon^{-1})/(\varepsilon^{-2} + x^2)$, $0 \le x \le 1$, from $i\varepsilon$ to the circle $|\tau - \frac{1}{2}| = \frac{1}{2}$ ($\Leftrightarrow \operatorname{Re}(\frac{1}{\tau} = 1)$ (call this Γ_2), then along this circle to the circle reflection of Γ_2 (call this Γ_3), then around that curve $\Gamma_4(x) = 1 - \overline{\Gamma_2(1-x)}$ near $\operatorname{Re}\tau = 1$ (call this Γ_4). Γ_4 is the reflection of Γ_2 in the line $\operatorname{Re}\tau = \frac{1}{2}$. Then up to $1 + i\varepsilon^{-1}$ (call this Γ_5), and across to $i\varepsilon^{-1}$ along $\operatorname{Im}\tau = \varepsilon^{-1}$ (call this Γ_6).

The image, $\lambda[\Gamma]$, (Figure 10.6.2), also has six pieces. $\lambda[\Gamma_1]$ runs from near 0 to near 1 along $\mathbb{R}$. It can, a priori, wander back and forth, but stays on $\mathbb{R}$. $\lambda[\Gamma_2]$ stays in a tiny neighborhood of $w = 1$. $\lambda[\Gamma_3]$ and $\lambda[\Gamma_5]$ stay along $\mathbb{R}$, running from just above 1 to near ∞ and from above $-\infty$ to below 0, and $\lambda[\Gamma_6]$ near $\lambda = 0$.

The key piece is $\lambda[\Gamma_4]$. $\Gamma_2(x)$ is picked so $\Gamma_2(x)^{-1} = x - i\varepsilon^{-1}$. By (10.6.3),

$$\lambda(\Gamma_2(x)^{-1}) = 16e^{-\varepsilon^{-1}}e^{i\pi x} + O(e^{-3\varepsilon^{-1}})$$

so, by (10.6.8), we have

$$\lambda(\Gamma_2(x)) = 1 - 16e^{-\varepsilon^{-1}}e^{-i\pi x} + O(e^{-3\varepsilon^{-1}})$$

Note that since λ is real on $\operatorname{Re}\tau = 0$, $\lambda(-\bar{\tau}) = \overline{\lambda(\tau)}$. Since $\Gamma_4(x) = 1 - \overline{\Gamma_2(1-x)}$, by (10.6.9),

$$\lambda(\Gamma_4(x)) = (16)^{-1}e^{\varepsilon^{-1}}e^{i\pi x} + (1 + O(e^{-2\varepsilon^{-1}})) \tag{10.6.12}$$

Then $\lambda(\Gamma)$ goes along $\mathbb{R}$ with two tiny approximate half-loops near 0 and 1, and a huge semicircle looping back in the upper half-plane near infinity.

If $\eta \in \mathbb{C}_+$ and $|\eta|$ is much smaller than $O(e^{\varepsilon^{-1}})$ and farther from 0 and 1 than $O(e^{-\varepsilon^{-1}})$, then $\lambda(\Gamma)$ will have winding number 1 about η and 0 about $\bar{\eta}$. By the argument principle, this implies (a).

(b), (c) Since λ is real on $\operatorname{Re} z = 0$,

$$\lambda(-\bar{\tau}) = \overline{\lambda(\tau)} \tag{10.6.13}$$

so λ is one-one on $-\widetilde{\mathcal{F}}^\sharp$, and since that set is open, $\lambda' \neq 0$ there. This set, together with $\mathcal{F}^\sharp$, covers all of $\widetilde{\mathcal{F}}$, except for the edges and for $\operatorname{Re}\tau = 0$.

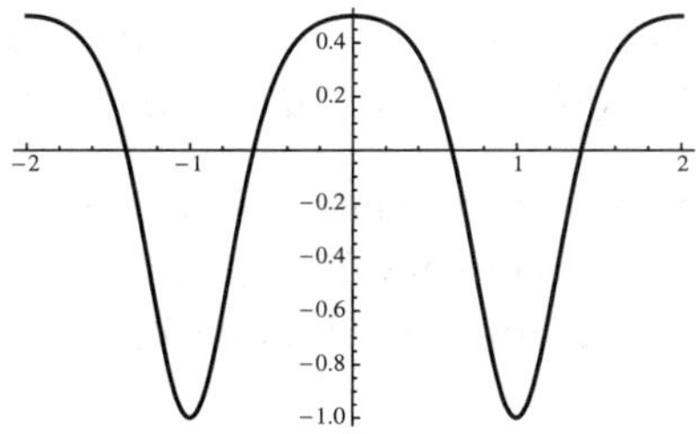

Figure 10.6.3. $\operatorname{Re}\lambda(x+i)$, $-2<x<2$.

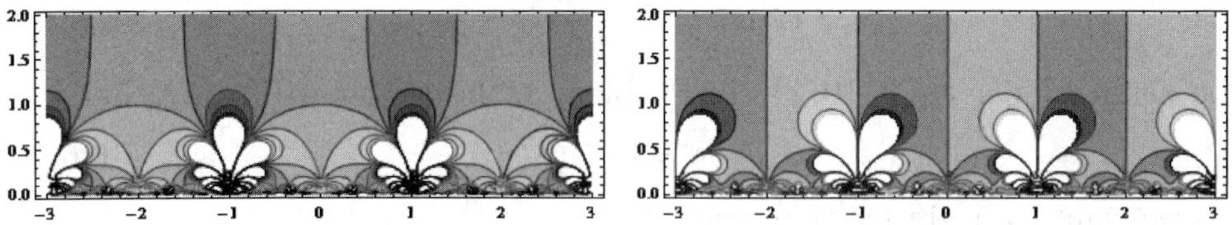

Figure 10.6.4. Re and Im of $\lambda(x+iy)$, $-3<x<3$, $0<y<2$.

λ is real on $\operatorname{Re}\tau=0$ with values in $(0,1)$, Since $\operatorname{Im}\tau>0$ on $0<\operatorname{Re}\tau<\varepsilon$ and $\operatorname{Im}\tau<0$ on $-\varepsilon<\operatorname{Re}\tau<0$, $\lambda'\neq 0$ on $\operatorname{Re}\tau=0$ (see Problem 2). Since $\frac{d\lambda(iy)}{dy}$ is real and nonvanishing, it has a single sign which, given the limiting values, must be negative. So λ is one-one on $\operatorname{Re}\tau=0$. By the covariance in Theorem 10.6.1, $\lambda\neq 0$ and λ is one-one on $\{\tau\mid\operatorname{Im}\tau>0,\operatorname{Re}\tau=1\}$ and on $\{\tau\mid\operatorname{Im}\tau>0,|\tau-\frac{1}{2}|=\frac{1}{2}\}$. This plus (a) proves that λ is a bijection of $\widetilde{\mathcal{F}}$ and $\mathbb{C}^\times$. By invariance under $\Gamma(2)$, we get $\lambda'\neq 0$ on all of $\mathbb{C}_+$.

(d) Theorem 10.6.1 implies $w=\gamma(z)\Rightarrow\lambda(w)=\lambda(z)$. On the other hand, if $w,z\in\mathbb{C}_+$ with $\lambda(w)=\lambda(z)$, there exists $\gamma,\tilde{\gamma}\in\Gamma(2)$ so $\gamma(w),\tilde{\gamma}(z)\in\widetilde{\mathcal{F}}$ and, by invariance of λ, $\lambda(\gamma(w))=\lambda(\tilde{\gamma}(z))$. Since λ is one-one on $\widetilde{\mathcal{F}}$, $\gamma(w)=\tilde{\gamma}(z)$, so $w=\tilde{\gamma}^{-1}\gamma(z)$. □

Figure 10.6.3 plots $\operatorname{Re}\lambda(x+i)$ and Figure 10.6.4 has contour plots of $\operatorname{Re}\lambda(x+iy)$ and $\operatorname{Im}\lambda(x+iy)$.

We can immediately obtain the first two and the fourth and fifth results advertised at the start of this section.

Theorem 10.6.4. *The map of* (10.6.2) *is the unique map of* $\mathcal{F}^\sharp$ *to* $\mathbb{D}$ *with limiting values* -1 *at* $\tau=\infty$, $(1-i)/(1+i)$ *at* $\tau=0$ *and* $+1$ *at* $\tau=1$.

Theorem 10.6.5. $\lambda\colon\mathbb{C}_+\to\mathbb{C}\setminus\{0,1\}$ *is a local bijection onto all of* $\mathbb{C}\setminus\{0,1\}$ *and* $\lambda(z)=w\Leftrightarrow\exists\gamma\in\Gamma(2)$, $z=\gamma(w)$.

Theorem 10.6.6. *For* $\tau\in\{iy\mid y\in(0,\infty)\}$, $e_1>e_2>e_3$, *and for any* $k^2\in(0,1)$, *there is a unique* $y\in(0,\infty)$, *so* $\lambda(iy)=k^2$.

Theorem 10.6.7. *For any $k^2 \in \mathbb{C}\setminus\{0,1\}$, there is a τ in $\mathbb{C}_+$ so $\lambda(\tau)=k^2$. All values of τ that solve this are related by an element of $\Gamma(2)$.*

Proofs. The only item not immediate from Theorem 10.6.3 is that $e_1 > e_2 > e_3$ on $i(0,\infty)$. By Problem 8 of Section 10.4, this is true for y large. Since e_j are real, nonequal, and continuous in y, this must persist for all $y \in (0,\infty)$. □

We now turn to analyzing the function J of (10.4.32). J can be written in terms of λ. Let

$$G(w)=\frac{(1-w+w^2)^3}{w^2(1-w)^2} \tag{10.6.14}$$

be defined for $w \in \mathbb{C}\setminus\{0,1\}$. In Problem 3, the reader will prove that

$$J(\tau)=\tfrac{4}{27}\,G(\lambda(\tau)) \tag{10.6.15}$$

so we want to begin by analyzing G.

Proposition 10.6.8. *The function G has the following properties:*

(a) $$G(w)=G\left(\frac{1}{w}\right)=G(1-w)=G\left(1-\frac{1}{w}\right) \tag{10.6.16}$$

$$=G((1-w)^{-1})=G(-w(1-w)^{-1}) \tag{10.6.17}$$

(b) *If $w \neq \frac{1}{2},-1,2,\omega_0,\omega_0^{-1}$, then $G'(w)\neq 0$.*

(c) *G maps $\mathbb{C}\setminus\{0,1\}$ onto all of $\mathbb{C}$, and for $\zeta \neq 0,\frac{27}{4}$, $G^{-1}(\{\zeta\})$ is six points. Moreover, $G^{-1}(\{\frac{27}{4}\})=\{\frac{1}{2},-1,2\}$, $G^{-1}(\{0\})=\{\omega_0,\omega_0^{-1}\}$.*

Remark. Recall $\omega_0=\frac{1}{2}+i\sqrt{3/4}$ and ω_0^{-1} are the two primitive sixth roots of unity.

Proof. (a) By direct calculation, $G(1/w)=G(w)=G(1-w)$. The FLTs $w\to 1/w$ and $w \to 1-w$ generate a group, $\mathbb{G}$, of six elements, and so G is invariant over that group as indicated.

(b), (c) $G(w)-\zeta=0$ is equivalent to a sixth-order polynomial, so the map is onto and is locally one-one unless $T(w)=w$ for some nonidentity T in $\mathbb{G}$. The solutions of these five equations in $\widehat{\mathbb{C}}$ and the values of $G(\omega)$ are below.

Map	$w=w^{-1}$	$w=1-w$	$w=1-\frac{1}{w}$	$w=(1-w)^{-1}$	$w=-w(1-w)^{-1}$
Solutions	± 1	$\frac{1}{2},\infty$	$\omega_0^{\pm 1}$	$\omega_0^{\pm 1}$	$2,0$
$G(\omega)$	$\infty,\frac{27}{4}$	$\frac{27}{4},\infty$	0	0	$\frac{27}{4},\infty$

Thus, G is six-one on $\widehat{\mathbb{C}}\setminus\{0,1,\infty,-1,2,\frac{1}{2},\omega_0,\omega_0^{-1}\}$, undefined (or ∞) at $0,1,\infty$, and $\frac{27}{4}$ (at $-1,2,\frac{1}{2}$) or 0 (at $\omega_0^{\pm 1}$). □

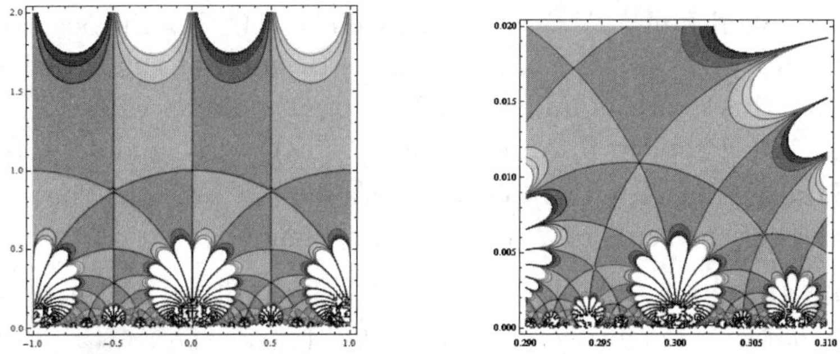

Figure 10.6.5. $\operatorname{Im} J[x+iy]$ for $-1 < x < 1$, $0 < y < 2$ and a blowup of a small piece.

Theorem 10.6.9. *J (defined on $\mathbb{C}_+$ by (10.4.32)) has the following properties:*

(a) *J is invariant under the action of $\mathbb{SL}(2,\mathbb{C})$ and $J(\tau) = J(\tilde{\tau})$ if and only if there is $A \in \mathbb{SL}(2,\mathbb{C})$ so $\tilde{\tau} = g_A(\tau)$.*
(b) *J maps $\mathbb{C}_+$ to all of $\mathbb{C}$.*
(c) *$J'(z) = 0$ only if $z = i$ or ω_0 or an image under the action of $\mathbb{SL}(2,\mathbb{C})$. $J(i) = 1$, $J(\omega_0) = 0$.*

$J(\tau)$ has a third-order zero at ω_0 and $J(\tau) - 1$ has a second-order zero at $\tau = i$.

Remarks. 1. The degeneracies and orders can be interpreted in terms of the sixth-order rotation symmetry of $\mathcal{J}_{1,\omega_0}$ and the fourth-order rotation symmetry of $\mathcal{J}_{1,i}$.

2. Figure 10.6.5 shows $\operatorname{Im} J(x+iy)$ for $-1 < x < 1$, $0 < y < 2$ and a blowup of the region near $x = 0.3 + i0$. Note that J is real on $\partial\mathcal{F}$ and its images, so these curves are contours in the contour plot of $\operatorname{Im} J$.

Proof. (a) Since λ is invariant under $\Gamma(2)$, by (10.6.15), J is invariant under $\Gamma(2)$. By Theorem 10.6.1 and Proposition 10.6.8, $J(-1/\tau) = J(\tau)$ and $J(\tau+1) = J(\tau)$. Since $\mathbb{SL}(2,\mathbb{Z})$ is generated by $\Gamma(2)$ (see Problem 4), $\pm\left(\begin{smallmatrix} 0 & 1 \\ -1 & 0 \end{smallmatrix}\right)$ and $\pm\left(\begin{smallmatrix} 1 & 1 \\ 1 & 0 \end{smallmatrix}\right)$, we have invariance under $\mathbb{SL}(2,\mathbb{Z})$.

Given that λ is one-one on $\widetilde{\mathcal{F}}$ and the six-one nature of G and six-fold intersection of $\mathbb{SL}(2,\mathbb{Z})$ orbits with $\widetilde{\mathcal{F}}$, we see that J is one-one on $\mathcal{F}$. This implies $J(\tau) = J(\tilde{\tau})$ if and only if $\tilde{\tau}$ is on the orbit of τ.

(b) This follows from the fact that λ is onto $\mathbb{C} \setminus \{0,1\}$ and that G maps $\mathbb{C} \setminus \{0,1\}$ to all of $\mathbb{C}$.

(c) By (10.6.15),

$$J'(\tau) = \tfrac{4}{27}\, G'(\lambda(\tau))\lambda'(\tau) \tag{10.6.18}$$

and λ' is nonvanishing, $J'(\tau) = 0 \Leftrightarrow G'(\lambda(\tau)) = 0$. By Proposition 10.6.8, this happens if $\lambda(\tau) \in \{\frac{1}{2}, -1, 2, \omega_0, \omega_0^{-1}\}$. Because of the invariance of G in (10.6.16), this is orbits under $\mathbb{SL}(2,\mathbb{C})$ of any points where $\lambda(\tau) = \frac{1}{2}$ or $\lambda(\tau) = \omega_0$. By Problem 5, $\lambda(i) = \frac{1}{2}$ and $\lambda(\omega_0) = \omega_0$, and so λ is one-one on $\widetilde{\mathcal{F}}$. These are the only $\Gamma(2)$ orbits where $\lambda(\tau)$ is $\frac{1}{2}$ or ω_0. The calculation of the orders of the zeros of J' is left to Problem 6. $\square$

Theorem 10.6.10. *For any g_2, g_3 in $\mathbb{C}^2$ with $g_2^3 - 27g_3^2 \neq 0$, there exists τ_1, τ_2 in $\mathbb{C}^2$ with $\mathrm{Im}(\tau_2/\tau_1) > 0$ so that $(\wp(z \mid \tau_1\tau_2), \wp'(z \mid \tau_1\tau_2))$ obey (10.4.40).*

Proof. Suppose first that $g_3 \neq 0$. Then $g_2^3/(g_2^3 - 27g_3^2) \neq 1$ so, by Theorem 10.6.9, there is τ solving $J(\tau) = g_2^3/(g_2^3 - 27g_3^2)$ and $g_1(1,\tau) \neq 0$, since $J(\tau) \neq 1$. Picking $\tilde{\tau}_1 = \mu$, $\tilde{\tau}_2 = \mu\tau$, we can arrange $\tilde{g}_3 = g_3$. Since $J(\tilde{\tau}) = J(\tau)$, we see $\tilde{g}_2^3 = g_2^3$, so $g_2 = \omega\tilde{g}_2$ for ω a cube root of unity. Since $g_2(\mu\tilde{\tau}_1, \mu\tilde{\tau}_2) = \mu^{-4}\tilde{g}_2$,

$$g_3(\mu\tilde{\tau}_1, \mu\tilde{\tau}_2) = \mu^{-6}\tilde{g}_3 \tag{10.6.19}$$

we pick $\mu = \omega^{-1}$. Then $g_2(\mu\tilde{\tau}_1, \mu\tilde{\tau}_2) = \mu^{-4}\tilde{g}_2 = \omega\tilde{g}_2 = g_2$ and $g_3(\mu\tau_1, \mu\tau_2) = \mu^{-6}\tilde{g}_3 = \tilde{g}_3 = g_3$,

If $g_3 = 0$, take $\tau_1 = \mu$, $\tau_2 = i\mu$ so, by (10.6.19), $g_3(\mu, i\mu) = -g_3(i\mu, -\mu) = -g_3(\mu, i\mu)$ since $S_{2k}(\tau_1,\tau_2) = S_{2k}(\pm\tau_1, \pm\tau_2) = S_{2k}(\tau_2,\tau_1)$. Thus, $g_3(\mu, i\mu) = 0$ and $g_2 \neq 0$. On the other hand, $S_4(1,i) \neq 0$ and thus we can pick $\mu = (g_1(1,i)/g_2)^{1/4}$ to get the desired value for g_2. $\square$

Notes and Historical Remarks. The term *modular function* is sometimes applied to any meromorphic function on $\mathbb{C}_+$ invariant under the action of the *modular group* $\mathbb{SL}(2,\mathbb{C})$ or a large subgroup like $\Gamma(2)$ and which obeys $|f(\tau)| \le Ce^{-K|\mathrm{Im}\,\tau|}$ if $\mathrm{Im}\,\tau > T$ for some C, K, T. λ is then called the λ-invariant or λ-modular function and J, the j-invariant or Klein's J invariant, after work of Klein [**309, 310**]. A *modular form* of weight k has $f((a\tau + b)/(c\tau + d)) = (c\tau + d)^k f(\tau)$ for some k. There are many books on modular forms [**140, 257, 322, 386**], most on their connections to number theory; see Section 13.1 of Part 2B.

Jacobi proved a remarkable explicit formula for $\Delta(\tau)$ in terms of $\tilde{q} = e^{2\pi i\tau}$ ($\tilde{q} = q^2$ for our usual q), namely,

$$\Delta(\tau) = (2\pi)^{12}\tilde{q}\prod_{n\ge 1}(1 - \tilde{q}^n)^{24}$$

See Apostol [**18**], Serre [**518**], and Silverman [**521**] for proofs.

It turns out $j(\tau) = 1728J(\tau)$ has remarkable properties. $1728 = 12^3$ is picked in part to make the leading coefficient in the series in $e^{2\pi i\tau}$ (leading as $\tau \to i\infty$) 1, but makes all the other Fourier coefficients integral and many other objects integral and is of number-theoretic and group-theoretic significance.

$(i_1 \dots i_\ell)$, with all i_j distinct, is shorthand for the permutation that takes $i_1 \to i_2, i_2 \to i_3, \dots, i_\ell \to i_1$ and leaves all other indices fixed.

Problems

1. Verify the table before Theorem 10.6.2 of maps $\tau \to \tilde{\tau}$ that implement the permutations on $e_j(\tau)$ (up to overall scaling) and the induced maps on λ.

2. If f is analytic in $\mathbb{D}_\varepsilon(0)$ with $\pm\operatorname{Im} f(z) \geq 0$ if $\pm\operatorname{Im} z > 0$, prove that $f'(z) \neq 0$ in $(-\varepsilon, \varepsilon)$.

3. (a) Prove that (with $\lambda = (e_2 - e_3)/(e_1 - e_3)$) that $(1+\lambda+\lambda^2)(e_1 - e_3)^2 = \frac{3}{4}g_2$.

   (b) Using (10.4.74), verify that $[\lambda(1-\lambda)(e_1 - e_3)^3]^2 = \Delta/16$.

   (c) With g given by (10.6.14) and $J(\tau)$ by (10.4.77), prove that (10.6.15) holds.

4. Prove that $\mathbb{SL}(2, \mathbb{Z})$ is generated by $\Gamma(2) \cup \{\pm\left(\begin{smallmatrix} 0 & 1 \\ -1 & 0 \end{smallmatrix}\right), \pm\left(\begin{smallmatrix} 1 & 1 \\ 1 & 0 \end{smallmatrix}\right)\}$.

5. (a) If $\tau = i$, prove that $\wp(iz \mid i) = -\wp(z \mid i)$ and that $e_3 = -e_1$, $e_2 = 0$, and then that $\lambda(i) = \frac{1}{2}$.

   (b) If $\tau = \omega_0$, prove that $\wp(\omega_0 z \mid \omega_0) = \omega_0^{-2}\wp(z \mid \omega_0)$ and that $e_3 = \omega_0^{-2} e_1$, $e_2 = \omega_0^{-4} e_1$, and $\lambda(\omega_0) = \omega_0$.

6. (a) Prove that near $\tau = i$, J is two-one so that $J'(i) = 0$, $J''(i) \neq 0$.

   (b) Prove that, near $\tau = \omega_0$, J is three-one so that $J'(\omega_0) = J''(\omega_0) = 0$, $J'''(\omega_0) \neq 0$.

7. (a) Let $f(\tau)$, defined and analytic for $\tau \in \mathbb{C}_+$, obey $f(\tau + \ell) = f(\tau)$ for some integral ℓ. Prove that there exists $K > 0$ so $|f(\tau)| \leq Ce^{K|\operatorname{Im}\tau|}$ if and only if f has a convergent expansion (uniformly on compact subsets of $\mathbb{C}_+$)

$$f(\tau) = \sum_{n=-m}^{\infty} a_n e^{2\pi i n\tau/\ell}$$

   for $m < \infty$.

   (b) A modular function is a function meromorphic on $\mathbb{C}_+$, invariant under $\mathbb{SL}(2, \mathbb{Z})$, with $|f(\tau)| \leq Ce^{K|\operatorname{Im}\tau|}$ for some $K < \infty$ and all τ with $|\operatorname{Im}\tau| > T$ for some T. Prove that every rational function of $J(\tau)$ is a modular function and vice-versa.

8. A modular form of weight k is like a modular function except invariance under $\mathbb{SL}(2,\mathbb{Z})$ is replaced by $f((a\tau+b)/(c\tau+d)) = (c\tau+d)^k f(\tau)$. Prove that $S_{2\ell}(1,\tau)$ is a modular form of weight 2ℓ.

10.7. The Equivalence Problem for Complex Tori

In Section 7.1, we induced complex tori, $\mathcal{J}_{\tau_1,\tau_2}$ for $\text{Im}(\tau_2/\tau_1) > 0$. In this section, we'll provide necessary and sufficient conditions for $\mathcal{J}_{\tilde{\tau}_1,\tilde{\tau}_2}$ and $\mathcal{J}_{\tau_1,\tau_2}$ to be conformally equivalent and, coincidentally, do the same for classifying cubic curves. Here are the main theorems:

Theorem 10.7.1. *Let $\tau_1,\tau_2,\tilde{\tau}_1,\tilde{\tau}_2$, all in $\mathbb{C}^\times$, be such that $\tau \equiv \tau_2/\tau_1$, $\tilde{\tau} \equiv \tilde{\tau}_2/\tilde{\tau}_1$ are both in $\mathbb{C}_+$. Then the following are equivalent:*

(1) *$\mathcal{J}_{\tau_1,\tau_2}$ and $\mathcal{J}_{\tilde{\tau}_1,\tilde{\tau}_2}$ are conformally equivalent.*
(2) *For some $\lambda \in \mathbb{C}^\times$, $\mathcal{L}_{\tilde{\tau}_1,\tilde{\tau}_2} = \lambda\mathcal{L}_{\tau_1,\tau_2}$.*
(3) *For some $A \in \mathbb{SL}(2,\mathbb{Z})$, $\tilde{\tau} = g_A(\tau)$.*
(4) *$J(\tilde{\tau}) = J(\tau)$* (10.7.1)

Theorem 10.7.2. *The elliptic curves generated by $y^2 = 4(x^3 - \tilde{g}_2 x - \tilde{g}_3$ and $y^2 = 4x^3 - g_2 x - g_3$ (where $\tilde{g}_2^3 - 27\tilde{g}_3^2 \neq 0 \neq g_2^3 - 27 g_3^2$) are conformally equivalent if and only if*

$$\frac{\tilde{g}_2^3}{\tilde{g}_2^3 - 27\tilde{g}_3^2} = \frac{g_2^3}{g_2^3 - 27g_3^2} \tag{10.7.2}$$

Proof of (2) ⇔ (3) ⇔ (4) ⇒ (1) in Theorem 10.7.1. If the lattices are the same, then there is $\left(\begin{smallmatrix} a & b \\ c & d \end{smallmatrix}\right) \in \mathbb{SL}(2,\mathbb{C})$ so

$$\tilde{\tau}_1 = a\lambda\tau_1 + b\lambda\tau_2 \tag{10.7.3}$$

$$\tilde{\tau}_2 = c\lambda\tau_1 + d\lambda\tau_2 \tag{10.7.4}$$

Thus, $\tilde{\tau} = g_A(\tau)$. Conversely, if $\tilde{\tau} = g_A(\tau)$, we can pick λ so (10.7.3) holds, and then (10.7.4) will hold, and so $\mathcal{L}_{\tilde{\tau}_1,\tilde{\tau}_2} = \lambda\mathcal{L}_{\tau_1,\tau_2}$.

(3) ⇔ (4) is immediate from Theorem 10.6.9. If (2) holds, $f([z]) = [\lambda^{-1}z]$ defines an analytic bijection of $\mathbb{C}/\mathcal{L}_{\tilde{\tau}_1,\tilde{\tau}_2}$ to $\mathbb{C}/\mathcal{L}_{\tau_1,\tau_2}$. □

We'll first provide two proofs of (1) ⇒ (2): one using uniqueness of universal covering spaces and the other, the elliptic modular function.

First proof of (1) ⇒ (2) in Theorem 10.7.1. Let $\pi\colon \mathbb{C} \to \mathcal{J}_{\tau_1,\tau_2}$ and $\tilde{\pi}\colon \mathbb{C} \to \mathcal{J}_{\tilde{\tau}_1,\tilde{\tau}_2}$ be the image of $z \in \mathbb{C}$ into the equivalence class in $\mathbb{C}/\mathcal{L}$. Let f be an analytic bijection of $\mathcal{J}_{\tau_1,\tau_2}$ to $\mathcal{J}_{\tilde{\tau}_1,\tilde{\tau}_2}$. Then $f \circ \pi$ is an analytic map of $\mathbb{C}$ to $\mathcal{J}_{\tilde{\tau}_1,\tilde{\tau}_2}$, which is a covering map since given $\zeta \in \mathcal{J}_{\tilde{\tau}_1,\tilde{\tau}_2}$, pick U as a neighborhood of $f^{-1}(\zeta)$ so $\pi^{-1}([U])$ is a collection of disjoint sets, each mapped one-one by π. Then $(f \circ \pi)^{-1}(f[U])$ is the necessary neighborhood of ζ.

By the uniqueness of the universal covering space (see Theorem 1.7.3), there is $g\colon \mathbb{C} \to \mathbb{C}$, a bicontinuous bijection, so that

$$\tilde{\pi} \circ g = f \circ \pi \tag{10.7.5}$$

Since analyticity is local and $\tilde{\pi}$ is the local analytic bijection, g must be analytic. Thus, by Theorem 7.3.4, for some λ and β,

$$g(z) = \lambda z + \beta \tag{10.7.6}$$

Then

$$\tilde{\pi}(\beta) = \tilde{\pi}(g(0)) = f(\pi(0)) = f(0)$$

and for $\omega \in \mathcal{L}_{\tau_1,\tau_2}$,

$$\tilde{\pi}(\beta + \lambda\omega) = \tilde{\pi}(g(\omega)) = f(\pi(\omega)) = f(0)$$

so $\lambda\omega \in \mathcal{L}_{\tilde{\tau}_1,\tilde{\tau}_2}$.

Conversely, if $\lambda\omega \in \mathcal{L}_{\tilde{\tau}_1,\tilde{\tau}_2}$, then

$$f(\pi(\omega)) = \tilde{\pi}(g(0)) = \tilde{\pi}(\beta + \lambda\omega) = \tilde{\pi}(\beta) = f(0)$$

so $\pi(\omega) = 0$, that is, $\omega \in \mathcal{L}_{\tau_1,\tau_2}$. We have thus proven that $\mathcal{L}_{\tilde{\tau}_1,\tilde{\tau}_2} = \lambda\mathcal{L}_{\tau_1,\tau_2}$. □

Second Proof of (1) ⇒ (2) in Theorem 10.7.2. Let $f\colon \mathcal{J}_{\tau_1,\tau_2} \to \mathcal{J}_{\tilde{\tau}_1,\tilde{\tau}_2}$ be an analytic bijection. Let $\wp(z \mid \tilde{\tau}_1\tilde{\tau}_2) \equiv \tilde{\wp}(z)$ be the Weierstrass $\wp$-function as a function on $\mathcal{J}_{\tilde{\tau}_1,\tilde{\tau}_2}$. Then $g(z) = \tilde{\wp}(f(z) - f(0))$ is a meromorphic function on $\mathcal{J}_{\tau_1,\tau_2}$ with second-order poles precisely at 0 and no other poles. It must have the form $g(z) = a\wp(z \mid \tau_1\tau_2) + b$ for suitable a and b. Thus, g has double values (i.e., $g(z) - g(z_0)$ has a single double zero) precisely at $\tau_1/2, \tau_2/2, (\tau_1+\tau_2)/2$, so $f(\tilde{\tau}_1/2) - f(0)$, $f(\tilde{\tau}_2/2) - f(0)$, $f(\tilde{\tau}_3/2) - f(0)$ must be a permutation of $\tau_1/2, \tau_2/2, \tau_3/2$ and $\{a\tilde{e}_j + b\}_{j=1}^3$ are a permutation of $\{e_j\}_{j=1}^3$.

Thus, $\lambda(\tau) = \lambda(\tilde{g}(\tilde{\tau}))$ for $\tilde{g}$, one of six FLTs in the $\tau \to \tilde{\tau}$ listed above Theorem 10.6.2, so $J(\tilde{\tau}) = J(\tau)$. By Theorem 10.6.9, (3) holds, and so (2) holds. □

Proof of Theorem 10.7.2. This is immediate from Corollary 10.4.6 and Theorems 10.6.9 and 10.7.1. □

Notes and Historical Remarks. Once one has the full uniformization theorem, it can also be proven that any Riemann surface which is topologically a torus is conformally equivalent to some $\mathcal{J}_{\tau_1,\tau_2}$. For the cover is, a priori, either $\mathbb{C}$ or $\mathbb{D}$. If $\mathbb{D}$, there must be a subgroup of elements of $\mathbb{A}\mathrm{ut}(\mathbb{D})$ with no fixed points isomorphic to $\mathbb{Z}^2$, but this does not happen (see Problem 1). Thus, the cover is $\mathbb{C}$, and the deck transformations are all translations. Thus, the image under the deck transformations of $0 \in \mathbb{C}$ must be a lattice $\mathcal{L}_{\tau_1,\tau_2}$ and the surface is then $\mathcal{J}_{\tau_1,\tau_2}$.

Problems

1. Let $f, g, \in \mathbb{A}\mathrm{ut}(\mathbb{C}_+)$ each be hyperbolic or parabolic and commuting.

   (a) Prove they have the same fixed points which, up to conjugacy, we can suppose are ∞ or 0 and ∞.

   (b) If the group generated by f and g is discrete, prove that there is h so $f = h^n$, $g = h^m$ for some integral n, m. (*Hint*: Discrete subgroups of $\mathbb{R}$ have a generator.)

   (c) Prove that $\mathbb{A}\mathrm{ut}(\mathbb{D})$ has no subgroups isomorphic to $\mathbb{Z}^2$.

2. Prove that for any τ_1, τ_2 with $\tau_1 \neq 0 \neq \tau_2$ and $\tau_2/\tau_1 \notin \mathbb{R}$, we have that $\mathbb{A}\mathrm{ut}(\mathcal{J}_{\tau_1,\tau_2})$ is naturally associated to $\mathbb{SL}(2, \mathbb{Z})$.

Chapter 11

Selected Additional Topics

> Raymond Edward Alan Christopher Paley was killed by an avalanche on April 7, 1933, while skiing in the vicinity of Banff, Alberta. Although only twenty-six years of age, he was already recognized as the ablest of the group of young English mathematicians who have been inspired by the genius of G. H. Hardy and J. E. Littlewood. In a group notable for its brilliant technique, no one had developed this technique to a higher degree than Paley. Nevertheless he should not be thought of primarily as a technician, for with this ability he combined creative power of the first order. As he himself was wont to say, technique without "rugger tactics" will not get one far, and these rugger tactics he practiced to a degree that was characteristic of his forthright and vigorous nature... In view of the very short time which he had been on this continent, the impression which Paley had made on American mathematicians is remarkable in the extreme."
>
> —*Norbert Wiener*, obituary of Raymond Paley [**593**]

Big Notions and Theorems: Paley–Wiener Theorem, Sheaf of Germs of Analytic Functions, Global Analytic Functions, Riemann Surface of an Analytic Function, Monodromy Theorem, Monodromy Group, Permanence of Relation, Cauchy's ODE Theorem, Picard's Little Theorem, Montel's Three-Value Theorem, Picard's Big Theorem, Zalcman's Lemma, Schottky Theorem, Bloch Theorem, Landau Theorem, Hartogs' Theorem, Bochner's Tube Theorem, $\mathbb{A}\mathrm{ut}(\mathbb{D}^2) \not\cong \mathbb{A}\mathrm{ut}(\mathbb{B}^2)$, Abel Map, Jacobi Variety, Abel's Theorem

Unlike all previous chapters which have had a central theme, this chapter is a collection of isolated topics—the only connections are that both Sections 11.3 and 11.4 prove Picard's theorem, albeit by different methods, and that Section 11.3 relies on a result from Section 11.2. That said, these are all important parts of a comprehensive look at complex analysis.

Section 11.1 discusses an idea associated with the names of Paley and Wiener that analyticity of the Fourier transform is equivalent to exponential decay of the original function. Section 11.2 codifies a theme we've hinted at several times before—the notion of all possible analytic continuations of an analytic function. Much of this section is language (we'll define a suitable sheaf), but there are two especially important theorems (the monodromy and Cauchy ODE theorems). The first is that if f can be analytically continued along any curve in a region $\Omega \subset \mathbb{C}$, then f has the same continuation along homotopic curves. In particular, if Ω is simply connected, then f is single-valued. The example of $\Omega = \mathbb{C}^\times$ and $f(z) = \sqrt{z}$ shows that simple connectivity is critical for this. We note that unlike Chapter 4, homotopy, not homology, is critical here; see Section 1.7. The second big result concerns solvability of linear ODEs with analytic coefficients throughout the region of analyticity. This is a theme we'll continue in Chapter 14 of Part 2B.

Sections 11.3 and 11.4 turn to the proof of Picard's theorems—Section 11.3 has the approach using the elliptic modular function, an approach close to Picard's original idea. Bonus Section 11.4 concerns the approach associated with Schottky and Bloch that uses some kind of a priori bound on the variation of a function. We note that Section 12.4 of Part 2B will have a third proof of Picard's theorem due to Ahlfors and Robinson, using ideas from Riemann geometry. All these approaches essentially rely on Section 11.3 for getting the great Picard theorem from Montel's three-value theorem, Theorem 6.2.10.

Besides these three proofs of Picard's little theorem, we have two more: Chapter 17 of Part 2B discusses Nevanlinna theory which generalizes Picard's theorem to give much more information about how often different values are taken. Section 3.3 of Part 3 has a "real variable" proof relying on harmonic and subharmonic functions. These five distinct proofs illustrate different things—every other proof I know of Picard's theorems are variants of these five, except for a proof of Davis [**131**] that relies on properties of typical Brownian motion curves in the plane.

This is a book on functions of one complex variable. It is clear that for real variables, there are very important differences between one and two or more variables because of the obvious geometric differences between $\mathbb{R}$ and $\mathbb{R}^n$ for $n \geq 2$. It seems less clear for complex variables. What Section 11.5 shows is that two big results from one complex variable—that any region is a domain of holomorphy (Corollary 9.5.4) and that all simply connected strict subregions are conformally equivalent (Theorem 8.1.1)—do not extend to $\mathbb{C}^2$. Bonus Section 11.6 describes (but does not prove) the general form of Abel's theorems for compact Riemann surfaces.

11.1. The Paley–Wiener Strategy

This section will assume the reader has some familiarity with the basics of Fourier transforms, such as in our presentation in Chapter 6 of Part 1. We normalize the Fourier transform, $\widehat{f}$, and inverse transform, $\check{f}$, by (we define for functions on $\mathbb{R}^\nu$ but focus here mainly on $\mathbb{R}$),

$$\widehat{f}(k) = (2\pi)^{-\nu/2} \int e^{-ik\cdot x} f(x)\, d^\nu x \tag{11.1.1}$$

$$\check{f}(x) = (2\pi)^{-\nu/2} \int e^{ik\cdot x} f(k)\, d^\nu k \tag{11.1.2}$$

While others (and we) refer to "the Paley–Wiener theorem," it is more a collection of theorems based on a single conception: analyticity of the Fourier transform is equivalent to exponential decay of the original function. This comes in two guises: decay at a fixed exponential rate is connected to analyticity of the transform in a strip and compact support (an extreme form of decay!) is equivalent to the transform being entire of order 1.

We'll illustrate with four theorems chosen so that there are equivalences. In general, equivalences only occur for three spaces, $L^2, \mathcal{S}, \mathcal{S}'$, and we'll look precisely at those three in one dimension, leaving $\mathbb{R}^\nu$, $\nu > 1$, to the Problems. Here are the results:

Theorem 11.1.1. *Let $f \in L^2(\mathbb{R}, dx)$. If $a > 0$ and for all $0 < b < a$, $e^{b|x|}f(x) \in L^2$, then the Fourier transform, $g \equiv \widehat{f}$, has an analytic continuation to the strip $\{z \mid |\operatorname{Im} z| < a\}$ that obeys*

$$\sup_{|y|\le b} \int_{-\infty}^{\infty} |g(x+iy)|^2\, dx < \infty \tag{11.1.3}$$

for all $b < a$. Conversely, if g is an analytic function on that strip and obeys (11.1.3), then $e^{b|x|}\check{g} \in L^2$ for all $b < a$.

Theorem 11.1.2. *Let $f \in L^2(\mathbb{R}, dx)$ with $\operatorname{supp}(f) \subset [-a, a]$. Then $g = \widehat{f}$ is an entire analytic function obeying*

$$|g(z)| \le Ce^{a|\operatorname{Im} z|} \tag{11.1.4}$$

and

$$\int_{-\infty}^{\infty} |g(x)|^2\, dx < \infty \tag{11.1.5}$$

Conversely, if g is an entire function obeying (11.1.4) and (11.1.5), then $g = \widehat{f}$ with $f \in L^2$ and $\operatorname{supp}(f) \subset [-a, a]$.

Remark. Actually, the proof shows more is true. One has $g(\,\cdot\,+ia)\in L^2$ for all a real and the converse only requires for all $\varepsilon>0$ and some $N(\varepsilon)$, $|g(z)|\le C_\varepsilon(1+|z|)^{N(\varepsilon)}e^{(a+\varepsilon)[\operatorname{Im} z]}$.

Theorem 11.1.3. *Let $f\in\mathcal{S}(\mathbb{R})$ with $\operatorname{supp}(f)\subset[-a,a]$. Then the Fourier transform, $g=\widehat{f}$, is an entire function so that for all n, there is C_n with*

$$|g(x+iy)|\le C_n(1+|x|+|y|)^{-n}\exp(a|y|) \tag{11.1.6}$$

Conversely, if g is an entire function that obeys (11.1.6), then $g\in\mathcal{S}$ and $\operatorname{supp}(\check{g})\subset[a,a]$.

Theorem 11.1.4. *Let $T\in\mathcal{S}'(\mathbb{R})$ with $\operatorname{supp}(T)\subset[-a,a]$. Then its Fourier transform, $S=\widehat{T}$, is an entire function, and for some N obeys*

$$|S(x+iy)|\le C(1+|x|+|y|)^N\exp(a|y|) \tag{11.1.7}$$

Conversely, if S is an entire function obeying

$$|S(x+iy)|\le C_\varepsilon(1+|x|+|y|)^N\exp((a+\varepsilon)|y|) \tag{11.1.8}$$

for N fixed and each $\varepsilon>0$, then there is a distribution T with $\widehat{T}=S$ and $\operatorname{supp}(T)\subset[-a,a]$.

The proofs are all straightforward.

Proof of Theorem 11.1.1. Given $b<a$, fix c so $0<b<c<a$. For z with $|\operatorname{Im} z|<c$, let $\varphi_z(x)=e^{-c|x|}e^{izx}/(2\pi)^{1/2}$. Then $\varphi_z\in L^2$ for all such z and is an analytic L^2-valued function of z (Problem 1).

By hypothesis, $e^{c|x|}f\in L^2$, so

$$g(z)=\langle\varphi_{\bar z},e^{c|x|}f\rangle \tag{11.1.9}$$

defines an analytic function of $z\in\{z\mid |\operatorname{Im} z|<c\}$. If, for $|\beta|<c$, β real,

$$f_\beta(x)=e^{\beta x}f(x) \tag{11.1.10}$$

then, by (11.1.1),

$$\widehat{f_\beta}(k)=g(i\beta+k) \tag{11.1.11}$$

In particular, $\widehat{f}(k)$ has an analytic continuation to $\{z\mid |\operatorname{Im} z|<a\}$.

Since $\|f_\beta\|_{L^2}\le\|e^{c|\cdot|}f\|_{L^2}$, we see, by the Plancherel theorem (see Theorem 6.2.14 in Part 1), that (11.1.3) holds.

Conversely, suppose that g is analytic in $\{z\mid |\operatorname{Im} z|<a\}$ and that (11.1.3) holds. For w in the strip, let

$$g_w(k)=g(k+w) \tag{11.1.12}$$

Then g_w is an L^2-valued analytic function (Problem 1). For w real,

$$\check{g}_w(x)=e^{-iw\cdot x}\check{g}(x) \tag{11.1.13}$$

If $h \in L^2$ has compact support,

$$\langle h, \check{g}_w \rangle = \int \overline{h(x)}\, e^{-iw\cdot x} \check{g}(x) \tag{11.1.14}$$

first for w real and then for all w in the strip since both sides are analytic in w.

Let $\chi_{[-R,R]}$ be the characteristic function of $[-R, R]$. Picking $w = \pm ib$, with $0 < b < a$ and $h = e^{-iw\cdot x}\check{g}(x)\chi_{[-R,R]}$, we get, by the Plancherel theorem, that

$$\|e^{\pm bx}\chi_{[-R,R]}\check{g}\| \le \|\check{g}_{\pm ib}\| = \|g_{\pm ib}\| = \left(\int |g(k \pm ib)|^2\, dk\right)^{1/2} \tag{11.1.15}$$

Since $e^{b|x|} \le e^{bx} + e^{-bx}$, we see that $e^{b|x|}\check{g} \in L^2$. $\square$

Proof of Theorem 11.1.3. Suppose first that f has compact support. In (11.1.1), k can be made complex and the integral still converges. By Theorem 3.1.6, $\widehat{f}$ defined by the integral is an entire function of k.

For k real, one has (see (6.2.43) of Part 1) for $n = 0, 1, 2, \dots,$

$$(ik)^n \widehat{f}(k) = (2\pi)^{-1/2} \int e^{-ikx} \frac{d^n f}{dx^n}(x)\, dx \tag{11.1.16}$$

and this holds for all k in $\mathbb{C}$ since both sides are entire functions of k. Thus,

$$|k|^n |\widehat{f}(k)| \le (2\pi)^{-1/2} e^{a|\mathrm{Im}\, k|} \left\| \frac{d^n}{dx^n} f \right\|_\infty \tag{11.1.17}$$

This implies (11.1.6).

Conversely, suppose g is entire and obeys (11.1.6). By a Cauchy estimate, for any m and k real,

$$\left| \frac{d^m g}{dk^m}(k) \right| \le \sup_{|w|\le 1} |g(k + w)| \le C_n(\tfrac{1}{2}|k| - 1)^{-n} e^a \tag{11.1.18}$$

so $g \in \mathcal{S}$.

Next, for any $\kappa \in \mathbb{R}$,

$$\begin{aligned} \check{g}(x) &= (2\pi)^{-1/2} \int_{-\infty}^{\infty} e^{ikx} f(k)\, dk \\ &= (2\pi)^{-1/2} \int_{-\infty}^{\infty} e^{i(k+i\kappa)x} f(k + i\kappa)\, dk \end{aligned} \tag{11.1.19}$$

where the shift of contour is justified by a use of the Cauchy theorem (Problem 2). Thus, for any κ,

$$\begin{aligned} |e^{\kappa x}\check{g}(x)| &\le e^{a|\kappa|}(2\pi)^{-1/2} C_2 \int_{-\infty}^{\infty} \frac{1}{(|k| + |\kappa| + 1)^2}\, dk \\ &\le D e^{a|\kappa|} \end{aligned} \tag{11.1.20}$$

so

$$|\check{g}(x)| \leq De^{a|\kappa|-\kappa x} \tag{11.1.21}$$

If $|x| > a$, taking $\kappa \to +\infty$ (if $x > 0$) or $\kappa \to -\infty$ (if $x < 0$) shows $\check{g}(x) = 0$, that is, $\check{g}$ is supported in $[-a, a]$. □

Finally, we turn to the result for distributions. This requires a number of facts about distributions which can be found, for example, in Chapter 6 of Part 1:

(a) For $T \in \mathcal{S}'(\mathbb{R}^\nu)$, one defines $\widehat{T}$ by

$$\widehat{T}(f) = T(\widehat{f}) \tag{11.1.22}$$

for all $f \in \mathcal{S}(\mathbb{R}^\nu)$.

(b) If $f \in \mathcal{S}(\mathbb{R}^\nu)$, $T \in \mathcal{S}'(\mathbb{R}^\nu)$, then $f * T$ obeys

$$\widehat{f * T} = (2\pi)^{\nu/2} \widehat{f}\,\widehat{T} \tag{11.1.23}$$

Here $f * T$ is the distribution given by a function equal to

$$(f * T)(x) = T(\widetilde{f}_{-x}) \tag{11.1.24}$$

where

$$\widetilde{f}_{-x}(y) = f(x - y) \tag{11.1.25}$$

(c) (See Problem 3.) If $T \in \mathcal{S}'(\mathbb{R}^\nu)$ has compact support and $f \in \mathcal{S}(\mathbb{R}^\nu)$ also has compact support, then $f*T \in \mathcal{S}'(\mathbb{R}^\nu)$ also has compact support, and

$$\operatorname{supp}(f * T) \subset \operatorname{supp}(f) + \operatorname{supp}(T) \tag{11.1.26}$$

(d) If $f \in \mathcal{S}$ and $T \in \mathcal{S}'(\mathbb{R}^\nu)$, then $fT \in \mathcal{S}'(\mathbb{R}^\nu)$ and

$$\widehat{fT} = (2\pi)^{-\nu/2} \widehat{f} * \widehat{T} \tag{11.1.27}$$

Proof of Theorem 11.1.4. We will prove a slightly weaker result, that if $\operatorname{supp}(T) \subset [-a, a]$, then (11.1.8) holds, leaving the proof of (11.1.7) to the Problems (see Problems 4 and 5). Fix $\varepsilon > 0$ and choose $\varphi_\varepsilon \in C_0^\infty(\mathbb{R})$ supported in $[a - \varepsilon, a + \varepsilon]$ with $\varphi_\varepsilon(x) = 1$ for $x \in [-a - \frac{\varepsilon}{2}, a + \frac{\varepsilon}{2}]$. Then $T = \varphi_\varepsilon T$, so we claim

$$\widehat{T}(k) = (2\pi)^{-1/2} T(\varphi_\varepsilon e^{-ik\cdot}) \tag{11.1.28}$$

This follows from (11.1.27), (11.1.24), and (11.1.22). Since

$$\begin{aligned} \widehat{T}(k) &= (2\pi)^{-1/2} (\widehat{\varphi}_\varepsilon * \widehat{T})(k) \\ &= \widehat{T}(\widetilde{\widehat{\varphi}}_{\varepsilon,-k}) \\ &= T((\widetilde{\widehat{\varphi}}_{\varepsilon,-k})\widehat{\ }) \end{aligned} \tag{11.1.29}$$

Next, note that

$$\begin{aligned}(\widetilde{\widehat{\varphi}_{\varepsilon,-k}})(\ell) &= \widehat{\varphi}_\varepsilon(k-\ell) \\ &= (2\pi)^{-1/2}\int e^{-i(k-\ell)x}\varphi_\varepsilon(x)\,dx \\ &= (e^{-ik\cdot}\varphi_\varepsilon)^\wedge(\ell) \end{aligned} \tag{11.1.30}$$

Thus, (11.1.28) follows from (11.1.29) and $\widehat{\check{g}} = g$.

Since $\varphi_\varepsilon \in C_0^\infty(\mathbb{R})$, $\varphi_\varepsilon e^{-ik\cdot} \in \mathcal{S}$ for all $k \in \mathbb{C}$ and lets us define $\widehat{T}(k)$ for all $k \in \mathbb{C}$. Moreover, by an easy argument (see Problem 6), this function is analytic in k.

Since T has compact support, there exist N and $C_0, \dots, C_N$ so that

$$|T(f)| \le \sum_{j=0}^N C_j \|f^{(j)}\|_\infty \tag{11.1.31}$$

Thus, (11.1.8) follows from (11.1.28) and

$$\|(\varphi_\varepsilon e^{-ik\cdot})^{(j)}\|_\infty \le C_{j,\varepsilon}(1+|k|)^j e^{|\mathrm{Im}\,k|(a+\varepsilon)} \tag{11.1.32}$$

For the converse, let S be an entire function obeying (11.1.8). Then $S \restriction \mathbb{R}$ is a polynomially bounded function, and so a tempered distribution. Thus, $S = \widehat{T}$ for a tempered distribution T. For $\delta > 0$, let $j_\delta(x) = \delta^{-1} j(x/\delta)$ be an approximate identity. Then, j_δ is entire, and by Theorem 11.1.3,

$$|\widehat{j_\delta}(k)| \le C_n(1+|k|)^{-n} e^{\delta|\mathrm{Im}\,k|} \tag{11.1.33}$$

so, by (11.1.23),

$$|\widehat{j_\delta * T}(k)| \le CC_n(1+|k|)^{-n+N} e^{(\delta+\varepsilon+a)|\mathrm{Im}\,k|} \tag{11.1.34}$$

By Theorem 11.1.3 again, $j_\delta * T \in \mathcal{S}$ and

$$\mathrm{supp}\,(j_\delta * T) \subset [-a-\varepsilon-\delta, a+\varepsilon+\delta] \tag{11.1.35}$$

Since ε and δ are arbitrary, and $j_\delta * T \to T$ as $\delta \downarrow 0$, we see that $\mathrm{supp}\,(T) \subset [-a, a]$. □

Proof of Theorem 11.1.2. Suppose first $f \in L^2$ with $\mathrm{supp}\,(f) \subset [-a, a]$. By the Plancherel theorem, g obeys (11.1.4). By Theorem 3.1.6, g is entire, and clearly,

$$|g(z)| \le e^{a|\mathrm{Im}\,z|} \int_{-a}^a |f(x)|\,dx \tag{11.1.36}$$

For the converse, if g obeys (11.1.4), f is a distribution supported in $[-a, a]$. By the Plancherel theorem and (11.1.4), $f \in L^2$. □

Notes and Historical Remarks. The development and exploitation of the notion of equivalence of falloff to analyticity of the Fourier transform is due to Paley and Wiener [**425**]. The extension to distributions, sometimes called the Paley–Wiener–Schwartz theorem, is due to Laurent Schwartz [**508**].

The space of entire functions g obeying (11.1.4) and (11.1.5) is called the Paley–Wiener space, PW_a. It is discussed in Section 6.6 of Part 1; see especially Theorem 6.6.16 and the Notes on the relevant work of Hardy.

Raymond Paley (1907–33), a British student of Littlewood, was killed in an avalanche while skiing in Canada at age only 26. Remarkably, he already had three major pieces of work to his name. Not only the work with Wiener, but also work with Littlewood, known as Littlewood–Paley theory, on when multiplying the Fourier transform by a function and inverse transforming preserves L^p, and work with Zygmund on natural boundaries of functions with random Taylor coefficients.

When one extends the results discussed here to dimension $\nu \geq 2$ (see Problems 7 and 8), a key fact is that the proper analog of an interval is typically a convex set.

If $f \in L^1(\mathbb{R})$ is supported on $[0, \infty)$, its Fourier transform $\widehat{f}(k)$, given by (11.1.1), clearly can be defined by a convergent integral if $\operatorname{Im} k < 0$. It is not hard to see that $\widehat{f}$ is analytic in $\mathbb{C}_-$ and continuous up to $\mathbb{R}$. In general, this leads to results on analyticity in half-spaces with some kind of boundary value connected to half-line support. In higher dimensions, cones replace half-spaces and tubes over conic half-planes. One place to read about this further is [**41**].

Problems

1. (a) For z with $|\operatorname{Im} z| < c$, $f_z(x) = e^{-c|x|}e^{izx}$ lies in L^2. Prove that $z \mapsto f_z$ is L^2-analytic in the sense that it has a power series about any point converging in L^2-norm.

   Remark. Do directly or use Theorem 3.1.12.

   (b) Let $\mathcal{H}$ be a separable Hilbert space. An $\mathcal{H}$-valued function, f, on a region $\Omega \subset \mathbb{C}$ is called analytic if it has a norm-convergent Taylor expansion near any $z_0 \in \mathcal{H}$ and *weakly analytic* if for every compact subset $K \subset \Omega$, $\sup_{z\in K}\|f(z)\| < \infty$ and if $z \mapsto \langle\psi, f(z)\rangle$ is analytic for each $\psi \in \mathcal{H}$. Prove that weak analyticity implies analyticity. (*Hint*: Pick a basis $\{\psi_n\}_{n=1}^\infty$. Prove first that for any ε, there are δ and N so that $\sum_{j=N+1}^\infty|\langle\psi_j, f(z)\rangle|^2 < \varepsilon$ for $z \in \mathbb{D}_\delta(0)$, then norm uniform convergence of the Fourier expansion and then norm analyticity; see also Theorem 3.1.12.)

(c) If g is analytic in $\mathcal{R} \equiv \{z \mid |\mathrm{Im}\, z| < a\}$ and obeys (11.1.3) for all $b \in (0, a)$, and if $g_w(k) = g(k+w)$ for $k \in \mathbb{R}$, prove that $w \mapsto g_w$ is weakly analytic, and so norm-analytic.

2. Let h be analytic in $\{z \mid |\mathrm{Im}\, z| < a\}$ so that for some C, $|f(x+iy)| \le C(1+|x|)^{-2}$ for all $|y| < a$. Prove that $\lim_{R\to\infty} \int_{-R}^{R} f(x+iy)\, dx$ is y-independent for $|y| < a$.

3. Using the definition of $f * T$ and of compact support for distributions (see Section 6.2 of Part 1), prove (11.1.26).

4. The problem will provide one way of proving (11.1.7) for $T \in \mathcal{S}'$ with $\mathrm{supp}\,(T) \subset [-a, a]$. The key intermediate step will be to show that for some signed measure μ supported on $(-\infty, a]$ and some ℓ, $T = d^\ell \mu / dx^\ell$.

(a) If G is supported on $[-a, a]$, show that there is a signed measure μ_0 and ℓ so that

$$T = \frac{d^\ell \mu_0}{dx^\ell} \tag{11.1.37}$$

(*Hint*: First show that there is a k with $|T(f)| \le C \sum_{j=0}^{k} \|f^{(j)}\|_\infty$.)

(b) If (11.1.37) holds, prove there is a polynomial P with $\deg(P) = \ell$ so that $\mu_0 - P(x)\, dx$ has support $(-\infty, a]$.

(c) If f is C^∞ and has support $(-a, \infty)$ with $f \equiv 1$ on $[a, \infty)$, prove that $\frac{d}{dx^2}((1-f)P)$ is a measure supported on $(-\infty, a]$.

(d) Conclude that there is a measure μ so that $\mathrm{supp}(\mu) \subset (-\infty, a]$ and $T = d^\ell \mu / dx^\ell$.

(e) Conclude that with $S = \widehat{T}$ for $y > 0$, we have (11.1.7) and then that, by symmetry, we have the result for all y.

5. (a) Prove there exists a sequence φ_n of functions in C_0^∞ with $\mathrm{supp}(\varphi_n) \subset [-a - \frac{1}{n}, a + \frac{1}{n}]$, $\varphi_n \equiv 1$ on $[-a, a]$ and obeying $\|d^\ell \varphi_n / dx^\ell\|_\infty \le C_\ell n^\ell$.

(b) Prove (11.1.7) for $T \in \mathcal{S}'(\mathbb{R})$ with $\mathrm{supp}\,(T) \subset [-a, a]$. (*Hint*: Use (11.1.28) with φ_ε replaced by φ_n if $n \le |y| < n+1$.)

6. (a) If $\varphi \in C_0^\infty(\mathbb{R})$ and $\varphi_k(x) = e^{-ikx}\varphi(x)$, prove directly that $k \to \varphi_k$ is analytic in the $\mathcal{S}$-topology (in that there is a power series converging in the metric of $\mathcal{S}$.)

(b) Prove this analyticity using Theorem 11.1.3.

7. Prove a multivariable version of Theorem 11.1.1. If $U \subset \mathbb{R}^\nu$ is open with $0 \in U$ and

$$\int |e^{a\cdot x} f(x)|^2 \, d^\nu x < \infty \tag{11.1.38}$$

for some $f \in L^2(\mathbb{R}^\nu, d^\nu x)$ and all $a \in U$, then $\widehat{f}(z)$ has an analytic continuation to $\{z = u + iv \mid v \in U,\, u \in \mathbb{R}^\nu\}$ with

$$\sup_{v\in K} \int |\widehat{f}(u+iv)|^2 \, d^\nu u < \infty \tag{11.1.39}$$

for all compact $K \subset U$.

Conversely, if $\widehat{f}$ is analytic in that set and (11.1.39) holds, then (11.1.38) holds for all $a \in U$.

8. Prove $\mathbb{R}^\nu$ versions of Theorems 11.1.3 and 11.1.4.

9. The purpose of this problem is for you to prove the following theorem: Suppose f is an entire function obeying
   (i) $f(0) = 1$, $|f(x)| \le 1$ for x real.
   (ii) $\int_{-\infty}^{\infty} |f(x)|^2 \, dx = 1$.
   (iii) All the zeros of f are real and can be labeled $\{z_j\}_{j=1}^{\infty} \cup \{z_j\}_{j=-\infty}^{-1}$ so that $|z_j| \ge |j|$ and $jz_j > 0$.
   (iv) $|f(z)| \le Ce^{D|z|}$.
   (v) f is real on $\mathbb{R}$.

   Then

$$f(z) = \frac{\sin(\pi z)}{\pi z} \tag{11.1.40}$$

   (a) Prove that for y real, $|f(iy)| \le \sinh(\pi y)/\pi y$. (*Hint*: What does Hadamard say?)

   (b) Prove that $|f(z)| \le Ce^{\pi|\operatorname{Im} z|}$ (*Hint*: Use Phragmén–Lindelöf on $e^{i\pi z} f(z)$ in the sector $0 \le \arg(z) \le \pi/2$.)

   (c) Prove that $\hat{f}(k)$ has support in $[-\pi, \pi]$ and that

$$\int \left| \hat{f}(k) - \frac{1}{(2\pi)^{1/2}} \chi_{[-\pi,\pi]}(k) \right|^2 dk = 0$$

   and then that equation (11.1.40) holds.

   Remark. This theorem is from [**27**], based in part on ideas from Lubinsky [**368**].

11.2. Global Analytic Functions

We've discussed the notion of analytic continuation of a function several times. Here we'll pursue the idea of all possible continuations of a function. The section will primarily present a language for this notion that is exceedingly useful and conceptually simple once one gets over the unusual notion of a fiber which is an infinite-dimensional space on which one puts the discrete topology! The one deep theorem will be the monodromy theorem—while we'll give a stand-alone proof of it, as we'll explain in the

Notes, it is essentially a variant of the fundamental theorem on covering maps (Theorem 1.7.2).

Definition. A *germ* of an analytic function is a point z_0 in $\mathbb{C}$ and a convergent Taylor series $\sum_{n=0}^{\infty} a_n(z-z_0)^n$ at z_0. If $\mathfrak{g}$ is a germ, $\pi(\mathfrak{g}) = z_0$ defines π, and $\rho(\mathfrak{g})$ is the radius of convergence of the power series. An analytic function, f, on $\Omega \subset \mathbb{C}$, defines a family of germs, one for each $z_0 \in \Omega$, via its Taylor series. We call it the set of germs induced by f.

Equivalently, a germ is an equivalence class of functions, g, analytic in a neighborhood of z_0 with $g \sim h$ if they agree near z_0. We'll write $g \in \mathfrak{g}$ to indicate this relation.

It is the topology that we put on the set on all germs that turns out to make the notion so valuable:

Definition. Let $\Omega \subset \mathbb{C}$ be a region. The *sheaf of germs* of analytic functions over Ω, $\mathcal{S}(\Omega)$, is the set of all germs, $\mathfrak{g}$, with $\pi(\mathfrak{g}) \in \Omega$. For each germ, $\mathfrak{g}$, and $r < \min(\rho(\mathfrak{g}), \text{dist}(z_0, \mathbb{C} \setminus \Omega))$, let $U_r(\mathfrak{g})$ be the set of all germs induced by $f(z) = \sum_{n=0}^{\infty} a_n(z-z_0)^n$ restricted to $\mathbb{D}_r(z_0)$. The topology on $\mathcal{S}(\Omega)$ is the one with $\{U_r(\mathfrak{g})\}$ as base.

Notice that the induced topology on $\{\mathfrak{g} \mid \pi(\mathfrak{g}) = z_0\}$ is discrete!

Definition. Let $\Omega \subset \mathbb{C}$ be a region and $\gamma\colon [0,1] \to \Omega$ a continuous curve. Let $\mathfrak{g}$ be a germ with $\pi(\mathfrak{g}) = \gamma(0)$. We say $\mathfrak{g}$ can be *analytically continued* along γ if and only if there is a continuous function, $\tilde{\gamma}\colon [0,1] \to \mathcal{S}(\Omega)$ so that $\pi \circ \tilde{\gamma} = \gamma$.

The reader should convince themselves that, if it exists, $\tilde{\gamma}$ is unique and that this captures the idea discussed after Corollary 2.3.9. The germ at $\tilde{\gamma}(1)$ will be denoted $\tau_\gamma(\mathfrak{g})$. Figure 11.2.1 is a schematic indication of what analytic continuation means.

Definition. A *global analytic function* is an arcwise connected component of $\mathcal{S}(\Omega)$. Given a germ $\mathfrak{g}$, the global analytic function for $\Omega = \mathbb{C}$ that contains $\mathfrak{g}$ is called the *Riemann surface of the function* defined by $\mathfrak{g}$.

The reader should check (Problem 1) that each global analytic function is separable. If $\mathcal{R}$ is a global analytic function and $\{\mathfrak{g}_n\}$ is a dense set of germs in $\mathcal{R}$, then $\pi\colon U_r \to \mathbb{D}_r(z_0)$ for $z_0 = \pi(\mathfrak{g}_n)$, r rational with $r < \rho(\mathfrak{g}_n)$ defines a family of maps which sets up a Riemann surface structure on $\mathcal{R}$ (Problem 2), so $\mathcal{R}$ is a Riemann surface, and the name is justified. The reader can confirm that for other choices of dense set, one gets a compatible Riemann surface structure, and that for something like $\log(z)$ $(z \in \mathbb{C}^\times)$, this is the same as the naive notion discussed in Example 7.1.6. For $\sqrt{z}$, this definition does not include $z = 0$, but see the Notes.

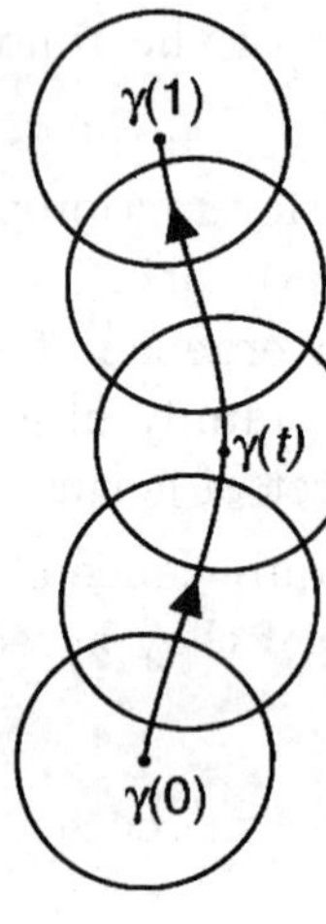

Figure 11.2.1. Analytic continuation (from `http://commons.wikimedia.org/wiki/File:Analytic_continuation_along_a_curve.png`).

The language makes it easy to approach the fundamental fact that continuation along homotopic curves, for which one can continue along every curve in the homotopy, yields the same function.

Theorem 11.2.1 (The Monodromy Theorem). *Let $\mathfrak{g}$ be a germ with $\pi(\mathfrak{g}) \in \Omega$, a region in $\mathbb{C}$. Suppose that $\mathfrak{g}$ can be continued along any curve in Ω. Then homotopic curves produce the same germs at their endpoints, that is, if $\Gamma\colon [0,1] \times [0,1] \to \Omega$ is continuous and $\gamma(0,s) = z_0 = \pi(\mathfrak{g})$, and $\gamma(1,s) = z_1$ is s-independent, then*

$$\tau_{\gamma(\cdot,0)}(\mathfrak{g}) = \tau_{\gamma(\cdot,1)}(\mathfrak{g}) \tag{11.2.1}$$

In particular, if Ω is simply connected, then there is a single-valued function, f, on Ω whose germ at z_0 is $\mathfrak{g}$.

Remark. The hypothesis that one can continue along any curve implies that $\pi\colon \mathcal{R} \to \Omega$ is a covering map, and this theorem is the fundamental theorem on covering maps; see the discussion in the Notes.

Proof. By connectedness of $[0,1]$, it suffices to prove that $\tau_{\gamma(\cdot,s)}(\mathfrak{g})$ is locally constant, that is, for every s_0, there is an ε so

$$\tau_{\gamma(\cdot,s)}(\mathfrak{g}) = \tau_{\gamma(\cdot,s_0)}(\mathfrak{g}) \tag{11.2.2}$$

for all $s \in [0,1]$ with $|s - s_0| < \varepsilon$. The strategy for this follows the proof of Theorem 2.6.5.

Given s_0, use the paving lemma (Proposition 2.2.7) to find disks $U_1, \dots, U_\ell$ and $0 = t_0 < t_1 < \cdots < t_n = 1$ so $\gamma(t, s_0) \in U_j$ if $t \in [t_{j-1}, t_j]$.

Then find ε so $\gamma(t,s) \in U_j$ if $|s-s_0| < \varepsilon$ and $t \in [t_{j-1}, t_j]$. Since $\mathfrak{g}$ can be continued from z_0 along $\gamma(\,\cdot\,, s_0)$ and then throughout U_j, the Taylor series for the continuation at the center of U_j gives the germ at every point in U_j.

It follows inductively in U_j that the continuation along $\gamma(\,\cdot\,, s_0)$ from z_0 to $\gamma(t_j, s_0)$ and then from $\gamma(t_j, s_0)$ by the line to $\gamma(t_j, s)$ agrees with continuation along $\gamma(\,\cdot\,, s)$ to $\gamma(t_j, s_0)$. Thus, (11.2.2) holds. □

The following is trivial to prove but important in applying and understanding analytic continuation.

Theorem 11.2.2 (Permanence of Relation). *Let $\Omega \subset \mathbb{C}$ be a region. Let $P(w_0, \dots, w_n, z)$ be a polynomial in $w_0, \dots, w_n$ with coefficients analytic in $z \in \Omega$. Let f be analytic in some $\mathbb{D}_\delta(z_0) \subset \Omega$ and suppose in that disk,*

$$P(f(z), f'(z), \dots, f^{(n)}(z), z) = 0 \tag{11.2.3}$$

Suppose f can be analytically continued along some curve γ in Ω and let $\tilde{f}$ be the continuation in some $\mathbb{D}_{\tilde{\gamma}}(\gamma(1))$. Then (11.2.3) holds for $\tilde{f}$ when $z \in \mathbb{D}_\delta(\gamma(1))$.

Proof. Cover γ with disks $U_1, \dots, U_n$ as in the last proof and note that for any analytic function, g, on U_j, $P(g(z), g'(z), \dots, z)$ is analytic on U_j. Thus, (11.2.3) continues disk by disk to U_n. □

Finally, we want to define the representations of the homotopy group that is sometimes possible in special but natural situations. In Problem 3, you will prove the following:

Theorem 11.2.3 (Cauchy's ODE Theorem). *Let $a_0(z), \dots, a_{n-1}(z)$ be analytic in some $\mathbb{D}_\delta(z_0)$. Then for any $(\lambda_0, \dots, \lambda_{n-1}) \in \mathbb{C}^{n-1}$, there exists a unique analytic function on $\mathbb{D}_\delta(z_0)$, $f_{\vec{\lambda}}(z)$, that obeys*

$$f^{(n)}(z) + \sum_{j=0}^{n-1} a_j(z) f^{(j)}(z) = 0 \tag{11.2.4}$$

$$f^{(j)}(z_0) = \lambda_j \qquad j = 0, \dots, n-1 \tag{11.2.5}$$

This sets up a one-one correspondence between $\mathbb{C}^n$ and all solutions of (11.2.4).

Theorem 11.2.4. *Let Ω be a region in $\mathbb{C}$ and $a_0(z), \dots, a_{n-1}(z)$ elements of $\mathfrak{A}(\Omega)$. Pick $z_0 \in \Omega$. Then for any closed curve in Ω with $\gamma(0) = \gamma(1) = z_0$ and any solution, f, of (11.2.4) near z_0, there is an analytic continuation of f along γ. $\gamma \mapsto \tau_\gamma(f)$ only depends on the homotopy class in γ. The map $(\lambda_0, \dots, \lambda_{n-1}) \to (g(0), \dots, g^{(n-1)}(0))$ where $g = \tau_\gamma(f_{\lambda_0, \dots, \lambda_{n-1}})$ is a* *map (linear in* $(\lambda_0,\dots,\lambda_{n-1})$*),* $\mathcal{M}_{[\gamma]}(\lambda_0,\dots,\lambda_{n-1})$ *and is a representation of* $\pi_1(\Omega,z_0)$ *in that*

$$\mathcal{M}_{[\gamma]}\mathcal{M}_{[\gamma']}=\mathcal{M}_{[\gamma,\gamma']} \tag{11.2.6}$$

Proof. Cover γ with disks lying in Ω, using the paving lemma (Proposition 2.2.7). By Theorem 11.2.3, we can continue along successive disks to get the continuation. Homotopy independence follows from Theorem 11.2.2 and linearity of $\mathcal{M}_{[\gamma]}$ from linearity of the differential equation. (11.2.6) follows from the definition of product in the homotopy group and the definition of analytic continuation. □

The group of maps $\{\mathcal{M}_{[\gamma]}\}$ is called the *monodromy group.* We'll say more about it in Chapter 14 of Part 2B. In particular, if Ω is simply connected, solutions define single-valued functions on all of Ω.

Theorem 11.2.5. *Let* Ω *be a simply connected region in* $\mathbb{C}$ *and* $a_0(z),\dots,a_n(z)$ *elements of* $\mathfrak{A}(\Omega)$*. For each* $(\lambda_0,\dots,\lambda_{n-1})\in\mathbb{C}^n$*, the unique solution of* (11.2.4)/(11.2.5) *near* z_0 *extends to an element of* $\mathfrak{A}(\Omega)$*.*

The study of what happens near an isolated singularity of the a_j's will be a subject of Section 14.2 of Part 2B.

Notes and Historical Remarks. Starting with Riemann's dissertation [**480**], the notion of Riemann surface was tied to the notion of Riemann surface of a function. The language of sheaves is convenient for our definition here—it and, in particular, its cohomology was developed for use in algebraic geometry and in several complex variables around 1950. The pioneers were Leray and H. Cartan, especially as part of Séminaire Henri Cartan, starting in 1950/51 [**99**]. Significant developments were codified in a book of Godement [**211**], a student of Cartan.

Monodromy was first introduced in the context of analytic continuation of hypergeometric functions—solutions of a certain ODE with analytic coefficients on $\mathbb{C}\setminus\{0,1\}$; see Sections 14.1 and 14.4 of Part 2B. The key innovation was in a paper of Riemann [**483**] who thought abstractly, instead of the rather concrete earlier work of Kummer. Hermite [**255**] had the idea of associating continuation around algebraic singularities with matrices, but it was Riemann who looked at products of such matrices. Further developments of monodromy groups were associated with Frobenius, Fuchs, Klein, Poincaré, and Schwarz. It was Weyl in his pathbreaking book on Riemann surfaces [**590**] that first went beyond monodromy groups and discussed what is essentially the general monodromy theorem.

An especially interesting case of monodromy is where an initial germ can be continued analytically along a curve in Ω, a region, typically $\mathbb{C}$ with a finite set of points removed. In that case, one can connect the removed points by "branch cuts" and get the Riemann surface of the function as a multisheeted object over Ω. In this situation, the elements of the monodromy group permute the sheets.

If $\mathfrak{g}$ is a germ near $z = 0$ which can be continued along each ray from 0 to any $z_1 \in \mathbb{D}_\delta$, then it is easy to see, using Corollary 2.3.9 and Theorem 3.1.1, that the radius of convergence of $\mathfrak{g}$ is at least δ. This shows that if $\mathfrak{g}$ is a germ in $\mathcal{S}(\Omega)$ which can be continued along any curve in Ω and $\mathfrak{g}_1$ is a $\tau_\gamma([\mathfrak{g}])$ and $z_1 = \pi(\mathfrak{g}_n)$, then $\rho(\mathfrak{g}_1)$ is at least $\delta = \operatorname{dist}(z_1, \mathbb{C} \setminus \Omega)$. Therefore, the map from all germs obtained from $\mathfrak{g}$ by continuation is a covering space for Ω with π as covering map. This shows that Theorem 11.2.1 can be proven from the fundamental theorem on covering maps (see Theorem 1.7.2).

For algebraic singularities, that is, if Ω is missing an isolated point, z_0, and f returns to its original germ after n loops, one can show that one can add a single point to the Riemann surface with $\pi(p) = z_0$ and $(z - z_0)^{1/n}$ as a local coordinate. π is no longer a covering map, but it is an analytic function between Riemann surfaces.

Cauchy apparently included some results on ODEs in his 1825–26 École Polytechnique lectures which were not widely available. In 1835, he wrote a paper [**105**] that summarized these results.

Problems

1. (a) Prove that if a germ, $\mathfrak{g}$, can be continued along a curve, γ, there is another curve, $\tilde{\gamma}$, which is a piecewise linear curve, with each segment of rational length and each slope a rational multiple of 2π, so that if $z_1 = \gamma(1)$, $\tilde{z}_1 = \tilde{\gamma}(1)$, then z_1 is in the circle of convergence of $\tilde{\mathfrak{g}} \equiv \tau_{\tilde{\gamma}}(\mathfrak{g})$ and $\tau_\gamma(\mathfrak{g})$ is the germ at z_1 of the function defined by the Taylor series of $\tilde{\mathfrak{g}}$.

   (b) Prove that the connected components of $\mathcal{S}(\Omega)$ defined by an initial germ, $\mathfrak{g}$, is a separable metric space.

2. Verify that the maps, π, on the collection $\mathcal{U}_r$, define an analytic structure on $\mathcal{R}$.

3. This problem will prove Theorem 11.2.3. For notational simplicity, take $z_0 = 0$.

   (a) Prove that the differential equation (11.2.4)/(11.2.5) is equivalent to a first-order vector-valued equation

$$y'(z) = A(z)y(z), \qquad y(z_0) = y^{(0)} \tag{11.2.7}$$

   where y is an n-component vector and A an $n \times n$ matrix with matrix elements also analytic in $\mathbb{D}_\delta(0)$. (*Hint*: $y(z) = (f(z), \dots, f^{(n-1)}(z_0))$.)

(b) At the level of formal power series, if

$$A(z) = \sum_{n=0}^{\infty} A_n z^n, \qquad y(z) = \sum_{n=0}^{\infty} y_n z^n \tag{11.2.8}$$

prove that (11.2.7) is equivalent to

$$y_{n+1} = (n+1)^{-1} \sum_{j=0}^{n} A_j y_{n-j}, \qquad y_0 = y^{(0)} \tag{11.2.9}$$

and that if the solution, $\{y_n\}_{n=0}^{\infty}$, of (11.2.10) defines a convergent power series in all of $\mathbb{D}_\delta(0)$, then Theorem 11.2.3 is proven.

(c) Pick $0 < \rho < r < \delta$ and suppose $\|A_n\| \le ar^{-n}$ for all n. Let $x_n = y_n \rho^{-n}$. Prove that

$$|x_{n+1}| \le (n+1)^{-1} a\rho \left(1 - \frac{\rho}{r}\right)^{-1} \max_{0 \le j \le n} |x_j| \tag{11.2.10}$$

and conclude that $\sup_n |x_n| < \infty$, so that the power series for y_n converges for $|z| < \rho$.

(d) Complete the proof of Theorem 11.2.3.

11.3. Picard's Theorem via the Elliptic Modular Function

Here are two spectacular theorems that we'll prove in this section:

Theorem 11.3.1 (Picard's Little Theorem). *A nonconstant entire function takes every value with perhaps one exception.*

The example $f(z) = e^z$, which is never zero, shows there can be the one exceptional value.

Theorem 11.3.2 (Picard's Great Theorem). *In any neighborhood of an isolated essential singularity, a function f takes every value with at most one exception infinitely often.*

Remarks. 1. This is also known as Picard's Big Theorem.

2. This is a vast and deep generalization of the Casorati–Weierstrass theorem (Theorem 3.8.2); it was proven only eleven years later than Casorati's publication.

Notice that the great theorem implies the little theorem, since if $f(z)$ is entire and nonconstant, then either f is a nonconstant polynomial which takes every value by the fundamental theorem of algebra (Theorem 3.1.11) or else $f(1/z)$ has an essential singularity at zero. That said, we will first prove the little theorem below since the proof will depend on a stronger result we need for the great theorem. However, in our two later proofs of these theorems (see Section 11.4 and Section 12.4 of Part 2B), we'll only prove the great theorem. The key to our proof in this section is the elliptic modular function, as used in the following:

Proposition 11.3.3. *Let γ be a curve in $\mathbb{C}\setminus\{0,1\}$. Then there is a curve $\tilde\gamma$ in $\mathbb{C}_+$ so $\lambda\circ\tilde\gamma=\gamma$.*

Remarks. 1. λ is the elliptic modular function as constructed in Section 8.3 or Section 10.6.

2. This is, of course, just a consequence of the fact that λ is a covering map, but rather than use the theory of covering spaces, we use the properties of λ as stated in Theorem 8.3.1.

Proof. Cover γ with disks $\mathbb{D}_{\delta(z_0)}(\gamma(t))$ so λ has an inverse η_t on $\mathbb{D}_{\delta(t)}(\gamma(t))$. By the paving lemma (Proposition 2.2.7), we can find $t_0=0<t_1<\cdots<t_n=1$, so each $\gamma([t_{j-1},t_j])\subset U_j$, one of the $\mathbb{D}_{\delta(s_j)}(\gamma(s_j))$. Let $\eta_j\equiv\eta_{s_j}$. Define $\tilde\gamma$ inductively as follows. Set $\tilde\gamma$ on $[t_0,t_1]$ to $\eta_1(\gamma(t))$. Since $\lambda(\eta_1(\gamma(t_1)))=\lambda(\eta_2(\gamma(t_1)))$, there is $f_1\in\Gamma(2)$, so $f_1(\eta_2(\gamma(t_1)))=\eta_1(\gamma(t_1))$. Define $\tilde\gamma$ on $[t_1,t_2]$ as $f_1(\eta_2(\gamma(t)))$. Having defined $\tilde\gamma$ on $[t_{j-1},t_j]$ and $f_1,\dots,f_{j-1}$ inductively, find $f_j\in\Gamma(2)$ so $f_j(\eta_{j+1}(\gamma(t_j)))=f_{j-1}(\eta_j(\gamma(t_j)))$ and define $\tilde\gamma$ on $[t_j,t_{j+1}]$ as $f_j(\eta_{j+1}(\gamma(t)))$. □

Proposition 11.3.4. *Let $\Omega\subset\mathbb{C}$ be simply connected and let $f\colon\Omega\to\mathbb{C}\setminus\{0,1\}$ be analytic. Then there exists an analytic $g\colon\Omega\to\mathbb{C}_+$ so that*

$$\lambda(g(z))=f(z) \tag{11.3.1}$$

for all $z\in\Omega$.

Remark. Again, one can use the fundamental lifting theorem for covering spaces instead of the argument below.

Proof. Fix $z_0\in\Omega$. Let γ_0 be a path in Ω with $\gamma_0(0)=z_0$. Pick $w_0\in\mathbb{C}_+$ with $\lambda(w_0)=f(z_0)$. Let $\gamma(z)=f(\gamma_0(t))$ so γ is a curve in $\mathbb{C}\setminus\{0,1\}$. Let $\tilde\gamma$ be a curve in $\mathbb{C}_+$ with $\lambda(\tilde\gamma(t))=\gamma(t)$ as constructed in Proposition 11.3.3.

Since $\lambda(\tilde\gamma(0))=\gamma(0)=f(z_0)$, we have $\lambda(w_0)=\lambda(\tilde\gamma(0))$, so there is $\alpha\in\Gamma(2)$ with $\alpha(\tilde\gamma(0))=w_0$. λ is analytic in a neighborhood of $\tilde\gamma$, so if we define $g(\gamma_0(t))=\lambda(\alpha(\tilde\gamma(t)))$, g will be analytic in a neighborhood of γ_0. Thus, g defined near z_0 as that inverse under λ of $f(z)$ can be continued along any curve. By the monodromy theorem (Theorem 11.2.1), g can be defined on all of Ω as a single-valued function which obeys (11.3.1). □

Proof of Theorem 11.3.1. Suppose α and β are two values not taken. By replacing f by $(f(z)-\alpha)/(\beta-\alpha)^{-1}$, we can suppose $\alpha=0$, $\beta=1$.

Pick $\Omega=\mathbb{C}$ in Proposition 11.3.3 and so find $g\colon\Omega\to\mathbb{C}_+$ with $\lambda\circ g=f$. Since g is entire and $(g(z)+i)^{-1}$ has $|(g(z)+i)^{-1}|\le 1$, $(g(z)+i)^{-1}$ is constant and obviously nonzero, so $g(z)$ is constant. Thus, $f(z) = \lambda(g(z))$ is constant. $\square$

Proposition 11.3.5. *Let $\{f_n\}_{n=1}^\infty$ be a sequence of functions in $\mathfrak{A}(\Omega)$ so that each f_n takes values in $\mathbb{C} \setminus \{0,1\}$, and for some $z_0 \in \Omega$, $f_n(z_0)$ has a finite limit in $\mathbb{C} \setminus \{0,1\}$. Then we can pass to a subsequence, f_{n_j}, which converges to some f in $\mathfrak{A}(\Omega)$ uniformly on compact subsets of Ω.*

Proof. It is easy to pick open disks, $\{U_m\}_{m=1}^\infty$ in Ω which obey: (i) $z_0 \in U_1$; (ii) $\bigcup_{m=1}^\infty U_m = \Omega$; (iii) $U_{m+1} \cap \bigcup_{j=1}^m U_j \neq \emptyset$ (see Problem 1).

Let $\tilde{\lambda}\colon \mathbb{D} \to \mathbb{C} \setminus \{0,1\}$ by

$$\tilde{\lambda}(z) = \lambda\biggl(-i\biggl(\frac{z-1}{z+1}\biggr)\biggr) \tag{11.3.2}$$

Let $f_n(z_0) = w_n$ and let $w_n \to w_\infty$. Pick $y_n \in \mathbb{D}$ so $\tilde{\lambda}(y_n) = w_n$ and then $y_n \to y_\infty$, so $\tilde{\lambda}(y_\infty) = w_\infty$. Pulling back Proposition 11.3.4 to $\mathbb{D}$, we can find $g_n\colon U_0 \to \mathbb{D}$, so $g_n(z_0) = y_n$ and $\tilde{\lambda}(g_n(z)) = f_n(z)$.

Since the g_n are uniformly bounded on U_1, by Montel's theorem (Theorem 6.2.2), we can find g_∞ analytic in U_1 and a subsequence so $g_{n_j} \to g_\infty$ uniformly on compact subsets of U_1. Since $g_\infty(z_0) = y_\infty \in \mathbb{D}$, by the maximum principle, $\operatorname{Ran}(g_\infty) \subset \mathbb{D}$. By continuity of $\tilde{\lambda}$ on $\mathbb{D}$, $f_{n_j} \to f_\infty \equiv \tilde{\lambda} \circ g_\infty$ on compacts of U_1.

Since $U_1 \cap U_2 \neq \emptyset$, find z_1 in the intersection and apply the above argument to U_2. Since f_{n_j} converges uniformly on $U_1 \cap U_2$, so do the g_{n_j} constructed in this step. Thus, by the Vitali convergence theorem (Theorem 6.2.8), g_{n_j}, and so $f_{n_j} = \tilde{\lambda} \circ g_{n_k}$ converges uniformly on compacts of U_2, and so f_{n_j} converges on compacts of $U_1 \cup U_2$.

Iterating this argument completes the proof. $\square$

This last proposition is a special case of the Montel three-value theorem. While it alone can be used to prove the big Picard theorem, with one extra trick, we can get the full Montel three-value theorem (see the discussion after Theorem 6.2.10 for why it is called "three-value"):

Theorem 11.3.6 (Montel Three-Value Theorem). *Let $\alpha \neq \beta$ lie in $\mathbb{C}$. The set of functions $f \in \mathfrak{A}(\Omega)$ with* $\operatorname{Ran}(f) \subset \mathbb{C} \setminus \{\alpha, \beta\}$ *is a normal family.*

Proof. By an argument that follows that of Proposition 11.3.5, we need only prove this when Ω is a disk (Problem 2) and by replacing f_n by $(f_n - \alpha)/(\beta - \alpha)$, we can suppose $\alpha = 0$, $\beta = 1$. Thus, we are reduced to showing that any sequence of functions, f_n, in $\mathfrak{A}(\mathbb{D})$ with values in $\mathbb{C} \setminus \{0,1\}$ either has a limit in $\mathfrak{A}(\mathbb{D})$ or converges uniformly to ∞.

Since $\widehat{\mathbb{C}}$ is compact, by passing to an initial subsequence, we can suppose $f_n(0) \to w_\infty \in \widehat{\mathbb{C}}$. If $w_\infty \in \mathbb{C} \setminus \{0, 1\}$, then by Proposition 11.3.5, f_n has a convergent subsequence.

Suppose next that $w_\infty = 1$. Since f_n is never zero, we can let $g_n = \sqrt{f_n}$, taking the square root with $f_n(0)$ near -1. Then g_n has a convergent subsequence, and thus, so does $f_n = g_n^2$. Indeed, by Hurwitz's theorem (Corollary 6.4.2), the subsequence must converge to $f_\infty \equiv 1$.

If $w_\infty = 0$, $g_n = 1 - f_n$ has a subsequence converging, and then $f_\infty \equiv 0$ by the above. Finally, if $w_\infty = \infty$, $g_n = 1 - f_n^{-1}$ has a convergent subsequence and $f_\infty \equiv \infty$ by the above. □

Proof of Theorem 11.3.6 ⇒ Theorem 11.3.2. Define g_n on $\mathbb{A}_{\frac{1}{2},2}$ by $g_n(z) = f(4^{-n}\rho z)$ where ρ is such that f is analytic in $\mathbb{D}_{\frac{1}{2}\rho}(0) \setminus \{0\}$. By Theorem 11.3.6, there is a subsequence g_{n_j}, so either g_{n_j} converges uniformly on $\partial\mathbb{D}$ or $g_{n_j}^{-1}$ converges to 0. In particular, either

$$\sup_{\substack{|z|=1 \\ j}} |g_{n_j}(z)| = S < \infty \quad \text{or} \quad \lim_{\substack{|z|=1 \\ j\to\infty}} |g_{n_j}(z)^{-1}| = 0$$

In the first case, f is uniformly bounded on the circles of radius $4^{-n_j}\rho$. But, by the maximum principle, the maximum on each annulus $\mathbb{A}_{4^{-n_{j-1}}\rho, 4^{-n_j}\rho}$ occurs on one of these circles, so f is bounded near 0 and so has a removable, not essential, singularity.

In the second case, applying the argument to f^{-1}, we show (recall f is never 0) $\lim_{z\to 0} f(z)^{-1} = 0$, so 0 is a polar singularity, not an essential singularity. □

Notes and Historical Remarks. Picard's theorems were proven, one shortly after the other, in 1879 (announcements in [**431, 432**] and full paper in [**433**]). His proofs relied on the elliptic modular function, albeit without the language of normal families. For a modern version of the proof in language closer to Picard, see Ullrich [**559**] and Veech [**562**]. See the Notes to Section 6.2 for the history of Montel's three-value theorem.

Schiff [**503**] has extensive discussion of the history of normal families and various wrinkles in the proofs in this section.

Julia [**295**] has proven the following strengthening of the great Picard theorem:

Theorem 11.3.7 (Julia's Theorem). *Let f be analytic in some $\mathbb{D}_\rho \setminus \{0\}$ and have an essential singularity at zero. Then there exists a direction $e^{i\theta_0} \in \partial\mathbb{D}$ so that for every $\varepsilon > 0$, $f(z)$ takes infinitely often every value, with at most one exception on the set $\{z = re^{i\theta} \mid 0 < r < 1,\ |\theta - \theta_0| < \varepsilon\}$.*

For a proof that uses no more than Montel's three-value theorem, see Markushevich [**373**, Ch. III, p. 344], Saks–Zygmund [**500**, p. 352], or Schiff [**503**, p. 62].

Picard's theorem has evoked considerable interest in regard to extending it to other contexts. One extension involves projective curves, that is, maps of $\mathbb{C}$ to $\mathbb{CP}(n)$. $\mathbb{CP}(1) = \widehat{\mathbb{C}}$, so Picard's theorem deals with the case $n = 1$. A *hyperplane*, H, in $\mathbb{CP}(n)$ is parameterized by $a \in \mathbb{C}^{n+1} \setminus \{\infty\}$ via

$$H = \left\{ [(z_1, \dots, z_{n+1})] \,\middle|\, \sum_{j=1}^{n+1} a_j z_j = 0 \right\} \tag{11.3.3}$$

For $n = 1$, Picard's theorem says if H_1, H_2, H_3 are three distinct hyperplanes (hyperplanes are points) and $f\colon \mathbb{C} \to \mathbb{CP}[1]$ has $\operatorname{Ran}(f) \cap H_j = \emptyset$, then f is constant. Cartan [**95, 98**] has proven that if $\{H_j\}_{j=1}^{2n+1}$ are $2n+1$ hypersurfaces in $\mathbb{CP}[n]$ with

$$\bigcap_{j\in I} H_j = \emptyset \tag{11.3.4}$$

for any $I \subset \{1, \dots, 2n+1\}$ with $\#(I) = n+1$ and if $f\colon \mathbb{C} \to \mathbb{CP}[n]$ is holomorphic with $\operatorname{Ran}(f) \cap H_j = \emptyset$ for $j = 1, \dots, 2n+1$, then f is constant. See also [**166, 167**]. Cartan proved and used an extension of Nevanlinna theory. The reader will prove this result in Problem 2 of Section 3.3 of Part 3.

There is also a Picard-type theorem for so-called quasiregular maps on $\mathbb{R}^n$; see Rickman [**479**] and Eremenko–Lewis [**168**].

Emile Picard (1856–1941) was a French mathematician whose life was impacted by war. His father died during the 1870 siege of Paris in the Franco–Prussian war and his mother struggled to put her two sons through lycée. Both of his sons were killed on the front in the First World War (Picard's student, Hadamard, also lost two sons in that war).

Picard spent almost his entire career in Paris, although his wonderful theorems we discuss in this section were done during a two-year period when he taught in Toulouse. His advisor was Darboux and, in turn, his students include Bernstein, Hadamard, Julia, Painlevé, and Weil.

His signature theorems relied on the elliptic modular function used earlier by his teacher and father-in-law, Hermite, in his solution of quintics (although Picard did this work two years before he married Hermite's daughter). Picard is also noted for his work on ordinary differential equations, in particular, his existence theorem that relied on iteration of the integral equation.

Problems

1. Let $z_0 \in \Omega \subset \mathbb{C}$ be a region. Prove there exist open disks $\{U_m\}_{m=1}^\infty$ covering Ω so $z_0 \in U_1$ and so $U_{m+1} \cap \bigcup_{j=1}^m U_j \neq \emptyset$.

2. Let $\{f_n\}_{n=1}^\infty$ be a sequence of functions analytic on $\Omega \subset \mathbb{C}$. Let $z_0 \in \Omega$ and let $\{U_n\}_{n=1}^\infty$ be a cover as constructed in Problem 1 and so that any subsequence of the f_n has a subsubsequence converging in $\mathfrak{A}(U_n)$. Prove there is a subsequence converging in $\mathfrak{A}(\Omega)$.

11.4. Bonus Section: Zalcman's Lemma and Picard's Theorem

In the twenty years following 1896, there appeared a number of approaches to Picard's theorems that replaced elliptic functions with rather explicit bounds on families of functions. Here we'll present a rather recent twist on these ideas that avoids explicit bounds by using an elegant lemma of Zalcman. We'll heavily use bonus Section 6.5 via the use of the spherical metric, the spherical derivative, $f^\sharp$, of (6.5.25) and Marty's theorem (Theorem 6.5.6). Here is the key construct:

Proposition 11.4.1 (Zalcman's Lemma). *Let S be a family of meromorphic functions on a region, $\Omega \subset \mathbb{C}$, which is not normal. Then there exists $\{f_n\}_{n=1}^\infty \subset S$, $z_n \in \Omega$, $z_n \to z_\infty \in \Omega$, and nonnegative numbers $\rho_n \to 0$ so that if*

$$g_n(\zeta) = f_n(z_n + \rho_n \zeta) \tag{11.4.1}$$

then for some entire meromorphic function, g_∞, we have

(i) *$g_n \to g_\infty$ normally on each $\mathbb{D}_R$ with $R < \infty$.*
(ii) *$g_\infty^\sharp(0) = 1$ so, in particular, g_∞ is not constant.*
(iii) *$g_\infty^\sharp(\zeta) \le 1$ for all $\zeta \in \mathbb{C}$.*

Moreover, if $V \subset \widehat{\mathbb{C}}$ is such that $\mathrm{Ran}(f_n) \subset V$ *for all n, then* $\mathrm{Ran}(g_\infty) \subset V$.

Proof. The key to the proof is that

$$g_n^\sharp(\zeta) = \rho_n f_n^\sharp(z_n + \rho_n \zeta) \tag{11.4.2}$$

Once one has the idea of looking at $\max((1-|z|^2)f_n^\sharp(z))$, the details are straightforward—at least once one knows the result is true. Throughout, we pick subsequences but continue to use f_n for the subsequences.

By Marty's theorem and the assumed non-normality, there exists a compact $K \subset \Omega$ so that $\sup_{f\in S,\, z\in K}\|f^\sharp(z)\| = \infty$. Thus, there is $w_n \to w_\infty \in K$ and $f_n \in S$ so that

$$f_n^\sharp(w_n) \ge n \tag{11.4.3}$$

For simplicity of notation, suppose $w_\infty = 0$ and that $\overline{\mathbb{D}} \subset \Omega$. By translation and scaling, we can certainly arrange that. Define

$$M_n = \sup_{|z|\le 1} (1-|z|^2) f_n^\sharp(z) \tag{11.4.4}$$

By (11.4.3), $M_n \geq (1 - |w_n|^2)n \to \infty$ since $w_n \to 0$. Moreover, since the function of z, whose sup is taken, is continuous and vanishing on $\partial\mathbb{D}$, there is $z_n \in \mathbb{D}$ so $(1 - |z_n|^2)f_n^\sharp(z_n) = M_n$, and by passing to a subsequence, that $z_n \to z_\infty \in \overline{\mathbb{D}} \subset \Omega$.

Define ρ_n by

$$\rho_n = [f^\sharp(z_n)]^{-1} = M_n^{-1}(1 - |z_n|^2) \leq M_n^{-1} \to 0 \tag{11.4.5}$$

Moreover,

$$(1 - |z_n|)^{-1}\rho_n = M_n^{-1}(1 + |z_n|) \leq 2M_n^{-1} \tag{11.4.6}$$

Thus, for R fixed and n so large that $R < \frac{1}{2}M_n$, we have

$$|\zeta| \leq R \Rightarrow |z_n + \rho_n\zeta| < 1 \tag{11.4.7}$$

Define g_n by (11.4.1) on the set where $|z_n + \rho_n\zeta| < 1$. By (11.4.2),

$$|g_n^\sharp(0)| = \rho_n[f_n^\sharp(z_n)] = 1 \tag{11.4.8}$$

and

$$\begin{aligned}
|g_n^\sharp(\zeta)| &\leq \rho_n f_n^\sharp(z_n + \rho_n\zeta) \\
&\leq \rho_n M_n(1 - |z_n + \rho_n\zeta|^2)^{-1} \\
&= \frac{1 - |z_n|^2}{1 - |z_n + \rho_n\zeta|^2} \qquad (11.4.9) \\
&\leq \frac{1 + |z_n|}{1 + |z_n| + \rho_n|\zeta|}\,\frac{1 - |z_n|}{1 - |z_n| - \rho_n|\zeta|} \qquad (11.4.10) \\
&\leq \frac{1}{1 - 2|\zeta|M_n^{-1}} \qquad (11.4.11)
\end{aligned}$$

so long as $|\zeta| < \frac{1}{2}M_n$. In the above, we go from (11.4.9) to (11.4.10) using $|z_n+\rho_n\zeta|^2 \leq (|z_n|+\rho_n|\zeta|)^2$ and $(1-a^2) = (1-a)(1+a)$. We go from (11.4.10) to (11.4.11) since the first term is bounded by 1 and $\rho_n|\zeta|(1 - |z_n|)^{-1} \leq 2|\zeta|M_n^{-1}$ by (11.4.6).

Thus, we have proven for each fixed R,

$$\lim_{n\to\infty} \sup_{|\zeta|\leq R} g_n^\sharp(\zeta) = 1 \tag{11.4.12}$$

That means, by Marty's theorem, we can arrange g_∞ meromorphic on $\mathbb{C}$ and a subsequence, also denoted g_n, so that $g_n \to g_\infty$ uniformly on compacts of $\mathbb{C}$. Moreover, (ii) and (iii) hold by (11.4.8) and (11.4.12).

Since $g_\infty^\sharp(0) = 1$, $g_\infty^\sharp$ is not constant, so by Hurwitz's theorem, g_∞ cannot take any value not taken by any f_n. In particular, g_∞ does not take the value ∞ if all $f \in S$ are analytic. $\square$

Proof of Montel's Three-Value Theorem. By Problem 2 of Section 11.3, we need only prove the result for simply connected Ω. Fix α, β, γ distinct in $\widehat{\mathbb{C}}$ and let $\mathcal{M}_{\alpha,\beta,\gamma}(\Omega)$ be the set of all meromorphic functions (or $f \equiv \infty$) that never take the value α, β, or γ. We need to prove $\mathcal{M}_{\alpha,\beta,\gamma}(\Omega)$ is normal. If F is any element of $\mathbb{A}\mathrm{ut}(\widehat{\mathbb{C}})$, it is easy to see that $\mathcal{M}_{F(\alpha),F(\beta),F(\gamma)}(\Omega)$ is normal if and only if $\mathcal{M}_{\alpha,\beta,\gamma}(\Omega)$ is (Problem 1). So without loss, we can take $\alpha = 0$, $\beta = 1$, $\gamma = \infty$, that is, look at S, those $f \in \mathfrak{A}(\Omega)$ which are never 0 or 1. Let S_k be a set of functions in $\mathfrak{A}(\Omega)$ that never take the value 0 or any of 2^k-th roots of unity.

Fix k for a moment. If f_n is a sequence in S, pick $g_n = f_n^{1/2^k}$ where we take any of the analytic 2^k-th roots guaranteed by Theorem 2.6.1 and the assumption that Ω is simply connected. If g_{n_j} is convergent, so is $f_{n_j} = g_{n_j}^{2^k}$. So if S is not normal, so is each S_k.

If S_k is not normal, by Zalcman's lemma, there exist entire functions, $g_\infty^{(k)}$, obeying (i) $(g_\infty^{(k)})^\sharp(0) = 1$; (ii) $(g_\infty^{(k)})^\sharp(z) \leq 1$ for all z; (iii) $g_\infty^{(k)}$ does not take the value 0 or any 2^k-th root of unity.

By Marty's theorem, there is a subsequence converging to an entire function g_∞ with (i) $g_\infty^\sharp(0) = 1$ (so g_∞ is not constant); (ii) $g_\infty^\sharp(z) \leq 1$ for all z; (iii) by Hurwitz's theorem, g_∞ does not take any value $e^{i\theta}$ where $\theta/2\pi$ is a dyadic rational. By the open mapping theorem, $|g_\infty(z)| \neq 1$ for all z. Thus, by continuity, either $|g_\infty(z)| < 1$ for all z or $|g_\infty(z)^{-1}| < 1$ for all z. Either way, by Liouville's theorem, g_∞ is constant, inconsistent with $g_\infty^\sharp(0) = 1$. This contradiction shows that S is normal. $\square$

Once we have Montel's three-value theorem, we get Picard's theorems as in Section 11.3.

Notes and Historical Remarks. These Notes will discuss what have come to be called "elementary" proofs of the Picard theorems, a term that seems to go back to Borel [**65**]. "Elementary" does not mean simple, less involved, or more natural. For the prime number theorem, the phrase is used for proofs that do not use complex analysis. It is difficult to imagine proofs of Picard's theorems that don't use complex analysis (but see Section 3.3 of Part 3). The term seems to mean proofs that don't use "fancy stuff"—initially that meant not using elliptic functions. Somehow the Ahlfors–Robinson proof (see Sections 12.3 and 12.4 of Part 2B), which uses curvature in conformal Riemann metrics, doesn't qualify for "elementary"!

Traditionally, it applied to proofs of Landau, Schottky, and Bloch that we turn to below. Those traditional proofs are not especially simpler than the elliptic modular function or Ahlfors–Robinson proofs. The proof via Zalcman's lemma is their descendant and it is undoubtedly the shortest proof from first principles (although the Nevanlinna theory (see Section 17.3 of Part 2B) and the Eremenko–Sodin proof (see Section 3.3 of Part 3) are fairly direct). One can argue that the essence of the Picard theorem is that $\mathbb{C}\setminus\{0,1\}$ has a universal cover that supports bounded nonconstant functions and, in that sense, the proof that relies on elliptic modular functions is most elementary.

Zalcman's lemma is from a 1975 paper [**600**]. Not surprisingly, it generated considerable followup, summarized in Zalcman [**601**]. Earlier, Lohwater–Pommerenke [**365**] had used a similar construction but for single functions on $\mathbb{D}$ obeying $\sup_{|z|}(1-|z|^2)f^\sharp(z)<\infty$ (such functions are called *normal functions*) rather than families. The elegant argument we use to prove Montel's three-value theorem from Zalcman's lemma is from Ros [**491**]. The general idea of multiplying by $(1-|z|^2)$ used in the proof of Zalcman's lemma is due to Landau [**341**].

Zalcman's lemma can be used to prove other normality results, for example (see Problem 2), the following theorem of Carathéodory [**90**]: If S is a family of meromorphic functions on $\Omega\subset\mathbb{C}$, a region so that there are three values $\alpha(f),\beta(f),\gamma(f)$, not taken by f (but f-dependent) and

$$\inf_{f\in S}\,[\min(\sigma(\alpha(f),\beta(f)),\sigma(\beta(f),\gamma(f)),\sigma(\gamma(f),\alpha(f)))]>0 \tag{11.4.13}$$

then S is normal.

There is a general principle that has been used as a guide to the study of normality—often called *Bloch's principle.* The idea is that if a property forced on entire functions causes the function to be constant, then that property on a family of functions on a region forces the family to be normal. Schiff [**503**] presents an explicit version of this, proved by Zalcman who was motivated in part by trying to provide a basis for this principle as suggested by Abraham Robinson. The name "Bloch's principle" comes from the statement in his paper [**55**] (the original quote is in Latin): "There is nothing in the infinite which did not exist before in the finite."

The first proof of a Picard theorem without elliptic modular functions was by E. Borel [**65**] in 1896. Seventeen years after Picard's work, he found an "elementary" proof of the little Picard theorem. It was Schottky [**506**] in 1904, motivated in part by the earlier work of Borel and of Landau [**338**], who found the first "elementary" proof of the great Picard theorem; he showed:

Theorem 11.4.2 (Schottky's Theorem). *For each $a\in\mathbb{C}$ and $\theta\in(0,1)$, there is a constant $C(a,\theta)$ so that if $f\in\mathfrak{A}(\mathbb{D}_R)$ with $f(0)=a$ and* $\operatorname{Ran}(f)\subset\mathbb{C}\setminus\{0,1\}$, *then*

$$\sup_{|z|\le\theta R}|f(z)|\le C(a,\theta) \tag{11.4.14}$$

It is fairly easy to see (Problem 3) that this result is equivalent to Montel's three-value theorem, so it is not difficult to believe it led to a proof of the great Picard theorem.

A new element was introduced by Bloch [**54**] in 1924 (when he published an announcement). He proved the remarkable

Theorem 11.4.3 (Bloch's Theorem). *For any nonconstant analytic function f in a neighborhood of $\overline{\mathbb{D}}$, let $b(f)$ be the* sup *of the radii of all disks, $\mathbb{D}_{f(z_0)}(\rho)$, in* Ran$(f)$ *which are a bijective image of f restricted to a subdomain of $\mathbb{D}$. Then there is a lower bound, b, on $\{b(f) \mid f'(0) = 1\}$.*

We emphasize that while the normalization is at 0, the z_0 need not be $f(0)$ and, indeed, there is no uniform lower bound on disks about $f(0)$ which are bijective images. Bloch also relates this result to Picard's little theorem. b is called *Bloch's constant.* A related number is *Landau's constant* (named after Landau [**341**]), the inf of the order of the radius of largest disk in Ran(f), known to be strictly larger than Bloch's constant. There is considerable literature on these constants, including conjectured values; see Finch [**188**, Sec. 7.1].

For further discussion of Schottky's and Bloch's theorems, including proofs, see Conway [**120**], Remmert [**477**], and Titchmarsh [**556**].

André Bloch (1893–1949) had a tragic life. In 1917, while on leave from the French army, he murdered his brother, his aunt, and his uncle. All his papers were written while he was resident in an asylum for the criminally insane.

Problems

1. Let f_n be a sequence of meromorphic functions on Ω. Let F be a fixed element of $\mathbb{A}\mathrm{ut}(\widehat{\mathbb{C}})$. Prove $F \circ f_n$ has a normal limit (in the sense of Section 6.5) if and only if f_n does.

2. Let S be a family of functions on a region, Ω, so that each $f \in S$ is missing three values, $\alpha(f), \beta(f), \gamma(f)$, possibly f-dependent, but so that (11.4.13) holds. Prove the theorem of Carathéodory that S is normal. (*Hint*: Use Zalcman's lemma but show, by passing to a subsequence, you can arrange that $\alpha(g_n), \beta(g_n), \gamma(g_n)$ each have limits in $\widehat{\mathbb{C}}$ with the ρ-metric and these limits are unequal. Then use the little Picard theorem on the limit.)

3. (a) Use the Montel three-value theorem to prove Schottky's theorem (Theorem 11.4.2).

   (b) Conversely, assuming Schottky's theorem, prove the three-value theorem.

4. Let f be an entire function and suppose $f \circ f$ has no fixed points. You'll prove that $f(z) = z + b$ for some b.

   (a) Show $g(z) = (f(f(z)) - z)/(f(z) - z)$ is an entire function which never takes the value 0 or 1 and so is a nonzero constant c.

   (b) Prove that $f'(f(z))$ never takes the values c or 0 (*Hint*: Prove that $f'(z)[f'(f(z)) - c] = 1 - c$) and conclude that $f(z) = az + b$ for some a, b.

   (c) Prove that $f(z) = z + b$.

5. Prove that if f and g are entire functions and $e^f + e^g = 1$, then f and g are both constant.

6. Consider the family on $\mathbb{C}$, $S = \{f_n(z)\}_{n=0,1,2,\dots}$ where $f_n(z) = z^n$.

   (a) Prove that S is not normal.

   (b) Prove that in Zalcman's lemma, one can pick $z_n \equiv 1$, $\rho_n = 2/n$. What is g_∞? Check that this g_∞ obeys (ii) and (iii) of Zalcman's lemma.

11.5. Two Results in Several Complex Variables: Hartogs' Theorem and a Theorem of Poincaré

Here we want to focus on two results about several complex variables. In $\mathbb{C}$, a fundamental role is played by the unit disk $\mathbb{D}$. On $\mathbb{C}^n$, one issue is that there are (at least) two natural candidates to replace $\mathbb{D}$:

Definition. In $\mathbb{C}^n$, the *polydisk*, $\mathbb{D}^n$, is defined by

$$\mathbb{D}^n = \{z = (z_1, \dots, z_n) \in \mathbb{C}^n \mid \max_{j=1,\dots,n} |z_j| < 1\}$$

and, more generally, $\mathbb{D}_r^n(z_0)$ is defined by making the condition $\max_{j=1,\dots,n} |z_j - z_j^{(0)}| < r$. The *ball* is defined by

$$\mathbb{B}^n = \left\{ z \in (z_1, \dots, z_n) \in \mathbb{C}^n \;\middle|\; |z| = \left(\sum_{j=1}^n |z_j|^2 \right)^{1/2} < 1 \right\}$$

A region, $\Omega \subset \mathbb{C}^n$, is an open connected set and $f \colon \Omega \to \mathbb{C}$ is called *analytic* on Ω if there is a convergent multivariable power series about each $z^{(0)} \in \Omega$, that is, for some $r > 0$ and all $z \in \mathbb{D}_r(z^{(0)})$, we have

$$f(z) = \sum_{j_1,\dots,j_n=0}^{\infty} a_{j_1 \dots j_n} (z_1 - z_1^{(0)})^{j_1} \dots (z_n - z_n^{(0)})^{j_n}$$

converging uniformly on $\mathbb{D}_r(z^{(0)})$. An analytic or conformal equivalence is a bijection $\Omega_1 \to \Omega_2$ so that each component is analytic.

Many results from one variable extend (see Problems 9 and 10). We want to focus here on two significant differences:

Theorem 11.5.1 (Hartogs' Theorem). *If f is analytic on $\mathbb{D}_1^2(0) \setminus \overline{\mathbb{D}_{1/2}^2(0)}$, then f has a continuation to all of $\mathbb{D}_1^2(0)$.*

Theorem 11.5.2 (Poincaré's Theorem). *$\mathbb{B}^2$ and $\mathbb{D}^2$ are not conformally equivalent.*

Of course, the situation in $\mathbb{C}$ is very different. If $f \in \mathfrak{A}(\mathbb{D})$ has a natural boundary on $\partial\mathbb{D}$, then $g(z) = f(z) - f(\frac{1}{2}z^{-1})$ is analytic on $\mathbb{A}_{\frac{1}{2},1}$ and no bigger region, and while any two simply connected regions in $\mathbb{C}$ are conformally equivalent, both $\mathbb{B}^2$ and $\mathbb{D}^2$ are simply connected and topologically trivial by all meanings of that notion.

Proof of Theorem 11.5.1. For each $w \in \mathbb{D}$, let $g_w(z) = f(z,w)$ where g is defined on $\mathbb{D}$ for $\frac{1}{2} < |w| < 1$ and on $\mathbb{A}_{\frac{1}{2},1}$ for $|w| \le \frac{1}{2}$. Thus,

$$g_w(z) = \sum_{n=-\infty}^{\infty} a_n(w) z^n \tag{11.5.1}$$

converging for $z \in \mathbb{A}_{\frac{1}{2},1}$ and

$$a_n(w) = \frac{1}{2\pi i} \oint_{|z|=\frac{3}{4}} z^{-n-1} f(z,w)\, dz \tag{11.5.2}$$

By the Cauchy theorem,

$$n \le -1 \quad \text{and} \quad \tfrac{1}{2} < |w| < 1 \Rightarrow a_n(w) = 0 \tag{11.5.3}$$

but by (11.5.2), $a_n(w)$ is analytic for $w \in \mathbb{D}$. It follows that $a_n(w) = 0$ for $n < 1$, so (11.5.1) converges for all $z \in \mathbb{D}$, $w \in \mathbb{D}$, that is, f has a continuation to all of $\mathbb{D}_1^{(2)}(0)$. $\square$

Theorem 11.5.2 is somewhat more subtle. For $0 \in \Omega \subset \mathbb{C}^2$, define $\mathbb{A}\mathrm{ut}_0(\Omega)$ to be the analytic bijections, f, of Ω with $f(0) = 0$. We'll prove that $\mathbb{A}\mathrm{ut}_0(\mathbb{B}^2)$ and $\mathbb{A}\mathrm{ut}_0(\mathbb{D}^2)$ are very different and use that to prove the two regions cannot be conformally equivalent. As preparation,

Theorem 11.5.3 (Cartan's First Theorem). *Let $\Omega \subset \mathbb{C}^2$ be bounded with $0 \in \Omega$ and $f \in \mathbb{A}\mathrm{ut}_0(\Omega)$ obey*

$$Df(0) = \mathbb{1} \tag{11.5.4}$$

Then $f(z) \equiv z$.

Remark. By Df, we mean the 2×2 matrix

$$Df = \begin{pmatrix} \partial f_1/\partial z_1 & \partial f_1/\partial z_2 \\ \partial f_2/\partial z_1 & \partial f_2/\partial z_2 \end{pmatrix} \tag{11.5.5}$$

where f_1, f_2 are the components of f.

Proof. The one-variable result appears in Problem 8 of Section 7.4. This proof is essentially the same; see Problem 1. ☐

Theorem 11.5.4 (Cartan's Second Theorem). *Let $\Omega \subset \mathbb{C}^2$ be a region with $0 \in \Omega$ and the property that $z \in \Omega \Rightarrow e^{i\theta} z \in \Omega$ for all $e^{i\theta} \in \partial\mathbb{D}$. Let $f \in \mathbb{A}\mathrm{ut}_0(\Omega)$. Then for some 2×2 matrix A, $f(z) = Az$.*

Proof. Let $Q_\theta(z) = e^{i\theta} z$. By hypothesis, Q_θ is an automorphism of Ω that leaves 0 fixed. Moreover, $DQ_\theta = \left(\begin{smallmatrix} e^{i\theta} & 0 \\ 0 & e^{i\theta} \end{smallmatrix}\right)$ is a multiple of $\mathbb{1}$. Thus, $D(Q_\theta^{-1} f^{-1} Q_\theta f)(0) = (DQ_\theta)^{-1}(Df)^{-1}(DQ_\theta)(Df) = (Df)^{-1}Df = \mathbb{1}$. So by Theorem 11.5.3, $Q_\theta^{-1} f^{-1} Q_\theta f$ is the identity, that is,

$$Q_\theta f = f Q_\theta \tag{11.5.6}$$

Write $f(z) = (f_1(z), f_2(z))$ and for z near zero,

$$f_j(z) = \sum_{k,\ell=1}^{\infty} b_{k\ell}^{(j)} z_1^k z_2^\ell \tag{11.5.7}$$

Then, by (11.5.6), $e^{i\theta} b_{k\ell}^{(j)} = b_{k\ell}^{(j)} e^{i(k+\ell)\theta}$, which means $b_{k\ell}^{(j)} = 0$ for $(k+\ell) \neq 1$, that is, f is linear near zero. By analyticity, f is linear. ☐

Theorem 11.5.5. *We have that*

$$\mathbb{A}\mathrm{ut}_0(\mathbb{B}^2) = \{f(z) = Az \mid A \in \mathbb{U}(2), \text{ the } 2 \times 2 \text{ unitary matrices}\} \tag{11.5.8}$$

$$\mathbb{A}\mathrm{ut}_0(\mathbb{D}^2) = \left\{ f(z) = Az \,\middle|\, A = \begin{pmatrix} e^{i\theta_1} & 0 \\ 0 & e^{i\theta_2} \end{pmatrix} \text{ or } A = \begin{pmatrix} 0 & e^{i\theta_1} \\ e^{i\theta_2} & 0 \end{pmatrix} \right\} \tag{11.5.9}$$

Remark. In Problem 2, the reader will find $\mathbb{A}\mathrm{ut}(\mathbb{B}^2)$ and $\mathbb{A}\mathrm{ut}(\mathbb{D}^2)$.

Proof. Both domains are invariant under $z \to e^{i\theta} z$ so Theorem 11.5.4 applies. $\mathbb{B}^2$ is the unit ball in Euclidean norm, so (Problem 3) $A[\mathbb{B}^2] = \mathbb{B}^2$ if and only if $\|\mathbb{A}u\|_2 = \|u\|_2$, and that happens if and only if A is unitary. This proves (11.5.8).

Let A: $[\mathbb{D}^2] \subset \mathbb{D}^2$ for a linear map $A = \left(\begin{smallmatrix} a_{11} & a_{12} \\ a_{21} & a_{22} \end{smallmatrix}\right)$. $\mathbb{D}^2$ is the unit ball in $\|z\|_\infty = \max(|z_1|, |z_2|)$ so $\mathbb{A}[\mathbb{D}^2] \subset \mathbb{D}^2$ if and only if $\|Az\|_\infty \leq \|z\|_\infty$, which holds if and only if (see Problem 4),

$$|a_{11}| + |a_{12}| \leq 1, \qquad |a_{21}| + |a_{22}| \leq 1 \tag{11.5.10}$$

On the other hand, since $\mathbb{A}[\mathbb{D}^2] = \mathbb{D}^2$, we must have that $A[\partial\mathbb{D}^2] = \partial\mathbb{D}^2$, so $\|A\binom{1}{0}\|_\infty = 1$, that is,

$$\max(|a_{11}|, |a_{21}|) = 1 = \max(|a_{12}|, |a_{22}|) \tag{11.5.11}$$

These last two equations imply

$$|a_{11}| = |a_{22}| = 1; \ |a_{12}| = |a_{21}| = 0 \text{ or } |a_{12}| = |a_{21}| = 1; \ |a_{11}| = |a_{22}| = 0 \tag{11.5.12}$$

which proves (11.5.9). □

Proof of Theorem 11.5.2. For any $z^{(0)} \in \mathbb{D}^2$, let

$$f_{z^{(0)}}(z) = \left(\frac{z_1 - z_1^{(0)}}{1 - \bar{z}_1^{(0)} z_1}, \frac{z_2 - z_2^{(0)}}{1 - \bar{z}_2^{(0)} z_2} \right)$$

Then $f \in \mathbb{A}\mathrm{ut}(\mathbb{D}^2)$ and $f(z^{(0)}) = 0$. Thus, if $g \colon \mathbb{B}^2 \to \mathbb{D}^2$ is an analytic bijection, $h \equiv f_{g(0)} \circ g$ is an analytic bijection with $h(0) = 0$, so we need only show there is no such analytic bijection.

If h is such a bijection, $F \to hFh^{-1}$ maps $\mathbb{A}\mathrm{ut}_0(\mathbb{B}^2)$ bijectively to $\mathbb{A}\mathrm{ut}_0(\mathbb{D}^2)$. Since

$$D(hFh^{-1})(0) = (Dh)(0)DF(0)Dh(0)^{-1} \tag{11.5.13}$$

this sets up a continuous bijection of $\mathbb{U}(2)$ and $\{A \mid A = \left(\begin{smallmatrix} e^{i\theta_1} & 0 \\ 0 & e^{i\theta_2} \end{smallmatrix}\right)$ or $A = \left(\begin{smallmatrix} 0 & e^{i\theta_1} \\ e^{i\theta_2} & 0 \end{smallmatrix}\right)$. Since one of these sets is connected (Problem 5) and the other is not, this cannot be. This contradiction proves the sets are not analytically equivalent. □

Notes and Historical Remarks. Friedrich Hartogs (1874–1943) was born in Belgium to a prosperous German-Jewish merchant family and spent his youth and career in Germany. He was a student of Pringsheim in Munich. His great paper [**242**], which made him the father of several complex variables, was his Habilitationschrift. Hartogs was also the first to use subharmonic functions in one complex variable theory.

He started teaching in Munich after this work. He was cautious by nature and, in 1922, turned down a full professorship in the new private university in Frankfurt because of the hyperinflation and concerns that the position might be less secure. Finally, in 1927, he was made a professor, after pressure from his colleagues at Munich, Carathéodory, Perron, and Tietze.

With the rise of the Nazis, he was dismissed. In the late thirties, he spent some time in Dachau. His wife was Aryan so he transferred their home to her, but when the laws allowed confiscation of the property of spouses of Jews, he divorced his wife although they continued to live together. Finally, in 1943, concerned about his situation and the future, he committed suicide, like Hausdorff, via an overdose of barbiturates.

His long great paper [**242**] included not only Theorem 11.5.1 but also a theorem that if f is analytic in each variable with the others fixed, it is jointly analytic. This result is often called Hartogs' theorem, and Theorem 11.5.1 called Hartogs' ball theorem (often, even in English, Kügelsatz). The name comes from the fact that the same argument (Problem 7) proves that if f is analytic in a neighborhood of $\partial\mathbb{B}^2$, it has an analytic continuation to all of $\mathbb{B}^2$. In fact [**242**], Hartogs proved by the same method that

Theorem 11.5.6 (Hartogs). *Let Ω be a region in $\mathbb{C}^n$ for $n \geq 2$ and let $K \subset \Omega$ be compact. Then any function analytic in $\Omega \setminus K$ has an analytic continuation to Ω.*

Modern approaches to Hartogs' theorem and its extensions use ideas different from the one we present (which is Hartogs' original idea). One, due to Bochner [**58**] and Martinelli [**374**], depends on multivariable variants of the Pompeiu formula (2.7.7) and first proves that if K is compact and f analytic and bounded on $\mathbb{C}^2 \setminus K$, then f is constant on the unbounded component of $\mathbb{C}^2 \setminus K$. Another idea, due to Ehrenpreis [**159**], uses PDE methods, exploiting special properties of the CR equations in higher dimensions. Range [**469**] has more history and sketches of the ideas. There is a pedagogical discussion in Scheidemann [**502**].

One well-known theorem on analytic continuation in higher dimension, due to Bochner [**57**], concerns tubes, that is, sets of the form

$$\mathcal{J}(K) = \{z \in \mathbb{C}^n \mid \operatorname{Im} z \in K\} \text{ for some } K \subset \mathbb{R}^n \tag{11.5.14}$$

Theorem 11.5.7 (Bochner Tube Theorem). *If f is analytic in $\mathcal{J}(K)$ for some open K in $\mathbb{R}^n$, it has an analytic continuation to the interior of the convex hull of $\mathcal{J}(K)$.*

This result has connections to the Paley–Wiener theorem (see Problem 8).

Note that Theorem 11.5.1 says that all local singularities are removable. In general, zeros and also singularities are $(n-1)$ (complex)-dimension objects that are sometimes hypersurfaces but can be more complicated.

Poincaré's theorem is from [**443**] and Cartan's two theorems from [**96**]. What we called Cartan's first theorem is often called the Cartan uniqueness theorem.

Henri Cartan (1904–2008), who lived to be 104, was the son of Elie Cartan, who is noted for his work on Lie algebras, Lie groups, and differential geometry. Henri's brother served in the French resistance in the Second World War and was captured and executed by the Germans. Henri Cartan was the central figure in several complex variables from 1930 to 1970, with contributions to the more classical aspects but also to sheaf theory and methods of homology. His book with Eilenberg, *Homological Algebra*, was a milestone in more than analysis. He was a founding member of Bourbaki.

He was a student of Montel and, in turn, his students included P. Cartier, J. Cerf, J. Deny, A. Douady, R. Godement, J.-L. Koszul, J.-P. Serre, and R. Thom. The last two both finished in 1951 and both went on to win Fields Medals.

There are a variety of approaches to the study of several complex variables, and the classical considerations of this section are not typical—rather sheaf theories, PDE and integral equation techniques have been used. See [**36, 100, 215, 227, 268, 330, 403, 468, 497, 502**] for monograph presentations.

There is also a theory of analytic functions on infinite-dimensional spaces, going back at least to Volterra [**566**]. For a discussion of the modern view and a history of its development, see Dineen [**143**].

Problems

1. (a) For $z \in \mathbb{D}^n_\delta(0)$, suppose $f(z) = z + A_k(z) + O(|z|^{k+1})$, where $k \geq 2$ and A_k is homogeneous of degree k and nonzero. Prove inductively that $f^{[n]}(z) = z + nA_k(z) + O(|z|^{k+1})$.

   (b) If $\Omega \subset \mathbb{C}^n$ is bounded and $g\colon \Omega \to \Omega$ has $g(z) = \sum_{k=0}^\infty A_k(z)$ as its Taylor expansion at zero, where A_k is homogeneous of degree k, prove a priori bounds (k-dependent but g-independent) on $\|A_k\|$ depending only on Ω. (*Hint*: Cauchy estimates.)

   (c) Prove Cartan's first theorem.

2. (a) For any $w \in \mathbb{D}^2$, prove
$$f_w(z) = \left(\frac{z_1 - w_1}{1 - \bar{w}_1 z_1}, \frac{z_2 - w_2}{1 - \bar{w}_2 z_2}\right)$$
   is in $\mathbb{A}\mathrm{ut}(\mathbb{D}^2)$.

   (b) Prove that every g in $\mathbb{A}\mathrm{ut}(\mathbb{D}^2)$ can be uniquely written as $g(z) = A[F_w(z)]$ where $w \in \mathbb{D}^2$ and $A \in \mathbb{A}\mathrm{ut}_0(\mathbb{D}^2)$.

   (c) For any $a \in \mathbb{D}$, prove that
$$h_a(z) = \left(\frac{z_0 - a}{1 - \bar{a}z_0}, \frac{(1-|a|^2)^{1/2} z_2}{1 - \bar{a}z_1}\right)$$
   is in $\mathbb{A}\mathrm{ut}(\mathbb{B}^2)$.

   (d) Prove that any g in $\mathbb{A}\mathrm{ut}(\mathbb{B}^2)$ can be written $g(z) = Uh_a(Vz)$ for 2×2 unitary U, V and $a \in \mathbb{D}$.

   Remark. This decomposition is not unique.

3. Let A be a linear map on $\mathbb{C}^n$. Prove the following are equivalent:
   (1) $A[\mathbb{B}^n] = \mathbb{B}^n$
   (2) $\|Az\|_2 = \|z\|_2$ for all z.
   (3) $\langle Az, Aw\rangle = \langle z, w\rangle$ for all $z, w \in \mathbb{C}^n$ where $\langle z, w\rangle = \sum_{j=1}^n \bar{z}_j w_j$.

4. Prove that if $A = (a_{jk})$ is an $n \times n$ complex matrix, then $A[\mathbb{D}^n] \subset \mathbb{D}^n$ if and only if $\sup_j \sum_{k=1}^n |a_{jk}| \le 1$.

5. Prove that $\mathbb{U}[2]$ is connected. (*Hint*: See Problem 19 in Section 7.3.)

6. Extend Theorem 11.5.5 and Problem 2 to $\mathbb{C}^n$.

7. (a) Use the method we gave for the proof of Theorem 11.5.1 to show that if f is analytic in some $\mathbb{B}^2_{r_1}(0) \setminus \overline{\mathbb{B}^2_{r_2}(0)}$ for $r_2 < r_1$, then f has an analytic continuation to $\mathbb{B}^2_{r_1}(0)$.

   (b) Prove that if f is analytic in a neighborhood of $\partial\mathbb{B}^2$, it has a continuation to $\mathbb{B}^2$.

8. (a) Let g be in $L^2(\mathbb{R}^\nu)$. Prove that $\{a \in \mathbb{R}^\nu \mid \int |e^{a\cdot x} g(z)|^2 \, d^\nu x < \infty\}$ is convex.

   (b) Prove the special case of Theorem 11.5.7 for f's that are bounded and analytic on $\mathcal{J}(K)$. (*Hint*: Let $h(z) = (z - z_0)^{-\nu} f(z)$ for some $z_0 \neq K$ and use the Paley–Wiener theorem in both directions.)

   Remark. The idea here was used by Bochner [**56**] a year before he proved the more general Theorem 11.5.7.

9. (a) Prove that if Ω is a region in $\mathbb{C}^n$ and f, g are analytic in Ω and agree near some $z_0 \in \Omega$, they agree on all of Ω.

   (b) Let $n \geq 2$ and let f be analytic and bounded on $\mathbb{C}^n \setminus K$ for some compact K. Show first that for some R and each z_j, $f(z_1^{(0)}, \dots, z_{j-1}^{(0)}, z_j, z_{j+1}^{(0)}, \dots, z_n^{(0)})$ is constant as z_j is varied for each fixed $z_1^{(0)}, \dots, z_{j-1}^{(0)}, z_{j+1}^{(0)}, \dots, z_n^{(0)}) \in \mathbb{C}^{n-1}$ with $|z_k^{(0)}| > R$. Conclude that f is constant.

10. (a) Prove a maximum principle for analytic functions of n variables.

    (b) Prove an open mapping theorem for analytic functions of n variables.

11.6. Bonus Section: A First Glance at Compact Riemann Surfaces

Here we want to describe the basic facts about meromorphic functions on general compact Riemann surfaces. Because it would require considerable development of the needed machinery, there will not be any proofs and only a few indications of the constructions. Since we will not provide any proofs and we have been careful to only place in the Notes stated theorems we don't prove (even in the Problems), we'll present the results in this section as a series of "Facts." The Notes will have references on where to find detailed presentations and proofs. We include this section because it both indicates further directions and really illuminates the theorems of Section 10.3.

Fact 11.6.1. Every compact two-dimension oriented manifold (and therefore, in particular, every compact Riemann surface) is topologically equivalent (aka homeomorphic to) a sphere with ℓ handles. ℓ is called the *genus* of the surface. $\ell = 0$ is the sphere, $\ell = 1$ is the torus.

In case S is a compact Riemann surface associated to an irreducible algebraic function of total degree n with branch points $\{z_j\}_{j=1}^k$ of orders $\{n_j\}_{j=1}^k$, the genus is given by the Riemann–Hurwitz formula

$$\ell = 1 - n + \tfrac{1}{2}\sum_{j=1}^{k} n_j$$

Fact 11.6.2. Consider the sphere, X_ℓ, with ℓ handles. If $\{\alpha_j\}_{j=1}^\ell$ and $\{\beta_j\}_{j=1}^\ell$ are the curves obtained by going from the base point to the ℓ-th handle and looping around the handle once along the sphere (to get α) or once around the outside (to get β) (see Figure 11.6.1), then $\pi_1(X_\ell, x_0)$ is generated by the α's and β's with exactly one relation $(\alpha_1\beta_1\alpha_1^{-1}\beta_1^{-1})(\alpha_2\beta_2\alpha_2^{-1}\beta_2^{-1})\dots(\alpha_\ell\beta_\ell\alpha_\ell^{-1}\beta_\ell^{-1}) = 1$. The first homology group, $H_1(X^\ell, x_0)$, is thus $\mathbb{Z}^{2\ell}$, by Hurewicz's theorem (see Theorem 1.8.1).

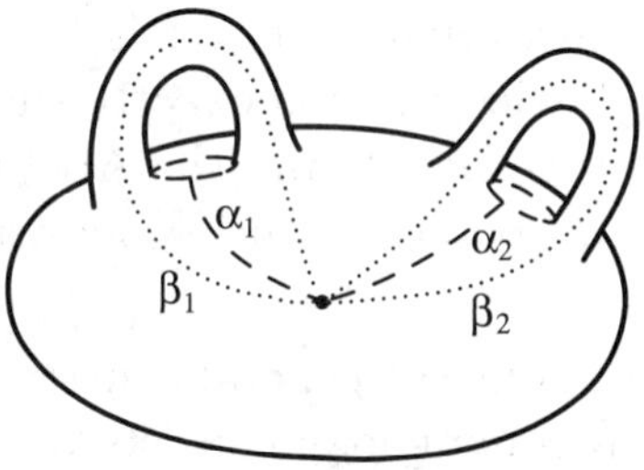

Figure 11.6.1. A homology basis.

Fact 11.6.3. Fix now a Riemann surface, X, of genus ℓ. In $\mathbb{C}^\ell$, there are 2ℓ points $\{\tau_j\}_{j=1}^{2\ell}$ which are linearly independent over $\mathbb{R}$ so that $\mathcal{L} = \{\sum_{k=1}^{2\ell} n_k\tau_k \mid (n_1, \dots, n_{2\ell}) \in \mathbb{Z}^{2\ell}\}$, called the *period lattice*, is a discrete subgroup of $\mathbb{C}^\ell$. The τ are determined by a choice of basis for $H_1(X^\ell, x_0)$, say $\{\alpha_j\}_{j=1}^\ell$ and $\{\beta_j\}_{j=1}^\ell$, and of ℓ other objects (a basis for the holomorphic differential forms). The quotient group, $V_X \equiv \mathbb{C}^\ell/\mathcal{L}$ is called the *Jacobi variety* of X. V_X is an abelian group for which we'll use additive notation.

Fact 11.6.4. There is a natural one-one map, $\mathfrak{A}$, of X into V_X, called the *Abel map*. It depends on a choice of base point, $p \in V_X$. If p_1 is another base point, the Abel maps $\mathfrak{A}_1$ (respectively, $\mathfrak{A}$) with p_1 (respectively, p) as base points, then $\mathfrak{A}_1 = \mathfrak{A} - \mathfrak{A}(p_1)$.

Fact 11.6.5. One has Abel's theorem: $\{q_j\}_{j=1}^{m_z}$ and $\{p_j\}_{j=1}^{m_p}$ are the zeros and poles (where no q_j is a p_k but p's or q's can be repeated to indicate the order of the zero or pole) of a meromorphic function, f, on X if and only if

(i) $m_z = m_p$ (11.6.1)

(ii) $\sum_{j=1}^{m_z} \mathfrak{A}(q_j) - \sum_{j=1}^{m_p} \mathfrak{A}(p_j) \in \mathcal{L}$ (11.6.2)

Fact 11.6.6. There exists a function, the Riemann theta function, $\Theta(z)$ on $\mathbb{C}^\ell$ which is entire and obeys

$$\Theta(z + \tau_j) = \exp(\alpha_j + \omega_j(z))\Theta(z) \tag{11.6.3}$$

Here ω_j are complex linear functions from $\mathbb{C}^\ell$ to $\mathbb{C}$. If (11.6.1) and (11.6.2) hold, then, up to a multiplicative constant, the meromorphic function f is given by

$$f(p) = \frac{\Theta(\mathfrak{A}(p) - \mathfrak{A}(q_1) - \tilde{\tau}) \prod_{j=2}^{m_1} \Theta(\mathfrak{A}(p) - \mathfrak{A}(q_j))}{\prod_{j=1}^{m_p} \Theta(\mathfrak{A}(p) - \mathfrak{A}(p_j))} \tag{11.6.4}$$

where $\tilde{\tau} \in \mathcal{L}$ is the element given on the left side of (11.6.2).

The reader will recognize the case $\ell = 1$ is a rewriting of the results of Sections 10.3–10.5. The lattice there is exactly the period lattice of $\mathcal{J}_{\tau_1,\tau_2}$, the Jacobi variety is the torus itself, and the Abel map (with base point 0) is the identity map. Θ can be either the Weierstrass σ-function or Jacobi theta function.

Here are a few words about the general constructions of $\mathcal{L}$, V, $\mathfrak{A}$, and Θ. $H^1(X)$, the first deRham cohomology consists of differential one-forms, g, with $dg = 0$, where two g's are equivalent if $g_1 - g_2 = df$ for a C^∞ function in X. By general principles (deRham's theorem; see the Notes to Section 1.8), $H^1(X)$ is a complex vector space of dimension 2ℓ. Inside this space, there are the forms whose expansion in $dz_j, d\bar{z}_j$ has only dz_j terms. Such forms are called *holomorphic one-forms.* These are of dimension ℓ.

Pick a basis $g_1, \dots, g_\ell$ of holomorphic one-forms. Then the periods τ_j are given by

$$(\tau_{2j-1})_k = \int_{\alpha_j} g_k \qquad (\tau_{2j})_k = \int_{\beta_j} g_k \tag{11.6.5}$$

and the Abel map is given by

$$(\mathfrak{A}(z))_k = \int_{z_0}^{z} g_k \tag{11.6.6}$$

This integral is along a smooth curve in X from z_0 to z. As complex numbers, the integrals in (11.6.6) are dependent on the choice of path, but the ambiguity is given by an element of $\mathcal{L}$, so $\mathfrak{A}$ is well-defined as a map with values in $\mathbb{C}^\ell/\mathcal{L}$. The Θ-function is a suitable quadratic exponential.

For $\ell = 1$, the only holomorphic form is dz. α_1, β_1 are maps which, lifted to $\mathbb{C}^\ell$ (from $\mathcal{J}_{\tau_1,\tau_2}$), are the curves on the τ_1 and τ_2 axes from 0 to the neighboring lattice points. So (11.6.5) indeed give the the periods. $\mathfrak{A}(z)$ is just z mod $\mathcal{L}$.

Notes and Historical Remarks. In modern guise, what we called Abel's theorem is the main result of Abel [**2**] on higher-dimensional abelian functions. When X is the Riemann surface of $w^2 - P(z) = 0$, the holomorphic one-forms are essentially $q(z)w^{-1}\,dz$, where $q(z)$ is a polynomial of degree at most $\ell - 1$. At zeros of w, this form is holomorphic since w is the proper local coordinate $w\,dw = P'(z)\,dz$, so $q(z)w^{-1}\,dz = q(z)(P'(z))^{-1}\,dw$ near this zero of z. Because p has degree at most $\ell - 1$ and $w \sim z^{(\ell+1)}$ near infinity, $q(z)w^{-1}\,dz$ is holomorphic in z^{-1} coordinates near $\infty_\pm$. The theta functions are those of Riemann [**484**].

For books with proofs of the results described here, see Donaldson [**149**], Farkas–Kra [**187**], Griffiths–Harris [**222**], and Miranda [**382**]. Other books on the subject include [**33, 194, 208, 226, 292, 400, 402, 533**].

Bibliography

[1] N. H. Abel, *Untersuchungen über die Reihe:* $1+\frac{m}{1}x+\frac{m(m-1)}{1\cdot 2}x^2+\frac{m(m-1)(m-2)}{1\cdot 2\cdot 3}x^3+\ldots$ *u.s.w.*, J. Reine Angew. Math. **1** (1826), 311–339. (Cited on 59, 61.)

[2] N. H. Abel, *Recherches sur les fonctions elliptiques*, J. Reine Angew. Math. **2** (1827), 101–181. (Cited on 498, 589.)

[3] W. Abikoff, *The uniformization theorem*, Amer. Math. Monthly **88** (1981), 574–592. (Cited on 367.)

[4] M. J. Ablowitz and A. S. Fokas, *Complex Variables: Introduction and Applications*, 2nd edition, Cambridge Texts in Applied Mathematics, Cambridge University Press, Cambridge, 2003; first edition, 1997. (Cited on 351.)

[5] S. Agmon, *Functions of exponential type in an angle and singularities of Taylor series*, Trans. Amer. Math. Soc. **70** (1951), 492–508. (Cited on 58, 242.)

[6] L. V. Ahlfors, *Bounded analytic functions*, Duke Math. J. **14** (1947), 1–11. (Cited on 378.)

[7] L. V. Ahlfors, *Development of the theory of conformal mapping and Riemann surfaces through a century*, in Contributions to the Theory of Riemann Surfaces, pp. 3–13, Annals of Mathematics Studies, Princeton University Press, Princeton, NJ, 1953. (Cited on 309, 314.)

[8] L. V. Ahlfors, *Conformal Invariants: Topics in Geometric Function Theory*, McGraw–Hill Series in Higher Mathematics, McGraw–Hill, New York–Düsseldorf-Johannesburg, 1973. (Cited on 324.)

[9] L. V. Ahlfors, *Complex Analysis. An Introduction to the Theory of Analytic Functions of One Complex Variable*, 3rd edition, McGraw–Hill, New York, 1979; first edition, 1953. (Cited on 142, 149, 362.)

[10] N. I. Akhiezer, *Elements of the Theory of Elliptic Functions*, Translations of Mathematical Monographs, American Mathematical Society, Providence, RI, 1990; Russian original, 1948. (Cited on 477.)

[11] A. Alexander, *Duel at Dawn. Heroes, martyrs, and the rise of modern mathematics*, New Histories of Science, Technology, and Medicine, Harvard University Press, Cambridge, MA, 2010. (Cited on 37, 499.)

[12] D. Alpay, *The Schur Algorithm, Reproducing Kernel Spaces and System Theory*, SMF/AMS Texts and Monographs, American Mathematical Society, Providence, RI; Société Mathématique de France, Paris, 2001. (Cited on 305.)

[13] G. E. Andrews, *A simple proof of Jacobi's triple product identity*, Proc. Amer. Math. Soc. **16** (1965), 333–334. (Cited on 535.)

[14] G. E. Andrews, *Generalized Frobenius partitions*, Mem. Amer. Math. Soc. **301**, (1984), 1–44. (Cited on 535.)

[15] G. E. Andrews, R. Askey, and R. Roy, *Special Functions*, Encyclopedia of Mathematics and Its Applications, Cambridge University Press, Cambridge, 1999. (Cited on 419, 421, 534, 535.)

[16] T. M. Apostol, *Another elementary proof of Euler's formula for $\zeta(2n)$*, Amer. Math. Monthly **80** (1973), 425–431. (Cited on 393.)

[17] T. M. Apostol, *A proof that Euler missed: evaluating $\zeta(2)$ the easy way*, Math. Intelligencer **5** (1983), 59–60. (Cited on 393.)

[18] T. M. Apostol, *Modular Functions and Dirichlet Series in Number Theory*, 2nd edition, Graduate Texts in Mathematics, Springer-Verlag, New York, 1990; first edition, 1976. (Cited on 550.)

[19] P. Appell, *Développements en série dans une aire limitée par des arcs de cercle*, Acta Math. **1** (1882), 145–152. (Cited on 156.)

[20] P. Appell, *Développements en série d'une fonction holomorphe dans une aire limitée par des arcs de cercle*, Math. Ann. **21** (1883), 118–124. (Cited on 156.)

[21] J. V. Armitage and W. F. Eberlein, *Elliptic Functions*, London Mathematical Society Student Texts, Cambridge University Press, Cambridge, 2006. (Cited on 477, 479, 536.)

[22] V. I. Arnol'd, *Mathematical Methods of Classical Mechanics*, corrected reprint of the 2nd (1989) edition, Graduate Texts in Mathematics, Springer-Verlag, New York, 1997; first edition of Russian original, 1976. (Cited on 479.)

[23] E. Artin, *The Gamma Function*, Athena Series: Selected Topics in Mathematics, Holt, Rinehart and Winston, New York–Toronto–London, 1964. (Cited on 419, 421, 423.)

[24] R. B. Ash, *Complex Variables*, Academic Press, New York–London, 1971. (Cited on 468.)

[25] R. B. Ash and W. P. Novinger, *Complex Variables*, 2nd edition, Dover Publicatoins, New York, 2007; available online at `http://www.math.uiuc.edu/~r-ash/CV.html`. (Cited on 150, 323.)

[26] D. Austin, *Trees, Teeth, and Time: The Mathematics of Clock Making*; available online at `http://www.ams.org/featurecolumn/archive/stern-brocot.html`, 2008. (Cited on 333.)

[27] A. Avila, Y. Last, and B. Simon, *Bulk universality and clock spacing of zeros for ergodic Jacobi matrices with absolutely continuous spectrum*, Analysis PDE **3** (2010), 81–108. (Cited on 564.)

[28] R. Ayoub, *The lemniscate and Fagnano's contributions to elliptic integrals*, Arch. Hist. Exact Sci. **29** (1984), 131–149. (Cited on 517.)

[29] P. Bachmann, *Die Analytische Zahlentheorie. Zahlentheorie. Pt. 2*, B. G. Teubner, Leipzig, 1894. (Cited on 12.)

[30] M. Bakonyi and T. Constantinescu, *Schur's Algorithm and Several Applications*, Pitman Research Notes in Mathematics Series, Longman Scientific & Technical,

Harlow; copublished in the United States with John Wiley, New York, 1992. (Cited on 305.)

[31] C. Bär, *Elementary Differential Geometry*, Cambridge University Press, Cambridge, 2010. (Cited on 21.)

[32] R. Beals and R. Wong, *Special Functions: A Graduate Text*, Cambridge Studies in Advanced Mathematics, Cambridge University Press, Cambridge, 2010. (Cited on 419, 536.)

[33] A. F. Beardon, *A Primer on Riemann Surfaces*, London Mathematical Society Lecture Notes, Cambridge University Press, Cambridge, 1984. (Cited on 589.)

[34] A. F. Beardon, *The Geometry of Discrete Groups*, corrected reprint of the 1983 original, Graduate Texts in Mathematics, Springer-Verlag, New York, 1995. (Cited on 335.)

[35] E. F. Beckenbach, *The stronger form of Cauchy's integral theorem*, Bull. Amer. Math. Soc. **49** (1943), 615–618. (Cited on 152, 194.)

[36] H. G. W. Begehr and A. Dzhuraev, *An Introduction to Several Complex Variables and Partial Differential Equations*, Pitman Monographs and Surveys in Pure and Applied Mathematics, Longman, Harlow, 1997. (Cited on 585.)

[37] E. T. Bell, *Men of Mathematics: The Lives and Achievements of the Great Mathematicians from Zeno to Poincaré*, Simon and Schuster, New York, 1986 (originally published in 1937). (Cited on 404.)

[38] S. R. Bell, *The Cauchy Transform, Potential Theory, and Conformal Mapping*, Studies in Advanced Mathematics, CRC Press, Boca Raton, FL, 1992. (Cited on 188, 323, 378.)

[39] B. Belhoste, *Augustin-Louis Cauchy. A biography*, translated from the French with a forward by Frank Ragland, Springer-Verlag, New York, 1991. (Cited on 39.)

[40] B. Belhoste, *Autour d'un mémoire in édit: la contribution d'Hermite au développement de la théorie des fonctions elliptiques*, Revue d'histoire des mathématiques **2**, 1–66. (Cited on 499.)

[41] E. J. Beltrami and M. R. Wohlers, *Distributions and the Boundary Values of Analytic Functions*, Academic Press, New York–London, 1966. (Cited on 562.)

[42] C. A. Berenstein and R. Gay, *Complex Variables. An introduction*, Graduate Texts in Mathematics, Springer-Verlag, New York, 1991. (Cited on 469.)

[43] Jakob Bernoulli, *Ars Conjectandi*, Impensis Thurnisiorum, fratrum, 1713; English translation available online at `http://www.sheynin.de/download/bernoulli.pdf`. (Cited on 437, 533.)

[44] L. Bers, *On rings of analytic functions*, Bull. Amer. Math. Soc. **54** (1948), 311–315. (Cited on 230.)

[45] L. Bers, *On Hilbert's 22nd problem*, in Mathematical Developments Arising From Hilbert Problems, pp. 559–609, Proc. Sympos. Pure Math., American Mathematical Society, Providence, RI, 1976. (Cited on 367.)

[46] L. Bers, *Quasiconformal mappings, with applications to differential equations, function theory and topology*, Bull. Amer. Math. Soc. **83** (1977), 1083–1100. (Cited on 38.)

[47] E. Betti, *La teorica delle funzioni ellittiche*, Ann. Mat. Pura Appl. **3** (1860), 65–159, 298–310; **4** (1861), 26–45, 57–70, 297–336. (Cited on 403, 519.)

[48] F. Beukers, J. A. C. Kolk, and E. Calabi, *Sums of generalized harmonic series and volumes*, Nieuw Arch. Wisk. (4) **11** (1993), 217–224. (Cited on 398.)

[49] L. Bieberbach, *Über einen Satz des Herrn Carathéodory*, Nachr. Königl. Ges. Wiss. Göttingen, Math.-phys. Klasse (1913), 552–560. (Cited on 294.)

[50] L. Bieberbach, *Conformal Mapping*, Chelsea Publishing, New York, 1953. (Cited on 350.)

[51] J. Binet, *Mémoire sur les intégrales définies eulériennes et sur leur application à la theorie des suites ainsi qu'à l'évaluation des fonctions des grandes nombres*, J. l'Ecole Polytechnique **16** (1838–39), 123–143. (Cited on 419, 447.)

[52] G. D. Birkhoff, *Démonstration d'un théorème élémentaire sur les fonctions entières* C. R. Acad. Sci. Paris **189** (1929), 473–475. (Cited on 161.)

[53] W. Blaschke, *Eine Erweiterung des Satzes von Vitali über Folgen analytischer Funktionen*, Leipz. Ber. **67** (1915), 194–200. (Cited on 455.)

[54] A. Bloch, *Les théorèmes de M. Valiron sur les fonctions entières et la théorie de l'uniformisation*, Ann. Fac. Sci. Toulouse (3) **17** (1925), 1–22. (Cited on 579.)

[55] A. Bloch, *La conception actuelle de la théorie des fonctions entières et méromorphes*, Enseign. Math. **25** (1926), 83–103. (Cited on 578.)

[56] S. Bochner, *Bounded analytic functions in several variables and multiple Laplace integrals*, Amer. J. Math. **59** (1937), 732–738. (Cited on 586.)

[57] S. Bochner, *A theorem on analytic continuation of functions in several variables*, Ann. Math. **39** (1938), 14–19. (Cited on 584.)

[58] S. Bochner, *Analytic and meromorphic continuation by means of Green's formula*, Ann. Math. **44** (1943), 652–673. (Cited on 584.)

[59] H. Bohr and E. Landau, *Beiträge zur Theorie der Riemannschen Zetafunktion*, Math. Ann. **74** (1913), 3–30. (Cited on 118.)

[60] H. Bohr and J. Mollerup, *Laerebog i Matematisk Analyse, Vol. III* [Textbook in Mathematical Analysis], J. Gjellerup, Copenhagen, 1922. (Cited on 420.)

[61] A. V. Bolsinov and A. T. Fomenko, *Integrable Hamiltonian Systems. Geometry, Topology, Classification*, Chapman & Hall/CRC, Boca Raton, FL, 2004. (Cited on 479.)

[62] R. Bombelli, *L'Algebra*, Bologna, 1572. (Cited on 4.)

[63] F. Bonahon, *Low-Dimensional Geometry: From Euclidean Surfaces to Hyperbolic Knots*, Student Mathematical Library, American Mathematical Society, Providence, RI; Institute for Advanced Study, Princeton, NJ, 2009. (Cited on 333.)

[64] C. W. Borchardt, *Leçons sur les fonctions doublement périodiques faites en 1847 par M. J. Liouville*, J. Reine Angew. Math. **88** (1880), 277–310. (Cited on 87, 497.)

[65] E. Borel, *Démonstration élémentaire d'un théoèreme de M. Picard sur les fonctions entières*, C. R. Acad. Sci. Paris **122** (1896), 1045–1048. (Cited on 577, 578.)

[66] E. Borel, *Sur les zéros des fonctions entières*, Acta Math. **20** (1897), 357–396. (Cited on 94, 182, 469.)

[67] E. Borel, *Leçons sur les fonctions entières*, Gauthier–Villars, Paris, 1900. (Cited on 469.)

[68] U. Bottazzini, *Complex Function Theory, 1780–1900*, in A History of Analysis, pp. 213–259, H. N. Jahnke, ed., History of Mathematics, American Mathematical Society, Providence, RI; London Mathematical Society, London, 2003. (Cited on 3.)

[69] U. Bottazzini and J. Gray, *Hidden Harmony–Geometric Fantasies. The Rise of Complex Function Theory*, Springer, New York, 2013. (Cited on 36, 55, 475, 517.)

[70] N. Bourbaki, *Éléments de mathématique. I: Les structures fondamentales de l'analyse*, Fascicule XI. Livre II: *Algèbre*, Chapitre 4: *Polynomes et fractions rationnelles*. Chapitre 5: *Corps commutatifs*. Deuxième édition, Actualités Scientifiques et Industrielles, No. 1102, Hermann, Paris, 1959. (Cited on 57.)

[71] F. Bowman, *Introduction to Elliptic Functions With Applications*, Dover Publications, New York, 1961. (Cited on 477.)

[72] J. Breuer and B. Simon, *Natural boundaries and spectral theory*, Adv. Math. **226** (2011), 4902–4920. (Cited on 58, 59, 241, 242.)

[73] C. Brezinski, *History of Continued Fractions and Padé Approximants*, Springer Series in Computational Mathematics, Springer-Verlag, Berlin, 1991. (Cited on 304.)

[74] E. Brieskorn and H. Knörrer, *Plane Algebraic Curves*, Birkhäuser Verlag, Basel, 1986. (Cited on 267.)

[75] Ch. Briot and J.-C. Bouquet, *Théorie des function doublement périodiques et, en particulier, des fonctions elliptique*, Mallet–Bachelier, Paris, 1859; 2nd edition, 1875. (Cited on 87, 130.)

[76] A. Brocot, *Calcul des rouages par approximation, nouvelle méthode*, Revue Chonométrique **3** (1861), 186–194. (Cited on 333.)

[77] L. E. J. Brouwer, *Beweis des Jordanschen Kurvensatzes*, Math. Ann. **69** (1910), 169–175. (Cited on 164.)

[78] A. Browder, *Introduction to Function Algebras*, W. A. Benjamin, New York–Amsterdam, 1969. (Cited on 157.)

[79] Y. Bugeaud, *Approximation by Algebraic Numbers*, Cambridge Tracts in Mathematics, Cambridge University Press, Cambridge, 2004. (Cited on 304.)

[80] R. B. Burckel, *An Introduction to Classical Complex Analysis, Vol. 1*, Pure and Applied Mathematics, Academic Press, New York–London, 1979. (Cited on 165, 323.)

[81] P. Bürgisser, M. Clausen, and M. A. Shokrollahi, *Algebraic Complexity Theory*, Grundlehren der Mathematischen Wissenschaften, Springer-Verlag, Berlin, 1997. (Cited on 112.)

[82] A. Burns, *Fractal Tilings*, The Mathematical Gazette, **78** (1994), 193–196. (Cited on 48.)

[83] P. L. Butzer and R. L. Stens, *The Euler–Maclaurin summation formula, the sampling theorem, and approximate integration over the real axis*, Linear Algebra Appl. **52/53** (1983), 141–155. (Cited on 443.)

[84] A.-P. Calderón, *Intermediate spaces and interpolation, the complex method*, Studia Math. **24** (1964), 113–190. (Cited on 177.)

[85] C. Carathéodory, *Über den Variabiletätsbereich der Koeffizienten den Potenzreihen, die gegebene Werte nicht annehmen*, Math. Ann. **64** (1907), 95–115. (Cited on 94, 239.)

[86] C. Carathéodory, *Über den Variabilitätsbereich der Fourier'schen Konstanten von positiven harmonischen Funktionen*, Rend. Circ. Mat. Palermo **32** (1911), 193–217. (Cited on 239.)

[87] C. Carathéodory, *Untersuchungen über die konformen Abbildungen von festen und veränderlichen Gebieten*, Math. Ann. **72** (1912) 107–144. (Cited on 117, 314, 315.)

[88] C. Carathéodory, *Über die gegenseitige Beziehung der Ränder bei der konformen Abbildung des Inneren einer Jordanschen Kurve auf einen Kreis*, Math. Ann. **73** (1913), 305–320. (Cited on 200, 323.)

[89] C. Carathéodory, *Über die Begrenzung einfach zusammenhängender Gebiete*, Math. Ann. **73** (1913), 323–370. (Cited on 324.)

[90] C. Carathéodory, *Theory of Functions of a Complex Variable. Vols. 1 and 2*, Chelsea Publishing, New York, 1954. Translation of *Funktionentheorie*, Birkhäuser, Basel, 1950. (Cited on 578.)

[91] C. Carathéodory and L. Fejér, *Remarques sur le théorème de M. Jensen*, C. R. Acad. Sci. Paris **145** (1908), 163–165. (Cited on 455.)

[92] C. Carathéodory and E. Landau, *Beiträge zur Konvergenz von Funktionenfolgen*, Berl. Ber. (1911), 587–613. (Cited on 238.)

[93] L. Carleson, *On a Class of Meromorphic Functions and Its Associated Exceptional Sets*, Thesis, University of Uppsala, 1950. (Cited on 194.)

[94] L. Carleson, *On null-sets for continuous analytic functions*, Ark. Mat. **1** (1951), 311–318. (Cited on 194.)

[95] H. Cartan, *Sur les systèmes de fonctions holomorphes à variétés linéaires lacunaires et leurs applications*, Ann. Sci. Ecole Norm. Sup. (3) **45** (1928), 255–346. (Cited on 574.)

[96] H. Cartan, *Les fonctions de deux variables complexes et le problème de la représentation analytique*, J. Math. Pures Appl. **10** (1931), 1–114. (Cited on 584.)

[97] H. Cartan, *Sur les fonctions de plusieurs variables complexes. L'itération des transformations intérieurs d'un domaine borné*, Math. Z. **35** (1932), 760–773. (Cited on 247.)

[98] H. Cartan, *Sur les zéros des combinaisons linéaires de p fonctions holomorphes données*, Mathematica (Cluj) **7** (1933), 80–103. (Cited on 574.)

[99] H. Cartan, *Séminaire Henri Cartan de l'Ecole Normale Supérieure, 1950/1951. Cohomologie des groupes, suite spectrale, faisceaux*, 2e éd., Secrétariat mathématique, 11 rue Pierre Curie, Paris, 1955. (Cited on 568.)

[100] H. Cartan, *Elementary Theory of Analytic Functions of One or Several Complex Variables*, Dover, New York, 1995; reprint of the 1973 edition, first edition of French original, 1961. (Cited on 57, 585.)

[101] F. Casorati, *Teoria delle funzioni di variabili complesse*, Pavia, 1868. (Cited on 128.)

[102] J. W. S. Cassels, *Lectures on Elliptic Curves*, London Mathematical Society Student Texts, Cambridge University Press, Cambridge, 1991. (Cited on 477.)

[103] A. L. Cauchy, *Cours d'analyse de l'École Royale polytechnique. I. Analyse algébrique*, Debure, Paris, 1821. (Cited on 49.)

[104] A. L. Cauchy, *Mémoire sur les intégrales définies, prises entre des limites imaginaires*, Chez de Bures Frères, Paris, 1825. (Cited on 39, 47.)

[105] A. L. Cauchy, *Mémoire sur l'intégration des équations différentielles*, paper written in Prague in 1835. (Cited on 569.)

[106] A. L. Cauchy, *Mémoire sur les fonctions alternées et sur les sommes alternées*, Exer. Anal. et Phys. Math. **2** (1841), 151–159. (Cited on 457.)

[107] A. L. Cauchy, *Mémoire sur quelques propositions fondamentales du calcul des résidus, et sur la théorie des intégrals singulières*, C. R. Acad. Sci. **19**, (1844), 1337–1344. (Cited on 87.)

[108] A. L. Cauchy, *Mémoire sur les variations intégrales des fonctions*, C. R. Acad. Sci. **40**, (1855), 651–658. (Cited on 100.)

[109] A. L. Cauchy, *Sur les compteurs logarithmiques*, C. R. Acad. Sci. **40**, (1855), 1009–1016. (Cited on 100.)

[110] K. Chandrasekharan, *Elliptic Functions*, Grundlehren der Mathematischen Wissenschaften, Springer-Verlag, Berlin, 1985. (Cited on 477.)

[111] R. Chapman, *Evaluating* $\zeta(2)$,available online at `http://www.secamlocal.ex.ac.uk/people/staff/rjchapma/etc/zeta2.pdf`. (Cited on 215.)

[112] J. Cheeger and D. G. Ebin, *Comparison Theorems in Riemannian Geometry*, revised reprint of the 1975 original, AMS Chelsea Publishing, Providence, RI, 2008. (Cited on 21.)

[113] M. S. Cheema, *Vector partitions and combinatorial identities*, Math. Comp. **18** (1964), 414–420. (Cited on 535.)

[114] J. Chen, *Theory of Real Functions*, Scientific Publishing, China, 1958. [Chinese] (Cited on 152.)

[115] Y. Choquet-Bruhat, C. DeWitt-Morette, and M. Dillard-Bleick, *Analysis, Manifolds and Physics*, 2nd edition, North–Holland, Amsterdam–New York, 1982. (Cited on 12.)

[116] E. Christoffel, *Sul problema delle temperature stazionarie e la rappresentazione diuna data superficie*, [On the problem of the stationary temperatures and the representation of a given surface], Brioschi Ann. (2) **1** (1867), 89–104. (Cited on 351.)

[117] J. A. Cima, A. L. Matheson, and W. T. Ross, *The Cauchy Transform*, Mathematical Surveys and Monographs, American Mathematical Society, Providence, RI, 2006. (Cited on 188.)

[118] A. Clebsch, *Über einen Satz von Steiner und einige Punkte der Theorie der Curven dritter Ordnung*, J. Reine Angew. Math. **63** (1864), 94–121. (Cited on 518.)

[119] T. Constantinescu, *Schur Parameters, Factorization and Dilation Problems*, Operator Theory: Advances and Applications, Birkhäuser Verlag, Basel, 1996. (Cited on 305.)

[120] J. B. Conway, *Functions of One Complex Variable*, 2nd edition, Graduate Texts in Mathematics, Springer-Verlag, New York–Berlin, 1978; first edition, 1973. (Cited on 323, 468, 579.)

[121] J. B. Conway, *Functions of One Complex Variable. II*, Graduate Texts in Mathematics, Springer-Verlag, New York, 1995. (Cited on 324.)

[122] E. T. Copson, *An Introduction to the Theory of Functions of a Complex Variable*, Oxford University Press, Oxford, 1946; original 1935 edition available online at `https://archive.org/details/TheoryOfTheFunctionsOfAComplexVariable`. (Cited on 214.)

[123] R. Cotes, *Logometria Auctore*, Phil. Trans. **29** (1714), 5–45. English translation in R. Gowing, *Roger Cotes–Natural philosopher*, Cambridge University Press, Cambridge, 1983. (Cited on 59.)

[124] R. Courant, *Über eine Eigenschaft der Abbildungsfunktionen bei konformer Abbildung*, Nachr. Kön. Ges. Wiss. Göttingen, Math.-phys. Klasse **1914**, 101–109. (Cited on 323.)

[125] D. A. Cox, *The arithmetic-geometric mean of Gauss*, Enseign. Math. (2) **30** (1984), 275–330. (Cited on 533.)

[126] W. Craig, *The trace formula for Schrödinger operators on the line*, Comm. Math. Phys. **126** (1989), 379–407. (Cited on 377.)

[127] D. Crowdy, *Schwarz–Christoffel mappings to unbounded multiply connected polygonal regions*, Math. Proc. Cambridge Philos. Soc. **142** (2007), 319–339. (Cited on 351.)

[128] J. d'Alembert, *Essai d'une nouvelle theorie de la resistance des fluids*, David, Paris, 1752. (Cited on 37.)

[129] G. Darboux, *Sur les développements en série des fonctions d'une seule variable.* Liouville J. **3** (1876), 291–312. (Cited on 438.)

[130] C. da Silva Dias, *Espacos vectoriais topológicos e sua applicação na teoria dos espaços funcionais analíticos* [Topological Vector Spaces and Their Application in the Theory of Analytic Functional Spaces], Thesis, University of São Paulo, 1951. (Cited on 230.)

[131] B. Davis, *Picard's theorem and Brownian motion*, Trans. Amer. Math. Soc. **213** (1975), 353–362. (Cited on 556.)

[132] P. J. Davis, *Leonhard Euler's integral: A historical profile of the gamma function*, Amer. Math. Monthly **66** (1959), 849–869. (Cited on 421.)

[133] P. J. Davis, *Interpolation and Approximation*, republication, with minor corrections, of the 1963 original, Dover Publications, New York, 1975. (Cited on 405.)

[134] A. de Moivre, *Miscellanea Analytica de Seriebus et Quadraturis*, J. Tonson and J. Watts, London, 1730. (Cited on 437.)

[135] A. de Moivre, *Miscellaneis Analyticis Supplementum*, J. Tonson and J. Watts, London, 1730. (Cited on 437.)

[136] A. Denjoy, *Sur les singularités discontinues des fonctions analytiques uniformes*, C. R. Acad. Sci. Paris **149** (1910), 386-388. (Cited on 194.)

[137] A. Denjoy, *Sur les polygones d'approximation d'une courbe rectifiable*, C. R. Acad. Sci. Paris **196** (1933), 29–32. (Cited on 152.)

[138] G. deRham, *Sur l'analysis situs des variétés à n dimensions*, Thése, J. Math. Pures Appl. **10** (1931), 115–200. (Cited on 26.)

[139] G. Desargues, *Brouillon project d'une atteinte aux evenemens des rencontres d'une cone avec un plan* [Proposed Draft of an Attempt to Deal with the Events of the Meeting of a Cone with a Plane], private publication, Paris, 1639. (Cited on 282.)

[140] F. Diamond and J. Shurman, *A First Course in Modular Forms*, Graduate Texts in Mathematics, Springer-Verlag, New York, 2005. (Cited on 550.)

[141] E. DiBenedetto, *Real Analysis*, Birkhäuser Advanced Texts: Basel Textbooks, Birkhäuser, Boston, 2002. (Cited on 12.)

[142] P. Dienes, *The Taylor Series: An Introduction to the Theory of Functions of a Complex Variable*, reprint of the 1931 edition, Dover, New York, 1957. (Cited on 57.)

[143] S. Dineen, *Complex Analysis on Infinite Dimensional Spaces*, Springer Monographs in Mathematics, Springer-Verlag, London, 1999. (Cited on 585.)

[144] A. Dinghas, *Einführung in die Cauchy–Weierstrass'sche Funktionentheorie*, B. I. Hochschultaschenbücher, Bibliographisches Institut, Mannheim–Zürich, 1968. (Cited on 57.)

[145] P. G. L. Dirichlet, *Über einen neuen Ausdruck zur Bestimmung der Dictigkeit einer unendlich dünnen Kugelschale, wenn der Werth des Potentials derselben in jedem Punkte ihrer Oberfläche gegeben ist*, Abh. Königlich. Preuss. Akad. Wiss. pp. 99–116, Berlin, 1850. (Cited on 315.)

[146] J. D. Dixon, *A brief proof of Cauchy's integral theorem*, Proc. Amer. Math. Soc. **29** (1971), 625–626. (Cited on 143.)

[147] G. Doetsch, *Über die obere Grenze des absoluten Betrages einer analytischen Funktion auf Geraden*, Math. Z. **8** (1920), 237–240. (Cited on 177.)

[148] D. Dominici, *Variations on a theme by James Stirling*, Note Mat. **28** (2008), 1–13. (Cited on 437.)

[149] S. Donaldson, *Riemann Surfaces*, Oxford Graduate Texts in Mathematics, Oxford University Press, Oxford, 2011. (Cited on 589.)

[150] T. A. Driscoll and L. N. Trefethen, *Schwarz–Christoffel Mapping*, Cambridge Monographs on Applied and Computational Mathematics, Cambridge University Press, Cambridge, 2002. (Cited on 350, 351.)

[151] D. S. Dummit and R. M. Foote, *Abstract Algebra*, 3rd edition, John Wiley & Sons, Hoboken, NJ, 2004, first edition, 1991. (Cited on 8.)

[152] N. Dunford, *Uniformity in linear spaces*, Trans. Amer. Math. Soc. **44** (1938), 305–356. (Cited on 88.)

[153] W. Dunham, *Euler and the fundamental theorem of algebra*, College Math. J. **22** (1991), 282–293; reprinted in [**154**]. (Cited on 88.)

[154] W. Dunham, *Euler: The Master of Us All*, The Dolciani Mathematical Expositions, Mathematical Association of America, Washington, DC, 1999. (Cited on 395, 599.)

[155] G. W. Dunnington, *Carl Friedrich Gauss: Titan of Science*, reprint of the 1955 original, Exposition Press, New York, with additional material by J. Gray and F.-E. Dohse, MAA Spectrum, Mathematical Association of America, Washington, DC, 2004. (Cited on 29.)

[156] P. Du Val, *Elliptic Functions and Elliptic Curves*, London Mathematical Society Lecture Note Series, Cambridge University Press, London–New York, 1973. (Cited on 477.)

[157] H.-D. Ebbinghaus, H. Hermes, F. Hirzebruch, et al., *Numbers*, Graduate Texts in Mathematics, Readings in Mathematics, Springer-Verlag, New York, 1991. (Cited on 397.)

[158] W. F. Eberlein, *On Euler's infinite product for the sine*, J. Math. Anal. Appl. **58** (1977), 147–151. (Cited on 397.)

[159] L. Ehrenpreis, *A new proof and an extension of Hartogs' theorem*, Bull. Amer. Math. Soc. **67** (1961), 507–509. (Cited on 584.)

[160] M. Eidelheit, *Zur Theorie der Systeme linearer Gleichungen*, Studia Math. **6** (1936), 139–148. (Cited on 405.)

[161] S. Eilenberg, *Singular homology theory*, Ann. Math. **45** (1944), 407–447. (Cited on 26.)

[162] S. Eilenberg and N. E. Steenrod, *Axiomatic approach to homology theory*, Proc. Nat. Acad. Sci. USA **31** (1945), 117–120. (Cited on 26.)

[163] G. Eisenstein, *Beiträge zur Theorie der elliptischen Functionen, VI. Genaue Untersuchung der unendlichen Doppelproducte, aus welchen die elliptischen Functionen als Quotienten zusammengesetzt sind*, J. Reine Angew. Math. **35** (1847), 153–274. (Cited on 517.)

[164] R. W. Emerson, Self-Reliance, an essay included in *Essays: First Series*, Houghton, Mifflin and Company, Boston, MA, 1883; available online at `http://en.wikisource.org/wiki/Essays:_First_Series/Self-Reliance` (self-reliance only) and `https://archive.org/details/essaysfirstseconx00emer` (1883 collection of First and Second Series). (Cited on 1.)

[165] D. B. A. Epstein, *Prime ends*, Proc. London Math. Soc. (3) **42** (1981), 385–414. (Cited on 324.)

[166] A. Eremenko, *Holomorphic curves omitting five planes in projective space*, Amer. J. Math. **118** (1996), 1141–1151. (Cited on 574.)

[167] A. Eremenko, *A Picard type theorem for holomorphic curves*, Periodica Math. Hungarica **38** (1999), 39–42. (Cited on 574.)

[168] A. Eremenko and J. Lewis, *Uniform limits of certain A-harmonic functions with applications to quasiregular mappings*, Ann. Acad. Sci. Fenn. Ser. A I Math. **16** (1991), 361–375. (Cited on 574.)

[169] T. Estermann, *Complex Numbers and Functions*, University of London, Athlone Press, London, 1962. (Cited on 101.)

[170] Euclid, *Euclid's Elements*, T. L. Heath (translator), Green Lion Press, Santa Fe, NM, 2002; available online at `http://farside.ph.utexas.edu/Books/Euclid/Elements.pdf`. (Cited on 306.)

[171] L. Euler, *De progressionibus transcendentibus seu quarum termini generales algebraice dari nequeunt*, Comm. Acad. Sci. Petropolitanae **5** (1730/1), 1738, 36–57 (based on presentation to the St. Petersburg Academy on November 28, 1729). English translation available at `http://home.sandiego.edu/~langton/eg.pdf`. (Cited on 419.)

[172] L. Euler, *De summis serierum reciprocarum*, [On the sums of series of reciprocals], Comm. Acad. Sci. Petropolitanae **7** (1734/35), 123–134. (Cited on 215, 393.)

[173] L. Euler, *Methodus universalis serierum convergentium summas quam proxime inveniendi*, Comm. Acad. Sci. Petropolitanae **8** (1736), 3–9; Opera Omnia, Vol. XIV, pp. 101–107. (Cited on 438.)

[174] L. Euler, *Methodus universalis series summandi ulterius promota*, Comm. Acad. Sci. Petropolitanae **8** (1736), 147–158; Opera Omnia, Vol. XIV, pp. 124–137. (Cited on 438.)

[175] L. Euler, *De fractionibus continuis dissertatio*, Comm. Acad. Sci. Petropolitanae **9** (1737), 98–137; translation by M. F. Wyman and B. F. Wyman as *An essay on continued fractions*, Math. Systems Theory **18** (1985), 295–328. (Cited on 255, 304.)

[176] L. Euler, *De productis ex infinitis factoribus ortis*, Comm. Acad. Sci. Petropolitanae **11** (1739), 3–31. (Cited on 419.)

[177] L. Euler, *Lettres à une princesse d'Allemagne sur divers sujets de physique et de philosophie*, Charpentier, Libraire-Éditeur, Paris, 1743. English translation available online at `http://www.archive.org/details/lettersofeuleron02eule`. (Cited on 395.)

[178] L. Euler, *Introductio in Analysin Infinitorum*, MM Bousquet, Lausanne, 1748. (Cited on 4, 59, 394, 395, 533.)

[179] L. Euler, *Recherches sur les racines imaginaires des équations*, Mémoires de l'Academie des Sciences de Berlin (1749), 222–288. (Cited on 87.)

[180] L. Euler, *Institutiones Calculi Differentialis*, in typographeo Petri Galeatii, 1755. (Cited on 438.)

[181] L. Euler, *Considerationes de traiectoriis orthogonalibus*, Novi Comment. Acad. Sci. Petropolitanae **14** (1770), 46–71; Opera Omnia, Vol. XXVIII, pp. 99–119). (Cited on 282.)

[182] G. Faber, *Über Potenzreihen mit unendlich vielen verschwindenden Koeffizienten*, Sitzungsber. Konigl. Bayer. Akad. Wiss., Math.-Phys. **36**, (1906), 581–583. (Cited on 58.)

[183] E. Fabry, *Sur les points singuliers d'une fonction donnée par son développement en série et l'impossibilité du prolongement analytique dans des cas très généraux*, Ann. Sci. Ecole Norm. Sup. (3) **13** (1896), 367–399. (Cited on 58.)

[184] C. G. Fagnano, *Metodo per misurare la lemniscata* [Method for measuring the lemniscates], Giornale dei letterati d'Italia **30** (1718), 87. (Cited on 516.)

[185] J. Farey, *On a curious property of vulgar fractions*, Philos. Magazine **47** (1816), 385–386. (Cited on 332.)

[186] H. M. Farkas and I. Kra, *Riemann Surfaces*, Graduate Texts in Mathematics, Springer, New York–Berlin, 1980. (Cited on 267.)

[187] H. M. Farkas and I. Kra, *Theta Constants, Riemann Surfaces and the Modular Group. An Introduction With Applications to Uniformization Theorems, Partition Identities and Combinatorial Number Theory*, Graduate Studies in Mathematics, American Mathematical Society, Providence, RI, 2001. (Cited on 533, 589.)

[188] S. R. Finch, *Mathematical Constants*, Encyclopedia of Mathematics and Its Applications, Cambridge University Press, Cambridge, 2003. (Cited on 579.)

[189] G. Fischer, *Plane Algebraic Curves*, Student Mathematical Library, American Mathematical Society, Providence, RI, 2001. (Cited on 267.)

[190] S. D. Fisher, *On Schwarz's lemma and inner functions*, Trans. Amer. Math. Soc. **138** (1969), 229–240. (Cited on 378.)

[191] L. R. Ford, *On the foundations of the theory of discontinuous groups of linear transformations*, Proc. Nat. Acad. Sci. USA **13** (1927), 286–289. (Cited on 282, 289.)

[192] L. R. Ford, *Fractions*, Amer. Math. Monthly **45** (1938), 586–601. (Cited on 304, 333.)

[193] L. R. Ford, *Automorphic Functions*, 2nd ed., Chelsea, New York, 1951. (Cited on 282, 289, 335.)

[194] O. Forster, *Lectures on Riemann Surfaces*, Graduate Texts in Mathematics, Springer-Verlag, New York, 1991; translated from the 1977 German original by Bruce Gilligan, reprint of the 1981 English translation. (Cited on 266, 267, 589.)

[195] J. Fredholm, *Om en speciell klass af singulära linier*, (Swedish) [On a special class of singular lines] Stockh. Öfv. (1890), 131–134. (Cited on 65.)

[196] B. Fuglede, *A sharpening of Wielandt's characterization of the gamma function*, Amer. Math. Monthly **115** (2008), 845–850. (Cited on 424.)

[197] W. Fulton, *Algebraic Topology. A First Course*, Graduate Texts in Mathematics, Springer-Verlag, New York, 1995. (Cited on 23.)

[198] T. W. Gamelin, *Uniform Algebras*, Prentice–Hall, Englewood Cliffs, NJ, 1969. (Cited on 157.)

[199] T. W. Gamelin, *Complex Analysis*, Undergraduate Texts in Mathematics, Springer-Verlag, New York, 2001. (Cited on 367.)

[200] F. R. Gantmacher, *The Theory of Matrices. Vol. 1*, reprint of the 1959 translation, AMS Chelsea Publishing, Providence, RI, 1998; Russian original, 1953 (Cited on 85.)

[201] S. J. Gardiner, *Harmonic Approximation*, London Mathematical Society Lecture Notes, Cambridge University Press, Cambridge, 1995. (Cited on 137.)

[202] L. Gårding, *Some Points of Analysis and Their History*, University Lecture Series, American Mathematical Society, Providence, RI; Higher Education Press, Beijing, 1997. (Cited on 167, 173.)

[203] L. Gårding and L. Hörmander, *Why is there no Nobel prize in mathematics?* Math. Intelligencer **7** (1985), 73–74. (Cited on 400.)

[204] J. Garnett, *Analytic Capacity and Measure*, Lecture Notes in Mathematics, Springer-Verlag, Berlin–New York, 1972. (Cited on 378.)

[205] C. F. Gauss, *Disquisitiones generales circa superficies curvas*, 1827; available online at `https://archive.org/details/disquisitionesg00gausgoog`; English translation: *General Investigations of Curved Surfaces*, Dover Publications, 2005. (Cited on 20.)

[206] C. F. Gauss, *Allgemeine Lehrsätze in Beziehung auf die im verkehrten Verhältnisse des Quadrats der Entfernung wirkenden Anziehungs- und Abstossungs-Kräfte*, Leipzig, 1839. (Cited on 315.)

[207] Ya. L. Geronimus, *On the trigonometric moment problem*, Ann. Math. **47** (1946), 742–761. (Cited on 306.)

[208] E. Girondo, and G. González-Diez, *Introduction to Compact Riemann Surfaces and Dessins d'Enfants*, London Mathematical Society Student Texts, Cambridge University Press, Cambridge, 2012. (Cited on 589.)

[209] J. W. L. Glaisher, *On some elliptic function and trigonometrical theorems*, Messenger of Math. **19** (1880), 92–97. (Cited on 536.)

[210] I. Glicksberg, *A remark on Rouché's theorem*, Amer. Math. Monthly **83** (1976), 186–187. (Cited on 101.)

[211] R. Godement, *Topologie algébrique et théorie des faisceaux*, Actualités Sci. Ind., Hermann, Paris 1958. (Cited on 568.)

[212] E. Goursat, *Démonstration du théorème de Cauchy. Extrait d'une lettre adressée à M. Hermite*, Acta Math. **4** (1884), 197–200. (Cited on 68.)

[213] E. Goursat, *Sur la définition générale des fonctions analytiques d'après Cauchy*, Trans. Amer. Math. Soc. **1** (1900), 14–16. (Cited on 68.)

[214] R. L. Graham, D. E. Knuth, and O. Patashnik, *Concrete Mathematics: A Foundation for Computer Science*, Addison–Wesley, Reading, MA, 1994. (Cited on 333.)

[215] H. Grauert and K. Fritzsche, *Several Complex Variables*, Graduate Texts in Mathematics, Springer-Verlag, New York–Heidelberg, 1976. (Cited on 585.)

[216] J. Gray, *On the history of the Riemann mapping theorem*, Rend. Circ. Mat. Palermo (2) **34** (1994), 47–94. (Cited on 314.)

[217] J. Gray, *Linear Differential Equations and Group Theory from Riemann to Poincaré*, reprint of the 2000 second edition, Modern Birkhäuser Classics, Birkhäuser, Boston, 2008, first edition, 1986. (Cited on 117, 335.)

[218] J. D. Gray and S. A. Morris, *When is a function that satisfies the Cauchy-Riemann equations analytic?*, Amer. Math. Monthly **85** (1978), 246–256. (Cited on 68.)

[219] G. Green, *An Essay on the Application of Mathematical Analysis to the Theories of Electricity and Magnetism*, Nottingham, 1828. (Cited on 315.)

[220] R. E. Greene and S. G. Krantz, *Function Theory of One Complex Variable*, 3rd edition, Graduate Studies in Mathematics, American Mathematical Society, Providence, RI, 2006; first edition, 1997. (Cited on 156, 468.)

[221] P. A. Griffiths, *Introduction to Algebraic Curves*, Translations of Mathematical Monographs, American Mathematical Society, Providence, RI, 1989. (Cited on 267.)

[222] P. Griffiths and J. Harris, *Principles of Algebraic Geometry*, Pure and Applied Mathematics, Wiley-Interscience, New York, 1978. (Cited on 267, 589.)

[223] A. Grothendieck, *Sur certains espaces de fonctions holomorphes. I*, J. Reine Angew. Math. **192** (1953), 35–64. (Cited on 230.)

[224] Chr. Gudermann, *Additamentum ad functionis* $\Gamma(a) = \int_0^\infty e^{-x} \cdot x^{a-1}\, \partial x$ *theoriam*, J. Reine Angew. Math. **29** (1845), 209–212. (Cited on 440.)

[225] H. Guggenheimer, *Differential Geometry*, corrected reprint of the 1963 edition, Dover Books on Advanced Mathematics, Dover Publications, New York, 1977. (Cited on 21.)

[226] R. C. Gunning, *Lectures on Riemann surfaces. Jacobi varieties*, Mathematical Notes, Princeton University Press, Princeton, NJ, 1973. (Cited on 589.)

[227] R. C. Gunning and H. Rossi, *Analytic Functions of Several Complex Variables*, Prentice–Hall, Englewood Cliffs, NJ, 1965. (Cited on 585.)

[228] J. Hadamard, *Sur le rayon de convergence des séries ordonnres suivant les puissances d'une variable*, C. R. Acad. Sci. Paris **106** (1888), 259–261. (Cited on 50.)

[229] J. Hadamard, *Essai sur l'étude des fonctions données par leur développement de Taylor*, J. Math. Pures Appl. (4) **8** (1892), 101–186. (Cited on 58.)

[230] J. Hadamard, *Étude sur les propriétés des fonctions entières et en particulier d'une fonction considérée par Riemann*, J. Math. Pures Appl. (4) **9** (1893), 171–215. (Cited on 468, 469.)

[231] J. Hadamard, *Sur l'expression du produit* $1.2.3\dots(n-1)$ *par une fonction entière*, Darboux Bull. (2) **19** (1895), 69–71. Available online at `http://www.luschny.de/math/factorial/hadamard/HadamardFactorial.pdf`. (Cited on 430.)

[232] L-S. Hahn and B. Epstein, *Classical Complex Analysis*, Jones & Bartlett Learning, Sudbury, MA, 1996. (Cited on 468.)

[233] E. Hairer and G. Wanner, *Analysis by Its History*, corrected reprint of the 1996 original, Undergraduate Texts in Mathematics, Readings in Mathematics, Springer, New York, 2008. (Cited on 305.)

[234] T. C. Hales, *Jordan's Proof of the Jordan Curve Theorem*, Studies in Logic, Grammar, and Rhetoric **10** (2007), 45–60. (Cited on 164.)

[235] H. Hanche-Olsen, *On Goursat's proof of Cauchy's integral theorem*, Amer. Math. Monthly **115** (2008), 648–652. (Cited on 75.)

[236] H. Hancock, *Lectures on the Theory of Elliptic Functions: Analysis*, Dover Publications, New York, 1958; an unaltered republication of the first edition, Wiley & Sons, New York, 1910. (Cited on 477, 536.)

[237] H. Hancock, *Elliptic Integrals*, Charleston, SC, Nabu Press, 2010; reprint of 1917 original, Wiley & Sons, New York, 1917. (Cited on 477, 536.)

[238] G. H. Hardy, *Orders of Infinity: The 'Infinitärcalcül' of Paul du Bois-Reymond*, Cambridge University Press, Cambridge, 1910; available online at `http://www.archive.org/details/ordersofinfinity00harduoft`. (Cited on 12.)

[239] G. H. Hardy, *Divergent Series*, Clarendon Press, Oxford, 1949; 2nd edition, AMS Chelsea, Providence, RI, 1991. (Cited on 12.)

[240] G. H. Hardy and E. M. Wright *An Introduction to the Theory of Nnumbers*, Sixth edition, revised by D. R. Heath–Brown and J. H. Silverman, with a foreword by Andrew Wiles, Oxford University Press, Oxford, 2008; First edition, 1938. (Cited on 333.)

[241] C. Haros, *Tables pour evaluer une fraction ordinaire avec autant de decimals qu'on voudra; et pour trover la fraction ordinaire la plus simple, et qui s'approche sensiblement d'une fraction decimale*, J. l'Ecole Polytechnique **4** (1802), 364–368. (Cited on 332.)

[242] F. Hartogs, *Zur Theorie der analytischen Funktionen mehrerer unabhängiger Veränderlichen, insbesondere über die Darstellung derselben durch Reihen welche nach Potentzen einer Veränderlichen fortschreiten*, Math. Ann. **62** (1906), 1–88. (Cited on 583, 584.)

[243] D. Hatch, *Hyperbolic planar tesselations*, available online at `http://www.plunk.org/~hatch/HyperbolicTesselations/`. (Cited on 335.)

[244] A. Hatcher, *Algebraic Topology*, Cambridge University Press, Cambridge, 2002; available online at `http://www.math.cornell.edu/~hatcher/AT/AT.pdf`. (Cited on 23, 24, 26, 142, 165.)

[245] J. Havil, *Gamma. Exploring Euler's Constant*, Princeton University Press, Princeton, NJ, 2003. (Cited on 420, 421.)

[246] S. Ya. Havinson, *Analytic capacity of sets, joint nontriviality of various classes of analytic functions and the Schwarz lemma in arbitrary domains* Amer. Math. Soc. Transl. (2) **43** (1964), 215–266. (Cited on 378.)

[247] B. Hayes, *Group Theory in the Bedroom, and Other Mathematical Diversions*, Hill and Wang, New York, 2008. (Cited on 333.)

[248] H. Heilbronn, *Zu dem Integralsatz von Cauchy*, Math. Z. **37** (1933), 37–38. (Cited on 152.)

[249] M. Heins, *On the number of* 1-1 *directly conformal maps which a multiply-connected plane region of finite connectivity* $p(> 2)$ *admits onto itself*, Bull. Amer. Math. Soc. **52** (1946), 454–457. (Cited on 292.)

[250] Y. Hellegouarch, *Invitation to the Mathematics of Fermat–Wiles*, Academic Press, San Diego, CA, 2002. (Cited on 533.)

[251] O. Helmer, *Divisibility properties of integral functions*, Duke Math. J. **6** (1940), 345–356. (Cited on 406.)

[252] P. Henrici, *Applied and Computational Complex Analysis, Vol. 1. Power Series—Integration—Conformal Mapping—Location of Zeros*, reprint of the 1974 original, Wiley Classics Library, John Wiley & Sons, New York, 1988. (Cited on 57.)

[253] D. Hensley, *Continued Fractions*, World Scientific Publishing, Hackensack, NJ, 2006. (Cited on 304.)

[254] G. Herglotz, *Über Potenzreihen mit positivem, reellem Teil im Einheitskreis*, Ber. Ver. Ges. wiss. Leipzig **63** (1911), 501–511. (Cited on 239.)

[255] Ch. Hermite, *Sur les fonctions algébriques*, C. R. Acad. Sci. Paris **32** (1851), 458–461. (Cited on 568.)

[256] Ch. Hermite, *Sur la résolution de l'équation du cinquième degré*, C. R. Acad. Sci. Paris **46** (1858), 508–515. (Cited on 479.)

[257] H. Hida, *Geometric Modular Forms and Elliptic Curves*, World Scientific Publishing, River Edge, NJ, 2000. (Cited on 550.)

[258] D. Hilbert, *Über die Entwickelung einer beliebigen analytischen Function einer Variabeln in eine unendliche, nach ganzen rationalen Functionen fortschreitende Reihe*, Nachr. Königl. Ges. Wiss. Göttingen, Math.-phys. Klasse **1897** (1897), 63–70. (Cited on 157.)

[259] D. Hilbert, *Mathematische Probleme*, Nachr. Kön. Ges. Wiss. Göttingen, Math.-phys. Klasse **1900**, 253–297; Arch. Math. Physik (3) **1** (1901), 44–63, 213-237; English translation available online at `http://aleph0.clarku.edu/~djoyce/hilbert/problems.html`. (Cited on 368.)

[260] D. Hilbert, *Über das Dirichlet'sche Princip*, Deutsche Math. Ver. **8** (1900), 184–188. (Cited on 238.)

[261] D. Hilbert, *Zur Theorie der konformen Abbildungen*, Nachr. Kön. Ges. Wiss. Göttingen, Math.-phys. Klasse **1909**, 314–323. (Cited on 368.)

[262] E. Hille, *Analysis. Vols. I, II*, reprint of the 1964 original with corrections, R. E. Krieger Publishing, Huntington, NY, 1979. (Cited on 157, 404.)

[263] E. Hille, *Ordinary Differential Equations in the Complex Domain*, reprint of the 1976 original, Dover Publications, Mineola, NY, 1997. (Cited on 12.)

[264] M. Hockman, *Continued fractions and the geometric decomposition of modular transformations*, Quaest. Math. **29** (2006), 427–446. (Cited on 333.)

[265] O. Hölder, *Über die Eigenschaft der Gammafunction keiner algebraischen Differentialgleichung zu genügen*, Math. Ann. **28** (1886), 1–13. (Cited on 419.)

[266] H. Hopf *Eine Verallgemeinerung der Euler–Poincaréschen Formel*, Nachr. Kön. Ges. Wiss. Göttingen, Math.-phys. Klasse **1928** (1928), 127–136. (Cited on 26.)

[267] H. Hopf and W. Rinow, *Über den Begriff der vollständigen differentialgeometrischen Fläche*, Comment. Math. Helv. **3** (1931), 209–225. (Cited on 20.)

[268] L. Hörmander, *An Introduction to Complex Analysis in Several Variables*, 3rd edition, North–Holland Mathematical Library, North–Holland, Amsterdam, 1990. (Cited on 585.)

[269] W. Hurewicz, *Beiträge zur Topologie der Deformationen. II: Homotopie- und Homologiegruppen*, Proc. Acad. Amsterdam **38** (1935), 521–528. (Cited on 26.)

[270] A. Hurwitz, *Über die Nullstellen der Bessel'schen Function*, Math. Ann. **33** (1889), 246–266. (Cited on 246.)

[271] A. Hurwitz, *Über die angenäherte Darstellung der Irrationalzahlen durch rationale Brüche*, Math. Ann. **39** (1891), 279–284. (Cited on 304.)

[272] A. Hurwitz and R. Courant, *Vorlesungen über allgemeine Funktionentheorie und elliptische Funktionen*, Interscience, New York, 1944. (Cited on 57.)

[273] D. Husemoller, *Elliptic Curves*, Graduate Texts in Mathematics, Springer-Verlag, New York, 1987. (Cited on 477, 518.)

[274] J. Igusa, *Theta Functions*, Die Grundlehren der mathematischen Wissenschaften, Springer-Verlag, New York–Heidelberg, 1972. (Cited on 534.)

[275] E. L. Ince, *Ordinary Differential Equations*, Dover Publications, New York, 1944; reprint of the English edition, Longmans, Green and Co., London, 1926. (Cited on 12.)

[276] A. E. Ingham, *The Distribution of Prime Numbers*, reprint of the 1932 original, with a foreword by R. C. Vaughan, Cambridge Mathematical Library, Cambridge University Press, Cambridge, 1990. (Cited on 214.)

[277] M. Iosifescu and C. Kraaikamp, *Metrical Theory of Continued Fractions*, Mathematics and Its Applications, Kluwer Academic Publishers, Dordrecht, 2002. (Cited on 304.)

[278] V. I. Ivanov and M. K. Trubetskov, *Handbook of Conformal Mapping with Computer-Aided Visualization*, CRC Press, Boca Raton, FL, 1995. (Cited on 350.)

[279] T. Iwaniec and G. Martin, *Geometric Function Theory and Non-linear Analysis*, Oxford Mathematical Monographs, The Clarendon Press, Oxford University Press, New York, 2001. (Cited on 128.)

[280] C. G. Jacobi, *Über den Ausdruck der verschiedenen Wurzeln einer Gleichung durch bestimmte Integrale*, J. Reine Angew. Math. **2** (1827), 1–8. (Cited on 498.)

[281] C. G. Jacobi, *Fundamenta nova theoriae functionum ellipticarum*, 1829; available online at `https://openlibrary.org/books/OL6506700M/Fundamenta_nova_theoriae_functionum_ellipticarum`. (Cited on 533, 534.)

[282] C. G. Jacobi, *Demonstratio formulae* $\int_0^1 w^{a-1}(1 - w)^{b-1}\partial w = \frac{\int^\infty e^{-x}x^{\alpha-1}\partial x \cdot \int_0^\infty e^{-x}x^{b-1}\partial w}{\int_0^\infty e^{-x}x^{a+b-1}\partial w} = \frac{\Gamma a \Gamma b}{\Gamma(a+b)}$, J. Reine Angew. Math. **11** (1834), 307. (Cited on 419, 450.)

[283] C. G. Jacobi, *De usu theoriae integralium ellipticorum et integralium Abelianorum in analysi Diophantea*, J. Reine Angew. Math. **13** (1835), 353–355. (Cited on 533.)

[284] J. L. Jensen, *Sur un nouvel et important théorème de la théorie des fonctions*, Acta Math. **22** (1899), 359–364. (Cited on 450.)

[285] J. L. Jensen, *Recherches sur la theorie des équations*, Acta Math. **36** (1913), 181–195. (Cited on 102.)

[286] R. Jentzsch, *Untersuchungen zur Theorie der Folgen analytischer Funktionen*, Acta Math. **41** (1916), 219–251. (Cited on 239.)

[287] G. A. Jones and D. Singerman, *Complex Functions. An Algebraic and Geometric Viewpoint*, Cambridge University Press, Cambridge, 1987. (Cited on 281.)

[288] G. A. Jones, D. Singerman, and K. Wicks, *The modular group and generalized Farey Graphs*, in Groups—St. Andrews 1989, Vol. 2, pp. 316–338, London Mathematical Society Lecture Notes, Cambridge University Press, Cambridge, 1991. (Cited on 333.)

[289] C. Jordan, *Cours d'analyse de l'Ecole Polytechnique; 3 Calcul Integral: équations différentielles*, Gauthier–Villars, Paris, 1909. The section on the Jordan Curve theorem is available at `http://www.maths.ed.ac.uk/~aar/jordan/jordan.pdf`; original publication 1887. (Cited on 164.)

[290] A. Joseph, A. Melnikov, and R. Rentschler (editors), *Studies in Memory of Issai Schur*, Progress in Mathematics, Birkhäuser, Boston, 2003. (Cited on 306.)

[291] J. Jost, *Riemannian Geometry and Geometric Analysis*, Third edition, Universitext, Springer-Verlag, Berlin, 2002. (Cited on 21.)

[292] J. Jost, *Compact Riemann Surfaces. An introduction to contemporary mathematics*, Third edition, Universitext, Springer-Verlag, Berlin, 2006. (Cited on 589.)

[293] R. Jost, *The General Theory of Quantized Fields*, Lectures in Applied Mathematics (Proc. Summer Seminar, Boulder, CO, 1960), Vol. IV, American Mathematical Society, Providence, RI, 1965. (Cited on 195.)

[294] N. Joukowski, *Über die Konturen der Tragflächen der Drachenflieger*, Zeitschrift für Flugtechnik und Motorluftschiffahart **1** (1910), 281–284. (Cited on 350.)

[295] G. Julia, *Leçons sur les fonctions uniformes à une point singulier essentiel isolé*, Gauthier–Villars, Paris, 1924. (Cited on 573.)

[296] J.-P. Kahane, *Some Random Series of Functions*, 2nd edition, Cambridge Studies in Advanced Mathematics, Cambridge University Press, Cambridge, 1985. (Cited on 58.)

[297] V. Karunakaran, *Complex Analysis*, 2nd edition, Alpha Science International, Harrow, 2005. (Cited on 214.)

[298] E. Kasner, *A new theory of polygenic (or non-monogenic) functions*, Science **66** (1927), 581–582. (Cited on 38.)

[299] T. Kato, *Perturbation Theory for Linear Operators*, 2nd edition, Grundlehren der Mathematischen Wissenschaften, Springer, Berlin-New York, 1976. (Cited on 131.)

[300] S. Katok, *Fuchsian Groups*, Chicago Lectures in Mathematics, University of Chicago Press, Chicago, 1992. (Cited on 335.)

[301] A. Ya. Khinchin, *Continued Fractions*, reprint of the 1964 translation, Dover Publications, Mineola, NY, 1997; first Russian edition 1936. (Cited on 304.)

[302] S. Khrushchev, *Schur's algorithm, orthogonal polynomials, and convergence of Wall's continued fractions in $L^2(\mathbb{T})$*, J. Approx. Theory **108** (2001), 161–248. (Cited on 306.)

[303] S. Khrushchev, *Classification theorems for general orthogonal polynomials on the unit circle*, J. Approx. Theory **116** (2002), 268–342. (Cited on 306.)

[304] S. Khrushchev, *Orthogonal polynomials: The first minutes*, in Spectral Theory and Mathematical Physics: A Festschrift in Honor of Barry Simon's 60th Birthday, pp. 875–905, Proc. Sympos. Pure Math., American Mathematical Society, Providence, RI, 2007. (Cited on 306.)

[305] S. Khrushchev, *Orthogonal Polynomials and Continued Fractions. From Euler's Point of View*, Encyclopedia of Mathematics and Its Applications, Cambridge University Press, Cambridge, 2008. (Cited on 306.)

[306] S-h. Kim and S. Östlund, *Simultaneous rational approximations in the study of dynamical systems*, Phys. Rev. A (3) **34** (1986), 3426–3434. (Cited on 333.)

[307] R. B. King, *Beyond the Quartic Equation*, reprint of the 1996 original, Modern Birkhäuser Classics, Birkhäuser, Boston, 2009. (Cited on 479.)

[308] F. Klein, *Über binäre Formen mit linearen Transformationen in sich selbst*, Math. Ann. **9** (1875), 183–208. (Cited on 283.)

[309] F. Klein, *Sull' equazioni dell' icosaedro nella risoluzione delle equazioni del quinto grado [per funzioni ellittiche]*, Rend. Istituto Lombardo (2) **10** (1877), 253–255. (Cited on 550.)

[310] F. Klein, *Über die Transformation der elliptischen Funktionen und die Auflösung der Gleichungen fünften Grades*, Math. Ann. **14** (1878), 111–172. (Cited on 550.)

[311] F. Klein, *Neue Beiträge zur Riemann'schen Functionentheorie*, Math. Ann. **21** (1883), 141–218. (Cited on 368.)

[312] F. Klein, *Lectures on the Icosahedron and the Solution of Equations of the Fifth Degree*, 2nd and revised edition, Dover Publications, New York, 1956; a republication of the English translation of the second revised edition, Kegan–Paul, London, 1914. Original German edition 1884; original English translation by above authors, 1888. (Cited on 283, 480.)

[313] F. Klein, *On Riemann's Theory of Algebraic Functions and Their Integrals: A Supplement to the Usual Treatises*, Dover Publications, New York, 1963; unaltered reprinting of the original translation first published in 1893. (Cited on 266.)

[314] F. Klein, *Development of Mathematics in the 19th Century*, Lie Groups: History, Frontiers and Applications, Math Sci Press, Brookline, MA, 1979. (Cited on 476.)

[315] A. W. Knapp, *Elliptic Curves*, Mathematical Notes, Princeton University Press, Princeton, NJ, 1992. (Cited on 477.)

[316] K. Knopp, *Theory of Functions. I. Elements of the General Theory of Analytic Functions*, Dover Publications, New York, 1945. (Cited on 420.)

[317] K. Knopp, *Theory of Functions. II. Applications and Continuation of the General Theory*, Dover Publications, New York, 1947. (Cited on 420.)

[318] K. Knopp, *Infinite Sequences and Series*, Dover Publications, New York, 1956. (Cited on 12.)

[319] K. Knopp, *Theory of Functions, Parts I and II*, Dover Publications, New York, 1996. (Cited on 420.)

[320] S. Kobayashi and K. Nomizu, *Foundations of Differential Geometry. Vol. II*, reprint of the 1969 original, Wiley Classics Library, John Wiley & Sons, New York, 1996. (Cited on 21.)

[321] H. Kober, *Dictionary of Conformal Representations*, Dover Publications, New York, 1952. (Cited on 350.)

[322] N. Koblitz, *Introduction to Elliptic Curves and Modular Forms*, 2nd edition, Graduate Texts in Mathematics, Springer-Verlag, New York, 1993. (Cited on 477, 550.)

[323] H. Koch, *Introduction to Classical Mathematics. I. From the Quadratic Reciprocity Law to the Uniformization theorem*, translated and revised from the 1986 German original, Mathematics and Its Applications, Kluwer, Dordrecht, 1991. (Cited on 368.)

[324] P. Koebe, *Über die Uniformisierung beliebiger analytischer Kurven, Erste Mitteilung, Zweite Mitteilung*, Nachr. Kön. Ges. Wiss. Göttingen, Math.-phys. Klasse (1907), 191–210; 633–669. (Cited on 314, 368.)

[325] P. Koebe, *Über die Uniformisierung beliebiger analytischer Kurven, Dritte Mitteilung*, Nachr. Kön. Ges. Wiss. Göttingen, Math.-phys. Klasse (1908), 337–358. (Cited on 238, 314.)

[326] P. Koebe, *Über die Uniformisierung der algebraischen Kurven. I*, Math. Ann. **67** (1909), 145–224. (Cited on 314.)

[327] P. Koebe, *Über eine neue Methode der konformen Abbildung und Uniformisierung*, Nachr. Kön. Ges. Wiss. Göttingen, Math.-phys. Klasse **1912**, 844–848. (Cited on 314.)

[328] G. Köthe, *Dualität in der Funktionentheorie*, J. Reine Angew. Math. **191** (1953), 30–49. (Cited on 230.)

[329] R. Kozhan, *Szegő asymptotics for matrix-valued measures with countably many bound states*, J. Approx. Theory **162** (2010), 1211–1224. (Cited on 456.)

[330] S. G. Krantz, *Function Theory of Several Complex Variables*, reprint of the 1992 edition, AMS Chelsea Publishing, Providence, RI, 2001. (Cited on 585.)

[331] S. G. Krantz, *Geometric Function Theory: Explorations in Complex Analysis*, Birkhäuser, Boston, 2006. (Cited on 323, 362.)

[332] W. Kühnel, *Differential Geometry: Curves—Surfaces—Manifolds*, Student Mathematical Library, American Mathematical Society, Providence, RI, 2002. (Cited on 21.)

[333] K. Kunugi, *Theory of Functions of a Complex Variable*, Iwanami Shoten, Publishers, Tokyo, Japan, 1958. [Chinese] (Cited on 152.)

[334] P. K. Kythe, *Computational Conformal Mapping*, Birkhäuser, Boston, 1998. (Cited on 350.)

[335] M. Laczkovich, *On Lambert's proof of the irrationality of* π, Amer. Math. Monthly **104** (1997), 439–443. (Cited on 305.)

[336] E. N. Laguerre, *Sur la determination de genre d'une fonction transcendente entière*, C. R. Acad. Sci. Paris **94** (1882), 635–638. (Cited on 469, 474.)

[337] J. H. Lambert, *Mémoire sur quelques proprietes remarquables des quantites transcendantes circulaires et logarithmiques*, Mémoires de l'Academie des Sciences de Berlin **17** (1761), 265–322. (Cited on 305.)

[338] E. Landau, *Über eine Verallgemeinerung des Picardschen Satzes*, Berl. Ber. (1904), 1118–1133. (Cited on 12, 578.)

[339] E. Landau, *On a familiar theorem of the theory of functions*, Bull. Amer. Math. Soc. **12** (1906), 155–156. (Cited on 128.)

[340] E. Landau, *Über die Zetafunktion und die Hadamardsche Theorie der ganzen Funktionen*, Math. Z. **26** (1927), 170–175. (Cited on 468.)

[341] E. Landau, *Über die Blochsche Konstante und zwei verwandte Weltkonstanten*, Math. Z. **30** (1929), 608–634. (Cited on 578, 579.)

[342] E. Landau and D. Gaier, *Darstellung und Begründung einiger neuerer Ergebnisse der Funktionentheorie*, 3rd edition, Springer-Verlag, Berlin, 1986. (Cited on 450.)

[343] N. S. Landkof, *Foundations of Modern Potential Theory*, Die Grundlehren der mathematischen Wissenschaften, Springer-Verlag, Berlin-New York, 1972. (Cited on 324.)

[344] S. Lang, *Elliptic Functions*, 2nd edition, Graduate Texts in Mathematics, Springer-Verlag, New York, 1987. (Cited on 477.)

[345] S. Lang, *Linear Algebra*, reprint of the 3rd edition, Undergraduate Texts in Mathematics, Springer-Verlag, New York, 1989. (Cited on 8.)

[346] S. Lang, *Real and Functional Analysis*, 3rd edition, Graduate Texts in Mathematics, Springer-Verlag, New York, 1993. (Cited on 12.)

[347] T. D. Lee and C. N. Yang, *Statistical theory of equations of state and phase transitions, I. Theory of condensation*, Phys. Rev. (2) **87** (1952), 404–409. (Cited on 239, 240.)

[348] T. D. Lee and C. N. Yang, *Statistical theory of equations of state and phase transitions. II. Lattice gas and Ising model*, Phys. Rev. (2) **87** (1952), 410–419. (Cited on 239, 240.)

[349] A.-M. Legendre, *Théorie Des Nombres*, Du Pratt, Paris, 1798. (Cited on 304.)

[350] A.-M. Legendre, *Exercices de Calcul intégral*, Paris, three volumes, Courcier, Paris, 1811–1817. (Cited on 498.)

[351] A.-M. Legendre, *Traité des Fonctions Elliptiques*, Paris, three volumes 1825–1832. (Cited on 498, 517.)

[352] G. M. Leibowitz, *Lectures on Complex Function Algebras*, Scott, Foresman and Co., Glenview, IL, 1970. (Cited on 157.)

[353] F. Lemmermeyer, *Reciprocity Laws. From Euler to Eisenstein*, Springer Monographs in Mathematics, Springer-Verlag, Berlin, 2000. (Cited on 479.)

[354] G. Leoni, *A First Course in Sobolev Spaces*, Graduate Studies in Mathematics, American Mathematical Society, Providence, RI, 2009. (Cited on 165.)

[355] E. Lindelöf, *Sur l'application de la méthode des approximations successives aux équations différentielles ordinaires du premier ordre*, C. R. Acad. Sci. Paris **114** (1894), 454–457. (Cited on 12.)

[356] E. Lindelöf, *Le calcul des résidus et ses applications à la théorie des fonctions*, Paris, Gauthier-Villars, 1905; available online at `http://www.gutenberg.org/ebooks/29781`. (Cited on 214.)

[357] E. Lindelöf, *Démonstration nouvelle d'un théorème fondamental sur les suites de fonctions monogènes*, Bull. Soc. Math. France **41** (1913), 171–178. (Cited on 239.)

[358] E. Lindelöf, *Sur la représentation conforme*, C. R. Acad. Sci. Paris **158** (1914), 245–247. (Cited on 323.)

[359] E. Lindelöf, *Sur un principe général de l'analyse et ses applications à la théorie de la représentation conforme*, Acta Soc. Fennicae **46** (1915), 35 S. (Cited on 177, 323.)

[360] J.-L. Lions, *Théorèmes de trace et d'interpolation. I*, Ann. Scuola Norm. Sup. Pisa (3) **13** (1959), 389–403. (Cited on 177.)

[361] J.-L. Lions, *Théorèmes de trace et d'interpolation. II*, Ann. Scuola Norm. Sup. Pisa (3) **14** (1960), 317–331. (Cited on 177.)

[362] J.-L. Lions, *Théorèmes de trace et d'interpolation. III*, J. Math. Pures Appl. (9) **42** (1963), 195–203. (Cited on 177.)

[363] J. Liouville, *Remarques de M. Liouville*, C. R. Acad. Sci. Paris **19** (1844), 1261–1263. (Cited on 87.)

[364] M. Livio, *The Equation That Couldn't Be Solved: How Mathematical Genius Discovered the Language of Symmetry*, Simon & Schuster, New York, 2005. (Cited on 499.)

[365] A. J. Lohwater and Ch. Pommerenke, *On normal meromorphic functions*, Ann. Acad. Sci. Fenn. A (1973), 12 pp. (Cited on 578.)

[366] H. Loomann, *Über die Cauchy-Riemannschen Differentialgleichungen*, Gött. Nachr. (1923), 97–108. (Cited on 68.)

[367] L. H. Loomis and S. Sternberg, *Advanced Calculus*, revised edition, Jones and Bartlett Publishers, Boston, 1990, first edition, 1968. (Cited on 12, 17.)

[368] D. S. Lubinsky, *Universality limits in the bulk for arbitrary measures on compact sets*, J. Anal. Math. **106** (2008), 373–394. (Cited on 564.)

[369] D. H. Luecking and L. A. Rubel, *Complex Analysis. A Functional Analysis Approach*, Universitext, Springer-Verlag, New York, 1984. (Cited on 229, 230, 233, 405.)

[370] J. Lützen, *Joseph Liouville 1809–1882: Master of Pure and Applied Mathematics*, Studies in the History of Mathematics and Physical Sciences, Springer-Verlag, New York, 1990. (Cited on 87, 499.)

[371] C. Maclaurin, *Treatise of Fluxions*, T. W. and T. Ruddimans, Edinburgh, 1742; available online at `http://echo.mpiwg-berlin.mpg.de/MPIWG:8W2EVQQN`. (Cited on 438.)

[372] A. I. Markushevich, *Theory of Functions of a Complex Variable. Vol. I, II, III*, 2nd English edition, Chelsea Publishing, New York, 1977; Russian originial, 1950. (Cited on 323, 502.)

[373] A. I. Markushevich, *Analytic Function Theory*, in Mathematics of the 19th Century: Vol. II: Geometry, Analytic Function Theory (A. N. Kolmogorov and A. P. Yushkevich, eds.), Birkhäuser, Basel, 1996. (Cited on 517, 574.)

[374] E. Martinelli, *Sopra una dimostrazione di R. Fueter per un teorema di Hartogs*, Comment. Math. Helv. **15** (1942/43), 340–349. (Cited on 584.)

[375] F. Marty, *Recherches sur la répartition des valeurs d'une fonction méromorphe*, Ann. Fac. Sci. Toulouse Math. (3) **23** (1931), 183–261. (Cited on 252.)

[376] J. N. Mather, *Topological proofs of some purely topological consequences of Carathéodory's theory of prime ends*, in Selected Studies: Physics-Astrophysics, Mathematics, History of Science, pp. 225–255, North–Holland, Amsterdam–New York, 1982. (Cited on 324.)

[377] V. Maz'ya and T. Shaposhnikova, *Jacques Hadamard, A Universal Mathematician*, History of Mathematics, American Mathematical Society, Providence, RI; London Mathematical Society, London, 1998. (Cited on 470.)

[378] H. McKean and V. Moll, *Elliptic Curves*, Cambridge University Press, Cambridge, 1997. (Cited on 477.)

[379] D. Menchoff, *Les conditions de monogeneite*, R. Bussière, Paris, 1936. (Cited on 68.)

[380] S. N. Mergelyan, *Uniform approximations of functions of a complex variable*, Uspehi Matem. Nauk (N.S.) **7** (1952), 31–122; English translation in Trans. Amer. Math. Soc. (1954), 99 pp. (Cited on 156.)

[381] D. Minda, *The image of the Ahlfors function*, Proc. Amer. Math. Soc. **83** (1981), 751–756. (Cited on 378.)

[382] R. Miranda, *Algebraic Curves and Riemann Surfaces*, Graduate Studies in Mathematics, American Mathematical Society, Providence, RI, 1995. (Cited on 267, 589.)

[383] D. S. Mitrinović and J. D. Kečkić, *The Cauchy Method of Residues. Theory and Applications*, Mathematics and Its Applications (East European Series), D. Reidel Publishing, Dordrecht, 1984. (Cited on 214.)

[384] G. Mittag-Leffler, *En metod att analytiskt framställa en funktion af rational karakter, hvilken blir oändlig alltid och endast uti vissa föreskrifna oändlighetspunkter, hvilkas konstanter äro påförhand angifna*, Öfversigt af Kongl. Vetenskaps-Akad. Förhandlingar Stockholm, (1876) 3–16. (Cited on 400.)

[385] G. Mittag-Leffler, *Sur la représentation analytique des fonctions monogènes uniformes d'une variable indépendante*, Acta Math. **4** (1884), 1–79. (Cited on 409.)

[386] T. Miyake, *Modular Forms*, reprint of the 1st 1989 English edition, Springer Monographs in Mathematics, Springer-Verlag, Berlin, 2006. (Cited on 550.)

[387] A. F. Möbius, *Über eine neue Verwandtschaft zwischen ebenen Figuren*, Ber. Verh. Königl. Sächs. Ges. Wiss. Math.-Phys. Kl. **5** (1852), 14–24; J. Reine Angew. Math. **52** (1856), 218–228. (Cited on 282.)

[388] A. F. Möbius, *Die Theorie der Kreisverwandtschaft in rein geometrischer Darstellung*, Abh. Königl. Sächs. Ges. Wiss. Math.-Phys. Kl. **2** (1855), 529–595. (Cited on 282.)

[389] P. Montel, *Sur les suites infinies de fonctions*, Ann. Sci. Ecole Norm. Sup. (3) **24** (1907), 233–334. (Cited on 238.)

[390] P. Montel, *Sur les familles de fonctions analytiques qui admettent des valeurs exceptionnelles dans un domaine*, Ann. Sci. Ec. Norm. Sup. (3) **29** (1912), 487–535. (Cited on 238.)

[391] P. Montel, *Sur les différentielles totales et les fonctions monogènes*, C. R. Acad. Sci. Paris **156** (1913), 1820–1822. (Cited on 68.)

[392] P. Montel, *Leçons sur les familles normales de fonctions analytiques et leurs applications. Recueillies et rédigées par J. Barbotte*, Gauthier–Villars, Paris, 1927; reissued in 1974 by Chelsea, New York. (Cited on 238.)

[393] C. D. Moore, *An Introduction to Continued Fractions*, The National Council of Teachers of Mathematics, Washington, DC, 1964. (Cited on 304.)

[394] L. J. Mordell, *On the rational solutions of the indeterminate equations of the third and fourth degrees*, Cambr. Phil. Soc. Proc. **21** (1922), 179–192. (Cited on 518.)

[395] L. J. Mordell, *On power series with the circle of convergence as a line of essential singularities*, J. London Math. Soc. **2** (1927), 146–148. (Cited on 58.)

[396] G. Morera, *Un teorema fondamentale nella teorica delle funzioni di una variabile complessa*, Rend. Reale Istituto Lombardo di scienze e lettere (2) **19** (1896), 304–307. (Cited on 87.)

[397] D. Mumford, C. Series, and D. Wright, *Indra's Pearls. The Vision of Felix Klein*, Cambridge University Press, New York, 2002. (Cited on 281, 335.)

[398] C. Müntz, *Über den Approximationssatz von Weierstrass*, in H. A. Schwarz Festschrift, pp. 303–312, Mathematische Abhandlungen, Berlin, Springer, 1914. (Cited on 456, 458.)

[399] P. J. Nahin, *An Imaginary Tale. The Story of $\sqrt{-1}$*, reprint of the 1998 edition, Princeton University Press, Princeton, NJ, 2007. (Cited on 17.)

[400] T. Napier and M. Ramachandran, *An Introduction to Riemann Surfaces*, Cornerstones, Birkhäuser/Springer, New York, 2011. (Cited on 589.)

[401] R. Narasimhan, *Analysis on Real and Complex Manifolds*, Advanced Studies in Pure Mathematics, Masson & Cie, Paris; North-Holland, Amsterdam 1968. (Cited on 17.)

[402] R. Narasimhan, *Compact Riemann Surfaces*, Lectures in Mathematics ETH Zürich, Birkhäuser Verlag, Basel, 1992. (Cited on 589.)

[403] R. Narasimhan, *Several Complex Variables*, reprint of the 1971 original, Chicago Lectures in Mathematics, University of Chicago Press, Chicago, 1995. (Cited on 585.)

[404] R. Narasimhan and Y. Nievergelt, *Complex Analysis in One Variable*, Second edition, Birkhäuser Boston, Inc., Boston, MA, 2001; first edition, 1985. (Cited on 68.)

[405] Z. Nehari, *Conformal Mapping*, reprinting of the 1952 edition, Dover Publications, New York, 1975. (Cited on 350, 362.)

[406] E. Neuenschwander, *Studies in the history of complex function theory. I. The Casorati–Weierstrass theorem*, Hist. Math. **5** (1978), 139–166. (Cited on 87, 128.)

[407] C. A. Neumann, *Vorlesungen über Riemann's Theorie der Abel'schen Integrale*, 2nd ed., Teubner, Leipzig 1884; first edition 1865. (Cited on 131.)

[408] R. Nevanlinna, *Zur Theorie der meromorphen Funktionen*, Acta Math. **46** (1925), 1–99. (Cited on 451.)

[409] E. H. Neville, *Jacobian Elliptic Functions*, Oxford University Press, Oxford, 1944. (Cited on 477.)

[410] F. W. Newman, *On $F(a)$ especially when a is negative*, Cambridge and Dublin Math. J. **3** (1848), 57–60. (Cited on 419.)

[411] E. Noether, *Ableitung der Elementarteilertheorie aus der Gruppentheorie*, Jahresber. Deutsch. Math.-Verein. **34** (1926), 104. (Cited on 26.)

[412] K. Ogura, *On a certain transcendental integral function in the theory of interpolation*, Tôhoku Math. J. **17** (1920), 64–72. (Cited on 221.)

[413] M. Ohtsuka, *Dirichlet Problem, Extremal Length and Prime Ends*, van Nostrand Reinhold, New York, 1970. (Cited on 323.)

[414] K. Oldham, J. Myland, and J. Spanier, *An Atlas of Functions. With Equator, the Atlas Function Calculator*, 2nd edition, Springer, New York, 2009. (Cited on 214.)

[415] C. D. Olds, *Continued Fractions*, Random House, New York, 1963. (Cited on 304.)

[416] B. Osgood, *Old and new on the Schwarzian derivative*, in Quasiconformal Mappings and Analysis (Ann Arbor, MI, 1995), pp. 275–308, Springer, New York, 1998. (Cited on 117.)

[417] W. F. Osgood, *Some points in the elements of the theory of functions*, Bull. Amer. Math. Soc. **2** (1896), 296–302. (Cited on 87.)

[418] W. F. Osgood, *On the existence of the Green's function for the most general simply connected plane region*, Trans. Amer. Math. Soc. **1** (1900), 310–314. (Cited on 314.)

[419] W. F. Osgood, *Note on the functions defined by infinite series whose terms are analytic functions of a complex variable; with corresponding theorems for definite integrals*, Ann. Math. (2) **3** (1901–1902), 25–34. (Cited on 128, 238.)

[420] W. F. Osgood, *Functions of Real Variables; Functions of a Complex Variable*, Chelsea Publishing, New York, 1958; originally two volumes published in 1936 by National University Press of Peking. (Cited on 79.)

[421] W. F. Osgood and E. H. Taylor, *Conformal transformations on the boundaries of their regions of definition*, Trans. Amer. Math. Soc. **14** (1913), 277–298. (Cited on 323.)

[422] A. Ostrowski, *Mathematische Miszellen. XV: Zur konformen Abbildung einfach zusammenhängender Gebiete*, Jahresber. Deutsch. Math.-Verein. **38** (1929), 168–182. (Cited on 315.)

[423] P. Painlevé, *Thèse d'Analyse. Sur les lignes singulières des fonctions analytiques*, Thesis, Paris, 1887. (Cited on 323.)

[424] H. Pajot, *Analytic Capacity, Rectifiability, Menger Curvature and the Cauchy Integral*, Lecture Notes in Mathematics, Springer-Verlag, Berlin, 2002. (Cited on 128, 378.)

[425] R. E. A. C. Paley and N. Wiener, *Fourier Transforms in the Complex Domain*, reprint of the 1934 original, American Mathematical Society Colloquium Publications, American Mathematical Society, Providence, RI, 1987. (Cited on 562.)

[426] R. E. A. C. Paley and A. Zygmund, *A note on analytic functions in the unit circle*, Proc. Cambridge Philos. Soc. **28** (1932), 266–272. (Cited on 58.)

[427] P. Pesic, *Abel's Proof: An Essay on the Sources and Meaning of Mathematical Unsolvability*, MIT Press, Cambridge, MA, 2003. (Cited on 499.)

[428] P. Petersen, *Riemannian Geometry*, Graduate Texts in Mathematics, Springer-Verlag, New York, 1998. (Cited on 21.)

[429] E. Phragmén, *Sur une extension d'un théorème classique de la théorie des fonctions*, Acta Math. **28** (1904), 351–368. (Cited on 173.)

[430] E. Phragmén and E. Lindelöf, *Sur une extension d'un principe classique de l'analyse et sur quelques propriétés des fonctions monogènes dans le voisinage d'un point singulier*, Acta Math. **31** (1908), 381–406. (Cited on 173.)

[431] E. Picard, *Sur les fonctions analytiques uniformes dans le voisinage d'un point singulier essentiel*, C. R. Acad. Sci. Paris **88**, (1879) 745–747. (Cited on 573.)

[432] E. Picard, *Sur une propriété des fonctions entières*, C. R. Acad. Sci. Paris **88** (1879), 1024–1027. (Cited on 573.)

[433] E. Picard, *Mémoire sur les fonctions entières*, Ann. Sci. Ecole Norm. Sup. (2) **9** (1880), 145–166. (Cited on 573.)

[434] E. Picard, *Mémoire sur théorie des équations aux dérivés partielles et la méthode des approximations successives*, J. Math. Pures Appl. **6** (1890), 145–210. (Cited on 12.)

[435] E. Picard, *Sur l'application des méthodes d'approximations successives à l'étude de certaines équations différentielles ordinaires*, J. Math. Pures Appl. **9** (1893), 217–271. (Cited on 12.)

[436] Plato, *The Republic*, circa 380 BC, available online at `http://classics.mit.edu/Plato/republic.3.ii.html`. (Cited on 1.)

[437] H. Poincaré, *Sur les transcendentes entières*, C. R. Acad. Sci. Paris **95** (1882), 23–26. (Cited on 469.)

[438] H. Poincaré, *Théorie des groupes fuchsiens*, Acta Math. **1** (1882), 1–62. (Cited on 282, 292.)

[439] H. Poincaré, *Mémoire sur les groupes kleinéens*, Acta Math. **3** (1883), 49–92. (Cited on 282.)

[440] H. Poincaré, *Sur un théorème de la théorie générale des fonctions*, Bull. Soc. Math. France **11** (1883), 112–125. (Cited on 368.)

[441] H. Poincaré, *Sur les fonctions entières*, Bull. Soc. Math. France **11** (1883), 136–144. (Cited on 469.)

[442] H. Poincaré, *Analysis situs*, J. l'Ecole Polytechnique (2) **1** (1895), 1–121. (Cited on 25.)

[443] H. Poincaré, *Les fonctions analytiques de deux variables et la representation conforme*, Rend. Circ. Mat. Palermo **23** (1907), 185–220. (Cited on 584.)

[444] H. Poincaré, *Sur l'uniformisation des fonctions analytiques*, Acta Math. **31** (1908), 1–63. (Cited on 368.)

[445] S. D. Poisson, *Suite du Memoire sur les intégrals definies et sur la sommation des series*, J. l'Ecole Polytechnique **19** (1823), 404–509. (Cited on 419.)

[446] A. Polishchuk, *Abelian Varieties, Theta Functions and the Fourier Transform*, Cambridge Tracts in Mathematics, Cambridge University Press, Cambridge, 2003. (Cited on 534.)

[447] G. Pólya, *On an integral function of an integral function*, J. London Math. Soc. **1** (1926), 12–15. (Cited on 468.)

[448] G. Pólya, *Remarks on computing the probability integral in one and two dimensions*, in Proc. Berkeley Sympos. on Mathematical Statistics and Probability, 1945, 1946, pp. 63–78, University of California Press, Berkeley and Los Angeles, 1949. (Cited on 214.)

[449] G. Pólya, *Induction And Analogy In Mathematics; Volume I of Mathematics and Plausible Reasoning*, Princeton University Press, Princeton, 1954; available online at `https://archive.org/details/Induction_And_Analogy_In_Mathematics_1_` (Cited on 387.)

[450] C. Pommerenke, *Über die analytische Kapazität*, Arch. Math. (Basel) **11** (1960), 270–277. (Cited on 378.)

[451] C. Pommerenke, *Univalent Functions*, Studia Mathematica/Mathematische Lehrbücher, Vandenhoeck & Ruprecht, Göttingen, 1975. (Cited on 323.)

[452] C. Pommerenke, *Boundary Behaviour of Conformal Maps*, Grundlehren der Mathematischen Wissenschaften, Springer-Verlag, Berlin, 1992. (Cited on 323, 324.)

[453] D. Pompeiu, *Sur la continuité des fonctions de variables complexes*, Ann. Fac. Sci. Toulouse (2) **7** (1905), 265–315. (Cited on 194.)

[454] D. Pompeiu, *Sur une classe de fonctions d'une variable complexe*, Rend. Circ. Mat. Palermo **33** (1912), 108–113. (Cited on 78.)

[455] M. B. Porter, *Concerning series of analytic functions*, Ann. Math. **6** (1904–1905), 190–192. (Cited on 238.)

[456] M. B. Porter, *On the polynomial convergents of a power series*, Ann. Math. **8** (1906), 189–192. (Cited on 58.)

[457] V. P. Potapov, *The multiplicative structure of J-contractive matrix functions*, Trans. Amer. Math. Soc. **15** (1960), 131–243; Russian original in Trudy Moskov. Math. Obšč. **4** (1955), 125–236. (Cited on 456.)

[458] A. Pressley, *Elementary Differential Geometry*, 2nd edition, Springer Undergraduate Mathematics Series, Springer-Verlag, London, 2010. (Cited on 21.)

[459] A. Pringsheim, *Über Functionen, welche in gewissen Punkten endliche Differentialquotienten jeder endlichen Ordnung, aber keine Taylor'sche Reihenentwickelung besitzen*, Math. Ann. **44** (1894), 41–56. (Cited on 63.)

[460] A. Pringsheim, *Über den Goursat'schen Beweis des Cauchy'schen Integralsatzes*, Trans. Amer. Math. Soc. **2** (1901), 413–421. (Cited on 68.)

[461] A. Pringsheim, *Über die Weierstraßsche Produktdarstellung ganzer transzendenter Funktionen und über bedingt konvergente unendliche Produkte*, Sitzungsber. Konigl. Bayer. Akad. Wiss., Math.-Phys. (1915), 387–400. (Cited on 405.)

[462] F. E. Prym, *Zur Theorie der Gammafunction*, J. Reine Angew. Math. **82** (1877), 165–172. (Cited on 430.)

[463] V. A. Puiseux, *Recherches sur les fonctions algébriques*, J. Math. Pures Appl. **15** (1850), 365–480. (Cited on 113, 266.)

[464] H. Rademacher, *Higher Mathematics From an Elementary Point of View*, Birkhäuser, Boston, 1983. (Cited on 304, 333.)

[465] H. Rademacher and E. Grosswald, *Dedekind Sums*, The Carus Mathematical Monographs, The Mathematical Association of America, Washington, DC, 1972. (Cited on 222.)

[466] T. Radó, *Über die Fundamentalabbildungen schlichter Gebiete*, Acta Sci. Math. Szeged **1** (1923), 240–251. (Cited on 315, 356.)

[467] T. Radó, *Über den Begriff der Riemannschen Fläche*, Acta Sci. Math. Szeged **2** (1925), 101–121. (Cited on 266.)

[468] R. M. Range, *Holomorphic Functions and Integral Representations in Several Complex Variables*, Graduate Texts in Mathematics, Springer-Verlag, New York, 1986. (Cited on 585.)

[469] R. Range, *Extension phenomena in multidimensional complex analysis: Correction of the historical record*, Math. Intelligencer **24** (2002), 4–12. (Cited on 584.)

[470] A. Ranicki, *Jordan curve theorem*, available online at `http://www.maths.ed.ac.uk/~aar/jordan/index.htm` (Cited on 164.)

[471] T. Ransford, *Potential Theory in the Complex Plane*, London Mathematical Society Student Texts, Cambridge University Press, Cambridge, 1995. (Cited on 324.)

[472] H. E. Rauch and H. M. Farkas, *Theta Functions With Applications to Riemann Surfaces*, Williams & Wilkins, Baltimore, MD, 1974. (Cited on 534.)

[473] H. E. Rauch and A. Lebowitz, *Elliptic Functions, Theta Functions, and Riemann Surfaces*, Williams & Wilkins, Baltimore, MD, 1973. (Cited on 477.)

[474] C. Reid, *Courant in Göttingen and New York. The Story of an Improbable Mathematician*, Springer-Verlag, New York–Heidelberg, 1976. (Cited on 369.)

[475] R. Remmert, *Theory of Complex Functions*, Graduate Texts in Mathematics, Springer-Verlag, New York, 1991. (Cited on 87.)

[476] R. Remmert, *Wielandt's theorem about the Γ-function*, Amer. Math. Monthly **103** (1996), 214–220. (Cited on 420.)

[477] R. Remmert, *Classical Topics in Complex Function Theory*, Graduate Texts in Mathematics, Springer-Verlag, New York, 1998. (Cited on 58, 159, 227, 315, 405, 420, 579.)

[478] A. Rice, *In search of the "birthday" of elliptic functions*, Math. Intelligencer **30** (2008), 48–56. (Cited on 477.)

[479] S. Rickman, *On the number of omitted values of entire quasiregular mappings*, J. Anal. Math. **37** (1980), 100–117. (Cited on 574.)

[480] G. F. B. Riemann, *Grundlagen für eine allgemeine Theorie der Functionen einer veränderlichen complexen Grösse*, inaugural dissertation (1851), Göttingen; Werke, 5–43. Available online in German at `http://www.maths.tcd.ie/pub/HistMath/People/Riemann/Grund/`. (Cited on 37, 38, 265, 314, 315, 568.)

[481] G. F. B. Riemann, *Über die Darstellbarkeit einer Function durch eine trigonometrische Reihe*, Habilitationsschrift, Universität Göttingen, 1854; Published posthumously in Abh. Königlich. Gesellschaft der Wissenschaften zu Göttingen **13**, 1867. Available online in German at `http://www.maths.tcd.ie/pub/HistMath/People/Riemann/Trig/`. (Cited on 316.)

[482] G. F. B. Riemann, *Über die Hypothesen, welche der Geometrie zu Grunde liegen*, written in 1854;. Published posthumously in Abh. Königlic. Gesellschaft der Wissenschaften zu Göttingen **13** (1867). Available online in German and in translation at `http://www.maths.tcd.ie/pub/HistMath/People/Riemann/Geom/`. (Cited on 21, 316.)

[483] G. F. B. Riemann, *Beiträge zur Theorie der durch die Gauss'sche Reihe $F(\alpha, \beta, \gamma, x)$ darstellbaren Functionen*, Abh. Königlich. Gesellschaft der Wissenschaften zu Göttingen **7** (1857); Available in German at `http://www.maths.tcd.ie/pub/HistMath/People/Riemann/PFunct/`. (Cited on 316, 568.)

[484] G. F. B. Riemann, *Theorie der Abel'schen Functionen*, J. Reine Angew. Math. **54** (1857), 101–155; Available online in German at `http://www.maths.tcd.ie/pub/HistMath/People/Riemann/AbelFn/`. (Cited on 266, 315, 534, 589.)

[485] G. F. B. Riemann, *Über die Anzahl der Primzahlen unter einer gegebenen Grösse*, Monatsber. Berlin Akad. (1858/60), 671–680; Available online in German and in translation at `http://www.maths.tcd.ie/pub/HistMath/People/Riemann/Zeta/`. (Cited on 316.)

[486] M. Riesz, *Sätze über Potenzreihen*, Ark. Mat., Astr. och Fys. **11** (1916), 1–16. (Cited on 242.)

[487] M. Riesz, *Sur les fonctions conjuguées*, Math. Z. **27** (1928), 218–244. (Cited on 177.)

[488] M. Riesz, *Sur les maxima des formes bilinéaires et sur les fonctionelles linéaires*, Acta Math. **49** (1927), 465–497. (Cited on 177.)

[489] E. Röding, *Über die Wertannahme der Ahlfors Funktion in beliebigen Gebieten*, Manuscripta Math. **20** (1977), 133–140. (Cited on 378.)

[490] W. W. Rogosinski, *On the order of the derivatives of a function analytic in an angle*, J. London Math. Soc. **20** (1945), 100–109. (Cited on 182.)

[491] A. Ros, *The Gauss map of minimal surfaces*, in Differential Geometry (Valencia, 2001), pp. 235–252, World Scientific, River Edge, NJ, 2002. (Cited on 578.)

[492] H. A. Rothe, *Systematisches Lehrbuch der Arithmetic*, Johann Ambrosius Barth, Leipzig, 1811. (Cited on 534.)

[493] E. Rouché, *Mémoire sur la série de Lagrange*, J. l'Ecole Polytechnique **22** (1868), 193–224. (Cited on 100.)

[494] L. A. Rubel, *How to use Runge's theorem*, Enseign. Math. **22** (1976), 185–190. (Cited on 161.)

[495] W. Rudin, *Lectures on the Edge-of-the-Wedge Theorem*, Conference Board of the Mathematical Sciences Regional Conference Series in Mathematics, American Mathematical Society, Providence, RI, 1971. (Cited on 194, 195.)

[496] W. Rudin, *Real and Complex Analysis*, 3rd edition, McGraw–Hill, New York, 1987. (Cited on 454.)

[497] W. Rudin, *Function Theory in the Unit Ball of $\mathbb{C}^n$*, reprint of the 1980 edition, Classics in Mathematics, Springer-Verlag, Berlin, 2008. (Cited on 585.)

[498] C. Runge, *Zur Theorie der eindeutigen analytischen Functionen*, Acta Math. **6** (1885), 229–244. (Cited on 156, 409.)

[499] L. Saalschütz, *Bemerkungen über die Gammafunctionen mit negativen Argumenten*, Z. Math. Phys. **32** (1887), 246–250. (Cited on 430.)

[500] S. Saks and A. Zygmund, *Analytic Functions*, 3rd edition, Elsevier, Amsterdam–London–New York; PWN—Polish Scientific Publishers, Warsaw, 1971 (Polish original first edition in 1938). (Cited on 149, 157, 238, 574.)

[501] Z. Sasvári, *An elementary proof of Binet's formula for the gamma function*, Amer. Math. Monthly **106** (1999), 156–158. (Cited on 447.)

[502] V. Scheidemann, *Introduction to Complex Analysis in Several Variables*, Birkhäuser Verlag, Basel, 2005. (Cited on 584, 585.)

[503] J. L. Schiff, *Normal Families*, Universitext, Springer-Verlag, New York, 1993. (Cited on 573, 574, 578.)

[504] R. Schinzinger and P. A. A. Laura, *Conformal Mapping. Methods and Applications*, revised edition of the 1991 original, Dover Publications, Mineola, NY, 2003. (Cited on 350.)

[505] O. Schlömilch, *Einiges über die Eulerischen Integrale der zweiten Art*, Arch. Math. Phys. **4** (1843), 167–174. (Cited on 419.)

[506] F. Schottky, *Über den Picard'schen Satz und die Borel'schen Ungleichungen*, Sitzungsber. Preuss. Akad. Wiss. **2** (1904), 1244–1262. (Cited on 578.)

[507] I. Schur, *Über Potenzreihen, die im Innern des Einheitskreises beschränkt sind, I, II*, J. Reine Angew. Math. **147** (1917), 205–232; **148** (1918), 122–145. English translation in I. Schur. Methods in Operator Theory and Signal Processing (edited by I. Gohberg), pp. 31–59, 66–88, Operator Theory: Advances and Applications, Birkhäuser, Basel, 1986. (Cited on 239, 305.)

[508] L. Schwartz, *Transformation de Laplace des distributions*, Comm. Sém. Math. Univ. Lund **1952**, Tome Suppl. (1952), 196–206. (Cited on 562.)

[509] H. A. Schwarz, *Über einige Abbildungsaufgaben*, J. Reine Angew. Math. **70** (1869), 105–120. (Cited on 351.)

[510] H. A. Schwarz, *Conforme Abbildung der Oberfläche eines Tetraeders auf die Oberfläche einer Kugel*, J. Reine Angew. Math. **70** (1869), 121–136. (Cited on 351.)

[511] H. A. Schwarz, *Zur Integration der partiellen Differential-Gleichung* $\frac{\partial^2 u}{\partial x^2} + \frac{\partial^2 u}{\partial y^2} = 0$, J. Reine Angew. Math. **74** (1872), 218–253. (Cited on 181.)

[512] H. A. Schwarz, *Über diejenigen Fälle, in welchen die Gaussische hypergeometrische Reihe eine algebraische Function ihres vierten Elementes darstellt*, J. Reine Angew. Math. **75** (1873), 292–335. (Cited on 283.)

[513] H. Schwerdtfeger, *Geometry of Complex Numbers. Circle Geometry, Moebius Transformation, Non-Euclidean Geometry*, Dover Books on Advanced Mathematics, Dover Publications, New York, 1979; a corrected reprinting of the 1962 edition: *Geometry of Complex Numbers*, Mathematical Expositions, University of Toronto Press, Toronto, 1962. (Cited on 281.)

[514] S. L. Segal, *Nine Introductions in Complex Analysis*, revised edition, North-Holland Mathematics Studies, Elsevier, Amsterdam, 2008; first edition, 1981. (Cited on 468, 502.)

[515] W. Seidel and J. L. Walsh, *On approximation by euclidean and non-euclidean translations of an analytic function*, Bull. Amer. Math. Soc. **47** (1941), 916–920. (Cited on 161.)

[516] A. Selberg, *Remarks on a multiple integral*, Norsk Mat. Tidsskr. **26** (1944), 71–78. [Norwegian] (Cited on 419.)

[517] C. Series, *The modular surface and continued fractions*, J. London Math. Soc. (2) **31** (1985), 69–80. (Cited on 333.)

[518] J.-P. Serre, *A Course in Arithmetic*, Graduate Texts in Mathematics, Springer-Verlag, New York–Heidelberg, 1973. (Cited on 550.)

[519] J. Serrin and H. Zou, *Cauchy–Liouville and universal boundedness theorems for quasilinear elliptic equations and inequalities*, Acta Math. **189** (2002), 79–142. (Cited on 87.)

[520] C. L. Siegel, *Topics in Complex Function Theory. Vol. I: Elliptic Functions and Uniformization Theory*, Interscience Tracts in Pure and Applied Mathematics, Wiley-Interscience, New York–London–Sydney, 1969. (Cited on 477.)

[521] J. H. Silverman, *Advanced Topics in the Arithmetic of Elliptic Curves*, Graduate Texts in Mathematics, Springer-Verlag, New York, 1994. (Cited on 550.)

[522] J. H. Silverman, *The Arithmetic of Elliptic Curves*, 2nd edition, Graduate Texts in Mathematics, Springer, Dordrecht, 2009. (Cited on 518.)

[523] B. Simon, *Representations of Finite and Compact Groups*, Graduate Studies in Mathematics, American Mathematical Society, Providence, RI, 1996. (Cited on 286.)

[524] B. Simon, *Orthogonal Polynomials on the Unit Circle, Part 1: Classical Theory*, AMS Colloquium Series, American Mathematical Society, Providence, RI, 2005. (Cited on 239, 306, 535.)

[525] B. Simon, *Szegő's Theorem and Its Descendants: Spectral Theory for L^2 Perturbations of Orthogonal Polynomials*, M. B. Porter Lectures, Princeton University Press, Princeton, NJ, 2011. (Cited on 289, 335, 350.)

[526] F. Smithies, *Cauchy and the Creation of Complex Function Theory*, Cambridge University Press, Cambridge, 1997. (Cited on 39.)

[527] Yu. V. Sokhotskii, *Theory of Integral Residues With Some Applications*, St. Petersburg, 1868. [Russian] (Cited on 128.)

[528] H. K. Sørensen, *Niels Henrik Abel: Transformation and Continuity of Mathematics in the 1820s* in Mathematics throughout the Ages, pp. 178–185, Prometheus, Prague, 2001. (Cited on 499.)

[529] H. K. Sørensen, *Abel and his Mathematics in Contexts*, NTM Intern. J. of History & Ethics of Natural Sciences Technology & Medicine **10** (2002), 137–155. (Cited on 499.)

[530] M. Spivak, *Calculus on Manifolds. A Modern Approach to Classical Theorems of Advanced Calculus*, W. A. Benjamin, New York–Amsterdam, 1965. (Cited on 17.)

[531] M. Spivak, *A Comprehensive Introduction to Differential Geometry. Vol. III*, 2nd edition, Publish or Perish, Wilmington, DE, 1979. (Cited on 21.)

[532] M. Z. Spivey, *The Euler–Maclaurin formula and sums of powers*, Math. Magazine **79** (2006), 61–65. (Cited on 444.)

[533] G. Springer, *Introduction to Riemann Surfaces*, 2nd edition, AMS Chelsea Publishing, American Mathematical Society, Providence, RI, 2002. (Cited on 267, 589.)

[534] K. N. Srinivasa Rao, *A contour for the Poisson integral*, Elem. Math. **27** (1972), 88–90. (Cited on 214.)

[535] G. K. Srinivasan, *The gamma function: an eclectic tour*, Amer. Math. Monthly **114** (2007), 297–315. (Cited on 421, 426.)

[536] E. M. Stein, *Interpolation of linear operators*, Trans. Amer. Math. Soc. **83** (1956), 482–492. (Cited on 177.)

[537] H. Steinhaus, *Über die Wahrscheinlichkeit dafür, daß der Konvergenzkreis einer Potenzreihe ihre natürliche Grenze ist*, Math. Z. **31** (1930), 408–416. (Cited on 58.)

[538] M. A. Stern, *Über eine zahlentheoretische Funktion*, J. Reine Angew. Math. **55** (1858), 193–220; available online at `http://gdz.sub.uni-goettingen.de/no_cache/en/dms/load/img/?IDDOC=268546`. (Cited on 333.)

[539] S. Sternberg, *Lectures on Differential Geometry*, 2nd edition, Chelsea Publishing, New York, 1983; first edition, 1964. (Cited on 21.)

[540] T. Stieltjes, *Sur le développement de* $\log\Gamma(a)$, J. Math. Pures Appl. (4) **5** (1889), 425–444. (Cited on 440.)

[541] T. Stieltjes, *Note sur l'intégrale* $\int_0^\infty e^{-u^2}\,du$, Nouv. Ann. de Math. (3) **9** (1890), 479–480. (Cited on 222.)

[542] T. Stieltjes, *Recherches sur les fractions continues*, Ann. Fac. Sci. Toulouse (1) **8** (1894), J1–J122; **9** (1895), A5–A47. (Cited on 238.)

[543] J. Stillwell, *Mathematics and Its History*, 3rd edition, Undergraduate Texts in Mathematics, Springer-Verlag, New York, 2010; first edition, 1989. (Cited on 518.)

[544] J. Stirling, *Methodus Differentialis*, Whiston and White, London, 1730. [James Stirling's *Methodus Differentialis*. An Annotated Translation of Stirling's Text, by I. Tweddle, Sources and Studies in the History of Mathematics and Physical Sciences, Springer-Verlag, London, 2003.] (Cited on 437.)

[545] E. L. Stout, *The Theory of Uniform Algebras*, Bogden & Quigley, Tarrytown-on-Hudson, NY, 1971. (Cited on 157.)

[546] R. F. Streater and A. S. Wightman, *PCT, Spin and Statistics, and All That*, 2nd edition, Mathematical Physics Monograph Series, Benjamin/Cummings, Reading, Mass.-London-Amsterdam, 1978; first edition, 1964. (Cited on 195.)

[547] A. Stubhaug, *Gösta Mittag-Leffler. A man of conviction*, Translated from the 2007 Norwegian original by Tiina Nunnally, Springer-Verlag, Berlin, 2010. (Cited on 400.)

[548] O. Szász, *Über die Approximation stetiger Funktionen durch lineare Aggregate von Potenzen*, Math. Ann. **77** (1916), 482–496. (Cited on 456, 458.)

[549] G. Szegő, *Tschebyscheffsche Polynome und nichtfortsetzbare Potenzreihen*, Math. Ann. **87** (1922), 90–111. (Cited on 58.)

[550] G. Szegő, *Orthogonal Polynomials*, Amer. Math. Soc. Colloq. Publ., American Mathematical Society, Providence, RI, 1939; 3rd edition, 1967. (Cited on 350.)

[551] S. Takakazu, *Katsuyo sampo*, [Essentials of Mathematics], 1712; available online at `http://mathdl.maa.org/mathDL/46/?pa=content&sa=viewDocument&nodeId=2591&bodyId=3549`. (Cited on 437.)

[552] B. Taylor, *Methodus incrementorum directa et inversa*, Impensis Gulielmi Innys, London, 1715; translation available online at `www.17centurymaths.com/contents/taylorscontents.html`. (Cited on 57.)

[553] N. M. Temme, *Chapter 7. Error Functions, Dawson's and Fresnel Integrals*, in NIST Digital Library of Mathematical Functions; available online at `http://dlmf.nist.gov/7`. (Cited on 214.)

[554] O. Thorin, *An extension of a convexity theorem due to M. Riesz*, Fys. Sallsk. Forh. **8** (1938), no. 14. (Cited on 177.)

[555] O. Thorin, *Convexity theorems generalizing those of M. Riesz and Hadamard*, Thesis; Ak. Avh. Lund, 1948. (Cited on 177.)

[556] E. C. Titchmarsh, *The Theory of Functions*, The Clarendon Press, Oxford, 1932; first edition, 1932; available online at `https://archive.org/details/TheTheoryOfFunctions`. (Cited on 214, 468, 579.)

[557] L. E. Turner, *The Mittag–Leffler Theorem: The Origin, Evolution, and Reception of a Mathematical Result, 1876–1884*, Simon Fraser University master's thesis, 2007; available online at `http://www.math.sfu.ca/~tarchi/turnermsc2007.pdf`. (Cited on 381.)

[558] K. Ueno, *An Introduction to Algebraic Geometry*, Translations of Mathematical Monographs, American Mathematical Society, Providence, RI, 1997. (Cited on 267.)

[559] D. C. Ullrich, *Complex Made Simple*, Graduate Studies in Mathematics, American Mathematical Society, Providence, RI, 2008. (Cited on 333, 573.)

[560] V. S. Varadarajan, *Euler Through Time: A New Look at Old Themes*, American Mathematical Society, Providence, RI, 2006. (Cited on 395.)

[561] O. Veblen, *Theory of plane curves in non-metrical analysis situs*, Trans. Amer. Math. Soc. **6** (1905), 83–98. (Cited on 164.)

[562] W. A. Veech, *A Second Course in Complex Analysis*, W. A. Benjamin, New York–Amsterdam, 1967. (Cited on 315, 333, 573.)

[563] L. Vietoris, *Über den höheren Zusammenhang kompakter Räume und eine Klasse von zusammenhangstreuen Abbildungen*, Math. Ann. **97** (1927), 454–472. (Cited on 26.)

[564] G. Vitali, *Sopra le serie de funzioni analitiche*, Rend. Ist. Lombardo (2) **36** (1903), 772–774; Ann. Mat. Pura Appl. (3) **10** (1904), 65–82. (Cited on 238.)

[565] G. Vivanti, *Sulle serie di potenze*, Rivista di Mat. **3** (1893) 111–114. (Cited on 63.)

[566] V. Volterra, *Sopra le funzioni che dipendono da altre funzioni, I, II, III*, Rend. Accad. Lincei (4) **3** (1887), 97–105, 141–146, 153–158. (Cited on 585.)

[567] T. von Kármán and E. Trefftz, *Potentialströmung um gegebene Tragflächenquerschnitte*, Z. für Flugtechnik und Motorluftschiffahart **9**, 1918. (Cited on 350.)

[568] H. von Koch, *Sur une courbe continue sans tangente, obtenue par une construction géométrique élémentaire*, Ark. Mat., Astr. och Fys. **1** (1904), 681–702. (Cited on 48.)

[569] R. Výborný, *On the use of differentiable homotopy in the proof of Cauchy's theorem*, Amer. Math. Monthly **86** (1979) 380–382. (Cited on 75.)

[570] H. S. Wall, *Continued fractions and bounded analytic functions*, Bull. Amer. Math. Soc. **50** (1944), 110–119. (Cited on 305.)

[571] H. S. Wall, *Analytic Theory of Continued Fractions*, Van Nostrand, New York, 1948; AMS Chelsea, Providence, RI, 2000. (Cited on 304.)

[572] J. Wallis, *Arithmetica Infinitorum*, Oxford, 1656; English translation: J. A. Stedall, *John Wallis: The Arithmetic of Infinitesimals*, Sources and Studies in the History of Mathematics and Physical Sciences, Springer-Verlag, New York, 2004. (Cited on 304.)

[573] J. L. Walsh, *On the location of the roots of the derivative of a polynomial*, Ann. Math. **22** (1920), 128–144. (Cited on 102.)

[574] J. L. Walsh, *The Cauchy–Goursat theorem for rectifiable Jordan curves*, Proc. Nat. Acad. Sci. USA **19** (1933), 540–541. (Cited on 152.)

[575] J. L. Walsh, *Interpolation and Approximation by Rational Functions in the Complex Domain*, Colloquium Publications, American Mathematical Society, New York, 1935. (Cited on 157.)

[576] F. W. Warner, *Foundations of Differentiable Manifolds and Lie Groups*, Scott, Foresman and Co., Glenview, IL–London, 1971. (Cited on 17.)

[577] L. C. Washington, *Elliptic Curves. Number Theory and Cryptography*, 2nd edition, Discrete Mathematics and Its Applications, Chapman & Hall/CRC, Boca Raton, FL, 2008. (Cited on 477.)

[578] R. H. Wasserman, *Tensors and Manifolds. With Applications to Mechanics and Relativity*, The Clarendon Press, Oxford University Press, New York, 1992. (Cited on 17.)

[579] G. N. Watson, *Complex Integration and Cauchy's Theorem*, reprint of Cambridge Tracts in Mathematics and Mathematical Physics, Hafner Publishing, New York, 1960; the original edition appeared in 1914, Cambridge University Press, London. (Cited on 214.)

[580] G. N. Watson, *The Marquis and the Land-Agent; A Tale of the Eighteenth Century*, Math. Gazette **17**, (1933), 5–17. This was the retiring presidential address of the British Mathematical Association. (Cited on 516.)

[581] J. H. M. Wedderburn, *On matrices whose coefficients are functions of a single variable*, Trans. Amer. Math. Soc. **16** (1915), 328–332. (Cited on 401.)

[582] K. Weierstrass, *Darstellung einer analytischen Function einer complexen Veränderlichen, deren absoluter Betrag zwischen zwei gegebenen Grenzen liegt*, Munster, 1841; only published in Math. Werke **1** (1894), 51–66. (Cited on 123.)

[583] K. Weierstrass, *Zur Theorie der Abelschen Functionen*, J. Reine Angew. Math. **47** (1854), 289–306. (Cited on 404, 498.)

[584] K. Weierstrass, *Über das sogenannte Dirichlet'sche Prinzip*, Gelesen in der Königl. Akademie der Wissenschaften am 14.7.1870; Math. Werke **2** (1895), 49–54. (Cited on 314.)

[585] K. Weierstrass, *Zur Theorie der eindeutigen analytischen Functionen*, Berlin Abh. (1876), 11–60; Math. Werke **2** (1897), 77–124. (Cited on 128, 403, 419.)

[586] K. Weierstrass, *Über einen Funktionentheoretischen Satz des Herrn G. Mittag-Leffler*, Monatsber. Berlin Akad. **5** (1880), 189–195. (Cited on 400.)

[587] K. Weierstrass, *Zur Functionenlehre*, Monatsber. Berlin Akad. (1880), 719–743. (Cited on 64.)

[588] K. Weierstrass, *Formeln und Lehrsätze zum Gebrauche der Elliptischen Functionen: Nach Vorlesungen und Aufzeichnunge*, edited by H. A. Schwarz, Dieterichsche Universitäts-Buchdruckerei, Göttingen; 2nd ed., Springer, Berlin 1893. Available online at `http://books.google.com/books?id=Pz8LAAAAYAAJ&dq=Formeln+und+Lehrsaetze&source=gbs_navlinks_s`; French translation as *Formules et propositions pour l'emploi des fonctions elliptiques d'après des leçons et des notes manuscrites de K. Weierstrass*, Gauthier-Villars, Paris 1893. (Cited on 518.)

[589] A. Weil, *Elliptic Functions According to Eisenstein and Kronecker*, reprint of the 1976 original, Classics in Mathematics, Springer-Verlag, Berlin, 1999. (Cited on 477, 518.)

[590] H. Weyl, *The Concept of a Riemann Surface*, Dover Publications, 2009; reprint of the 1955 edition published by Addison-Wesley, Reading, MA. Original: *Die Idee der Riemannschen Fläche*, B. G. Teubner, Leipzig, 1913. (Cited on 47, 266, 368, 568.)

[591] H. Weyl, *Symmetry*, reprint of the 1952 original, Princeton Science Library, Princeton University Press, Princeton, NJ, 1989. (Cited on 283.)

[592] E. T. Whittaker and G. N. Watson, *A Course of Modern Analysis. An Introduction to the General Theory of Infinite Processes and of Analytic Functions; with an Account of the Principal Transcendental Functions*, reprint of the fourth (1927) edition, Cambridge Mathematical Library, Cambridge University Press, Cambridge, 1996. (Cited on 57, 438.)

[593] N. Wiener, *R. E. A. C. Paley — in memoriam*, Bull. Amer. Math. Soc. **39** (1933), 476. (Cited on 555.)

[594] W. Wirtinger, *Zur formalen Theorie der Funktionen von mehr komplexen Veränderlichen*, Math. Ann. **97** (1927), 357–375. (Cited on 37.)

[595] E. M. Wright, *An enumerative proof of an identity of Jacobi*, J. London Math. Soc. **40** (1965), 55–57. (Cited on 535.)

[596] A. Yamada, *A remark on the image of the Ahlfors function*, Proc. Amer. Math. Soc. **88** (1983), 639–642. (Cited on 378.)

[597] A. Yamada, *Ahlfors functions on Denjoy domains*, Proc. Amer. Math. Soc. **115** (1992), 757–763. (Cited on 378.)

[598] B. H. Yandell, *The Honors Class. Hilbert's Problems and Their Solvers*, A K Peters, Natick, MA, 2002. (Cited on 368, 369.)

[599] R. M. Young, *Excursions in Calculus, An interplay of the continuous and the discrete*, The Dolciani Mathematical Expositions, Mathematical Association of America, Washington, DC, 1992. (Cited on 393.)

[600] L. Zalcman, *A heuristic principle in complex function theory*, Amer. Math. Monthly **82** (1975), 813–817. (Cited on 578.)

[601] L. Zalcman, *Normal families: New perspectives*, Bull. Amer. Math. Soc. **35** (1998), 215–230. (Cited on 578.)

[602] S.-Q. Zheng, B.-N. Guo, and F. Qi, *A concise proof for properties of three functions involving the exponential function*, Appl. Math. E-Notes **9** (2009), 177–183. (Cited on 447.)

Symbol Index

Subject Index

Author Index

Index of Capsule Biographies

图字：01-2022-1696 号

基础复分析：分析综合教程
Jichu Fufenxi: Fenxi Zonghe Jiaocheng

图书在版编目 (CIP) 数据

基础复分析：分析综合教程．第 2A 部分 = Basic Complex Analysis: A Comprehensive Course in Analysis, Part 2A：英文 /（美）巴里·西蒙 (Barry Simon) 著．— 影印本．— 北京：高等教育出版社，2023.3
ISBN 978-7-04-059300-6
Ⅰ．①基… Ⅱ．①巴… Ⅲ．①复分析—英文 Ⅳ．① O174.5
中国版本图书馆 CIP 数据核字 (2022) 第 160710 号

策划编辑 李 鹏　　责任编辑 李 鹏
封面设计 张申申　　责任印制 刁 毅

出版发行 高等教育出版社
社址 北京市西城区德外大街 4 号
邮政编码 100120
购书热线 010-58581118
咨询电话 400-810-0598
网址 http://www.hep.edu.cn
http://www.hep.com.cn
网上订购 http://www.hepmall.com.cn
http://www.hepmall.com
http://www.hepmall.cn
印刷 山东临沂新华印刷物流集团有限责任公司

开本 787mm × 1092mm 1/16
印张 42
字数 1060 千字
版次 2023 年 3 月第 1 版
印次 2023 年 3 月第 1 次印刷
定价 269.00 元

[物 料 号 59300-00]

美国数学会经典影印系列

1 **Lars V. Ahlfors**, Lectures on Quasiconformal Mappings, Second Edition 9787040470109

2 **Dmitri Burago, Yuri Burago, Sergei Ivanov**, A Course in Metric Geometry 9787040469080

3 **Tobias Holck Colding, William P. Minicozzi II**, A Course in Minimal Surfaces 9787040469110

4 **Javier Duoandikoetxea**, Fourier Analysis 9787040469011

5 **John P. D'Angelo**, An Introduction to Complex Analysis and Geometry 9787040469981

6 **Y. Eliashberg, N. Mishachev**, Introduction to the *h*-Principle 9787040469028

7 **Lawrence C. Evans**, Partial Differential Equations, Second Edition 9787040469356

8 **Robert E. Greene, Steven G. Krantz**, Function Theory of One Complex Variable, Third Edition 9787040469073

9 **Thomas A. Ivey, J. M. Landsberg**, Cartan for Beginners: Differential Geometry via Moving Frames and Exterior Differential Systems 9787040469172

10 **Jens Carsten Jantzen**, Representations of Algebraic Groups, Second Edition 9787040470086

11 **A. A. Kirillov**, Lectures on the Orbit Method 9787040469103

12 **Jean-Marie De Koninck, Armel Mercier**, 1001 Problems in Classical Number Theory 9787040469998

13 **Peter D. Lax, Lawrence Zalcman**, Complex Proofs of Real Theorems 9787040470000

14 **David A. Levin, Yuval Peres, Elizabeth L. Wilmer**, Markov Chains and Mixing Times 9787040469943

15 **Dusa McDuff, Dietmar Salamon**, *J*-holomorphic Curves and Symplectic Topology 9787040469936

16 **John von Neumann**, Invariant Measures 9787040469974

17 **R. Clark Robinson**, An Introduction to Dynamical Systems: Continuous and Discrete, Second Edition 9787040470093

18 **Terence Tao**, An Epsilon of Room, I: Real Analysis: pages from year three of a mathematical blog 9787040469004

19 **Terence Tao**, An Epsilon of Room, II: pages from year three of a mathematical blog 9787040468991

20 **Terence Tao**, An Introduction to Measure Theory 9787040469059

21 **Terence Tao**, Higher Order Fourier Analysis 9787040469097

22 **Terence Tao**, Poincaré's Legacies, Part I: pages from year two of a mathematical blog 9787040469950

23 **Terence Tao**, Poincaré's Legacies, Part II: pages from year two of a mathematical blog 9787040469967

24 **Cédric Villani**, Topics in Optimal Transportation 9787040469219

25 **R. J. Williams**, Introduction to the Mathematics of Finance 9787040469127

26 **T. Y. Lam**, Introduction to Quadratic Forms over Fields 9787040469196

27 **Jens Carsten Jantzen**, Lectures on Quantum Groups

28 **Henryk Iwaniec**, Topics in Classical Automorphic Forms

29 **Sigurdur Helgason**, Differential Geometry, Lie Groups, and Symmetric Spaces

30 **John B.Conway**, A Course in Operator Theory

31 **James E. Humphreys**, Representations of Semisimple Lie Algebras in the BGG Category *O*

32 **Nathanial P. Brown, Narutaka Ozawa**, C*-Algebras and Finite-Dimensional Approximations

33 **Hiraku Nakajima**, Lectures on Hilbert Schemes of Points on Surfaces

34 **S. P. Novikov**, **I. A. Taimanov**, **Translated by Dmitry Chibisov**, Modern Geometric Structures and Fields

35 **Luis Caffarelli, Sandro Salsa**, A Geometric Approach to Free Boundary Problems

36 **Paul H. Rabinowitz**, Minimax Methods in Critical Point Theory with Applications to Differential Equations

37 **Fan R. K. Chung**, Spectral Graph Theory

38 **Susan Montgomery**, Hopf Algebras and Their Actions on Rings

39 **C. T. C. Wall, Edited by A. A. Ranicki**, Surgery on Compact Manifolds, Second Edition

40 **Frank Sottile**, Real Solutions to Equations from Geometry

41 **Bernd Sturmfels**, Gröbner Bases and Convex Polytopes

42 **Terence Tao**, Nonlinear Dispersive Equations: Local and Global Analysis

43 **David A. Cox, John B. Little, Henry K. Schenck**, Toric Varieties

44 **Luca Capogna, Carlos E. Kenig, Loredana Lanzani**, Harmonic Measure: Geometric and Analytic Points of View

45 **Luis A. Caffarelli, Xavier Cabré**, Fully Nonlinear Elliptic Equations

46 **Teresa Crespo, Zbigniew Hajto,** Algebraic Groups and Differential Galois Theory

47 **Barbara Fantechi, Lothar Göttsche, Luc Illusie, Steven L. Kleiman, Nitin Nitsure, Angelo Vistoli,** Fundamental Algebraic Geometry: Grothendieck's FGA Explained

48 **Shinichi Mochizuki,** Foundations of *p*-adic Teichmüller Theory

49 **Manfred Leopold Einsiedler, David Alexandre Ellwood, Alex Eskin, Dmitry Kleinbock, Elon Lindenstrauss, Gregory Margulis, Stefano Marmi, Jean-Christophe Yoccoz,** Homogeneous Flows, Moduli Spaces and Arithmetic

50 **David A. Ellwood, Emma Previato,** Grassmannians, Moduli Spaces and Vector Bundles

51 **Jeffery McNeal, Mircea Mustaţă,** Analytic and Algebraic Geometry: Common Problems, Different Methods

52 **V. Kumar Murty,** Algebraic Curves and Cryptography

53 **James Arthur, James W. Cogdell, Steve Gelbart, David Goldberg, Dinakar Ramakrishnan, Jiu-Kang Yu,** On Certain *L*-Functions

54 **Rick Miranda,** Algebraic Curves and Riemann Surfaces 9787040469066

55 **Hershel M. Farkas, Irwin Kra,** Theta Constants, Riemann Surfaces and the Modular Group 9787040469042

56 **Fritz John,** Nonlinear Wave Equations, Formation of Singularities 9787040517262

57 **Henryk Iwaniec, Emmanuel Kowalski,** Analytic Number Theory 9787040517231

58 **Jan Malý, William P. Ziemer,** Fine Regularity of Solutions of Elliptic Partial Differential Equations 9787040517248

59 **Jin Hong, Seok-Jin Kang,** Introduction to Quantum Groups and Crystal Bases 9787040517255

60 **V. I. Arnold,** Topological Invariants of Plane Curves and Caustics 9787040517057

61 **Dusa McDuff, Dietmar Salamon,** *J*-Holomorphic Curves and Quantum Cohomology 9787040517040

62 **James Eells, Luc Lemaire,** Selected Topics in Harmonic Maps 9787040517279

63 **Yuri I. Manin,** Frobenius Manifolds, Quantum Cohomology, and Moduli Spaces 9787040517026

64 **Bernd Sturmfels,** Solving Systems of Polynomial Equations 9787040517033

65 **Liviu I. Nicolaescu,** Notes on Seiberg-Witten Theory 9787040469035

66 **I. G. Macdonald,** Symmetric Functions and Orthogonal Polynomials 9787040535938

67 **Craig Huneke,** Tight Closure and Its Applications 9787040535945

68 **Victor Guillemin, Viktor Ginzburg, Yael Karshon,** Moment Maps, Cobordisms, and Hamiltonian Group Actions 9787040535952

69 **Gerald B. Folland,** Fourier Analysis and Its Applications 9787040535969

70 **Luis Barreira, Yakov B. Pesin,** Lyapunov Exponents and Smooth Ergodic Theory 9787040534962

71 **Gilles Pisier,** Factorization of Linear Operators and Geometry of Banach Spaces 9787040535037

72 **Richard Elman, Nikita Karpenko, Alexander Merkurjev,** The Algebraic and Geometric Theory of Quadratic Forms 9787040534955

73 **William Fulton,** Introduction to Intersection Theory in Algebraic Geometry 9787040534870

74 **S. K. Lando,** Lectures on Generating Functions 9787040535006

75 **Frank D. Grosshans, Gian-Carlo Rota, Joel A. Stein,** Invariant Theory and Superalgebras 9787040534979

76 **Lawrence C. Evans,** Weak Convergence Methods for Nonlinear Partial Differential Equations 9787040534993

77 **Harish-Chandra,** Admissible Invariant Distributions on Reductive *p*-adic Groups 9787040535020

78 **Marc Levine,** Mixed Motives 9787040535013

79 **Pavel I. Etingof, Igor B. Frenkel, Alexander A. Kirillov, Jr.,** Lectures on Representation Theory and Knizhnik-Zamolodchikov Equations 9787040534986

80 **Nicholas M. Katz, Peter Sarnak,** Random Matrices, Frobenius Eigenvalues, and Monodromy 9787040534948

81 **Kenneth R. Davidson,** C*-Algebras by Example 9787040534924

82 **Max-Albert Knus, Alexander Merkurjev, Markus Rost, Jean-Pierre Tignol,** The Book of Involutions 9787040534931

83 **Judy L. Walker,** Codes and Curves 9 787040 534917

84 **J. Amorós, M. Burger, K. Corlette, D. Kotschick, D. Toledo,** Fundamental Groups of Compact Kähler Manifolds 9 787040 536331

85 **V. I. Arnold,** Experimental Mathematics 9 787040 556261

86 **Leslie Jane Federer Vaaler, James W. Daniel,** Mathematical Interest Theory, Second Edition 9 787040 556278

87 **Paul R. Halmos,** Linear Algebra Problem Book 9 787040 556285

88 **Luis Barreira, Claudia Valls,** Ordinary Differential Equations: Qualitative Theory 9 787040 556315

89 **Terence Tao,** Topics in Random Matrix Theory 9 787040 556476

90 **Gerald Teschl,** Ordinary Differential Equations and Dynamical Systems 9 787040 556483

91 **Lawrence C. Evans,** An Introduction to Stochastic Differential Equations 9 787040 556490

92 **Gerald Teschl,** Mathematical Methods in Quantum Mechanics: With Applications to Schrödinger Operators, Second Edition 9 787040 556506

93 **András Vasy,** Partial Differential Equations: An Accessible Route through Theory and Applications 9 787040 569773

94 **Weinan E, Tiejun Li, Eric Vanden-Eijnden,** Applied Stochastic Analysis 9 787040 556391

95 **David R. Mazur,** Combinatorics: A Guided Tour 9 787040 556339

96 **David Bressoud,** A Radical Approach to Real Analysis 9 787040 556322

97 **H. Iwaniec,** Lectures on the Riemann Zeta Function 9 787040 556308

98 **Terence Tao,** Hilbert's Fifth Problem and Related Topics 9 787040 556292

99 **David Lannes,** The Water Waves Problem: Mathematical Analysis and Asymptotics 9 787040 556346

100 **Seán Dineen,** Probability Theory in Finance: A Mathematical Guide to the Black-Scholes Formula, Second Edition 9 787040 556353

101 **Steven G. Krantz,** Complex Analysis: The Geometric Viewpoint, Second Edition 9 787040 556360

102 **A. Borel, N. Wallach,** Continuous Cohomology, Discrete Subgroups, and Representations of Reductive Groups, Second Edition 9 787040 556377

103 **H. S. M. Coxeter,** Non-Euclidean Geometry, Sixth Edition 9 787040 556384

104 **Kurt O. Friedrichs**, Mathematical Methods of Electromagnetic Theory 9 787040 569810

105 **Jeffrey Rauch,** Hyperbolic Partial Differential Equations and Geometric Optics 9 787040 569803

106 **Alberto Bressan,** Lecture Notes on Functional Analysis: With Applications to Linear Partial Differential Equations 9 787040 569773

107 **Wilhelm Schlag,** A Course in Complex Analysis and Riemann Surfaces 9 787040 569797

108 **Emmanuel Kowalski,** An Introduction to the Representation Theory of Groups 9 787040 569780

109 **Steven G. Krantz,** A Guide to Complex Variables 9 787040 570229

110 **Rajendra Bhatia,** Fourier Series 9 787040 570236

111 **Fernando Q. Gouvêa,** A Guide to Groups, Rings, and Fields 9 787040 569995

112 **Steven G. Krantz,** A Guide to Functional Analysis

113 **Alberto Torchinsky,** Problems in Real and Functional Analysis

114 **Fritz Colonius, Wolfgang Kliemann,** Dynamical Systems and Linear Algebra

115 **Jerome K. Percus, Stephen Childress,** Mathematical Models in Developmental Biology

116 **Steven G. Krantz,** A Panorama of Harmonic Analysis

117 **Ralph P. Boas, Revised by Harold P. Boas,** Invitation to Complex Analysis, Second Edition

118 **William H. Meeks III, Joaquín Pérez,** A Survey on Classical Minimal Surface Theory

119 **Peter Duren,** Invitation to Classical Analysis

120 **Helene Shapiro,** Linear Algebra and Matrices: Topics for a Second Course

121 **Ralph P. Boas, Revised and updated by Harold P. Boas,** A Primer of Real Functions, Fourth Edition

122 **John P. D'Angelo,** Inequalities from Complex Analysis

123 **Terence Tao,** Expansion in Finite Simple Groups of Lie Type

124 **Goro Shimura,** Arithmetic and Analytic Theories of Quadratic Forms and Clifford Groups

125 **Victor M. Buchstaber, Taras E. Panov,** Toric Topology

126 **Benson Farb, Richard Hain, Eduard Looijenga,** Moduli Spaces of Riemann Surfaces

127 **Charles Livingston,** Knot Theory

128 **Mladen Bestvina, Michah Sageev, Karen Vogtmann,** Geometric Group Theory

129 **Edward Packel,** The Mathematics of Games and Gambling, Second Edition

130 **Kiran S. Kedlaya, Bjorn Poonen, Ravi Vakil,** The William Lowell Putnam Mathematical Competition 1985–2000: Problems, Solutions, and Commentary

131 **Ida Kantor, Jiří Matoušek, Robert Šámal,** Mathematics++: Selected Topics Beyond the Basic Courses

132 **John Roe,** Winding Around: The Winding Number in Topology, Geometry, and Analysis

133 **Wolfgang Kühnel, Translated by Bruce Hunt,** Differential Geometry: Curves–Surfaces–Manifolds, Third Edition

134 **Daniel W. Stroock,** Mathematics of Probability

135 **Jennifer Schultens,** Introduction to 3-Manifolds

136 **Gábor Székelyhidi,** An Introduction to Extremal Kähler Metrics

137 **Diane Maclagan, Bernd Sturmfels,** Introduction to Tropical Geometry

138 **Barry Simon,** Real Analysis : A Comprehensive Course in Analysis, Part 1

139 **Barry Simon,** Basic Complex Analysis: A Comprehensive Course in Analysis, Part 2A

140 **Barry Simon,** Advanced Complex Analysis: A Comprehensive Course in Analysis, Part 2B

141 **Barry Simon,** Harmonic Analysis: A Comprehensive Course in Analysis, Part 3

142 **Barry Simon,** Operator Theory: A Comprehensive Course in Analysis, Part 4